Arithmetic Operations:

$$ab+ac = a(b+c)$$

$$\frac{a}{b} + \frac{c}{d} = \frac{ad+bc}{bd}$$

$$\frac{a+b}{c} = \frac{a}{c} + \frac{b}{c}$$

$$\frac{\left(\frac{a}{b}\right)}{\left(\frac{c}{d}\right)} = \frac{ad}{bc}$$

$$a\left(\frac{b}{c}\right) = \frac{ab}{c}$$

$$\frac{a-b}{c-d} = \frac{b-a}{d-c}$$

$$\frac{ab+ac}{a} = b+c, \, a \neq 0$$

$$\frac{\left(\frac{a}{b}\right)}{c} = \frac{a}{bc}$$

$$\frac{a}{\left(\frac{b}{c}\right)} = \frac{ac}{b}$$

Exponents and Radicals:

$$a^0 = 1, \, a \neq 0$$

$$\frac{a^x}{a^y} = a^{x-y}$$

$$\left(\frac{a}{b}\right)^x = \frac{a^x}{b^x}$$

$$\sqrt[n]{a^m} = a^{m/n} = (\sqrt[n]{a})^m$$

$$a^{-x} = \frac{1}{a^x}$$

$$(a^x)^y = a^{xy}$$

$$\sqrt{a} = a^{1/2}$$

$$\sqrt[n]{ab} = \sqrt[n]{a}\,\sqrt[n]{b}$$

$$a^x a^y = a^{x+y}$$

$$(ab)^x = a^x b^x$$

$$\sqrt[n]{a} = a^{1/n}$$

$$\sqrt[n]{\left(\frac{a}{b}\right)} = \frac{\sqrt[n]{a}}{\sqrt[n]{b}}$$

Algebraic Errors to Avoid:

$$\frac{a}{x+b} \neq \frac{a}{x} + \frac{a}{b}$$ (To see this error, let $a = b = x = 1$.)

$$\sqrt{x^2+a^2} \neq x + a$$ (To see this error, let $x = 3$ and $a = 4$.)

$$a-b(x-1) \neq a-bx-b$$ (Remember to distribute negative signs. The equation should be $a-b(x-1) = a-bx+b$.)

$$\frac{\left(\frac{x}{a}\right)}{b} \neq \frac{bx}{a}$$ (To divide fractions, invert and multiply. The equation should be

$$\frac{\frac{x}{a}}{b} = \frac{\frac{x}{a}}{\frac{b}{1}} = \left(\frac{x}{a}\right)\left(\frac{1}{b}\right) = \frac{x}{ab}.)$$

$$\sqrt{-x^2+a^2} \neq -\sqrt{x^2-a^2}$$ (We can't factor a negative sign outside of the square root.)

$$\frac{a+bx}{a} \neq 1+bx$$ (This is one of many examples of incorrect cancellation. The equation should be $\frac{a+bx}{a} = \frac{a}{a} + \frac{bx}{a} = 1 + \frac{bx}{a}$.)

$$\frac{1}{x^{1/2}-x^{1/3}} \neq x^{-1/2}-x^{-1/3}$$ (This error is a sophisticated version of the first error.)

$$(x^2)^3 \neq x^5$$ (The equation should be $(x^2)^3 = x^2 x^2 x^2 = x^6$.)

Conversion Table:

1 centimeter = 0.394 inches	1 joule = 0.738 foot-pounds	1 mile = 1.609 kilometers
1 meter = 39.370 inches	1 gram = 0.035 ounces	1 gallon = 3.785 liters
= 3.281 feet	1 kilogram = 2.205 pounds	1 pound = 4.448 newtons
1 kilometer = 0.621 miles	1 inch = 2.540 centimeters	1 foot-lb = 1.356 joules
1 liter = 0.264 gallons	1 foot = 30.480 centimeters	1 ounce = 28.350 grams
1 newton = 0.225 pounds	= 0.305 meters	1 pound = 0.454 kilograms

Also available for student purchase

Look for the supplements to this text to give you learning tips and extra practice.

Start with the **Study and Solutions Guide** (28306-1). This convenient guide gives you important information and extra help, including:

• Detailed solutions to many odd-numbered exercises, worked in the same style you'll find in this text.
• Summaries of each section's important concepts
• Chapter self-tests
• Study strategies

You can also look for **College Algebra Tutor** software for your personal computer. This easy-to-use software contains tutorials keyed to every section of your textbook. The software is available for IBM PC-compatible computers on 5-1/4" disks (28591-9) or 3-1/2" disks (28590-0), or for Macintosh computers (34095-2). It provides:

• Section-by-section coverage, giving support for every topic
• Extra practice in working with problems and concepts
• Diagnostic feedback to help you correct errors
• Glossaries and randomly-generated warm ups, and chapter post-tests

Check for these outstanding supplements in your college bookstore. If you can't find them, ask your textbook department to order them. Or call D. C. Heath toll-free: (800) 334-3284. Give yourself the edge in mathematics!

D. C. Heath and Company
125 Spring Street
Lexington, MA 02173

Please note: A $2.00 shipping and handling charge will be added to orders
placed directly through D. C. Heath for UPS ground transportation.

College Algebra

THIRD EDITION

with Technology Updates

Roland E. Larson

Robert P. Hostetler

The Pennsylvania State University
The Behrend College

with the assistance of
David E. Heyd
The Pennsylvania State University
The Behrend College

D.C. Heath and Company
Lexington, Massachusetts   Toronto

Address editorial correspondence to:

D. C. Heath and Company
125 Spring Street
Lexington, MA 02173

Acquisitions Editor: Ann Marie Jones
Senior Development Editor: Cathy Cantin
Production Editor: Sarah Doyle
Designer: Sally Steele
Art Editor: Sally Steele
Production Coordinator: Lisa Merrill
Technical Art: Folium
Cover: Lance Hidy

Published simultaneously in Canada.

Printed in the United States of America.

International Standard Book Number 0-669-39828-4

10   9   8   7   6   5   4   3   2   1

Preface

Success in college-level mathematics courses begins with a good understanding of algebra, and one goal of *College Algebra, Third Edition with Technology Updates* is to help students develop this understanding. Another goal is to show students how algebra can be used as a modeling language for real-life problems. Although we review some of the basic concepts of algebra, we assume that most students in this course have completed two years of high school algebra.

New to the Third Edition

Many users of the second edition of the text have given us suggestions for improving the text. We appreciate this type of input very much and have incorporated most of the suggestions into the Third Edition. Every section in the text was revised or considered for revision, and some sections were completely rewritten. Many new examples and exercises were added throughout the text. In addition, examples, exercises, and applications were revised and updated to expand the students' opportunities to practice their algebra skills. There is increased emphasis on the proper use of scientific and graphics calculators and graphing utilities in problem solving. Many examples and exercises now contain real data with credited source lines. Discussion problems were added to every section of the text. Each chapter now includes a two-page technology feature, and an introduction to graphics calculators, graphing calculator programs, and exercises requiring the use of technology were added to the appendix. New cumulative tests offer additional review after every three chapters.

Features

The features of this text are designed to help students develop their algebra and problem-solving skills, as well as acquire an understanding of mathematical concepts. To do this, the text has several key features.

Graphics

The ability to visualize a problem is a critical part of a student's ability to solve the problem. To encourage the development of this skill, the text has many figures in examples and exercise sets and in answers to odd-numbered exercises in the back of the text. Various types of graphics show geometric representations, including graphs of functions, geometric figures, symmetry, displays of statistical information, and screen outputs from graphing technology. All graphs are computer-generated for accuracy.

Applications

Numerous pertinent applications, many new to the Third Edition, are integrated throughout every section of the text, both as solved examples and as exercises. This encourages students to use and review their problem-solving skills. The text applications are current, and students learn to apply the process of mathematical modeling to real-world situations in many areas, such as business, economics, biology, engineering, chemistry, and physics. Many applications in the text use real-world data, and source lines are included to help motivate student interest.

Examples

For the Third Edition, many examples were revised and several new ones, including real-life applications, were added. Each was carefully chosen to illustrate a particular concept or problem-solving technique.

Discussion Problems

New to the Third Edition, the discussion problems offer students the opportunity for thinking, reasoning, and communicating about mathematics in different ways. Individually or in teams, for in-class discussion, writing assignments, or class presentations, students are encouraged to draw new conclusions about the concepts presented. The problem might ask for further explanation, synthesis, experimentation, or extension of the section concepts. Discussion problems appear at the end of each text section.

Exercise Sets

The exercise sets were extensively revised for the Third Edition. Many sets include a group of exercises that provide the graphs of functions involved. More exploratory and conceptual questions and questions involving geometry were added to the exercises to enhance their effectiveness. Each exercise set is carefully graded in difficulty to allow students to gain confidence as they progress. Exercise sets, including warm-up exercises, appear at the end of each text section. Review exercises are included at the end of each chapter, and now cumulative tests are included to review what students have learned from the three preceding chapters. The opportunity to use calculators is available with several topics to show patterns, experiment, calculate, or create graphic models.

Problem Solving Using Technology

Every chapter contains an optional feature that shows how graphics calculators and computer graphing utilities can be used to solve applications. New to the Third Edition, these features enhance and expand the range of problem-solving techniques using real data, computer-generated art, and graphing-technology screen output to simulate real-life problem-solving situations. Many of the problems discussed in these features are previews of classic problems in calculus, helping students develop an intuitive foundation for further study. In addition, there are opportunities throughout the text to use graphing technology in problems and applications in section exercises.

Enhanced Presentation

The Third Edition incorporates the use of additional colors to strengthen the text as a pedagogical tool. Color is used consistently to aid both reading and reference. For instance, definitions are highlighted by tan boxes, and equation side comments are given in red. Color in the art helps students visualize relationships.

These and other features of the text are described in greater detail on the following pages.

Features of the Text

Chapter Opener

Each chapter begins with a list of the topics to be covered and a brief overview. This provides a survey of the contents of the chapter, showing students how the topics fit into the overall development of algebra. Each section begins with a list of important topics covered in that section.

Definitions

All of the important rules, formulas, and definitions are boxed for emphasis. Each is also titled for easy reference.

Algebra of Calculus

Special emphasis has been given to algebraic skills that are needed in calculus. In addition to the material in Section 1.7 shown here, many other examples in the Third Edition discuss algebraic techniques that are used in calculus. These examples are clearly identified.

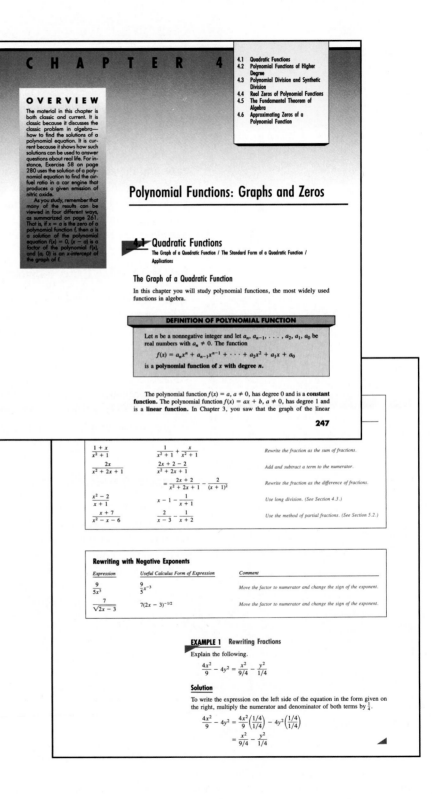

C H A P T E R   4

4.1 Quadratic Functions
4.2 Polynomial Functions of Higher Degree
4.3 Polynomial Division and Synthetic Division
4.4 Real Zeros of Polynomial Functions
4.5 The Fundamental Theorem of Algebra
4.6 Approximating Zeros of a Polynomial Function

OVERVIEW
The material in this chapter is both classic and current. It is classic because it discusses the classic problem in algebra—how to find the solutions of a polynomial equation. It is current because it shows how such solutions can be used to answer questions about real life. For instance, Exercise 58 on page 280 uses the solution of a polynomial equation to find the air-fuel ratio in a car engine that produces a given emission of nitric oxide.
As you study, remember that many of the results can be viewed in four different ways, as summarized on page 261. That is, if $x = a$ is the zero of a polynomial function f, then a is a solution of the polynomial equation $f(x) = 0$, $(x - a)$ is a factor of the polynomial $f(x)$, and $(a, 0)$ is an x-intercept of the graph of f.

Polynomial Functions: Graphs and Zeros

4.1 Quadratic Functions
The Graph of a Quadratic Function / The Standard Form of a Quadratic Function / Applications

The Graph of a Quadratic Function

In this chapter you will study polynomial functions, the most widely used functions in algebra.

DEFINITION OF POLYNOMIAL FUNCTION

Let n be a nonnegative integer and let $a_n, a_{n-1}, \ldots, a_2, a_1, a_0$ be real numbers with $a_n \neq 0$. The function

$$f(x) = a_n x^n + a_{n-1} x^{n-1} + \cdots + a_2 x^2 + a_1 x + a_0$$

is a **polynomial function of** x **with degree** n.

The polynomial function $f(x) = a$, $a \neq 0$, has degree 0 and is a **constant function**. The polynomial function $f(x) = ax + b$, $a \neq 0$, has degree 1 and is a **linear function**. In Chapter 3, you saw that the graph of the linear

247

$\dfrac{1+x}{x^2+1}$	$\dfrac{1}{x^2+1} + \dfrac{x}{x^2+1}$	*Rewrite the fraction as the sum of fractions.*
$\dfrac{2x}{x^2+2x+1}$	$\dfrac{2x+2-2}{x^2+2x+1}$	*Add and subtract a term to the numerator.*
	$= \dfrac{2x+2}{x^2+2x+1} - \dfrac{2}{(x+1)^2}$	*Rewrite the fraction as the difference of fractions.*
$\dfrac{x^2-2}{x+1}$	$x - 1 - \dfrac{1}{x+1}$	*Use long division. (See Section 4.3.)*
$\dfrac{x+7}{x^2-x-6}$	$\dfrac{2}{x-3} - \dfrac{1}{x+2}$	*Use the method of partial fractions. (See Section 5.2.)*

Rewriting with Negative Exponents

Expression	Useful Calculus Form of Expression	Comment
$\dfrac{9}{5x^3}$	$\dfrac{9}{5}x^{-3}$	*Move the factor to numerator and change the sign of the exponent.*
$\dfrac{7}{\sqrt{2x-3}}$	$7(2x-3)^{-1/2}$	*Move the factor to numerator and change the sign of the exponent.*

EXAMPLE 1   Rewriting Fractions

Explain the following.

$$\frac{4x^2}{9} - 4y^2 = \frac{x^2}{9/4} - \frac{y^2}{1/4}$$

Solution

To write the expression on the left side of the equation in the form given on the right, multiply the numerator and denominator of both terms by $\frac{1}{4}$.

$$\frac{4x^2}{9} - 4y^2 = \frac{4x^2}{9}\left(\frac{1/4}{1/4}\right) - 4y^2\left(\frac{1/4}{1/4}\right)$$

$$= \frac{x^2}{9/4} - \frac{y^2}{1/4}$$

Examples

The Third Edition contains more than 600 text examples. They are titled for easy reference, and many include side comments that explain or justify steps in the solution. Students are encouraged to check their solutions.

Remarks

Special instructional notes to students appear with definitions, theorems, rules, and examples. Anticipating students' needs, they give additional insight, help avoid common errors, or describe generalizations.

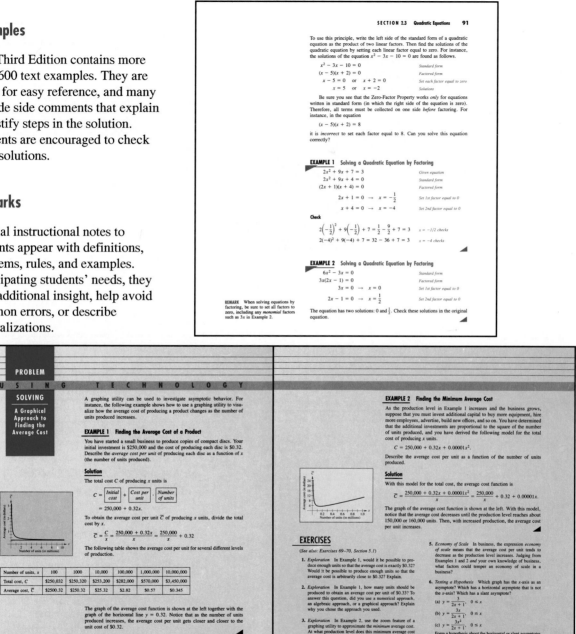

Problem Solving Using Technology

This optional feature in each chapter examines applications using graphing technology. As the basis for discussion, class demonstration, or student assignment, a variety of problems are explored, many of which use real data. Many of the problems discussed in these features are previews of classic problems in calculus.

Problem Solving

A consistent strategy for solving problems is emphasized throughout: analyze the problem, create a verbal model, construct an algebraic model, solve, and check the answer in the statement of the original problem. This problem-solving process has wide applicability and can be used with analytical, graphical, and numerical approaches to problem solving.

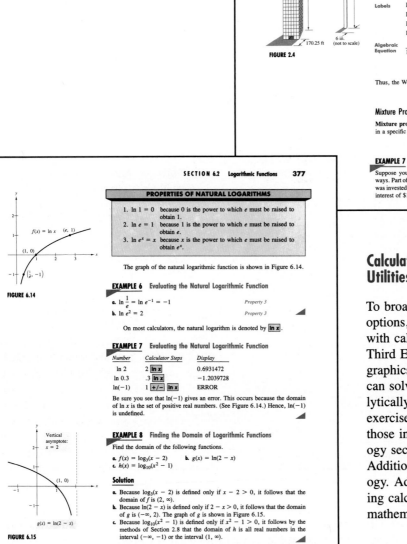

SECTION 2.2   Linear Equations and Modeling   81

In the next example, we use this technique to solve a problem involving similar triangles.

EXAMPLE 6   An Application Involving Similar Triangles

In order to measure the height of the twin towers of the World Trade Center (in New York City), suppose you use the following scheme. You measure the shadow cast by one of the buildings and find it to be 170.25 feet, as shown in Figure 2.4. Then you measure the shadow cast by a 4-foot post and find that its shadow is 6 inches long. Use this information to determine the height of the building.

Solution

To solve this problem, you use a result from geometry that tells you that the ratios of corresponding sides of similar triangles are equal.

Verbal Model $\dfrac{\text{Height of building}}{\text{Length of shadow}} = \dfrac{\text{Height of post}}{\text{Length of shadow}}$

Labels   Height of building = x (measured in feet)
Length of building's shadow = 170.25 feet
Height of stake = 4 feet = 48 inches
Length of stake's shadow = 6 inches

Algebraic Equation $\dfrac{x}{170.25} = \dfrac{48}{6}$
$x = 8 \cdot 170.25$
$x = 1362$ feet

Thus, the World Trade Center is about 1362 feet high.

x ft
48 in.
6 in. (not to scale)
170.25 ft
FIGURE 2.4

Mixture Problems

Mixture problems involve two different unknown quantities that are *mixed* in a specific way. Watch for a *hidden product* in the verbal model.

EXAMPLE 7   A Simple Interest Problem

Suppose you received $10,000 from an inheritance and invested it in two ways. Part of the money was invested at $9\frac{1}{2}\%$ simple interest and the remainder was invested at 11%. After one year, the two investments returned a combined interest of $1,038.50. How much did you invest in each type of account?

SECTION 6.2   Logarithmic Functions   377

PROPERTIES OF NATURAL LOGARITHMS

1. $\ln 1 = 0$ because 0 is the power to which e must be raised to obtain 1.
2. $\ln e = 1$ because 1 is the power to which e must be raised to obtain e.
3. $\ln e^x = x$ because x is the power to which e must be raised to obtain e^x.

The graph of the natural logarithmic function is shown in Figure 6.14.

$f(x) = \ln x$ $(e, 1)$
$(1, 0)$
$\left(\frac{1}{e}, -1\right)$
FIGURE 6.14

EXAMPLE 6   Evaluating the Natural Logarithmic Function

a. $\ln \dfrac{1}{e} = \ln e^{-1} = -1$ Property 3

b. $\ln e^2 = 2$ Property 3

On most calculators, the natural logarithm is denoted by [ln x].

EXAMPLE 7   Evaluating the Natural Logarithmic Function

Number	Calculator Steps	Display
ln 2	2 [ln x]	0.6931472
ln 0.3	.3 [ln x]	-1.2039728
ln(-1)	1 [+/-] [ln x]	ERROR

Be sure you see that ln(-1) gives an error. This occurs because the domain of ln x is the set of positive real numbers. (See Figure 6.14.) Hence, ln(-1) is undefined.

EXAMPLE 8   Finding the Domain of Logarithmic Functions

Find the domain of the following functions.

a. $f(x) = \log_3(x - 2)$ **b.** $g(x) = \ln(2 - x)$
c. $h(x) = \log_{10}(x^2 - 1)$

Solution

a. Because $\log_3(x - 2)$ is defined only if $x - 2 > 0$, it follows that the domain of f is $(2, \infty)$.
b. Because $\ln(2 - x)$ is defined only if $2 - x > 0$, it follows that the domain of g is $(-\infty, 2)$. The graph of g is shown in Figure 6.15.
c. Because $\log_{10}(x^2 - 1)$ is defined only if $x^2 - 1 > 0$, it follows by the methods of Section 2.8 that the domain of h is all real numbers in the interval $(-\infty, -1)$ or the interval $(1, \infty)$.

Vertical asymptote: $x = 2$
$(1, 0)$
$g(x) = \ln(2 - x)$
FIGURE 6.15

Calculators and Computer Graphing Utilities

To broaden the range of teaching and learning options, hints and instructions for working with calculators occur in many places in the Third Edition. If your students have access to graphics calculators or graphing utilities, they can solve exercises both graphically and analytically beginning with Chapter 3. Some exercises require a graphing utility, including those in the Problem Solving Using Technology section in each chapter and Appendix G, Additional Problem Solving Using Technology. Additionally, Appendix F contains graphing calculator programs for exploring key mathematical concepts.

Discussion Problems

A discussion problem appears at the end of each section. Each one encourages students to think, reason, and write about mathematics, individually or in groups. Examining the mathematics in a different way from that presented in the section, these problems emphasize synthesis and experimentation.

Warm Ups

Each section (except Sections 1.1 and 1.7) contains a set of 10 warm-up exercises for students to review and practice the previously learned skills that are necessary to master the new skills and concepts presented in the section. All warm-up exercises are answered in the back of the text.

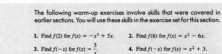

SECTION 3.5   Exercises   213

DISCUSSION PROBLEM
Increasing and Decreasing Functions

Use your school's library or some other reference source to find examples of three different functions that represent data between 1980 and 1990. Find one that decreased during the decade, one that increased, and one that was constant. For instance, the value of the dollar decreased, the population of the United States increased, and the land size of the United States remained constant. Can you find three other examples? Present your results graphically.

WARM UP

The following warm-up exercises involve skills that were covered in earlier sections. You will use these skills in the exercise set for this section.

1. Find $f(2)$ for $f(x) = -x^3 + 5x$.   2. Find $f(6)$ for $f(x) = x^2 - 6x$.

3. Find $f(-x)$ for $f(x) = \dfrac{3}{x}$.   4. Find $f(-x)$ for $f(x) = x^2 + 3$.

In Exercises 5 and 6, solve for x.

5. $x^3 - 16x = 0$ 6. $2x^2 - 3x + 1 = 0$

In Exercises 7–10, find the domain of the function.

7. $g(x) = \dfrac{4}{x - 4}$ 8. $f(x) = \dfrac{2x}{x^2 - 9x + 20}$

9. $h(t) = \sqrt[4]{5 - 3t}$ 10. $f(t) = t^3 + 3t - 5$

EXERCISES for Section 3.5

In Exercises 1–6, find the domain and range of the function.

1. $f(x) = \sqrt{x - 1}$ 2. $f(x) = 4 - x^2$ 3. $f(x) = \sqrt{x^2 - 4}$ 4. $f(x) = |x - 2|$

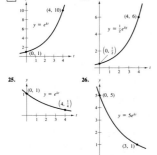

SECTION 6.5   Exercises   405

In Exercises 17–22, complete the table for the given radioactive isotope.

Isotope	Half-life (years)	Initial quantity	Amount after 1000 years	Amount after 10,000 years
17. Ra^{226}	1,620	10 g		
18. Ra^{226}	1,620		1.5 g	
19. C^{14}	5,730			2 g
20. C^{14}	5,730	3 g		
21. Pu^{230}	24,360		2.1 g	
22. Pu^{230}	24,360			0.4 g

In Exercises 23–26, find the constant k such that the exponential function $y = Ce^{kt}$ passes through the given points on the graph.

23.
24.
25.
26.

27. *Population*   The population P of a city is given by
$P = 105,300e^{0.015t}$
where t is the time in years, with $t = 0$ corresponding to 1990. According to this model, in what year will the city have a population of 150,000?

28. *Population*   The population P of a city is given by
$P = 240,360e^{0.012t}$

where t is the time in years, with $t = 0$ corresponding to 1990. According to this model, in what year will the city have a population of 250,000?

29. *Population*   The population P of a city is given by
$P = 2500e^{kt}$
where t is the time in years, with $t = 0$ corresponding to the year 1990. In 1945, the population was 1350. Find the value of k and use this result to predict the population in the year 2010.

30. *Population*   The population P of a city is given by
$P = 140,500e^{kt}$
where t is the time in years, with $t = 0$ corresponding to the year 1990. In 1960, the population was 100,250. Find the value of k and use this result to predict the population in the year 2000.

31. *Population*   The population of Dhaka, Bangladesh was 4.22 million in 1990, and its projected population for the year 2000 is 6.49 million. (*Source:* U.S. Bureau of the Census) Find the exponential growth model $y = Ce^{kt}$ for the population growth of Dhaka by letting $t = 0$ correspond to 1990. Use the model to predict the population of the city in 2010.

32. *Population*   The population of Houston, Texas was 2.30 million in 1990, and its projected population for the year 2000 is 2.65 million. (*Source:* U.S. Bureau of the Census) Find the exponential growth model $y = Ce^{kt}$ for the population growth of Houston by letting $t = 0$ correspond to 1990. Use the model to predict the population of the city in 2010.

33. *Bacteria Growth*   The number of bacteria N in a culture is given by the model
$N = 100e^{kt}$
where t is the time in hours, with $t = 0$ corresponding to the time when $N = 100$. When $t = 5$, $N = 300$. How long will it take the population to double in size?

34. *Bacteria Growth*   The number of bacteria N in a culture is given by the model
$N = 250e^{kt}$
where t is the time in hours, with $t = 0$ corresponding to the time when $N = 250$. When $t = 10$, $N = 280$. How long will it take the population to double in size?

35. *Radioactive Decay*   The half-life of radioactive radium (Ra^{226}) is 1620 years. What percentage of a present amount of radioactive radium will remain after 100 years?

Exercises

The nearly 6000 exercises include computational, conceptual, exploratory, and applied problems. These are designed to build competence, skill, and understanding. Each exercise set is graded in difficulty to allow students to gain confidence as they progress. Some exercises require the use of a graphing utility. Answers to odd-numbered exercises are in the back of the text. Boxed numbers indicate exercises that are solved in detail in the *Study and Solutions Guide*.

Applications

Real-world applications are integrated throughout the text both in examples and in exercises. This offers students insight about the usefulness of algebra, develops strategies for solving problems, and emphasizes the relevance of the mathematics. Many of the applications use current real data, and all are titled for reference.

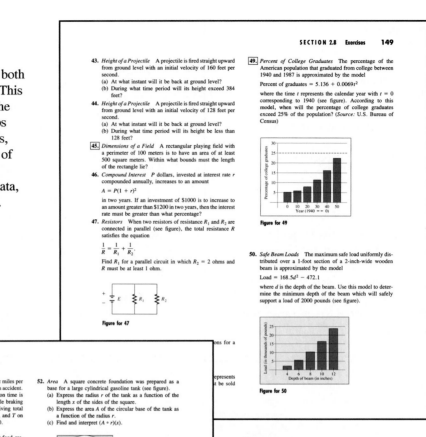

SECTION 2.8   Exercises   **149**

43. *Height of a Projectile*   A projectile is fired straight upward from ground level with an initial velocity of 160 feet per second.
 (a) At what instant will it be back at ground level?
 (b) During what time period will its height exceed 384 feet?

44. *Height of a Projectile*   A projectile is fired straight upward from ground level with an initial velocity of 128 feet per second.
 (a) At what instant will it be back at ground level?
 (b) During what time period will its height be less than 128 feet?

45. *Dimensions of a Field*   A rectangular playing field with a perimeter of 100 meters is to have an area of at least 500 square meters. Within what bounds must the length of the rectangle lie?

46. *Compound Interest* P dollars, invested at interest rate r compounded annually, increases to an amount

$$A = P(1 + r)^2$$

in two years. If an investment of $1000 is to increase to an amount greater than $1200 in two years, then the interest rate must be greater than what percentage?

47. *Resistors*   When two resistors of resistance R_1 and R_2 are connected in parallel (see figure), the total resistance R satisfies the equation

$$\frac{1}{R} = \frac{1}{R_1} + \frac{1}{R_2}.$$

Find R_1 for a parallel circuit in which $R_2 = 2$ ohms and R must be at least 1 ohm.

Figure for 47

49. *Percent of College Graduates*   The percentage of the American population that graduated from college between 1940 and 1987 is approximated by the model

Percent of graduates $= 5.136 + 0.0069t^2$

where the time t represents the calendar year with $t = 0$ corresponding to 1940 (see figure). According to this model, when will the percentage of college graduates exceed 25% of the population? (*Source:* U.S. Bureau of Census)

Figure for 49

50. *Safe Beam Loads*   The maximum safe load uniformly distributed over a 1-foot section of a 2-inch-wide wooden beam is approximated by the model

Load $= 168.5d^2 - 472.1$

where d is the depth of the beam. Use this model to determine the minimum depth of the beam which will safely support a load of 2000 pounds (see figure).

Figure for 50

224   CHAPTER 3   Functions and Graphs

49. *Stopping Distance*   While traveling in a car at x miles per hour, you are required to stop quickly to avoid an accident. The distance the car travels during your reaction time is given by $R(x) = \frac{3}{4}x$. The distance traveled while braking is given by $B(x) = \frac{1}{15}x^2$. Find the function giving total stopping distance T. Graph the functions R, B, and T on the same set of coordinate axes for $0 \le x \le 60$.

50. *Comparing Sales*   Suppose you own two fast-food restaurants in town. From 1985 to 1990, the sales for one restaurant have been decreasing according to the function

$R_1 = 500 - 0.8t^2$, $t = 5, 6, 7, 8, 9, 10$

where R_1 represents the sales for the first restaurant (in thousands of dollars) and t represents the calendar year with $t = 5$ corresponding to 1985. During the same six-year period, the sales for the second restaurant have been increasing according to the function

$R_2 = 250 + 0.78t$, $t = 5, 6, 7, 8, 9, 10$.

Write a function that represents the total sales for the two restaurants. Use the *stacked bar graph* in the figure, which represents the total sales during the six-year period, to determine whether the total sales have been increasing or decreasing.

Figure for 50

51. *Ripples*   A pebble is dropped into a calm pond, causing ripples in the form of concentric circles (see figure). The radius (in feet) of the outer ripple is given by $r(t) = 0.6t$, where t is the time in seconds after the pebble strikes the water. The area of the circle is given by the function $A(r) = \pi r^2$. Find and interpret $(A \circ r)(t)$.

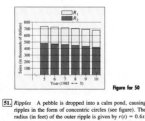

Figure for 51

52. *Area*   A square concrete foundation was prepared as a base for a large cylindrical gasoline tank (see figure).
 (a) Express the radius r of the tank as a function of the length x of the sides of the square.
 (b) Express the area A of the circular base of the tank as a function of the radius r.
 (c) Find and interpret $(A \circ r)(x)$.

Figure for 52

53. *Cost*   The weekly cost of producing x units in a manufacturing process is given by the function $C(x) = 60x + 750$. The number of units produced in t hours is given by $x(t) = 50t$. Find and interpret $(C \circ x)(t)$.

54. *Air Traffic Control*   An air traffic controller spots two planes at the same altitude flying toward the same point (see figure). Their flight paths form a right angle at point P. One plane is 150 miles from point P and is moving at 450 miles per hour. The second plane is 200 miles from point P and is moving at 450 miles per hour. Write the distance s between the planes as a function of time t.

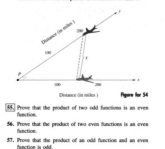

Figure for 54

55. Prove that the product of two odd functions is an even function.

56. Prove that the product of two even functions is an even function.

57. Prove that the product of an odd function and an even function is odd.

Geometry

Geometric formulas and concepts are reviewed throughout the text. For easy reference, common formulas are given inside the back cover.

Graphics

The ability to visualize problems is a critical skill that students need in order to solve them. To encourage the development of this skill and to reinforce concepts, the text has over 1000 figures, with every graph computer-generated for accuracy.

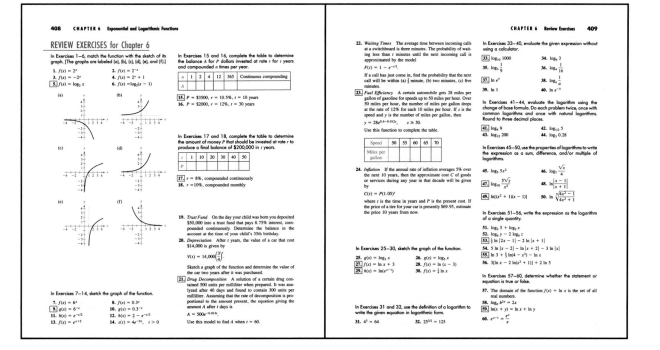

Review Exercises

A set of review exercises at the end of each chapter gives students an opportunity for additional practice. The review exercises include both computational and applied problems covering a wide range of topics.

Cumulative Tests

Cumulative tests appear after Chapters 3, 6, and 9. These tests help students judge their mastery of previously covered concepts. They also help students maintain the knowledge-base they have been building throughout the text, preparing them for other exams and future courses.

Supplements

College Algebra, Third Edition by Larson and Hostetler is accompanied by a comprehensive supplements package for maximum teaching effectiveness and efficiency.

Instructor's Annotated Edition

Complete Solutions Guide

Study and Solutions Guide by Dianna L. Zook, Indiana University–Purdue University at Fort Wayne

Test Item File and Resource Guide by Meredythe M. Burrows, The Pennsylvania State University, The Behrend College

College Algebra Videotapes by Dana Mosely

College Algebra Tutor by Timothy R. Larson, Paula M. Sibeto, and John R. Musser

Test-Generating Software

Transparency Package

Bestgrapher Software by George Best

Computer Activities for Precalculus Software
by Technology Training Associates

Precalculus Experiments with the Casio Graphics Calculator
by Lawrence G. Gilligan, OMI College of Applied Science,
University of Cincinnati

Precalculus Experiments with the TI-81 Graphics Calculator
by Lawrence G. Gilligan, OMI College of Applied Science,
University of Cincinnati

The Algebra of Calculus
by Eric Braude

Derive* software by Soft Warehouse

This complete supplements package offers ancillary materials for students,
for instructors, and for classroom resources. Most items are keyed directly to
the textbook for easy use. For the convenience of software users, a technical
support telephone number is available with all D. C. Heath software
products: (617) 860-1218. The components of this comprehensive teaching
and learning package are outlined on the following pages.

* *Derive* is available to adopters at a discounted price.

INSTRUCTORS

Complete Solutions Guide
Solutions to all warm ups, text exercises, discussion problems, technology features, and cumulative tests

Instructor's Annotated Edition
• Answers to all warm ups, exercises, and cumulative tests
• Teaching strategies
• Additional examples and exercises

Test Item File and Resource Guide
• Printed test bank
• Over 2000 test items
• Open-ended and multiple-choice test items
• Available as test-generating software
• Sample tests
• Survey of assessment methods

STUDENTS

Study and Solutions Guide
• Solutions to selected odd-numbered text exercises
• Solutions match methods of text
• Summaries of key concepts in each text chapter
• Self-tests
• Study strategies

The Algebra of Calculus
• Review of problems using algebra that students will encounter in calculus
• Over 200 examples
• Exercise sets

Precalulus Experiments with the TI-81 Graphics Calculator
• More than 20 labs
• Worktext format: includes numerous screen displays
• Examples, exercises, and cumulative exercise sets

Precalculus Experiments with the Casio Graphics Calculator
• More than 20 labs
• Worktext format: includes numerous screen displays
• Examples, exercises, and cumulative exercise set

CLASSROOM RESOURCES

Instructor's Annotated Edition
• Answers to all warm ups, exercises, and cumulative tests
• Teaching strategies

Transparency Package
• 50 color transparencies
• Color-coded by text topic

Software		VIDEOTAPES

Computerized Testing Software
- Test-generating software
- Over 2000 test items
- Also available as a printed test item file

Derive

Bestgrapher
- Function grapher
- Screen simultaneously displays equation, graph, and table of values
- Some features anticipate calculus
- Includes zoom and print features for use on assignments

Tutor
- Interactive tutorial software follows text section by section
- Diagnostic feedback
- Additional practice
- Chapter self-tests
- Glossary
- Guided exercises provide step-by-step solutions; can be useful for class demonstration

Derive

Bestgrapher
- Function grapher
- Screen simultaneously displays equation, graph, and table of values
- Some features anticipate calculus
- Includes zoom and print features for use on assignments

Computer Activities for Precalculus
- Function grapher
- Directed tutorial on selected topics

Videotapes
- Comprehensive coverage
- Computer-generated animation
- For media/resource centers
- Additional explanation of concepts, sample problems, and applications

Tutor
- Interactive tutorial software follows text section by section
- Diagnostic feedback
- Additional practice
- Chapter self-tests
- Glossary
- Guided exercises provide step-by-step solutions; can be useful for class demonstration

Derive

Bestgrapher
- Function grapher
- Screen simultaneously displays equation, graph, and table of values
- Includes zoom

Computer Activities for Precalculus
- Function grapher
- Directed tutorial

Videotapes
- Comprehensive coverage
- Computer-generated animation
- Additional explanation of concepts, sample problems, and applications

INTEGRATED LEARNING PACKAGE

Computer Activities for Precalculus

```
           4

 -15  ⌐ ⌐ ⌐ ⌐ ⌐ ⌐ ⌐ ⌐ ⌐ ⌐ ⌐   15
          -4

f(x) = log(2,x)
a = □
Enter a positive value for the base a
other than 1.

   Answer   Notation   Help   Menu
```

Instructor's Annotated Edition

equation is in logarithmic form and
the second is in exponential form.

$$3 = \log_2 8 \quad \text{if and only if} \quad 2^3 = 8.$$

Base

Logarithm is an exponent

EXAMPLE 1 Evaluating Logarithms

The logarithmic function is one of
the most difficult for students to
understand. Remind students that a
logarithm is an exponent. Converting back and forth from logarithmic
form to exponential form supports
this concept.

a. $\log_2 32 = 5$ because $2^5 = 32$.

b. $\log_3 27 = 3$ because $3^3 = 27$.

c. $\log_4 2 = \frac{1}{2}$ because $4^{1/2} = \sqrt{4} = 2$.

d. $\log_{10} \frac{1}{100} = -2$ because $10^{-2} = \frac{1}{10^2} = \frac{1}{100}$.

e. $\log_3 1 = 0$ because $3^0 = 1$.

f. $\log_2 2 = 1$ because $2^1 = 2$.

Precalculus Experiments with the TI-81 Graphics Calculator

In questions 1 through 4, graph the given logarithmic function and state its domain and range.

1. $y = \log(x + 1)$ 2. $y = 4 + \ln x$

Domain: _____   Domain: _____

Range: _____   Range: _____

Precalculus Experiments with the Casio Graphics Calculator

Procedure 1. We graph the common logarithm, $y = \log x$ in Figure 8.1. Note the Casio range
settings and the domain and range. (The domain and range correspond, of course, to the
range and domain of the function $y = 10^x$.)

```
Range
Xmin:  -1.
 max:  25.
 scl:   5.
Ymin:  -2.
 max:   2.
 scl:   0.5
```

Domain: $x > 0$
Range: **R**
Intercept: $(1, 0)$
Vertical asymptote: $x = 0$

FIGURE 8.1. $y = \log x$

Transparency Package

Graph of $y = \log_a x$, $a > 1$
$y = \log_a x$
$(1, 0)$
- Domain: $(0, \infty)$
- Range: $(-\infty, \infty)$
- Intercept: $(1, 0)$
- Increasing
- y-axis is a vertical asymptote ($\log_a x \to -\infty$ as $x \to 0^+$)
- Continuous
- Reflection of graph of $y = a^x$ about the line $y = x$

Graph of $y = \ln x$
$y = \ln x$
$(1, 0)$
- Domain: $(0, \infty)$
- Range: $(-\infty, \infty)$
- Intercept: $(1, 0)$
- Increasing
- y-axis is a vertical asymptote ($\ln x \to -\infty$ as $x \to 0^+$)
- Continuous
- Reflection of graph of $y = e^x$ about the line $y = x$

Exponential and Logarithmic Functions—19   Graphs of Logarithmic Functions

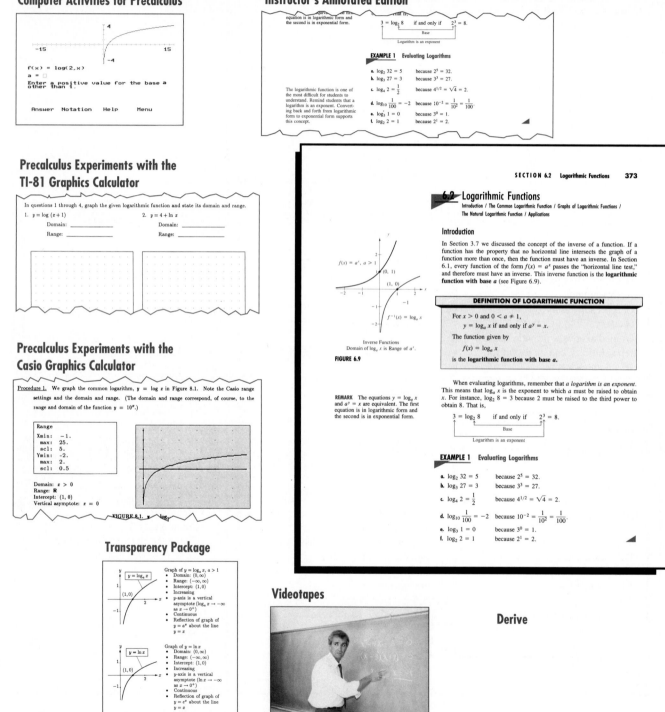

SECTION 6.2   Logarithmic Functions   **373**

6.2 Logarithmic Functions

Introduction / The Common Logarithmic Function / Graphs of Logarithmic Functions / The Natural Logarithmic Function / Applications

Introduction

In Section 3.7 we discussed the concept of the inverse of a function. If a function has the property that no horizontal line intersects the graph of a function more than once, then the function must have an inverse. In Section 6.1, every function of the form $f(x) = a^x$ passes the "horizontal line test," and therefore must have an inverse. This inverse function is the **logarithmic function with base** a (see Figure 6.9).

$f(x) = a^x$, $a > 1$
$(0, 1)$
$(1, 0)$
$f^{-1}(x) = \log_a x$

Inverse Functions
Domain of $\log_a x$ is Range of a^x.

FIGURE 6.9

DEFINITION OF LOGARITHMIC FUNCTION

For $x > 0$ and $0 < a \neq 1$,

$$y = \log_a x \text{ if and only if } a^y = x.$$

The function given by

$$f(x) = \log_a x$$

is the **logarithmic function with base** a.

REMARK The equations $y = \log_a x$ and $a^y = x$ are equivalent. The first equation is in logarithmic form and the second is in exponential form.

When evaluating logarithms, remember that *a logarithm is an exponent*. This means that $\log_a x$ is the exponent to which a must be raised to obtain x. For instance, $\log_2 8 = 3$ because 2 must be raised to the third power to obtain 8. That is,

$$3 = \log_2 8 \quad \text{if and only if} \quad 2^3 = 8.$$

Base

Logarithm is an exponent

EXAMPLE 1 Evaluating Logarithms

a. $\log_2 32 = 5$ because $2^5 = 32$.

b. $\log_3 27 = 3$ because $3^3 = 27$.

c. $\log_4 2 = \frac{1}{2}$ because $4^{1/2} = \sqrt{4} = 2$.

d. $\log_{10} \frac{1}{100} = -2$ because $10^{-2} = \frac{1}{10^2} = \frac{1}{100}$.

e. $\log_3 1 = 0$ because $3^0 = 1$.

f. $\log_2 2 = 1$ because $2^1 = 2$.

Videotapes

Derive

Complete Solutions Guide

r	0.005	0.010	0.015	0.020	0.025	0.030
t	138.6	69.3	46.2	34.7	27.7	23.1

60. $t = \dfrac{\ln K}{0.095}$

(a)

K	1	2	4	6	8	10	12
t	0	7.3	14.6	18.9	21.9	24.2	26.2

(b)

61. $y = 80.4 - 11 \ln x = 80.4 - 11 \ln 300$

$y \approx 17.66$ cubic feet per minute

62. (a) $\frac{450}{30} = 15$ cubic feet per minute

(b) 380 cubic feet

Total air require $= 380(40) = 15,400$

Study and Solutions Guide

(b) $\log_a a = 1$

(c) $\log_a a^x = x$

■ You should know the definition of the natural logarithmic function.

$\log_e x = \ln x, \quad x > 0$

■ You should know the properties of the natural logarithmic function.

(a) $\ln 1 = 0$

(b) $\ln e = 1$

(c) $\ln e^x = x$

■ You should be able to graph logarithmic functions.

Solutions to Selected Exercises

5. Evaluate $\log_{16} 4$ without using a calculator.

Solution:

$\log_{16} 4 = \log_{16} \sqrt{16} = \log_{16} 16^{1/2} = \frac{1}{2}$

9. Evaluate $\log_{10} 0.01$ without using a calculator.

Solution:

Test Item File and Resource Guide

29. Sketch the graph: $f(x) = 1 + \log_5 x$

1–O–Ans:

30. Sketch the graph: $y = \ln(1 - x)$

2–O–Ans:

31. Students in an algebra class were given an exam and then tested monthly with an equivalent exam. average score for the class was given by the human memory model

$f(t) = 85 - 16 \log_{10}(t + 1), \quad 0 \le t \le 12$

where t is the time in months. What is the average score after 3 months?

(a) 77   (b) 67   (c) 75

(d) 63   (e) None of these

2–M–Ans: c

32. Students in an algebra class were given an exam and then tested monthly with an equivalent exam. average score for the class was given by the human memory model

Tutor

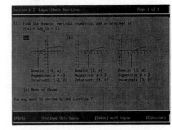

Computerized Testing Software

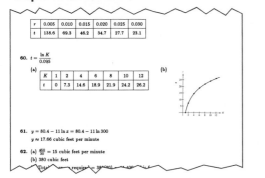

380   **CHAPTER 6**   **Exponential and Logarithmic Functions**

In Exercises 39–44, use the graph of $y = \ln x$ to match the given function to its graph. [The graphs are labeled (a), (b), (c), (d), (e), and (f).]

39. $f(x) = \ln x + 2$ **40.** $f(x) = -\ln x$

41. $f(x) = -\ln(x + 2)$ **42.** $f(x) = \ln(x - 1)$

43. $f(x) = \ln(1 - x)$ **44.** $f(x) = -\ln(-x)$

(a)   (b)

(c)   (d)

(e)   (f)

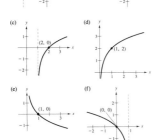

In Exercises 45–56, find the domain, vertical asymptote, and x-intercept of the logarithmic function and sketch its graph.

45. $f(x) = \log_4 x$ **46.** $g(x) = \log_6 x$

47. $h(x) = \log_4(x - 3)$ **48.** $f(x) = -\log_6(x + 2)$

49. $y = -\log_3 x + 2$ **50.** $y = \log_5(x - 1) + 4$

51. $y = \log_{10}\left(\dfrac{x}{5}\right)$ **52.** $y = \log_{10}(-x)$

53. $f(x) = \ln(x - 2)$ **54.** $h(x) = \ln(x + 1)$

55. $g(x) = \ln(-x)$ **56.** $f(x) = \ln(3 - x)$

57. *Human Memory Model*   Students in a mathematics class were given an exam and then tested monthly with an equivalent exam. The average score for the class was given by the human memory model

$f(t) = 80 - 17 \log_{10}(t + 1), \quad 0 \le t \le 12$

where t is the time in months.

(a) What was the average score on the original exam $(t = 0)$?

(b) What was the average score after 4 months?

(c) What was the average score after 10 months?

58. *Population Growth*   The population of a town will double in

$t = \dfrac{10 \ln 2}{\ln 67 - \ln 50}$

years. Find t.

59. *World Population Growth*   The time in years for the world population to double if it is increasing at a continuous rate of r is given by

$t = \dfrac{\ln 2}{r}.$

Complete the table.

r	0.005	0.010	0.015	0.020	0.025	0.030
t						

60. *Investment Time*   A principal P invested at $9\frac{1}{2}\%$ and compounded continuously increases to an amount K times the original principal after t years, where t is given by

$t = \dfrac{\ln K}{0.095}.$

(a) Complete the table.

K	1	2	4	6	8	10	12
t							

(b) Use the table in part (a) to graph this function.

Ventilation Rates   In Exercises 61 and 62, use the model

$y = 80.4 - 11 \ln x$

which approximates the minimum required ventilation rate in terms of the air space per child in a public school classroom. In the model, x is the air space per child in cubic feet and y is the ventilation rate in cubic feet per minute (see figure).

Acknowledgments

We would like to thank the many people who have helped us prepare the text and supplements package. Their encouragement, criticisms, and suggestions have been invaluable to us.

Third Edition Reviewers

Dennis Alber, Palm Beach Community College; Randall Allbritton, Daytona Beach Community College; Paul J. Allen, University of Alabama; Catherine C. Aust, Clayton State College; Judith Barclay, Cuesta College; Derek Bloomfield, Orange County Community College; Gene Clegg, Johnson County Community College; Maria Cossio, LaGuardia Community College; Margaret D. Dolgas, University of Delaware; Gregory Dotseth, University of Northern Iowa; Don A. Edwards, Houston Community College; Iris B. Fetta, Clemson University; Dewey Furness, Ricks College; James A. Gauthier, Louisiana State University; Nancy Henry, Indiana University at Kokomo; Marlene F. Hubbard, University of Arizona; Norma F. James, New Mexico State University; Moana Karsteter, Tallahassee Community College; Gary Ling, City College of San Francisco; Sheila D. McNicholas, University of Illinois at Chicago; Karla Neal, Louisiana State University; Mary Ellen O'Leary, University of South Carolina; Michael Perkowski, University of Missouri; Beverly B. Phillips, Thomas Nelson Community College; Beverly M. Reed, Kent State University; Mary Ellen Rivers, Grand Valley State University; Patricia B. D. Shure, University of Michigan; Burla J. Sims, University of Arkansas at Little Rock; Michael Smith, Auburn University; Theresa Stalder, University of Illinois at Chicago; Greg St. George, University of Montana; Arthur Szylewicz, Moorpark College; Mohan Tikoo, Southeast Missouri State University; Marvel D. Townsend, University of Florida.

Reviewers, First and Second Editions

Hollie Baker, Norfolk State University; Derek Bloomfield, Orange County Community College; Ben P. Bockstage, Broward Community College; Daniel D. Bonar, Denison University; John E. Bruha, University of Northern Iowa; Richard Cutts, University of Wisconsin–Stout; H. Eugene Hall, DeKalb Community College; Randal Hoppens, Blinn College; E. John Hornsby, Jr., University of New Orleans; William B. Jones, University of Colorado; Jimmie D. Lawson, Louisiana State University; Peter J. Livorsi, Oakton Community College; Wade T. Macey, Appalachian State University; Jerome L. Paul, University of Cincinnati; Marilyn Schiermeier, North Carolina State University; George W. Schultz, St. Petersburg Junior College; Edith Silver, Mercer County Community College; Shirley C. Sorensen, University of Maryland; Charles Stone, DeKalb Community College; Bruce Williamson, University of Wisconsin–River Falls.

Survey Respondents, Second Edition

Holli Adams, Portland Community College; Marion Baumler, Niagara County Community College; Diane Blansett, Delta State University; Derek Bloomfield, Orange County Community College; Daniel D. Bonar, Denison University; John E. Bruha, University of Northern Iowa; William L. Campbell, University of Wisconsin–Platteville; John Caraluzzo, Orange County Community College; William E. Chatfield, University of Wisconsin–Platteville; Robert P. Finley, Mississippi State University; August J. Garver, University of Missouri–Rolla; Sue Goodman, University of North Carolina; Louis Hoelzle, Bucks County Community College; Randal Hoppens, Blinn College; Moana Karsteter, Tallahassee Community College; Robert C. Limburg, St. Louis Community College at Florissant Valley; Peter J. Livorsi, Oakton Community College; John Locker, University of North Alabama; Wade T. Macey, Appalachian State University; J. Kent Minichiello, Howard University; Terry Mullen, Carroll College; Richard Nation, Palomar College; William Paul, Appalachian State University; Richard A. Quint, Ventura College; Charles T. Scarborough, Mississippi State University; Shannon Schumann, University of Wyoming; Arthur E. Schwartz, Mercer County Community College; Joseph Sharp , West Georgia College; Burla J. Sims, University of Arkansas at Little Rock; James R. Smith, Appalachian State University; B. Louise Whisler, San Bernardino Valley College; Bruce Williamson, University of Wisconsin–River Falls.

A special thanks to all the people at D. C. Heath and Company who worked with us in the development and production of the text, especially Ann Marie Jones, Mathematics Acquisitions Editor; Cathy Cantin, Senior Developmental Editor; Sarah Doyle, Production Editor; Elizabeth Gale, Editorial Assistant; Sally Steele, Designer and Art Editor; Carolyn Johnson, Editorial Associate; Mike O'Dea, Production Manager; and Lisa Merrill, Production Supervisor.

Several other people worked on this project. David E. Heyd assisted us in writing the text and solved the exercises; Dianna L. Zook wrote the *Study and Solutions Guide*; Meredythe Burrows wrote the *Test Item File* and checked the manuscript for accuracy; and Helen Medley checked the manuscript for accuracy. The following people also worked on the project: Richard J. Bambauer, Linda M. Bollinger, Laurie A. Brooks, Patti Jo Campbell, Jordan C. Feidler, Linda L. Kifer, Deanna G. Larson, Patricia S. Larson, Timothy R. Larson, Amy L. Marshall, John R. Musser, R. Scott O'Neil, Louis R. Rieger, Paula M. Sibeto, and Evelyn A. Wedzikowski.

On a personal level, we are grateful to our wives, Deanna Gilbert Larson and Eloise Hostetler, for their love, patience, and support. Also, a special thanks goes to R. Scott O'Neil.

If you have suggestions for improving the text, please feel free to write to us. Over the past two decades, we have received many useful comments from both instructors and students, and we value these very much.

Roland E. Larson
Robert P. Hostetler

The Larson and Hostetler Precalculus Series

College Algebra, Third Edition

This text is designed for a one-term course covering standard topics such as algebraic functions and their graphs, exponential and logarithmic functions, systems of equations, matrices, determinants, sequences, series, and probability.

Trigonometry, Third Edition

This text is for use in a one-term course covering the trigonometric functions and their graphs, exponential and logarithmic functions, and analytic geometry (including polar coordinates and parametric equations).

Algebra and Trigonometry, Third Edition

This book combines the contents of the two texts mentioned above (with the exception of polar coordinates and parametric equations). It is comprehensive enough for a two-term course or may be used selectively in a one-term course.

Precalculus, Third Edition

With this book, students cover the algebraic, exponential, logarithmic, and trigonometric functions and their graphs, as well as analytic geometry in preparation for a course in calculus. This text may be used in a one- or two-term course.

Also available :

College Algebra: Concepts and Models by Larson, Hostetler, and Munn

College Algebra: A Graphing Approach by Larson, Hostetler, and Edwards

Precalculus: A Graphing Approach by Larson, Hostetler, and Edwards

Algebra and Trigonometry: A Graphing Approach by Larson, Hostetler, and Edwards

Precalculus Functions and Graphs: A Graphing Approach by Larson, Hostetler, and Edwards

Precalculus with Limits: A Graphing Approach by Larson, Hostetler, and Edwards

Contents

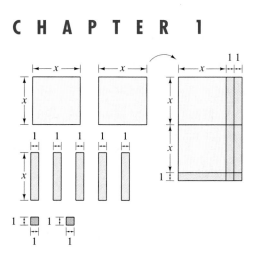
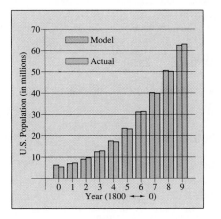

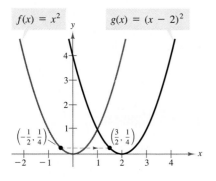

$f(x) = x^2$ $g(x) = (x - 2)^2$

Horizontal shift to the right: 2 units

C H A P T E R 3

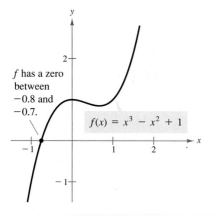

f has a zero between -0.8 and -0.7.

$f(x) = x^3 - x^2 + 1$

C H A P T E R 4

CHAPTER 5

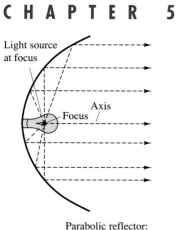

Light source at focus

Axis

Focus

Parabolic reflector:
Light is reflected
in parallel rays.

CHAPTER 6

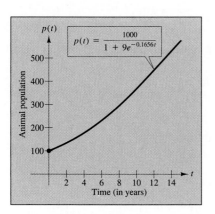

$$p(t) = \frac{1000}{1 + 9e^{-0.1656t}}$$

C H A P T E R   7

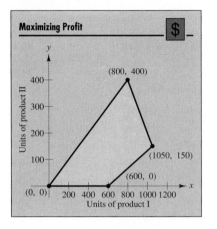

C H A P T E R   8

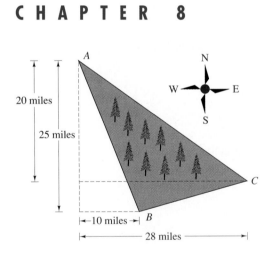

CHAPTER 9

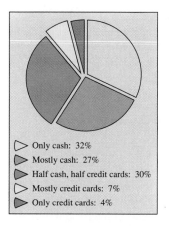

▷ Only cash: 32%
▷ Mostly cash: 27%
▷ Half cash, half credit cards: 30%
▷ Mostly credit cards: 7%
▷ Only credit cards: 4%

Introduction to Calculators

This text includes several examples and exercises that require the use of a calculator, thus enabling you to relate concepts that are closer to the way they are used in the workplace and in real-life applications. As each new calculator application is encountered, it is accompanied with instructions for efficient calculator use. These instructions, however, are general in nature and may not agree precisely with the steps required by your calculator.

Scientific Calculators

For use with this text, a scientific calculator with the following features will be helpful.

1. At least 8-digit display
2. Four arithmetic operations: $+$, $-$, $\times$, $\div$
3. Change sign key or Negation key: $+/-$ or $(-)$
4. Memory key and Recall key: STO , RCL
5. Parentheses: $($, $)$
6. Exponent key: y^x , e^x , $\wedge$
7. Logarithm key: $\ln x$, $\log x$
8. Pi and Degree-Radian conversion: π , DRG
9. Inverse, reciprocal, square root: INV , $1/x$, $\sqrt{2}$

One of the basic differences in calculators is their order of operations. Some calculators use an order of operations called RPN (for Reverse Polish Notation). In this text, however, all calculator steps are given using algebraic order of operations. For example, the calculation

$$4.69[5 + 2(6.87 - 3.042)]$$

can be performed with the following steps.

4.69 $\times$ $($ 5 $+$ 2 $\times$ $($ 6.87 $-$ 3.042 $)$ $)$ $=$

This yields the value of 59.35664. Without parentheses, you could enter the expression from the inside out with the sequence

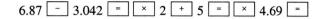

6.87 $\boxed{-}$ 3.042 $\boxed{=}$ $\boxed{\times}$ 2 $\boxed{+}$ 5 $\boxed{=}$ $\boxed{\times}$ 4.69 $\boxed{=}$

to obtain the same result.

Graphing Calculators

A graphing calculator expands the features of a scientific calculator to include graphs of functions and programming. If you have access to such a calculator, you should consult its user's manual to see how to enter and evaluate expressions. You should also practice using its graphing features. To help you become familiar with the graphing features, we have included Appendix A, Graphing Utilities. Be sure to read the examples in Appendix A and work the exercises. Answers to odd-numbered exercises in Appendix A are given in the back of the book.

Rounding Numbers

For all their usefulness, calculators do have a problem representing numbers because they are limited to a finite number of digits. For instance, what does your calculator display when you compute 2 ÷ 3? Some calculators simply truncate (drop) the digits that exceed their display range and display .66666666. Others will round the number and display .66666667. Although the second display is more accurate, both of these decimal representations of 2/3 contain a rounding error.

When rounding decimals, we suggest the following guidelines.

1. Determine the number of digits of accuracy you want to keep. The digit in the last position you keep is the **rounding digit,** and the digit in the first position you discard is the **decision digit.**

2. If the decision digit is 5 or greater, round up by adding 1 to the rounding digit.

3. If the decision digit is 4 or less, round down by leaving the rounding digit unchanged.

Here are some examples. Note that you round down in the first example because the decision digit is 4 or less, and you round up in the other two examples because the decison digit is 5 or greater.

Number	*Rounded to three decimal places*	
a. $\sqrt{2}$ = 1.4142136	1.414	*Round down*
b. π = 3.1415927	3.142	*Round up*
c. $\frac{7}{9}$ = 0.7777777	0.778	*Round up*

One of the best ways to minimize error due to rounding is to leave numbers in your calculator until your calculations are complete. If you want to save a number for future use, store it in your calculator's memory.

Remember that once you (or your calculator) have rounded a number, a round-off error has been introduced. For instance, if a number is rounded to $x \approx 27.3$, then the actual value of x can lie anywhere between 27.25 and 27.35. That is, $27.25 \le x < 27.35$.

Problem Solving Using a Calculator

Here are some guidelines to consider when using any type of calculator in problem solving.

1. Be sure you understand the operation of your own calculator. You need to be skilled at entering expressions in a way that will guarantee that your calculator is performing the operations correctly.

2. Focus first on analyzing the problem. After you have developed a strategy, you may be able to use your calculator to help implement the strategy. Write down your steps in an organized way to clearly outline the strategy used and the results.

3. Most problems can be solved in a variety of ways. If you choose to solve a problem using a table, try checking the solution with an analytic (or algebraic) approach. Or, if you choose to solve a problem using algebra, try checking the solution with a graphing approach.

4. If you have access to a graphing calculator or other graphing utility, you will find many uses for it beginning with Chapter 3.

5. After obtaining a solution with a calculator, be sure to ask yourself if the solution is reasonable (within the context of the problem).

6. To lessen the chance of errors, clear the calculator display (and check the settings) before beginning a new problem.

C H A P T E R 1

OVERVIEW

The rules of algebra reviewed in this chapter are generalizations of arithmetic rules. For instance, the rule for adding fractional expressions is a generalization of the rule for adding numerical fractions.

Rules for operations are an important feature of algebra. The *most* important feature of algebra, however, is that it is a language that can be used to answer questions about real life. For instance, Example 8 on page 34 uses algebra to model the volume of an open box. In the discussion problem that follows Example 8, you are asked to experiment with the model to find the maximum possible volume of the box.

Pay special attention to the terminology and notation used in the chapter. The meaning of such words as *term*, *factor*, and *expression* must be clear.

Review of Fundamental Concepts of Algebra

1.1 The Real Number System

The Set of Real Numbers / The Real Number Line / Ordering the Real Numbers / The Absolute Value of a Real Number / The Distance Between Two Real Numbers

The Set of Real Numbers

Real numbers are used to describe quantities like age, miles per gallon, container size, population, and so on. The set of real numbers contains subsets with which you need to be familiar. For instance, the numbers

$$\ldots, -3, -2, -1, 0, 1, 2, 3, \ldots$$

are **integers.** A real number is **rational** if it can be written as the ratio p/q of two integers, where $q \neq 0$. For instance, the numbers

$$\frac{1}{3} = 0.3333 \ldots, \quad \frac{1}{8} = 0.125, \quad \text{and} \quad \frac{125}{111} = 1.126126 \ldots$$

1

are rational. The decimal representation of a rational number either repeats (as in 3.1454545. . .) or terminates (as in $\frac{1}{2} = 0.5$). Real numbers that cannot be written as the ratio of two integers are **irrational.** For instance, the numbers

$$\sqrt{2} \approx 1.4142136 \quad \text{and} \quad \pi \approx 3.1415927$$

are irrational. (The symbol $\approx$ means "approximately equal to.") The decimal representation of an irrational number neither terminates nor repeats. For example, 3.202002 . . . is an irrational number. Several subsets of real numbers are shown in Figure 1.1.

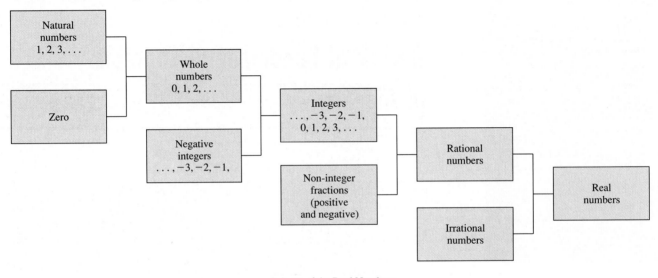

Subsets of the Real Numbers

FIGURE 1.1

The Real Number Line

The **real number line** is a visual model of the set of real numbers. The point that corresponds to 0 is the **origin.** Points to the right of the origin correspond to positive numbers, and points to the left of the origin correspond to negative numbers, as shown in Figure 1.2. The term **nonnegative** describes a number that is either positive or zero.

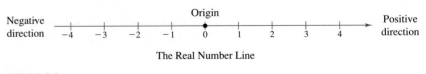

The Real Number Line

FIGURE 1.2

One-to-One Correspondence

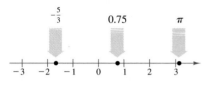

Every real number corresponds to exactly one point on the real number line.

Every point on the real number line corresponds to exactly one real number.

FIGURE 1.3

Each point on the real number line corresponds to one and only one real number and *each real number corresponds to one and only one point on the real number line.* The number associated with a point on the real number line is called the **coordinate** of the point because of the **one-to-one correspondence** between real numbers and points on the real number line. However, it usually isn't necessary to distinguish between a point and its coordinate. (See Figure 1.3.)

Ordering the Real Numbers

One important property of real numbers is that they are **ordered.**

> ### DEFINITION OF ORDER ON THE REAL NUMBER LINE
>
> If a and b are real numbers, then a is **less than** b if $b - a$ is positive. We denote this order by the **inequality**
>
> $$a < b.$$
>
> This relationship can also be described by saying that b is **greater than** a and by writing $b > a$. The inequality $a \leq b$ means that a is **less than or equal to** b and the inequality $b \geq a$ means that b is **greater than or equal to** a. The symbols $<, >, \leq$, and $\geq$ are called **inequality symbols.**

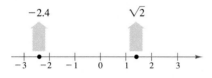

$a < b$ if and only if a lies to the left of b.

FIGURE 1.4

Geometrically, this definition implies that $a < b$ if and only if a lies to the *left* of b on the real number line, as shown in Figure 1.4.

EXAMPLE 1 Interpreting Inequalities

a. The inequality $x \leq 2$ denotes all real numbers less than or equal to 2, as shown in Figure 1.5(a).

b. The inequality $-2 \leq x < 3$ means that $x \geq -2$ *and* $x < 3$. This "double" inequality denotes all real numbers between -2 and 3, including -2 but *not* including 3, as shown in Figure 1.5(b).

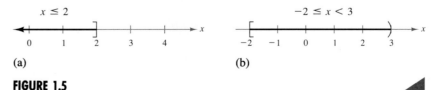

(a)   (b)

FIGURE 1.5

EXAMPLE 2   Using Inequalities to Represent Sets of Real Numbers

Use inequality notation to describe each of the following.

a. c is nonnegative.   **b.** d is negative and greater than -3.

Solution

a. "c is nonnegative" means that c is greater than or equal to zero, which can be written as $c \geq 0$.

b. "d is negative" can be written as $d < 0$, and "d is greater than -3" can be written as $-3 < d$. Combining these two inequalities produces the *double inequality* $-3 < d < 0$. ◢

The **Law of Trichotomy** states that for any two real numbers a and b, *precisely* one of three relationships is possible.

$$a = b, \quad a < b, \quad \text{or} \quad a > b \qquad \qquad \textit{Law of Trichotomy}$$

The Absolute Value of a Real Number

The **absolute value** of a real number is its value, disregarding its sign.

DEFINITION OF ABSOLUTE VALUE

If a is a real number, then the **absolute value** of a is

$$|a| = \begin{cases} a, & \text{if } a \geq 0 \\ -a, & \text{if } a < 0. \end{cases}$$

REMARK   The absolute value of a real number is either positive or zero. Moreover, 0 is the only real number whose absolute value is zero. Thus, $|0| = 0$.

Be sure you see that the absolute value of a real number is never negative. For instance, if $a = -5$, then $|-5| = -(-5) = 5$.

EXAMPLE 3   Evaluating the Absolute Value of a Number

Evaluate the fraction $\dfrac{|x|}{x}$ for (a) $x > 0$ and (b) $x < 0$.

Solution

a. If $x > 0$, then $|x| = x$ and $\dfrac{|x|}{x} = \dfrac{x}{x} = 1$.

b. If $x < 0$, then $|x| = -x$ and $\dfrac{|x|}{x} = \dfrac{-x}{x} = -1$. ◢

Try to formulate verbal descriptions of the following properties—they are easier to remember that way. For instance, the third property states that the absolute value of a product of two numbers is equal to the product of the absolute values of the two numbers.

PROPERTIES OF ABSOLUTE VALUES

Let a and b be real numbers.

1. $|a| \geq 0$
2. $|-a| = |a|$
3. $|ab| = |a|\,|b|$
4. $\left|\dfrac{a}{b}\right| = \dfrac{|a|}{|b|}, \quad b \neq 0$

The Distance Between Two Real Numbers

Absolute value can be used to define the distance between two numbers on the real number line. For instance, the distance between -3 and 4 is $|4 - (-3)| = |7| = 7$, as shown in Figure 1.6.

DISTANCE BETWEEN TWO POINTS ON THE REAL LINE

Let a and b be real numbers. The **distance between a and b** is

$$d(a, b) = |b - a| = |a - b|.$$

The distance between -3 and 4 is 7.

FIGURE 1.6

REMARK From Example 4(a), we can reason that the distance between any real number x and the origin is $|x - 0| = |x|$.

EXAMPLE 4 Distance and Absolute Value

a. The distance between -4 and the origin is

$$d(-4, 0) = |-4 - 0| = |-4| = 4.$$

b. The statement "the distance between c and -2 is at least 7" can be written

$$d(c, -2) = |c - (-2)| = |c + 2| \geq 7.$$

DISCUSSION

PROBLEM

Decimal Approximations of Irrational Numbers

At the beginning of this section we noted that $\sqrt{2}$ is not a rational number. There are, however, rational numbers whose squares are very close to 2. For instance, if you square the rational number

$$\frac{140}{99}$$

you obtain 1.9998. Can you find another rational number whose square is even closer to 2? Write a short paragraph explaining how you obtained this number.

EXERCISES for Section 1.1

In Exercises 1–6, determine which numbers are (a) natural numbers, (b) integers, (c) rational numbers, and (d) irrational numbers.

1. $-9, -\frac{7}{2}, 5, \frac{2}{3}, \sqrt{2}, 0, 1$
2. $\sqrt{5}, -7, -\frac{7}{3}, 0, 3.12, \frac{5}{4}$
3. $2.01, 0.666\ldots, -13, 0.010110111\ldots$
4. $2.3030030003\ldots, 0.7575, -4.63, \sqrt{10}$
* 5. $-\pi, -\frac{1}{3}, \frac{6}{3}, \frac{1}{2}\sqrt{2}, -7.5$
6. $25, -17, -\frac{12}{5}, \sqrt{9}, 3.12, \frac{1}{2}\pi$

In Exercises 7–12, plot the real numbers on the real number line. Then place the appropriate inequality sign ($<$ or $>$) between them.

7. $\frac{3}{2}, 7$
8. $-3.5, 1$
9. $-4, -8$
10. $1, \frac{16}{3}$
11. $\frac{5}{6}, \frac{2}{3}$
12. $-\frac{8}{7}, -\frac{3}{7}$

In Exercises 13–22, verbally describe the subset of real numbers represented by the inequality. Then sketch the subset on the real number line.

13. $x \le 5$
14. $x \ge -2$
15. $x < 0$
16. $x > 3$
17. $x \ge 4$
18. $x < 2$
19. $-2 < x < 2$
20. $0 \le x \le 5$
21. $-1 \le x < 0$
22. $0 < x \le 6$

In Exercises 23–30, use inequality notation to describe the set.

23. x is negative.
24. z is at least 10.
25. y is no more than 25.
26. y is greater than 5 and less than or equal to 12.
27. The person's age, A, is at least 30.
28. The yield, Y, is no more than 45 bushels per acre.
29. The annual rate of inflation, r, is expected to be at least 3.5%, but no more than 6%.
30. The price, p, of unleaded gasoline is not expected to go above \$1.35 per gallon during the coming year.

*A boxed number indicates that a detailed solution can be found in the *Study and Solutions Guide*.

In Exercises 31–40, evaluate the expression.

31. $|-10|$
32. $|0|$
33. $|3 - \pi|$
34. $|4 - \pi|$
35. $\dfrac{-5}{|-5|}$
36. $-3 - |-3|$
37. $-3|-3|$
38. $|-1| - |-2|$
39. $-|16.25| + 20$
40. $2|33|$

In Exercises 41–46, place the correct symbol ($<$, $>$, or $=$) between the numbers.

41. $|-3| \; \blacksquare \; -|-3|$
42. $|-4| \; \blacksquare \; |4|$
43. $-5 \; \blacksquare \; -|5|$
44. $-|-6| \; \blacksquare \; |-6|$
45. $-|-2| \; \blacksquare \; -|2|$
46. $-(-2) \; \blacksquare \; -2$

In Exercises 47–54, find the distance between a and b.

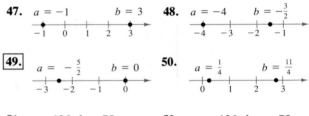

47. $a = -1$, $b = 3$
48. $a = -4$, $b = -\frac{3}{2}$
49. $a = -\frac{5}{2}$, $b = 0$
50. $a = \frac{1}{4}$, $b = \frac{11}{4}$

51. $a = 126, b = 75$
52. $a = -126, b = -75$
53. $a = 9.34, b = -5.65$
54. $a = \frac{16}{5}, b = \frac{112}{75}$

In Exercises 55–60, use absolute value notation to describe the situation.

55. The distance between x and 5 is no more than 3.
56. The distance between x and -10 is at least 6.
57. While traveling, you remember passing milepost 7, then milepost 18. How far did you travel during that time period?
58. While traveling, you remember passing milepost 103, then milepost 86. How far did you travel during that time period?
59. y is at least six units from 0.
60. y is at most two units from a.

Budget Variance In Exercises 61–64, the accounting department of a company is checking to see whether the actual expenses of a department differ from the budgeted expenses by more than \$500 or by more than 5%. Complete the missing parts of the table, and determine whether the actual expense passes the "budget variance test."

		Budgeted Expense, b	Actual Expense, a	$\lvert a - b \rvert$	0.05b
61.	Wages	$112,700.00	$113,356.52	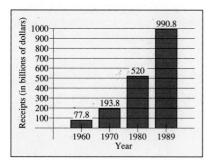	
62.	Utilities	$9,400.00	$9,772.59		
63.	Taxes	$37,640.00	$37,335.80		
64.	Insurance	$2,575.00	$2,613.15		

Federal Deficit In Exercises 65–68, the bar graph shows the receipts of the federal government (in billions of dollars) for selected years from 1960 through 1989. In each exercise you are given the expenses of the federal government. Find the absolute value of the surplus or deficit for the year. (*Source:* U.S. Treasury Department)

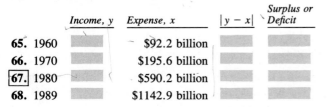

Figure for 65–68

		Income, y	Expense, x	$\lvert y - x \rvert$	Surplus or Deficit
65.	1960		$92.2 billion		
66.	1970		$195.6 billion		
67.	1980		$590.2 billion		
68.	1989		$1142.9 billion		

In Exercises 69 and 70, use a calculator to order the numbers from smallest to largest.

69. $\frac{7071}{5000}$, $\frac{584}{413}$, $\sqrt{2}$, $\frac{47}{33}$, $\frac{127}{90}$

70. $\frac{26}{15}$, $\sqrt{3}$, 1.7320, $\frac{381}{220}$, $\sqrt{10} - \sqrt{2}$

In Exercises 71–74, use a calculator to find the decimal form of the rational number. If it is a nonterminating decimal, write the repeating pattern.

71. $\frac{5}{8}$ **72.** $\frac{1}{3}$

73. $\frac{41}{333}$ **74.** $\frac{6}{11}$

75. Use a calculator to complete the table. Observe that $5/n$ increases without bound as n approaches zero.

n	10	1	0.5	0.01	0.0001	0.000001
$5/n$						

In Exercises 76–80, determine whether the statement is true or false.

76. The reciprocal of a nonzero integer is an integer.

77. The reciprocal of a nonzero rational number is rational.

78. Every integer is rational.

79. Every real number is either rational or irrational.

80. The absolute value of a real number is positive.

1.2 The Basic Rules of Algebra

Algebraic Expressions / Basic Rules of Algebra / Equations / Exponents / Scientific Notation

Algebraic Expressions

One characteristic of algebra is the use of letters (or combinations of letters) to represent numbers. The letters are **variables,** and combinations of letters and numbers are **algebraic expressions.** Here are a few examples.

$$5x, \qquad 2x - 3, \qquad \frac{4}{x^2 + 2}, \qquad 7x + y$$

> ### DEFINITION OF AN ALGEBRAIC EXPRESSION
>
> A collection of letters (**variables**) and real numbers (**constants**) combined using the operations of addition, subtraction, multiplication, division, and exponentiation is an **algebraic expression.**

The **terms** of an algebraic expression are those parts that are separated by *addition*. For example,

$$x^2 - 5x + 8 = x^2 + (-5x) + 8$$

has three terms: x^2 and $-5x$ are the **variable terms** and 8 is the **constant term.** The numerical factor of a variable term is the **coefficient** of the variable term. For instance, the coefficient of $-5x$ is -5 and the coefficient of x^2 is 1.

To **evaluate** an algebraic expression, substitute numerical values for each of the variables in the expression. Here are two examples.

Expression	Value of Variable	Substitute	Value of Expression
$-3x + 5$	$x = 3$	$-3(3) + 5$	$-9 + 5 = -4$
$3x^2 + 2x - 1$	$x = -1$	$3(-1)^2 + 2(-1) - 1$	$3 - 2 - 1 = 0$

Basic Rules of Algebra

There are four arithmetic operations with real numbers: **addition, multiplication, subtraction,** and **division,** denoted by the symbols $+$, $\times$ or $\cdot$, $-$, and $\div$. Of these, addition and multiplication are the two primary operations. Subtraction and division are the inverse operations of addition and multiplication.

Subtraction	Division
$a - b = a + (-b)$	If $b \neq 0$, then $a \div b = a\left(\dfrac{1}{b}\right) = \dfrac{a}{b}$.

In these definitions, $-b$ is the **additive inverse** (or opposite) of b, and $1/b$ is the **multiplicative inverse** (or reciprocal) of b. In place of $a \div b$, we often use the fraction symbol a/b. In this fractional form, a is the **numerator** of the fraction and b is the **denominator.**

Be sure you see that the following properties, which we call the **basic rules of algebra,** are true for variables and algebraic expressions as well as for real numbers.

BASIC RULES OF ALGEBRA

Let a, b, and c be *real numbers, variables,* or *algebraic expressions.*

Property	Example
Commutative Property of Addition: $a + b = b + a$	$4x + x^2 = x^2 + 4x$
Commutative Property of Multiplication: $ab = ba$	$(4 - x)x^2 = x^2(4 - x)$
Associative Property of Addition: $(a + b) + c = a + (b + c)$	$(-x + 5) + 2x^2 = -x + (5 + 2x^2)$
Associative Property of Multiplication: $(ab)c = a(bc)$	$(2x \cdot 3y)(8) = (2x)(3y \cdot 8)$
Distributive Property: $a(b + c) = ab + ac$ and $(a + b)c = ac + bc$	$3x(5 + 2x) = 3x \cdot 5 + 3x \cdot 2x$ and $(y + 8)y = y \cdot y + 8 \cdot y$
Additive Identity Property: $a + 0 = a$	$5y^2 + 0 = 5y^2$
Multiplicative Identity Property: $a \cdot 1 = 1 \cdot a = a$	$(4x^2)(1) = (1)(4x^2) = 4x^2$
Additive Inverse Property: $a + (-a) = 0$	$5x^3 + (-5x^3) = 0$
Multiplicative Inverse Property: $a \cdot \dfrac{1}{a} = 1, \quad a \neq 0$	$(x^2 + 4)\left(\dfrac{1}{x^2 + 4}\right) = 1$

REMARK Since subtraction is defined as "adding the opposite," the Distributive Properties are also true for subtraction. For instance, the "subtraction form" of $a(b + c) = ab + ac$ is

$$a(b - c) = a[b + (-c)] = ab + a(-c) = ab - ac. \quad \blacktriangleleft$$

The following are summaries of the basic properties of negation, zero, and fractions. We suggest that you not only learn a verbal description of each property, but that you also try to gain an *intuitive feeling* for the validity of each.

PROPERTIES OF NEGATION

Let a and b be real numbers, variables, or algebraic expressions.

Property	Example
1. $(-1)a = -a$	$(-1)7 = -7$
2. $-(-a) = a$	$-(-6) = 6$
3. $(-a)b = -(ab) = a(-b)$	$(-5)3 = -(5 \cdot 3) = 5(-3)$
4. $(-a)(-b) = ab$	$(-2)(-6) = 12$
5. $-(a + b) = (-a) + (-b)$	$-(3 + 8) = (-3) + (-8)$

Be sure you see the difference between the *opposite of a number* and a *negative number.* If a is already negative, then its opposite, $-a$, is positive. For instance, if $a = -5$, then $-a = -(-5) = 5$.

PROPERTIES OF ZERO

Let a and b be real numbers, variables, or algebraic expressions.

1. $a + 0 = a$ and $a - 0 = a$
2. $a \cdot 0 = 0$
3. $\dfrac{0}{a} = 0$, $a \neq 0$

4. $\dfrac{a}{0}$ is undefined.
5. Zero-Factor Property:
   If $ab = 0$, then $a = 0$ or $b = 0$.

The "or" in the Zero-Factor Property includes the possibilities that either or both factors may be zero. This is an **inclusive or**, and it is the way the word "or" is generally used in mathematics.

PROPERTIES OF FRACTIONS

Let a, b, c, and d be real numbers, variables, or algebraic expressions such that $b \neq 0$ and $d \neq 0$.

1. Equivalent Fractions:
   $\dfrac{a}{b} = \dfrac{c}{d}$ if and only if $ad = bc$.

2. Rules of Signs: $-\dfrac{a}{b} = \dfrac{-a}{b} = \dfrac{a}{-b}$ and $\dfrac{-a}{-b} = \dfrac{a}{b}$

3. Generate Equivalent Fractions: $\dfrac{a}{b} = \dfrac{ac}{bc}$, $c \neq 0$

4. Add or Subtract with Like Denominators:
   $\dfrac{a}{b} \pm \dfrac{c}{b} = \dfrac{a \pm c}{b}$

5. Add or Subtract with Unlike Denominators:
   $\dfrac{a}{b} \pm \dfrac{c}{d} = \dfrac{ad \pm bc}{bd}$

6. Multiply Fractions: $\dfrac{a}{b} \cdot \dfrac{c}{d} = \dfrac{ac}{bd}$

7. Divide Fractions:
   $\dfrac{a}{b} \div \dfrac{c}{d} = \dfrac{a}{b} \cdot \dfrac{d}{c} = \dfrac{ad}{bc}$, $c \neq 0$

In Property 1 (equivalent fractions) the phrase "if and only if" implies two statements. One statement is: If $a/b = c/d$, then $ad = bc$. The other statement is: If $ad = bc$, where $b \neq 0$ and $d \neq 0$, then $a/b = c/d$.

EXAMPLE 1   Properties of Zero and Properties of Fractions

a. $x - \dfrac{0}{5} = x - 0 = x$ *Properties 3 and 1 of zero*

b. $\dfrac{x}{5} = \dfrac{3 \cdot x}{3 \cdot 5} = \dfrac{3x}{15}$ *Generate equivalent fractions*

c. $\dfrac{x}{3} + \dfrac{2x}{5} = \dfrac{5 \cdot x + 3 \cdot 2x}{15}$ *Add fractions with unlike denominators*

d. $\dfrac{7}{x} \div \dfrac{3}{2} = \dfrac{7}{x} \cdot \dfrac{2}{3} = \dfrac{14}{3x}$ *Divide fractions* ◢

If a, b, and c are integers such that $ab = c$, then a and b are **factors** or **divisors** of c. For example, 2 and 3 are factors of 6. A **prime number** is a positive integer that has exactly two factors: itself and 1. For example, 2, 3, 5, 7, and 11 are prime numbers. The numbers 4, 6, 8, 9, and 10 are **composite** because they can be written as the product of two or more prime numbers. The number 1 is neither prime nor composite. The **Fundamental Theorem of Arithmetic** states that every positive integer greater than 1 can be written as the product of prime numbers in precisely one way (disregarding order). For instance, the *prime factorization* of 24 is $24 = 2 \cdot 2 \cdot 2 \cdot 3$.

When adding or subtracting fractions with unlike denominators, you have two options. You could use Property 5 of fractions as in Example 1(c). Alternatively, you could use Property 4 of fractions by rewriting both fractions so that they have the same denominator. We call this the **least common denominator** (LCD) method. For adding or subtracting *two* fractions, Property 5 is often more convenient. For *three or more* fractions, the LCD method is usually preferred.

EXAMPLE 2   The LCD Method of Adding or Subtracting Fractions

Evaluate the following.

$$\frac{2}{15} - \frac{5}{9} + \frac{4}{5}$$

Solution

By prime factoring the denominators ($15 = 3 \cdot 5, 9 = 3 \cdot 3$, and $5 = 5$) you can see that the least common denominator is $3 \cdot 3 \cdot 5 = 45$.

$$\frac{2}{15} - \frac{5}{9} + \frac{4}{5} = \frac{2(3)}{15(3)} - \frac{5(5)}{9(5)} + \frac{4(9)}{5(9)} = \frac{6 - 25 + 36}{45} = \frac{17}{45}$$

◢

Equations

An **equation** is a statement of equality between two expressions. Thus, the statement

$$a + b = c + d$$

means that the expressions $a + b$ and $c + d$ represent the same number.

PROPERTIES OF EQUALITY

Let a, b, and c be real numbers, variables, or algebraic expressions.

1. Reflexive: $a = a$
2. Symmetric: If $a = b$, then $b = a$.
3. Transitive: If $a = b$ and $b = c$, then $a = c$.
4. Substitution Principle: If $a = b$, then a can be replaced by b in any expression involving a.

Two important consequences of the Substitution Principle are the following rules.

1. If $a = b$, then $a + c = b + c$.   *Add c to both sides*
2. If $a = b$, then $ac = bc$.   *Multiply both sides by c*

The first rule allows you to add the same number to both sides of an equation. The second rule allows you to multiply both sides of an equation by the same number. The converses of these two rules are the **Cancellation Laws** for addition and multiplication.

1. If $a + c = b + c$, then $a = b$.   *Subtract c from both sides*
2. If $ac = bc$ and $c \neq 0$, then $a = b$.   *Divide both sides by c*

Exponents

Repeated *multiplications* can be written in **exponential form.**

Repeated Multiplication	Exponential Form
$7 \cdot 7$	7^2
$a \cdot a \cdot a \cdot a \cdot a$	a^5
$(-4)(-4)(-4)$	$(-4)^3$
$(2x)(2x)(2x)(2x)$	$(2x)^4$

EXPONENTIAL NOTATION

Let a be a real number, variable, or algebraic expression, and let n be a positive integer. Then

$$a^n = \underbrace{a \cdot a \cdot a \cdots a}_{n \text{ factors}}$$

where n is the **exponent** and a is the **base.** The expression a^n is read "a to the nth **power.**"

It is important to recognize the différence between expressions such as $(-2)^4$ and -2^4. In $(-2)^4$, the parentheses indicate that the exponent applies to the negative sign as well as to the 2, but in $-2^4 = -(2^4)$, the exponent applies only to the 2.

When multiplying exponential expressions with the same base, you *add* exponents.

$$a^m \cdot a^n = a^{m+n} \qquad \textit{Add exponents when multiplying}$$

When dividing exponential expressions, you *subtract* exponents. That is,

$$\frac{a^m}{a^n} = a^{m-n}, \qquad a \neq 0. \quad \textit{Subtract exponents when dividing}$$

There are two special cases involving division of exponential expressions. If $m = n$, then

$$\frac{a^n}{a^n} = a^{n-n} = a^0 = 1, \qquad a \neq 0$$

and we say that *any nonzero number raised to the zero power is 1*. If n is a positive integer, then

$$\frac{1}{a^n} = a^{-n}, \qquad a \neq 0.$$

PROPERTIES OF EXPONENTS

Let a and b be real numbers, variables, or algebraic expressions, and let m and n be integers. (Assume all denominators and bases are nonzero.)

Property	*Example*
1. $a^m a^n = a^{m+n}$	$3^2 \cdot 3^4 = 3^{2+4} = 3^6$
2. $\dfrac{a^m}{a^n} = a^{m-n}$	$\dfrac{x^7}{x^4} = x^{7-4} = x^3$
3. $a^{-n} = \dfrac{1}{a^n} = \left(\dfrac{1}{a}\right)^n$	$y^{-4} = \dfrac{1}{y^4} = \left(\dfrac{1}{y}\right)^4$
4. $a^0 = 1, \quad a \neq 0$	$(x^2 + 1)^0 = 1$
5. $(ab)^m = a^m b^m$	$(5x)^3 = 5^3 x^3 = 125 x^3$
6. $(a^m)^n = a^{mn}$	$(y^3)^{-4} = y^{3(-4)} = y^{-12} = \dfrac{1}{y^{12}}$
7. $\left(\dfrac{a}{b}\right)^m = \dfrac{a^m}{b^m}$	$\left(\dfrac{2}{x}\right)^3 = \dfrac{2^3}{x^3} = \dfrac{8}{x^3}$
8. $\lvert a^2 \rvert = \lvert a \rvert^2 = a^2$	$\lvert (-2)^2 \rvert = \lvert -2 \rvert^2 = (-2)^2 = 4$

These properties of exponents apply for *all* integers *m* and *n*, not just positive ones. For instance, by Property 2,

$$\frac{3^4}{3^{-5}} = 3^{4-(-5)} = 3^{4+5} = 3^9.$$

EXAMPLE 3   Using Properties of Exponents

a. $(-3ab^4)(4ab^{-3}) = -12(a)(a)(b^4)(b^{-3}) = -12a^2b$

b. $(2xy^2)^3 = 2^3(x)^3(y^2)^3 = 8x^3y^6$ *Apply exponent to coefficient*

c. $3a(-4a^2)^0 = 3a(1) = 3a, \quad a \neq 0$

d. $\left(\dfrac{5x^3}{y}\right)^2 = \dfrac{5^2(x^3)^2}{y^2} = \dfrac{25x^6}{y^2}$ *Apply exponent to coefficient*

EXAMPLE 4   Rewriting with Positive Exponents

a. $x^{-1} = \dfrac{1}{x}$ *Property 3: $a^{-n} = \dfrac{1}{a^n}$*

b. $\dfrac{1}{3x^{-2}} = \dfrac{1(x^2)}{3} = \dfrac{x^2}{3}$ *−2 exponent does not apply to 3*

c. $\dfrac{12a^3b^{-4}}{4a^{-2}b} = \dfrac{12a^3 \cdot a^2}{4b \cdot b^4} = \dfrac{3a^5}{b^5}$

d. $\left(\dfrac{3x^2}{y}\right)^{-2} = \dfrac{3^{-2}(x^2)^{-2}}{y^{-2}} = \dfrac{3^{-2}x^{-4}}{y^{-2}} = \dfrac{y^2}{3^2x^4} = \dfrac{y^2}{9x^4}$

Rarely in algebra is there only one way to solve a problem. Don't be concerned if the steps you use to solve a problem are not exactly the same as the steps presented in this text. The important thing is to use steps that you understand *and*, of course, are justified by the rules of algebra. For instance, you might prefer the following steps for Example 4(d).

$$\left(\frac{3x^2}{y}\right)^{-2} = \left(\frac{y}{3x^2}\right)^2 = \frac{y^2}{9x^4}$$

Note how Property 3 is used in the first step of this solution. The fraction form of Property 3 is

$$\left(\frac{a}{b}\right)^{-m} = \left(\frac{b}{a}\right)^{m}.$$

EXAMPLE 5 **Using a Calculator to Raise a Number to a Power**

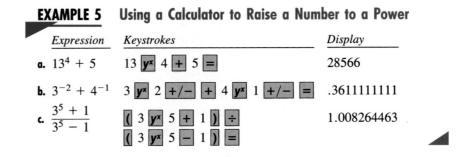

Expression	Keystrokes	Display
a. $13^4 + 5$	13 y^x 4 $+$ 5 $=$	28566
b. $3^{-2} + 4^{-1}$	3 y^x 2 $+/-$ $+$ 4 y^x 1 $+/-$ $=$	.3611111111
c. $\dfrac{3^5 + 1}{3^5 - 1}$	(3 y^x 5 $+$ 1) $\div$ (3 y^x 5 $-$ 1) $=$	1.008264463

Scientific Notation

Exponents provide an efficient way of writing and computing with very large (or very small) numbers. For instance, a drop of water contains more than 33 billion billion molecules—that is, 33 followed by 18 zeros.

$$33,000,000,000,000,000,000$$

It is convenient to write such numbers in **scientific notation.** This notation has the form $c \times 10^n$, where $1 \le c < 10$ and n is an integer. Thus, the number of molecules in a drop of water can be written in scientific notation as

$$3.3 \times 10,000,000,000,000,000,000 = 3.3 \times 10^{19}.$$

The *positive* exponent 19 indicates that the number is large (10 or more) and that the decimal point has been moved 19 places. A *negative* exponent in scientific notation indicates that the number is *small* (less than 1). For instance, the mass (in grams) of one electron is approximately

$$9.0 \times 10^{-28} = 0.0000000000000000000000000009.$$

28 decimal places

EXAMPLE 6 Scientific Notation

a. $1.345 \times 10^2 = 134.5$

Two places

b. $0.0000782 = 7.82 \times 10^{-5}$

Five places

c. $9.36 \times 10^{-6} = 0.00000936$

Six places

d. $836,100,000.0 = 8.361 \times 10^8$

Eight places

Most scientific calculators automatically switch to scientific notation when they are showing large (or small) numbers that exceed the display range. Try multiplying $86,500,000 \times 6,000$. If your calculator follows standard conventions, its display should be 5.19 11 or 5.19 E 11 . This means that $c = 5.19$ and the exponent of 10 is $n = 11$, which implies that the number is 5.19×10^{11}.

EXAMPLE 7   Using Scientific Notation with a Calculator

Use a calculator to evaluate $65{,}000 \times 3{,}400{,}000{,}000$.

Solution

Since $65{,}000 = 6.5 \times 10^4$ and $3{,}400{,}000{,}000 = 3.4 \times 10^9$, you can multiply the two numbers using the following calculator steps.

6.5 | EE | 4 | × | 3.4 | EE | 9 | = | Display: | 2.21   E   14 |

The product of the two numbers is

$$(6.5 \times 10^4)(3.4 \times 10^9) = 2.21 \times 10^{14} = 221{,}000{,}000{,}000{,}000.$$

DISCUSSION
PROBLEM
Calculators and
Order of
Operations

When entering expressions into a calculator, you must be aware of the calculator's built-in order of operations. To override the built-in order of operations, you can insert parentheses in appropriate places. For instance, if you want to evaluate

$$\frac{3+5}{2}$$

you *cannot* use the following keystroke sequence.

3 | + | 5 | ÷ | 2 | = |

Which arithmetic expression does this keystroke sequence represent? Write a keystroke sequence for each of the following expressions.

1. $\dfrac{3+5}{2}$ 2. $\dfrac{3(3+5)}{2}$ 3. $3(3) + \dfrac{5}{2}$

WARM UP

The following warm-up exercises involve skills that were covered in earlier sections. You will use these skills in the exercise set for this section.

In Exercises 1–4, place the correct inequality symbol ($<$ or $>$) between the numbers.

1. -4 ▧ -2 2. 0 ▧ -3

3. $\sqrt{3}$ ▧ 1.73 4. $-\pi$ ▧ -3

In Exercises 5–8, find the distance between the numbers.

5. $4, 6$ 6. $-2, 2$

7. $0, -5$ 8. $-1, 3$

In Exercises 9 and 10, evaluate the expression.

9. $|-7| + |7|$ 10. $-|8 - 10|$

EXERCISES for Section 1.2

In Exercises 1–6, identify the terms of the expression.

1. $7x + 4$

2. $-5 + 3x$

3. $x^2 - 4x + 8$

4. $3x^2 - 8x - 11$

5. $4x^3 + x - 5$

6. $3x^4 + 3x^3$

In Exercises 7–12, evaluate the expression for the values of x. (If not possible, state the reason.)

Expression	Values	
7. $4x - 6$	(a) $x = -1$	(b) $x = 0$
8. $9 - 7x$	(a) $x = -3$	(b) $x = 3$
9. $x^2 - 3x + 4$	(a) $x = -2$	(b) $x = 2$
10. $-x^2 + 5x - 4$	(a) $x = -1$	(b) $x = 1$
11. $\dfrac{x + 1}{x - 1}$	(a) $x = 1$	(b) $x = -1$
12. $\dfrac{x}{x + 2}$	(a) $x = 2$	(b) $x = -2$

In Exercises 13–22, identify the rule(s) of algebra illustrated by the equation.

13. $x + 9 = 9 + x$

14. $(x + 3) - (x + 3) = 0$

15. $\dfrac{1}{h + 6}(h + 6) = 1, \quad h \neq -6$

16. $2\left(\frac{1}{2}\right) = 1$

17. $2(x + 3) = 2x + 6$

18. $(z - 2) + 0 = z - 2$

19. $1 \cdot (1 + x) = 1 + x$

20. $x + (y + 10) = (x + y) + 10$

21. $x(3y) = (x \cdot 3)y = (3x)y$

22. $\frac{1}{7}(7 \cdot 12) = \left(\frac{1}{7} \cdot 7\right)12 = 1 \cdot 12 = 12$

In Exercises 23–26, evaluate the expression. (If not possible, state the reason.)

23. $\dfrac{81 - (90 - 9)}{5}$

24. $10(23 - 30 + 7)$

25. $\dfrac{8 - 8}{-9 + (6 + 3)}$

26. $15 - \dfrac{3 - 3}{5}$

In Exercises 27–36, perform the indicated operations. (Write fractional answers in reduced form.)

27. $(4 - 7)(-2)$

28. $\dfrac{27 - 35}{4}$

29. $\frac{3}{16} + \frac{5}{16}$

30. $\frac{6}{7} - \frac{4}{7}$

31. $\frac{5}{8} - \frac{5}{12} + \frac{1}{6}$

32. $\frac{10}{11} + \frac{6}{33} - \frac{13}{66}$

33. $\frac{4}{5} \cdot \frac{1}{2} \cdot \frac{3}{4}$

34. $\frac{11}{16} \div \frac{3}{4}$

35. $12 \div \frac{1}{4}$

36. $\left(\frac{3}{5} \div 3\right) - \left(6 \cdot \frac{4}{8}\right)$

In Exercises 37–40, use a calculator to evaluate the expression. (Round your answer to two decimal places.)

37. $-3 + \frac{3}{7}$

38. $3\left(-\frac{5}{12} + \frac{3}{8}\right)$

39. $\dfrac{11.46 - 5.37}{3.91}$

40. $\dfrac{(1/5)(-8 - 9)}{-1/3}$

In Exercises 41–48, evaluate the expression.

41. $\dfrac{5^5}{5^2}$

42. $3 \cdot 3^3$

43. $(3^3)^2$

44. -3^2

45. $(2^3 \cdot 3^2)^2$

46. $\left(-\frac{3}{5}\right)^3\left(\frac{5}{3}\right)^2$

47. $\dfrac{4 \cdot 3^{-2}}{2^{-2} \cdot 3^{-1}}$

48. $(-2)^0$

In Exercises 49–52, evaluate the expression for the given value of x.

Expression	Value
49. $-3x^3$	2
50. $7x^{-2}$	4
51. $6x^0 - (6x)^0$	10
52. $5(-x)^3$	3

In Exercises 53–80, simplify the expression.

53. $(-5z)^3$

54. $(3x)^2$

55. $5x^4(x^2)$

56. $(4x^3)^2$

57. $6y^2(2y^4)^2$

58. $(-z)^3(3z^4)$

59. $\dfrac{3x^5}{x^3}$

60. $\dfrac{25y^8}{10y^4}$

61. $\dfrac{7x^2}{x^3}$

62. $\dfrac{r^4}{r^6}$

63. $\dfrac{12(x + y)^3}{9(x + y)}$

64. $\left(\dfrac{4}{y}\right)^3 \left(\dfrac{3}{y}\right)^4$

65. $(x + 5)^0$, $x \neq -5$

66. $(2x^5)^0$, $x \neq 0$

67. $(2x^2)^{-2}$

68. $(z + 2)^{-3}(z + 2)^{-1}$

69. $(-2x^2)^3(4x^3)^{-1}$

70. $(4y^{-2})(8y^4)$

71. $\left(\dfrac{x}{10}\right)^{-1}$

72. $\left(\dfrac{x^{-3}y^4}{5}\right)^{-3}$

73. $(4a^{-2}b^3)^{-3}$

74. $[(x^2y^{-2})^{-1}]^{-1}$

75. $\left(\dfrac{5x^2}{y^{-2}}\right)^{-4}$

76. $(5x^2y^4z^6)^3(5x^2y^4z^6)^{-3}$

77. $3^n \cdot 3^{2n}$

78. $\dfrac{x^2 \cdot x^n}{x^3 \cdot x^n}$

79. $\left(\dfrac{a^{-2}}{b^{-2}}\right)\left(\dfrac{b}{a}\right)^3$

80. $\left(\dfrac{a^{-3}}{b^{-3}}\right)\left(\dfrac{a}{b}\right)^3$

In Exercises 81–84, write the number in scientific notation.

81. Land Area of Earth: 57,500,000 square miles

82. Light Year: 9,461,000,000,000,000 kilometers

83. Relative Density of Hydrogen: 0.0000899 gram per cm³

84. One Micron (Millionth of Meter): 0.00003937 inch

In Exercises 85–88, write the number in decimal form.

85. U.S. Daily Coca-Cola Consumption: 5.24×10^8 servings

86. Interior Temperature of Sun: 1.3×10^7 degrees Celsius

87. Charge of Electron: 4.8×10^{-10} electrostatic units

88. Width of Human Hair: 9.0×10^{-4} meters

In Exercises 89 and 90, use a calculator to evaluate the expression. (Round to three decimal places.)

89. (a) $750\left(1 + \dfrac{0.11}{365}\right)^{800}$

   (b) $\dfrac{67,000,000 + 93,000,000}{0.0052}$

90. (a) $(9.3 \times 10^6)^3(6.1 \times 10^{-4})^4$

   (b) $\dfrac{(2.414 \times 10^4)^6}{(1.68 \times 10^5)^5}$

91. *Speed of Light*   The speed of light is 11,160,000 miles per minute. The distance from the sun to the earth is 93,000,000 miles. Find the time for light to travel from the sun to the earth.

92. *Balance in an Account*   The balance A after t years in an account earning an annual interest rate of r compounded n times per year is

$$A = P\left(1 + \dfrac{r}{n}\right)^{nt}$$

where P is the original deposit. Complete the table for $500 deposited in an account earning 12% compounded daily. (Note that $r = 0.12$ implies an interest rate of 12%.)

Number of Years	5	10	20	30	40	50
Balance						

93. *Highway Fatalities*   There were 45,555 fatal accidents in the United States in 1989. Find the number for each of the categories indicated in the figure. (*Source:* National Highway Traffic Safety Administration)

94. *Municipal Landfills*   There were 179.6 million tons of municipal waste generated in 1988. Find the number of tons for each of the categories indicated in the figure. (*Source:* U.S. Environmental Protection Agency)

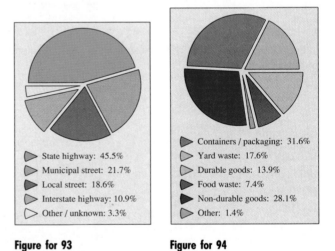

State highway: 45.5%
Municipal street: 21.7%
Local street: 18.6%
Interstate highway: 10.9%
Other / unknown: 3.3%

Containers / packaging: 31.6%
Yard waste: 17.6%
Durable goods: 13.9%
Food waste: 7.4%
Non-durable goods: 28.1%
Other: 1.4%

Figure for 93   **Figure for 94**

95. *Calculator Keystrokes*   Write the expression that corresponds to the keystrokes.

5 × (2.7 − 9.4) =

96. *Calculator Keystrokes*   Write the expression that corresponds to the keystrokes.

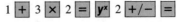

1 + 3 × 2 = y^x 2 +/− =

1.3 Radicals and Rational Exponents

Radicals and Properties of Radicals / Simplifying Radicals /
Rationalizing Denominators and Numerators / Rational Exponents / Radicals and Calculators

Radicals and Properties of Radicals

You already know how to square a number—raise it to the second power by using the number *twice* as a factor. For instance, 5 squared is $5 \cdot 5 = 5^2 = 25$. Conversely, a **square root of a number** is one of its two equal factors. For example, 5 is a square root of 25 because 5 is one of the two equal factors of 25. In a similar way, a **cube root** of a number is one of its three equal factors. Consider the following examples.

Number	Equal Factors	Root
$25 = (-5)^2$	$(-5)(-5)$	-5 (square root)
$-64 = (-4)^3$	$(-4)(-4)(-4)$	-4 (cube root)
$81 = 3^4$	$3 \cdot 3 \cdot 3 \cdot 3$	3 (fourth root)

DEFINITION OF *n*TH ROOT OF A NUMBER

Let a and b be real numbers and let $n \geq 2$ be a positive integer. If

$$a = b^n$$

then b is an ***n*th root of a.** If $n = 2$, then the root is a **square root.** If $n = 3$, then the root is a **cube root.**

Some numbers have more than one *n*th root. For example, both 5 and -5 are square roots of 25. The **principal *n*th root** of a number is defined as follows.

PRINCIPAL *n*TH ROOT OF A NUMBER

Let a be a real number that has at least one *n*th root. The **principal *n*th root of a** is the *n*th root that has the same sign as a. It is denoted by a **radical symbol**

$$\sqrt[n]{a}. \quad \textit{Principal nth root}$$

The positive integer n is the **index** of the radical, and the number a is the **radicand.** If $n = 2$, we omit the index and write $\sqrt{a}$ rather than $\sqrt[2]{a}$.

REMARK   The plural of index is *indices*.

> **EXAMPLE 1**   **Evaluating Expressions Involving Radicals**
>
> **a.** $\sqrt{49} = 7$ because $7^2 = 49$.
>
> **b.** $-\sqrt{49} = -7$ because $7^2 = 49$.
>
> **c.** $\sqrt[3]{\dfrac{125}{64}} = \dfrac{5}{4}$ because $\left(\dfrac{5}{4}\right)^3 = \dfrac{5^3}{4^3} = \dfrac{125}{64}$.
>
> **d.** $\sqrt[5]{-32} = -2$ because $(-2)^5 = -32$.
>
> **e.** $\sqrt[4]{-81}$ is not a real number because there is no real number that can be raised to the fourth power to produce -81.

Here are some generalizations about the nth roots of a real number.

1. If a is a positive real number and n is a positive *even* integer, then a has exactly two (real) nth roots: $\sqrt[n]{a}$ and $-\sqrt[n]{a}$. (See Examples 1(a) and 1(b).)
2. If a is any real number and n is an *odd* integer, then a has only one (real) nth root: $\sqrt[n]{a}$. (See Examples 1(c) and 1(d).)
3. If a is a negative real number and n is an *even* integer, then a has no (real) nth root. (See Example 1(e).)
4. $\sqrt[n]{0} = 0$.

Integers such as 1, 4, 9, 16, 25, and 36 are called **perfect squares** because they have integer square roots. Similarly, integers such as 1, 8, 27, 64, and 125 are called **perfect cubes** because they have integer cube roots.

PROPERTIES OF RADICALS

Let a and b be real numbers, variables, or algebraic expressions such that the indicated roots are real numbers, and let m and n be positive integers.

Property	Example
1. $\sqrt[n]{a^m} = (\sqrt[n]{a})^m$	$\sqrt[3]{8^2} = (\sqrt[3]{8})^2 = (2)^2 = 4$
2. $\sqrt[n]{a} \cdot \sqrt[n]{b} = \sqrt[n]{ab}$	$\sqrt{5} \cdot \sqrt{7} = \sqrt{5 \cdot 7} = \sqrt{35}$
3. $\dfrac{\sqrt[n]{a}}{\sqrt[n]{b}} = \sqrt[n]{\dfrac{a}{b}}, \quad b \neq 0$	$\dfrac{\sqrt[4]{27}}{\sqrt[4]{9}} = \sqrt[4]{\dfrac{27}{9}} = \sqrt[4]{3}$
4. $\sqrt[m]{\sqrt[n]{a}} = \sqrt[mn]{a}$	$\sqrt[3]{\sqrt{10}} = \sqrt[6]{10}$
5. $(\sqrt[n]{a})^n = a$	$(\sqrt{3})^2 = 3$
6. For n even, $\sqrt[n]{a^n} = |a|$.	$\sqrt{(-12)^2} = |-12| = 12$
For n odd, $\sqrt[n]{a^n} = a$.	$\sqrt[3]{(-12)^3} = -12$

REMARK   A common special case of Property 6 is $\sqrt{a^2} = |a|$.

Simplifying Radicals

An expression involving radicals is in **simplest form** when the following conditions are satisfied.

1. All possible factors have been removed from the radical.
2. All fractions have radical-free denominators (accomplished by a process called *rationalizing the denominator*).
3. The index of the radical is reduced.

   To simplify a radical, we factor the radicand into factors whose exponents are multiples of the index. The roots of these factors are written outside the radical and the "leftover" factors make up the new radicand.

EXAMPLE 2   Simplifying Even Roots

a.

$$\underset{\substack{\text{Perfect} \\ \text{4th power}}}{} \quad \underset{\substack{\text{Leftover} \\ \text{factor}}}{}$$

$$\sqrt[4]{48} = \sqrt[4]{16 \cdot 3} = \sqrt[4]{2^4 \cdot 3} = 2\sqrt[4]{3}$$

b.

$$\underset{\substack{\text{Perfect} \\ \text{squares}}}{} \quad \underset{\substack{\text{Leftover} \\ \text{factor}}}{}$$

$$\sqrt{48x^5y^2} = \sqrt{16x^4 \cdot y^2 \cdot 3x} \qquad \textit{Find largest square factors}$$
$$= \sqrt{(4x^2)^2 \cdot y^2 \cdot 3x}$$
$$= 4x^2 \, |y| \sqrt{3x} \qquad \textit{Find root of perfect squares}$$

c. $\sqrt[4]{(5x)^4} = |5x| = 5|x|$

EXAMPLE 3   Simplifying Odd Roots

a.

$$\underset{\substack{\text{Perfect} \\ \text{cube}}}{} \quad \underset{\substack{\text{Leftover} \\ \text{factor}}}{}$$

$$\sqrt[3]{24} = \sqrt[3]{8 \cdot 3} = \sqrt[3]{2^3 \cdot 3} = 2\sqrt[3]{3}$$

b.

$$\underset{\substack{\text{Perfect} \\ \text{cube}}}{} \quad \underset{\substack{\text{Leftover} \\ \text{factors}}}{}$$

$$\sqrt[3]{24a^4} = \sqrt[3]{8a^3 \cdot 3a} \qquad \textit{Find largest cube factor}$$
$$= \sqrt[3]{(2a)^3 \cdot 3a}$$
$$= 2a\sqrt[3]{3a} \qquad \textit{Find root of perfect cube}$$

c. $\sqrt[3]{-40x^6} = \sqrt[3]{(-8x^6) \cdot 5} = \sqrt[3]{(-2x^2)^3 \cdot 5} = -2x^2\sqrt[3]{5}$

Rationalizing Denominators and Numerators

To rationalize a denominator or numerator of the form $a - b\sqrt{m}$ (or $a + b\sqrt{m}$), multiply both numerator and denominator by a **conjugate**: $a + b\sqrt{m}$ and $a - b\sqrt{m}$ are conjugates of each other. If $a = 0$, then the rationalizing factor for $\sqrt{m}$ is itself, $\sqrt{m}$.

EXAMPLE 4   Rationalizing Single-Term Denominators

a. $\dfrac{5}{2\sqrt{3}} = \dfrac{5}{2\sqrt{3}} \cdot \dfrac{\sqrt{3}}{\sqrt{3}} = \dfrac{5\sqrt{3}}{2(3)} = \dfrac{5\sqrt{3}}{6}$

b. $\dfrac{2}{\sqrt[3]{5}} = \dfrac{2}{\sqrt[3]{5}} \cdot \dfrac{\sqrt[3]{5^2}}{\sqrt[3]{5^2}} = \dfrac{2\sqrt[3]{(5)^2}}{\sqrt[3]{5^3}} = \dfrac{2\sqrt[3]{25}}{5}$

EXAMPLE 5   Rationalizing a Denominator with Two Terms

$$\dfrac{2}{3 + \sqrt{7}} = \dfrac{2}{3 + \sqrt{7}} \cdot \dfrac{3 - \sqrt{7}}{3 - \sqrt{7}}$$ *Multiply numerator and denominator by conjugate*

$$= \dfrac{2(3 - \sqrt{7})}{(3)^2 - (\sqrt{7})^2}$$

$$= \dfrac{2(3 - \sqrt{7})}{9 - 7}$$

$$= \dfrac{2(3 - \sqrt{7})}{2}$$ *Divide like factors*

$$= 3 - \sqrt{7}$$

EXAMPLE 6   Rationalizing the Numerator

$$\dfrac{\sqrt{5} - \sqrt{7}}{2} = \dfrac{\sqrt{5} - \sqrt{7}}{2} \cdot \dfrac{\sqrt{5} + \sqrt{7}}{\sqrt{5} + \sqrt{7}}$$ *Multiply numerator and denominator by conjugate*

$$= \dfrac{5 - 7}{2(\sqrt{5} + \sqrt{7})}$$

$$= \dfrac{-2}{2(\sqrt{5} + \sqrt{7})}$$

$$= \dfrac{-1}{\sqrt{5} + \sqrt{7}}$$ *Simplify*

Do not confuse an expression like $\sqrt{5} + \sqrt{7}$ with the expression $\sqrt{5 + 7}$. In general,

$$\sqrt{x + y} \quad \text{DOES NOT EQUAL} \quad \sqrt{x} + \sqrt{y}.$$

Similarly,

$$\sqrt{x^2 + y^2} \quad \text{DOES NOT EQUAL} \quad x + y.$$

Rational Exponents

Up to this point, our work with exponents has been restricted to *integer* exponents. In the following definition, note how radicals are used to define **rational exponents.**

DEFINITION OF RATIONAL EXPONENTS

If a is a real number and n is a positive integer such that the principal nth root of a exists, then we define $a^{1/n}$ to be

$$a^{1/n} = \sqrt[n]{a}.$$

Moreover, if m is a positive integer that has no common factor with n, then

$$a^{m/n} = (a^{1/n})^m = (\sqrt[n]{a})^m \quad \text{and} \quad a^{m/n} = (a^m)^{1/n} = \sqrt[n]{a^m}.$$

The numerator of a rational exponent denotes the *power* to which the base is raised, and the denominator denotes the *index* or the *root* to be taken, as shown below.

$$b^{m/n} = (\sqrt[n]{b})^m = \sqrt[n]{b^m}$$

where the numerator is the Power and the index is the Index.

When working with rational exponents, the properties of integer exponents still apply. For instance,

$$2^{1/2}2^{1/3} = 2^{(1/2)+(1/3)} = 2^{5/6}.$$

EXAMPLE 7   **Changing from Radical to Exponential Form**

a. $\sqrt{3} = 3^{1/2}$

b. $\sqrt{(3xy)^5} = \sqrt[2]{(3xy)^5} = (3xy)^{(5/2)}$

c. $2x\sqrt[4]{x^3} = (2x)(x^{3/4}) = 2x^{1+(3/4)} = 2x^{7/4}$

EXAMPLE 8   Changing from Exponential to Radical Form

a. $(x^2 + y^2)^{3/2} = (\sqrt{x^2 + y^2})^3 = \sqrt{(x^2 + y^2)^3}$

b. $2y^{3/4}z^{1/4} = 2(y^3z)^{1/4} = 2\sqrt[4]{y^3z}$

c. $a^{-3/2} = \dfrac{1}{a^{3/2}} = \dfrac{1}{\sqrt{a^3}}$

d. $x^{0.2} = x^{1/5} = \sqrt[5]{x}$

REMARK Rational exponents can be tricky, and you must remember that the expression $b^{m/n}$ is not defined unless $\sqrt[n]{b}$ is a real number. This restriction produces some unusual-looking results. For instance, the number $(-8)^{1/3}$ is defined because $\sqrt[3]{-8} = -2$, but the number $(-8)^{2/6}$ is undefined because $\sqrt[6]{-8}$ is not a real number.

Rational exponents are particularly useful for evaluating roots of numbers on a calculator, for reducing the index of a radical, and for simplifying expressions encountered in calculus.

EXAMPLE 9   Simplifying with Rational Exponents

a. $(27)^{2/6} = (27)^{1/3} = \sqrt[3]{27} = 3$

b. $(-32)^{-4/5} = (\sqrt[5]{(-32)})^{-4} = (-2)^{-4} = \dfrac{1}{(-2)^4} = \dfrac{1}{16}$

c. $(-5x^{5/3})(3x^{-3/4}) = -15x^{(5/3)-(3/4)} = -15x^{11/12}, \; x \neq 0$

EXAMPLE 10   Reducing the Index of a Radical

a. $\sqrt[9]{a^3} = a^{3/9}$ *Rewrite with rational exponents*

     $= a^{1/3}$ *Reduce exponent*

     $= \sqrt[3]{a}$ *Rewrite in radical form*

b. $\sqrt[3]{\sqrt{125}} = (125^{1/2})^{1/3} = 125^{1/6} = \sqrt[6]{125} = \sqrt[6]{(5)^3} = 5^{3/6} = 5^{1/2} = \sqrt{5}$

EXAMPLE 11   Simplifying Algebraic Expressions

a. $(2x - 1)^{4/3}(2x - 1)^{-1/3} = (2x - 1)^{(4/3)-(1/3)}$

                             $= (2x - 1)^1$

                             $= 2x - 1, \quad x \neq \dfrac{1}{2}$

b. $\dfrac{x - 1}{(x - 1)^{-1/2}} \cdot \dfrac{\sqrt{x - 1}}{\sqrt{x - 1}} = \dfrac{(x - 1)^{3/2}}{(x - 1)^0} = (x - 1)^{3/2}, \quad x \neq 1$

Radical expressions can be combined (added or subtracted) if they are **like radicals**—that is, if they have the same index and radicand. For instance, $2\sqrt{3x}$, $-\sqrt{3x}$, and $\sqrt{3x}/2$ are like radicals but $\sqrt[3]{3x}$ and $2\sqrt{3x}$ are not like radicals. To determine whether two radicals are like radicals, you should first simplify each radical.

EXAMPLE 12 **Combining Radicals**

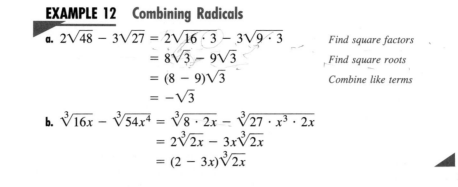

a. $2\sqrt{48} - 3\sqrt{27} = 2\sqrt{16 \cdot 3} - 3\sqrt{9 \cdot 3}$ *Find square factors*

$\qquad\qquad\qquad = 8\sqrt{3} - 9\sqrt{3}$ *Find square roots*

$\qquad\qquad\qquad = (8 - 9)\sqrt{3}$ *Combine like terms*

$\qquad\qquad\qquad = -\sqrt{3}$

b. $\sqrt[3]{16x} - \sqrt[3]{54x^4} = \sqrt[3]{8 \cdot 2x} - \sqrt[3]{27 \cdot x^3 \cdot 2x}$

$\qquad\qquad\qquad = 2\sqrt[3]{2x} - 3x\sqrt[3]{2x}$

$\qquad\qquad\qquad = (2 - 3x)\sqrt[3]{2x}$

Radicals and Calculators

There are two basic methods of evaluating radicals on most calculators. For square roots, use the *square root key* $\boxed{\sqrt{}}$. For other roots, first convert the radical to exponential form and then use the *exponential key* $\boxed{y^x}$. You may be able to find other ways to evaluate radicals on your calculator.

EXAMPLE 13 **Evaluating Radicals with a Calculator**

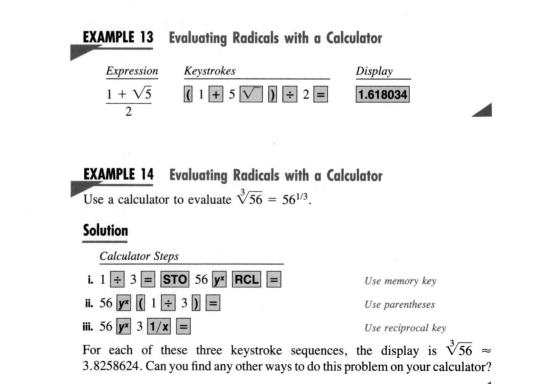

Expression	*Keystrokes*	*Display*
$\dfrac{1 + \sqrt{5}}{2}$	$\boxed{(}$ 1 $\boxed{+}$ 5 $\boxed{\sqrt{}}$ $\boxed{)}$ $\boxed{\div}$ 2 $\boxed{=}$	$\boxed{\textbf{1.618034}}$

EXAMPLE 14 **Evaluating Radicals with a Calculator**

Use a calculator to evaluate $\sqrt[3]{56} = 56^{1/3}$.

Solution

Calculator Steps

i. 1 $\boxed{\div}$ 3 $\boxed{=}$ $\boxed{\text{STO}}$ 56 $\boxed{y^x}$ $\boxed{\text{RCL}}$ $\boxed{=}$ *Use memory key*

ii. 56 $\boxed{y^x}$ $\boxed{(}$ 1 $\boxed{\div}$ 3 $\boxed{)}$ $\boxed{=}$ *Use parentheses*

iii. 56 $\boxed{y^x}$ 3 $\boxed{1/x}$ $\boxed{=}$ *Use reciprocal key*

For each of these three keystroke sequences, the display is $\sqrt[3]{56} \approx$ 3.8258624. Can you find any other ways to do this problem on your calculator?

Johannes Kepler (1571–1630), the well-known German astronomer, discovered a relationship between the average distance of a planet from the sun and the time (or period) it takes the planet to orbit the sun. At the time, people knew that planets that are closer to the sun take less time to complete an orbit than planets that are farther from the sun, as indicated in Figure 1.7. What Kepler discovered was that the distance and period are related by an exact mathematical formula. The following table shows the average distance (in astronomical units) and period (in years) for the six planets that are closest to the sun. By completing the table, can you rediscover Kepler's relationship? Discuss your conclusions.

Planet	Mercury	Venus	Earth	Mars	Jupiter	Saturn
Average Distance, x	0.387	0.723	1.0	1.523	5.203	9.541
$\sqrt{x}$						
Period, y	0.241	0.615	1.0	1.881	11.861	29.457
$\sqrt[3]{y}$						

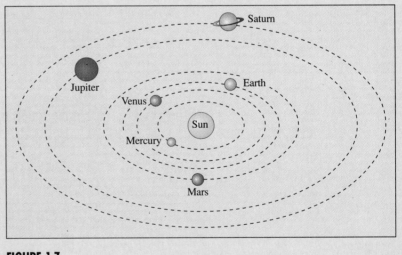

FIGURE 1.7

WARM UP

The following warm-up exercises involve skills that were covered in earlier sections. You will use these skills in the exercise set for this section.

In Exercises 1–10, simplify the expression.

1. $\left(\frac{1}{3}\right)\left(\frac{2}{3}\right)^2$

2. $3(-4)^2$

3. $(-2x)^3$

4. $(-2x^3)(-3x^4)$

5. $(7x^5)(4x)$

6. $(5x^4)(25x^2)^{-1}, \quad x \neq 0$

7. $\dfrac{12z^6}{4z^2}, \quad z \neq 0$

8. $\left(\dfrac{2x}{5}\right)^2\left(\dfrac{2x}{5}\right)^{-4}, \quad x \neq 0$

9. $\left(\dfrac{3y^2}{x}\right)^0, \quad x \neq 0, y \neq 0$

10. $[(x+2)^2(x+2)^3]^2$

EXERCISES for Section 1.3

In Exercises 1–12, fill in the missing description.

Radical Form	*Rational Exponent Form*
1. $\sqrt{9} = 3$	
2. $\sqrt[3]{64} = 4$	
3. ▨	$32^{1/5} = 2$
4. ▨	$-(144^{1/2}) = -12$
5. ▨	$196^{1/2} = 14$
6. $\sqrt[3]{614.125} = 8.5$	
7. $\sqrt[3]{-216} = -6$	
8. ▨	$(-243)^{1/5} = -3$
9. ▨	$27^{2/3} = 9$
10. $(\sqrt[4]{81})^3 = 27$	
11. $\sqrt[4]{81^3} = 27$	
12. ▨	$16^{5/4} = 32$

In Exercises 13–30, evaluate the expression. (Do not use a calculator.)

13. $\sqrt{9}$

14. $\sqrt{49}$

15. $\sqrt[3]{8}$

16. $\sqrt[3]{\dfrac{27}{8}}$

17. $-\sqrt[3]{-27}$

18. $\sqrt[3]{0}$

19. $\dfrac{4}{\sqrt{64}}$

20. $\dfrac{\sqrt[4]{81}}{3}$

21. $(\sqrt[3]{-125})^3$

22. $\sqrt[4]{562^4}$

23. $36^{3/2}$

24. $27^{1/3}$

25. $32^{-3/5}$

26. $100^{-3/2}$

27. $\left(\frac{16}{81}\right)^{-3/4}$

28. $\left(\frac{9}{4}\right)^{-1/2}$

29. $\left(-\frac{1}{64}\right)^{-1/3}$

30. $\left(-\frac{125}{27}\right)^{-1/3}$

In Exercises 31–42, simplify by removing all possible factors from the radical.

31. $\sqrt{8}$

32. $\sqrt[3]{\dfrac{16}{27}}$

33. $\sqrt{9 \times 10^{-4}}$

34. $\sqrt{4.5 \times 10^9}$

35. $\sqrt{72x^3}$

36. $\sqrt{54xy^4}$

37. $\sqrt{\dfrac{18x^2}{z^3}}$

38. $\sqrt{\dfrac{32a^4}{b^2}}$

39. $\sqrt[3]{16x^5}$

40. $\sqrt[4]{(3x^2)^4}$

41. $\sqrt{75x^2y^{-4}}$

42. $\sqrt[5]{96x^5}$

In Exercises 43–50, rationalize the denominator. Then simplify your answer.

43. $\dfrac{1}{\sqrt{3}}$

44. $\dfrac{5}{\sqrt{10}}$

45. $\dfrac{8}{\sqrt[3]{2}}$

46. $\dfrac{5}{\sqrt[3]{(5x)^2}}$

47. $\dfrac{2x}{5 - \sqrt{3}}$

48. $\dfrac{5}{\sqrt{14} - 2}$

49. $\dfrac{3}{\sqrt{5} + \sqrt{6}}$

50. $\dfrac{5}{2\sqrt{10} - 5}$

In Exercises 51–56, rationalize the numerator. Then simplify your answer.

51. $\dfrac{\sqrt{8}}{2}$

52. $\dfrac{\sqrt{2}}{3}$

53. $\dfrac{\sqrt{5} + \sqrt{3}}{3}$

54. $\dfrac{\sqrt{3} - \sqrt{2}}{2}$

55. $\dfrac{\sqrt{7} - 3}{4}$

56. $\dfrac{2\sqrt{3} + \sqrt{3}}{3}$

In Exercises 57–60, reduce the index of the radical.

57. $\sqrt[4]{3^2}$

58. $\sqrt[6]{x^3}$

59. $\sqrt[6]{(x + 1)^4}$

60. $\sqrt[4]{(3x^2)^4}$

In Exercises 61–64, write a single radical. Then simplify your answer.

61. $\sqrt{\sqrt{32}}$

62. $\sqrt{\sqrt{243(x + 1)}}$

63. $\sqrt{\sqrt[4]{2x}}$

64. $\sqrt{\sqrt[3]{10a^7b}}$

In Exercises 65–70, simplify the expression.

65. $5\sqrt{x} - 3\sqrt{x}$

66. $3\sqrt{x + 1} + 10\sqrt{x + 1}$

67. $2\sqrt{50} + 12\sqrt{8}$

68. $4\sqrt{27} - \sqrt{75}$

69. $-2\sqrt{9y} + 10\sqrt{y}$

70. $7\sqrt{80x} - 2\sqrt{125x}$

In Exercises 71–76, perform the indicated operations and simplify.

71. $5^{4/3} \cdot 5^{8/3}$

72. $\dfrac{8^{12/5}}{8^{2/5}}$

73. $\dfrac{(2x^2)^{3/2}}{2^{1/2}x^4}$

74. $\dfrac{x^{4/3}y^{2/3}}{(xy)^{1/3}}$

75. $\dfrac{x^{-3} \cdot x^{1/2}}{x^{3/2} \cdot x^{-1}}, \quad x > 0$

76. $\dfrac{5^{-1/2} \cdot 5x^{5/2}}{(5x)^{3/2}}, \quad x > 0$

In Exercises 77–82, use a calculator to approximate the number. Round to three decimal places.

77. $\sqrt{57}$

78. $\sqrt[3]{45^2}$

79. $\sqrt[6]{125}$

80. $(15.25)^{-1.4}$

81. $\sqrt{75 + 3\sqrt{8}}$

82. $(2.65 \times 10^{-4})^{1/3}$

In Exercises 83–88, fill in the blank with $<$, $=$, or $>$.

83. $\sqrt{5} + \sqrt{3}$ ▇ $\sqrt{5 + 3}$

84. $\sqrt{3} - \sqrt{2}$ ▇ $\sqrt{3 - 2}$

85. 5 ▇ $\sqrt{3^2 + 2^2}$

86. 5 ▇ $\sqrt{3^2 + 4^2}$

87. $\sqrt{3} \cdot \sqrt[4]{3}$ ▇ $\sqrt[3]{3}$

88. $\sqrt{\dfrac{3}{11}}$ ▇ $\dfrac{\sqrt{3}}{\sqrt{11}}$

Declining Balances Depreciation In Exercises 89 and 90, find the annual depreciation rate r. To find the annual depreciation rate by the **declining balances method**, use the formula

$$r = 1 - \left(\frac{S}{C}\right)^{1/n}$$

where n is the useful life of the item (in years), S is the salvage value (in dollars), and C is the original cost (in dollars).

89.

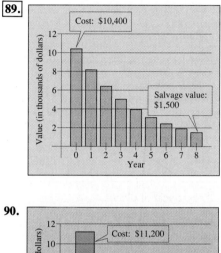

90.

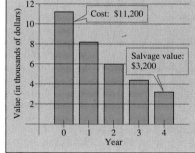

91. *Dimensions of a Cube* Find the dimensions of a cube that has a volume of 13,824 cubic inches (see figure).

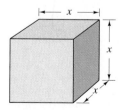

Figure for 91

92. *Funnel Time* A funnel is filled with water to a height of h (see figure). The time t it takes for the funnel to empty is

$$t = 0.03[12^{5/2} - (12 - h)^{5/2}], \qquad 0 \le h \le 12.$$

Find t for $h = 7$ centimeters.

Figure for 92

93. *Period of a Pendulum* The period T in seconds of a pendulum is

$$T = 2\pi\sqrt{\frac{L}{32}}$$

where L is the length of the pendulum in feet (see figure). Find the period of a pendulum whose length is 2 feet.

Figure for 93

94. *Particles in a Stream* A stream of water moving at the rate of v feet per second can carry particles of size $0.03\sqrt{v}$ inches. Find the size particle that can be carried by a stream flowing at the rate of $\frac{3}{4}$ feet per second.

95. *Calculator Experiment* Enter any positive real number in your calculator and repeatedly take the square root. What real number does the display appear to be approaching?

96. *Calculator Experiment* Square the real number $2/\sqrt{5}$ and note that the radical is eliminated from the denominator. Is this equivalent to rationalizing the denominator? Why or why not?

97. List all possible unit digits of the square of a positive integer. Use that list to determine whether $\sqrt{5233}$ is an integer.

1.4 Polynomials and Special Products

Polynomials / Operations with Polynomials / Special Products / Applications

Polynomials

The most common kind of algebraic expression is the **polynomial.** Some examples are

$$2x + 5, \quad 3x^4 - 7x^2 + 2x + 4, \quad \text{and} \quad 5x^2y^2 - xy + 3.$$

The first two are *polynomials in x* and the third one is a *polynomial in x and y.* The terms of a polynomial in x have the form ax^k, where a is called the **coefficient** and k the **degree** of the term. For instance, the third degree polynomial

$$2x^3 - 5x^2 + 1 = 2x^3 + (-5)x^2 + (0)x + 1$$

has coefficients 2, -5, 0, and 1.

DEFINITION OF A POLYNOMIAL IN x

Let $a_0, a_1, a_2, \ldots, a_n$ be *real numbers* and let n be a *nonnegative integer.* A **polynomial in x** is an expression of the form

$$a_n x^n + a_{n-1} x^{n-1} + \cdots + a_1 x + a_0$$

where $a_n \neq 0$. The polynomial is of **degree n,** a_n is the **leading coefficient,** and a_0 is the **constant term.**

REMARK Polynomials with one, two, or three terms are called **monomials, binomials,** or **trinomials,** respectively.

In **standard form,** a polynomial is written with descending powers of x.

EXAMPLE 1 Writing Polynomials in Standard Form

Polynomial	Standard Form	Degree
a. $4x^2 - 5x^7 - 2 + 3x$	$-5x^7 + 4x^2 + 3x - 2$	7
b. $4 - 9x^2$	$-9x^2 + 4$	2
c. 8	$8 \ (8 = 8x^0)$	0

A polynomial that has all zero coefficients is called the **zero polynomial,** denoted by 0. We do not assign a degree to the zero polynomial. Expressions such as $\sqrt{x^2 - 3x}$ and $x^2 + 5x^{-1}$ are not polynomials.

For polynomials in more than one variable, the degree of a *term* is the sum of the exponents of the variables in the term. The degree of the *polynomial* is the highest degree of its terms. For instance, the polynomial $5x^3y - x^2y^2 + 2xy - 5$ has two terms of degree 4, one term of degree 2, and one term of degree 0. The degree of the polynomial is 4.

Operations with Polynomials

You can **add** and **subtract** polynomials in much the same way you add and subtract real numbers. Simply add or subtract the *like terms* (terms having the same variables to the same powers) by adding their coefficients. For instance, $-3xy^2$ and $5xy^2$ are like terms and their sum is

$$-3xy^2 + 5xy^2 = (-3 + 5)xy^2 = 2xy^2.$$

EXAMPLE 2   Sums and Differences of Polynomials

a. $(5x^3 - 7x^2 - 3) + (x^3 + 2x^2 - x + 8)$

$\qquad = (5x^3 + x^3) + (2x^2 - 7x^2) - x + (8 - 3)$ *Group like terms*

$\qquad = 6x^3 - 5x^2 - x + 5$ *Combine like terms*

b. $(7x^4 - x^2 - 4x + 2) - (3x^4 - 4x^2 + 3x)$

$\qquad = 7x^4 - x^2 - 4x + 2 - 3x^4 + 4x^2 - 3x$

$\qquad = (7x^4 - 3x^4) + (4x^2 - x^2) + (-3x - 4x) + 2$ *Group like terms*

$\qquad = 4x^4 + 3x^2 - 7x + 2$ *Combine like terms*

REMARK   A common mistake is to fail to change the sign of *each* term inside parentheses preceded by a negative sign. For instance, note that

$$-(3x^4 - 4x^2 + 3x) = -3x^4 + 4x^2 - 3x$$

and

$$-(3x^4 - 4x^2 + 3x) \neq -3x^4 - 4x^2 + 3x.$$ *Common mistake*

To find the **product** of two polynomials, use the left and right Distributive Properties. For example, if you treat $(5x + 7)$ as a single quantity, you can multiply $(3x - 2)$ by $(5x + 7)$ as follows.

$$(3x - 2)(5x + 7) = 3x(5x + 7) - 2(5x + 7)$$

$$= (3x)(5x) + (3x)(7) - (2)(5x) - (2)(7)$$

$$= 15x^2 + 21x - 10x - 14$$

Product of **F**irst terms	Product of **O**uter terms	Product of **I**nner terms	Product of **L**ast terms

$$= 15x^2 + 11x - 14$$

With practice, you should be able to multiply two binomials without writing all of the above steps. In fact, the four products in the boxes above suggest a single step: the **FOIL Method.**

When multiplying two polynomials, be sure to multiply *each* term of one polynomial by *each* term of the other. A vertical arrangement is helpful.

EXAMPLE 3   Using a Vertical Arrangement to Multiply Polynomials

Multiply $(x^2 - 2x + 2)$ by $(x^2 + 2x + 2)$.

Solution

$$
\begin{array}{ll}
x^2 - 2x + 2 & \textit{Standard form} \\
x^2 + 2x + 2 & \textit{Standard form} \\
\hline
x^4 - 2x^3 + 2x^2 & \leftarrow x^2(x^2 - 2x + 2) \\
\quad\quad 2x^3 - 4x^2 + 4x & \leftarrow 2x(x^2 - 2x + 2) \\
\quad\quad\quad\quad 2x^2 - 4x + 4 & \leftarrow 2(x^2 - 2x + 2) \\
\hline
x^4 + 0x^3 + 0x^2 - 0x + 4 = x^4 + 4 &
\end{array}
$$

Thus, $(x^2 - 2x + 2)(x^2 + 2x + 2) = x^4 + 4$.

Special Products

SPECIAL PRODUCTS

Let u and v be real numbers, variables, or algebraic expressions.

Special Product	*Example*
Sum and Difference of Same Terms	
$(u + v)(u - v) = u^2 - v^2$	$(x + 4)(x - 4) = x^2 - 4^2 = x^2 - 16$
Square of a Binomial	
$(u + v)^2 = u^2 + 2uv + v^2$	$(x + 3)^2 = x^2 + 2(x)(3) + 3^2 = x^2 + 6x + 9$
$(u - v)^2 = u^2 - 2uv + v^2$	$(3x - 2)^2 = (3x)^2 - 2(3x)(2) + 2^2 = 9x^2 - 12x + 4$
Cube of a Binomial	
$(u + v)^3 = u^3 + 3u^2v + 3uv^2 + v^3$	$(x + 2)^3 = x^3 + 3x^2(2) + 3x(2^2) + 2^3$
	$= x^3 + 6x^2 + 12x + 8$
$(u - v)^3 = u^3 - 3u^2v + 3uv^2 - v^3$	$(x - 1)^3 = x^3 - 3x^2(1) + 3x(1^2) - 1^3$
	$= x^3 - 3x^2 + 3x - 1$

EXAMPLE 4   Sum and Difference of Same Two Terms

Find the product of $(5x + 9)$ and $(5x - 9)$.

Solution

The product of a sum and a difference of the *same* two terms has no middle term and it takes the form $(u + v)(u - v) = u^2 - v^2$.

$$(5x + 9)(5x - 9) = (5x)^2 - 9^2 = 25x^2 - 81$$

EXAMPLE 5   Square of a Binomial

Find $(6x - 5)^2$.

Solution

The square of a binomial has the form $(u - v)^2 = u^2 - 2uv + v^2$.

$$(6x - 5)^2 = (6x)^2 - 2(6x)(5) + 5^2 = 36x^2 - 60x + 25$$

EXAMPLE 6   Cube of a Binomial

Find $(3x + 2)^3$.

Solution

The cube of a binomial has the form $(u + v)^3 = u^3 + 3u^2v + 3uv^2 + v^3$. Note the *decrease* of powers of u and the *increase* of powers of v.

$$(3x + 2)^3 = (3x)^3 + 3(3x)^2(2) + 3(3x)(2)^2 + 2^3$$
$$= 27x^3 + 54x^2 + 36x + 8$$

Occasionally, the formulas for special products can be extended to cover products of two trinomials.

EXAMPLE 7   The Product of Two Trinomials

Find the product of $(x + y - 2)$ and $(x + y + 2)$.

Solution

By grouping $x + y$ in parentheses, you can write

$$(x + y - 2)(x + y + 2) = [(x + y) - 2][(x + y) + 2]$$
$$= (x + y)^2 - 2^2$$
$$= x^2 + 2xy + y^2 - 4.$$

Applications

A polynomial is the most common type of mathematical model used to represent real-world situations, and you will encounter many of these in the text. Example 8 gives you some idea of how polynomials can be used as models.

EXAMPLE 8   An Application: Volume of a Box

An open box is made by cutting squares out of the corners of a piece of metal that is 16 inches by 20 inches, as shown in Figure 1.8. The edge of each cutout square is x inches. What is the volume of the box? Find the volume when $x = 1$, $x = 2$, and $x = 3$.

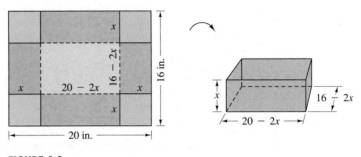

FIGURE 1.8

Solution

The volume of a rectangular box is equal to the product of its length, width, and height. From the figure, the length is $20 - 2x$, the width is $16 - 2x$, and the height is x. Thus, the volume of the box is

$$\text{Volume} = (20 - 2x)(16 - 2x)(x) = 320x - 72x^2 + 4x^3.$$

When $x = 1$ inch, the volume of the box is

$$\text{Volume} = 320(1) - 72(1^2) + 4(1^3) = 252 \text{ cubic inches.}$$

When $x = 2$ inches, the volume of the box is

$$\text{Volume} = 320(2) - 72(2^2) + 4(2^3) = 384 \text{ cubic inches.}$$

When $x = 3$ inches, the volume of the box is

$$\text{Volume} = 320(3) - 72(3^2) + 4(3^3) = 420 \text{ cubic inches.}$$

DISCUSSION PROBLEM

A Mathematical Experiment

In Example 8, you found the volume of the open metal box was given by

$$\text{Volume} = 320x - 72x^2 + 4x^3.$$

Suppose you want to create a box that has as much volume as possible. From Example 8, you know that by cutting 1-, 2-, and 3-inch squares from the corners, you can create boxes whose volumes are 252, 384, and 420 cubic inches, respectively. Try several other values of x to see if you can find the size of the square that should be cut from the corners to produce a box that has a maximum volume. (*Hint:* The answer is not a whole number.)

WARM UP

The following warm-up exercises involve skills that were covered in earlier sections. You will use these skills in the exercise set for this section.

In Exercises 1–10, perform the indicated operations.

1. $(7x^2)(6x)$

2. $(10z^3)(-2z^{-1})$

3. $(-3x^2)^3$

4. $-3(x^2)^3$

5. $\dfrac{27z^5}{12z^2}$

6. $\sqrt{24} \cdot \sqrt{2}$

7. $\left(\dfrac{2x}{3}\right)^{-2}$

8. $16^{3/4}$

9. $\dfrac{4}{\sqrt{8}}$

10. $\sqrt[3]{-27x^3}$

EXERCISES for Section 1.4

In Exercises 1–6, find the degree and leading coefficient of the polynomial.

1. $2x^2 - x + 1$

2. $-3x^4 + 2x^2 - 5$

3. $x^5 - 1$

4. 3

5. $4x^5 + 6x^4 - x - 1$

6. $2x$

In Exercises 7–12, determine whether the expression is a polynomial. If it is, write the polynomial in standard form.

7. $2x - 3x^3 + 8$

8. $2x^3 + x - 3x^{-1}$

9. $\dfrac{3x + 4}{x}$

10. $\dfrac{x^2 + 2x - 3}{2}$

11. $y^2 - y^4 + y^3$

12. $\sqrt{y^2 - y^4}$

In Exercises 13–26, perform the indicated operations and write the result in standard form.

13. $(6x + 5) - (8x + 15)$
14. $(2x^2 + 1) - (x^2 - 2x + 1)$
15. $-(x^3 - 2) + (4x^3 - 2x)$
16. $-(5x^2 - 1) - (-3x^2 + 5)$
17. $(15x^2 - 6) - (-8x^3 - 14x^2 - 17)$
18. $(15x^4 - 18x - 19) - (13x^4 - 5x + 15)$
19. $5z - [3z - (10z + 8)]$
20. $(y^3 + 1) - [(y^2 + 1) + (3y - 7)]$
21. $3x(x^2 - 2x + 1)$
22. $y^2(4y^2 + 2y - 3)$
23. $-5z(3z - 1)$
24. $-4x(3 - x^3)$
25. $(-2x)(-3x)(5x + 2)$
26. $(1 - x^3)(4x)$

In Exercises 27–60, find the product. (*Note:* Some of the expressions are *not* polynomials, but the formulas can still be used.)

27. $(x + 3)(x + 4)$
28. $(x - 5)(x + 10)$
29. $(3x - 5)(2x + 1)$
30. $(7x - 2)(4x - 3)$
31. $(2x + 3)^2$
32. $(4x + 5)^2$
33. $(2x - 5y)^2$
34. $(5 - 8x)^2$
35. $[(x - 3) + y]^2$
36. $[(x + 1) - y]^2$
37. $(x + 10)(x - 10)$
38. $(2x + 3)(2x - 3)$
39. $(x + 2y)(x - 2y)$
40. $(2x + 3y)(2x - 3y)$
41. $(m - 3 + n)(m - 3 - n)$
42. $(x + y + 1)(x + y - 1)$
43. $(2r^2 - 5)(2r^2 + 5)$
44. $(3a^3 - 4b^2)(3a^3 + 4b^2)$
45. $(x + 1)^3$
46. $(x - 2)^3$

47. $(2x - y)^3$
48. $(3x + 2y)^3$
49. $(\sqrt{x} + \sqrt{y})(\sqrt{x} - \sqrt{y})$
50. $(5 + \sqrt{x})(5 - \sqrt{x})$
51. $(4x^3 - 3)^2$
52. $(8x + 3)^2$
53. $(x^2 + 9)(x^2 - x - 4)$
54. $(x - 2)(x^2 + 2x + 4)$
55. $(x^2 - x + 1)(x^2 + x + 1)$
56. $(x^2 + 3x - 2)(x^2 - 3x - 2)$
57. $5x(x + 1) - 3x(x + 1)$
58. $(2x - 1)(x + 3) + 3(x + 3)$
59. $(x + \sqrt{5})(x - \sqrt{5})(x + 4)$
60. $(x + y)(x - y)(x^2 + y^2)$

61. *Compound Interest* After two years, an investment of $500 compounded annually at an interest rate r will yield an amount of

    $500(1 + r)^2$.

    Write this polynomial in standard form.

62. *Compound Interest* After three years, an investment of $1200 compounded annually at an interest rate r will yield an amount of

    $1200(1 + r)^3$.

    Write this polynomial in standard form.

63. *Volume of a Box* An open box is made by cutting squares out of the corners of a piece of metal that is 18 inches by 26 inches (see figure). The edge of each cut-out square is x inches. What is the volume of the box? Find the volume when $x = 1$, $x = 2$, and $x = 3$.

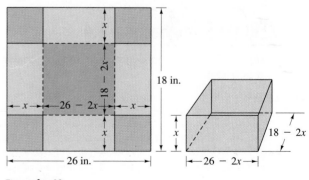

Figure for 63

64. *Volume of a Box*   A closed box is constructed by cutting along the solid lines and folding along the broken lines of the rectangular piece of metal shown in the figure. The length and width of the rectangle are 45 inches and 15 inches, respectively. Find the volume of the box in terms of x. Find the volume when $x = 3$, $x = 5$, and $x = 7$.

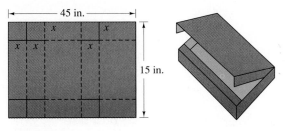

Figure for 64

65. *Area of a Region*   Find the area of the red shaded region in the figure. Write your result as a polynomial in standard form.

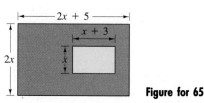

Figure for 65

66. *Floor Space*   Find a polynomial that represents the total number of square feet for the floor plan shown in the figure.

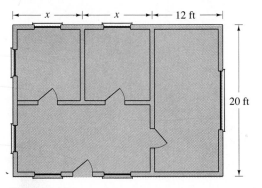

Figure for 66

67. *Stopping Distance*   The stopping distance of an automobile is the distance traveled during the driver's reaction time plus the distance traveled after the brakes are applied. In an experiment, these distances were measured (in feet) when the automobile was traveling at a speed of x miles per hour (see figure). The distance traveled during the

reaction time is $R = 1.1x$, and the braking distance is $B = 0.14x^2 - 4.43x + 58.40$. Determine the polynomial that represents the total stopping distance. Use this polynomial to estimate the total stopping distance when $x = 30$, $x = 40$, and $x = 55$.

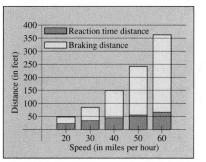

Figure for 67

68. *Safe Beam Load*   A uniformly distributed load is placed on a 1-inch-wide steel beam. When the span of the beam is x feet and its depth is 6 inches, the safe load is approximated by

$$S_6 = (0.06x^2 - 2.42x + 38.71)^2.$$

When the depth is 8 inches, the safe load is approximated by

$$S_8 = (0.08x^2 - 3.30x + 51.93)^2.$$

Estimate the difference in the safe loads of these two beams when the span is 10 feet (see figure).

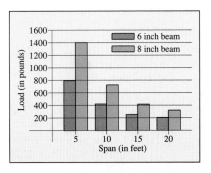

Figure for 68

69. Find the degree of the product of two polynomials of degrees m and n.

70. Find the degree of the sum of two polynomials of degrees m and n if $m < n$.

71. A student's homework paper included the following.

$$(x - 3)^2 = x^2 + 9$$

Write a paragraph fully explaining the error and giving the correct method for squaring a binomial.

SOLVING

A Numerical Approach to Maximizing a Volume

Many mathematical results are discovered experimentally by calculating examples and looking for patterns. Prior to the 1950s, this mode of discovery was very time-consuming because the calculations had to be done by hand. The introduction of computer and calculator technology has removed the drudgery of calculation.

Using technology to conduct mathematical experiments usually involves creating an algebraic **model** to represent the quantity under question. For instance, the following example shows how to create a model for the volume of a rectangular box.

EXAMPLE 1   Creating a Model for the Volume of a Box

Consider a rectangular box with a square base and a surface area of 216 square inches. Let x represent the length (in inches) of each side of the base. Use the variable x to write a model, or expression, for the volume of the box.

Solution

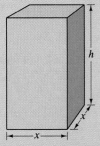

You can begin by writing a model for the height h (in inches) in terms of x.

$$\boxed{\begin{array}{c}\textit{Surface}\\\textit{area}\end{array}} = \boxed{\begin{array}{c}\textit{Area of}\\\textit{base}\end{array}} + \boxed{\begin{array}{c}\textit{Area of}\\\textit{top}\end{array}} + 4\boxed{\begin{array}{c}\textit{Area of}\\\textit{side}\end{array}}$$

$$216 = x^2 + x^2 + 4xh$$

$$216 = 2x^2 + 4xh$$

$$216 - 2x^2 = 4xh$$

$$\frac{216 - 2x^2}{4x} = h$$

$$\frac{54}{x} - \frac{x}{2} = h$$

Now, having written the height in terms of x, you can write the following model for the volume in terms of x.

$$\boxed{\begin{array}{c}\textit{Volume}\\\textit{of box}\end{array}} = \boxed{\begin{array}{c}\textit{Length}\\\textit{of box}\end{array}} \cdot \boxed{\begin{array}{c}\textit{Width}\\\textit{of box}\end{array}} \cdot \boxed{\begin{array}{c}\textit{Height}\\\textit{of box}\end{array}}$$

$$V = (x)(x)\left(\frac{54}{x} - \frac{x}{2}\right)$$

$$= 54x - \frac{1}{2}x^3, \quad 0 < x \le \sqrt{108} \quad \blacktriangleleft$$

Now, suppose you wanted to answer the following question. "Of all rectangular boxes with square bases and surface areas of 216 square inches, which has the largest volume?" You can use the model created in Example 1 to experimentally answer the question.

EXAMPLE 2 Finding the Maximum Volume of a Box

Of all rectangular boxes with a square base and a surface area of 216 square inches, which has the largest volume?

Solution

To answer the question experimentally, you can calculate several volumes. For instance, you could let x vary between 1.0 inches and 10 inches and calculate the resulting volume.

Base, x	Height	Surface Area	Volume of Box
1.0	53.5	216.0	53.5
1.5	35.3	216.0	79.3
2.0	26.0	216.0	104.0
2.5	20.4	216.0	127.2
3.0	16.5	216.0	148.5
3.5	13.7	216.0	167.6
4.0	11.5	216.0	184.0
4.5	9.8	216.0	197.4
5.0	8.3	216.0	207.5
5.5	7.1	216.0	213.8
6.0	6.0	216.0	216.0
6.5	5.1	216.0	213.7
7.0	4.2	216.0	206.5
7.5	3.5	216.0	194.1
8.0	2.8	216.0	176.0
8.5	2.1	216.0	151.9
9.0	1.5	216.0	121.5
9.5	0.9	216.0	84.3
10.0	0.4	216.0	40.0

From the results of the experiment, it appears that the *cube* (the box whose dimensions are 6 by 6 by 6) has the greatest volume. (In Chapter 4 on page 269, we look at a graphical approach to solving this problem.) ◢

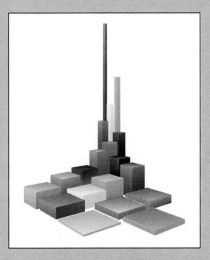

The diagram shows the 19 different boxes whose volumes were calculated in Example 2.

EXERCISES

(See also: Discussion Problem, page 35; Exercises 63–64, Section 1.4)

1. *Exploration* In Example 2, what happens to the height of the boxes as x gets closer and closer to 0? Of all boxes with a square base and a surface area of 216 square inches, is there a tallest?

2. *Exploration* In Example 2, what happens to the height of the boxes as x gets closer and closer to $\sqrt{108}$? Is there a shortest box that has a square base and a surface area of 216 square inches?

3. *Exploration* Complete the table. Why does this table lend further support to the conclusion obtained in Example 2?

Base, x	5.9	5.99	5.999	6.1	6.01	6.001
Volume, V	?	?	?	?	?	?

4. Of all rectangular boxes with a surface area of 216 square inches and a base that is x inches by $2x$ inches, which has the maximum volume?

39

1.5 Factoring

Introduction / Polynomials with Common Factors / Factoring Special Polynomial Forms / Trinomials with Binomial Factors / Factoring by Grouping

Introduction

The process of writing a polynomial as a product is called **factoring.** It is an important tool for solving equations and for reducing fractional expressions.

Unless noted otherwise, we will limit our discussion of factoring to polynomials whose factors have integer coefficients. If a polynomial cannot be factored using integer coefficients, then it is called **prime** or **irreducible over the integers.** For instance, the polynomial $x^2 - 3$ is irreducible over the integers. [Over the *real numbers,* this polynomial can be factored as $x^2 - 3 = (x + \sqrt{3})(x - \sqrt{3})$.]

A polynomial is said to be **completely factored** when each of its factors is prime. For instance,

$$x^3 - x^2 + 4x - 4 = (x - 1)(x^2 + 4)$$

is completely factored, but

$$x^3 - x^2 - 4x + 4 = (x - 1)(x^2 - 4)$$

is not completely factored. Its complete factorization would be

$$x^3 - x^2 - 4x + 4 = (x - 1)(x + 2)(x - 2).$$

Polynomials with Common Factors

We start with polynomials that can be written as the product of a monomial and another polynomial. The technique used here is the distributive property, $a(b + c) = ab + ac$, in the *reverse* direction.

$$ab + ac = a(b + c) \qquad \text{\textit{a is a common factor}}$$

Removing (factoring out) a common factor is the first step in completely factoring polynomials.

EXAMPLE 1   Removing Common Factors

a. $6x^3 - 4x = 2x(3x^2) - 2x(2) = 2x(3x^2 - 2)$

b. $(x - 2)(2x) + (x - 2)(3) = (x - 2)(2x + 3)$

Factoring Special Polynomial Forms

FACTORING SPECIAL POLYNOMIAL FORMS

Factored Form   *Example*

Difference of Two Squares

$$u^2 - v^2 = (u + v)(u - v)$$

$$9x^2 - 4 = (3x)^2 - 2^2$$
$$= (3x + 2)(3x - 2)$$

Perfect Square Trinomial

$$u^2 + 2uv + v^2 = (u + v)^2$$

$$x^2 + 6x + 9 = x^2 + 2(x)(3) + 3^2$$
$$= (x + 3)^2$$

$$u^2 - 2uv + v^2 = (u - v)^2$$

$$x^2 - 6x + 9 = x^2 - 2(x)(3) + 3^2$$
$$= (x - 3)^2$$

Sum or Difference of Two Cubes

$$u^3 + v^3 = (u + v)(u^2 - uv + v^2)$$

$$x^3 + 8 = x^3 + 2^3$$
$$= (x + 2)(x^2 - 2x + 4)$$

$$u^3 - v^3 = (u - v)(u^2 + uv + v^2)$$

$$27x^3 - 1 = (3x)^3 - 1^3$$
$$= (3x - 1)(9x^2 + 3x + 1)$$

One of the easiest special polynomial forms to factor is the difference of two squares. Think of the form as follows.

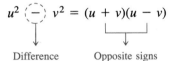

Factors are a conjugate pair

Difference   Opposite signs

To recognize perfect square terms, look for coefficients that are squares of integers and variables raised to *even powers*.

REMARK In Example 2, note that the first step in factoring a polynomial is to check for common factors. Once the common factor is removed, it is often possible to recognize patterns that were not immediately obvious.

EXAMPLE 2 Removing a Common Factor First

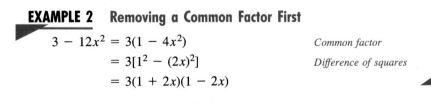

$$3 - 12x^2 = 3(1 - 4x^2)$$ *Common factor*
$$= 3[1^2 - (2x)^2]$$ *Difference of squares*
$$= 3(1 + 2x)(1 - 2x)$$

EXAMPLE 3 Factoring the Difference of Two Squares

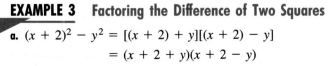

a. $(x + 2)^2 - y^2 = [(x + 2) + y][(x + 2) - y]$
$$= (x + 2 + y)(x + 2 - y)$$
$$= (x + y + 2)(x - y + 2)$$

b. Apply the difference of two squares formula twice.

$$16x^4 - 81 = (4x^2)^2 - 9^2$$
$$= (4x^2 + 9)(4x^2 - 9) \qquad \textit{First application}$$
$$= (4x^2 + 9)[(2x)^2 - 3^2]$$
$$= (4x^2 + 9)(2x + 3)(2x - 3) \quad \textit{Second application}$$

A perfect square trinomial is the square of a binomial, and it has the following form.

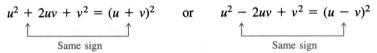

$$u^2 + 2uv + v^2 = (u + v)^2 \qquad \text{or} \qquad u^2 - 2uv + v^2 = (u - v)^2$$

Same sign   Same sign

Note that the first and last terms are squares and the middle term is twice the product of u and v.

EXAMPLE 4   Factoring Perfect Square Trinomials

a. $16x^2 + 8x + 1 = (4x)^2 + 2(4x)(1) + 1^2$
$$= (4x + 1)^2$$

b. $x^2 - 10x + 25 = x^2 - 2(x)(5) + 5^2$
$$= (x - 5)^2$$

The next two formulas show sums and differences of cubes. Pay special attention to the signs of the terms.

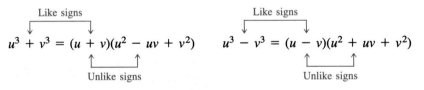

Like signs   Like signs

$$u^3 + v^3 = (u + v)(u^2 - uv + v^2) \qquad u^3 - v^3 = (u - v)(u^2 + uv + v^2)$$

Unlike signs   Unlike signs

EXAMPLE 5   Factoring the Difference of Cubes

$$x^3 - 27 = x^3 - 3^3 = (x - 3)(x^2 + 3x + 9)$$

EXAMPLE 6   Factoring the Sum of Cubes

$$3x^3 + 192 = 3(x^3 + 64)$$
$$= 3(x^3 + 4^3)$$
$$= 3(x + 4)(x^2 - 4x + 16)$$

Trinomials with Binomial Factors

To factor a trinomial of the form $ax^2 + bx + c$, use the following pattern.

$$\underset{\text{Factors of } c}{\overset{\text{Factors of } a}{ax^2 + bx + c = (\blacksquare x + \blacksquare)(\blacksquare x + \blacksquare)}}$$

The goal is to find a combination of factors of a and c so that the outer and inner products add up to the middle term bx. For instance, in the trinomial $6x^2 + 17x + 5$, you can write

$$\overset{\text{F} \quad \text{O} \quad \text{I} \quad \text{L}}{(2x + 5)(3x + 1) = 6x^2 + 2x + 15x + 5 = 6x^2 + 17x + 5.}$$

Note that the outer (O) and inner (I) products add up to $17x$.

EXAMPLE 7   Factoring a Trinomial: Leading Coefficient Is 1

Factor the trinomial $x^2 - 7x + 12$.

Solution

The possible factorizations are

$$(x - 2)(x - 6), \quad (x - 1)(x - 12), \quad \text{and} \quad (x - 3)(x - 4).$$

Testing the middle term, you will find the correct factorization to be

$$x^2 - 7x + 12 = (x - 3)(x - 4).$$

EXAMPLE 8   Factoring a Trinomial: Leading Coefficient Is Not 1

Factor the trinomial $2x^2 + x - 15$.

Solution

The eight possible factorizations are as follows.

$$(2x - 1)(x + 15) \quad (2x + 1)(x - 15)$$
$$(2x - 3)(x + 5) \quad (2x + 3)(x - 5)$$
$$(2x - 5)(x + 3) \quad (2x + 5)(x - 3)$$
$$(2x - 15)(x + 1) \quad (2x + 15)(x - 1)$$

Testing the middle term, you will find the correct factorization to be

$$2x^2 + x - 15 = (2x - 5)(x + 3).$$

Factoring by Grouping

Sometimes polynomials with more than three terms can be factored by a method called **factoring by grouping.** It is not always obvious which terms to group, and sometimes several different groupings will work.

EXAMPLE 9   Factoring by Grouping

$$\begin{aligned}
x^3 - 2x^2 - 3x + 6 &= (x^3 - 2x^2) - (3x - 6) & \text{\textit{Group terms}} \\
&= x^2(x - 2) - 3(x - 2) & \text{\textit{Factor groups}} \\
&= (x - 2)(x^2 - 3) & \text{\textit{Common factor}}
\end{aligned}$$

As general guidelines for completely factoring polynomials, consider, in order, the following factorizations.

Guidelines for Factoring Polynomials

1. Factor out any common factors.
2. Factor according to one of the special polynomial forms.
3. Factor as $ax^2 + bx + c = (mx + r)(nx + s)$.
4. Factor by grouping.

Factoring a trinomial can involve quite a bit of trial and error. An alternative technique that some people like to use is factoring by grouping.

EXAMPLE 10   Factoring a Trinomial by Grouping

Use factoring by grouping to factor the following trinomial.

$$2x^2 + 5x - 3$$

Solution

In the trinomial $2x^2 + 5x - 3$, we have $a = 2$ and $c = -3$, which implies that the product ac is -6. Now, since -6 factors as $(6)(-1)$ and $6 - 1 = 5 = b$, we rewrite the middle term as $5x = 6x - x$. This produces the following.

$$\begin{aligned}
2x^2 + 5x - 3 &= 2x^2 + 6x - x - 3 & \text{\textit{Rewrite middle term}} \\
&= (2x^2 + 6x) - (x + 3) & \text{\textit{Group terms}} \\
&= 2x(x + 3) - (x + 3) & \text{\textit{Factor groups}} \\
&= (x + 3)(2x - 1) & \text{\textit{Distributive Property}}
\end{aligned}$$

Therefore, the given trinomial factors as

$$2x^2 + 5x - 3 = (x + 3)(2x - 1).$$

Figure 1.9 shows two cubes: a large cube whose volume is a^3 and a smaller cube whose volume is b^3. If the smaller cube is removed from the larger, the remaining solid has a volume of $a^3 - b^3$ and is composed of three rectangular boxes, labeled Box 1, Box 2, and Box 3. Find the volume of each box and describe how these results are related to the following special product formula.

$$a^3 - b^3 = (a - b)(a^2 + ab + b^2)$$
$$= (a - b)a^2 + (a - b)ab + (a - b)b^2$$

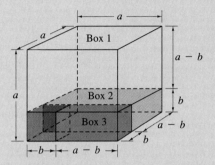

FIGURE 1.9

WARM UP

The following warm-up exercises involve skills that were covered in earlier sections. You will use these skills in the exercise set for this section.

In Exercises 1–10, find the product.

1. $3x(5x - 2)$ **2.** $-2y(y + 1)$

3. $(2x + 3)^2$ **4.** $(3x - 8)^2$

5. $(2x - 3)(x + 8)$ **6.** $(4 - 5z)(1 + z)$

7. $(2y + 1)(2y - 1)$ **8.** $(x + a)(x - a)$

9. $(x + 4)^3$ **10.** $(2x - 3)^3$

EXERCISES for Section 1.5

In Exercises 1–6, factor out the common factor.

1. $3x + 6$ **2.** $5y - 30$

3. $2x^3 - 6x$ **4.** $4x^3 - 6x^2 + 12x$

5. $(x - 1)^2 + 6(x - 1)$ **6.** $3x(x + 2) - 4(x + 2)$

In Exercises 7–12, factor the difference of two squares.

7. $x^2 - 36$ **8.** $x^2 - \frac{1}{4}$

9. $16y^2 - 9$ **10.** $49 - 9y^2$

11. $(x - 1)^2 - 4$ **12.** $25 - (z + 5)^2$

In Exercises 13–18, factor the perfect square trinomial.

13. $x^2 - 4x + 4$

14. $x^2 + 10x + 25$

15. $4t^2 + 4t + 1$

16. $9x^2 - 12x + 4$

17. $25y^2 - 10y + 1$

18. $z^2 + z + \frac{1}{4}$

In Exercises 19–32, factor the trinomial.

19. $x^2 + x - 2$

20. $x^2 + 5x + 6$

21. $s^2 - 5s + 6$

22. $t^2 - t - 6$

23. $y^2 + y - 20$

24. $z^2 - 5z - 24$

25. $x^2 - 30x + 200$

26. $x^2 - 13x + 42$

27. $3x^2 - 5x + 2$

28. $2x^2 - x - 1$

29. $9z^2 - 3z - 2$

30. $12x^2 + 7x + 1$

31. $5x^2 + 26x + 5$

32. $5u^2 + 13u - 6$

In Exercises 33–38, factor the sum or difference of cubes.

33. $x^3 - 8$

34. $x^3 - 27$

35. $y^3 + 64$

36. $z^3 + 125$

37. $8t^3 - 1$

38. $27x^3 + 8$

In Exercises 39–44, factor by grouping.

39. $x^3 - x^2 + 2x - 2$

40. $x^3 + 5x^2 - 5x - 25$

41. $2x^3 - x^2 - 6x + 3$

42. $5x^3 - 10x^2 + 3x - 6$

43. $6 + 2x - 3x^3 - x^4$

44. $x^5 + 2x^3 + x^2 + 2$

In Exercises 45–50, factor the trinomial by grouping.

45. $3x^2 + 10x + 8$

46. $2x^2 + 9x + 9$

47. $6x^2 + x - 2$

48. $6x^2 - x - 15$

49. $15x^2 - 11x + 2$

50. $12x^2 - 13x + 1$

In Exercises 51–80, completely factor the expression and express the answer with no negative exponents.

51. $x^3 - 4x^2$

52. $6x^2 - 54$

53. $1 - 4x + 4x^2$

54. $9x^2 - 6x + 1$

55. $-2x^2 - 4x + 2x^3$

56. $2y^3 - 7y^2 - 15y$

57. $9x^2 + 10x + 1$

58. $13x + 6 + 5x^2$

59. $3x^3 + x^2 + 15x + 5$

60. $5 - x + 5x^2 - x^3$

61. $x^4 - 4x^3 + x^2 - 4x$

62. $3u - 2u^2 + 6 - u^3$

63. $4 - (x + 4)^2$

64. $(t - 1)^2 - 49$

65. $(x^2 + 1)^2 - 4x^2$

66. $(x^2 + 8)^2 - 36x^2$

67. $2t^3 - 16$

68. $5x^3 + 40$

69. $4x(2x - 1) + (2x - 1)^2$

70. $5(3 - 4x)^2 - 8(3 - 4x)(5x - 1)$

71. $2(x + 1)(x - 3)^2 - 3(x + 1)^2(x - 3)$

72. $7(3x + 2)^2(1 - x)^2 + (3x + 2)(1 - x)^3$

73. $7x(2)(x^2 + 1)(2x) - (x^2 + 1)^2(7)$

74. $3(x - 2)^2(x + 1)^4 + (x - 2)^3(4)(x + 1)^3$

75. $2x(x - 5)^4 - x^2(4)(x - 5)^3$

76. $5(x^6 + 1)^4(6x^5)(3x + 2)^3 + 3(3x + 2)^2(3)(x^6 + 1)^5$

77. $\dfrac{x^2}{2}(x^2 + 1)^4 - (x^2 + 1)^5$

78. $5w^{-3}(9w + 1)^4(9) + (9w + 1)^5(3w^{-2})$

79. $x^2(x^2 + 1)^{-5} - (x^2 + 1)^{-4}$

80. $2x(x - 5)^{-3} - 4x^2(x - 5)^{-4}$

In Exercises 81 and 82, find two integers c such that the trinomial can be factored. (There are many correct answers.)

81. $2x^2 + 5x + c$

82. $3x^2 - 10x + c$

In Exercises 83–86, match the "geometric factoring model" with the correct factoring formula. [The models are labeled (a), (b), (c), and (d).]

83. $a^2 - b^2 = (a + b)(a - b)$

84. $a^2 + 2ab + b^2 = (a + b)^2$

85. $a^2 + 2a + 1 = (a + 1)^2$

86. $ab + a + b + 1 = (a + 1)(b + 1)$

(a)

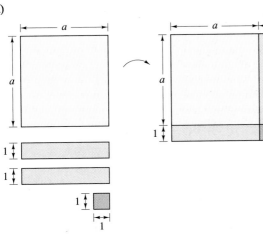

(b)

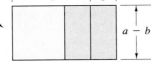

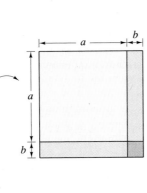

(c)

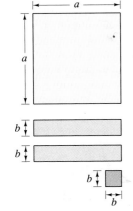

(d)

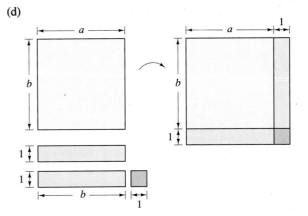

In Exercises 87–90, make a "geometric factoring model" to represent the given factorization. For instance, a factoring model for $2x^2 + 5x + 2 = (2x + 1)(x + 2)$ is shown in the accompanying figure.

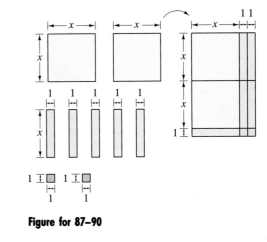

Figure for 87–90

87. $3x^2 + 7x + 2 = (3x + 1)(x + 2)$

88. $x^2 + 4x + 3 = (x + 3)(x + 1)$

89. $2x^2 + 7x + 3 = (2x + 1)(x + 3)$

90. $x^2 + 3x + 2 = (x + 2)(x + 1)$

91. *Volume*   The cylindrical shell shown in the figure has a
volume of $V = \pi R^2 h - \pi r^2 h$.
(a) Factor the expression for the volume.
(b) From the result of part (a) show that the volume is
$V = 2\pi(\text{average radius})(\text{thickness of the shell})h$.

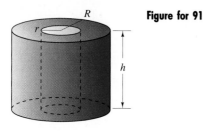

Figure for 91

92. The rate of change of an autocatalytic chemical reaction
is $kQx - kx^2$, where Q is the amount of the original substance, x is the amount of substance formed, and k is a
constant of proportionality. Factor this expression.

1.6 Fractional Expressions

Domain of an Algebraic Expression / Simplifying Rational Expressions /
Operations with Rational Expressions / Compound Fractions

Domain of an Algebraic Expression

The set of real numbers for which an expression is defined is the **domain** of
the expression. We will not usually state the domain of a given algebraic
expression, but it will be implied. Two algebraic expressions are said to be
equivalent if they yield the same value for all numbers in their domain. For
instance, the expressions $[(x + 1) + (x + 2)]$ and $2x + 3$ are equivalent.

EXAMPLE 1   Finding the Domain of an Algebraic Expression

a. The domain of the polynomial

$$2x^3 + 3x + 4$$

is the set of all real numbers. In fact, the domain of any polynomial is the
set of all real numbers (unless the domain is specifically restricted).

b. The domain of the radical expression

$$\sqrt{x - 2}$$

is the set of real numbers greater than or equal to 2, because the square
root of a negative number is not a real number.

c. The domain of the expression

$$\frac{x + 2}{x - 3}$$

is the set of all real numbers except $x = 3$, which would produce an
undefined division by zero.

The quotient of two algebraic expressions is a **fractional expression.** Moreover, the quotient of two *polynomials* such as

$$\frac{1}{x}, \qquad \frac{2x - 1}{x + 1}, \qquad \text{or} \qquad \frac{x^2 - 1}{x^2 + 1}$$

is a **rational expression.** Recall that a fraction is in reduced form if its numerator and denominator have no factors in common aside from ± 1. To write a fraction in reduced form, apply the following rule.

$$\frac{a \cdot c}{b \cdot c} = \frac{a}{b}, \qquad b \neq 0 \quad \text{and} \quad c \neq 0$$

The key to success in simplifying rational expressions lies in your ability to *factor* polynomials.

EXAMPLE 2   Reducing a Rational Expression

Write the following rational expression in reduced form.

$$\frac{x^2 + 4x - 12}{3x - 6}$$

Solution

Factoring both numerator and denominator, and then reducing, produces the following.

$$\frac{x^2 + 4x - 12}{3x - 6} = \frac{(x + 6)(x - 2)}{3(x - 2)} \qquad \text{\textit{Factor completely}}$$

$$= \frac{x + 6}{3}, \quad x \neq 2 \qquad \text{\textit{Reduce}}$$

Note that the original expression is undefined when $x = 2$ (because division by zero is undefined). To make sure that the reduced expression is *equivalent* to the original expression, you must restrict the domain of the reduced expression by excluding the value $x = 2$.

REMARK   In Example 2, do not make the mistake of trying to reduce further by dividing *terms*.

$$\frac{x + 6}{3} \qquad \text{DOES NOT EQUAL} \qquad \frac{x + 6}{3} \quad \text{or} \quad x + 2.$$

Remember that to reduce fractions, divide common *factors,* not terms.

Simplifying Rational Expressions

When simplifying rational expressions, be sure to factor each polynomial completely before concluding that the numerator and denominator have no factors in common. Moreover, changing the sign of a factor may allow further reduction, as seen in part (b) of the next example.

EXAMPLE 3   Reducing Rational Expressions

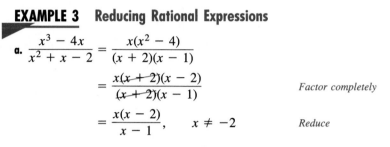

a. $\dfrac{x^3 - 4x}{x^2 + x - 2} = \dfrac{x(x^2 - 4)}{(x + 2)(x - 1)}$

$\qquad\qquad = \dfrac{x(x + 2)(x - 2)}{(x + 2)(x - 1)}$ \hfill *Factor completely*

$\qquad\qquad = \dfrac{x(x - 2)}{x - 1}, \qquad x \neq -2$ \hfill *Reduce*

Note that the implied domain excludes $x = 1$ in both the original and the reduced expressions.

b. $\dfrac{12 + x - x^2}{2x^2 - 9x + 4} = \dfrac{(4 - x)(3 + x)}{(2x - 1)(x - 4)}$ \hfill *Factor completely*

$\qquad\qquad = \dfrac{-(x - 4)(3 + x)}{(2x - 1)(x - 4)}$ \hfill $(4 - x) = -(x - 4)$

$\qquad\qquad = -\dfrac{3 + x}{2x - 1}, \qquad x \neq 4$ \hfill *Reduce*

Operations with Rational Expressions

To multiply or divide rational expressions, use the properties of fractions (discussed in Section 1.2). Recall that to divide fractions we invert the divisor and multiply.

EXAMPLE 4   Multiplying Rational Expressions

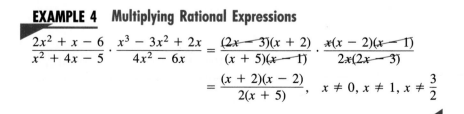

$\dfrac{2x^2 + x - 6}{x^2 + 4x - 5} \cdot \dfrac{x^3 - 3x^2 + 2x}{4x^2 - 6x} = \dfrac{(2x - 3)(x + 2)}{(x + 5)(x - 1)} \cdot \dfrac{x(x - 2)(x - 1)}{2x(2x - 3)}$

$\qquad\qquad\qquad = \dfrac{(x + 2)(x - 2)}{2(x + 5)}, \qquad x \neq 0, x \neq 1, x \neq \dfrac{3}{2}$

EXAMPLE 5   Dividing Rational Expressions

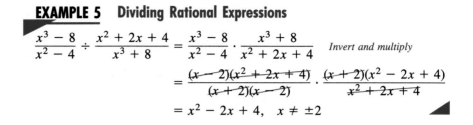

$$\frac{x^3 - 8}{x^2 - 4} \div \frac{x^2 + 2x + 4}{x^3 + 8} = \frac{x^3 - 8}{x^2 - 4} \cdot \frac{x^3 + 8}{x^2 + 2x + 4} \qquad \textit{Invert and multiply}$$

$$= \frac{(x - 2)(x^2 + 2x + 4)}{(x + 2)(x - 2)} \cdot \frac{(x + 2)(x^2 - 2x + 4)}{x^2 + 2x + 4}$$

$$= x^2 - 2x + 4, \quad x \neq \pm 2 \qquad \blacktriangleleft$$

To add or subtract rational expressions, use the familiar LCD (least common denominator) method or the basic definition

$$\frac{a}{b} \pm \frac{c}{d} = \frac{ad \pm bc}{bd}, \quad b \neq 0 \text{ and } d \neq 0.$$

This definition provides an efficient way of adding or subtracting *two* fractions that have no common factors in their denominators.

EXAMPLE 6   Subtracting Rational Expressions Using the Basic Definition

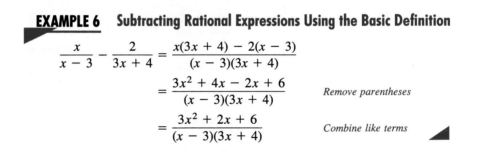

$$\frac{x}{x - 3} - \frac{2}{3x + 4} = \frac{x(3x + 4) - 2(x - 3)}{(x - 3)(3x + 4)}$$

$$= \frac{3x^2 + 4x - 2x + 6}{(x - 3)(3x + 4)} \qquad \textit{Remove parentheses}$$

$$= \frac{3x^2 + 2x + 6}{(x - 3)(3x + 4)} \qquad \textit{Combine like terms} \qquad \blacktriangleleft$$

For three or more fractions, or for fractions with a repeated factor in the denominator, the LCD method works well. Recall that the LCD of several fractions consists of the product of all prime factors in the denominators, with each factor given the highest power of its occurrence in any denominator.

EXAMPLE 7   The LCD Method for Combining Rational Expressions

Perform the given operations and simplify.

$$\frac{3}{x - 1} - \frac{2}{x} + \frac{x + 3}{x^2 - 1}$$

Solution

Using the factored denominators $(x - 1)$, x, and $(x + 1)(x - 1)$, you can see that the LCD is $x(x + 1)(x - 1)$.

$$\frac{3}{x - 1} - \frac{2}{x} + \frac{x + 3}{(x + 1)(x - 1)}$$

$$= \frac{3(x)(x + 1)}{x(x + 1)(x - 1)} - \frac{2(x + 1)(x - 1)}{x(x + 1)(x - 1)} + \frac{(x + 3)(x)}{x(x + 1)(x - 1)}$$

$$= \frac{3(x)(x + 1) - 2(x + 1)(x - 1) + (x + 3)(x)}{x(x + 1)(x - 1)}$$

$$= \frac{3x^2 + 3x - 2x^2 + 2 + x^2 + 3x}{x(x + 1)(x - 1)}$$

$$= \frac{2x^2 + 6x + 2}{x(x + 1)(x - 1)} = \frac{2(x^2 + 3x + 1)}{x(x + 1)(x - 1)}$$

REMARK   Sometimes the numerator of the answer has a factor in common with the denominator. In such cases the answer should be reduced.

Compound Fractions

So far in this section we have limited our operations to simple fractional expressions. Fractional expressions with fractions in the numerator, denominator, or both, are called **compound fractions.** A compound fraction can be simplified by first combining both its numerator and its denominator into single fractions, then inverting the denominator and multiplying.

EXAMPLE 8   Simplifying a Compound Fraction

$$\frac{\left(\dfrac{2}{x} - 3\right)}{\left(1 - \dfrac{1}{x - 1}\right)} = \frac{\left(\dfrac{2 - 3(x)}{x}\right)}{\left(\dfrac{1(x - 1) - 1}{x - 1}\right)} \qquad \textit{Combine fractions}$$

$$= \frac{\left(\dfrac{2 - 3x}{x}\right)}{\left(\dfrac{x - 2}{x - 1}\right)} \qquad \textit{Simplify}$$

$$= \frac{2 - 3x}{x} \cdot \frac{x - 1}{x - 2} \qquad \textit{Invert and multiply}$$

$$= \frac{(2 - 3x)(x - 1)}{x(x - 2)}, \quad x \neq 1$$

Another way to simplify a compound fraction is to multiply each term in its numerator and denominator by the LCD of all fractions in both its numerator and denominator. Each product is then reduced to obtain a single fraction.

EXAMPLE 9   Simplifying a Compound Fraction by Multiplying by the LCD

Use the LCD to simplify the following compound fraction.

$$\frac{\left(\dfrac{1}{x^2} - \dfrac{1}{y^2}\right)}{\left(\dfrac{1}{x} + \dfrac{1}{y}\right)}$$

Solution

For the four fractions in the numerator and denominator, the LCD is x^2y^2. Multiplying each term of the numerator and denominator by this LCD yields the following.

$$\frac{\left(\dfrac{1}{x^2} - \dfrac{1}{y^2}\right)x^2y^2}{\left(\dfrac{1}{x} + \dfrac{1}{y}\right)x^2y^2} = \frac{\left(\dfrac{1}{x^2}\right)x^2y^2 - \left(\dfrac{1}{y^2}\right)x^2y^2}{\left(\dfrac{1}{x}\right)x^2y^2 + \left(\dfrac{1}{y}\right)x^2y^2}$$

$$= \frac{y^2 - x^2}{xy^2 + x^2y} = \frac{(y - x)(y + x)}{xy(y + x)} = \frac{y - x}{xy}$$ ◢

The next two examples illustrate some methods for simplifying expressions involving radicals and negative exponents. (These types of expressions occur frequently in calculus.)

EXAMPLE 10   Simplifying an Expression with Negative Exponents

Simplify the expression

$$x(1 - 2x)^{-3/2} + (1 - 2x)^{-1/2}.$$

Solution

By rewriting the given expression with positive exponents, we obtain

$$\frac{x}{(1 - 2x)^{3/2}} + \frac{1}{(1 - 2x)^{1/2}}$$

which could then be combined by the LCD method. However, by first removing the common factor with the *smaller exponent*, the process is simpler.

$$x(1 - 2x)^{-3/2} + (1 - 2x)^{-1/2} = (1 - 2x)^{-3/2}[x + (1 - 2x)^{(-1/2)-(-3/2)}]$$

$$= (1 - 2x)^{-3/2}[x + (1 - 2x)^1]$$

$$= \frac{1 - x}{(1 - 2x)^{3/2}}$$

Note that when factoring, we subtract exponents.   ◢

> **EXAMPLE 11**   **Simplifying a Compound Fraction: LCD Method**

Simplify the expression

$$\frac{\sqrt{4 - x^2} + \dfrac{x^2}{\sqrt{4 - x^2}}}{4 - x^2}.$$

Solution

$$\frac{\sqrt{4 - x^2} + \dfrac{x^2}{\sqrt{4 - x^2}}}{4 - x^2} = \frac{\sqrt{4 - x^2} + \dfrac{x^2}{\sqrt{4 - x^2}}}{4 - x^2} \cdot \frac{\sqrt{4 - x^2}}{\sqrt{4 - x^2}}$$

$$= \frac{(4 - x^2) + x^2}{(4 - x^2)^{3/2}}$$

$$= \frac{4}{\sqrt{(4 - x^2)^3}}$$

DISCUSSION

PROBLEM

Comparing Domains of Two Expressions

Complete the following table by evaluating the expressions

$$\frac{x^2 - 3x + 2}{x - 2} \qquad \text{and} \qquad x - 1$$

at the indicated values of x. Write a short paragraph describing the equivalence or nonequivalence of the two expressions.

x	-3	-2	-1	0	1	2	3
$\dfrac{x^2 - 3x + 2}{x - 2}$							
$x - 1$							

WARM UP

The following warm-up exercises involve skills that were covered in earlier sections. You will use these skills in the exercise set for this section.

In Exercises 1–10, completely factor the polynomials.

1. $5x^2 - 15x^3$ **2.** $16x^2 - 9$ **3.** $9x^2 - 6x + 1$

4. $9 + 12y + 4y^2$ **5.** $z^2 + 4z + 3$ **6.** $x^2 - 15x + 50$

7. $3 + 8x - 3x^2$ **8.** $3x^2 - 46x + 15$ **9.** $s^3 + s^2 - 4s - 4$

10. $y^3 + 64$

EXERCISES for Section 1.6

In Exercises 1–10, find the domain.

1. $3x^2 - 4x + 7$

2. $2x^2 + 5x - 2$

3. $4x^3 + 5x + 3, \quad x \geq 0$

4. $6x^2 + 7x - 9, \quad x > 0$

5. $\dfrac{1}{x - 2}$

6. $\dfrac{x + 1}{2x + 1}$

7. $\dfrac{x - 1}{x^2 - 4x}$

8. $\dfrac{2x + 1}{x^2 - 9}$

9. $\sqrt{x + 1}$

10. $\dfrac{1}{\sqrt{x + 1}}$

In Exercises 11–16, find the factor that makes the two fractions equivalent.

11. $\dfrac{5}{2x} = \dfrac{5(\rule{1cm}{0.4pt})}{6x^2}$

12. $\dfrac{3}{4} = \dfrac{3(\rule{1cm}{0.4pt})}{4(x + 1)}$

13. $\dfrac{x + 1}{x} = \dfrac{(x + 1)(\rule{1cm}{0.4pt})}{x(x - 2)}$

14. $\dfrac{3y - 4}{y + 1} = \dfrac{(3y - 4)(\rule{1cm}{0.4pt})}{y^2 - 1}$

15. $\dfrac{3x}{x - 3} = \dfrac{3x(\rule{1cm}{0.4pt})}{x^2 - x - 6}$

16. $\dfrac{1 - z}{z^2} = \dfrac{(1 - z)(\rule{1cm}{0.4pt})}{z^3 + z^2}$

In Exercises 17–30, write in reduced form.

17. $\dfrac{15x^2}{10x}$

18. $\dfrac{18y^2}{60y^5}$

19. $\dfrac{3xy}{xy + x}$

20. $\dfrac{9x^2 + 9x}{2x + 2}$

21. $\dfrac{x - 5}{10 - 2x}$

22. $\dfrac{x^2 - 25}{5 - x}$

23. $\dfrac{x^3 + 5x^2 + 6x}{x^2 - 4}$

24. $\dfrac{x^2 + 8x - 20}{x^2 + 11x + 10}$

25. $\dfrac{y^2 - 7y + 12}{y^2 + 3y - 18}$

26. $\dfrac{3 - x}{x^2 - 5x + 6}$

27. $\dfrac{2 - x + 2x^2 - x^3}{x - 2}$

28. $\dfrac{x^2 - 9}{x^3 + x^2 - 9x - 9}$

29. $\dfrac{z^3 - 8}{z^2 + 2z + 4}$

30. $\dfrac{y^3 - 2y^2 - 3y}{y^3 + 1}$

In Exercises 31–56, perform the indicated operations and simplify.

31. $\dfrac{5}{x - 1} \cdot \dfrac{x - 1}{25(x - 2)}$

32. $\dfrac{(x + 5)(x - 3)}{x + 2} \cdot \dfrac{1}{(x + 5)(x + 2)}$

33. $\dfrac{(x - 9)(x + 7)}{x + 1} \cdot \dfrac{x}{9 - x}$

34. $\dfrac{x + 13}{x^3(3 - x)} \cdot \dfrac{x(x - 3)}{5}$

35. $\dfrac{r}{r - 1} \cdot \dfrac{r^2 - 1}{r^2}$

36. $\dfrac{4y - 16}{5y + 15} \cdot \dfrac{2y + 6}{4 - y}$

37. $\dfrac{t^2 - t - 6}{t^2 + 6t + 9} \cdot \dfrac{t + 3}{t^2 - 4}$

38. $\dfrac{y^3 - 8}{2y^3} \cdot \dfrac{4y}{y^2 - 5y + 6}$

39. $\dfrac{x^2 + xy - 2y^2}{x^3 + x^2y} \cdot \dfrac{x}{x^2 + 3xy + 2y^2}$

40. $\dfrac{x^3 - 1}{x + 1} \cdot \dfrac{x^2 + 1}{x^2 - 1}$

41. $\dfrac{3(x + y)}{4} \div \dfrac{x + y}{2}$

42. $\dfrac{x + 2}{5(x - 3)} \div \dfrac{x - 2}{5(x - 3)}$

43. $\dfrac{\left(\dfrac{x^2}{(x + 1)^2}\right)}{\left(\dfrac{x}{(x + 1)^3}\right)}$

44. $\dfrac{\left(\dfrac{x^2 - 1}{x}\right)}{\left(\dfrac{(x - 1)^2}{x}\right)}$

45. $\dfrac{5}{x - 1} + \dfrac{x}{x - 1}$

46. $\dfrac{2x - 1}{x + 3} + \dfrac{1 - x}{x + 3}$

47. $6 - \dfrac{5}{x + 3}$

48. $\dfrac{3}{x - 1} - 5$

49. $\dfrac{3}{x - 2} + \dfrac{5}{2 - x}$

50. $\dfrac{2x}{x - 5} - \dfrac{5}{5 - x}$

51. $\dfrac{2}{x^2 - 4} - \dfrac{1}{x^2 - 3x + 2}$

52. $\dfrac{x}{x^2 + x - 2} - \dfrac{1}{x + 2}$

53. $\dfrac{1}{x^2 - x - 2} - \dfrac{x}{x^2 - 5x + 6}$

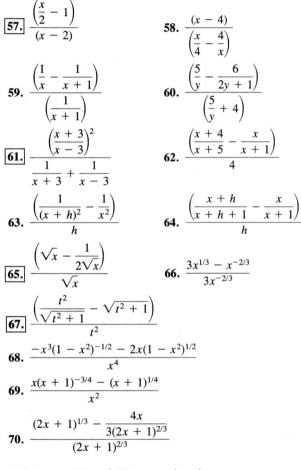

54. $\dfrac{2}{x^2 - x - 2} + \dfrac{10}{x^2 + 2x - 8}$

55. $-\dfrac{1}{x} + \dfrac{2}{x^2 + 1} + \dfrac{1}{x^3 + x}$

56. $\dfrac{2}{x + 1} + \dfrac{2}{x - 1} + \dfrac{1}{x^2 - 1}$

In Exercises 57–70, simplify the compound fraction.

57. $\dfrac{\left(\dfrac{x}{2} - 1\right)}{(x - 2)}$

58. $\dfrac{(x - 4)}{\left(\dfrac{x}{4} - \dfrac{4}{x}\right)}$

59. $\dfrac{\left(\dfrac{1}{x} - \dfrac{1}{x + 1}\right)}{\left(\dfrac{1}{x + 1}\right)}$

60. $\dfrac{\left(\dfrac{5}{y} - \dfrac{6}{2y + 1}\right)}{\left(\dfrac{5}{y} + 4\right)}$

61. $\dfrac{\left(\dfrac{x + 3}{x - 3}\right)^2}{\dfrac{1}{x + 3} + \dfrac{1}{x - 3}}$

62. $\dfrac{\left(\dfrac{x + 4}{x + 5} - \dfrac{x}{x + 1}\right)}{4}$

63. $\dfrac{\left(\dfrac{1}{(x + h)^2} - \dfrac{1}{x^2}\right)}{h}$

64. $\dfrac{\left(\dfrac{x + h}{x + h + 1} - \dfrac{x}{x + 1}\right)}{h}$

65. $\dfrac{\left(\sqrt{x} - \dfrac{1}{2\sqrt{x}}\right)}{\sqrt{x}}$

66. $\dfrac{3x^{1/3} - x^{-2/3}}{3x^{-2/3}}$

67. $\dfrac{\left(\dfrac{t^2}{\sqrt{t^2 + 1}} - \sqrt{t^2 + 1}\right)}{t^2}$

68. $\dfrac{-x^3(1 - x^2)^{-1/2} - 2x(1 - x^2)^{1/2}}{x^4}$

69. $\dfrac{x(x + 1)^{-3/4} - (x + 1)^{1/4}}{x^2}$

70. $\dfrac{(2x + 1)^{1/3} - \dfrac{4x}{3(2x + 1)^{2/3}}}{(2x + 1)^{2/3}}$

In Exercises 71 and 72, rationalize the numerator.

71. $\dfrac{\sqrt{x + 2} - \sqrt{x}}{2}$

72. $\dfrac{\sqrt{z - 3} - \sqrt{z}}{3}$

73. *Rate*   A photocopier copies at a rate of 16 pages per minute.
   (a) Find the time required to copy one page.
   (b) Find the time required to copy x pages.
   (c) Find the time required to copy 60 pages.

74. *Rate*   After two people work together for t hours on a common task, the fractional part of the job done by each of the two workers is $t/3$ and $t/5$. What fractional part of the task has been completed?

75. *Average*   Find the average of $x/3$ and $2x/5$.

76. *Partition into Equal Parts*   Find three real numbers that divide the real number line between $x/3$ and $3x/4$ into four equal parts.

Monthly Payment   In Exercises 77 and 78, use the following formula, which gives the approximate annual percentage rate r of a monthly installment loan

$$r = \dfrac{\left(\dfrac{24(NM - P)}{N}\right)}{\left(P + \dfrac{NM}{12}\right)}$$

where N is the total number of payments, M is the monthly payment, and P is the amount financed.

77. (a) Approximate the annual percentage rate for a four-year car loan of $15,000 that has monthly payments of $400.
   (b) Simplify the expression for the annual percentage rate r, and then rework part (a).

78. (a) Approximate the annual percentage rate for a five-year car loan of $18,000 that has monthly payments of $400.
   (b) Simplify the expression for the annual percentage rate r, and then rework part (a).

79. *Electronics*   When two resistors are connected in parallel, the total resistance is given by

$$\dfrac{1}{\dfrac{1}{R_1} + \dfrac{1}{R_2}}.$$

Simplify this compound fraction.

80. *Refrigeration*   When food (at room temperature) is placed in a refrigerator, the time required for the food to cool depends on the amount of food, the air circulation in the refrigerator, the original temperature of the food, and the temperature of the refrigerator. Consider the following model, which gives the temperature of food that is 75°F and is placed in a 40°F refrigerator

$$T = 10\left(\dfrac{4t^2 + 16t + 75}{t^2 + 4t + 10}\right)$$

where T is the temperature in degrees Fahrenheit and t is the time in hours. Sketch a bar graph showing the temperature of the food when $t = 0, 1, 2, 3, 4,$ and 5 hours.

1.7 Algebraic Errors and Some Algebra of Calculus

Algebraic Errors to Avoid / Some Algebra of Calculus

Algebraic Errors to Avoid

Before wrapping up our review of the fundamental concepts of algebra, we list some common algebraic errors. Many of these errors are made because they seem to be the *easiest* things to do.

Errors Involving Parentheses

Potential Error		Correct Form	Comment
$a - (x - b)$ DOES NOT EQUAL	$a - x - b$	$a - (x - b) = a - x + b$	*Change all signs when distributing negative sign through parentheses.*
$(a + b)^2$ DOES NOT EQUAL	$a^2 + b^2$	$(a + b)^2 = a^2 + 2ab + b^2$	*Remember the middle term when squaring binomials.*
$\left(\dfrac{1}{2}a\right)\left(\dfrac{1}{2}b\right)$ DOES NOT EQUAL	$\dfrac{1}{2}(ab)$	$\left(\dfrac{1}{2}a\right)\left(\dfrac{1}{2}b\right) = \dfrac{1}{4}(ab) = \dfrac{ab}{4}$	$\frac{1}{2}$ *occurs twice as a factor.*
$(3x + 6)^2$ DOES NOT EQUAL	$3(x + 2)^2$	$(3x + 6)^2 = [3(x + 2)]^2$ $= 3^2(x + 2)^2$	*When factoring, apply exponents to all factors.*
$6(3x + 4)^5$ DOES NOT EQUAL	$(18x + 24)^5$	No simpler form	*Exponent 5 does not apply to 6.*

Errors Involving Fractions

Potential Error		Correct Form	Comment
$\dfrac{a}{x + b}$ DOES NOT EQUAL	$\dfrac{a}{x} + \dfrac{a}{b}$	Leave as $\dfrac{a}{x + b}$	*Do not add denominators when adding fractions.*
$\dfrac{\left(\dfrac{x}{a}\right)}{b}$ DOES NOT EQUAL	$\dfrac{bx}{a}$	$\dfrac{\left(\dfrac{x}{a}\right)}{b} = \left(\dfrac{x}{a}\right)\left(\dfrac{1}{b}\right) = \dfrac{x}{ab}$	*Multiply by the reciprocal when dividing fractions.*
$\dfrac{1}{a} + \dfrac{1}{b}$ DOES NOT EQUAL	$\dfrac{1}{a + b}$	$\dfrac{1}{a} + \dfrac{1}{b} = \dfrac{a + b}{ab}$	*Use the definition for adding fractions.*
$\dfrac{1}{3x}$ DOES NOT EQUAL	$\dfrac{1}{3}x$	$\dfrac{1}{3x} = \dfrac{1}{3} \cdot \dfrac{1}{x}$	*Use the definition for multiplying fractions.*
$(1/3)x$ DOES NOT EQUAL	$\dfrac{1}{3x}$	$(1/3)x = \dfrac{1}{3} \cdot x = \dfrac{x}{3}$	*Be careful when using a slash to denote division.*
$(1/x) + 2$ DOES NOT EQUAL	$\dfrac{1}{x + 2}$	$(1/x) + 2 = \dfrac{1}{x} + 2 = \dfrac{1 + 2x}{x}$	*Be careful when using a slash to denote division.*

Errors Involving Exponents and Radicals

Potential Error		Correct Form	Comment
$(x^2)^3$	DOES NOT EQUAL x^5	$(x^2)^3 = x^{2 \cdot 3} = x^6$	Multiply exponents when raising an exponential form to a power.
$x^2 \cdot x^3$	DOES NOT EQUAL x^6	$x^2 \cdot x^3 = x^{2+3} = x^5$	Add exponents when multiplying exponential forms with like bases.
$-2x^3$	DOES NOT EQUAL $(-2x)^3$	$-2x^3 = -2(x^3)$	Exponents have priority over multiplication.
$3x^{-5}$	DOES NOT EQUAL $\dfrac{1}{3x^5}$	$3x^{-5} = \dfrac{3}{x^5}$	Negative exponent does not apply to coefficient.
$\dfrac{1}{x^{1/2} - x^{1/3}}$	DOES NOT EQUAL $x^{-1/2} - x^{-1/3}$	Leave as $\dfrac{1}{x^{1/2} - x^{1/3}}$	Do not move term-by-term from denominator to numerator.
$\sqrt{5x}$	DOES NOT EQUAL $5\sqrt{x}$	$\sqrt{5x} = \sqrt{5}\sqrt{x}$	Radicals apply to every factor inside the radical.
$\sqrt{x^2 + a^2}$	DOES NOT EQUAL $x + a$	Leave as $\sqrt{x^2 + a^2}$	Do not apply radicals term-by-term.
$\sqrt{-x^2 + a^2}$	DOES NOT EQUAL $-\sqrt{x^2 - a^2}$	Leave as $\sqrt{-x^2 + a^2}$	Do not factor negative signs out of square roots.

Errors Involving Cancellation

Potential Error		Correct Form	Comment
$\dfrac{a + bx}{a}$	DOES NOT EQUAL $1 + bx$	$\dfrac{a + bx}{a} = \dfrac{a}{a} + \dfrac{bx}{a} = 1 + \dfrac{b}{a}x$	Reduce common factors, not common terms.
$\dfrac{a + ax}{a}$	DOES NOT EQUAL $a + x$	$\dfrac{a + ax}{a} = \dfrac{a(1 + x)}{a} = 1 + x$	Factor before reducing.
$1 + \dfrac{x}{2x}$	DOES NOT EQUAL $1 + \dfrac{1}{x}$	$1 + \dfrac{x}{2x} = 1 + \dfrac{1}{2} = \dfrac{3}{2}$	Reduce common factors.

Some Algebra of Calculus

In calculus it is often necessary to take a simplified algebraic expression and "unsimplify" it. See the following list, taken from a standard calculus text.

Unusual Factoring

Expression	Useful Calculus Form of Expression	Comment
$\dfrac{5x^4}{8}$	$\dfrac{5}{8}x^4$	*Write with fractional coefficient.*
$\dfrac{x^2 + 3x}{-6}$	$-\dfrac{1}{6}(x^2 + 3x)$	*Write with fractional coefficient.*
$2x^2 - x - 3$	$2\left(x^2 - \dfrac{x}{2} - \dfrac{3}{2}\right)$	*Factor out the leading coefficient.*
$\dfrac{x}{2}(x + 1)^{-1/2} + (x + 1)^{1/2}$	$\dfrac{(x + 1)^{-1/2}}{2}[x + 2(x + 1)]$	*Factor out factor with least power.*

Inserting Factors or Terms

Expression	Useful Calculus Form of Expression	Comment
$(2x - 1)^3$	$\dfrac{1}{2}(2x - 1)^3(2)$	*Multiply and divide by 2.*
$7x^2(4x^3 - 5)^{1/2}$	$\dfrac{7}{12}(4x^3 - 5)^{1/2}(12x^2)$	*Multiply and divide by 12.*
$\dfrac{4x^2}{9} - 4y^2 = 1$	$\dfrac{x^2}{9/4} - \dfrac{y^2}{1/4} = 1$	*Write with fractional denominators.*
$\dfrac{x}{x + 1}$	$\dfrac{x + 1 - 1}{x + 1} = 1 - \dfrac{1}{x + 1}$	*Add and subtract the same term.*

Writing a Fraction as a Sum

Expression	*Useful Calculus Form of Expression*	*Comment*
$\dfrac{x + 2x^2 + 1}{\sqrt{x}}$	$x^{1/2} + 2x^{3/2} + x^{-1/2}$	*Divide each term by $x^{1/2}$.*
$\dfrac{1 + x}{x^2 + 1}$	$\dfrac{1}{x^2 + 1} + \dfrac{x}{x^2 + 1}$	*Rewrite the fraction as the sum of fractions.*
$\dfrac{2x}{x^2 + 2x + 1}$	$\dfrac{2x + 2 - 2}{x^2 + 2x + 1}$	*Add and subtract a term to the numerator.*
	$= \dfrac{2x + 2}{x^2 + 2x + 1} - \dfrac{2}{(x + 1)^2}$	*Rewrite the fraction as the difference of fractions.*
$\dfrac{x^2 - 2}{x + 1}$	$x - 1 - \dfrac{1}{x + 1}$	*Use long division. (See Section 4.3.)*
$\dfrac{x + 7}{x^2 - x - 6}$	$\dfrac{2}{x - 3} - \dfrac{1}{x + 2}$	*Use the method of partial fractions. (See Section 5.2.)*

Rewriting with Negative Exponents

Expression	*Useful Calculus Form of Expression*	*Comment*
$\dfrac{9}{5x^3}$	$\dfrac{9}{5}x^{-3}$	*Move the factor to numerator and change the sign of the exponent.*
$\dfrac{7}{\sqrt{2x - 3}}$	$7(2x - 3)^{-1/2}$	*Move the factor to numerator and change the sign of the exponent.*

EXAMPLE 1 Rewriting Fractions

Explain the following.

$$\frac{4x^2}{9} - 4y^2 = \frac{x^2}{9/4} - \frac{y^2}{1/4}$$

Solution

To write the expression on the left side of the equation in the form given on the right, multiply the numerator and denominator of both terms by $\frac{1}{4}$.

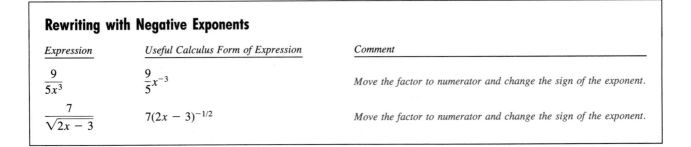

$$\frac{4x^2}{9} - 4y^2 = \frac{4x^2}{9}\left(\frac{1/4}{1/4}\right) - 4y^2\left(\frac{1/4}{1/4}\right)$$

$$= \frac{x^2}{9/4} - \frac{y^2}{1/4}$$

EXAMPLE 2   Rewriting with Negative Exponents

Rewrite the expression using negative exponents.

$$\frac{2}{5x^3} - \frac{1}{\sqrt{x}} + \frac{3}{16x^2}$$

Solution

$$\frac{2}{5x^3} - \frac{1}{\sqrt{x}} + \frac{3}{16x^2} = \frac{2}{5x^3} - \frac{1}{x^{1/2}} + \frac{3}{(4x)^2}$$

$$= \frac{2}{5}x^{-3} - x^{-1/2} + 3(4x)^{-2}$$

EXAMPLE 3   Factors Involving Negative Exponents

Factor $x(x + 1)^{-1/2} + (x + 1)^{1/2}$.

Solution

When multiplying factors with like bases, you add exponents. When factoring, you are undoing multiplication, and so you *subtract* exponents.

$$x(x + 1)^{-1/2} + (x + 1)^{1/2} = (x + 1)^{-1/2}[x(x + 1)^0 + (x + 1)^1]$$

$$= (x + 1)^{-1/2}[x + (x + 1)]$$

$$= (x + 1)^{-1/2}(2x + 1)$$

EXAMPLE 4   Writing a Fraction as a Sum of Terms

Rewrite the fraction as the sum of three terms.

$$\frac{x + 2x^2 + 1}{\sqrt{x}}$$

Solution

$$\frac{x + 2x^2 + 1}{\sqrt{x}} = \frac{x}{x^{1/2}} + \frac{2x^2}{x^{1/2}} + \frac{1}{x^{1/2}}$$

$$= x^{1/2} + 2x^{3/2} + x^{-1/2}$$

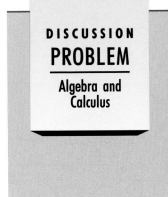

DISCUSSION
PROBLEM
Algebra and
Calculus

Suppose you are taking a course in calculus, and for one of the homework problems you obtain the following answer.

$$\frac{1}{10}(2x - 1)^{5/2} + \frac{1}{6}(2x - 1)^{3/2}$$

The answer in the back of the book is

$$\frac{1}{15}(2x - 1)^{3/2}(3x + 1).$$

Are these two answers equivalent? If so, show how the second answer can be obtained from the first.

EXERCISES for Section 1.7

In Exercises 1–24, find and correct any errors.

1. $2x - (3y + 4) = 2x - 3y + 4$

2. $\dfrac{4}{16x - (2x + 1)} = \dfrac{4}{14x + 1}$

3. $5z + 3(x - 2) = 5z + 3x - 2$

4. $x(yz) = (xy)(xz)$

5. $-\dfrac{x - 3}{x - 1} = \dfrac{3 - x}{1 - x}$

6. $\dfrac{x - 1}{(5 - x)(-x)} = \dfrac{1 - x}{x(5 - x)}$

7. $a\left(\dfrac{x}{y}\right) = \dfrac{ax}{ay}$

8. $(5z)(6z) = 30z$

9. $(4x)^2 = 4x^2$

10. $\left(\dfrac{x}{y}\right)^3 = \dfrac{x^3}{y}$

11. $\sqrt{x + 9} = \sqrt{x} + 3$

12. $\sqrt{25 - x^2} = 5 - x$

13. $\dfrac{6x + y}{6x - y} = \dfrac{x + y}{x - y}$

14. $\dfrac{2x^2 + 1}{5x} = \dfrac{2x + 1}{5}$

15. $\dfrac{1}{x + y^{-1}} = \dfrac{y}{x + 1}$

16. $\dfrac{1}{a^{-1} + b^{-1}} = \left(\dfrac{1}{a + b}\right)^{-1}$

17. $x(2x - 1)^2 = (2x^2 - x)^2$

18. $x(x + 5)^{1/2} = (x^2 + 5x)^{1/2}$

19. $\sqrt[3]{x^3 + 7x^2} = x^2\sqrt[3]{x + 7}$

20. $(3x^2 - 6x)^3 = 3x(x - 2)^3$

21. $\dfrac{3}{x} + \dfrac{4}{y} = \dfrac{7}{x + y}$

22. $\dfrac{7 + 5(x + 3)}{x + 3} = 12$

23. $\dfrac{1}{2y} = \left(\dfrac{1}{2}\right)y$

24. $\dfrac{2x + 3x^2}{4x} = \dfrac{2 + 3x^2}{4}$

In Exercises 25–44, insert the required factor in the parentheses.

25. $\dfrac{3x + 2}{5} = \dfrac{1}{5}(\rule{1cm}{0.3cm})$

26. $\frac{3}{4}x + \frac{1}{2} = \frac{1}{4}(\rule{1cm}{0.3cm})$

27. $\frac{1}{3}x^3 + 5 = (\rule{1cm}{0.3cm})(x^3 + 15)$

28. $\frac{5}{2}z^2 - \frac{1}{4}z + 2 = (\rule{1cm}{0.3cm})(10z^2 - z + 8)$

29. $x(1 - 2x^2)^3 = (\rule{1cm}{0.3cm})(1 - 2x^2)^3(-4x)$

30. $5x\sqrt[3]{1 + x^2} = (\rule{1cm}{0.3cm})\sqrt[3]{1 + x^2}(2x)$

31. $\dfrac{1}{\sqrt{x}(1 + \sqrt{x})^2} = (\rule{1cm}{0.3cm})\dfrac{1}{(1 + \sqrt{x})^2}\left(\dfrac{1}{2\sqrt{x}}\right)$

32. $\dfrac{4x + 6}{(x^2 + 3x + 7)^3} = (\rule{1cm}{0.3cm})\dfrac{1}{(x^2 + 3x + 7)^3}(2x + 3)$

33. $\dfrac{x + 1}{(x^2 + 2x - 3)^2} = (\rule{1cm}{0.3cm})\dfrac{1}{(x^2 + 2x - 3)^2}(2x + 2)$

34. $\dfrac{1}{(x - 1)\sqrt{(x - 1)^4 - 4}} = \dfrac{(\rule{1cm}{0.3cm})}{(x - 1)^2\sqrt{(x - 1)^4 - 4}}$

35. $\dfrac{3}{x} + \dfrac{5}{2x^2} - \dfrac{3}{2}x = (\rule{1cm}{0.3cm})(6x + 5 - 3x^3)$

36. $\dfrac{(x - 1)^2}{169} + (y + 5)^2 = \dfrac{(x - 1)^3}{169(\rule{1cm}{0.3cm})} + (y + 5)^2$

37. $\dfrac{x^2}{1/12} - \dfrac{y^2}{2/3} = \dfrac{12x^2}{(\rule{1cm}{0.3cm})} - \dfrac{3y^2}{(\rule{1cm}{0.3cm})}$

38. $\dfrac{x^2}{4/9} + \dfrac{y^2}{7/8} = \dfrac{9x^2}{(\rule{1cm}{0.3cm})} + \dfrac{8y^2}{(\rule{1cm}{0.3cm})}$

39. $\sqrt{x} + (\sqrt{x})^3 = \sqrt{x}(\rule{1cm}{0.3cm})$

40. $(1 - 3x)^{4/3} - 4x(1 - 3x)^{1/3} = (1 - 3x)^{1/3}(\rule{1cm}{0.3cm})$

41. $\dfrac{x^2}{\sqrt{x^2+1}} - \sqrt{x^2+1} = \dfrac{1}{\sqrt{x^2+1}}(\rule{1cm}{0.3cm})$

42. $\dfrac{1}{2\sqrt{x}} + 5x^{3/2} - 10x^{5/2} = \dfrac{1}{2\sqrt{x}}(\rule{1cm}{0.3cm})$

43. $\dfrac{1}{10}(2x+1)^{5/2} - \dfrac{1}{6}(2x+1)^{3/2} = \dfrac{(2x+1)^{3/2}}{15}(\rule{1cm}{0.3cm})$

44. $\dfrac{3}{7}(t+1)^{7/3} - \dfrac{3}{4}(t+1)^{4/3} = \dfrac{3(t+1)^{4/3}}{28}(\rule{1cm}{0.3cm})$

In Exercises 45–50, write the fraction as a sum of two or more terms.

45. $\dfrac{16 - 5x - x^2}{x}$

46. $\dfrac{x^3 - 5x^2 + 4}{x^2}$

47. $\dfrac{4x^3 - 7x^2 + 1}{x^{1/3}}$

48. $\dfrac{2x^5 - 3x^3 + 5x - 1}{x^{3/2}}$

49. $\dfrac{3 - 5x^2 - x^4}{\sqrt{x}}$

50. $\dfrac{x^3 - 5x^4}{3x^2}$

In Exercises 51–58, simplify the expression.

51. $\dfrac{-2(x^2-3)^{-3}(2x)(6x+1)^3 - 3(6x+1)^2(6)(x^2-3)^{-2}}{[(6x+1)^3]^2}$

52. $\dfrac{(3x+2)^5(-3)(x^2+1)^{-4}(2x) - (x^2+1)^{-3}(5)(3x+2)^4(3)}{[(3x+2)^5]^2}$

53. $\dfrac{(6x+1)^3(27x^2+2) - (9x^3+2x)(3)(6x+1)^2(6)}{[(6x+1)^3]^2}$

54. $\dfrac{(4x^2+9)^{1/2}(2) - (2x+3)(\frac{1}{2})(4x^2+9)^{-1/2}(8x)}{[(4x^2+9)^{1/2}]^2}$

55. $\dfrac{(3x+2)^{3/4}(2x+3)^{-2/3}(2) - (2x+3)^{1/3}(3x+2)^{-1/4}(3)}{[(3x+2)^{3/4}]^2}$

56. $\dfrac{\sqrt{2x-1} - \dfrac{x+2}{\sqrt{2x-1}}}{2x-1}$

57. $\dfrac{2(3x-1)^{1/3} - (2x+1)(\frac{1}{3})(3x-1)^{-2/3}(3)}{(3x-1)^{2/3}}$

58. $\dfrac{(x+1)(\frac{1}{2})(2x-3x^2)^{-1/2}(2-6x) - (2x-3x^2)^{1/2}}{(x+1)^2}$

REVIEW EXERCISES for Chapter 1

In Exercises 1 and 2, determine which numbers in the given set are (a) natural numbers, (b) integers, (c) rational numbers, and (d) irrational numbers.

1. $11, -14, -\frac{8}{9}, \frac{5}{2}, \sqrt{6}, 0.4$

2. $\sqrt{15}, -22, -\frac{10}{3}, 0, 5.2, \frac{3}{7}$

In Exercises 3 and 4, plot the two real numbers on the real number line and place the appropriate inequality sign (< or >) between them.

3. $-4, -3$

4. $\frac{1}{2}, \frac{1}{3}$

In Exercises 5 and 6, give a verbal description of the subset of real numbers that is represented by the inequality, and sketch the subset on the real number line.

5. $x \le 7$

6. $x > 1$

In Exercises 7 and 8, write the given expression without using absolute value signs.

7. $-|-14|$

8. $|-4 - 2|$

In Exercises 9–12, use absolute value notation to describe the given expression.

9. The distance between x and 7 is at least 4.

10. The distance between x and 25 is no more than 10.

11. The distance between y and -30 is less than 5.

12. The distance between y and $\frac{1}{2}$ is more than 2.

In Exercises 13–24, perform the indicated operations without the aid of a calculator.

13. $|-3| + 4(-2) - 6$

14. $(16 - 8) \div 4$

15. $\sqrt{5} \cdot \sqrt{125}$

16. $\dfrac{\sqrt{72}}{\sqrt{2}}$

17. $6[4 - 2(6 + 8)]$

18. $-4[16 - 3(7 - 10)]$

19. $\left(\dfrac{3^2}{5^2}\right)^{-3}$

20. $6^{-4}(-3)^5$

21. $2(-27)^{2/3}$

22. $\left(\dfrac{25}{16}\right)^{-1/2}$

23. $(3 \times 10^4)^2$

24. $\dfrac{1}{(4 \times 10^{-2})^3}$

In Exercises 25 and 26, write the number in scientific notation.

25. *Daily U.S. Consumption of Dunkin' Donuts:* 2,740,000
26. *Number of Meters in One Foot:* 0.3048

In Exercises 27 and 28, write the number in decimal form.

27. *Distance Between Sun and Jupiter:* 4.833×10^8 miles
28. *Ratio of Day to Year:* 2.74×10^{-3}

In Exercises 29 and 30, use a calculator to evaluate the given expression. (Round your answer to three decimal places.)

29. (a) $1800(1 + 0.08)^{24}$
   (b) $0.0024(7,658,400)$

30. (a) $50,000\left(1 + \dfrac{0.075}{12}\right)^{48}$
   (b) $\dfrac{28,000,000 + 34,000,000}{87,000,000}$

In Exercises 31–44, describe the error and then make the necessary correction.

31. $10(4 \cdot 7) = 40 \cdot 70$
32. $\left(\frac{1}{3}x\right)\left(\frac{1}{3}y\right) = \frac{1}{3}xy$
33. $4\left(\frac{3}{7}\right) = \frac{12}{28}$
34. $\frac{2}{9} \times \frac{4}{9} = \frac{8}{9}$
35. $\dfrac{x - 1}{1 - x} = 1$
36. $\dfrac{\frac{15}{16}}{\frac{2}{3}} = \dfrac{5}{8}$
37. $(2x)^4 = 2x^4$
38. $(-x)^6 = -x^6$
39. $(3^4)^4 = 3^8$
40. $-x^2(-x^2 + 3) = x^4 + 3x^2$
41. $\sqrt{3^2 + 4^2} = 3 + 4$
42. $(5 + 8)^2 = 5^2 + 8^2$
43. $\sqrt{10x} = 10\sqrt{x}$
44. $\sqrt{7x}\sqrt[3]{2} = \sqrt{14x}$

In Exercises 45 and 46, fill in the missing description.

Radical Form	*Rational Exponent Form*
45. $\sqrt{16} = 4$	
46. ▨	$16^{1/4} = 2$

In Exercises 47 and 48, simplify by removing all possible factors from the radical.

47. $\sqrt{4x^4}$
48. $\sqrt[3]{\dfrac{2x^3}{27}}$

In Exercises 49 and 50, simplify the expression.

49. $\sqrt{50} - \sqrt{18}$
50. $\sqrt{8x^3} + \sqrt{2x}$

In Exercises 51 and 52, rewrite the expression by rationalizing the denominator. Simplify your answer.

51. $\dfrac{1}{2 - \sqrt{3}}$
52. $\dfrac{1}{\sqrt{x} - 1}$

In Exercises 53–58, perform the required multiplication and simplify your answer.

53. $(2x - 3)^2$
54. $(3\sqrt{5} + 2)(3\sqrt{5} - 2)$
55. $(x^2 - 2x + 1)(x^3 - 1)$
56. $(x^3 - 3x)(2x^2 + 3x + 5)$
57. $(y^2 - y)(y^2 + 1)(y^2 + y + 1)$
58. $\left(x - \dfrac{1}{x}\right)(x + 2)$

In Exercises 59–64, factor completely.

59. $x^3 - x$
60. $x(x - 3) + 4(x - 3)$
61. $2x^2 + 21x + 10$
62. $3x^2 + 14x + 8$
63. $x^3 - x^2 + 2x - 2$
64. $x^3 - 1$

In Exercises 65–68, insert the missing factor.

65. $\frac{3}{4}x^2 - \frac{5}{6}x + 4 = \frac{1}{12}(\blacksquare)$
66. $\frac{2}{3}x^4 - \frac{3}{8}x^3 + \frac{5}{6}x^2 = \frac{1}{24}x^2(\blacksquare)$
67. $\dfrac{t}{\sqrt{t + 1}} - \sqrt{t + 1} = \dfrac{1}{\sqrt{t + 1}}(\blacksquare)$
68. $2x(x^2 - 3)^{1/3} - 5(x^2 - 3)^{4/3} = (x^2 - 3)^{1/3}(\blacksquare)$

In Exercises 69–78, perform the required operation and simplify the answer.

69. $\dfrac{x^2 - 4}{x^4 - 2x^2 - 8} \cdot \dfrac{x^2 + 2}{x^2}$

70. $\dfrac{2x - 1}{x + 1} \cdot \dfrac{x^2 - 1}{2x^2 - 7x + 3}$

71. $\dfrac{x^2(5x - 6)}{2x + 3} \div \dfrac{5x}{2x + 3}$

72. $\dfrac{4x - 6}{(x - 1)^2} \div \dfrac{2x^2 - 3x}{x^2 + 2x - 3}$

73. $x - 1 + \dfrac{1}{x + 2} + \dfrac{1}{x - 1}$

74. $2x + \dfrac{3}{2(x - 4)} - \dfrac{1}{2(x + 2)}$

75. $\dfrac{1}{x} - \dfrac{x - 1}{x^2 + 1}$

76. $\dfrac{1}{x - 1} + \dfrac{1 - x}{x^2 + x + 1}$

77. $\dfrac{1}{x - 2} + \dfrac{1}{(x - 2)^2} + \dfrac{1}{x + 2}$

78. $\dfrac{1}{L}\left(\dfrac{1}{y} - \dfrac{1}{L - y}\right)$, where L is a constant

In Exercises 79–82, simplify the given compound fraction.

79. $\dfrac{\left(\dfrac{1}{x} - \dfrac{1}{y}\right)}{(x^2 - y^2)}$

80. $\dfrac{\left(\dfrac{1}{x} - \dfrac{1}{y}\right)}{\left(\dfrac{1}{x} + \dfrac{1}{y}\right)}$

81. $\dfrac{\left(\dfrac{3a}{(a^2/x) - 1}\right)}{\left(\dfrac{a}{x} - 1\right)}$

82. $\dfrac{\left(\dfrac{1}{2x - 3} - \dfrac{1}{2x + 3}\right)}{\left(\dfrac{1}{2x} - \dfrac{1}{2x + 3}\right)}$

83. Use a calculator to complete the following table.

	1	10	10^2	10^4	10^6	10^{10}

What number is $5/\sqrt{n}$ approaching as n increases without bound?

84. Calculate $\sqrt[5]{107}\,\sqrt[5]{1145}$ in two ways. First, use the key-stroke sequence

107 $\boxed{y^x}$ .2 $\boxed{\times}$ 1145 $\boxed{y^x}$ .2 $\boxed{=}$.

Second, use the sequence

107 $\boxed{\times}$ 1145 $\boxed{=}$ $\boxed{y^x}$ .2 $\boxed{=}$.

Why do these two methods give the same result?

85. Let m and n be any two integers. Then $2m$ and $2n$ are even integers and $(2m + 1)$ and $(2n + 1)$ are odd integers.
(a) Prove that the sum of two even integers is even.
(b) Prove that the sum of two odd integers is even.
(c) Prove that the product of an even integer and any integer is even.

86. *Weight of Copper Tubing* One foot of $\frac{1}{8}$-inch copper wire weighs 4 ounces. What is the weight of $\frac{5}{8}$ mile of this tubing? [*Note:* 1 mile = 5280 feet.]

87. *Surface Area* The inside and outside radii of a thrust washer are r inches and R inches, respectively (see figure). Find an algebraic expression for the surface area of one side of the washer. Factor the expression if possible.

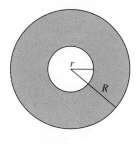

Figure for 87

C H A P T E R 2

OVERVIEW

The primary focus of the first chapter was *simplifying expressions.* Here, the primary focus is *solving equations* and *inequalities.* The chapter is organized according to equation and inequality types—linear, quadratic, and other types.

In some cases, you are asked to *use* models to answer questions. For instance, Example 7 on page 72 asks you to use a linear model to estimate when weekly earnings in the United States will average $500. In other cases, you are asked to first *create* a model and then use it. For instance, Example 9 on page 96 asks you to create a quadratic model to answer questions about construction safety.

As you study, do not overlook the importance of *checking* solutions in the original equation or problem statement.

Algebraic Equations and Inequalities

2.1 Linear Equations

Equations and Solutions of Equations / Linear Equations in One Variable / Equations Involving Fractional Expressions / Applications

Equations and Solutions of Equations

In Chapter 1 we reviewed the fundamentals of algebra. We now *use* these fundamentals to solve problems that can be expressed in the form of equations or inequalities. Such problems are common in science, business, industry, and government.

An **equation** is a statement that two algebraic expressions are equal. For example, $3x - 5 = 7$, $x^2 - x - 6 = 0$, and $\sqrt{2x} = 4$, are equations. To **solve** an equation in x means to find all values of x for which the equation is true. Such values are **solutions.** For instance, $x = 4$ is a solution of the equation $3x - 5 = 7$, because $3(4) - 5 = 7$ is a true statement.

The solutions of an equation depend upon the kinds of numbers being considered. For instance, in the set of rational numbers the equation $x^2 = 10$ has no solution because there is no rational number whose square is 10. However, in the set of real numbers this equation has two solutions, $\sqrt{10}$ and $-\sqrt{10}$, because $(\sqrt{10})^2 = 10$ and $(-\sqrt{10})^2 = 10$.

An equation that is true for *every* real number in the domain of the variable is an **identity.** For example, $x^2 - 9 = (x + 3)(x - 3)$ is an identity

67

because it is a true statement for any real value of x, and $x/3x^2 = 1/3x$, where $x \neq 0$, is an identity because it is true for any nonzero real value of x.

An equation that is true for just *some* (or even none) of the real numbers in the domain of the variable is a **conditional equation.** For example, the equation $x^2 - 9 = 0$ is conditional because $x = 3$ and $x = -3$ are the only values in the domain that satisfy the equation. Learning to solve conditional equations is the primary focus of this chapter.

Linear Equations in One Variable

The most common type of conditional equation is a **linear equation.**

DEFINITION OF LINEAR EQUATION

A **linear equation** in one variable x is an equation that can be written in the standard form

$$ax + b = 0$$

where a and b are real numbers with $a \neq 0$.

Linear equations have exactly one solution. To see this, consider the following steps. (Remember that $a \neq 0$.)

$$ax + b = 0 \qquad \textit{Given equation}$$

$$ax = -b \qquad \textit{Subtract b from both sides}$$

$$x = -\frac{b}{a} \qquad \textit{Divide both sides by a}$$

To solve a conditional equation in x, isolate x on one side of the equation by a sequence of **equivalent** (and usually simpler) equations, each having the same solution(s) as the original equation. The operations that yield equivalent equations come from the Substitution Principle and the simplification techniques studied in Chapter 1.

Generating Equivalent Equations

An equation can be transformed into an *equivalent equation* by one or more of the following steps.

	Given Equation	*Equivalent Equation*
1. Remove symbols of grouping, combine like terms, or reduce fractions on one or both sides of the equation.	$2x - x = 4$	$x = 4$
2. Add (or subtract) the same quantity to *both* sides of the equation.	$x + 1 = 6$	$x = 5$
3. Multiply (or divide) *both* sides of the equation by the same *nonzero* quantity.	$2x = 6$	$x = 3$
4. Interchange the two sides of the equation.	$2 = x$	$x = 2$

EXAMPLE 1 Solving a Linear Equation

$3x - 6 = 0$	*Given equation*
$3x = 6$	*Add 6 to both sides*
$x = 2$	*Divide both sides by 3*

After solving an equation, you should **check each solution** in the *original* equation. In Example 1, check that 2 is a solution by substituting 2 for x in the original equation.

$$3(2) - 6 = 6 - 6 = 0 \qquad \textit{Check solution}$$

Some linear equations have no solution because all the x-terms subtract out and a contradictory (false) statement such as $0 = 5$ or $12 = 7$ is obtained. Watch for this type of linear equation in the exercises.

EXAMPLE 2 Solving a Linear Equation

$6(x - 1) + 4 = 3(7x + 1)$	*Given equation*
$6x - 6 + 4 = 21x + 3$	*Remove parentheses*
$6x - 2 = 21x + 3$	*Simplify*
$-15x = 5$	*Add 2 and subtract 21x*
$x = -\dfrac{1}{3}$	*Divide by −15*

Check Check this solution by substitution in the original equation.

$6(x - 1) + 4 = 3(7x + 1)$	*Given equation*
$6\left(-\dfrac{1}{3} - 1\right) + 4 \stackrel{?}{=} 3\left[7\left(-\dfrac{1}{3}\right) + 1\right]$	*Replace x by −1/3*
$6\left(-\dfrac{4}{3}\right) + 4 \stackrel{?}{=} 3\left[-\dfrac{7}{3} + 1\right]$	*Add fractions*
$-\dfrac{24}{3} + 4 \stackrel{?}{=} -\dfrac{21}{3} + 3$	
$-8 + 4 \stackrel{?}{=} -7 + 3$	*Simplify*
$-4 = -4$	*Solution checks*

Students sometimes tell us that a solution looks easy when we work it out in class, but that they don't see where to begin when trying it alone. Keep in mind that no one—not even great mathematicians—can expect to look at every mathematical problem and immediately know where to begin. Many problems involve some trial and error before a solution is found. To make algebra work for you, you must put in a lot of time, you must expect to try solution methods that end up not working, and you must learn from both your successes and your failures.

Equations Involving Fractional Expressions

To solve an equation involving fractional expressions, we find the least common denominator of all terms in the equation and multiply every term by this LCD. This procedure clears the equation of fractions.

EXAMPLE 3   Solving an Equation Involving Fractional Expressions

Solve the equation for x.

$$\frac{x}{3} + \frac{3x}{4} = 2$$

Solution

The least common denominator is 12.

$$\frac{x}{3} + \frac{3x}{4} = 2 \qquad\qquad \textit{Given equation}$$

$$(12)\frac{x}{3} + (12)\frac{3x}{4} = (12)2 \qquad\qquad \textit{Multiply by the LCD}$$

$$4x + 9x = 24 \qquad\qquad \textit{Reduce and multiply}$$

$$13x = 24 \qquad\qquad \textit{Combine like terms}$$

$$x = \frac{24}{13} \qquad\qquad \textit{Divide by 13}$$

The equation has one solution: $\frac{24}{13}$. Check this solution in the original equation.

     When multiplying or dividing an equation by a *variable* quantity, it is possible to introduce an **extraneous** solution that does not satisfy the original equation.

EXAMPLE 4   An Equation with an Extraneous Solution

Solve the equation for x.

$$\frac{1}{x - 2} = \frac{3}{x + 2} - \frac{6x}{x^2 - 4}$$

Solution

The LCD is $x^2 - 4$ or $(x + 2)(x - 2)$. Multiply every term by this LCD and reduce.

$$\frac{1}{x-2}(x+2)(x-2) = \frac{3}{x+2}(x+2)(x-2) - \frac{6x}{x^2-4}(x+2)(x-2)$$

$$x + 2 = 3(x - 2) - 6x, \quad x \neq \pm 2$$

$$x + 2 = 3x - 6 - 6x$$

$$4x = -8$$

$$x = -2$$

In the original equation, $x = -2$ yields a denominator of zero; so, $x = -2$ is an extraneous solution, and the equation has *no solution*. ◣

An equation with a *single fraction* on each side can be cleared of denominators by **cross multiplying,** which is equivalent to multiplying by the LCD and then reducing. For instance, in the equation

$$\frac{2}{x-3} = \frac{3}{x+1}$$

the LCD is $(x - 3)(x + 1)$. Multiply both sides of the equation by this LCD.

$$\frac{2}{x-3}(x-3)(x+1) = \frac{3}{x+1}(x-3)(x+1)$$

$$2(x + 1) = 3(x - 3), \quad x \neq -1, x \neq 3$$

By comparing this equation to the original, you can see that the original numerators and denominators were "cross multiplied." That is, the left numerator was multiplied by the right denominator and the right numerator was multiplied by the left denominator.

EXAMPLE 5 Cross Multiplying to Solve an Equation

Solve for y.

$$\frac{3y-2}{2y+1} = \frac{6y-9}{4y+3}$$

Solution

Since this equation has a single fraction on each side, you can cross multiply.

$$\frac{3y-2}{2y+1} = \frac{6y-9}{4y+3} \qquad \textit{Given equation}$$

$$(3y - 2)(4y + 3) = (6y - 9)(2y + 1), \quad y \neq -\frac{1}{2}, -\frac{3}{4} \qquad \textit{Cross multiply}$$

$$12y^2 + y - 6 = 12y^2 - 12y - 9 \qquad \textit{Combine like terms}$$

$$13y = -3 \qquad \textit{Divide by 13}$$

$$y = -\frac{3}{13}$$

Check that $y = -\frac{3}{13}$ is a solution of the original equation. ◣

Up to this point, we have carefully chosen examples in which the calculations are simple. This is rather artificial, since real-world problems frequently involve numbers that are not simple integers or fractions. In such cases a calculator is useful.

EXAMPLE 6 Using a Calculator to Solve an Equation

Solve for x.

$$\frac{1}{9.38} - \frac{3}{x} = \frac{5}{0.3714}$$

Solution

Round-off error will be minimized if you solve for x before doing any calculations. In this case the LCD is $(9.38)(0.3714)(x)$.

$$\frac{1}{9.38} - \frac{3}{x} = \frac{5}{0.3714}$$

$$(9.38)(0.3714)(x)\left(\frac{1}{9.38} - \frac{3}{x}\right) = (9.38)(0.3714)(x)\left(\frac{5}{0.3714}\right)$$

$$0.3714x - 3(9.38)(0.3714) = (9.38)(5)(x), \quad x \neq 0$$

$$0.3714x - 5(9.38)x = 3(9.38)(0.3714)$$

$$[0.3714 - 5(9.38)]x = 3(9.38)(0.3714)$$

$$x = \frac{3(9.38)(0.3714)}{0.3714 - 5(9.38)}$$

$$x \approx -0.225 \qquad \textit{Round to three places}$$

REMARK Because of round-off error, a check of a decimal solution may not yield exactly the same values for both sides of the original equation. The difference, however, will usually be quite small.

Applications

Of all the different mathematical models you will be studying, linear models are the most commonly used.

EXAMPLE 7 An Application: Weekly Earnings

The average weekly earnings for workers in the United States between 1980 and 1990 can be approximated by the linear equation

$$y = 14.5t + 270$$

where y represents the average weekly earnings in dollars and t represents the calendar year (with $t = 0$ corresponding to 1980). From Figure 2.1, you can see that the average weekly earnings increased in a linear pattern during the 1980s. Assuming that this pattern continues, find the year when the average weekly earnings will reach $500. (*Source:* U.S. Bureau of Labor Statistics)

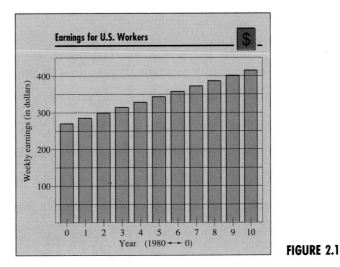

FIGURE 2.1

Solution

Let $y = 500$ and solve for t.

$y = 14.5t + 270$	*Given equation*
$500 = 14.5t + 270$	*Let y = 500*
$230 = 14.5t$	*Subtract 270 from both sides*
$t = \dfrac{230}{14.5}$	*Divide both sides by 14.5*
$t \approx 16$	*Solution*

Since $t = 0$ corresponds to 1980, you can conclude that $t = 16$ must correspond to 1996. Thus, from this model the average weekly earnings will reach $500 by 1996.

DISCUSSION PROBLEM

A Mathematical Fallacy

A mathematical **fallacy** is an argument that appears to prove something that we know is incorrect. For instance, the following argument appears to prove that $1 = 0$. Can you find the error in this argument?

$x = 1$	*Given equation*
$x - 1 = 0$	*Subtract 1 from both sides*
$x(x - 1) = 0$	*Multiply both sides by x*
$\dfrac{x(x - 1)}{x - 1} = \dfrac{0}{x - 1}$	*Divide both sides by x − 1*
$\dfrac{x(x - 1)}{x - 1} = 0$	*Reduce*
$x = 0$	*Solution*

WARM UP

The following warm-up exercises involve skills that were covered in earlier sections. You will use these skills in the exercise set for this section.

In Exercises 1–10, perform the indicated operations and simplify your answer.

1. $(2x - 4) - (5x + 6)$

2. $(3x - 5) + (2x - 7)$

3. $2(x + 1) - (x + 2)$

4. $-3(2x - 4) + 7(x + 2)$

5. $\dfrac{x}{3} + \dfrac{x}{5}$

6. $x - \dfrac{x}{4}$

7. $\dfrac{1}{x + 1} - \dfrac{1}{x}$

8. $\dfrac{2}{x} + \dfrac{3}{x}$

9. $\dfrac{4}{x} + \dfrac{3}{x - 2}$

10. $\dfrac{1}{x + 1} - \dfrac{1}{x - 1}$

EXERCISES for Section 2.1

In Exercises 1–10, determine whether the equation is an identity or a conditional equation.

1. $2(x - 1) = 2x - 2$

2. $3(x + 2) = 3x + 4$

3. $-2(x - 3) + 5 = -2x + 10$

4. $3(x + 2) - 5 = 3x + 1$

5. $4(x + 1) - 2x = 2(x + 2)$

6. $-7(x - 3) + 4x = 3(7 - x)$

7. $x^2 - 8x + 5 = (x - 4)^2 - 11$

8. $x^2 + 2(3x - 2) = x^2 + 6x - 4$

9. $3 + \dfrac{1}{x + 1} = \dfrac{4x}{x + 1}$

10. $\dfrac{5}{x} + \dfrac{3}{x} = 24$

In Exercises 11–18, determine whether the given value of x is a solution of the equation.

Equation	Values	
11. $5x - 3 = 3x + 5$	(a) $x = 0$	(b) $x = -5$
	(c) $x = 4$	(d) $x = 10$
12. $7 - 3x = 5x - 17$	(a) $x = -3$	(b) $x = 0$
	(c) $x = 8$	(d) $x = 3$
13. $3x^2 + 2x - 5 = 2x^2 - 2$	(a) $x = -3$	(b) $x = 1$
	(c) $x = 4$	(d) $x = -5$

Equation	Values	
14. $5x^3 + 2x - 3$ $= 4x^3 + 2x - 11$	(a) $x = 2$ (c) $x = 0$	(b) $x = -2$ (d) $x = 10$
15. $\dfrac{5}{2x} - \dfrac{4}{x} = 3$	(a) $x = -\frac{1}{2}$ (c) $x = 0$	(b) $x = 4$ (d) $x = \frac{1}{4}$
16. $3 + \dfrac{1}{x + 2} = 4$	(a) $x = -1$ (c) $x = 0$	(b) $x = -2$ (d) $x = 5$
17. $(x + 5)(x - 3) = 20$	(a) $x = 3$ (c) $x = 5$	(b) $x = -5$ (d) $x = -7$
18. $\sqrt[3]{x - 8} = 3$	(a) $x = 2$ (c) $x = 35$	(b) $x = -2$ (d) $x = 8$

In Exercises 19–56, solve the equation (if possible) and check your answer.

19. $x + 10 = 15$

20. $7 - x = 18$

21. $7 - 2x = 15$

22. $7x + 2 = 16$

23. $8x - 5 = 3x + 10$

24. $7x + 3 = 3x - 13$

25. $2(x + 5) - 7 = 3(x - 2)$

26. $2(13t - 15) + 3(t - 19) = 0$

27. $6[x - (2x + 3)] = 8 - 5x$

28. $8(x + 2) - 3(2x + 1) = 2(x + 5)$

29. $\dfrac{5x}{4} + \dfrac{1}{2} = x - \dfrac{1}{2}$

30. $\dfrac{x}{5} - \dfrac{x}{2} = 3$

31. $\dfrac{3}{2}(z + 5) - \dfrac{1}{4}(z + 24) = 0$

32. $\dfrac{3x}{2} + \dfrac{1}{4}(x - 2) = 10$

33. $0.25x + 0.75(10 - x) = 3$

34. $0.60x + 0.40(100 - x) = 50$

35. $x + 8 = 2(x - 2) - x$

36. $3(x + 3) = 5(1 - x) - 1$

37. $\dfrac{100 - 4u}{3} = \dfrac{5u + 6}{4} + 6$

38. $\dfrac{17 + y}{y} + \dfrac{32 + y}{y} = 100$

39. $\dfrac{5x - 4}{5x + 4} = \dfrac{2}{3}$

40. $\dfrac{10x + 3}{5x + 6} = \dfrac{1}{2}$

41. $10 - \dfrac{13}{x} = 4 + \dfrac{5}{x}$

42. $\dfrac{15}{x} - 4 = \dfrac{6}{x} + 3$

43. $\dfrac{1}{x - 3} + \dfrac{1}{x + 3} = \dfrac{10}{x^2 - 9}$

44. $\dfrac{1}{x - 2} + \dfrac{3}{x + 3} = \dfrac{4}{x^2 + x - 6}$

45. $\dfrac{x}{x + 4} + \dfrac{4}{x + 4} + 2 = 0$

46. $\dfrac{2}{(x - 4)(x - 2)} = \dfrac{1}{x - 4} + \dfrac{2}{x - 2}$

47. $\dfrac{7}{2x + 1} - \dfrac{8x}{2x - 1} = -4$

48. $\dfrac{4}{u - 1} + \dfrac{6}{3u + 1} = \dfrac{15}{3u + 1}$

49. $\dfrac{1}{x} + \dfrac{2}{x - 5} = 0$

50. $\dfrac{6}{x} - \dfrac{2}{x + 3} = \dfrac{3(x + 5)}{x(x + 3)}$

51. $\dfrac{3}{x(x - 3)} + \dfrac{4}{x} = \dfrac{1}{x - 3}$

52. $3 = 2 + \dfrac{2}{z + 2}$

53. $(x + 2)^2 + 5 = (x + 3)^2$

54. $(x + 1)^2 + 2(x - 2) = (x + 1)(x - 2)$

55. $(x + 2)^2 - x^2 = 4(x + 1)$

56. $(2x + 1)^2 = 4(x^2 + x + 1)$

In Exercises 57–60, solve for x.

57. $4(x + 1) - ax = x + 5$

58. $6x + ax = 2x + 5$

59. $4 - 2(x - 2b) = ax + 3$

60. $5 + ax = 12 - bx$

In Exercises 61–64, use your calculator to solve the equation for x and round to three decimal places.

61. $0.275x + 0.725(500 - x) = 300$

62. $2.763 - 4.5(2.1x - 5.1432) = 6.32x + 5$

63. $\dfrac{x}{0.6321} + \dfrac{x}{0.0692} = 1000$

64. $\dfrac{2}{7.398} - \dfrac{4.405}{x} = \dfrac{1}{x}$

In Exercises 65 and 66, find an equation of the form $ax + b = cx$ that has the given solution. (There are many correct answers.)

65. $x = 2$

66. $x = \dfrac{1}{3}$

In Exercises 67–70, evaluate the expression in two ways. (a) Calculate entirely on your calculator by storing intermediate results, and then round to two decimal places. (b) Round both the numerator and the denominator to two decimal places before dividing, and then round the final answer to two decimal places. Does the second method introduce an additional round-off error?

67. $\dfrac{1 + 0.73205}{1 - 0.73205}$

68. $\dfrac{1 + 0.86603}{1 - 0.86603}$

69. $\dfrac{3.33 + \dfrac{1.98}{0.74}}{4 + \dfrac{6.25}{3.15}}$

70. $\dfrac{1.73205 - 1.19195}{3 - (1.73205)(1.19195)}$

Human Height In Exercises 71 and 72, use the following information. The relationship between the length of an adult's thigh bone and the height of the adult can be approximated by the linear equations

$y = 0.432x - 10.44$ Female
$y = 0.449x - 12.15$ Male

where y is the length of the thigh bone in inches and x is the height in inches (see figure).

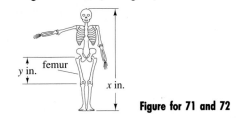

Figure for 71 and 72

71. An anthropologist discovers a thigh bone belonging to an adult human female. The bone is 16 inches long. Estimate the height of the female.

72. From the foot bones of an adult human male, an anthropologist estimates that the height of the male was 69 inches. Near where the foot bone was discovered, the anthropologist finds a male adult thigh bone 19 inches long. Is it possible that both bones came from the same person?

Negative Income Tax In Exercises 73–76, use the following information about a possible negative income tax for a family of two adults and two children. The plan would guarantee the poor a minimum income while encouraging families to increase their private income (see figure).

Family's earned income: $I = x$
Government payment: $G = 8{,}000 - \frac{1}{2}x, \quad 0 \le x \le 16{,}000$
Spendable income: $S = I + G$

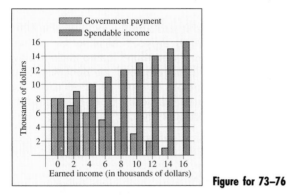

Figure for 73–76

73. Express the spendable income S in terms of x.

74. Government payment is $4600. Find the earned income x.

75. Spendable income is $11,800. Find the earned income x.

76. Spendable income is $10,500. Find the government payment G.

77. The surface area S of the rectangular solid in the figure is
$$S = 2(24) + 2(4x) + 2(6x).$$
Find the length of the box x if the surface area is 248 square inches.

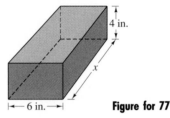

Figure for 77

78. A delivery company has a fleet of vans. The annual operating cost per van is
$$C = 0.32m + 2{,}500$$
where m is the number of miles traveled by a van in a year. What number of miles will yield an annual operating cost that is equal to $10,000?

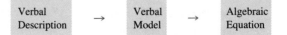

2.2 Linear Equations and Modeling

Introduction to Problem Solving / Using Mathematical Models to Solve Problems / Mixture Problems / Common Formulas

Introduction to Problem Solving

In this section you will learn how algebra can be used to solve problems that occur in real-life situations. This procedure is **mathematical modeling.**

A good approach to mathematical modeling is to use two stages. Begin by using the verbal description of the problem to form a *verbal model*. Then, after assigning labels to the unknown quantities in the verbal model, form a *mathematical model* or *algebraic equation*.

Verbal Description	→	Verbal Model	→	Algebraic Equation

When you are trying to construct a verbal model, it is helpful to look for a *hidden equality*—a statement that two algebraic expressions are equal. These two expressions might be explicitly stated as being equal, or they might be known to be equal (based on prior knowledge or experience).

EXAMPLE 1 **Using a Verbal Model to Construct an Algebraic Equation**

You have accepted a job for which your annual salary will be $24,740. This salary includes a year-end bonus of $500. If you are paid twice a month, what will be your gross pay for each paycheck?

Solution

Since there are 12 months in a year and you will be paid twice a month, it follows that you will receive 24 paychecks during the year. You can construct an algebraic equation for this problem. Begin with a verbal model, then assign labels, and finally form an algebraic equation.

Verbal Model   Income for year = 24 paychecks + Bonus

Labels   Income for year = $24,740

Amount of each paycheck = x

Bonus = $500

Algebraic Equation $24,740 = 24x + 500$

The algebraic equation for this problem is a *linear equation* in the variable x. Using the techniques discussed in Section 2.1, you can solve this equation for x. If you do that, you will find that the solution is $x = \$1010$.

To build an algebraic equation, it is important that you recognize the *vocabulary* of algebraic operations.

Translating Key Words and Phrases

Key Words and Phrases	Verbal Description	Algebraic Statement
Equality Equals, equal to, is, are, was, will be, represents	The sale price S is $10 less than the list price L.	$S = L - 10$
Addition Sum, plus, greater, increased by, more than, exceeds, total of	The sum of 5 and x Seven more than y	$5 + x$ $y + 7$
Subtraction Difference, minus, less, decreased by, subtracted from, reduced by, the remainder	The difference of 4 and b Three less than z	$4 - b$ $z - 3$
Multiplication Product, multiplied by, twice, times, percent of	Two times x	$2x$
Division Quotient, divided by, ratio, per	The ratio of x and 8	$\dfrac{x}{8}$

Using Mathematical Models to Solve Problems

Study the next several examples carefully. Your goal should be to develop a *general problem-solving strategy.*

EXAMPLE 2   Finding the Percent of a Raise

Suppose you accepted a job that pays $8 an hour. You are told that after a two-month probationary period, your hourly wage will be increased to $9 an hour. What percent raise will you receive after the two-month period?

Solution

Verbal Model   Raise = Percent · Old wage

Labels   Old wage = $8

New wage = $9

Raise = 9 − 8 = $1

Percent = r (in decimal form)

Algebraic Equation $1 = r \cdot 8$

$$\frac{1}{8} = r$$

$0.125 = r$

You will receive a raise of $0.125 = 12.5\%$.

EXAMPLE 3   Finding the Percent of a Benefits Package

Suppose your annual salary is $22,000. In addition to your salary, your employer provides the following benefits.

Social Security (employer's portion)	7.65% of salary	$1,683.00
Workers' compensation	0.5% of salary	110.00
Unemployment compensation	0.75% of salary	165.00
Medical insurance	$2,240 per year	2,240.00
Retirement contribution	6.5% of salary	1,430.00

The total of this benefits package represents what percent of your annual salary?

Solution

To begin, add the dollar amount of the benefits package and obtain a total of $5,628. Then set up the following model.

Verbal Model	Benefits package $=$ Percent $\cdot$ Salary

Labels	Salary $=$ \$22,000
	Benefits package $=$ \$5,628
	Percent $= r$ (in decimal form)

Algebraic Equation	$5{,}628 = r \cdot 22{,}000$

$$\frac{5{,}628}{22{,}000} = r$$

$$0.256 \approx r$$

Your benefits package is approximately $0.256 = 25.6\%$ of your salary.

EXAMPLE 4   Finding the Dimensions of a Room

A rectangular family room is twice as long as it is wide, and its perimeter is 84 feet. Find the dimensions of the family room.

Solution

For this problem, it helps to sketch a picture, as shown in Figure 2.2.

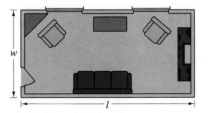

FIGURE 2.2

Verbal Model	$2 \cdot$ Length $+ \, 2 \cdot$ Width $=$ Perimeter

Labels	Perimeter $=$ 84 feet
	Width $= w$ (measured in feet)
	Length $= l = 2w$ (measured in feet)

Algebraic Equation	$2(2w) + 2w = 84$

$$4w + 2w = 84$$
$$6w = 84$$
$$w = 14 \text{ feet}$$
$$l = 2w = 28 \text{ feet}$$

The dimensions of the room are 14 feet by 28 feet.

EXAMPLE 5   A Distance Problem

A plane is flying nonstop from New York to San Francisco, a distance of about 2700 miles, as shown in Figure 2.3. After $1\frac{1}{2}$ hours in the air, the plane flies over Chicago (a distance of 800 miles from New York). How long will it take the plane to fly from New York to San Francisco? (Assume that the plane flies at a constant speed during the entire flight.)

FIGURE 2.3

Solution

To solve this problem, use the formula that relates distance, rate, and time.

Verbal Model   Distance = Rate · Time

Labels   Distance = 2700 miles

Time = t (in hours)

Rate = $\dfrac{\text{Distance to Chicago}}{\text{Time to Chicago}} = \dfrac{800}{1.5} = 533.33$ mph

Algebraic Equation $2700 = (533.33)t$

$\dfrac{2700}{533.33} = t$

$t \approx 5.06$ hours

Since 0.06 hours represents about 4 minutes, we conclude that the trip will take about 5 hours and 4 minutes.

Another way to solve the distance problem in Example 5 is to use the concept of **ratio and proportion.** To do this, we let x represent the time it takes to fly from New York to San Francisco, and we set up the following proportion.

$$\frac{\text{Time to Chicago}}{\text{Distance to Chicago}} = \frac{\text{Time to San Francisco}}{\text{Distance to San Francisco}}$$

$$\frac{1.5}{800} = \frac{x}{2700}$$

$$\frac{1.5}{800} \cdot 2700 = x$$

$$5.06 \approx x$$

In the next example, we use this technique to solve a problem involving similar triangles.

EXAMPLE 6   An Application Involving Similar Triangles

In order to measure the height of the twin towers of the World Trade Center (in New York City), suppose you use the following scheme. You measure the shadow cast by one of the buildings and find it to be 170.25 feet, as shown in Figure 2.4. Then you measure the shadow cast by a 4-foot post and find that its shadow is 6 inches long. Use this information to determine the height of the building.

Solution

To solve this problem, you use a result from geometry that tells you that the ratios of corresponding sides of similar triangles are equal.

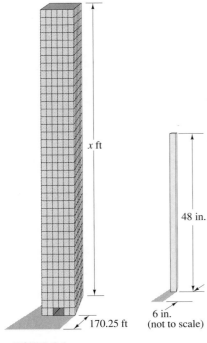

x ft

48 in.

6 in.
(not to scale)

 170.25 ft

FIGURE 2.4

Verbal Model $$\dfrac{\text{Height of building}}{\text{Length of shadow}} = \dfrac{\text{Height of post}}{\text{Length of shadow}}$$

Labels   Height of building $= x$ (measured in feet)

Length of building's shadow $= 170.25$ feet

Height of stake $= 4$ feet $= 48$ inches

Length of stake's shadow $= 6$ inches

Algebraic Equation $$\dfrac{x}{170.25} = \dfrac{48}{6}$$

$$x = 8 \cdot 170.25$$

$$x = 1362 \text{ feet}$$

Thus, the World Trade Center is about 1362 feet high.

Mixture Problems

Mixture problems involve two different unknown quantities that are *mixed* in a specific way. Watch for a *hidden product* in the verbal model.

EXAMPLE 7   A Simple Interest Problem

Suppose you received $10,000 from an inheritance and invested it in two ways. Part of the money was invested at $9\frac{1}{2}\%$ simple interest and the remainder was invested at 11%. After one year, the two investments returned a combined interest of $1,038.50. How much did you invest in each type of account?

Solution

Simple interest problems are based on the formula $I = Prt$, where I is the interest, P is the principal, r is the annual percentage rate (in decimal form), and t is the time in years.

Verbal Model

$$\dfrac{\text{Interest}}{\text{from } 9\frac{1}{2}\%} + \dfrac{\text{Interest}}{\text{from } 11\%} = \dfrac{\text{Total}}{\text{interest}}$$

Labels

Amount invested at $9\frac{1}{2}\% = x$ (in dollars)

Amount invested at $11\% = 10,000 - x$ (in dollars)

Interest from $9\frac{1}{2}\% = Prt = (x)(0.095)(1)$ (in dollars)

Interest from $11\% = Prt = (10,000 - x)(0.11)(1)$ (in dollars)

Total interest $= \$1,038.50$

Algebraic Equation

$$0.095x + 0.11(10,000 - x) = 1,038.5$$
$$0.095x + 1100 - 0.11x = 1,038.5$$
$$-0.015x = -61.5$$
$$x = \$4,100 \text{ at } 9\tfrac{1}{2}\%$$
$$10,000 - x = \$5,900 \text{ at } 11\%$$

Thus, the amount invested at $9\frac{1}{2}\%$ was $\$4,100$ and the amount invested at 11% was $\$5,900$.   ◢

In Example 7, did you recognize the hidden product in the two terms on the left side of the equation?

EXAMPLE 8   An Inventory Problem

A department store has $\$30,000$ of inventory in 12-inch and 19-inch color television sets. The profit on a 12-inch set is 22%, while the profit on a 19-inch set is 40%. If the profit for the entire stock is 35%, how much is invested in each type of television?

Solution

Verbal Model

$$\dfrac{\text{Profit from}}{\text{12-inch sets}} + \dfrac{\text{Profit from}}{\text{19-inch sets}} = \dfrac{\text{Total}}{\text{profit}}$$

Labels

Inventory of 12-inch sets $= x$ (in dollars)

Inventory of 19-inch sets $= 30,000 - x$ (in dollars)

Profit from 12-inch sets $= 0.22x$ (in dollars)

Profit from 19-inch sets $= 0.40(30,000 - x)$ (in dollars)

Total profit $= 0.35(30,000) = \$10,500$

Algebraic
Equation

$$0.22(x) + 0.40(30,000 - x) = 0.35(30,000)$$

$$0.22x + 12,000 - 0.4x = 10,500$$

$$-0.18x = -1,500$$

$$x \approx \$8,333.33 \quad \text{12-inch sets}$$

$$30,000 - x \approx \$21,666.67 \quad \text{19-inch sets}$$

Thus, the total inventory for 12-inch sets is approximately \$8333.33 and the total inventory for 19-inch sets is approximately \$21,666.67.

Common Formulas

Many common types of geometric, scientific, and investment problems use ready-made equations called **formulas.** Knowing these formulas will help you translate and solve a wide variety of real-life problems.

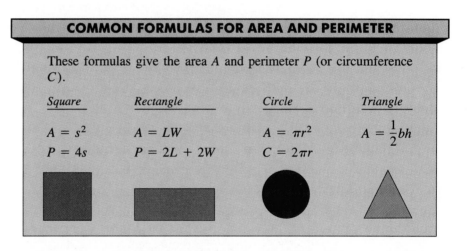

COMMON FORMULAS FOR AREA AND PERIMETER

These formulas give the area A and perimeter P (or circumference C).

Square	*Rectangle*	*Circle*	*Triangle*
$A = s^2$	$A = LW$	$A = \pi r^2$	$A = \dfrac{1}{2}bh$
$P = 4s$	$P = 2L + 2W$	$C = 2\pi r$	

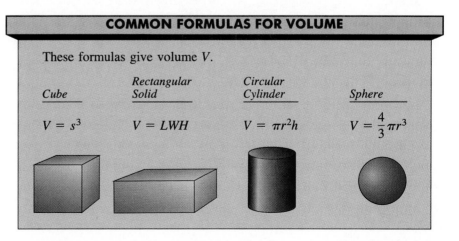

COMMON FORMULAS FOR VOLUME

These formulas give volume V.

Cube	*Rectangular Solid*	*Circular Cylinder*	*Sphere*
$V = s^3$	$V = LWH$	$V = \pi r^2 h$	$V = \dfrac{4}{3}\pi r^3$

MISCELLANEOUS COMMON FORMULAS

Temperature $F = \dfrac{9}{5}C + 32$

F = degrees Fahrenheit C = degrees Celsius

Simple Interest $I = Prt$

I = interest P = principal

r = annual interest rate t = time in years

Compound Interest $A = P\left(1 + \dfrac{r}{n}\right)^{nt}$

A = balance P = principal

r = annual interest rate n = compoundings per year

t = time in years

Distance $d = rt$

d = distance traveled r = rate

t = time

When working with applied problems you often need to rewrite one of the common formulas. For instance, the formula, $P = 2L + 2W$, for the perimeter of a rectangle, can be rewritten or solved for W in this way.

$$P = 2L + 2W \qquad \textit{Given formula}$$

$$P - 2L = 2W$$

$$\frac{P - 2L}{2} = W \qquad \textit{Solve for W}$$

EXAMPLE 9   Using a Formula

A cylindrical can has a volume of 300 cubic centimeters (cm³) and a radius of 3 centimeters (cm), as shown in Figure 2.5. Find the height of the can.

Solution

The formula for the *volume of a cylinder* is $V = \pi r^2 h$. To find the height of the can, solve for h.

$$h = \frac{V}{\pi r^2}$$

Then, using $V = 300$ cm³ and $r = 3$ cm, find the height.

$$h = \frac{300 \text{ cm}^3}{\pi (3 \text{ cm})^2}$$

$$= \frac{300 \text{ cm}^3}{9\pi \text{ cm}^2}$$

$$\approx 10.61 \text{ cm} \qquad \textit{Round to two decimal places}$$

←3 cm→

h

FIGURE 2.5

Applied problems in textbooks often give precisely the right amount of information that is necessary to solve a given problem. In real-life, however, we often must sort through the given information and discard any that is irrelevant to the problem. Such irrelevant information is called a **red herring.** Find the red herrings in the following problems.

1. You have accepted a job for which your annual salary will be $18,600. You will be paid once a month, which implies that your monthly salary will be $1,550. How much of your annual salary will be deducted for Social Security tax? (Assume that the Social Security tax rate is 7.65%.)
2. You are driving to a concert (a 240-mile trip). After traveling for one hour, you stop for a snack, which takes 20 minutes. You then continue driving for four more hours until you reach the theater where the concert is being held. At the end of the trip you notice that you used $10\frac{1}{2}$ gallons of gasoline (for the entire trip). How many miles per gallon did your car average on the trip?

WARM UP

The following warm-up exercises involve skills that were covered in earlier sections. You will use these skills in the exercise set for this section.

Solve the equations (if possible) and check your answers.

1. $3x - 42 = 0$ 2. $64 - 16x = 0$

3. $2 - 3x = 14 + x$ 4. $7 + 5x = 7x - 1$

5. $5[1 + 2(x + 3)] = 6 - 3(x - 1)$ 6. $2 - 5(x - 1) = 2[x + 10(x - 1)]$

7. $\dfrac{x}{3} + \dfrac{x}{2} = \dfrac{1}{3}$ 8. $\dfrac{2}{x} + \dfrac{2}{5} = 1$

9. $1 - \dfrac{2}{z} = \dfrac{z}{z + 3}$ 10. $\dfrac{x}{x + 1} - \dfrac{1}{2} = \dfrac{4}{3}$

EXERCISES for Section 2.2

In Exercises 1–10, write an algebraic equation for the verbal model.

1. The sum of two consecutive natural numbers
2. The product of two consecutive natural numbers
3. *Distance Traveled*   The distance traveled in t hours by a car traveling at 50 miles per hour
4. *Travel Time*   The travel time for a plane traveling at a rate of r miles per hour for 200 miles

5. *Acid Solution*   The amount of acid in x gallons of a 20% acid solution
6. *Discount*   The sale price for an item that is discounted 20% of its list price L
7. *Perimeter of a Rectangle*   The perimeter of a rectangle whose width is x and whose length is twice the width
8. *Area of a Triangle*   The area of a triangle whose base is 20 inches and whose height is h inches

9. *Total Cost*   The total cost of producing x units for which the fixed costs are $1200 and the cost per unit is $25

10. *Total Revenue*   The total revenue obtained by selling x units at $3.59 per unit

Number Problems   In Exercises 11–24, write a mathematical model and solve the problem.

11. The sum of two consecutive natural numbers is 525. Find the two numbers.

12. Find three consecutive natural numbers whose sum is 804.

13. One positive number is five times another number. The difference between the two numbers is 148. Find the numbers.

14. One positive number is one-fifth of another number. The difference between the two numbers is 76. Find the numbers.

15. Find two consecutive integers whose product is five less than the square of the smaller number.

16. Find two consecutive natural numbers such that the difference of their reciprocals is one-fourth the reciprocal of the smaller number.

17. What is 30% of 45?

18. What is 175% of 360?

19. What is 0.045% of 2,650,000?

20. 432 is what percent of 1600?

21. 459 is what percent of 340?

22. 12 is $\frac{1}{2}$% of what number?

23. 70 is 40% of what number?

24. 825 is 250% of what number?

25. *Federal Government Income*   The pie chart shows the sources of income for the federal government in 1989. The total income was 990,800,000,000 dollars. Find the income for each of the indicated categories. (Round your answers to the nearest billion dollars.) (*Source:* Office of Management and Budget)

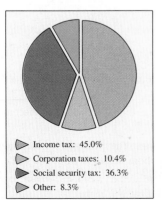

▷ Income tax: 45.0%
▷ Corporation taxes: 10.4%
▷ Social security tax: 36.3%
▷ Other: 8.3%

Figure for 25

26. *Federal Government Expenses*   The pie chart shows the types of expenses for the federal government in 1989. The total expenses were 1,142,869,000,000 dollars. Find the expense for each of the indicated categories. (Round your answers to the nearest billion dollars.) (*Source:* Office of Management and Budget)

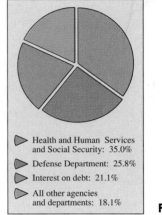

▷ Health and Human Services and Social Security: 35.0%
▷ Defense Department: 25.8%
▷ Interest on debt: 21.1%
▷ All other agencies and departments: 18.1%

Figure for 26

27. *Private Debt*   A family has annual loan payments equaling 58.6% of their annual income. During the year, their loan payments total $13,077.75. What is their income?

28. *Discount*   The price of a swimming pool has been discounted 16.5%. The sale price is $1210.75. Find the original list price of the pool.

29. *Monthly Profit*   The total profit for a company in February was 20% higher than it was in January. The total profit for the two months was $157,498. Find the profit for each month.

30. *Weekly Paycheck*   Your weekly paycheck is 15% *less* than your coworker's. The total of your paycheck and your coworker's paycheck is $645. Find the amount of each person's paycheck.

Then and Now   In Exercises 31–34, the prices of different items are given for 1940 and 1990. Find the percentage increase for each item. (*Source: USA Today*)

Item	1940	1990
31. Worker's annual earnings	$1,500.00	$27,000.00
32. A new car	$800.00	$16,400.00
33. A pound of ground beef	$0.15	$1.44
34. $2\frac{1}{2}$ acres on the Potomac	$7,500.00	$2,500,000.00

35. *Dimensions of a Room* A room is 1.5 times as long as it is wide, and its perimeter is 75 feet (see figure). Find the dimensions of the room.

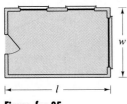

Figure for 35

36. *Dimensions of a Picture* A picture frame has a total perimeter of 3 feet (see figure). The width of the frame is 0.62 times its height. Find the dimensions of the frame.

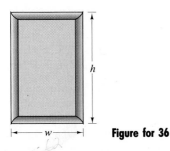

Figure for 36

37. *Course Grade* To get an A in a course a student must have an average of at least 90 on four tests of 100 points each. A student's scores on the first three tests were 87, 92, and 84. What must the student score on the fourth test to get an A in the course?

38. *Course Grade* Suppose you are taking a course that has four tests. The first three tests have 100 points each and the fourth test has 200 points. To get an A in the course, a student must have an average of at least 90% on the four tests. Your scores on the first three tests were 87, 92, and 84. What must you score on the fourth test to get an A in the course?

39. *Travel Time* Suppose you are driving on a freeway to a town 150 miles from your home. After 30 minutes you pass a freeway exit that you know is 25 miles from your home. Assuming that you continue at the same constant speed, how long will the entire trip take?

40. *Travel Time* On the first part of a 317-mile trip a salesman averaged 58 miles per hour. He averaged only 52 miles per hour on the last part of the trip because of the increased volume of traffic. Find the amount of time at each of the speeds if the total time was 5 hours and 45 minutes.

41. *Travel Time* Two cars start at one point and travel in the same direction at average speeds of 40 miles per hour and 55 miles per hour. How much time must elapse before the two cars are 5 miles apart?

42. *Catch Up Time* Students are traveling in two cars to a football game 135 miles away. The first car travels at an average speed of 45 miles per hour. The second car starts one-half hour after the first and travels at an average speed of 55 miles per hour. How long will it take the second car to catch up to the first car? Will the second car catch up to the first car before the first car arrives at the game?

43. *Travel Time* Two families meet at a park for a picnic. At the end of the day one family travels east at an average speed of 42 miles per hour and the other family travels west at an average speed of 50 miles per hour. Both families have approximately 160 miles to travel.
(a) Find the time it takes each family to get home.
(b) Find the time that will have elapsed when they are 100 miles apart.
(c) Find the distance the eastbound family has to travel after the westbound family has arrived home.

44. *Average Speed* A truck driver traveled at an average speed of 55 miles per hour on a 200-mile trip to pick up a load of freight. On the return trip (with the truck fully loaded), the average speed was 40 miles per hour. Find the average speed for the round trip.

45. *Wind Speed* An executive flew in the corporate jet to a meeting in a city 1500 miles away (see figure). After traveling the same amount of time on the return flight, the pilot mentioned that they still had 300 miles to go. If the air speed of the plane was 600 miles per hour, how fast was the wind blowing? (Assume that the wind direction was parallel to the flight path and constant all day.)

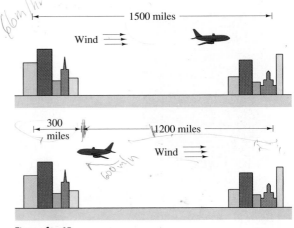

1500 miles

Wind

300 miles

1200 miles

Wind

Figure for 45

46. *Speed of Light*   Light travels at the speed of 3.0×10^8 meters per second. Find the time in minutes required for light to travel from the sun to the earth (a distance of 1.5×10^{11} meters).

47. *Radio Waves*   Radio waves travel at the speed of light, 3.0×10^8 meters per second. Find the time required for a radio wave to travel from mission control in Houston to NASA astronauts on the surface of the moon 3.86×10^8 meters away.

48. *Height of a Tree*   Suppose you want to measure the height of a tree. To do this, you measure the tree's shadow and find that it is 25 feet long. You also measure the shadow of a 5-foot lamp post and find that its shadow is 2 feet long (see figure). How tall is the tree?

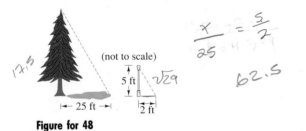

(not to scale)

5 ft

|← 25 ft →| 2 ft

Figure for 48

49. *Height of a Building*   Suppose you want to measure the height of a building. To do this, you measure the building's shadow and find that it is 50 feet long. You also measure the shadow of a 4-foot stake and find that its shadow is $3\frac{1}{2}$ feet long (see figure). How tall is the building?

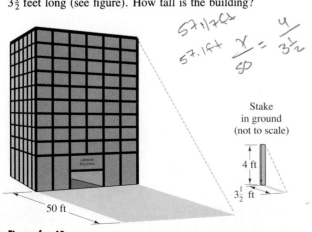

Stake in ground (not to scale)

4 ft

$3\frac{1}{2}$ ft

50 ft

Figure for 49

50. *Height of a Flagpole*   A person who is 6 feet tall walks away from a flagpole toward the tip of the shadow of the pole. When the person is 30 feet from the pole, the tip of the person's shadow and the shadow cast by the pole coincide at a point 5 feet in front of the person. Find the height of the pole.

51. *Walking Distance*   A person who is 6 feet tall walks away from a 50-foot silo and towards the tip of the silo's shadow. At a distance of 32 feet from the silo, the person's shadow begins to emerge beyond the silo's shadow. How much farther must the person walk to be completely out of the silo's shadow?

52. *Investment Mix*   Suppose you invest $12,000 in two funds paying $10\frac{1}{2}\%$ and 13% simple interest. The total annual interest is $1,447.50. How much is invested in each fund?

53. *Investment Mix*   Suppose you invest $25,000 in two funds paying 11% and $12\frac{1}{2}\%$ simple interest. The total annual interest is $2,975. How much is invested in each fund?

54. *Comparing Investment Returns*   Jack invested $12,000 in a fund paying $9\frac{1}{2}\%$ simple interest and $8,000 in a fund where the interest rate is variable. At the end of the year he was notified that his total interest for both funds was $2,054.40. Find the equivalent simple interest rate on the variable rate fund.

55. *Comparing Investment Returns*   Mary has $10,000 on deposit earning simple interest with the interest rate linked to the *prime rate*. Because of a drop in the prime rate, the rate on Mary's investment dropped by $1\frac{1}{2}\%$ for the last quarter of the year. Her annual earnings on the fund are $1,112.50. Find the interest rate for the first three quarters of the year and the last quarter.

56. *Mixture Problem*   Using the values from the table, determine the amounts of Solutions 1 and 2, respectively, needed to obtain the desired amount and concentration of the final mixture.

	Concentration			
	Solution 1	Solution 2	Final solution	Amount of final solution
(a)	10%	30%	25%	100 gal
(b)	25%	50%	30%	5 L
(c)	15%	45%	30%	10 qt
(d)	70%	90%	75%	25 gal

57. *Mixture Problem*   A 55-gallon barrel contains a mixture with a concentration of 40% (see figure). How much of this mixture must be withdrawn and replaced by 100% concentrate to bring the mixture up to 75% concentration?

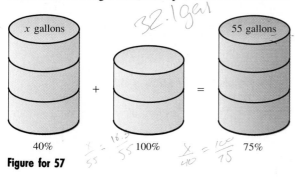

Figure for 57

58. *Mixture Problem*   A farmer mixed gasoline and oil to have 2 gallons of mixture for his two-cycle chain saw engine. This mixture was 32 parts gasoline and 1 part two-cycle oil. How much gasoline must be added to bring the mixture to 40 parts gasoline and 1 part oil?

59. *Mixture Problem*   A grocer mixes two kinds of nuts, costing $2.49 per pound and $3.89 per pound, respectively, to make 100 pounds of a mixture costing $3.19 per pound. How much of each kind of nut was put into the mixture?

60. *Production Limit*   A company has fixed costs of $10,000 per month and variable costs of $8.50 per unit manufactured. The company has $85,000 available to cover the monthly costs. How many units can the company manufacture? (*Fixed costs* occur regardless of the level of production. *Variable costs* depend on the level of production.)

61. *Production Limit*   A company has fixed costs of $10,000 per month and variable costs of $9.30 per unit manufactured. The company has $85,000 available to cover the monthly costs. How many units can the company manufacture?

62. *Water Depth*   A trough is 12 feet long, 3 feet deep, and 3 feet wide (see figure). Find the depth of the water when the trough contains 70 gallons. (1 gallon $\approx$ 0.13368 cubic feet)

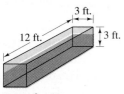

Figure for 62

Static Problems   In Exercises 63 and 64, suppose you have a uniform beam of length L with a fulcrum x feet from one end (see figure). If there are objects with weights W_1 and W_2 placed at opposite ends of the beam, then the beam will balance if

$$W_1 x = W_2(L - x).$$

Find x so the beam will balance.

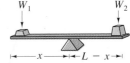

Figure for 63 and 64

63. Two children weighing 50 pounds and 75 pounds are going to play on a seesaw that is 10 feet long.

64. A person weighing 200 pounds is attempting to move a 550-pound rock with a bar that is 5 feet long.

In Exercises 65–84, solve for the indicated variable.

65. *Area of a Triangle*
  Solve for h: $A = \frac{1}{2}bh$.

66. *Volume of a Right Circular Cylinder*
  Solve for h: $V = \pi r^2 h$.

67. *Markup*
  Solve for C: $S = C + RC$.

68. *Discount*
  Solve for L: $S = L - RL$.

69. *Investment at Simple Interest*
  Solve for r: $A = P + Prt$.

70. *Investment at Compound Interest*
  Solve for P: $A = P\left(1 + \dfrac{r}{n}\right)^{nt}$.

71. *Area of a Trapezoid*
  Solve for b: $A = \frac{1}{2}(a + b)h$.

72. *Area of a Sector of a Circle*
  Solve for θ: $A = \dfrac{\pi r^2 \theta}{360}$.

73. *Volume of a Spherical Segment*
  Solve for r: $V = \frac{1}{3}\pi h^2(3r - h)$.

74. *Volume of an Oblate Spheroid*
  Solve for b: $V = \frac{4}{3}\pi a^2 b$.

75. *Thermal Expansion*
  Solve for α: $L = L_0[1 + \alpha(\Delta t)]$.

76. *Free-falling Body*

Solve for a: $h = v_0 t + \frac{1}{2}at^2$.

77. *Newton's Law of Universal Gravitation*

Solve for m_2: $F = \alpha \dfrac{m_1 m_2}{r^2}$.

78. *Heat Flow*

Solve for t_1: $H = \dfrac{KA(t_2 - t_1)}{L}$.

79. *Lensmaker's Equation*

Solve for R_1: $\dfrac{1}{f} = (n - 1)\left(\dfrac{1}{R_1} - \dfrac{1}{R_2}\right)$.

80. *Capacitance in Series Circuits*

Solve for C_1: $C = \dfrac{1}{\dfrac{1}{C_1} + \dfrac{1}{C_2}}$.

81. *Arithmetic Progression*

Solve for n: $L = a + (n - 1)d$.

82. *Arithmetic Progression*

Solve for a: $S = \dfrac{n}{2}[2a + (n - 1)d]$.

83. *Geometric Progression*

Solve for r: $S = \dfrac{rL - a}{r - 1}$.

84. *Prismoidal Formula*

Solve for S_1: $V = \frac{1}{6}H(S_0 + 4S_1 + S_2)$.

2.3 Quadratic Equations

Solving Quadratic Equations by Factoring / Solving Quadratic Equations by Extracting Square Roots / Solving Quadratic Equations by Completing the Square / Applications

Solving Quadratic Equations by Factoring

DEFINITION OF A QUADRATIC EQUATION

A **quadratic equation** in x is an equation that can be written in the standard form

$$ax^2 + bx + c = 0$$

where a, b, and c are real numbers with $a \neq 0$. A quadratic equation in x is also known as a **second-degree polynomial equation in x.**

We will discuss four methods for solving quadratic equations: factoring, extracting square roots, completing the square, and the Quadratic Formula. (The Quadratic Formula is discussed in the next section.) The first technique is based on the Zero-Factor Property given in Section 1.2.

If $ab = 0$, then $a = 0$ or $b = 0$.   *Zero-Factor Property*

To use this principle, write the left side of the standard form of a quadratic equation as the product of two linear factors. Then find the solutions of the quadratic equation by setting each linear factor equal to zero. For instance, the solutions of the equation $x^2 - 3x - 10 = 0$ are found as follows.

$x^2 - 3x - 10 = 0$	*Standard form*
$(x - 5)(x + 2) = 0$	*Factored form*
$x - 5 = 0$ or $x + 2 = 0$	*Set each factor equal to zero*
$x = 5$ or $x = -2$	*Solutions*

Be sure you see that the Zero-Factor Property works *only* for equations written in standard form (in which the right side of the equation is zero). Therefore, all terms must be collected on one side *before* factoring. For instance, in the equation

$$(x - 5)(x + 2) = 8$$

it is *incorrect* to set each factor equal to 8. Can you solve this equation correctly?

EXAMPLE 1   Solving a Quadratic Equation by Factoring

$2x^2 + 9x + 7 = 3$	*Given equation*
$2x^2 + 9x + 4 = 0$	*Standard form*
$(2x + 1)(x + 4) = 0$	*Factored form*
$2x + 1 = 0 \;\rightarrow\; x = -\dfrac{1}{2}$	*Set 1st factor equal to 0*
$x + 4 = 0 \;\rightarrow\; x = -4$	*Set 2nd factor equal to 0*

Check

$$2\left(-\frac{1}{2}\right)^2 + 9\left(-\frac{1}{2}\right) + 7 = \frac{1}{2} - \frac{9}{2} + 7 = 3 \qquad x = -1/2 \text{ checks}$$

$$2(-4)^2 + 9(-4) + 7 = 32 - 36 + 7 = 3 \qquad x = -4 \text{ checks}$$

EXAMPLE 2   Solving a Quadratic Equation by Factoring

$6x^2 - 3x = 0$	*Standard form*
$3x(2x - 1) = 0$	*Factored form*
$3x = 0 \;\rightarrow\; x = 0$	*Set 1st factor equal to 0*
$2x - 1 = 0 \;\rightarrow\; x = \dfrac{1}{2}$	*Set 2nd factor equal to 0*

REMARK   When solving equations by factoring, be sure to set all factors to zero, including any *monomial* factors such as $3x$ in Example 2.

The equation has two solutions: 0 and $\frac{1}{2}$. Check these solutions in the original equation.

If the two factors of a quadratic expression are the same, then the corresponding solution is a **double** or **repeated solution.**

EXAMPLE 3   A Quadratic Equation with a Repeated Solution

$$9x^2 - 6x + 1 = 0 \qquad \textit{Standard form}$$
$$(3x - 1)^2 = 0 \qquad \textit{Factored form}$$
$$3x - 1 = 0 \qquad \textit{Set repeated factor equal to 0}$$
$$x = \frac{1}{3} \qquad \textit{Solution}$$

This equation has only one solution: $\frac{1}{3}$. Check this solution in the original equation.

Solving Quadratic Equations by Extracting Square Roots

There is a shortcut for solving quadratic equations of the form

$$u^2 = d$$

where $d > 0$ and u is an algebraic expression. By factoring, you can see that this equation has two solutions.

$$u^2 = d \qquad \textit{Given equation}$$
$$u^2 - d = 0 \qquad \textit{Standard form}$$
$$(u + \sqrt{d})(u - \sqrt{d}) = 0 \qquad \textit{Factor}$$
$$u + \sqrt{d} = 0 \;\rightarrow\; u = -\sqrt{d} \quad \textit{Set 1st factor equal to 0}$$
$$u - \sqrt{d} = 0 \;\rightarrow\; u = \sqrt{d} \quad \textit{Set 2nd factor equal to 0}$$

Since the two solutions differ only in sign, you can write the solutions together, using a "plus or minus sign."

$$u = \pm\sqrt{d}$$

This form of the solution is read "u is equal to plus or minus the square root of d." Solving an equation of the form $u^2 = d$ without going through the steps of factoring is **extracting square roots.**

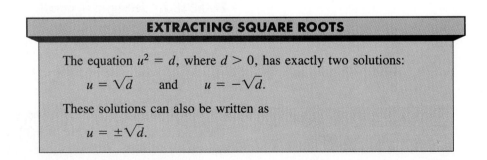

EXTRACTING SQUARE ROOTS

The equation $u^2 = d$, where $d > 0$, has exactly two solutions:

$$u = \sqrt{d} \qquad \text{and} \qquad u = -\sqrt{d}.$$

These solutions can also be written as

$$u = \pm\sqrt{d}.$$

EXAMPLE 4 **Solving a Quadratic Equation by Extracting Square Roots**

$4x^2 = 12$	*Given equation*
$x^2 = 3$	*Divide both sides by 4*
$x = \pm\sqrt{3}$	*Extract square roots*

The equation has two solutions: $\sqrt{3}$ and $-\sqrt{3}$. Note that $x^2 - 3 = 0$ factors as $(x + \sqrt{3})(x - \sqrt{3}) = 0$, which gives the same two solutions. Check these solutions in the original equation.

EXAMPLE 5 **Solving a Quadratic Equation by Extracting Square Roots**

Solve the quadratic equation by extracting the square roots.

$$(x - 3)^2 = 7$$

Solution

In this case, an extra step is needed *after* extracting square roots.

$(x - 3)^2 = 7$	*Given equation*
$x - 3 = \pm\sqrt{7}$	*Extract square roots*
$x = 3 \pm \sqrt{7}$	*Add 3 to both sides*

Check these solutions in the original equation.

Solving Quadratic Equations by Completing the Square

The equation in Example 5 was given in the form $(x - 3)^2 = 7$ so that we could find the solution by extracting square roots. Suppose, however, that the equation $(x - 3)^2 = 7$ had been given in the standard form

$$x^2 - 6x + 2 = 0. \qquad \textit{Standard form}$$

This equation is equivalent to the original and thus has the same two solutions, $x = 3 \pm \sqrt{7}$. However, the left side of the equation is not factorable, and we cannot find its solutions unless we can *reverse* the steps shown above.

COMPLETING THE SQUARE

To **complete the square** for the expression

$$x^2 + bx$$

we add $(b/2)^2$, which is the square of half the coefficient of x. Consequently,

$$x^2 + bx + \left(\frac{b}{2}\right)^2 = \left(x + \frac{b}{2}\right)^2.$$

When solving quadratic equations by completing the square, you must add $(b/2)^2$ to *both sides* to maintain equality.

EXAMPLE 6   Completing the Square: Leading Coefficient Is 1

Solve the equation $x^2 - 6x + 2 = 0$ by completing the square. Compare the solutions to those obtained in Example 5.

Solution

$x^2 - 6x + 2 = 0$	*Given equation*
$x^2 - 6x = -2$	*Subtract 2 from both sides*
$x^2 - 6x + 3^2 = -2 + 3^2$	*Add 3^2 to both sides*
$\underbrace{\qquad}_{(\text{half})^2}$	
$x^2 - 6x + 9 = 7$	*Simplify*
$(x - 3)^2 = 7$	*Perfect square trinomial*
$x - 3 = \pm\sqrt{7}$	*Extract square roots*
$x = 3 \pm \sqrt{7}$	*Solutions*

If the leading coefficient of a quadratic is not 1, you must divide both sides of the equation by this coefficient *before* completing the square. For instance, to complete the square for $3x^2 - 4x - 5 = 0$, first divide each term by the leading coefficient 3.

$$x^2 - \frac{4}{3}x - \frac{5}{3} = 0$$

Then proceed as in Example 6.

Completing the square has many uses. In the next section you will see how it is used to develop a general formula for solving quadratic equations. In Section 5.4 you will use it to write equations of conics. In Example 7 you will use it to rewrite algebraic expressions in forms that simplify calculus operations.

When rewriting algebraic expressions by completing the square, you must add *and subtract* the quantity $(b/2)^2$.

EXAMPLE 7   Completing the Square Within an Algebraic Expression

Rewrite the denominator as the sum or difference of two squares.

$$\frac{1}{x^2 - 2x - 3}$$

Solution

Complete the square of the denominator.

$$x^2 - 2x - 3 = \left(x^2 - 2x + 1^2\right) - 3 - 1^2 \qquad \textit{Add and subtract } 1^2$$
$$= (x^2 - 2x + 1) - 4 \qquad \textit{Group terms}$$
$$= (x - 1)^2 - 2^2 \qquad \textit{Difference of two squares}$$

Therefore, the original expression can be written

$$\frac{1}{x^2 - 2x - 3} = \frac{1}{(x - 1)^2 - 2^2}.$$

If the leading coefficient is negative, as in $1/(3 + 2x - x^2)$, first factor out the negative coefficient.

$$\frac{1}{-(x^2 - 2x - 3)}$$

Then proceed to complete the square as in Example 7.

REMARK Although the quadratic $x^2 - 2x - 3$ in Example 7 is factorable, some operations in calculus are simpler with the completed square form than with the factored form.

Applications

Quadratic equations often occur in problems dealing with area. Here is a simple example. "A square room has an area of 144 square feet. Find the dimensions of the room." To solve this problem, let x represent the length of each side of the room. Then, by solving the equation

$$x^2 = 144$$

you can conclude that each side of the room is 12 feet long. Note that although the equation $x^2 = 144$ has two solutions, -12 and 12, the negative solution makes no sense for this problem, so you choose the positive solution.

EXAMPLE 8   Finding the Dimensions of a Room

A bedroom is 3 feet longer than it is wide and has an area of 154 square feet, as shown in Figure 2.6. Find the dimensions of the room.

Solution

Verbal Model	$\dfrac{\text{Width}}{\text{of room}} \cdot \dfrac{\text{Length}}{\text{of room}} = \dfrac{\text{Area}}{\text{of room}}$
Labels	Area of room = 154 square feet
	Width of room = w feet
	Length of room = $(w + 3)$ feet
Algebraic Equation	$w(w + 3) = 154$
	$w^2 + 3w - 154 = 0$
	$(w - 11)(w + 14) = 0$
	$w - 11 = 0 \;\rightarrow\; w = 11$
	$w + 14 = 0 \;\rightarrow\; w = -14$

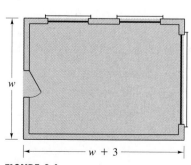

FIGURE 2.6

Choosing the positive value, you find that the width is 11 feet and the length is $w + 3 = 14$ feet. You can check this solution by observing that the length is 3 feet longer than the width *and* that the product of the length and width is 154 square feet.

Another common application of quadratic equations involves an object that is falling (or projected into the air). The general equation that gives the height of such an object is a **position equation,** and (on the earth's surface) it has the form

$$s = -16t^2 + v_0 t + s_0.$$

In this equation, s represents the height of the object (in feet), v_0 represents the original velocity of the object (in feet per second), s_0 represents the original height of the object (in feet), and t represents the time (in seconds).

EXAMPLE 9   Falling Time

A construction worker on the 24th floor of a building project (see Figure 2.7) accidentally drops a wrench and immediately yells: "Look out below!" Could a person at ground level hear this warning in time to get out of the way of the falling wrench?

Solution

Let's assume that each floor of the building is 10 feet high, so that the wrench is dropped from a height of 240 feet. Since sound travels at about 1100 feet per second, we assume that a person at ground level hears the warning within 1 second of the time the wrench is dropped. To set up a mathematical model for the height of the wrench, we use the position equation

$$s = -16t^2 + v_0 t + s_0.$$

Since the object is dropped (rather than thrown) we can conclude that the initial velocity is $v_0 = 0$. Moreover, since the initial height is $s_0 = 240$ feet, we have the following model.

$$s = -16t^2 + 240$$

After falling 1 second, the height of the wrench is $-16(1^2) + 240 = 214$. After falling 2 seconds, the height of the wrench is $-16(2^2) + 240 = 176$. To find the number of seconds it takes the wrench to hit the ground, let the height s be zero and solve the resulting equation for t.

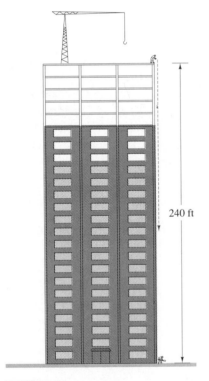

FIGURE 2.7

$s = -16t^2 + 240$	*Position equation*
$0 = -16t^2 + 240$	*Set height equal to 0*
$16t^2 = 240$	*Add $16t^2$ to both sides*
$t^2 = 15$	*Divide both sides by 16*
$t = \sqrt{15} \approx 3.87$	*Extract positive square root*

240 ft

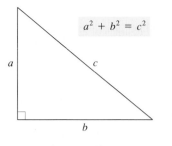

$a^2 + b^2 = c^2$

a c

b

Pythagorean Theorem

FIGURE 2.8

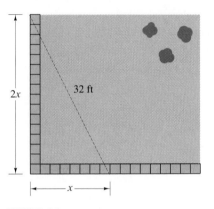

$2x$

32 ft

x

FIGURE 2.9

Thus, the wrench will take about 3.87 seconds to hit the ground. If the person hears the warning 1 second after the wrench is dropped, the person still has almost 3 seconds to get out of the way.

A third type of application that often involves a quadratic equation is one dealing with the hypotenuse of a right triangle. Recall from geometry that the sides of a right triangle are related by the **Pythagorean Theorem,** as indicated in Figure 2.8. This theorem states that if a and b are the lengths of the sides of the triangle and c is the length of the hypotenuse, then

$$a^2 + b^2 = c^2. \qquad \textit{Pythagorean Theorem}$$

EXAMPLE 10   Cutting Across the Lawn

Suppose your house is on a large corner lot. Several of the children in the neighborhood cut across your lawn, as shown in Figure 2.9. The distance across the lawn is 32 feet. Assuming that the picture in Figure 2.9 is drawn to scale, how many feet does a person save by walking across the lawn instead of walking on the sidewalk?

Solution

In Figure 2.9, we let x represent the length of the shorter part of the sidewalk. Using a ruler, we find that the length of the longer part of the sidewalk is twice the shorter, so we represent its length by $2x$. Now, using the Pythagorean Theorem, we have

$$\begin{aligned} x^2 + (2x)^2 &= 32^2 & \textit{Pythagorean Theorem} \\ 5x^2 &= 1024 & \textit{Add like terms} \\ x^2 &= 204.8 & \textit{Divide both sides by 5} \\ x &= \sqrt{204.8}. & \textit{Extract positive square root} \end{aligned}$$

Thus, the total distance on the sidewalk is

$$x + 2x = 3x = 3\sqrt{204.8} \approx 42.9 \text{ feet.}$$

Therefore, cutting across the lawn saves a person about 10.9 feet of walking distance.

DISCUSSION

PROBLEM

Comparing
Methods

In this section we discussed three different methods for solving a quadratic equation. Which of the three methods do you think works best for each of the following quadratic equations? Write a short paragraph explaining (or discuss) your choices.

1. $x^2 - 2x - 3 = 0$ 2. $x^2 - 4x - 3 = 0$ 3. $x^2 - 3 = 0$

WARM UP

The following warm-up exercises involve skills that were covered in earlier sections. You will use these skills in the exercise set for this section.

In Exercises 1–4, simplify the expressions.

1. $\sqrt{\frac{7}{50}}$ **2.** $\sqrt{32}$ **3.** $\sqrt{7^2 + 3 \cdot 7^2}$ **4.** $\sqrt{\frac{1}{4} + \frac{3}{8}}$

In Exercises 5–10, factor the algebraic expressions.

5. $3x^2 + 7x$ **6.** $4x^2 - 25$

7. $16 - (x - 11)^2$ **8.** $x^2 + 7x - 18$

9. $10x^2 + 13x - 3$ **10.** $6x^2 - 73x + 12$

EXERCISES for Section 2.3

In Exercises 1–10, write the quadratic equation in standard quadratic form.

1. $2x^2 = 3 - 5x$ **2.** $4x^2 - 2x = 9$

3. $x^2 = 25x$ **4.** $10x^2 = 90$

5. $(x - 3)^2 = 2$ **6.** $12 - 3(x + 7)^2 = 0$

7. $x(x + 2) = 3x^2 + 1$ **8.** $x^2 + 1 = \dfrac{x - 3}{2}$

9. $\dfrac{3x^2 - 10}{5} = 12x$ **10.** $x(x + 5) = 2(x + 5)$

In Exercises 11–22, solve the quadratic equation for x by factoring.

11. $6x^2 + 3x = 0$ **12.** $9x^2 - 1 = 0$

13. $x^2 - 2x - 8 = 0$ **14.** $x^2 - 10x + 9 = 0$

15. $x^2 + 10x + 25 = 0$ **16.** $16x^2 + 56x + 49 = 0$

17. $3 + 5x - 2x^2 = 0$ **18.** $2x^2 = 19x + 33$

19. $x^2 + 4x = 12$ **20.** $-x^2 + 8x = 12$

21. $x^2 + 2ax + a^2 = 0$ **22.** $(x + a)^2 - b^2 = 0$

In Exercises 23–34, solve the equation by extracting square roots. List both the exact answer *and* the decimal answer rounded to two decimal places.

23. $x^2 = 16$ **24.** $x^2 = 144$

25. $x^2 = 7$ **26.** $x^2 = 27$

27. $3x^2 = 36$ **28.** $9x^2 = 25$

29. $(x - 12)^2 = 18$ **30.** $(x + 13)^2 = 21$

31. $(x + 2)^2 = 12$ **32.** $(x - 5)^2 = 20$

33. $(x - 7)^2 = (x + 3)^2$ **34.** $(x + 5)^2 = (x + 4)^2$

In Exercises 35–44, solve the quadratic equation by completing the square.

35. $x^2 - 2x = 0$ **36.** $x^2 + 4x = 0$

37. $x^2 + 4x - 32 = 0$ **38.** $x^2 - 2x - 3 = 0$

39. $x^2 + 6x + 2 = 0$ **40.** $x^2 + 8x + 14 = 0$

41. $9x^2 - 18x + 3 = 0$ **42.** $9x^2 - 12x - 14 = 0$

43. $8 + 4x - x^2 = 0$ **44.** $4x^2 - 4x - 99 = 0$

In Exercises 45–68, solve the equation for x.

45. $x^2 = 64$ **46.** $7x^2 = 32$

47. $x^2 - 2x - 1 = 0$ **48.** $x^2 - 6x + 4 = 0$

49. $16x^2 - 9 = 0$ **50.** $11x^2 + 33x = 0$

51. $4x^2 - 12x + 9 = 0$ **52.** $x^2 - 14x + 49 = 0$

53. $(x + 3)^2 = 81$ **54.** $(x - 5)^2 = 8$

55. $4x = 4x^2 - 3$ **56.** $80 + 6x = 9x^2$

57. $50 + 5x = 3x^2$ **58.** $144 - 73x + 4x^2 = 0$

59. $12x = x^2 + 27$ **60.** $26x = 8x^2 + 15$

61. $x^2 - x - \frac{11}{4} = 0$ **62.** $x^2 + 3x - \frac{3}{4} = 0$

63. $50x^2 - 60x - 7 = 0$ **64.** $9x^2 + 12x + 3 = 0$

65. $(x + 3)^2 - 4 = 0$ **66.** $a^2x^2 - b^2 = 0$

67. $(x + 1)^2 = x^2$ **68.** $(x + 1)^2 = 4x^2$

In Exercises 69–72, complete the square for the quadratic portion of the algebraic expression.

69. $\dfrac{1}{x^2 - 4x - 12}$ **70.** $\dfrac{4}{4x^2 + 4x - 3}$

71. $\dfrac{1}{\sqrt{6x - x^2}}$ **72.** $\dfrac{1}{\sqrt{16 - 6x - x^2}}$

73. *Dimensions of a Building*   A one-story building is 14 feet longer than it is wide (see figure). It has 1632 square feet of floor space. Find the length and width of the building.

Figure for 73

74. *Dimensions of a Triangular Sign*   The height of a triangular sign is equal to its base. The area of the sign is 10 square feet. Find the base and height of the sign.

In Exercises 75 and 76, use the position equation in Example 9 as the model for the problem.

75. *CN Tower*   At 1821 feet tall, the CN Tower in Toronto, Ontario, is the world's tallest self-supporting structure. An object is dropped from the top of the tower. How long will it take for it to hit the ground?

76. *Warfare*   A bomber flying at 32,000 feet over level terrain drops a 500-pound bomb.
(a) How long will it take for the bomb to strike the ground?
(b) If the plane is flying at 600 miles per hour, how far will the bomb travel horizontally during its descent?

77. *An Isosceles Right Triangle*   The hypotenuse of an isosceles right triangle is 5 centimeters long. How long are its sides? (An isosceles triangle has two sides of equal length.)

78. *An Equilateral Triangle*   An equilateral triangle has a height of 10 inches. How long are each of its sides? (*Hint:* Use the height of the triangle to partition the triangle into two congruent right triangles.)

79. *Depth of a Submarine*   The sonar of a Navy cruiser detects a submarine that is 3000 feet from the cruiser. The angle between the water level and the submarine is 45° (see figure). How deep is the submarine?

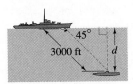

Figure for 79

80. *Height of a Kite*   You have a spool of 100 feet of string attached to a kite. When the kite has pulled all of the string from the spool you estimate that the angle of elevation to the kite is 45°. Approximate the height of the kite.

81. *Total Revenue*   The demand equation for a certain product is $p = 20 - 0.0002x$ where p is the price per unit and x is the number of units sold. The total revenue is given by

$$\text{Revenue} = xp = x(20 - 0.0002x).$$

How many units must be sold to produce a revenue of \$500,000?

82. *Consecutive Even Integers*   Find two consecutive positive even integers whose product is 288.

83. *Population of the U.S.*   The population of the United States from 1800 to 1890 can be approximated by the model

$$\text{Population} = 0.6942t^2 + 6.183$$

where the population is given in millions of people and the time t represents the calendar year with $t = 0$ corresponding to 1800, $t = 1$ corresponding to 1810, and so on. If this model had continued to be valid up through the present time, when would the population of the United States have reached 250 million? Judging from the given figure, would you say this model was a good representation of the population through 1890? Through 1990? (*Source:* U.S. Bureau of Census)

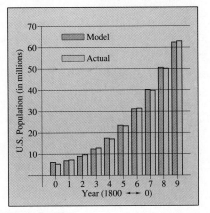

Figure for 83

84. If a and b are nonzero real numbers, determine the solutions of the equation $ax^2 + bx = 0$.

85. If a is a nonzero real number, determine the solutions of the equation $ax^2 - ax = 0$.

86. Find a quadratic equation with solutions $x = -4$ and $x = 6$. (There are many correct answers.)

SOLVING

A Numerical Approach to Finding a Falling Time

So far in this chapter, you have studied algebraic methods for solving equations. For many problems, you can gain further insight by using a numerical approach either instead of, or in addition to, an algebraic approach.

EXAMPLE 1   Finding the Falling Time for an Object

At 9:55 A.M. on Saturday, July 28, 1945, a terrible airplane accident occurred. A B-25 bomber crashed into the 78th and 79th floors of the Empire State Building in New York City. Hundreds of pieces of debris fell 975 feet to the streets below. How much time after hearing the crash did the people on the street have to get out of the way?

Solution

Assume that the debris *dropped* from a height of 975 feet. Using an initial velocity of $v_0 = 0$ and an initial height of $s_0 = 0$, you can write the following model for the height (in feet) of the falling debris: $s = -16t^2 + 975$. The table gives the heights at different times.

Time, t	0	1	2	3	4	5	6	7	7.8
Height, s	975	959	911	831	719	575	399	191	2

From the table, you can see that the debris took about 8 seconds to hit the ground. Because it would have taken about 1 second for the sound of the crash to reach the ground (sound travels at about 1100 feet per second), you can conclude that people had about 7 seconds to get out of the way of the falling debris.

This conclusion was reached numerically by constructing a table. You can reach the same conclusion algebraically by solving the equation $s = -16t^2 + 975$ for the time that gives a height of $s = 0$ in feet.

Fourteen people were killed in this accident. The toll would have been greater, but it was a drizzly Saturday morning—most of the building's 15,000 office workers were at home. (Reproduced from *Great Disasters*, © 1989, The Reader's Digest Association, Inc. Used by permission. Illustration by Dennis Lyall.)

$$-16t^2 + 975 = s \qquad \textit{Falling object model}$$
$$-16t^2 + 975 = 0 \qquad \textit{Substitute 0 for s}$$
$$975 = 16t^2 \qquad \textit{Add } 16t^2 \textit{ to both sides}$$
$$\frac{975}{16} = t^2 \qquad \textit{Divide both sides by 16}$$
$$\sqrt{\frac{975}{16}} = t \qquad \textit{Find the positive square root}$$
$$7.8 \approx t \qquad \textit{Use a calculator}$$

Once you have created a mathematical model and used it to solve a real-life problem, try rethinking your model to check how accurately it represents the real-life situation.

EXAMPLE 2   Adding Sophistication to a Model

In Example 1, because the airplane accident was accompanied by an explosion, some of the debris was most likely *propelled* downward. How does this consideration affect the time the people on the street had to get out of the way of the falling debris?

Solution

You could start by assuming that some of the debris was propelled straight down with initial speed of about 100 feet per second. Using $v_0 = -100$ and $s_0 = 975$, you can obtain the model $s = -16t^2 - 100t + 975$. The table gives the heights of the debris at several times.

Time, t	0	1	2	3	4	5	5.3
Height, s	975	859	711	531	319	75	−4.44

With this model, note that the falling time has decreased considerably. Again figuring that it took about 1 second to hear the crash, you can conclude that the people on the street had only about 4.3 seconds to get out of the way.

◢

EXERCISES

(See also Section 2.3, Example 9, Exercises 75–76)

1. *Real or False Accuracy?*   In Example 2, suppose you solved the equation

   $$0 = -16t^2 - 100t + 975$$

   algebraically to obtain

   $$t = \frac{25 - 5\sqrt{181}}{-8} \approx 5.284.$$

   Because sound traveling at 1100 feet per second takes 0.886 second to travel 975 feet, you could reason that the people had $5.284 - 0.886 = 4.398$ seconds to get out of the way. This solution appears to be more accurate than that obtained in Example 2. Is it?

2. *Exploration*   In Example 2, how long would the people on the street have had to get out of the way if some of the debris had been propelled with an initial velocity of 200 feet per second?

3. *Working Backward*   In the actual accident, all of the casualties were people who were in the building. What does this suggest about the initial velocity of any of the debris that was propelled downward?

4. *Exploration*   Use the falling object model and table given in Example 1.
   (a) What was the average speed of the debris during its first second of fall?
   (b) Was the debris falling faster during its next second of fall? Explain.
   (c) Calculate the height of the debris after 7.7 seconds of fall.
   (d) Use the result of part (c) to approximate the terminal speed of the debris.

5. Use the model given in Example 2 to approximate the terminal speed of the debris. Is the terminal speed given by this model greater than the terminal speed given by the model in Example 1?

6. In Example 2, we assumed that some of the debris was propelled straight down with an initial speed of 100 feet per second. With this assumption, it seems reasonable that other parts of the debris would have been propelled straight up with the same initial speed. Under these conditions, approximate the duration at which debris continued to hit the streets.

101

2.4 The Quadratic Formula

Development of the Quadratic Formula / The Discriminant /
Solving a Quadratic Equation by the Quadratic Formula / Applications

Development of the Quadratic Formula

Often in mathematics you are taught the long way of solving a problem first. Then, the longer method is used to develop shorter techniques. The long way stresses understanding and the short way stresses efficiency.

In Section 2.3 you learned the technique of completing the square to solve nonfactorable quadratic equations. This method can be thought of as a "long way" of solving a quadratic equation. When you use completing the square to solve a quadratic equation, you must complete the square for *each* equation separately. In the following development you complete the square *once* in a general setting to obtain the **Quadratic Formula**—a shortcut for solving a quadratic equation.

$$ax^2 + bx + c = 0 \qquad \text{\textit{Standard form, } } a \neq 0$$

$$ax^2 + bx = -c \qquad \text{\textit{Subtract c from both sides}}$$

$$x^2 + \frac{b}{a}x = -\frac{c}{a} \qquad \text{\textit{Divide both sides by a}}$$

$$x^2 + \frac{b}{a}x + \left(\frac{b}{2a}\right)^2 = -\frac{c}{a} + \left(\frac{b}{2a}\right)^2 \qquad \text{\textit{Complete the square}}$$

$$\underbrace{\phantom{x^2 + \frac{b}{a}x +}}_{(\text{half})^2}$$

$$\left(x + \frac{b}{2a}\right)^2 = \frac{b^2 - 4ac}{4a^2} \qquad \text{\textit{Simplify}}$$

$$x + \frac{b}{2a} = \pm\sqrt{\frac{b^2 - 4ac}{4a^2}} \qquad \text{\textit{Extract square roots}}$$

$$x = -\frac{b}{2a} \pm \frac{\sqrt{b^2 - 4ac}}{2|a|} \qquad \text{\textit{Solutions}}$$

Note that since $\pm 2|a|$ represents the same numbers as $\pm 2a$, you can omit the absolute value sign.

THE QUADRATIC FORMULA

The solutions of a quadratic equation in the standard form

$$ax^2 + bx + c = 0, \quad a \neq 0$$

are given by the **Quadratic Formula**

$$x = \frac{-b \pm \sqrt{b^2 - 4ac}}{2a}.$$

The Quadratic Formula is one of the most important formulas in algebra. You should learn the verbal statement of the Quadratic Formula: "Negative b, plus or minus the square root of b squared minus $4ac$, all divided by $2a$."

The Discriminant

In the Quadratic Formula, the quantity under the radical sign, $b^2 - 4ac$, is the **discriminant** of the quadratic expression $ax^2 + bx + c$. It can be used to determine the nature of the solutions of a quadratic equation.

SOLUTIONS OF A QUADRATIC EQUATION

The solutions of a quadratic equation $ax^2 + bx + c = 0$, $a \neq 0$, can be classified as follows.

1. If the discriminant $b^2 - 4ac$ is positive, then the quadratic equation has *two* different real solutions.
2. If the discriminant $b^2 - 4ac$ is zero, then the quadratic equation has *one* repeated solution.
3. If the discriminant $b^2 - 4ac$ is negative, then the quadratic equation has *no* real solution.

EXAMPLE 1   Using the Discriminant

REMARK If the discriminant of a quadratic equation is negative, as in case 3 above, then its square root is imaginary (not a real number), and the quadratic formula yields two complex solutions. We will study this case in Section 2.5.

Use the discriminant to determine the number of real solutions of the quadratic equations.

a. $4x^2 - 20x + 25 = 0$ **b.** $13x^2 + 7x + 1 = 0$ **c.** $5x^2 = 8x$

Solution

a. Since $a = 4$, $b = -20$, and $c = 25$, the discriminant is

$$b^2 - 4ac = 400 - 4(4)(25) = 400 - 400 = 0.$$

Therefore, there is *one* repeated real solution.

b. In this case, $a = 13$, $b = 7$, and $c = 1$. Thus, the discriminant is

$$b^2 - 4ac = 49 - 4(13)(1) = 49 - 52 = -3.$$

Since the discriminant is negative, the equation has no real solution.

c. In standard form, the equation is $5x^2 - 8x = 0$ with $a = 5$, $b = -8$, and $c = 0$. Thus, the discriminant is

$$b^2 - 4ac = 64 - 4(5)(0) = 64.$$

Since the discriminant is positive, the equation has two different real solutions.

Solving a Quadratic Equation by the Quadratic Formula

When using the Quadratic Formula, remember that *before* the formula can be applied, you must first write the quadratic equation in standard form.

EXAMPLE 2   **Using the Quadratic Formula: Two Distinct Solutions**

Use the Quadratic Formula to solve $x^2 + 3x = 9$.

Solution

$$x^2 + 3x = 9 \qquad \text{\textit{Given equation}}$$

$$x^2 + 3x - 9 = 0 \qquad \text{\textit{Standard form with}}$$
$$\textit{a = 1, b = 3, c = -9}$$

$$x = \frac{-b \pm \sqrt{b^2 - 4ac}}{2a} \qquad \text{\textit{Quadratic Formula}}$$

$$x = \frac{-3 \pm \sqrt{(3)^2 - 4(1)(-9)}}{2(1)} \qquad \text{\textit{Substitute}}$$

$$x = \frac{-3 \pm \sqrt{45}}{2}$$

$$x = \frac{-3 \pm 3\sqrt{5}}{2} \qquad \text{\textit{Solutions}}$$

Therefore, the equation has two solutions:

$$x = \frac{-3 + 3\sqrt{5}}{2} \qquad \text{and} \qquad x = \frac{-3 - 3\sqrt{5}}{2}.$$

Check these solutions in the original equation.

EXAMPLE 3   **Using the Quadratic Formula: One Repeated Solution**

Use the Quadratic Formula to solve $8x^2 - 24x + 18 = 0$.

Solution

Note that this equation has a common factor of 2. To simplify things, first divide both sides of the equation by 2.

$$8x^2 - 24x + 18 = 0 \qquad \text{\textit{Common factor of 2}}$$

$$4x^2 - 12x + 9 = 0 \qquad \text{\textit{Standard form with}}$$
$$\textit{a = 4, b = -12, c = 9}$$

$$x = \frac{-b \pm \sqrt{b^2 - 4ac}}{2a} \qquad \text{\textit{Quadratic Formula}}$$

$$x = \frac{-(-12) \pm \sqrt{(-12)^2 - 4(4)(9)}}{2(4)} \qquad \text{\textit{Substitute}}$$

$$x = \frac{12 \pm \sqrt{0}}{8} = \frac{3}{2} \qquad \text{\textit{Repeated solution}}$$

Therefore, this quadratic equation has only one solution: $\frac{3}{2}$. Check this solution in the original equation.

The discriminant in Example 3 is a perfect square (zero in this case), and you could have factored the quadratic as

$$4x^2 - 12x + 9 = (2x - 3)^2 = 0$$

to conclude that the solution is $x = \frac{3}{2}$. Since factoring is easier than applying the Quadratic Formula, try factoring first. If, however, factors cannot easily be found, then use the Quadratic Formula. For instance, try solving the quadratic equation $x^2 - x - 12 = 0$ in two ways—by factoring and by the Quadratic Formula—to see that you get the same solutions either way.

When using a calculator to evaluate the Quadratic Formula, you should get in the habit of using the memory key. This will save steps and minimize round-off error.

EXAMPLE 4   Using a Calculator to Evaluate the Quadratic Formula

Solve $16.3x^2 - 197.6x + 7.042 = 0$.

Solution

In this case, $a = 16.3$, $b = -197.6$, and $c = 7.042$.

$$x = \frac{-b \pm \sqrt{b^2 - 4ac}}{2a}$$

$$= \frac{-(-197.6) \pm \sqrt{(-197.6)^2 - 4(16.3)(7.042)}}{2(16.3)}$$

To evaluate these solutions, begin by calculating the square root of the discriminant.

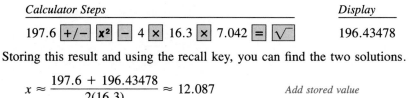

Calculator Steps	*Display*
197.6 $\boxed{+/-}$ $\boxed{x^2}$ $\boxed{-}$ 4 $\boxed{\times}$ 16.3 $\boxed{\times}$ 7.042 $\boxed{=}$ $\boxed{\sqrt{}}$	196.43478

Storing this result and using the recall key, you can find the two solutions.

$$x \approx \frac{197.6 + 196.43478}{2(16.3)} \approx 12.087 \qquad \textit{Add stored value}$$

$$x \approx \frac{197.6 - 196.43478}{2(16.3)} \approx 0.036 \qquad \textit{Subtract stored value}$$

Applications

In Section 2.3 we looked at four basic types of applications involving quadratic equations: area, falling bodies, Pythagorean Theorem, and quadratic models. The solution of each of these types of problems can involve the Quadratic Formula.

EXAMPLE 5 School Expenditures

From 1975 to 1989, the total school expenditures (elementary through university) in the United States closely followed the quadratic model

$$\text{Expenditures} = 459.5t^2 + 9170.3t + 107{,}298$$

where the expenditures are measured in millions of dollars and the time t represents the calendar year with $t = 0$ corresponding to 1975. These expenses are shown graphically in Figure 2.10. Assuming that this model continues to be valid, when will the total school expenditures reach $400,000,000,000 per year? (*Source:* U.S. National Center for Education Statistics)

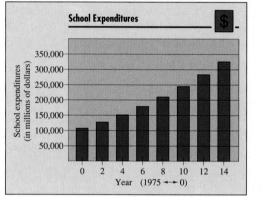

FIGURE 2.10

Solution

Since the expenditures are measured in millions of dollars, we need to solve the equation

$$459.5t^2 + 9170.3t + 107{,}298 = 400{,}000.$$

To begin, write the equation in standard form.

$$459.5t^2 + 9170.3t - 292{,}702 = 0$$

Then, apply the Quadratic Formula with $a = 459.5$, $b = 9170.3$, and $c = -292{,}702$.

$$t = \frac{-9170.3 \pm \sqrt{9170.3^2 - 4(459.5)(-292{,}702)}}{2(459.5)}$$

Choosing the positive solution, we find that

$$t = \frac{-9170.3 + \sqrt{9170.3^2 - 4(459.5)(-292{,}702)}}{2(459.5)} \approx 17.$$

Since $t = 0$ corresponds to 1975, it follows that $t = 17$ must correspond to 1992. Thus, from the model we conclude that the total school expenditures will have reached $400,000,000,000 by 1992.

The position equation given in Section 2.3, $s = -16t^2 + v_0t + s_0$, gives the height of a free-falling object on the surface of the earth. Because of the difference in gravitational force, this equation is different on other planets (and their moons). For instance, on the surface of the moon, the position equation is

$$s = -2.7t^2 + v_0t + s_0.$$

In this equation, s represents the height of the object (in feet), v_0 represents the original velocity of the object (in feet per second), s_0 represents the original height of the object (in feet), and t represents the time (in seconds).

FIGURE 2.11

EXAMPLE 6   Throwing an Object on the Moon

An astronaut standing on the surface of the moon throws a rock upward, as shown in Figure 2.11. The height of the rock is given by

$$s = -2.7t^2 + 27t + 6.$$

How long will it take for the rock to hit the surface of the moon? If the rock were thrown with the same initial velocity on the surface of earth, how long would it remain in the air?

Solution

Since s gives the height of the rock at any time t, you can find the time that the rock hits the surface of the moon by setting s equal to zero and solving for t.

$$-2.7t^2 + 27t + 6 = 0$$
$$t = \frac{-27 \pm \sqrt{27^2 - 4(-2.7)(6)}}{2(-2.7)}$$
$$\approx 10.2 \text{ seconds}$$

Note that we chose the positive solution since it makes no sense, in this problem, to choose a negative value for t. To find the time the rock would have remained in the air on earth, use the position equation

$$s = -16t^2 + 27t + 6.$$

After setting s equal to zero and solving for t, we find that the rock would have remained in the air for only 1.9 seconds. The reason that the rock remains aloft so much longer on the moon is that the moon's gravitational force is much weaker than the earth's.

DISCUSSION PROBLEM

A Revenue Function

From your experience as a consumer, you know that the demand for a product tends to decrease as the price increases. This relationship can be represented by a **demand equation** such as

$$p = 50 - 0.0001x$$

where p is the price of the item and x is the total demand in number of units. For instance, if the price of this item is \$50, then $x = 0$, which means that no one is willing to pay that much for the item. On the other hand, if the price is \$25, then $x = 250,000$, which means that a quarter of a million units of the product could be sold. The total revenue obtained by selling x units of this product is given by the following **revenue equation.**

$$R = \text{(number of units)(price per unit)}$$
$$= xp$$
$$= x(50 - 0.0001x)$$

From this model, would it be possible to obtain a revenue of \$50,000,000? Would it be possible to obtain a revenue of \$70,000,000? Discuss your answers.

WARM UP

The following warm-up exercises involve skills that were covered in earlier sections. You will use these skills in the exercise set for this section.

In Exercises 1–4, simplify.

1. $\sqrt{9 - 4(3)(-12)}$

2. $\sqrt{36 - 4(2)(3)}$

3. $\sqrt{12^2 - 4(3)(4)}$

4. $\sqrt{15^2 + 4(9)(12)}$

In Exercises 5–10, solve the quadratic equations by factoring.

5. $x^2 - x - 2 = 0$

6. $2x^2 + 3x - 9 = 0$

7. $x^2 - 4x = 5$

8. $2x^2 + 13x = 7$

9. $x^2 = 5x - 6$

10. $x(x - 3) = 4$

EXERCISES for Section 2.4

In Exercises 1–6, use the discriminant to determine the number of real solutions of the quadratic equation.

1. $4x^2 - 4x + 1 = 0$

2. $2x^2 - x - 1 = 0$

3. $2x^2 - 5x + 5 = 0$

4. $3x^2 - 6x + 3 = 0$

5. $\frac{1}{5}x^2 + \frac{6}{5}x - 8 = 0$

6. $\frac{1}{3}x^2 - 5x + 25 = 0$

In Exercises 7–28, use the Quadratic Formula to solve the equation.

7. $2x^2 + x - 1 = 0$

8. $2x^2 - x - 1 = 0$

9. $16x^2 + 8x - 3 = 0$

10. $25x^2 - 20x + 3 = 0$

11. $2 + 2x - x^2 = 0$

12. $x^2 - 10x + 22 = 0$

13. $x^2 + 14x + 44 = 0$ **14.** $6x = 4 - x^2$

15. $x^2 + 8x - 4 = 0$ **16.** $4x^2 - 4x - 4 = 0$

17. $12x - 9x^2 = -3$ **18.** $16x^2 + 22 = 40x$

19. $36x^2 + 24x - 7 = 0$ **20.** $3x + x^2 - 1 = 0$

21. $4x^2 + 4x = 7$ **22.** $16x^2 - 40x + 5 = 0$

23. $28x - 49x^2 = 4$ **24.** $9x^2 + 24x + 16 = 0$

25. $25h^2 + 80h + 61 = 0$ **26.** $8t = 5 + 2t^2$

27. $(y - 5)^2 = 2y$ **28.** $(z + 6)^2 = -2z$

In Exercises 29–32, use a calculator to solve the equation. Round your answers to three decimal places.

29. $5.1x^2 - 1.7x - 3.2 = 0$

30. $-0.005x^2 + 0.101x - 0.193 = 0$

31. $422x^2 - 506x - 347 = 0$

32. $2x^2 - 2.50x - 0.42 = 0$

In Exercises 33–42, solve the equation.

33. $3x + 4 = 2x - 7$ **34.** $x^2 - 2x + 5 = x^2 - 5$

35. $4x^2 - 15 = 25$ **36.** $4x^2 + 2x + 4 = 2x + 8$

37. $x^2 + 3x + 1 = 0$ **38.** $x^2 + 3x - 4 = 0$

39. $(x - 1)^2 = 9$ **40.** $2x^2 - 4x - 6 = 0$

41. $100x^2 - 400 = 0$

42. $2x^2 + 4x - 9 = 2(x - 1)^2$

Number Problems   In Exercises 43–46, solve the number problem *and* make up an applied problem that could be represented by this verbal model. (For example, an applied problem that could be represented by Exercise 43 is: "The perimeter of a one-story house is 200 feet, and the house has 2500 square feet of floor space. What are the length and width of the house?")

43. Find two numbers whose sum is 100 and whose product is 2500.

44. Find two consecutive positive integers whose product is 72.

45. Find two consecutive positive integers such that the sum of their squares is 113.

46. Find two consecutive even integers whose product is 440.

In Exercises 47–50, use the cost equation to find the number of units x that a manufacturer can produce for the given cost C. Round your answer to the nearest positive integer.

47. $C = 0.125x^2 + 20x + 500$ $C = \$14,000$

48. $C = 0.5x^2 + 15x + 5,000$ $C = \$11,500$

49. $C = 800 + 0.04x + 0.002x^2$ $C = \$1,680$

50. $C = 800 - 10x + \dfrac{x^2}{4}$ $C = \$896$

51. *Dimensions of a Corral*   A rancher has 200 feet of fencing to enclose two adjacent rectangular corrals (see figure). Find the dimensions that would create an enclosed area of 1400 square feet.

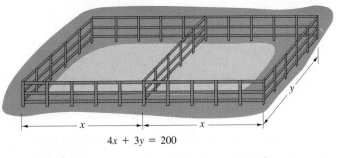

$$4x + 3y = 200$$

Figure for 51

52. *Seating Capacity*   A rectangular classroom seats 72 students. If the seats were rearranged with three more seats in each row, the classroom would have two fewer rows. Find the original number of seats in each row.

53. *Lawn Mowing*   Two people must mow a rectangular lawn 100 feet by 200 feet. Each wants to mow no more than half of the lawn. The first starts by mowing around the outside of the lawn. How wide a strip must the person mow on each of the four sides? If the mower has a 24-inch cut, approximate the required number of trips around the lawn.

54. *Lawn Mowing*   Two people must mow a rectangular lawn 100 feet by 200 feet. The first person agrees to mow three-fourths of the lawn and starts by mowing around the outside. How wide a strip must the person mow on each of the four sides? If the mower has a 24-inch cut, approximate the required number of trips around the lawn.

55. *Dimensions of a Box*   An open box with a square base is to be constructed from 108 square inches of material (see figure). What should the dimensions of the base be if the height of the box is to be 3 inches? (*Hint:* The surface area is given by $S = x^2 + 4xh$.)

56. *Dimensions of a Box*   An open box is to be made from a square piece of material by cutting 2-inch squares from each corner and turning up the sides (see figure). The volume of the finished box is to be 200 cubic inches. Find the size of the original piece of material.

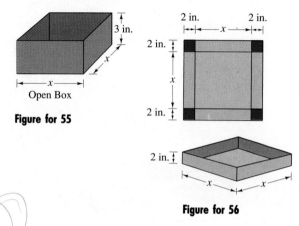

Figure for 55

Figure for 56

57. *Recreational Spending*   The total number of dollars spent on recreation in the United States from 1980 to 1988 can be approximated by the model

Spending $= 116.289 + 9.5067t + 0.84145t^2$

where the spending is measured in billions of dollars and the time t represents the calendar year with $t = 0$ corresponding to 1980. The figure shows the exact spending and the spending represented by the model. From this model, predict the year when total recreational spending will reach $300,000,000,000. (*Source:* U.S. Bureau of Economic Analysis)

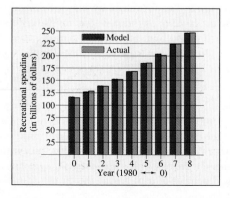

Figure for 57

58. *Oxygen Consumption*   The metabolic rate of ectothermic organisms increases with increasing temperatures within a certain range. The figure shows experimental data for oxygen consumption (microliters per gram per hour) of a beetle for certain temperatures. These data can be approximated by the model

Consumption $= 0.45x^2 - 1.65x + 50.75, \quad 10 \le x \le 25$

where x is the air temperature in degrees Centigrade.
(a) Find the air temperature if the oxygen consumption is 150 microliters per gram per hour.
(b) If the temperature is increased from 10° to 20°, the oxygen consumption is increased by approximately what factor?

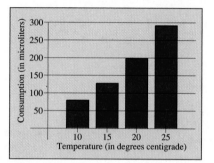

Figure for 58

59. *Flying Speed*   Two planes leave simultaneously from the same airport, one flying due east and the other due south (see figure). The eastbound plane is flying 50 miles per hour faster than the southbound plane. After three hours the planes are 2440 miles apart. Find the speed of each plane.

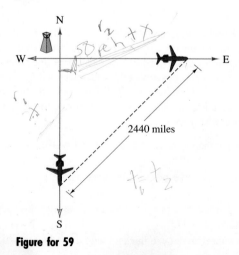

Figure for 59

60. *Flying Distance*   A small commuter airline flies to three cities whose locations form the vertices of a right triangle (see figure). The total flight distance (from City A to City B to City C and back to City A) is 1400 miles. It is 600 miles between the two cities that are farthest apart. Find the other two distances between cities.

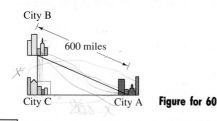

City B

600 miles

City C   City A   **Figure for 60**

61. *Distance from the Dock*   A windlass is used to tow a boat to the dock. The rope is attached to the boat at a point 15 feet below the level of the windlass (see figure). Find the distance from the boat to the dock when 75 feet of rope has been let out. $30\sqrt{6}$ ft ≈ 73.5 ft

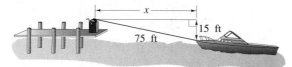

x

75 ft

15 ft

Figure for 61

62. *Total Revenue*   The demand equation for a certain product is $p = 60 - 0.0004x$ where p is the price per unit and x is the number of units sold. The total revenue for selling x units is given by

Revenue $= xp = x(60 - 0.0004x)$.

How many units must be sold to produce a revenue of $220,000?

2.5 Complex Numbers

The Imaginary Unit *i* / Operations with Complex Numbers /
Complex Conjugates and Division / Complex Solutions of Quadratic Equations

The Imaginary Unit *i*

When we introduced the Quadratic Formula in Section 2.4, we noted that some quadratic equations have no real solutions. For instance, the quadratic equation $x^2 + 1 = 0$ has no real solution because there is no real number x that can be squared to produce -1. To overcome this deficiency, mathematicians created an expanded system of numbers using the **imaginary unit *i*,** defined as

$$i = \sqrt{-1}$$

where $i^2 = -1$. Adding real numbers to real multiples of this imaginary unit gives the set of **complex numbers.** Each complex number can be written in the **standard form,** $a + bi$.

DEFINITION OF A COMPLEX NUMBER

For real numbers a and b, the number

$$a + bi$$

is a **complex number.** If $a = 0$ and $b \neq 0$, then the complex number bi is an **imaginary number.**

The set of real numbers is a subset of the set of complex numbers because every real number a can be written as a complex number using $b = 0$. That is, for every real number a, we can write $a = a + 0i$.

Two complex numbers $a + bi$ and $c + di$, written in standard form, are **equal** to each other

$$a + bi = c + di \qquad \textit{Equality of two complex numbers}$$

if and only if $a = c$ and $b = d$. For instance, $a + 3i = -2 + 3i$ only if $a = -2$.

Operations with Complex Numbers

To add (or subtract) two complex numbers, add (or subtract) the real and imaginary parts of the numbers separately.

ADDITION AND SUBTRACTION OF COMPLEX NUMBERS

If $a + bi$ and $c + di$ are two complex numbers written in standard form, then their sum and difference are defined as follows.

Sum $\qquad (a + bi) + (c + di) = (a + c) + (b + d)i$

Difference $\quad (a + bi) - (c + di) = (a - c) + (b - d)i$

The **additive identity** in the complex number system is zero (the same as in the real number system). Furthermore, the **additive inverse** of the complex number $a + bi$ is

$$-(a + bi) = -a - bi. \qquad\qquad \textit{Additive inverse}$$

Thus, we have

$$(a + bi) + (-a - bi) = 0 + 0i = 0.$$

EXAMPLE 1 Adding and Subtracting Complex Numbers

Write the sums and differences in standard form.

REMARK Note in Example 1(b) that the sum of two complex numbers can be a real number.

a. $(3 - i) + (2 + 3i)$
b. $2i + (-4 - 2i)$
c. $3 - (-2 + 3i) + (-5 + i)$

Solution

a.
$$
\begin{aligned}
(3 - i) + (2 + 3i) &= 3 - i + 2 + 3i && \text{\textit{Remove parentheses}} \\
&= 3 + 2 - i + 3i && \text{\textit{Group real and imaginary terms}} \\
&= (3 + 2) + (-1 + 3)i \\
&= 5 + 2i && \text{\textit{Standard form}}
\end{aligned}
$$

b.
$$
\begin{aligned}
2i + (-4 - 2i) &= 2i - 4 - 2i && \text{\textit{Remove parentheses}} \\
&= -4 + 2i - 2i && \text{\textit{Group real and imaginary terms}} \\
&= -4 && \text{\textit{Standard form}}
\end{aligned}
$$

c.
$$
\begin{aligned}
3 - (-2 + 3i) + (-5 + i) &= 3 + 2 - 3i - 5 + i \\
&= 3 + 2 - 5 - 3i + i \\
&= 0 - 2i \\
&= -2i
\end{aligned}
$$

Many of the properties of real numbers are valid for complex numbers as well. For example:

Associative Property of Addition and Multiplication,
Commutative Property of Addition and Multiplication, and
Distributive Property of Multiplication over Addition.

Notice how these properties are used when two complex numbers are multiplied.

$$
\begin{aligned}
(a + bi)(c + di) &= a(c + di) + bi(c + di) && \text{\textit{Distributive}} \\
&= ac + (ad)i + (bc)i + (bd)i^2 && \text{\textit{Distributive}} \\
&= ac + (ad)i + (bc)i + (bd)(-1) && \text{\textit{Definition of i}} \\
&= ac - bd + (ad)i + (bc)i && \text{\textit{Commutative}} \\
&= (ac - bd) + (ad + bc)i && \text{\textit{Associative}}
\end{aligned}
$$

Rather than trying to memorize this multiplication rule, we suggest that you just remember how the distributive property is used to multiply two complex numbers. The procedure is similar to multiplying two polynomials and combining like terms (as in the FOIL method).

EXAMPLE 2   Multiplying Complex Numbers

a. $(i)(-3i) = -3i^2 = -3(-1) = 3$

b. $(2 - i)(4 + 3i) = 8 + 6i - 4i - 3i^2$ *Binomial product*

$\qquad\qquad\qquad\quad = 8 + 6i - 4i - 3(-1)$ $i^2 = -1$

$\qquad\qquad\qquad\quad = 8 + 3 + 6i - 4i$ *Collect terms*

$\qquad\qquad\qquad\quad = 11 + 2i$ *Standard form*

c. $(3 + 2i)(3 - 2i) = 9 - 6i + 6i - 4i^2$ *Binomial product*

$\qquad\qquad\qquad\qquad = 9 - 4(-1)$ $i^2 = -1$

$\qquad\qquad\qquad\qquad = 9 + 4$ *Simplify*

$\qquad\qquad\qquad\qquad = 13$ *Standard form*

Complex Conjugates and Division

Notice in Example 2(c) that the product of two complex numbers can be a real number. This occurs with pairs of complex numbers of the form $a + bi$ and $a - bi$, called **complex conjugates.** In general, the product of two complex conjugates can be written as follows.

$$(a + bi)(a - bi) = a^2 - abi + abi - b^2i^2$$
$$= a^2 - b^2(-1)$$
$$= a^2 + b^2$$

Complex conjugates can be used to divide one complex number by another. That is, to find the quotient

$$\frac{a + bi}{c + di}, \qquad c \text{ and } d \text{ not both zero}$$

multiply the numerator and denominator by the conjugate of the denominator to obtain

$$\frac{a + bi}{c + di} = \frac{a + bi}{c + di}\left(\frac{c - di}{c - di}\right) = \frac{(ac + bd) + (bc - ad)i}{c^2 + d^2}.$$

EXAMPLE 3   Dividing Complex Numbers

Write the complex number in standard form.

$$\frac{2 + 3i}{4 - 2i}$$

Solution

$$\frac{2 + 3i}{4 - 2i} = \frac{2 + 3i}{4 - 2i}\left(\frac{4 + 2i}{4 + 2i}\right)$$

$$= \frac{8 + 4i + 12i + 6i^2}{16 - 4i^2}$$

$$= \frac{8 - 6 + 16i}{16 + 4}$$

$$= \frac{1}{20}(2 + 16i)$$

$$= \frac{1}{10} + \frac{4}{5}i$$

◢

Complex Solutions of Quadratic Equations

Using the Quadratic Formula to solve a quadratic equation often yields a result like $\sqrt{-3}$, which is not a real number. By factoring out $i = \sqrt{-1}$, you can write this number in standard form.

$$\sqrt{-3} = \sqrt{3(-1)} = \sqrt{3}\sqrt{-1} = \sqrt{3}i \quad \text{or} \quad i\sqrt{3}$$

We call $\sqrt{3}\,i$ the principal square root of -3.

REMARK When working with square roots of negative numbers, be sure to convert to standard form *before* multiplying.

PRINCIPAL SQUARE ROOT OF A NEGATIVE NUMBER

If a is a positive number, then the **principal square root** of the negative number $-a$ is defined as

$$\sqrt{-a} = \sqrt{a}\,i.$$

In this definition we are using the rule $\sqrt{ab} = \sqrt{a}\sqrt{b}$, for $a > 0$ and $b < 0$. This rule is not valid if *both* a and b are negative. For example,

$$\sqrt{-5}\sqrt{-5} = (\sqrt{5}i)(\sqrt{5}i) = 5i^2 = -5$$

whereas

$$\sqrt{(-5)(-5)} = \sqrt{25} = 5.$$

Consequently, $\sqrt{(-5)(-5)} \neq \sqrt{-5}\sqrt{-5}$.

EXAMPLE 4   **Writing Complex Numbers in Standard Form**

a. $\sqrt{-3}\sqrt{-12} = \sqrt{3}i\sqrt{12}i = \sqrt{36}i^2 = 6(-1) = -6$

b. $\sqrt{-48} - \sqrt{-27} = \sqrt{48}i - \sqrt{27}i = 4\sqrt{3}i - 3\sqrt{3}i = \sqrt{3}i$

c. $(-1 + \sqrt{-3})^2 = (-1 + \sqrt{3}i)^2$

$$= (-1)^2 - 2\sqrt{3}i + (\sqrt{3})^2(i^2)$$

$$= 1 - 2\sqrt{3}i + 3(-1)$$

$$= -2 - 2\sqrt{3}i$$

EXAMPLE 5   **Complex Solutions of a Quadratic Equation**

$$3x^2 - 2x + 5 = 0 \qquad \text{\textit{Given equation}}$$

$$x = \frac{-(-2) \pm \sqrt{(-2)^2 - 4(3)(5)}}{2(3)} \qquad \text{\textit{Quadratic Formula}}$$

$$= \frac{2 \pm \sqrt{-56}}{6}$$

$$= \frac{2 \pm 2\sqrt{14}i}{6}$$

$$= \frac{1}{3} \pm \frac{\sqrt{14}}{3}i$$

Thus, the given equation has two solutions:

$$\frac{1}{3} + \frac{\sqrt{14}}{3}i \qquad \text{and} \qquad \frac{1}{3} - \frac{\sqrt{14}}{3}i.$$

When working with polynomials of degree higher than 2, it may be necessary to raise i to the third or higher power. The pattern looks like this.

$$i^1 = i$$
$$i^2 = -1$$
$$i^3 = i^2 \cdot i = -i$$
$$i^4 = i^2 \cdot i^2 = (-1)(-1) = 1$$
$$i^5 = i^4 \cdot i = i$$

Since the pattern begins to repeat after the fourth power, you can compute the value of i^n for any natural number n. Simply factor out the multiples of 4 in the exponent and compute the remaining portion. For example,

$$i^{38} = i^{36} \cdot i^2 \qquad \text{or} \qquad i^{27} = i^{24} \cdot i^3$$

$$= (i^4)^9 \cdot i^2 \qquad\qquad = (i^4)^6 \cdot i^3$$

$$= (1)^9(-1) = -1 \qquad\qquad = (1)^6(-i) = -i.$$

DISCUSSION PROBLEM

Cube Roots of Unity

In the real number system, the equation $x^3 = 1$ has 1 as its only solution. However, in the complex number system, this equation has three solutions.

$$1, \quad \frac{-1 + \sqrt{3}i}{2}, \quad \text{and} \quad \frac{-1 - \sqrt{3}i}{2}$$

Try cubing each of these numbers to show that each number has the property that $x^3 = 1$. Then show how you can use the Quadratic Formula to *find* each number from the equation $x^3 = 1$.

WARM UP

The following warm-up exercises involve skills that were covered in earlier sections. You will use these skills in the exercise set for this section.

In Exercises 1–8, simplify the expression.

1. $\sqrt{12}$ **2.** $\sqrt{500}$

3. $\sqrt{20} - \sqrt{5}$ **4.** $\sqrt{27} - \sqrt{243}$

5. $\sqrt{24}\sqrt{6}$ **6.** $2\sqrt{18}\sqrt{32}$

7. $\dfrac{1}{\sqrt{3}}$ **8.** $\dfrac{2}{\sqrt{2}}$

In Exercises 9 and 10, solve the quadratic equation.

9. $x^2 + x - 1 = 0$ **10.** $x^2 + 2x - 1 = 0$

$3 - 3\sqrt{2}i$

EXERCISES for Section 2.5

In Exercises 1–4, find real numbers a and b so that the equation is true.

1. $a + bi = -10 + 6i$ **2.** $a + bi = 13 + 4i$

3. $(a - 1) + (b + 3)i = 5 + 8i$

4. $(a + 6) + 2bi = 6 - 5i$

In Exercises 5–16, write the complex number in standard form.

5. $4 + \sqrt{-9}$ **6.** $3 + \sqrt{-16}$

7. $2 - \sqrt{-27}$ **8.** $1 + \sqrt{-8}$

9. $\sqrt{-75}$ **10.** 45

11. $-6i + i^2$ **12.** $-4i^2 + 2i$

13. 8 **14.** $(\sqrt{-4})^2 - 5$

15. $\sqrt{-0.09}$ **16.** $\sqrt{-0.0004}$

In Exercises 17–26, perform the indicated addition or subtraction and write the result in standard form.

17. $(5 + i) + (6 - 2i)$ **18.** $(13 - 2i) + (-5 + 6i)$

19. $(8 - i) - (4 - i)$ **20.** $(3 + 2i) - (6 + 13i)$

21. $(-2 + \sqrt{-8}) + (5 - \sqrt{-50})$

22. $(8 + \sqrt{-18}) - (4 + 3\sqrt{2}i)$

23. $13i - (14 - 7i)$ **24.** $22 + (-5 + 8i) + 10i$

25. $-\left(\frac{3}{2} + \frac{5}{2}i\right) + \left(\frac{5}{3} + \frac{11}{3}i\right)$

26. $(1.6 + 3.2i) + (-5.8 + 4.3i)$

In Exercises 27–34, write the conjugate of the complex number and find the product of the number and its conjugate.

27. $5 + 3i$

28. $9 - 12i$

29. $-2 - \sqrt{5}i$

30. $-4 + \sqrt{2}i$

31. $20i$

32. $\sqrt{-15}$

33. $\sqrt{8}$

34. $-3 - \sqrt{-5}$

In Exercises 35–56, perform the specified operation and write the result in standard form.

35. $\sqrt{-6} \cdot \sqrt{-2}$

36. $\sqrt{-5} \cdot \sqrt{-10}$

37. $(\sqrt{-10})^2$

38. $(\sqrt{-75})^2$

39. $(1 + i)(3 - 2i)$

40. $(6 - 2i)(2 - 3i)$

41. $6i(5 - 2i)$

42. $-8i(9 + 4i)$

43. $(\sqrt{14} + \sqrt{10}i)(\sqrt{14} - \sqrt{10}i)$

44. $(3 + \sqrt{-5})(7 - \sqrt{-10})$

45. $(4 + 5i)^2$

46. $(2 - 3i)^3$

47. $(2 + 3i)^2 + (2 - 3i)^2$

48. $(1 - 2i)^2 - (1 + 2i)^2$

49. $\dfrac{4}{4 - 5i}$

50. $\dfrac{3}{1 - i}$

51. $\dfrac{2 + i}{2 - i}$

52. $\dfrac{8 - 7i}{1 - 2i}$

53. $\dfrac{6 - 7i}{i}$

54. $\dfrac{8 + 20i}{2i}$

55. $\dfrac{1}{(4 - 5i)^2}$

56. $\dfrac{(2 - 3i)(5i)}{2 + 3i}$

In Exercises 57–64, use the Quadratic Formula to solve the quadratic equation.

57. $x^2 - 2x + 2 = 0$

58. $x^2 + 6x + 10 = 0$

59. $4x^2 + 16x + 17 = 0$

60. $9x^2 - 6x + 37 = 0$

61. $4x^2 + 16x + 15 = 0$

62. $9x^2 - 6x - 35 = 0$

63. $16t^2 - 4t + 3 = 0$

64. $5s^2 + 6s + 3 = 0$

65. Write the first 16 positive integer powers of i (that is, i, i^2, i^3, ..., i^{16}), and express each as i, $-i$, 1, or -1.

66. Express each of the powers of i as i, $-i$, 1, or -1.
(a) i^{40} (b) i^{25}
(c) i^{50} (d) i^{67}

In Exercises 67–74, simplify the complex number and write it in standard form.

67. $-6i^3 + i^2$

68. $4i^2 - 2i^3$

69. $-5i^5$

70. $(-i)^3$

71. $(\sqrt{-75})^3$

72. $(\sqrt{-2})^6$

73. $\dfrac{1}{i^3}$

74. $\dfrac{1}{(2i)^3}$

75. Cube the complex numbers.
$$2, \quad -1 + \sqrt{3}i, \quad -1 - \sqrt{3}i$$

76. Raise the numbers to the fourth power.
$$2, \quad -2, \quad 2i, \quad -2i$$

77. Prove that the sum of a complex number $a + bi$ and its conjugate is a real number.

78. Prove that the difference of a complex number $a + bi$ and its conjugate is an imaginary number.

79. Prove that the product of a complex number $a + bi$ and its conjugate is a real number.

80. Prove that the conjugate of the product of two complex numbers $a_1 + b_1i$ and $a_2 + b_2i$ is the product of their conjugates.

81. Prove that the conjugate of the sum of two complex numbers $a_1 + b_1i$ and $a_2 + b_2i$ is the sum of their conjugates.

2.6 Other Types of Equations

Solving Polynomial Equations / Solving Equations Involving Radicals /
Solving Equations Involving Fractions or Absolute Values / Applications

Solving Polynomial Equations

We have only two basic methods for solving nonlinear equations—*factoring* and the *Quadratic Formula*. The main goal of this section is to learn to *rewrite* nonlinear equations in factorable or quadratic form. In this section, we will give only real solutions. Complex solutions will be discussed in detail in Section 4.5.

Because factoring polynomials is so crucial to this section, you may want to review the special forms in Section 1.5.

EXAMPLE 1   Solving a Polynomial Equation by Factoring

Solve $3x^4 = 48x^2$.

Solution

The basic approach here is to write the polynomial equation in standard form (with zero on the right side), factor the left side, and then set each factor equal to zero.

$$3x^4 = 48x^2 \qquad \text{\textit{Given equation}}$$
$$3x^4 - 48x^2 = 0 \qquad \text{\textit{Collect terms on left side}}$$
$$(3x^2)(x^2 - 16) = 0 \qquad \text{\textit{Common monomial factor}}$$
$$3(x^2)(x + 4)(x - 4) = 0 \qquad \text{\textit{Difference of two squares}}$$
$$x^2 = 0 \quad \rightarrow \quad x = 0 \qquad \text{\textit{Set 1st factor equal to 0}}$$
$$x + 4 = 0 \quad \rightarrow \quad x = -4 \qquad \text{\textit{Set 2nd factor equal to 0}}$$
$$x - 4 = 0 \quad \rightarrow \quad x = 4 \qquad \text{\textit{Set 3rd factor equal to 0}}$$

Thus, the equation has three solutions: 0, -4, and 4. Check each of these solutions in the original equation.

REMARK   A common mistake when factoring an equation like that given in Example 1 is to divide both sides of the equation by the variable factor x^2 before attempting to solve the equation. The solution $x = 0$ is lost. When using factoring to solve an equation, be sure to set each factor equal to zero. Don't divide both sides of an equation by a variable factor in an attempt to simplify the equation.

EXAMPLE 2   Solving a Polynomial Equation by Factoring

$$x^3 - 3x^2 - 3x + 9 = 0 \qquad \text{\textit{Given equation}}$$
$$x^2(x - 3) - 3(x - 3) = 0 \qquad \text{\textit{Group terms}}$$
$$(x - 3)(x^2 - 3) = 0 \qquad \text{\textit{Factor by grouping}}$$
$$x - 3 = 0 \quad \rightarrow \quad x = 3 \qquad \text{\textit{Set 1st factor equal to 0}}$$
$$x^2 - 3 = 0 \quad \rightarrow \quad x = \pm\sqrt{3} \qquad \text{\textit{Set 2nd factor equal to 0}}$$

Thus, the equation has three solutions: 3, $\sqrt{3}$, and $-\sqrt{3}$. Check these solutions in the original equation.

Occasionally, mathematical models involve equations that are of **quadratic type.** In general, an equation is of quadratic type if it can be written in the form

$$au^2 + bu + c = 0$$

where $a \neq 0$ and u is an algebraic expression. For instance, the equation

$$x^4 - 3x^2 + 2 = 0 \quad \rightarrow \quad \overbrace{(x^2)^2}^{u^2} - 3\overbrace{(x^2)}^{u} + 2 = 0$$

is of quadratic type with $u = x^2$. Once an equation is in this quadratic form, we can solve it by factoring or by the Quadratic Formula.

EXAMPLE 3   Solving an Equation of Quadratic Type

Solve $x^4 - 3x^2 + 2 = 0$.

Solution

This equation is of quadratic type with $u = x^2$. Factor the left side of the equation as the product of two second-degree polynomials.

$$x^4 - 3x^2 + 2 = 0 \qquad \textit{Given equation}$$

$$\overbrace{(x^2)^2}^{u^2} - 3\overbrace{(x^2)}^{u} + 2 = 0 \qquad \textit{Quadratic form}$$

$$(x^2 - 1)(x^2 - 2) = 0 \qquad \textit{Partially factor}$$

$$(x + 1)(x - 1)(x^2 - 2) = 0 \qquad \textit{Completely factor}$$

$$x + 1 = 0 \quad \rightarrow \quad x = -1 \qquad \textit{Set 1st factor equal to 0}$$

$$x - 1 = 0 \quad \rightarrow \quad x = 1 \qquad \textit{Set 2nd factor equal to 0}$$

$$x^2 - 2 = 0 \quad \rightarrow \quad x = \pm\sqrt{2} \qquad \textit{Set 3rd factor equal to 0}$$

Thus, the equation has four solutions: -1, 1, $\sqrt{2}$, and $-\sqrt{2}$. Check these solutions in the original equation.

Solving Equations Involving Radicals

The steps involved in solving the remaining equations in this section will often introduce *extraneous solutions*, as discussed in Section 2.1. Operations like squaring both sides of an equation, raising both sides of an equation to a rational power, and multiplying both sides by a variable quantity all have this potential danger. Thus, when you use any of these operations a check *of each tentative solution* is crucial.

EXAMPLE 4   An Equation Involving a Rational Exponent

Solve $4x^{2/3} - 16 = 0$.

Solution

The strategy used to solve this equation is to isolate the expression $x^{2/3}$ and raise both sides of the equation to the 3/2 power.

$$4x^{2/3} - 16 = 0 \qquad \qquad \text{\textit{Given equation}}$$
$$4x^{2/3} = 16 \qquad \qquad \text{\textit{Add 16 to both sides}}$$
$$x^{2/3} = 4 \qquad \qquad \text{\textit{Divide both sides by 4}}$$
$$(x^{2/3})^{3/2} = \pm(4)^{3/2} \qquad \qquad \text{\textit{Raise both sides to 3/2 power}}$$
$$x = \pm 4^{3/2} \qquad \qquad \text{\textit{x has exponent 1}}$$
$$x = \pm(\sqrt{4})^3 \qquad \qquad \text{\textit{Radical form}}$$
$$x = \pm(2)^3 = \pm 8 \qquad \qquad \text{\textit{Simplify}}$$

Check

$$4x^{2/3} - 16 = 0 \qquad \qquad \text{\textit{Given equation}}$$
$$4(-8)^{2/3} - 16 \stackrel{?}{=} 0 \qquad \qquad \text{\textit{Replace x by} } -8$$
$$4(-2)^2 - 16 \stackrel{?}{=} 0 \qquad \qquad \text{\textit{Property of exponents}}$$
$$16 - 16 = 0 \qquad \qquad \text{\textit{Solution checks}}$$

A check of $x = 8$ will also satisfy the equation.

◢

EXAMPLE 5   An Equation Involving a Radical

Solve $\sqrt{2x + 7} - x = 2$.

Solution

The basic approach here is to eliminate the square root by squaring both sides of the equation. To do this, first isolate the radical.

$$\sqrt{2x + 7} - x = 2 \qquad \qquad \text{\textit{Given equation}}$$
$$\sqrt{2x + 7} = x + 2 \qquad \qquad \text{\textit{Add x to both sides}}$$
$$2x + 7 = x^2 + 4x + 4 \qquad \qquad \text{\textit{Square both sides}}$$
$$0 = x^2 + 2x - 3 \qquad \qquad \text{\textit{Standard form}}$$
$$0 = (x + 3)(x - 1) \qquad \qquad \text{\textit{Factor}}$$
$$x + 3 = 0 \;\rightarrow\; x = -3 \qquad \qquad \text{\textit{Set 1st factor equal to 0}}$$
$$x - 1 = 0 \;\rightarrow\; x = 1 \qquad \qquad \text{\textit{Set 2nd factor equal to 0}}$$

Check

$$\sqrt{2(-3) + 7} + 3 = \sqrt{1} + 3 \neq 2 \qquad \qquad \text{\textit{-3 does not check}}$$
$$\sqrt{2(1) + 7} - 1 = \sqrt{9} - 1 = 2 \qquad \qquad \text{\textit{1 checks}}$$

Thus, the equation has only one solution: 1.

◢

For equations with two or more radicals, it may be necessary to repeat the "isolate a radical and square both sides" routine demonstrated in Example 5. Remember to include the middle term when squaring a binomial.

EXAMPLE 6   An Equation Involving Two Radicals

$$\sqrt{2x + 6} - \sqrt{x + 4} = 1 \qquad \text{\textit{Given equation}}$$
$$\sqrt{2x + 6} = 1 + \sqrt{x + 4} \qquad \text{\textit{Isolate radical}}$$
$$2x + 6 = 1 + 2\sqrt{x + 4} + (x + 4) \qquad \text{\textit{Square both sides}}$$
$$x + 1 = 2\sqrt{x + 4} \qquad \text{\textit{Isolate radical}}$$
$$x^2 + 2x + 1 = 4(x + 4) \qquad \text{\textit{Square both sides}}$$
$$x^2 - 2x - 15 = 0 \qquad \text{\textit{Standard form}}$$
$$(x - 5)(x + 3) = 0 \qquad \text{\textit{Factor}}$$
$$x - 5 = 0 \;\rightarrow\; x = 5 \qquad \text{\textit{Set 1st factor equal to 0}}$$
$$x + 3 = 0 \;\rightarrow\; x = -3 \qquad \text{\textit{Set 2nd factor equal to 0}}$$

Check

$$\sqrt{2(5) + 6} - \sqrt{5 + 4} = 4 - 3 = 1 \qquad \text{\textit{5 is a solution}}$$
$$\sqrt{2(-3) + 6} - \sqrt{-3 + 4} = 0 - 1 \neq 1 \qquad \text{\textit{−3 is not a solution}}$$

Thus, the only solution is $x = 5$.

EXAMPLE 7   An Equation Involving a Rational Exponent

Solve $(x^2 - x - 4)^{3/4} = 8$.

Solution

The left side of the equation is raised to the 3/4 power, so we begin by raising both sides to the *reciprocal* power, 4/3. Note that

$$8^{4/3} = (\sqrt[3]{8})^4 = 2^4 = 16.$$

$$(x^2 - x - 4)^{3/4} = 8 \qquad \text{\textit{Given equation}}$$
$$(x^2 - x - 4) = (8)^{4/3} \qquad \text{\textit{Raise both sides to 4/3 power}}$$
$$x^2 - x - 4 = 16 \qquad \text{\textit{Simplify}}$$
$$x^2 - x - 20 = 0 \qquad \text{\textit{Standard form}}$$
$$(x - 5)(x + 4) = 0 \qquad \text{\textit{Factor}}$$
$$x - 5 = 0 \;\rightarrow\; x = 5 \qquad \text{\textit{Set 1st factor equal to 0}}$$
$$x + 4 = 0 \;\rightarrow\; x = -4 \qquad \text{\textit{Set 2nd factor equal to 0}}$$

Thus, the equation has two solutions: 5 and −4. A check will show that both solutions are valid.

Solving Equations Involving Fractions or Absolute Values

To solve an equation involving fractions, we multiply both sides of the equation by the least common denominator of the fractions in the equation. This procedure will "clear an equation of fractions." For instance, in the equation

$$\frac{2}{x^2 + 1} + \frac{1}{x} = \frac{2}{x}$$

we can multiply both sides of the equation by $x(x^2 + 1)$ to obtain

$$(2x) + (x^2 + 1) = 2(x^2 + 1).$$

Try solving this equation. You should obtain one solution: $x = 1$.

EXAMPLE 8   An Equation Involving Fractions

$$\frac{2}{x} = \frac{3}{x - 2} - 1 \qquad \textit{Given equation}$$

$$x(x - 2)\frac{2}{x} = x(x - 2)\frac{3}{x - 2} - x(x - 2)(1) \quad \textit{Multiply by LCD}$$

$$2(x - 2) = 3x - x(x - 2), \qquad x \neq 0, 2$$

$$2x - 4 = -x^2 + 5x$$

$$x^2 - 3x - 4 = 0$$

$$(x - 4)(x + 1) = 0$$

$$x - 4 = 0 \quad \rightarrow \quad x = 4$$

$$x + 1 = 0 \quad \rightarrow \quad x = -1$$

Thus, the equation has two solutions: 4 and -1. A check will show that both solutions are valid.

To solve an equation involving an absolute value, consider the fact that the expression inside the absolute value can be positive or negative. This consideration results in *two* separate equations, each of which must be solved. For instance, the equation

$$|x - 2| = 3$$

results in the two equations

$$x - 2 = 3 \qquad \text{and} \qquad -(x - 2) = 3$$

which implies that the equation has two solutions: 5 and -1.

EXAMPLE 9 An Equation Involving Absolute Value

Solve $|x^2 - 3x| = -4x + 6$.

Solution

Since the variable expression inside the absolute value signs can be positive or negative, we must solve two equations.

First Equation

$$x^2 - 3x = -4x + 6 \qquad \text{\textit{Use positive expression}}$$

$$x^2 + x - 6 = 0 \qquad \text{\textit{Standard form}}$$

$$(x + 3)(x - 2) = 0 \qquad \text{\textit{Factor}}$$

$$x + 3 = 0 \ \rightarrow \ x = -3 \qquad \text{\textit{Set 1st factor equal to 0}}$$

$$x - 2 = 0 \ \rightarrow \ x = 2 \qquad \text{\textit{Set 2nd factor equal to 0}}$$

Second Equation

$$-(x^2 - 3x) = -4x + 6 \qquad \text{\textit{Use negative expression}}$$

$$x^2 - 7x + 6 = 0 \qquad \text{\textit{Standard form}}$$

$$(x - 1)(x - 6) = 0 \qquad \text{\textit{Factor}}$$

$$x - 1 = 0 \ \rightarrow \ x = 1 \qquad \text{\textit{Set 1st factor equal to 0}}$$

$$x - 6 = 0 \ \rightarrow \ x = 6 \qquad \text{\textit{Set 2nd factor equal to 0}}$$

Check

$$|(-3)^2 - 3(-3)| = -4(-3) + 6 \qquad \text{\textit{-3 checks}}$$

$$|2^2 - 3(2)| \neq -4(2) + 6 \qquad \text{\textit{2 does not check}}$$

$$|1^2 - 3(1)| = -4(1) + 6 \qquad \text{\textit{1 checks}}$$

$$|6^2 - 3(6)| \neq -4(6) + 6 \qquad \text{\textit{6 does not check}}$$

Thus, the equation has only two solutions: -3 and 1.

Applications

It would be impossible to categorize the many different types of applications that involve nonlinear and nonquadratic models. However, from the few examples and exercises that are given, we hope that you will gain some appreciation for the variety of applications that can occur.

EXAMPLE 10   Reduced Rates

A ski club chartered a bus for a ski trip at a cost of $480. In an attempt to lower the bus fare, the club invited nonmembers to go along. After five nonmembers joined the trip, the fare per skier decreased by $4.80. How many club members are going on the trip?

Solution

Verbal Model
$$\text{Cost per skier} \cdot \text{Number of skiers} = \text{Cost of trip}$$

Labels
Cost of trip = $480

Number of ski club members = x

Number of skiers = $x + 5$

Original cost per member = $\dfrac{480}{x}$ (in dollars)

Cost per skier = $\dfrac{480}{x} - 4.80$ (in dollars)

Algebraic Equation

$$\left(\frac{480}{x} - 4.80\right)(x + 5) = 480$$

$$\left(\frac{480 - 4.8x}{x}\right)(x + 5) = 480$$

$$(480 - 4.8x)(x + 5) = 480x, \qquad x \neq 0$$

$$480x - 4.8x^2 - 24x + 2400 = 480x$$

$$-4.8x^2 - 24x + 2400 = 0$$

$$x^2 + 5x - 500 = 0$$

$$(x + 25)(x - 20) = 0$$

$$x + 25 = 0 \;\; \rightarrow \;\; x = -25$$

$$x - 20 = 0 \;\; \rightarrow \;\; x = 20$$

Choosing the positive value of x, we find that the ski club has 20 members. Check this result in the original statement of the problem.

EXAMPLE 11   Market Research

The marketing department at a publishing firm is asked to determine the price of a book. The department determines that the demand for the book depends on the price of the book according to the formula

$$p = 40 - \sqrt{0.0001x + 1}$$

where p is the price per book in dollars and x is the number of books sold at the given price. For instance, in Figure 2.12, note that if the price were $39, then (according to the model) no one would be willing to buy the book. On the other hand, if the price were $17.60, five million copies could be sold. If the publisher set the price at $12.95, how many copies could be sold?

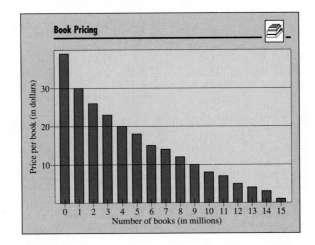

Book Pricing

Price per book (in dollars) / Number of books (in millions)

FIGURE 2.12

Solution

$$
\begin{aligned}
p &= 40 - \sqrt{0.0001x + 1} & &\textit{Given model} \\
12.95 &= 40 - \sqrt{0.0001x + 1} & &\textit{Set price at \$12.95} \\
\sqrt{0.0001x + 1} &= 27.05 & &\textit{Isolate radical} \\
0.0001x + 1 &= 731.7025 & &\textit{Square both sides} \\
0.0001x &= 730.7025 & &\textit{Subtract 1 from both sides} \\
x &= 7{,}307{,}025 & &\textit{Divide both sides by 0.0001}
\end{aligned}
$$

Thus, with a price of $12.95, the publisher could expect to sell about 7.3 million copies.

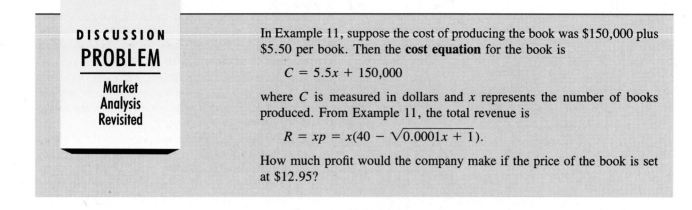

DISCUSSION PROBLEM

Market Analysis Revisited

In Example 11, suppose the cost of producing the book was $150,000 plus $5.50 per book. Then the **cost equation** for the book is

$$C = 5.5x + 150,000$$

where C is measured in dollars and x represents the number of books produced. From Example 11, the total revenue is

$$R = xp = x(40 - \sqrt{0.0001x + 1}).$$

How much profit would the company make if the price of the book is set at $12.95?

WARM UP

The following warm-up exercises involve skills that were covered in earlier sections. You will use these skills in the exercise set for this section.

In Exercises 1–10, find the real solution(s) of each equation.

1. $x^2 - 22x + 121 = 0$ 2. $x(x - 20) + 3(x - 20) = 0$

3. $(x + 20)^2 = 625$ 4. $5x^2 + x = 0$

5. $3x^2 + 4x - 4 = 0$ 6. $12x^2 + 8x - 55 = 0$

7. $x^2 + 4x - 5 = 0$ 8. $4x^2 + 4x - 15 = 0$

9. $x^2 - 3x + 1 = 0$ 10. $x^2 - 4x + 2 = 0$

EXERCISES for Section 2.6

In Exercises 1–70, find all solutions of the equation. Check your answers in the original equation.

1. $4x^4 - 18x^2 = 0$
2. $20x^3 - 125x = 0$
3. $x^3 - 2x^2 - 3x = 0$
4. $2x^4 - 15x^3 + 18x^2 = 0$
5. $x^4 - 81 = 0$
6. $x^6 - 64 = 0$
7. $5x^3 + 30x^2 + 45x = 0$
8. $9x^4 - 24x^3 + 16x^2 = 0$
9. $x^3 - 3x^2 - x + 3 = 0$
10. $x^3 + 2x^2 + 3x + 6 = 0$
11. $x^4 - x^3 + x - 1 = 0$
12. $x^4 + 2x^3 - 8x - 16 = 0$
13. $x^4 - 10x^2 + 9 = 0$
14. $x^4 - 29x^2 + 100 = 0$
15. $x^4 + 5x^2 - 36 = 0$
16. $x^4 - 4x^2 + 3 = 0$
17. $4x^4 - 65x^2 + 16 = 0$
18. $36t^4 + 29t^2 - 7 = 0$
19. $x^6 + 7x^3 - 8 = 0$
20. $x^6 + 3x^3 + 2 = 0$

21. $\dfrac{1}{t^2} + \dfrac{8}{t} + 15 = 0$

22. $6\left(\dfrac{s}{s+1}\right)^2 + 5\left(\dfrac{s}{s+1}\right) - 6 = 0$

23. $2x + 9\sqrt{x} - 5 = 0$
24. $6x - 7\sqrt{x} - 3 = 0$
25. $5 - 3x^{1/3} - 2x^{2/3} = 0$
26. $9t^{2/3} + 24t^{1/3} + 16 = 0$
27. $\sqrt{2x} - 10 = 0$
28. $4\sqrt{x} - 3 = 0$
29. $\sqrt{x - 10} - 4 = 0$
30. $\sqrt{5 - x} - 3 = 0$
31. $\sqrt[3]{2x + 5} + 3 = 0$
32. $\sqrt[3]{3x + 1} - 5 = 0$
33. $x = \sqrt{11x - 30}$
34. $2x - \sqrt{15 - 4x} = 0$
35. $-\sqrt{26 - 11x} + 4 = x$
36. $x + \sqrt{31 - 9x} = 5$
37. $\sqrt{x + 1} - 3x = 1$
38. $\sqrt{x + 5} = \sqrt{x - 5}$
39. $\sqrt{x} + \sqrt{x - 20} = 10$
40. $\sqrt{x} - \sqrt{x - 5} = 1$
41. $\sqrt{x + 5} + \sqrt{x - 5} = 10$
42. $2\sqrt{x + 1} - \sqrt{2x + 3} = 1$

43. $\sqrt{7x + 36} - \sqrt{5x + 16} = 2$

44. $3\sqrt{x} - \dfrac{4}{\sqrt{x}} = 4$

45. $(x - 5)^{2/3} = 16$

46. $(x + 3)^{4/3} = 16$

47. $(x + 3)^{3/4} = 27$

48. $(x^2 + 2)^{5/2} = 32$

49. $(x^2 - 5)^{2/3} = 16$

50. $(x^2 - x - 22)^{4/3} = 16$

51. $3x(x - 1)^{1/2} + 2(x - 1)^{3/2} = 0$

52. $4x^2(x - 1)^{1/3} + 6x(x - 1)^{4/3} = 0$

53. $\dfrac{20 - x}{x} = x$

54. $\dfrac{4}{x} - \dfrac{5}{3} = \dfrac{x}{6}$

55. $\dfrac{1}{x} - \dfrac{1}{x + 1} = 3$

56. $\dfrac{x}{x^2 - 4} + \dfrac{1}{x + 2} = 3$

57. $x = \dfrac{3}{x} + \dfrac{1}{2}$

58. $4x + 1 = \dfrac{3}{x}$

59. $\dfrac{1}{x} = \dfrac{4}{x - 1} + 1$

60. $x + \dfrac{9}{x + 1} = 5$

61. $\dfrac{4}{x + 1} - \dfrac{3}{x + 2} = 1$

62. $\dfrac{x + 1}{3} - \dfrac{x + 1}{x + 2} = 0$

63. $|x + 1| = 2$

64. $|x - 2| = 3$

65. $|2x - 1| = 5$

66. $|3x + 2| = 7$

67. $|x| = x^2 + x - 3$

68. $|x^2 + 6x| = 3x + 18$

69. $|x - 10| = x^2 - 10x$

70. $|x + 1| = x^2 - 5$

In Exercises 71–74, use a calculator to find the real solutions of the equation. Round to three decimal places.

71. $3.2x^4 - 1.5x^2 - 2.1 = 0$

72. $7.08x^6 + 4.15x^3 - 9.6 = 0$

73. $1.8x - 6\sqrt{x} - 5.6 = 0$

74. $4x^{2/3} + 8x^{1/3} + 3.6 = 0$

75. *Sharing the Cost* A college charters a bus for $1700 to take a group of students to a museum. When six more students join the trip, the cost per student drops by $7.50. How many students were in the original group?

76. *Sharing the Cost* Three students are planning to rent an apartment for a year and share equally in the monthly rent. By adding a fourth person to the group, each person could save $75 per month. How much is the monthly rent?

77. *Airspeed* An airline runs a commuter flight between two cities that are 720 miles apart. If the average speed of the plane is increased by 40 miles per hour, the travel time is decreased by 12 minutes. What airspeed is required to obtain this decrease in travel time?

78. *Average Speed* A family drove 1080 miles to their vacation lodge. Because of increased traffic density, their average speed on the return trip was decreased by 6 miles per hour and the trip took $2\frac{1}{2}$ hours longer. Determine their average speed on the way to the lodge.

79. *Compound Interest* A deposit of $2500 reaches a balance of $3544.06 after five years. The interest in the account is compounded monthly. What is the annual percentage rate for this investment?

80. *Compound Interest* A sales representative for a mutual fund company describes a "guaranteed investment fund" that if you deposit $10,000 in the fund you will be guaranteed a return of at least $25,000 after 20 years. (Assume the interest is compounded quarterly.)
(a) What is the annual percentage rate if the investment only meets the minimum guaranteed amount?
(b) If after 20 years you received $35,000, what annual percentage rate did you receive?

81. *Saturated Steam* The temperature (in degrees Fahrenheit) of saturated steam increases as pressure increases (see figure). This relationship is approximated by the model

$$\text{Temperature} = 75.82 - 2.11x + 43.51\sqrt{x}, \quad 5 \le x \le 40$$

where x is the absolute pressure in pounds per square inch.
(a) The temperature of steam at sea level ($x = 14.696$) is 212°. Evaluate the model at this pressure.
(b) Use the model to approximate the pressure if the temperature of the steam is 240°.

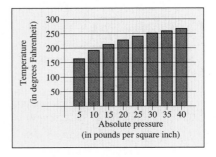

Figure for 81

82. *Airline Passengers* An airline offers daily flights between Chicago and Denver. The total monthly cost of these flights is

$$C = \sqrt{0.2x + 1}$$

where C is measured in millions of dollars and x is measured in thousands of passengers. The total cost of the flights for a certain month is 2.5 million dollars. How many passengers flew that month?

83. *Market Research* The demand equation for a certain product is

$$p = 40 - \sqrt{0.01x + 1}$$

where x is the number of units demanded per day and p is the price per unit. Find the demand if the price is set at $37.55.

84. *Market Research* The demand equation for a certain product is

$$p = 30 - \sqrt{0.0001x + 1}$$

where x is the number of units demanded per day and p is the price per unit. Find the demand if the price is set at $25.30.

85. *Power Line* A power station is on one side of a river that is $\frac{1}{2}$ mile wide. A factory is six miles downstream on the other side of the river. It costs $18 per foot to run power lines overland and $24 per foot to run them underwater. The total cost of the project is $616,877.27. Find the length x as labeled in the figure.

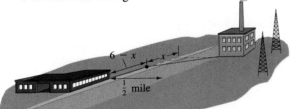

Figure for 85

86. *Baseball Diamond* A baseball diamond has the shape of a square where the distance from home plate to second base is approximately $127\frac{1}{2}$ feet. Approximate the distance between each of the bases.

87. *Surface Area* The surface area of a cone is $S = \pi r \sqrt{r^2 + h^2}$. Solve this equation for h.

88. *Inductance* An equation describing a circuit containing inductance i and capacitance C is

$$i = \pm\sqrt{\frac{1}{LC}}\sqrt{Q^2 - q}.$$

Solve this equation for Q.

In Exercises 89 and 90, consider an equation of the form $x + \sqrt{x - a} = b$ where a and b are constants.

89. Find a and b if the solution to the equation is $x = 20$. (There are many correct answers.)

90. Write a short paragraph listing the steps required in solving the equation for x.

2.7 Linear Inequalities

Inequalities and Intervals on the Real Number Line / Properties of Inequalities / Solving a Linear Inequality / Inequalities Involving Absolute Value / Applications

Inequalities and Intervals on the Real Number Line

Simple inequalities were introduced in Section 1.1 to *order* the real numbers. There, inequality symbols $<$, $\leq$, $>$, and $\geq$ were used to compare two numbers and to denote subsets of real numbers. For instance, the simple inequality

$$x \geq 3$$

denotes all real numbers x that are greater than or equal to 3.

In this section we expand our work with inequalities to include more involved statements such as

$$5x - 7 < 3x + 9 \qquad \text{and} \qquad -3 \leq 6x - 1 < 3.$$

As with an equation, we **solve an inequality** in the variable x by finding all values of x for which the inequality is true. These values are **solutions** and the solutions **satisfy** the inequality. The set of all real numbers that are solutions of an inequality is the **solution set** of the inequality.

The set of all points on the real number line that represent the solution set is the **graph** of the inequality. Graphs of many types of inequalities consist of intervals on the real number line.

BOUNDED INTERVALS ON THE REAL NUMBER LINE

Let a and b be real numbers such that $a < b$. The following intervals on the real number line are **bounded intervals.** The numbers a and b are the **endpoints** of each interval.

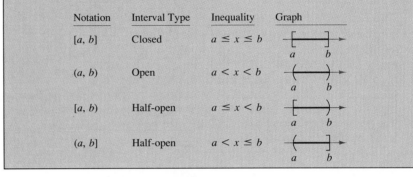

Notation	Interval Type	Inequality	Graph
$[a, b]$	Closed	$a \leq x \leq b$	
(a, b)	Open	$a < x < b$	
$[a, b)$	Half-open	$a \leq x < b$	
$(a, b]$	Half-open	$a < x \leq b$	

Note that a *closed interval* contains both of its endpoints, a *half-open interval* contains only one of its endpoints, and an *open interval* contains neither of its endpoints. Often, the solution of an inequality is an interval on the real line that is **unbounded.** For instance, the interval consisting of all positive numbers is unbounded.

UNBOUNDED INTERVALS ON THE REAL NUMBER LINE

Let a and b be real numbers. The following intervals on the real number line are **unbounded intervals.**

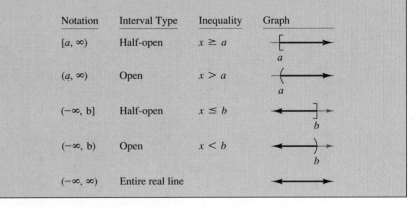

Notation	Interval Type	Inequality	Graph
$[a, \infty)$	Half-open	$x \geq a$	
(a, ∞)	Open	$x > a$	
$(-\infty, b]$	Half-open	$x \leq b$	
$(-\infty, b)$	Open	$x < b$	
$(-\infty, \infty)$	Entire real line		

The symbols ∞ (**positive infinity**) and $-\infty$ (**negative infinity**) do not represent real numbers. They are simply convenient symbols used to describe the unboundedness of an interval such as $(1, \infty)$.

EXAMPLE 1 Intervals and Inequalities

Write an inequality to represent each of the intervals and state whether the interval is bounded or unbounded.

a. $(-3, 5]$ **b.** $(-3, \infty)$ **c.** $[0, 2]$

Solution

a. $(-3, 5]$ corresponds to $-3 < x \le 5$. *Bounded*
b. $(-3, \infty)$ corresponds to $-3 < x$. *Unbounded*
c. $[0, 2]$ corresponds to $0 \le x \le 2$. *Bounded*

Properties of Inequalities

The procedures for solving linear inequalities in one variable are much like those for solving linear equations. To isolate the variable, we use **properties of inequalities.** These properties are similar to the properties of equality, but there are two important exceptions. When both sides of an inequality are multiplied or divided by a negative number, the direction of the inequality symbol must be reversed.

$$-2 < 5 \qquad \text{\textit{Given inequality}}$$
$$(-3)(-2) > (-3)(5) \qquad \text{\textit{Multiply both sides by} } -3$$
$$\text{\textit{and reverse the inequality}}$$
$$6 > -15$$

Two inequalities that have the same solution set are **equivalent.** The following list describes operations that can be used to create equivalent inequalities.

PROPERTIES OF INEQUALITIES

Let a, b, c, and d be real numbers.

Property	*Example*
1. Transitive Property: $a < b$ and $b < c \rightarrow a < c$	Since $-2 < 5$ and $5 < 7$, it follows that $-2 < 7$.
2. Addition of Inequalities: $a < b$ and $c < d \rightarrow a + c < b + d$.	Since $2 < 4$ and $3 < 5$, it follows that $2 + 3 < 4 + 5$.
3. Addition of a Constant: $a < b \rightarrow a + c < b + c$	Since $-3 < 7$, it follows that $-3 + 2 < 7 + 2$.
4. Multiplying by a Constant: (i) For $c > 0$, $\quad a < b \rightarrow ac < bc$	Since $5 > 0$ and $3 < 9$, it follows that $3(5) < 9(5)$.
(ii) For $c < 0$, $\quad a < b \rightarrow ac > bc$	Since $-5 < 0$ and $3 < 9$, it follows that $3(-5) > 9(-5)$.

Solving a Linear Inequality

The simplest type of inequality to solve is a **linear inequality** in a single variable. Each of the inequalities

$$2x + 3 > 4, \qquad 3x - 4 \le 2x + 5, \qquad \text{and} \qquad 3 \ge -5x + 4$$

is a linear inequality in the variable x because the (implied) exponent of x in each inequality is 1.

REMARK Each of the properties of inequalities is true if the symbol $<$ is replaced by $\le$.

As you read through the examples, pay special attention to the steps in which the inequality symbol is reversed. Remember that when you multiply or divide an inequality by a negative number, you must reverse the inequality symbol.

EXAMPLE 2 Solving a Linear Inequality

Solve the linear inequality and sketch the graph of its solution set.

$$5x - 7 > 3x + 9$$

Solution

REMARK The five inequalities forming the solution steps of Example 2 are all **equivalent** in the sense that each has the same solution set. Moreover, the solution set can be denoted in three ways—by an inequality, by an interval, or by a graph.

$5x - 7 > 3x + 9$	*Given inequality*
$5x > 3x + 16$	*Add 7 to both sides*
$5x - 3x > 16$	*Subtract 3x from both sides*
$2x > 16$	*Combine terms*
$x > 8$ (or $8 < x$)	*Divide both sides by 2*

Thus, the solution set consists of all real numbers that are greater than 8. The interval notation for this solution set is $(8, \infty)$. The graph of this solution set is shown in Figure 2.13.

Solution interval: $(8, \infty)$

FIGURE 2.13

Checking the solution set of an inequality is not as simple as checking the solution of an equation. (There are usually too many x-values to substitute back into the original inequality.) We can, however, get an indication of the validity of a solution set by substituting a few convenient values of x. For instance, in Example 2, we found the solution of $5x - 7 > 3x + 9$ to be $x > 8$. Check to see that $x = 9$ satisfies the original inequality, whereas $x = 7$ does not.

EXAMPLE 3 Solving a Linear Inequality

Solve the linear inequality and sketch the graph of its solution set.

$$1 - \frac{3x}{2} \ge x - 4$$

Solution

$$1 - \frac{3x}{2} \geq x - 4 \qquad \textit{Given inequality}$$

$$2 - 3x \geq 2x - 8 \qquad \textit{Multiply both sides by LCD}$$

$$-3x \geq 2x - 10 \qquad \textit{Subtract 2 from both sides}$$

$$-5x \geq -10 \qquad \textit{Subtract 2x from both sides}$$

$$x \leq 2 \qquad \textit{Divide both sides by -5 and reverse inequality}$$

Thus, the solution set consists of all real numbers that are less than or equal to 2. The interval notation for this solution set is $(-\infty, 2]$. The graph of this solution set is shown in Figure 2.14.

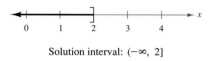

Solution interval: $(-\infty, 2]$

FIGURE 2.14

Sometimes it is convenient to write two inequalities as a **double inequality.** For instance, you can write the two inequalities $-4 \leq 5x - 2$ and $5x - 2 < 7$ more simply as

$$-4 \leq 5x - 2 < 7.$$

This form allows you to solve the two given inequalities together.

EXAMPLE 4   Solving a Double Inequality

Solve the double inequality and sketch the graph of its solution set.

$$-3 \leq 6x - 1 < 3$$

Solution

To solve this double inequality, isolate x as the middle term.

$$-3 \leq 6x - 1 < 3 \qquad \textit{Given inequality}$$

$$-2 \leq 6x < 4 \qquad \textit{Add 1 to all three parts}$$

$$-\frac{1}{3} \leq x < \frac{2}{3} \qquad \textit{Divide by 6 and reduce}$$

Solution interval: $\left[-\frac{1}{3}, \frac{2}{3}\right)$

FIGURE 2.15

Thus, the solution set consists of all real numbers that are greater than or equal to $-\frac{1}{3}$ and less than $\frac{2}{3}$. The interval notation for this solution set is $\left[-\frac{1}{3}, \frac{2}{3}\right)$. The graph of this solution set is shown in Figure 2.15.

The double inequality in Example 4 could have been solved in two parts.

$$-3 \le 6x - 1 \quad \text{and} \quad 6x - 1 < 3$$
$$-2 \le 6x \qquad\qquad\qquad 6x < 4$$
$$-\frac{1}{3} \le x \qquad\qquad\qquad x < \frac{2}{3}$$

The solution set consists of all real numbers that satisfy *both* inequalities. In other words, the solution set is the set of all values of x for which $-\frac{1}{3} \le x < \frac{2}{3}$.

When combining two inequalities to form a double inequality, be sure that the inequalities satisfy the Transitive Property. For instance, it is *incorrect* to combine the inequalities $3 < x$ and $x \le -1$ as $3 < x \le -1$. This "inequality" is obviously wrong because 3 is not less than -1.

Inequalities Involving Absolute Value

To see how to solve inequalities involving absolute values, consider the following comparisons.

Equation or Inequality	*Geometric Solution*	*Graph*
$\lvert x \rvert = 2$	Values of x that lie 2 units from 0	$x = -2$ and $x = 2$
$\lvert x \rvert < 2$	Values of x that lie *less than* 2 units from 0	$-2 < x < 2$
$\lvert x \rvert > 2$	Values of x that lie *more than* 2 units from 0	$x < -2$ or $x > 2$

SOLVING AN ABSOLUTE VALUE INEQUALITY

Let x be a variable or an algebraic expression and let a be a real number such that $a \ge 0$.

1. The solutions of $\lvert x \rvert < a$ are all values of x that lie between $-a$ and a. That is,

   $\lvert x \rvert < a$ if and only if $-a < x < a$.

2. The solutions of $\lvert x \rvert > a$ are all values of x that are less than $-a$ or greater than a. That is,

   $\lvert x \rvert > a$ if and only if $x < -a$ or $x > a$.

These two rules are also valid if $<$ is replaced by $\le$ and $>$ is replaced by $\ge$.

EXAMPLE 5   Solving an Absolute Value Inequality

Solve the inequality and sketch the graph of its solution set.

$$|x - 5| < 2$$

Solution

$\|x - 5\| < 2$	*Given inequality*
$-2 < x - 5 < 2$	*Equivalent inequalities*
$-2 + 5 < x - 5 + 5 < 2 + 5$	*Add 5 to all three parts*
$3 < x < 7$	*Solution set*

Thus, the solution set consists of all real numbers that are greater than 3 and less than 7. The interval notation for this solution set is (3, 7). The graph of this solution set is shown in Figure 2.16.

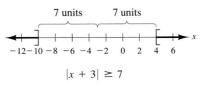

$$|x - 5| < 2$$

FIGURE 2.16

EXAMPLE 6   Solving an Absolute Value Inequality

Solve the inequality and sketch the graph of its solution set.

$$|x + 3| \geq 7$$

Solution

$\|x + 3\| \geq 7$			*Given inequality*
$x + 3 \leq -7$	or	$x + 3 \geq 7$	*Equivalent inequalities*
$x + 3 - 3 \leq -7 - 3$		$x + 3 - 3 \geq 7 - 3$	*Subtract 3 from both sides*
$x \leq -10$		$x \geq 4$	*Solution set*

Thus, the solution set consists of all real numbers that are less than or equal to -10 *or* greater than or equal to 4. The interval notation for this solution set is $(-\infty, -10] \cup [4, \infty)$. The symbol $\cup$ is a **union** symbol, used to denote the combining of two sets. The graph of this solution set is shown in Figure 2.17.

$$|x + 3| \geq 7$$

FIGURE 2.17

EXAMPLE 7   An Inequality Involving Absolute Values

$\left\|2 - \dfrac{x}{3}\right\| < 0.01$	*Given inequality*
$-0.01 < 2 - \dfrac{x}{3} < 0.01$	*Equivalent inequalities*
$-2.01 < \dfrac{-x}{3} < -1.99$	*Subtract 2 from both sides*
$6.03 > x > 5.97$	*Multiply by -3 and reverse both inequalities*

Therefore, the solution interval is (5.97, 6.03).

Applications

Linear inequalities in real-life problems involve statements with phrases such as "at least," "no more than," "minimum value," and so on.

EXAMPLE 8 Comparative Shopping

A subcompact car can be rented from Company A for $180 per week with no extra charge for mileage. A similar car can be rented from Company B for $100 per week plus 20 cents for each mile driven. How many miles must a person drive in a week to make the rental fee for Company A less than that for Company B?

Solution

Verbal Model

$$\text{Weekly cost for Company B} > \text{Weekly cost for Company A}$$

Labels

m = number of miles driven in one week

Weekly cost for Company A = $180

Weekly cost for Company B = $100 + 0.20m$ (in dollars)

Inequality

$$100 + 0.2m > 180$$
$$0.2m > 80$$
$$m > 400 \text{ miles}$$

Thus, the car from Company A is cheaper if the person plans to drive more than 400 miles in a week.

EXAMPLE 9 Accuracy of a Measurement

Suppose you go to a candy store to buy chocolates that cost $9.89 per pound. The scale used in the store has a state seal of approval that indicates the scale is accurate to within half an ounce. According to the scale, your purchase weighs one-half pound and costs $4.95. How much might you have been undercharged or overcharged due to an error in the scale?

Solution

To solve this problem, let x represent the *true* weight of the candy. Since the state seal indicates that the scale is accurate to within half an ounce (or $\frac{1}{32}$ of a pound), you can conclude that the difference between the exact weight (x) and the scale weight $\left(\frac{1}{2}\right)$ is less than or equal to $\frac{1}{32}$ of a pound. That is,

$$\left| x - \frac{1}{2} \right| \leq \frac{1}{32}.$$

Solve this inequality.

$$-\frac{1}{32} \le x - \frac{1}{2} \le \frac{1}{32}$$

$$\frac{15}{32} \le x \le \frac{17}{32}$$

$$0.46875 \le x \le 0.53125$$

In other words, your "one-half" pound of candy could have weighed as little as 0.46875 pounds (which would have cost $4.64) or as much as 0.53125 pounds (which would have cost $5.25). Thus, you could have been undercharged by as much as

$$5.25 - 4.95 = \$0.30$$

or overcharged by as much as

$$4.95 - 4.64 = \$0.31.$$

DISCUSSION
PROBLEM

Verbal Statements of Properties

You may find that it is easier to learn a verbal statement than to learn a mathematical formula. For instance, you can remember the Pythagorean Theorem, $a^2 + b^2 = c^2$, as "the sum of the squares of the two sides is equal to the square of the hypotenuse."

At the beginning of this section, four properties of inequalities were given. Translate each of these mathematical statements into *verbal statements*.

WARM UP

The following warm-up exercises involve skills that were covered in earlier sections. You will use these skills in the exercise set for this section.

In Exercises 1–4, determine which of the two numbers is larger.

1. $-\frac{1}{2}$, -7

2. $-\frac{1}{3}$, $-\frac{1}{6}$

3. $-\pi$, -3

4. -6, $\frac{13}{2}$

In Exercises 5–8, use inequality notation to describe the statement.

5. x is nonnegative.

6. z is strictly between -3 and 10.

7. P is no more than 2.

8. W is at least 200.

In Exercises 9 and 10, evaluate the expression for the given values of x.

9. $|x - 10|$, $x = 12$, $x = 3$

10. $|2x - 3|$, $x = \frac{3}{2}$, $x = 1$

EXERCISES for Section 2.7

In Exercises 1–4, write an inequality to represent the interval and state whether the interval is bounded or unbounded.

1. $[-1, 3]$ **2.** $(4, 10]$

3. $(10, \infty)$ **4.** $[-6, \infty)$

In Exercises 5–8, determine whether the values of x are solutions of the inequality.

Inequality	Values	
5. $5x - 12 > 0$	(a) $x = 3$	(b) $x = -3$
	(c) $x = \frac{5}{2}$	(d) $x = \frac{3}{2}$
6. $x + 1 < \frac{2x}{3}$	(a) $x = 0$	(b) $x = 4$
	(c) $x = -4$	(d) $x = -3$
7. $0 \le \frac{x - 2}{4} < 2$	(a) $x = 4$	(b) $x = 10$
	(c) $x = 0$	(d) $x = \frac{7}{2}$
8. $-1 < \frac{3 - x}{2} \le 1$	(a) $x = 0$	(b) $x = \sqrt{5}$
	(c) $x = 1$	(d) $x = 5$

In Exercises 9–16, match the inequality with its graph. [The graphs are labeled (a), (b), (c), (d), (e), (f), (g), and (h).]

9. $x < 4$ **10.** $x \ge 6$

11. $-2 < x \le 5$ **12.** $0 \le x \le \frac{7}{2}$

13. $|x| < 4$ **14.** $|x| > 3$

15. $|x - 5| > 2$ **16.** $|x + 6| < 3$

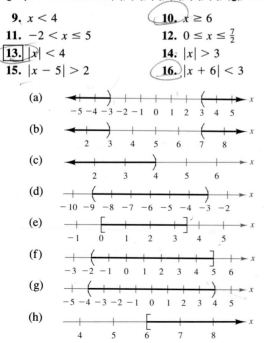

In Exercises 17–50, solve the inequality and sketch the solution on the real number line.

17. $4x < 12$ **18.** $2x > 3$

19. $-10x < 40$ **20.** $-6x > 15$

21. $x - 5 \ge 7$ **22.** $x + 7 \le 12$

23. $4(x + 1) < 2x + 3$ **24.** $2x + 7 < 3$

25. $2x - 1 \ge 0$ **26.** $3x + 1 \ge 2$

27. $4 - 2x < 3$ **28.** $6x - 4 \le 2$

29. $1 < 2x + 3 < 9$ **30.** $-8 \le 1 - 3(x - 2) < 13$

31. $-4 < \dfrac{2x - 3}{3} < 4$ **32.** $0 \le \dfrac{x + 3}{2} < 5$

33. $\dfrac{3}{4} > x + 1 > \dfrac{1}{4}$ **34.** $-1 < -\dfrac{x}{3} < 1$

35. $|x| < 5$ **36.** $|2x| < 6$

37. $\left|\dfrac{x}{2}\right| > 3$ **38.** $|5x| > 10$

39. $|x - 20| \le 4$ **40.** $|x - 7| < 6$

41. $|x - 20| \ge 4$ **42.** $|x + 14| + 3 > 17$

43. $\left|\dfrac{x - 3}{2}\right| \ge 5$ **44.** $|1 - 2x| < 5$

45. $|9 - 2x| - 2 < -1$ **46.** $\left|1 - \dfrac{2x}{3}\right| < 1$

47. $2|x + 10| \ge 9$ **48.** $3|4 - 5x| \le 9$

49. $|x - 5| < 0$ **50.** $|x - 5| \ge 0$

In Exercises 51–56, find the interval(s) on the real number line for which the radicand is nonnegative (greater than or equal to zero).

51. $\sqrt{x - 5}$ **52.** $\sqrt{x - 10}$

53. $\sqrt{x + 3}$ **54.** $\sqrt[4]{6x + 15}$

55. $\sqrt[4]{7 - 2x}$ **56.** $\sqrt{3 - x}$

In Exercises 57–64, use absolute value notation to define each interval (or pair of intervals) on the real line.

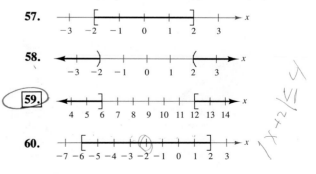

61. All real numbers within 10 units of 12

62. All real numbers at least 5 units from 8

63. All real numbers whose distances from -3 are more than 5

64. All real numbers whose distances from -6 are no more than 7

65. *Comparative Shopping* Suppose you can rent a midsize car from Company A for $250 per week with no extra charge for mileage. A similar car can be rented from Company B for $150 per week plus $0.25 cents for each mile driven. How many miles must you drive in a week to make the rental fee for Company B *greater than* that for Company A?

66. *Comparative Shopping* Suppose your department sends its copying to the photocopy center of your company. The center bills your department $0.10 per page. You have investigated the possibility of buying a departmental copier for $3000. With your own copier the cost per page would be $0.03. The expected life of the copier is four years. How many copies must you make in the four-year period to justify purchasing the copier?

67. *Simple Interest* In order for an investment of $1000 to grow to *more than* $1250 in two years, what must the interest rate be? $[A = P(1 + rt)]$

68. *Simple Interest* In order for an investment of $750 to grow to *more than* $1050 in two years, what must the interest rate be?

69. *Break-even Analysis* The revenue for selling x units of a product is

$$R = 115.95x.$$

The cost of producing x units is

$$C = 95x + 750.$$

In order to obtain a profit, the revenue must be *greater than* the cost. For what values of x will this product return a profit?

70. *Annual Operating Cost* A utility company has a fleet of vans. The annual operating cost per van is

$$C = 0.32m + 2,300$$

where m is the number of miles traveled by a van in a year. What number of miles will yield an annual operating cost that is less than $10,000?

71. *Daily Sales* A doughnut shop sells a dozen doughnuts for $2.95. Beyond the fixed costs (rent, utilities, and insurance) of $150 per day, it costs $1.45 for enough materials (flour, sugar, and so on) and labor to produce a dozen doughnuts. If the daily profit varies between $50 and $200, between what levels (in dozens) do the daily sales vary?

72. *Teachers' Salaries* The average salary for elementary and secondary teachers in the United States from 1980 to 1989 is approximated by the model

Salary $= 16.116 + 1.496t$

where the salary is given in thousands of dollars and the time t represents the calendar year with $t = 0$ corresponding to 1980 (see figure). Assuming the model is correct, when will the average salary *exceed* $35,000? (*Source:* National Education Association)

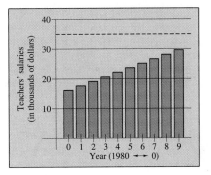

Figure for 72

73. *Accuracy of Measurement* The side of a square is measured as 10.4 inches with a possible error of $\frac{1}{16}$ inch. Determine the interval containing the area of the square.

74. *Accuracy of Measurement* Suppose you buy a bag of oranges that cost $0.95 per pound. The weight that is listed on the bag is 4.65 pounds. If the scale that weighed the bag is only accurate to within 1 ounce, how much money might you have been undercharged or overcharged?

75. *Height* The heights, h, of two-thirds of the members of a certain population satisfy the inequality

$$\left| \frac{h - 68.5}{2.7} \right| \leq 1$$

where h is measured in inches. Determine the interval on the real line in which these heights lie.

76. *Relative Humidity* A certain electronic device is to be operated in an environment with relative humidity h in the interval defined by

$$|h - 50| \leq 30.$$

What are the minimum and maximum relative humidities for the operation of this device?

77. Given two real numbers a and b, such that $a > b > 0$, prove that

$$\frac{1}{a} < \frac{1}{b}.$$

2.8 Other Types of Inequalities

Polynomial Inequalities / Rational Inequalities / Applications

Polynomial Inequalities

To solve a polynomial inequality like

$$x^2 - 2x - 3 < 0$$

we use the fact that a polynomial can change signs only at its zeros (the x-values that make the polynomial zero). Between two consecutive zeros a polynomial must be entirely positive or entirely negative. This means that when the real zeros of a polynomial are put in order, they divide the real line into intervals in which the polynomial has no sign changes. These zeros are the **critical numbers** of the inequality, and the resulting intervals are the **test intervals** for the inequality. For example, the polynomial

$$x^2 - 2x - 3 = (x + 1)(x - 3)$$

has two zeros, $x = -1$ and $x = 3$, and these zeros divide the real line into three test intervals:

$$(-\infty, -1), \quad (-1, 3), \quad \text{and} \quad (3, \infty).$$

Thus, to solve the inequality $x^2 - 2x - 3 < 0$, we need only to test one value from each of these test intervals.

Finding Test Intervals for a Polynomial

To determine the intervals on which the values of a polynomial are entirely negative or entirely positive, use the following steps.

1. Find all real zeros of the polynomial, and arrange the zeros in increasing order (from smallest to largest). The zeros of a polynomial are its **critical numbers.**
2. Use the critical numbers of the polynomial to determine its **test intervals.**
3. Choose one representative x-value in each test interval and evaluate the polynomial at that value. If the value of the polynomial is negative, then the polynomial will have negative values for *every* x-value in the interval. If the value of the polynomial is positive, then the polynomial will have positive values for *every* x-value in the interval.

EXAMPLE 1   Solving a Polynomial Inequality

Determine the interval(s) on which the inequality $x^2 - x - 6 < 0$ is true.

Solution

By factoring the quadratic as

$$x^2 - x - 6 = (x + 2)(x - 3)$$

we see that the critical numbers occur at $x = -2$ and $x = 3$. Therefore, the test intervals for the quadratic are

$$(-\infty, -2), \qquad (-2, 3), \qquad \text{and} \qquad (3, \infty). \quad \textit{Test intervals}$$

In each test interval, we choose a representative x-value and evaluate the polynomial, as shown in Table 2.1.

TABLE 2.1

Test interval	Representative x-value	Value of polynomial	Conclusion
$(-\infty, -2)$	$x = -3$	$(-3)^2 - (-3) - 6 = 6$	Polynomial is positive
$(-2, 3)$	$x = 0$	$(0)^2 - (0) - 6 = -6$	Polynomial is negative
$(3, \infty)$	$x = 4$	$(4)^2 - (4) - 6 = 6$	Polynomial is positive

Therefore, the polynomial has negative values for every x in the interval $(-2, 3)$. This result is shown graphically in Figure 2.18.

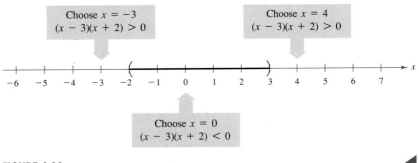

FIGURE 2.18

To determine the test intervals for a polynomial inequality, the inequality must first be written in standard form (with the polynomial on the left and zero on the right).

EXAMPLE 2   Solving a Polynomial Inequality

$$2x^3 + 5x^2 > 12x \qquad \textit{Given inequality}$$

$$2x^3 + 5x^2 - 12x > 0 \qquad \textit{Write in standard form}$$

$$x(x + 4)(2x - 3) > 0 \qquad \textit{Factor}$$

Critical Numbers $x = -4, \, x = 0, \, x = \frac{3}{2}$

Test Intervals $(-\infty, -4), \, (-4, 0), \, \left(0, \frac{3}{2}\right), \, \left(\frac{3}{2}, \infty\right)$

Test   Is $x(x + 4)(2x - 3) > 0$?

After testing these intervals, as shown in Figure 2.19, we see that the polynomial $2x^3 + 5x^2 - 12x$ is positive in the interval $(-4, 0)$ and in the interval $\left(\frac{3}{2}, \infty\right)$. Therefore, the solution set of the inequality is

$$(-4, 0) \cup \left(\frac{3}{2}, \infty\right). \qquad \textit{Solution set}$$

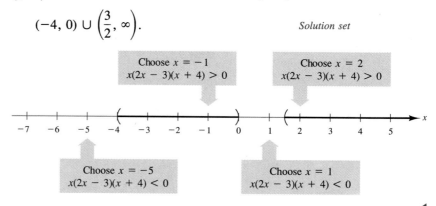

FIGURE 2.19

EXAMPLE 3   Using the Quadratic Formula

Solve the inequality $x^2 + 4x + 1 \leq 0$.

Solution

By the Quadratic Formula, the zeros of $x^2 + 4x + 1$ are

$$x = \frac{-4 \pm \sqrt{16 - 4}}{2} = \frac{-4 \pm 2\sqrt{3}}{2} = -2 \pm \sqrt{3}.$$

Now, having found the zeros of the polynomial $x^2 + 4x + 1$, proceed as usual.

Critical Numbers $x = -2 - \sqrt{3}, \, x = -2 + \sqrt{3}$

Test Intervals $(-\infty, -2 - \sqrt{3}), \, (-2 - \sqrt{3}, -2 + \sqrt{3}), \, (-2 + \sqrt{3}, \infty)$

Test   Is $x^2 + 4x + 1 \leq 0$?

After testing these intervals, as shown in Figure 2.20, we see that the polynomial $x^2 + 4x + 1$ is less than or equal to zero in the *closed* interval $[-2 - \sqrt{3}, -2 + \sqrt{3}]$. Therefore, the solution set of the inequality is

$$[-2 - \sqrt{3}, -2 + \sqrt{3}].$$ *Solution set*

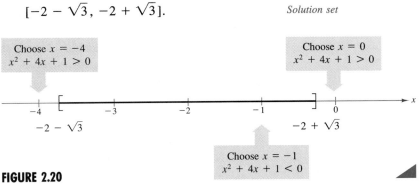

FIGURE 2.20

When solving a polynomial inequality, be sure you have accounted for the particular type of inequality symbol given in the inequality. For instance, in Example 3, note that the solution is a closed interval because the original inequality contained a "less than or equal to" symbol. If the original inequality had been $x^2 + 4x + 1 < 0$, the solution would have been the *open* interval $(-2 - \sqrt{3}, -2 + \sqrt{3})$.

EXAMPLE 4 Unusual Solution Sets

Solve the inequalities to verify that the solution set is correct.

a. The solution set of the quadratic inequality

$$x^2 + 2x + 4 > 0$$

consists of the entire set of real numbers, $(-\infty, \infty)$. In other words, the quadratic $x^2 + 2x + 4$ is positive for every real value of x. (Note that this quadratic inequality has *no* critical numbers. In such a case, there is only one test interval—the entire real line.)

b. The solution set of the quadratic inequality

$$x^2 + 2x + 1 \le 0$$

consists of the single real number $\{-1\}$.

c. The solution set of the quadratic inequality

$$x^2 + 3x + 5 < 0$$

is empty. In other words, the quadratic $x^2 + 3x + 5$ is not less than zero for any value of x.

d. The solution set of the quadratic inequality

$$x^2 - 4x + 4 > 0$$

consists of all real numbers *except* the number 2. In interval notation, this solution set can be written as $(-\infty, 2) \cup (2, \infty)$.

Rational Inequalities

The concepts of critical numbers and test intervals can be extended to inequalities involving rational expressions. To do this, we use the fact that the value of a rational expression can change sign only at its *zeros* (the x-values for which its numerator is zero) and its *undefined values* (the x-values for which its denominator is zero). These two types of numbers make up the **critical numbers** of a rational inequality. For instance, the critical numbers of the inequality

$$\frac{x - 1}{(x - 2)(x + 3)} < 0$$

are $x = 1$ (the numerator is zero), and $x = 2$ and $x = -3$ (the denominator is zero). From these three critical numbers we see that the given inequality has *four* test intervals:

$$(-\infty, -3), \quad (-3, 1), \quad (1, 2), \quad \text{and} \quad (2, \infty).$$

EXAMPLE 5 Solving a Rational Inequality

$$\frac{2x - 7}{x - 5} \leq 3 \qquad \text{\textit{Given inequality}}$$

$$\frac{2x - 7}{x - 5} - 3 \leq 0 \qquad \text{\textit{Standard form}}$$

$$\frac{2x - 7 - 3x + 15}{x - 5} \leq 0 \qquad \text{\textit{Add fractions}}$$

$$\frac{-x + 8}{x - 5} \leq 0 \qquad \text{\textit{Simplify}}$$

Now, in standard form we see that the critical numbers are 5 and 8.

Critical Numbers $\quad x = 5, x = 8 \qquad$ *Denominator is zero when $x = 5$*

Test Intervals $\quad (-\infty, 5), (5, 8), (8, \infty)$

Test $\qquad$ Is $\dfrac{-x + 8}{x - 5} \leq 0?$

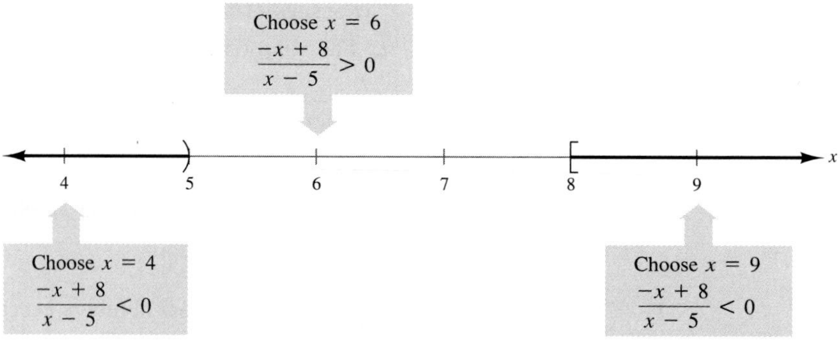

Choose $x = 6$
$$\frac{-x + 8}{x - 5} > 0$$

Choose $x = 4$
$$\frac{-x + 8}{x - 5} < 0$$

Choose $x = 9$
$$\frac{-x + 8}{x - 5} < 0$$

FIGURE 2.21

After testing these intervals, as shown in Figure 2.21, we see that the rational expression $(-x + 8)/(x - 5)$ is negative in the open intervals $(-\infty, 5)$ and $(8, \infty)$. Moreover, since $(-x + 8)/(x - 5) = 0$ when $x = 8$, we conclude that the solution set of the inequality is

$(-\infty, 5) \cup [8, \infty).$ *Solution set*

EXAMPLE 6   An Inequality Involving Fractions

$$\frac{x}{x - 2} > \frac{1}{x + 3}$$ *Given inequality*

$$\frac{x}{x - 2} - \frac{1}{x + 3} > 0$$ *Standard form*

$$\frac{x^2 + 3x - x + 2}{(x - 2)(x + 3)} > 0$$ *Combine fractions*

$$\frac{x^2 + 2x + 2}{(x - 2)(x + 3)} > 0$$ *Simplify*

The discriminant $b^2 - 4ac = 4 - 8 = -4 < 0$ indicates that the numerator has no zeros. Therefore, $x = 2$ and $x = -3$ are the only critical numbers.

Critical Numbers $x = -3, x = 2$

Test Intervals $(-\infty, -3), (-3, 2), (2, \infty)$

Test   Is $\dfrac{x^2 + 2x + 2}{(x - 2)(x + 3)} > 0$?

After testing these intervals, as shown in Figure 2.22, we see that the rational expression $(x^2 + 2x + 2)/[(x - 2)(x + 3)]$ is positive in the open intervals $(-\infty, -3)$ and $(2, \infty)$. Therefore, the solution set of the inequality is

$(-\infty, -3) \cup (2, \infty).$ *Solution set*

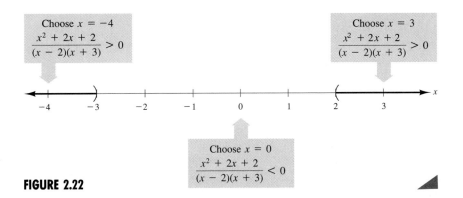

FIGURE 2.22

Applications

EXAMPLE 7   Finding the Domain of an Expression

Find the domain of the expression $\sqrt{64 - 4x^2}$.

Solution

Remember that the domain of an expression is the set of all x-values for which the expression is defined (has real values). Since $\sqrt{64 - 4x^2}$ is defined only if $64 - 4x^2$ is nonnegative, the domain is given by $64 - 4x^2 \geq 0$.

$64 - 4x^2 \geq 0$	*Standard form*
$16 - x^2 \geq 0$	*Divide both sides by 4*
$(4 - x)(4 + x) \geq 0$	*Factor*

Thus, the inequality has two critical numbers: -4 and 4. We use these two numbers to test the inequality.

Critical Numbers $x = -4, \, x = 4$

Test Intervals $(-\infty, -4), \, (-4, 4), \, (4, \infty)$

Test   Is $(4 - x)(4 + x) \geq 0$?

A test shows that $64 - 4x^2 \geq 0$ in the *closed interval* $[-4, 4]$. Thus, the domain of the expression $\sqrt{64 - 4x^2}$ is the interval $[-4, 4]$.

EXAMPLE 8   The Height of a Projectile

A projectile is fired straight upward from ground level with an initial velocity of 384 feet per second. During what time period will its height exceed 2000 feet? (See Figure 2.23.)

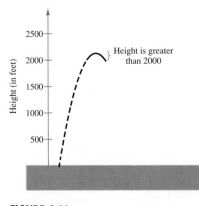

FIGURE 2.23

Solution

Recall from Section 2.4 that the position of an object moving in a vertical path is given by $s = -16t^2 + v_0t + s_0$, where s is the height in feet and t is the time in seconds. In this case, $s_0 = 0$ and $v_0 = 384$. Thus, we need to solve the inequality $-16t^2 + 384t > 2000$.

$$-16t^2 + 384t > 2000 \qquad \text{\small \textit{Given inequality}}$$

$$t^2 - 24t < -125 \qquad \text{\small \textit{Divide by }} -16 \text{ \small \textit{and reverse}} \atop \text{\small \textit{inequality}}$$

$$t^2 - 24t + 125 < 0 \qquad \text{\small \textit{Standard form}}$$

By the Quadratic Formula, the critical numbers are

$$t = \frac{24 \pm \sqrt{76}}{2} = \frac{24 \pm 2\sqrt{19}}{2} = 12 \pm \sqrt{19} \approx 7.64 \text{ or } 16.36.$$

A test will verify that the height of the projectile will exceed 2000 feet during the time interval

$$7.64 \text{ sec} < t < 16.36 \text{ sec}.$$

DISCUSSION

PROBLEM

A
Computer
Experiment

If you have access to a computer that has the BASIC language, try running the following program.

```
10 DEF FNY(X)=X^2-9
20 FOR I=1 TO 80
30 X=(I-38)/6
40 IF FNY(X)<0 THEN PRINT "-";
50 IF FNY(X)=0 THEN PRINT "0";
60 IF FNY(X)>0 THEN PRINT "+";
70 NEXT
80 PRINT: FOR I=-6 TO 6: PRINT "-|----";:
   NEXT: PRINT "-|";
90 FOR I=-6 TO 6: PRINT USING "##";I;:
   PRINT " ";: NEXT: PRINT " 7"
100 END
```

The printout from this program is shown in Figure 2.24. Notice that the program test evaluates the expression $x^2 - 9$ at several values between -6 and 7. If the value of $x^2 - 9$ is negative, a minus sign is printed above the number line; if the value of $x^2 - 9$ is zero, a zero is printed; and if the value of $x - 9$ is positive, a plus sign is printed.

FIGURE 2.24

The following warm-up exercises involve skills that were covered in earlier sections. You will use these skills in the exercise set for this section.

In Exercises 1–10, solve the inequality.

1. $-\dfrac{y}{3} > 2$

2. $-6z < 27$

3. $-3 \le 2x + 3 < 5$

4. $-3x + 5 \ge 20$

5. $10 > 4 - 3(x + 1)$

6. $3 < 1 + 2(x - 4) < 7$

7. $2|x| \le 7$

8. $|x - 3| > 1$

9. $|x + 4| > 2$

10. $|2 - x| \le 4$

EXERCISES for Section 2.8

In Exercises 1–34, solve the inequality and give the answer in interval notation.

1. $x^2 \le 9$

2. $x^2 < 5$

3. $x^2 > 4$

4. $(x - 3)^2 \ge 1$

5. $(x + 2)^2 < 25$

6. $(x + 6)^2 \le 8$

7. $x^2 + 4x + 4 \ge 9$

8. $x^2 - 6x + 9 < 16$

9. $x^2 + x < 6$

10. $x^2 + 2x > 3$

11. $3(x - 1)(x + 1) > 0$

12. $6(x + 2)(x - 1) < 0$

13. $x^2 + 2x - 3 < 0$

14. $x^2 - 4x - 1 > 0$

15. $4x^3 - 6x^2 < 0$

16. $4x^3 - 12x^2 > 0$

17. $x^3 - 4x \ge 0$

18. $2x^3 - x^4 \le 0$

19. $(x - 1)^2(x + 2)^3 \ge 0$

20. $x^4(x - 3) \le 0$

21. $\dfrac{1}{x} - x > 0$

22. $\dfrac{1}{x} - 4 < 0$

23. $\dfrac{x + 6}{x + 1} - 2 < 0$

24. $\dfrac{x + 12}{x + 2} - 3 \ge 0$

25. $\dfrac{3x - 5}{x - 5} > 4$

26. $\dfrac{5 + 7x}{1 + 2x} < 4$

27. $\dfrac{4}{x + 5} > \dfrac{1}{2x + 3}$

28. $\dfrac{5}{x - 6} > \dfrac{3}{x + 2}$

29. $\dfrac{1}{x - 3} \le \dfrac{9}{4x + 3}$

30. $\dfrac{1}{x} \ge \dfrac{1}{x + 3}$

31. $\dfrac{x^2 + 2x}{x^2 - 9} \le 0$

32. $\dfrac{x^2 + x - 6}{x} \ge 0$

33. $\dfrac{3}{x - 1} - \dfrac{2}{x + 1} < 1$

34. $\dfrac{3x}{x - 1} \le \dfrac{x}{x + 4}$

In Exercises 35–40, find the domain of x in the expression.

35. $\sqrt[4]{4 - x^2}$

36. $\sqrt{x^2 - 4}$

37. $\sqrt{x^2 - 7x + 12}$

38. $\sqrt{144 - 9x^2}$

39. $\sqrt{12 - x - x^2}$

40. $\sqrt{\dfrac{x}{x^2 - 9}}$

In Exercises 41 and 42, use a calculator to solve the inequality. (Round each number in your answer to two decimal places.)

41. $-0.5x^2 + 12.5x + 1.6 > 0$

42. $\dfrac{2}{3.1x - 3.7} > 5.8$

43. *Height of a Projectile* A projectile is fired straight upward from ground level with an initial velocity of 160 feet per second.
(a) At what instant will it be back at ground level?
(b) During what time period will its height exceed 384 feet?

44. *Height of a Projectile* A projectile is fired straight upward from ground level with an initial velocity of 128 feet per second.
(a) At what instant will it be back at ground level?
(b) During what time period will its height be less than 128 feet?

45. *Dimensions of a Field* A rectangular playing field with a perimeter of 100 meters is to have an area of at least 500 square meters. Within what bounds must the length of the rectangle lie?

46. *Compound Interest* P dollars, invested at interest rate r compounded annually, increases to an amount

$$A = P(1 + r)^2$$

in two years. If an investment of $1000 is to increase to an amount greater than $1200 in two years, then the interest rate must be greater than what percentage?

47. *Resistors* When two resistors of resistance R_1 and R_2 are connected in parallel (see figure), the total resistance R satisfies the equation

$$\frac{1}{R} = \frac{1}{R_1} + \frac{1}{R_2}.$$

Find R_1 for a parallel circuit in which $R_2 = 2$ ohms and R must be at least 1 ohm.

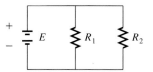

Figure for 47

48. *Company Profits* The revenue and cost equations for a product are

$$R = x(50 - 0.0002x)$$
$$C = 12x + 150,000$$

where R and C are measured in dollars and x represents the number of units sold. How many units must be sold to obtain a profit of at least $1,650,000?

49. *Percent of College Graduates* The percentage of the American population that graduated from college between 1940 and 1987 is approximated by the model

Percent of graduates $= 5.136 + 0.0069t^2$

where the time t represents the calendar year with $t = 0$ corresponding to 1940 (see figure). According to this model, when will the percentage of college graduates exceed 25% of the population? (*Source:* U.S. Bureau of Census)

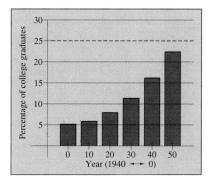

Figure for 49

50. *Safe Beam Loads* The maximum safe load uniformly distributed over a 1-foot section of a 2-inch-wide wooden beam is approximated by the model

Load $= 168.5d^2 - 472.1$

where d is the depth of the beam. Use this model to determine the minimum depth of the beam which will safely support a load of 2000 pounds (see figure).

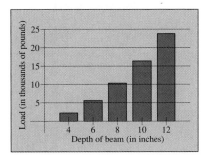

Figure for 50

REVIEW EXERCISES for Chapter 2

In Exercises 1 and 2, determine whether the equation is an identity or conditional equation.

1. $6 - (x - 2)^2 = 2 + 4x - x^2$

2. $3(x - 2) + 2x = 2(x + 3)$

In Exercises 3 and 4, determine whether the value of x is a solution of the equation.

Equation	Values	
3. $3x^2 + 7x + 5 = x^2 + 9$	(a) $x = 0$	(b) $x = -4$
	(c) $x = \frac{1}{2}$	(d) $x = -1$
4. $6 + \dfrac{3}{x - 4} = 5$	(a) $x = 4$	(b) $x = 0$
	(c) $x = -2$	(d) $x = 1$

In Exercises 5–42, solve the equation (if possible) and check your answer.

5. $3x - 2(x + 5) = 10$
6. $4x + 2(7 - x) = 5$
7. $4(x + 3) - 3 = 2(4 - 3x) - 4$
8. $\frac{1}{2}(x - 3) - 2(x + 1) = 5$

9. $3\left(1 - \dfrac{1}{5t}\right) = 0$
10. $\dfrac{1}{x - 2} = 3$

11. $6x = 3x^2$
12. $3x^2 + 1 = 0$

13. $6x^2 = 5x + 4$
14. $15 + x - 2x^2 = 0$

15. $(x + 4)^2 = 18$
16. $16x^2 = 25$

17. $x^2 - 12x + 30 = 0$
18. $x^2 + 6x - 3 = 0$

19. $5x^4 - 12x^3 = 0$
20. $4x^3 - 6x = 0$

21. $4t^3 - 12t^2 + 8t = 0$
22. $12t^3 - 84t^2 + 120t = 0$

23. $2 - x^{-2} = 0$
24. $2 + 8x^{-2} = 0$

25. $\dfrac{4}{(x - 4)^2} = 1$
26. $\dfrac{1}{(t + 1)^2} = 1$

27. $\dfrac{1}{x} + \dfrac{1}{x + 1} = 2$
28. $\dfrac{4}{x - 3} - \dfrac{4}{x} = 1$

29. $\sqrt{x + 4} = 3$
30. $\sqrt{x - 2} - 8 = 0$

31. $2\sqrt{x} - 5 = 0$
32. $\sqrt{3x - 2} = 4 - x$

33. $\sqrt{2x + 3} + \sqrt{x - 2} = 2$
34. $5\sqrt{x} - \sqrt{x - 1} = 6$

35. $(x - 1)^{2/3} - 25 = 0$
36. $(x + 2)^{3/4} = 27$

37. $(x + 4)^{1/2} + 5x(x + 4)^{3/2} = 0$
38. $8x^2(x^2 - 4)^{1/3} + (x^2 - 4)^{4/3} = 0$

39. $|x - 5| = 10$
40. $|2x + 3| = 7$
41. $|x^2 - 3| = 2x$
42. $|x^2 - 6| = x$

In Exercises 43–46, solve the equation for the indicated variable.

43. Solve for r: $V = \frac{1}{3}\pi r^2 h$

44. Solve for X: $Z = \sqrt{R^2 - X^2}$

45. Solve for p: $L = \dfrac{k}{3\pi r^2 p}$

46. Solve for v: $E = 2kw\left(\dfrac{v}{2}\right)^2$

In Exercises 47–56, solve the inequality.

47. $\frac{1}{2}(3 - x) > \frac{1}{3}(2 - 3x)$
48. $\dfrac{x}{5} - 6 \leq -\dfrac{x}{2} + 6$

49. $x^2 - 4 \leq 0$
50. $x^2 - 2x \geq 3$

51. $\dfrac{x - 5}{3 - x} < 0$
52. $\dfrac{2}{x + 1} \leq \dfrac{3}{x - 1}$

53. $|x - 2| < 1$
54. $|x| \leq 4$
55. $\left|x - \frac{3}{2}\right| \geq \frac{3}{2}$
56. $|x + 3| > 4$

In Exercises 57 and 58, find the domain of the expression by finding the interval(s) on the real number line for which the radicand is nonnegative.

57. $\sqrt{2x - 10}$
58. $\sqrt{x(x - 4)}$

In Exercises 59–70, perform the indicated operations and write the result in standard form.

59. $(7 + 5i) + (-4 + 2i)$

60. $-(6 - 2i) + (-8 + 3i)$

61. $\left(\dfrac{\sqrt{2}}{2} - \dfrac{\sqrt{2}}{2}i\right) - \left(\dfrac{\sqrt{2}}{2} + \dfrac{\sqrt{2}}{2}i\right)$

62. $(13 - 8i) - 5i$

63. $5i(13 - 8i)$
64. $(1 + 6i)(5 - 2i)$

65. $(10 - 8i)(2 - 3i)$
66. $i(6 + i)(3 - 2i)$

67. $\dfrac{6 + i}{i}$
68. $\dfrac{3 + 2i}{5 + i}$

69. $\dfrac{4}{-3i}$
70. $\dfrac{1}{(2 + i)^4}$

71. *Monthly Profit* In October, a company's total profit was 12% more than it was in September. The total profit for the two months was $689,000. Find the profit for each month.

72. *Discount Rate* The price of a television set has been discounted $85. The sale price is $340. What percent is the discount?

73. *Mixture Problem* A car radiator contains 10 quarts of a 30% antifreeze solution. How many quarts will have to be replaced with pure antifreeze if the resulting solution is to be 50% antifreeze?

74. *Starting Position* A fitness center has two running tracks around a rectangular playing floor. The tracks are 3 feet wide and form semicircles at the narrow end of the rectangular floor (see figure). Determine the distance between the starting positions if two runners must run the same distance to the finish line in one lap around the track.

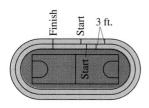

Figure for 74

75. *Depth of an Underwater Cable* The sonar of a ship locates a cable 2000 feet from the ship (see figure). The angle between the water level and the cable is 45 degrees. How deep is the cable?

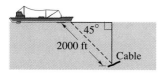

Figure for 75

76. *Average Speed* You drove 56 miles on a service call for your company and it took 10 minutes longer than the return trip when you drove an average of 8 miles per hour faster. What was your average speed on the return trip?

77. *Cost Sharing* A group of farmers agree to share equally in the cost of a $48,000 piece of machinery. If they could find two more farmers to join the group, each person's share of the cost would decrease by $4,000. How many farmers are presently in the group?

78. *Venture Capital* An individual is planning to start a small business that will require $90,000 before any income can be generated. Since it is difficult to borrow for new ventures, the individual wants a group of friends to divide the cost equally for a future share of the profit. Some are willing, but three more are needed in order for the per-person price to be $2,500 less. How many investors are needed?

79. *Market Research* The demand equation for a product is

$$p = 42 - \sqrt{0.001x + 2}$$

where x is the number of units demanded per day and p is the price per unit. Find the demand if the price is set at $29.95.

80. *Compound Interest* P dollars invested at interest rate r compounded annually for five years increases to an amount

$$A = P(1 + r)^5.$$

If an investment of $1000 is to increase to an amount greater than $1400 in five years, then the interest rate must be greater than what percentage?

81. *Break-even Analysis* The revenue for selling x units of a product is

$$R = 125.95x.$$

The cost of producing x units is

$$C = 92x + 1200.$$

In order to obtain a profit, the revenue must be greater than the cost. For what values of x will this product return a profit?

82. *Pendulum* The period of a pendulum is

$$T = 2\pi\sqrt{\frac{L}{32}}$$

where T is the time in seconds and L is the length of the pendulum in feet. If the period is to be at least 2 seconds, determine the minimum length of the pendulum.

C H A P T E R 3

OVERVIEW

In this chapter, you will learn how a graph can be used to visualize relationships between two variables. For instance, in Exercises 59 and 60 on page 164, a graph shows the changing cost of a Super Bowl ad from 1967 through 1991.

You will also learn that many of the equations used to model real-life relationships between two variables are *functions.* For instance, in Exercise 49 on page 224, the stopping distance of a car is a function of the speed of the car.

As you study the chapter, you should know that algebra is *not* a static branch of mathematics—it changes. Modern algebra employs technology, estimation, problem-solving strategies, functions, and their graphs to provide a deeper understanding of traditional algebraic concepts.

Functions and Graphs

3.1 The Cartesian Plane

Introduction / The Distance Between Two Points in the Plane / The Midpoint Formula

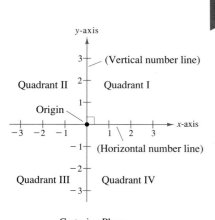

Cartesian Plane

FIGURE 3.1

Introduction

Just as we can represent real numbers by points on the real line, we can represent ordered pairs of real numbers by points in a plane. This plane is a **rectangular coordinate system** or the **Cartesian plane,** after the French mathematician René Descartes (1596–1650).

The Cartesian plane is formed by two real number lines intersecting at right angles, as shown in Figure 3.1. The horizontal number line is usually called the **x-axis** and the vertical number line is usually called the **y-axis.** (The plural of axis is *axes*.) The point of intersection of the two axes is the **origin.** The axes separate the plane into four regions called **quadrants.**

Each point in the plane corresponds to an **ordered pair** (x, y) of real numbers x and y, called **coordinates** of the point. The first number (**x-**

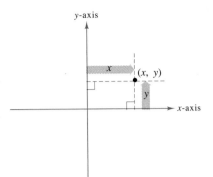

FIGURE 3.2

coordinate) tells how far to the left or right the point is from the vertical axis, and the second number (**y-coordinate**) tells how far up or down the point is from the horizontal axis, as shown in Figure 3.2.

$$(x, y)$$

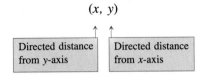

REMARK   It is customary to use the notation (x, y) to denote both a point in the plane and an open interval on the real number line. The nature of a specific problem will show which of the two we are talking about.   ◢

EXAMPLE 1   Plotting Points in the Cartesian Plane

Plot the points $(-1, 2)$, $(3, 4)$, $(0, 0)$, $(3, 0)$, and $(-2, -3)$ in the Cartesian plane.

Solution

To plot the point $(-1, 2)$ we envision a vertical line through -1 on the x-axis and a horizontal line through 2 on the y-axis. The intersection of these two lines is the point $(-1, 2)$, as shown in Figure 3.3. The other four points can be plotted in a similar way.

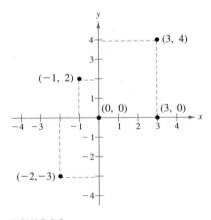

FIGURE 3.3   ◢

Shifting points in the Cartesian plane is a technique used in computer graphics.

EXAMPLE 2   **Shifting Points in the Plane**

The triangle shown in Figure 3.4(a) has vertices at the points $(-1, 2)$, $(1, -4)$, and $(2, 3)$. Shift the triangle three units to the right and two units up and find the vertices of the shifted triangle.

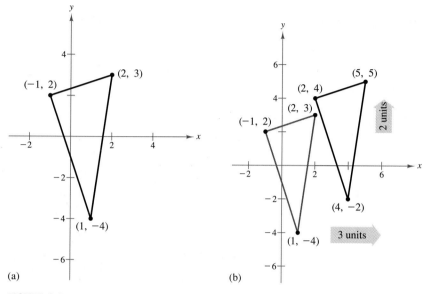

FIGURE 3.4

Solution

To shift the vertices three units to the right, we add 3 to each of the x-coordinates. To shift the vertices two units up, we add 2 to each of the y-coordinates. The coordinates of the shifted vertices are as follows.

Original Vertices	Shifted Vertices
$(-1, 2)$	$(-1 + 3, 2 + 2) = (2, 4)$
$(1, -4)$	$(1 + 3, -4 + 2) = (4, -2)$
$(2, 3)$	$(2 + 3, 3 + 2) = (5, 5)$

The original and shifted triangles are shown in Figure 3.4(b).

The rectangular coordinate system allows you to visualize relationships between variables x and y. Today, Descartes's ideas are commonly used in virtually every scientific and business-related field.

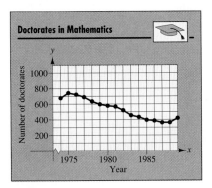

FIGURE 3.5

EXAMPLE 3   Number of Doctor's Degrees in Mathematics

The number of doctor's degrees in mathematics granted to United States citizens by universities in the United States from 1974 to 1989 is given in Table 3.1. Plot these points in a rectangular coordinate system. (*Source: American Mathematical Society*)

TABLE 3.1

Year	1974	1975	1976	1977	1978	1979	1980	1981
Degrees	677	741	722	689	634	596	578	567

Year	1982	1983	1984	1985	1986	1987	1988	1989
Degrees	519	455	433	396	386	362	363	419

Solution

The points are shown in Figure 3.5. Note that the break in the *x*-axis indicates that we have omitted the numbers between 0 and 1973.

The Distance Between Two Points in the Plane

We know from Section 1.1 that the distance d between two points a and b on the real number line is simply $d = |b - a|$. The same "absolute value rule" is used to find the distance between two points that lie on the same *vertical* or *horizontal* line in the plane.

EXAMPLE 4   Finding Horizontal and Vertical Distances

a. Find the distance between the points $(1, -1)$, and $(1, 4)$.
b. Find the distance between the points $(-3, -1)$ and $(1, -1)$.

Solution

a. Because the *x*-coordinates are equal, we envision a vertical line through the points $(1, -1)$ and $(1, 4)$, as shown in Figure 3.6. The distance between these two points is the absolute value of the difference of their *y*-coordinates. That is,

$$\text{Vertical distance} = |4 - (-1)| = 5. \qquad \textit{Subtract y-coordinates}$$

b. Because the *y*-coordinates are equal, we envision a horizontal line through the points $(-3, -1)$ and $(1, -1)$, as shown in Figure 3.6. The distance between these two points is the absolute value of the difference of their *x*-coordinates. That is,

$$\text{Horizontal distance} = |1 - (-3)| = 4. \qquad \textit{Subtract x-coordinates}$$

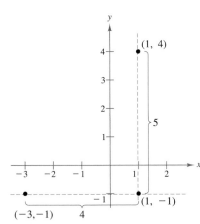

FIGURE 3.6

The technique shown in Example 4 can be used to develop a general formula for finding the distance between two points in a plane. This formula will work for any two points, even if they do not lie on the same vertical or horizontal line. To develop the formula, use the Pythagorean Theorem, which states that for a right triangle, the hypotenuse c and sides a and b are related by the formula $a^2 + b^2 = c^2$, as shown in Figure 3.7. (The converse is also true: If $a^2 + b^2 = c^2$, then the triangle is a right triangle.)

To develop a general formula for the distance between two points, let (x_1, y_1) and (x_2, y_2) represent two points in the plane (that do not lie on the same horizontal or vertical line). With these two points, a right triangle can be formed, as shown in Figure 3.8. Note that the third vertex of the triangle is (x_1, y_2). Since (x_1, y_1) and (x_1, y_2) lie on the same vertical line, the length of the vertical side of the triangle is $|y_2 - y_1|$. Similarly, the length of the horizontal side is $|x_2 - x_1|$. Thus, by the Pythagorean Theorem, the distance between (x_1, y_1) and (x_2, y_2) is

$$d^2 = |x_2 - x_1|^2 + |y_2 - y_1|^2.$$

Since the distance d must be positive, we choose the positive square root and write

$$d = \sqrt{|x_2 - x_1|^2 + |y_2 - y_1|^2}.$$

Finally, replacing $|x_2 - x_1|^2$ and $|y_2 - y_1|^2$ by the equivalent expressions $(x_2 - x_1)^2$ and $(y_2 - y_1)^2$ gives the following formula for the distance between two points in a rectangular coordinate plane.

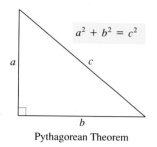

$$a^2 + b^2 = c^2$$

Pythagorean Theorem

FIGURE 3.7

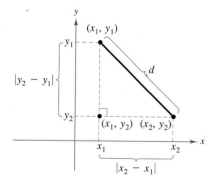

Distance Between Two Points

FIGURE 3.8

THE DISTANCE FORMULA

The distance d between the two points (x_1, y_1) and (x_2, y_2) in the coordinate plane is

$$d = \sqrt{(x_2 - x_1)^2 + (y_2 - y_1)^2}.$$

REMARK   Note that for the special case in which the two points lie on the same vertical or horizontal line, the Distance Formula still works. For instance, applying the Distance Formula to the points $(1, -1)$ and $(1, 4)$ produces

$$d = \sqrt{(1 - 1)^2 + [4 - (-1)]^2} = \sqrt{5^2} = 5,$$

which is the same result obtained in Example 4.

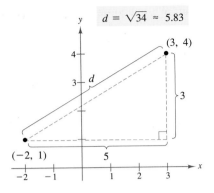

$d = \sqrt{34} \approx 5.83$

FIGURE 3.9

EXAMPLE 5 Finding the Distance Between Two Points

Find the distance between the points $(-2, 1)$ and $(3, 4)$.

Solution

Letting $(x_1, y_1) = (-2, 1)$ and $(x_2, y_2) = (3, 4)$, apply the Distance Formula to obtain

$$d = \sqrt{[3 - (-2)]^2 + (4 - 1)^2}$$
$$= \sqrt{5^2 + 3^2} = \sqrt{25 + 9} = \sqrt{34} \approx 5.83.$$

See Figure 3.9.

In Example 5, the figure provided was not essential to the solution of the problem. *Nevertheless*, we recommend that you include graphs with your problem solutions.

EXAMPLE 6 An Application of the Distance Formula

Show that the points $(2, 1)$, $(4, 0)$, and $(5, 7)$ are vertices of a right triangle.

Solution

The three points are plotted in Figure 3.10. Using the Distance Formula, you can find the lengths of the three sides of the triangle.

$$d_1 = \sqrt{(5 - 2)^2 + (7 - 1)^2} = \sqrt{9 + 36} = \sqrt{45}$$
$$d_2 = \sqrt{(4 - 2)^2 + (0 - 1)^2} = \sqrt{4 + 1} = \sqrt{5}$$
$$d_3 = \sqrt{(5 - 4)^2 + (7 - 0)^2} = \sqrt{1 + 49} = \sqrt{50}$$

Since $d_1^2 + d_2^2 = 45 + 5 = 50 = d_3^2$, you can conclude from the Pythagorean Theorem that the triangle is a right triangle.

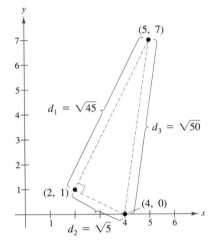

FIGURE 3.10

EXAMPLE 7 An Application of the Distance Formula

In a football game, a quarterback throws a pass from the 5-yard line, 20 yards from the sideline. The pass is caught by a wide receiver on the 45-yard line, 50 yards from the same sideline, as shown in Figure 3.11. How long was the pass?

Solution

$$d = \sqrt{(50 - 20)^2 + (45 - 5)^2} \qquad \textit{Distance Formula}$$
$$= \sqrt{900 + 1600}$$
$$= \sqrt{2500}$$
$$= 50 \text{ yards}$$

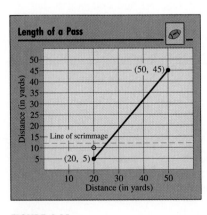

FIGURE 3.11

REMARK In Example 7, the scale along the goal line does not normally appear on a football field. However, when you use coordinate geometry to solve real-life problems, you are free to place the coordinate system in any way that is convenient in the problem.

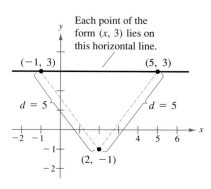

Each point of the form $(x, 3)$ lies on this horizontal line.

$(-1, 3)$ $(5, 3)$

$d = 5$ $d = 5$

$(2, -1)$

FIGURE 3.12

EXAMPLE 8   Finding Points at a Specified Distance from a Given Point

Find x so that the distance between $(x, 3)$ and $(2, -1)$ is 5.

Solution

$$\sqrt{(x - 2)^2 + (3 + 1)^2} = 5 \qquad \textit{Distance Formula}$$

$$(x^2 - 4x + 4) + 16 = 25 \qquad \textit{Square both sides}$$

$$x^2 - 4x - 5 = 0 \qquad \textit{Standard form}$$

$$(x - 5)(x + 1) = 0 \qquad \textit{Factor}$$

$$x - 5 = 0 \quad \rightarrow \quad x = 5 \qquad \textit{Set 1st factor equal to 0}$$

$$x + 1 = 0 \quad \rightarrow \quad x = -1 \qquad \textit{Set 2nd factor equal to 0}$$

There are two solutions: Each of the points $(5, 3)$ and $(-1, 3)$ lies five units from the point $(2, -1)$, as shown in Figure 3.12.

The Midpoint Formula

Next you will consider a formula for finding the midpoint of a line segment joining two points. The coordinates of the midpoint are the average values of the corresponding coordinates of the two endpoints.

THE MIDPOINT FORMULA

The **midpoint** of the line segment joining the points (x_1, y_1) and (x_2, y_2) in the coordinate plane is

$$\left(\frac{x_1 + x_2}{2}, \frac{y_1 + y_2}{2} \right).$$

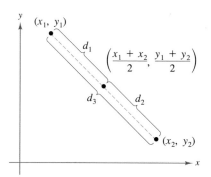

(x_1, y_1)

d_1

$\left(\frac{x_1 + x_2}{2}, \frac{y_1 + y_2}{2} \right)$

d_3 d_2

(x_2, y_2)

Midpoint Formula

FIGURE 3.13

Proof

Using Figure 3.13, you must show that

$$d_1 = d_2 \qquad \text{and} \qquad d_1 + d_2 = d_3.$$

By the Distance Formula, you obtain

$$d_1 = \sqrt{\left(\frac{x_1 + x_2}{2} - x_1 \right)^2 + \left(\frac{y_1 + y_2}{2} - y_1 \right)^2} = \frac{1}{2}\sqrt{(x_2 - x_1)^2 + (y_2 - y_1)^2}$$

$$d_2 = \sqrt{\left(x_2 - \frac{x_1 + x_2}{2} \right)^2 + \left(y_2 - \frac{y_1 + y_2}{2} \right)^2} = \frac{1}{2}\sqrt{(x_2 - x_1)^2 + (y_2 - y_1)^2}$$

$$d_3 = \sqrt{(x_2 - x_1)^2 + (y_2 - y_1)^2}.$$

Thus, it follows that $d_1 = d_2$ and $d_1 + d_2 = d_3$.

EXAMPLE 9 Finding the Midpoint of a Line Segment

Find the midpoint of the line segment joining the points $(-5, -3)$ and $(9, 3)$.

Solution

Figure 3.14 shows the two given points and their midpoint. By the Midpoint Formula,

$$\text{Midpoint} = \left(\frac{-5 + 9}{2}, \frac{-3 + 3}{2}\right) = (2, 0).$$

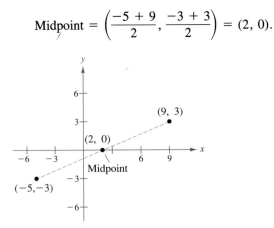

FIGURE 3.14

Although you have used the Midpoint Formula in the context of geometry, it can also be applied to problems that don't appear to be geometrical.

EXAMPLE 10 An Application: Retail Sales

A business had annual retail sales of \$240,000 in 1986 and \$312,000 in 1992. Find the sales for 1989, assuming that the annual increase in sales followed a *linear* pattern.

Solution

To make the computations simpler, let $t = 0$ represent the year 1986 and $t = 6$ represent the year 1992. Then, the retail sales in 1000s of dollars for 1986 and 1992 are represented by the points

$(0, 240)$ and $(6, 312)$.

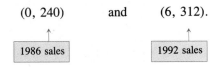

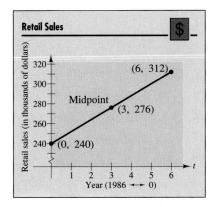

FIGURE 3.15

Since 1989 is midway between 1986 and 1992, and since the growth pattern is linear, you can use the Midpoint Formula to find the 1989 sales.

$$\text{Midpoint} = \left(\frac{0 + 6}{2}, \frac{240 + 312}{2} \right) = (3, 276)$$

1989 sales

Thus, the 1989 sales were approximately $276,000, as indicated in Figure 3.15.

DISCUSSION
PROBLEM

A
Misleading
Graph

While graphs can help you visualize relationships between two variables, they can also mislead people. The graphs shown in Figure 3.16 represent the *same* data points. Which of the two graphs is misleading, and why?

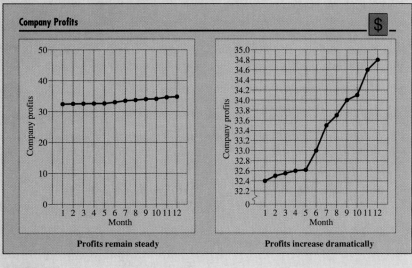

FIGURE 3.16

WARM UP

The following warm-up exercises involve skills that were covered in earlier sections. You will use these skills in the exercise set for this section.

In Exercises 1–6, simplify the expression.

1. $\sqrt{(2-6)^2 + [1-(-2)]^2}$ **2.** $\sqrt{(1-4)^2 + (-2-1)^2}$

3. $\dfrac{4+(-2)}{2}$ **4.** $\dfrac{-1+(-3)}{2}$

5. $\sqrt{18} + \sqrt{45}$ **6.** $\sqrt{12} + \sqrt{44}$

In Exercises 7–10, solve for x or y.

7. $\sqrt{(4-x)^2 + (5-2)^2} = \sqrt{58}$ **8.** $\sqrt{(8-6)^2 + (y-5)^2} = 2\sqrt{5}$

9. $\dfrac{x+3}{2} = 7$ **10.** $\dfrac{-2+y}{2} = 1$

EXERCISES for Section 3.1

In Exercises 1–4, sketch the polygon with the indicated vertices.

1. Triangle: $(-1, 1)$, $(2, -1)$, $(3, 4)$
2. Triangle: $(0, 3)$, $(-1, -2)$, $(4, 8)$
3. Square: $(2, 4)$, $(5, 1)$, $(2, -2)$, $(-1, 1)$
4. Parallelogram: $(5, 2)$, $(7, 0)$, $(1, -2)$, $(-1, 0)$

In Exercises 5 and 6, the figure is shifted to a new position in the plane. Find the coordinates of the vertices of the figure in its *new* position.

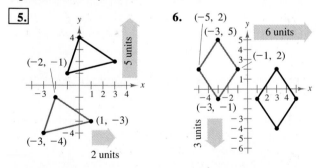

In Exercises 7–10, find the distance between the points. (*Note:* In each case the two points lie on the same horizontal or vertical line.)

7. $(6, -3)$, $(6, 5)$ **8.** $(1, 4)$, $(8, 4)$
9. $(-3, -1)$, $(2, -1)$ **10.** $(-3, -4)$, $(-3, 6)$

In Exercises 11–14, (a) find the length of the two sides of the right triangle and use the Pythagorean Theorem to find the length of the hypotenuse, and (b) use the Distance Formula to find the length of the hypotenuse of the triangle.

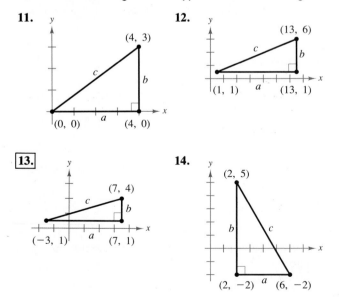

In Exercises 15–26, (a) plot the points, (b) find the distance between the points, and (c) find the midpoint of the line segment joining the points.

15. (1, 1), (9, 7) **16.** (1, 12), (6, 0)
17. (−4, 10), (4, −5) **18.** (−7, −4), (2, 8)
19. (−1, 2), (5, 4) **20.** (2, 10), (10, 2)
21. $\left(\frac{1}{2}, 1\right), \left(-\frac{5}{2}, \frac{4}{3}\right)$ **22.** $\left(-\frac{1}{3}, -\frac{1}{3}\right), \left(-\frac{1}{6}, -\frac{1}{2}\right)$
23. (6.2, 5.4), (−3.7, 1.8) **24.** (−16.8, 12.3), (5.6, 4.9)
25. (−36, −18), (48, −72)
26. (1.451, 3.051), (5.906, 11.360)

In Exercises 27 and 28, use the Midpoint Formula to estimate the sales of a company for 1991. Assume the sales followed a linear pattern.

27.

Year	1989	1993
Sales	$520,000	$740,000

28.

Year	1989	1993
Sales	$4,200,000	$5,650,000

In Exercises 29–32, show that the points form the vertices of the indicated polygon.

29. Right triangle: (4, 0), (2, 1), (−1, −5)
30. Isosceles triangle: (1, −3), (3, 2), (−2, 4)
31. Rhombus: (0, 0), (1, 2), (2, 1), (3, 3)
(A rhombus is a parallelogram whose sides are all of the same length.)
32. Parallelogram: (0, 1), (3, 7), (4, 4), (1, −2)

In Exercises 33 and 34, find x so that the distance between the points is 13.

33. (1, 2), (x, −10) **34.** (−8, 0), (x, 5)

In Exercises 35 and 36, find y so that the distance between the points is 17.

35. (0, 0), (8, y) **36.** (−8, 4), (7, y)

In Exercises 37 and 38, find a relationship between x and y so that (x, y) is equidistant from the two points.

37. (4, −1), (−2, 3) **38.** $\left(3, \frac{5}{2}\right), (−7, 1)$
$3x - 2y - y = 0$

In Exercises 39–48, determine the quadrant(s) in which (x, y) is located so that the conditions are satisfied.

39. $x > 0$ and $y < 0$ **40.** $x < 0$ and $y < 0$
41. $x > 0$ and $y > 0$ **42.** $x < 0$ and $y > 0$
43. $x = -4$ and $y > 0$ **44.** $x > 2$ and $y = 3$
45. $y < -5$ **46.** $x > 4$
47. $(x, -y)$ is in the second quadrant.
48. $(-x, y)$ is in the fourth quadrant.

49. A line segment has (x_1, y_1) as one endpoint and (x_m, y_m) as its midpoint. Find the other endpoint (x_2, y_2) of the line segment in terms of $x_1, y_1, x_m,$ and y_m.
50. Use the result of Exercise 49 to find the coordinates of one endpoint of a line segment if the coordinates of the other endpoint and midpoint are, respectively,
(a) (1, −2), (4, −1) (b) (−5, 11), (2, 4).
51. Use the Midpoint Formula twice to find the three points that divide the line segment joining (x_1, y_1) and (x_2, y_2) into four parts.
52. Use the result of Exercise 51 to find the points that divide the line segment joining the points into four equal parts.
(a) (1, −2), (4, −1) (b) (−2, −3), (0, 0)
53. *Football Pass* In a football game, a quarterback throws a pass from the 15-yard line, 10 yards from the sideline. The pass is caught on the 40-yard line, 45 yards from the same sideline. How long was the pass? (Assume the pass and the reception are on the same side of midfield.)
54. *Flying Distance* A plane flies in a straight line to a city that is 100 miles east and 150 miles north of the point of departure. How far did it fly?

In Exercises 55 and 56, plot the points whose coordinates are given in the table.

55. *Normal Temperatures* The normal temperature y (Fahrenheit) for Duluth, Minnesota for each month of the year is given in the table. The months are numbered 1 through 12, with 1 corresponding to January. (*Source:* NOAA)

x	1	2	3	4	5	6	7	8	9	10	11	12
y	6	12	23	38	50	59	65	63	54	44	28	14

56. *Stock Price*   The price y per common share of stock on December 31 for the years 1984 through 1989 is given in the table. The time in years is given by x. (*Source:* Ameritech Annual Report for 1989)

x	1984	1985	1986	1987	1988	1989
y	$25.50	$35.50	$44.125	$42.25	$47.875	$68.00

Milk Prices   In Exercises 57 and 58, refer to the figure. (*Source:* U.S. Department of Agriculture and the National Milk Producers Federation)

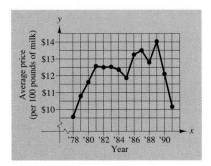

Figure for 57 and 58

57. What is the highest price of milk shown in the graph? When did this occur?

58. Find the percentage drop in the price of milk from the highest price shown in the graph to the price paid to farmers in January 1991.

TV Advertising   In Exercises 59 and 60, refer to the figure. (*Source:* Nielson Media Research)

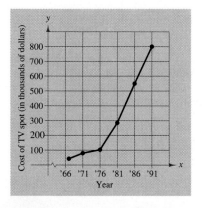

Figure for 59 and 60

59. Find the percentage increase in the cost of a 30-second spot from Super Bowl I to Super Bowl XXV.

60. Estimate the increase in the cost of a 30-second spot (a) from Super Bowl V to Super Bowl XV, and (b) from Super Bowl XV to Super Bowl XXV.

61. Plot the points $(2, 1)$, $(-3, 5)$, and $(7, -3)$ on the rectangular coordinate system. Now plot the corresponding points when the sign of the x-coordinate is reversed. What can you infer about the result of the location of a point when the sign of the x-coordinate is changed?

62. Plot the points $(2, 1)$, $(-3, 5)$, and $(7, -3)$ on the rectangular coordinate system. Now plot the corresponding points when the sign of the y-coordinate is reversed. What can you infer about the result of the location of a point when the sign of the y-coordinate is changed?

63. Prove that the diagonals of the parallelogram in the figure bisect each other.

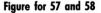

Figure for 63

3.2 Graphs of Equations

The Graph of an Equation / Intercepts of a Graph / Symmetry / The Equation of a Circle

The Graph of an Equation

News magazines often show graphs comparing the rate of inflation, the federal deficit, wholesale prices, or the unemployment rate to the time of year. Industrial firms and businesses use graphs to report their monthly production and sales statistics. Such graphs provide a simple geometric picture of the way one quantity changes with respect to another.

Frequently, the relationship between two quantities is expressed in the form of an equation. In this section, we introduce the basic procedure for determining the geometric picture associated with an algebraic equation.

For an equation in variables x and y, a point (a, b) is a **solution point** if the substitution of $x = a$ and $y = b$ satisfies the equation. Most equations have *infinitely* many solution points. For example, the equation $3x + y = 5$ has solution points $(0, 5)$, $(1, 2)$, $(2, -1)$, $(3, -4)$, and so on. The set of all solution points of a given equation is the **graph** of the equation.

The Point-Plotting Method of Graphing

To sketch the graph of an equation by point plotting, use the following method.

1. If possible, rewrite the equation so that one of the variables is isolated on the left side of the equation.
2. Make up a table of several solution points.
3. Plot these points in the coordinate plane.
4. Connect the points with a smooth curve.

EXAMPLE 1   Sketching the Graph of an Equation

Sketch a graph of the equation $3x + y = 5$.

Solution

Isolate the variable y.

$$y = 5 - 3x$$

Use negative, zero, and positive values for x to obtain a table of values (solution points).

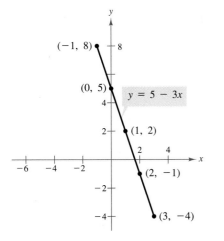

x	-1	0	1	2	3
$y = 5 - 3x$	8	5	2	-1	-4

Next, plot these points and connect them as shown in Figure 3.17. It appears that the graph is a straight line. (You will study lines extensively in Section 3.3.)

FIGURE 3.17

FIGURE 3.18

Step 4 of the point-plotting method can be difficult. For instance, how would you connect the four points in Figure 3.18? Without further information about the equation, any one of the three graphs in Figure 3.19 would be reasonable.

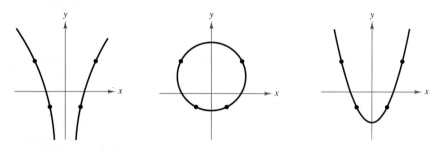

FIGURE 3.19

With too few solution points, you could grossly misrepresent the graph of an equation. Just how many points should be plotted? For straight-line graphs, two points are sufficient. For more complicated graphs, you need many more points, enough to reveal the essential behavior of the graph. A programmable calculator is useful for determining the many solution points needed for an accurate graph.

Plot several points.

FIGURE 3.20

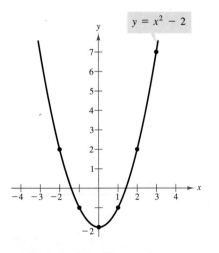

Connect points with a smooth curve.

FIGURE 3.21

EXAMPLE 2   Sketching the Graph of an Equation

Sketch the graph of the equation $y = x^2 - 2$.

Solution

First, make a table of values.

x	-2	-1	0	1	2	3
$y = x^2 - 2$	2	-1	-2	-1	2	7

Next, plot the corresponding solution points, as shown in Figure 3.20. Finally, connect the points with a smooth curve, as shown in Figure 3.21.

Intercepts of a Graph

Two types of points that are especially useful when sketching a graph are those for which either the y-coordinate or the x-coordinate is zero. These points are called **intercepts** because they are points at which the graph intersects the x- or y-axis.

REMARK Sometimes it is convenient to denote the x-intercept as simply the x-coordinate, a, of the point $(a, 0)$ rather than the point itself. The same is true with the y-intercept. Unless it is necessary to make a distinction, we will use "intercept" to mean either the point or the coordinate.

> ### DEFINITION OF INTERCEPTS
>
> 1. The point $(a, 0)$ is an **x-intercept** of the graph of an equation if it is a solution point of the equation. To find the x-intercepts, let y be zero and solve the equation for x.
> 2. The point $(0, b)$ is a **y-intercept** of the graph of an equation if it is a solution point of the equation. To find the y-intercepts, let x be zero and solve the equation for y.

Of course, it is possible that a particular graph will have no intercepts or several intercepts. For instance, consider the three graphs in Figure 3.22.

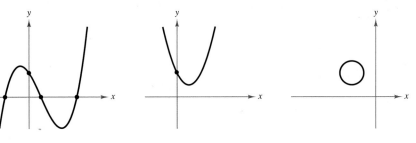

Three x-Intercepts
One y-Intercept

No x-Intercept
One y-Intercept

No Intercepts

FIGURE 3.22

EXAMPLE 3   Finding the x- and y-Intercepts of a Graph

Find the x- and y-intercepts of the graph of $x = y^2 - 3$.

Solution

To find the x-intercept, let $y = 0$. This produces $x = -3$, which implies that the graph has one x-intercept, which occurs at the point

$(-3, 0)$.   *x-intercept*

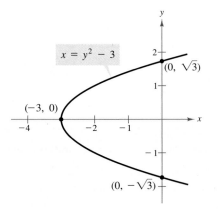

FIGURE 3.23

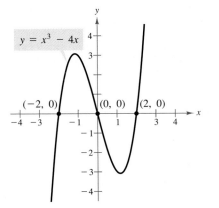

FIGURE 3.24

To find the y-intercept, let $x = 0$. This produces $0 = y^2 - 3$, which has two solutions: $y = \pm\sqrt{3}$. Thus, the equation has two y-intercepts, which occur at the points

$$(0, \sqrt{3}) \quad \text{and} \quad (0, -\sqrt{3}). \qquad \textit{y-intercepts}$$

See Figure 3.23.

EXAMPLE 4   Finding the *x*- and *y*-Intercepts of a Graph

Find the x- and y-intercepts of the graph of $y = x^3 - 4x$.

Solution

To find the x-intercepts, let $y = 0$.

$$0 = x^3 - 4x = x(x^2 - 4)$$

Set each factor equal to zero and solve for x to obtain $x = 0$ and $x = \pm 2$. Thus, the equation has three x-intercepts, which occur at the points

$$(0, 0), \quad (2, 0), \quad \text{and} \quad (-2, 0). \qquad \textit{x-intercepts}$$

Let $x = 0$ to obtain $y = 0$, which tells us that the y-intercept is

$$(0, 0). \qquad \textit{y-intercept}$$

See Figure 3.24.

Symmetry

The graphs shown in Figures 3.21, 3.23, and 3.24 each have a type of **symmetry** with respect to one of the coordinate axes or with respect to the origin.

Figure 3.21	$y = x^2 - 2$	*y-axis symmetry*
Figure 3.23	$x = y^2 - 3$	*x-axis symmetry*
Figure 3.24	$y = x^3 - 4x$	*Origin symmetry*

Symmetry with respect to the x-axis means that if the Cartesian plane were folded along the x-axis, the portion of the graph above the x-axis would coincide with the portion below the x-axis. Symmetry with respect to the y-axis or to the origin is described in a similar manner, as shown in Figure 3.25.

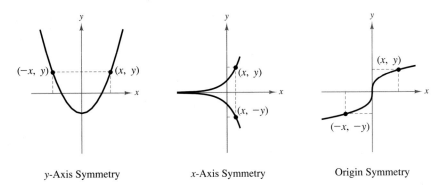

y-Axis Symmetry   x-Axis Symmetry   Origin Symmetry

FIGURE 3.25

Knowing the symmetry of a graph *before* attempting to sketch it is helpful, because then you need only half as many solution points to sketch the graph. There are three basic types of symmetry. (See Exercises 29–32.) A graph is **symmetric with respect to the y-axis** if, whenever (x, y) is on the graph, $(-x, y)$ is also on the graph. A graph is **symmetric with respect to the x-axis** if, whenever (x, y) is on the graph, $(x, -y)$ is also on the graph. A graph is **symmetric with respect to the origin** if, whenever (x, y) is on the graph, $(-x, -y)$ is also on the graph.

The graph of $y = x^2 - 2$ is symmetric with respect to the y-axis because the point $(-x, y)$ satisfies the equation.

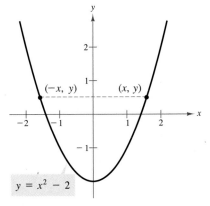

$y = x^2 - 2$

y-Axis Symmetry

FIGURE 3.26

$y = x^2 - 2$	*Given equation*
$y = (-x)^2 - 2$	*Substitute $(-x, y)$ for (x, y)*
$y = x^2 - 2$	*Replacement yields equivalent equation*

See Figure 3.26.

The graph of $y = x^3 - 4x$ (see Figure 3.24) is symmetric with respect to the origin because the point $(-x, -y)$ satisfies the equation.

$y = x^3 - 4x$	*Given equation*
$-y = (-x)^3 - 4(-x)$	*Substitute $(-x, -y)$ for (x, y)*
$-y = -x^3 + 4x$	*Equation is equivalent to given equation*

A similar test can be made for symmetry with respect to the x-axis.

Tests for Symmetry

1. The graph of an equation is symmetric with respect to the *y-axis* if replacing x with $-x$ yields an equivalent equation.
2. The graph of an equation is symmetric with respect to the *x-axis* if replacing y with $-y$ yields an equivalent equation.
3. The graph of an equation is symmetric with respect to the *origin* if replacing x with $-x$ and y with $-y$ yields an equivalent equation.

EXAMPLE 5　Using Intercepts and Symmetry as Sketching Aids

Use intercepts and symmetry to sketch the graph of $x - y^2 = 1$.

Solution

Letting $x = 0$, you see that $-y^2 = 1$ or $y^2 = -1$ has no real solutions. Hence, there are no y-intercepts. Let $y = 0$ to obtain $x = 1$. Thus, the x-intercept is $(1, 0)$.

y-intercept: None

x-intercept: $(1, 0)$

Of the three tests for symmetry, the only one that is satisfied by this equation is the test for x-axis symmetry.

$x - y^2 = 1$	*Given equation*
$x - (-y)^2 = 1$	*Replace y with $-y$*
$x - y^2 = 1$	*Replacement yields equivalent equation*

Thus, the graph is symmetric with respect to the x-axis. Using symmetry, you need only to find solution points above the x-axis and then reflect them to obtain the desired graph, as shown in Figure 3.27.

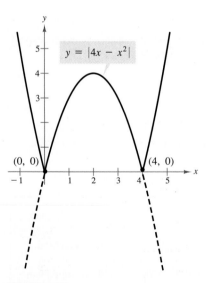

FIGURE 3.27

y	0	1	2
$x = y^2 + 1$	1	2	5

EXAMPLE 6　Sketching the Graph of an Equation

Sketch the graph of $y = |4x - x^2|$.

Solution

Intercepts: Letting $x = 0$ yields $y = 0$, which means that $(0, 0)$ is a y-intercept. Letting $y = 0$ yields $x = 0$ and $x = 4$, which means that $(0, 0)$ and $(4, 0)$ are x-intercepts.

Symmetry: This equation fails all three tests for symmetry and consequently its graph is not symmetric with respect to either axis or the origin.

The absolute value sign indicates that y is always nonnegative.

x	-1	0	1	2	3	4	5
$y = |4x - x^2|$	5	0	3	4	3	0	5

The graph is shown in Figure 3.28. The dotted portion of the figure shows how the graph would differ if no absolute value signs were used.

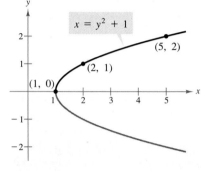

FIGURE 3.28

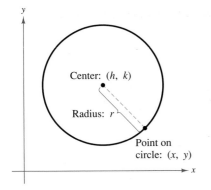

FIGURE 3.29

The Equation of a Circle

Thus far you have studied the point-plotting method and two additional concepts (intercepts and symmetry) that can be used to streamline the graphing procedure. Another graphing aid is *equation recognition*, the ability to recognize the general shape of a graph simply by looking at its equation. A circle is one type of graph that is easily recognized.

Figure 3.29 shows a circle of radius r with center at the point (h, k). The point (x, y) is on this circle if and only if its distance from the center (h, k) is r. This means that a **circle** in the plane consists of all points (x, y) that are a given positive distance r from a fixed point (h, k). Using the Distance Formula, you can express this relationship by saying that the point (x, y) lies on the circle if and only if

$$\sqrt{(x - h)^2 + (y - k)^2} = r.$$

You can square both sides of this equation to obtain the **standard form of the equation of a circle.**

STANDARD FORM OF THE EQUATION OF A CIRCLE

The **standard form of the equation of a circle** is

$$(x - h)^2 + (y - k)^2 = r^2.$$

The point (h, k) is the **center** of the circle and the positive number r is the **radius** of the circle.

REMARK The standard form of the equation of a circle whose center is the *origin* is simply $x^2 + y^2 = r^2$.

EXAMPLE 7 Finding an Equation of a Circle

The point $(3, 4)$ lies on a circle whose center is at $(-1, 2)$, as shown in Figure 3.30. Find an equation for the circle.

Solution

The radius r of the circle is the distance between $(-1, 2)$ and $(3, 4)$.

$$r = \sqrt{[3 - (-1)]^2 + (4 - 2)^2}$$
$$= \sqrt{16 + 4}$$
$$= \sqrt{20}$$

Thus, the center of the circle is $(h, k) = (-1, 2)$ and the radius is $r = \sqrt{20}$. Now, write the standard form of the equation of the circle.

$$(x - h)^2 + (y - k)^2 = r^2 \qquad \textit{Standard form}$$
$$[x - (-1)]^2 + (y - 2)^2 = (\sqrt{20})^2 \qquad \textit{Let } h = -1, k = 2, \textit{ and } r = \sqrt{20}$$
$$(x + 1)^2 + (y - 2)^2 = 20 \qquad \textit{Equation of circle}$$

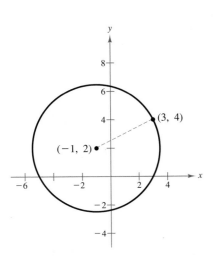

FIGURE 3.30

Removing the parentheses in the standard equation in Example 7 yields

$$(x + 1)^2 + (y - 2)^2 = 20 \qquad \textit{Standard form}$$
$$x^2 + 2x + 1 + y^2 - 4y + 4 = 20 \qquad \textit{Expand terms}$$
$$x^2 + y^2 + 2x - 4y - 15 = 0. \qquad \textit{General form}$$

The last equation is in the **general form of the equation of a circle.**

$$Ax^2 + Ay^2 + Dx + Ey + F = 0, \qquad A \neq 0$$

The general form of the equation of a circle is less useful than the standard form. For instance, it is not immediately apparent from the general equation of the circle in Example 7

$$x^2 + y^2 + 2x - 4y - 15 = 0$$

that the center is $(-1, 2)$ and the radius is $\sqrt{20}$. To graph the equation of a circle, it is best to write the equation in standard form. You can do this by **completing the square.**

EXAMPLE 8   Completing the Square to Sketch a Circle

Identify the center and radius of the circle given by the equation and sketch the circle.

$$4x^2 + 4y^2 + 20x - 16y + 37 = 0$$

Solution

To write the given equation in standard form, you must complete the square for both the x-terms *and* the y-terms.

$$4x^2 + 4y^2 + 20x - 16y + 37 = 0 \qquad \textit{General form}$$

$$x^2 + y^2 + 5x - 4y + \frac{37}{4} = 0 \qquad \textit{Divide by 4}$$

$$(x^2 + 5x + \quad) + (y^2 - 4y + \quad) = -\frac{37}{4} \qquad \textit{Group terms}$$

$$\left(x^2 + 5x + \left(\frac{5}{2}\right)^2\right) + (y^2 - 4y + 2^2) = -\frac{37}{4} + \frac{25}{4} + 4 \qquad \textit{Complete square}$$

$$\underbrace{\qquad}_{(\text{half})^2} \qquad \underbrace{\qquad}_{(\text{half})^2}$$

$$\left(x + \frac{5}{2}\right)^2 + (y - 2)^2 = 1 \qquad \textit{Standard form}$$

Thus, the center of the circle is $\left(-\frac{5}{2}, 2\right)$ and the radius of the circle is 1. Use this information to sketch the circle shown in Figure 3.31.

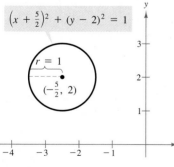

FIGURE 3.31

DISCUSSION PROBLEM

Two Unusual Types of Equations

The general equation $Ax^2 + Ay^2 + Dx + Ey + F = 0$ may not always represent a circle. Such an equation will have no solution points if the procedure of completing the square yields the *impossible* result

$$(x - h)^2 + (y - k)^2 = \text{(negative number)}.$$

Moreover, the general equation $Ax^2 + Ay^2 + Dx + Ey + F = 0$ will have exactly one solution point if the procedure of completing the square yields the result $(x - h)^2 + (y - k)^2 = 0$. The point (h, k) is the only solution point for this equation.

Which of the following equations has no solution and which has exactly one solution? Discuss your reasons.

1. $x^2 + y^2 + 1 = 0$ 2. $x^2 + y^2 = 0$

WARM UP

The following warm-up exercises involve skills that were covered in earlier sections. You will use these skills in the exercise set for this section.

In Exercises 1 and 2, solve for y in terms of x.

1. $3x - 5y = 2$ 2. $x^2 - 4x + 2y - 5 = 0$

In Exercises 3–6, solve for x.

3. $x^2 - 4x + 4 = 0$ 4. $(x - 1)(x + 5) = 0$

5. $x^3 - 9x = 0$ 6. $x^4 - 8x^2 + 16 = 0$

In Exercises 7–10, simplify the equations.

7. $-y = (-x)^3 + 4(-x)$ 8. $(-x)^2 + (-y)^2 = 4$

9. $y = 4(-x)^2 + 8$ 10. $(-y)^2 = 3(-x) + 4$

EXERCISES for Section 3.2

In Exercises 1–6, determine whether the indicated points lie on the graph of the equation.

Equation	Points	
1. $y = \sqrt{x + 4}$	(a) $(0, 2)$	(b) $(5, 3)$
2. $y = x^2 - 3x + 2$	(a) $(2, 0)$	(b) $(-2, 8)$
3. $2x - y - 3 = 0$	(a) $(1, 2)$	(b) $(1, -1)$
4. $x^2 + y^2 = 20$	(a) $(3, -2)$	(b) $(-4, 2)$
5. $x^2y - x^2 + 4y = 0$	(a) $\left(1, \frac{1}{5}\right)$	(b) $\left(2, \frac{1}{2}\right)$
6. $y = \dfrac{1}{x^2 + 1}$	(a) $(0, 0)$	(b) $(3, 0.1)$

In Exercises 7–10, find the constant C such that the ordered pair is a solution point of the equation.

7. $y = x^2 + C$, $(2, 6)$
8. $y = Cx^3$, $(-4, 8)$
9. $y = C\sqrt{x + 1}$, $(3, 8)$
10. $x + C(y + 2) = 0$, $(4, 3)$

In Exercises 11 and 12, complete the table. Use the solution points to sketch the graph of the equation.

11. $2x + y = 3$

x	-4			2	4
y		7	3		
(x, y)					

12. $y = 4 - x^2$

x		-1		2	
y	0		4		-5
(x, y)					

In Exercises 13–20, find the x- and y-intercepts of the graph of the equation.

13. $y = x - 5$ **14.** $y = (x - 1)(x - 3)$
15. $y = x^2 + x - 2$ **16.** $y = 4 - x^2$
17. $y = x\sqrt{x + 2}$ **18.** $xy = 4$
19. $xy - 2y - x + 1 = 0$ **20.** $x^2y - x^2 + 4y = 0$

In Exercises 21–28, check for symmetry with respect to both axes and the origin.

21. $x^2 - y = 0$ **22.** $xy^2 + 10 = 0$
23. $x - y^2 = 0$ **24.** $y = \sqrt{9 - x^2}$
25. $y = x^3$ **26.** $xy = 4$
27. $y = \dfrac{x}{x^2 + 1}$ **28.** $y = x^4 - x^2 + 3$

In Exercises 29–32, use symmetry to sketch the complete graph of the equation.

29.

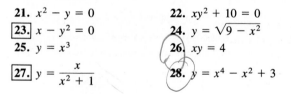

$y = -x^2 + 4$

y-Axis Symmetry

30.

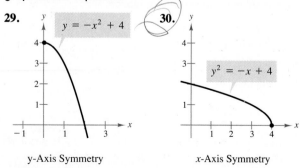

$y^2 = -x + 4$

x-Axis Symmetry

31.

$y = -x^3 + x$

Origin Symmetry

32.

$y = |x| - 2$

y-Axis Symmetry

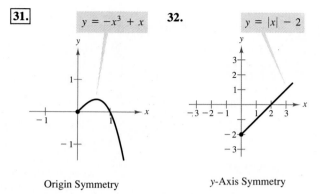

In Exercises 33–38, match the equation with its graph. [The graphs are labeled (a), (b), (c), (d), (e), and (f).]

33. $y = 4 - x$ **34.** $y = x^2 + 2x$
35. $y = \sqrt{4 - x^2}$ **36.** $y = \sqrt{x}$
37. $y = x^3 - x$ **38.** $y = |x| - 2$

(a)   (b)

(c)   (d)

(e)   (f)

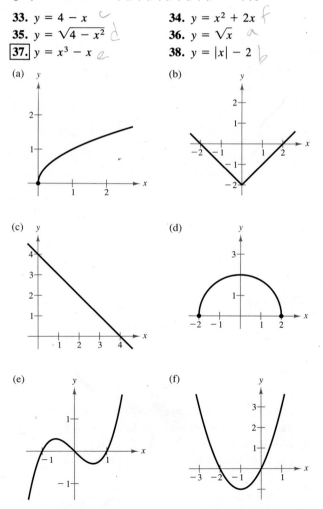

In Exercises 39–58, sketch the graph of the equation. Identify any intercepts and test for symmetry.

39. $y = -3x + 2$
40. $y = 2x - 3$
41. $y = 1 - x^2$
42. $y = x^2 - 1$
43. $y = x^2 - 4x + 3$
44. $y = -x^2 - 4x$
45. $y = x^3 + 2$
46. $y = x^3 - 1$
47. $y = x(x - 2)^2$
48. $y = \dfrac{4}{x^2 + 1}$
49. $y = \sqrt{x - 3}$
50. $y = \sqrt{1 - x}$
51. $y = \sqrt[3]{x}$
52. $y = \sqrt[3]{x + 1}$
53. $y = |x - 2|$
54. $y = 4 - |x|$
55. $x = y^2 - 1$
56. $x = y^2 - 4$
57. $x^2 + y^2 = 4$
58. $x^2 + y^2 = 16$

In Exercises 59–66, find the standard form of the equation of the specified circle.

59. Center: $(0, 0)$; radius: 3
60. Center: $(0, 0)$; radius: 5
61. Center: $(2, -1)$; radius: 4
62. Center: $\left(0, \frac{1}{3}\right)$; radius: $\frac{1}{3}$
63. Center: $(-1, 2)$; passing through: $(0, 0)$
64. Center: $(3, -2)$; passing through: $(-1, 1)$
65. Endpoints of a diameter: $(0, 0)$, $(6, 8)$
66. Endpoints of a diameter: $(-4, -1)$, $(4, 1)$

In Exercises 67–74, find the center and radius, and sketch the graph of the equation.

67. $x^2 + y^2 - 2x + 6y + 6 = 0$
68. $x^2 + y^2 - 2x + 6y - 15 = 0$
69. $x^2 + y^2 - 2x + 6y + 10 = 0$
70. $3x^2 + 3y^2 - 6y - 1 = 0$
71. $2x^2 + 2y^2 - 2x - 2y - 3 = 0$
72. $4x^2 + 4y^2 - 4x + 2y - 1 = 0$
73. $16x^2 + 16y^2 + 16x + 40y - 7 = 0$
74. $x^2 + y^2 - 4x + 2y + 3 = 0$

75. *Depreciation*   A manufacturing plant purchases a new molding machine for \$225,000. The depreciated value y after t years is

$$y = 225{,}000 - 20{,}000t, \qquad 0 \le t \le 8.$$

Sketch the graph of the equation over the given interval for t.

76. *Dimensions of a Rectangle*   A rectangle of length l and width w has a perimeter of 12 meters.
(a) Show that the width of the rectangle is $w = 6 - l$ and its area is $A = l(6 - l)$.
(b) Sketch the graph of the equation for the area.
(c) From the graph of part (b), estimate the dimensions of the rectangle that yield maximum area.

In Exercises 77 and 78, (a) sketch a graph to compare the given data and the model for that data, (b) use the model to predict y for the year 1994, and (c) for the year 2000.

77. *Federal Debt*   The table gives the per capita federal debt for the United States for selected years from 1950 to 1990. (*Source:* U.S. Treasury Department)

Year	1950	1960	1970
Per capita debt	\$1688	\$1572	\$1807

Year	1980	1985	1990
Per capita debt	\$3981	\$7614	\$12,848

A mathematical model for the per capita debt during this period is

$$y = 0.40t^3 - 9.42t^2 + 1053.24$$

where y represents the per capita debt and t is the time in years with $t = 0$ corresponding to 1950.

78. *Life Expectancy*   The following table gives the life expectancy of a child (at birth) for selected years from 1920 to 1989. (*Source:* Department of Health and Human Services)

Year	1920	1930	1940	1950
Life expectancy	54.1	59.7	62.9	68.2

Year	1960	1970	1980	1989
Life expectancy	69.7	70.8	73.7	75.2

A mathematical model for the life expectancy during this period is

$$y = \frac{t + 66.94}{0.01t + 1}$$

where y represents the life expectancy and t represents the time in years with $t = 0$ corresponding to 1950.

79. *Earnings Per Share* The earnings per share for Eli Lilly Corporation from 1980 to 1986 can be approximated by the mathematical model

$$y = 1.097t + 0.15, \qquad 0 \le t \le 6$$

where y is the earnings and t represents the calendar year with $t = 0$ corresponding to 1980. Sketch the graph of this equation. (*Source:* NYSE Stock Reports)

80. *Copper Wire* The resistance y in ohms of 1000 feet of solid copper wire at 77 degrees Fahrenheit can be approximated by the mathematical model

$$y = \frac{10{,}770}{x^2} - 0.37, \qquad 5 \le x \le 100$$

where x is the diameter of the wire in mils (0.001 in.). Use the model to estimate the resistance when $x = 50$. (*Source:* American Wire Gage)

81. Find a and b if the x-intercept of the graph of $y = \sqrt{ax + b}$ is (5, 0). (The answer is not unique.)

82. Find a and b if the graph of $y = ax^2 + bx^3$ is symmetric to (a) the y-axis and (b) the origin. (The answer is not unique.)

3.3 Lines in the Plane

The Slope of a Line / The Point-Slope Form of the Equation of a Line / Sketching Graphs of Lines / Parallel and Perpendicular Lines

The Slope of a Line

In this section, we study lines and their equations. Throughout this text, we use the term **line** to mean a *straight* line.

The **slope** of a nonvertical line represents the number of units a line rises or falls vertically for each unit of horizontal change from left to right. For instance, consider the two points (x_1, y_1) and (x_2, y_2) on the line shown in Figure 3.32. As we move from left to right along this line, a change of $(y_2 - y_1)$ units in the vertical direction corresponds to a change of $(x_2 - x_1)$ units in the horizontal direction. That is,

$$y_2 - y_1 = \text{the change in } y$$

and

$$x_2 - x_1 = \text{the change in } x.$$

The slope of the line is given by the ratio of these two changes.

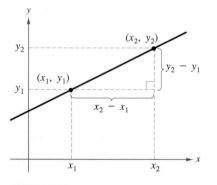

FIGURE 3.32

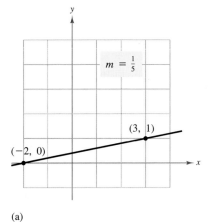

(a)

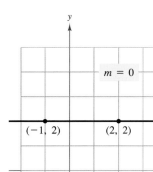

(b)

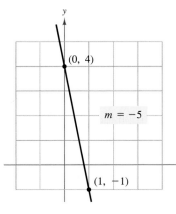

(c)

FIGURE 3.33

DEFINITION OF THE SLOPE OF A LINE

The **slope** m of the nonvertical line passing through the points (x_1, y_1) and (x_2, y_2) is

$$m = \frac{y_2 - y_1}{x_2 - x_1} = \frac{\text{change in } y}{\text{change in } x}$$

where $x_1 \neq x_2$.

When this formula is used, the *order of subtraction* is important. Given two points on a line, you are free to label either one of them as (x_1, y_1), and the other as (x_2, y_2). However, once this is done, you must form the numerator and denominator using the same order of subtraction.

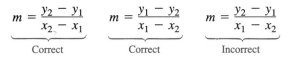

$$m = \frac{y_2 - y_1}{x_2 - x_1} \qquad m = \frac{y_1 - y_2}{x_1 - x_2} \qquad m = \frac{y_2 - y_1}{x_1 - x_2}$$

$\qquad\quad$ Correct $\qquad\qquad\quad$ Correct $\qquad\qquad$ Incorrect

EXAMPLE 1   **Finding the Slope of a Line Passing Through Two Points**

Find the slopes of the lines passing through the pairs of points.

a. $(-2, 0)$ and $(3, 1)$
b. $(-1, 2)$ and $(2, 2)$
c. $(0, 4)$ and $(1, -1)$

Solution

a. $m = \dfrac{y_2 - y_1}{x_2 - x_1}$ $\leftarrow$ *Difference in y-values*
$\qquad\qquad\qquad$ $\leftarrow$ *Difference in x-values*

$\qquad = \dfrac{1 - 0}{3 - (-2)}$

$\qquad = \dfrac{1}{3 + 2}$

$\qquad = \dfrac{1}{5}$

b. $m = \dfrac{2 - 2}{2 - (-1)} = \dfrac{0}{3} = 0$

c. $m = \dfrac{-1 - 4}{1 - 0} = \dfrac{-5}{1} = -5$

The graphs of the three lines are shown in Figure 3.33.

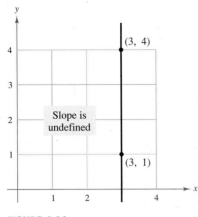

FIGURE 3.34

Note that the definition of slope does not apply to vertical lines. For instance, consider the points (3, 4) and (3, 1) on the vertical line shown in Figure 3.34. Applying the formula for slope,

$$m = \frac{4 - 1}{3 - 3}.$$ *Undefined division by zero*

Because division by zero is not defined, we do not define the slope of a vertical line.

From the slopes of the lines shown in Figures 3.33 and 3.34, you can make the following generalizations about the slope of a line.

1. A line with positive slope ($m > 0$) *rises* from left to right.
2. A line with negative slope ($m < 0$) *falls* from left to right.
3. A line with zero slope ($m = 0$) is *horizontal*.
4. A line with undefined slope is *vertical*.

Any two points on a line can be used to calculate its slope. This can be verified from the similar triangles shown in Figure 3.35. Recall that the ratios of corresponding sides of similar triangles are equal.

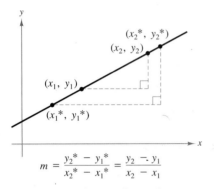

$$m = \frac{y_2{}^* - y_1{}^*}{x_2{}^* - x_1{}^*} = \frac{y_2 - y_1}{x_2 - x_1}$$

Any two points on a line can be used to determine the slope of the line.

FIGURE 3.35

The Point-Slope Form of the Equation of a Line

If you know the slope of a line *and* the coordinates of one point on the line, then you can find an equation for the line. For instance, in Figure 3.36, let (x_1, y_1) be a given point on the line whose slope is m. If (x, y) is any *other* point on the line, then it follows that

$$\frac{y - y_1}{x - x_1} = m.$$

This equation in variables x and y can be rewritten in the form

$$y - y_1 = m(x - x_1)$$

which is the **point-slope form** of the equation of a line.

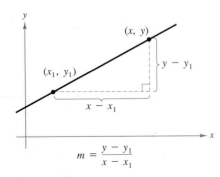

$$m = \frac{y - y_1}{x - x_1}$$

FIGURE 3.36

POINT-SLOPE FORM OF THE EQUATION OF A LINE

The **point-slope** form of the equation of the line that passes through the point (x_1, y_1) and has a slope of m is

$$y - y_1 = m(x - x_1).$$

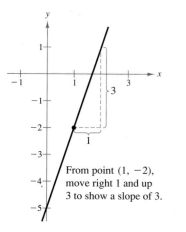

FIGURE 3.37

EXAMPLE 2 The Point-Slope Form of the Equation of a Line

Find an equation of the line that passes through the point $(1, -2)$ and has a slope of 3.

Solution

$$y - y_1 = m(x - x_1) \qquad \textit{Point-slope form}$$

$$y - (-2) = 3(x - 1) \qquad \textit{Substitute } y_1 = -2, \ x_1 = 1, \text{ and } m = 3$$

$$y + 2 = 3x - 3$$

$$y = 3x - 5 \qquad \textit{Equation of line}$$

The graph of this line is shown in Figure 3.37. ◀

The point-slope form can be used to find the equation of a line passing through two points (x_1, y_1) and (x_2, y_2). First, use the formula for the slope of the line passing through two points.

$$m = \frac{y_2 - y_1}{x_2 - x_1}$$

Then, once you know the slope, use the point-slope form to obtain the equation

$$y - y_1 = \frac{y_2 - y_1}{x_2 - x_1}(x - x_1).$$

This is sometimes called the **two-point form** of the equation of a line.

EXAMPLE 3 A Linear Model for Sales Prediction

During the first two quarters of the year, a company had total sales of $3.4 million and $3.7 million, respectively.

a. Write a linear equation giving the total sales y in terms of the quarter x.
b. Use the equation to predict the total sales during the fourth quarter.

Solution

a. In Figure 3.38 let $(1, 3.4)$ and $(2, 3.7)$ be two points on the line representing the total sales. The slope of the line passing through these two points is

$$m = \frac{3.7 - 3.4}{2 - 1} = 0.3.$$

By the point-slope form, the equation of the line is as follows.

$$y - y_1 = m(x - x_1)$$

$$y - 3.4 = 0.3(x - 1)$$

$$y = 0.3x - 0.3 + 3.4$$

$$y = 0.3x + 3.1$$

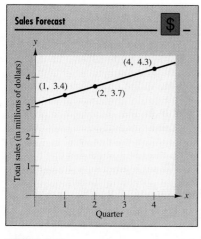

FIGURE 3.38

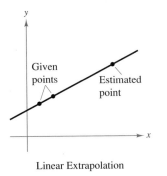

Linear Extrapolation

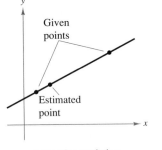

Linear Interpolation

FIGURE 3.39

b. Using the equation from part (a), you can estimate the fourth-quarter sales $(x = 4)$.

$$y = 0.3(4) + 3.1 = 4.3 \text{ million dollars}$$

The approximation method illustrated in Example 3 is **linear extrapolation.** Note in Figure 3.39 that for linear extrapolation, the estimated point lies to the right of the given points. When the estimated point lies *between* two given points, the procedure is **linear interpolation.**

Sketching Graphs of Lines

Many problems in coordinate (or analytic) geometry can be classified in two basic categories.

1. Given a graph (or parts of it), find its equation.
2. Given an equation, find its graph.

For lines, the first problem is solved easily by using the point-slope form. This formula, however, is not particularly useful for solving the second type of problem. The form that is better suited to graphing linear equations is the **slope-intercept form** of the equation of a line. To derive the slope-intercept form, we write the following.

$$y - y_1 = m(x - x_1) \qquad \textit{Point-slope form}$$
$$y = mx - mx_1 + y_1 \qquad \textit{Distributive Property}$$
$$y = mx + b \qquad b = -mx_1 + y_1, \textit{ a constant}$$

SLOPE-INTERCEPT FORM OF THE EQUATION OF A LINE

The graph of the equation

$$y = mx + b$$

is a line whose slope is m and y-intercept is $(0, b)$.

EXAMPLE 4   Using the Slope-Intercept Form

Sketch the graph of each of the linear equations.

a. $y = 2x + 1$ **b.** $y = 2$ **c.** $x + y = 2$

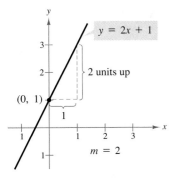

(a) When m is positive, the line rises.

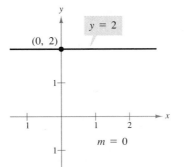

(b) When m is zero, the line is horizontal.

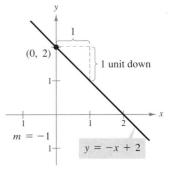

(c) When m is negative, the line falls.

FIGURE 3.40

Solution

a. Since $b = 1$, the y-intercept is $(0, 1)$. Moreover, since the slope is $m = 2$, this line *rises* two units for each unit the line moves to the right, as shown in Figure 3.40(a).

b. By writing the equation $y = 2$ in the form $y = (0)x + 2$, you see that the y-intercept is $(0, 2)$ and the slope is zero. A zero slope implies that the line is horizontal, as shown in Figure 3.40(b).

c. By writing the equation $x + y = 2$ in slope-intercept form, $y = -x + 2$, you see that the y-intercept is $(0, 2)$. Moreover, since the slope is $m = -1$, this line *falls* one unit for each unit the line moves to the right, as shown in Figure 3.40(c).

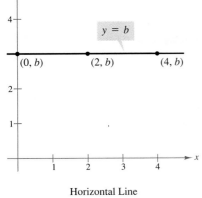

Horizontal Line

FIGURE 3.41

From the slope-intercept form of the equation of a line, you see that a horizontal line ($m = 0$) has an equation of the form

$$y = (0)x + b \qquad \text{or} \qquad y = b. \qquad \textit{Horizontal line}$$

This is consistent with the fact that each point on a horizontal line through $(0, b)$ has a y-coordinate of b, as shown in Figure 3.41.

Similarly, each point on a vertical line through $(a, 0)$ has an x-coordinate of a, as shown in Figure 3.42. Hence, a vertical line has an equation of the form

$$x = a. \qquad \textit{Vertical line}$$

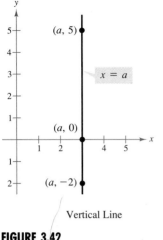

FIGURE 3.42

This equation cannot be written in the slope-intercept form because the slope of a vertical line is undefined. However, *every* line has an equation that can be written in the **general form**

$$Ax + By + C = 0 \qquad \textit{General form}$$

where A and B are not *both* zero. If $A = 0$ (and $B \neq 0$), the equation can be reduced to the form $y = b$, a horizontal line. If $B = 0$ (and $A \neq 0$), the general equation can be reduced to the form $x = a$, a vertical line.

SUMMARY OF EQUATIONS OF LINES

1. General form: $Ax + By + C = 0$
2. Vertical line: $x = a$
3. Horizontal line: $y = b$
4. Slope-intercept form: $y = mx + b$
5. Point-slope form: $y - y_1 = m(x - x_1)$

Parallel and Perpendicular Lines

PARALLEL LINES

Two distinct nonvertical lines are **parallel** if and only if their slopes are equal.

EXAMPLE 5   Equations of Parallel Lines

Find an equation of the line that passes through the point $(2, -1)$ and is parallel to the line $2x - 3y = 5$, as shown in Figure 3.43.

Solution

Write the given equation in slope-intercept form.

$$2x - 3y = 5 \qquad \textit{Given equation}$$
$$3y = 2x - 5$$
$$y = \frac{2}{3}x - \frac{5}{3} \qquad \textit{Slope-intercept form}$$

Therefore, the given line has a slope of $m = \frac{2}{3}$. Since any line parallel to the given line must also have a slope of $\frac{2}{3}$, the required line through $(2, -1)$ has the following equation.

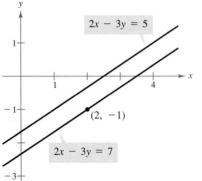

FIGURE 3.43

$$y - (-1) = \frac{2}{3}(x - 2)$$ *Point-slope form*

$$y = \frac{2}{3}x - \frac{4}{3} - 1$$

$$y = \frac{2}{3}x - \frac{7}{3}$$ *Slope-intercept form*

Notice the similarity between the slope-intercept form of the original equation and the slope-intercept form of the parallel equation.   ◢

PERPENDICULAR LINES

Two nonvertical lines are **perpendicular** if and only if their slopes are negative reciprocals of each other. That is,

$$m_1 = -\frac{1}{m_2}.$$

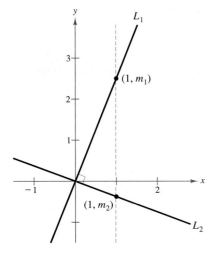

FIGURE 3.44

Proof

Recall that the phrase "if and only if" is a way of stating two rules in one. One rule says, "If two nonvertical lines are perpendicular, then their slopes must be negative reciprocals." The other rule is the converse, which says, "If two lines have slopes that are negative reciprocals, they must be perpendicular." We will prove the first of these two rules.

Assume that you are given two nonvertical perpendicular lines L_1 and L_2 with slopes m_1 and m_2. For simplicity's sake, let these two lines intersect at the origin, as shown in Figure 3.44. The vertical line $x = 1$ will intersect L_1 and L_2 at the respective points $(1, m_1)$ and $(1, m_2)$. Since L_1 and L_2 are perpendicular, the triangle formed by these two points and the origin is a right triangle. Thus, you can apply the Pythagorean Theorem and conclude that

$$\left(\begin{array}{c}\text{Distance between}\\(0, 0) \text{ and } (1, m_1)\end{array}\right)^2 + \left(\begin{array}{c}\text{Distance between}\\(0, 0) \text{ and } (1, m_2)\end{array}\right)^2 = \left(\begin{array}{c}\text{Distance between}\\(1, m_1) \text{ and } (1, m_2)\end{array}\right)^2.$$

Using the Distance Formula,

$$(\sqrt{1 + m_1{}^2})^2 + (\sqrt{1 + m_2{}^2})^2 = (\sqrt{0^2 + (m_1 - m_2)^2})^2$$
$$1 + m_1{}^2 + 1 + m_2{}^2 = (m_1 - m_2)^2$$
$$2 + m_1{}^2 + m_2{}^2 = m_1{}^2 - 2m_1m_2 + m_2{}^2$$
$$2 = -2m_1m_2$$
$$-1 = m_1m_2$$
$$-\frac{1}{m_2} = m_1.$$ ◢

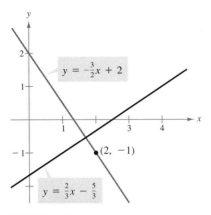

FIGURE 3.45

EXAMPLE 6   Equations of Perpendicular Lines

Find an equation of the line that passes through the point $(2, -1)$ and is perpendicular to the line $2x - 3y = 5$.

Solution

By writing the given line in the form $y = \frac{2}{3}x - \frac{5}{3}$ you see that the line has a slope of $\frac{2}{3}$. Hence, any line that is perpendicular to this line must have a slope of $-\frac{3}{2}$ (because $-\frac{3}{2}$ is the negative reciprocal of $\frac{2}{3}$). Therefore, the required line through the point $(2, -1)$ has the following equation:

$$y - (-1) = -\frac{3}{2}(x - 2) \qquad \textit{Point-slope form}$$

$$y = -\frac{3}{2}x + 3 - 1$$

$$y = -\frac{3}{2}x + 2. \qquad \textit{Slope-intercept form}$$

The graphs of both equations are shown in Figure 3.45.

DISCUSSION

PROBLEM

Linear Interpolation

Linear interpolation can be used to approximate the x-intercept of the graph of an equation. For example, consider the graph of $y = x^3 + x + 1$ as shown in Figure 3.46. When $x = -0.69$ the value of y is negative, and when $x = -0.68$ the value of y is positive. This implies that the graph must have an x-intercept whose x-coordinate is between -0.69 and -0.68. Write a short paragraph describing how you could use linear interpolation with the following two points on the graph of $y = x^3 + x + 1$ to approximate the x-intercept of the graph.

$$(-0.69, -0.018509) \qquad \text{and} \qquad (-0.68, 0.005568)$$

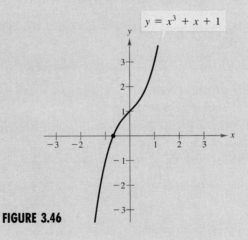

FIGURE 3.46

WARM UP

The following warm-up exercises involve skills that were covered in earlier sections. You will use these skills in the exercise set for this section.

In Exercises 1–4, simplify the expressions.

1. $\dfrac{4 - (-5)}{-3 - (-1)}$

2. $\dfrac{-5 - 8}{0 - (-3)}$

3. Find $\dfrac{-1}{m}$ for $m = \dfrac{4}{5}$.

4. Find $\dfrac{-1}{m}$ for $m = -2$.

In Exercises 5–10, solve for y in terms of x.

5. $2x - 3y = 5$

6. $4x + 2y = 0$

7. $y - (-4) = 3[x - (-1)]$

8. $y - 7 = \frac{2}{3}(x - 3)$

9. $y - (-1) = \dfrac{3 - (-1)}{2 - 4}(x - 4)$

10. $y - 5 = \dfrac{3 - 5}{0 - 2}(x - 2)$

EXERCISES for Section 3.3

In Exercises 1–6, estimate the slope of the line from its graph.

1.

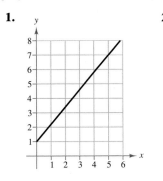

2.

3.

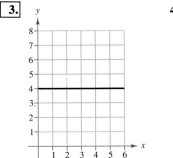

4.

5.

6.

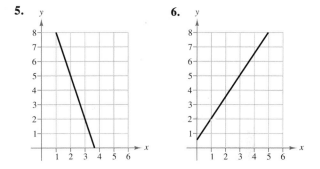

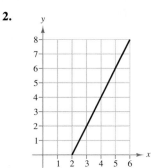

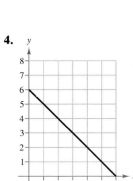

In Exercises 7 and 8, sketch the graph of the lines through the given point with the indicated slope. Make the sketches on the same set of coordinate axes.

	Point	*Slopes*
7.	$(2, 3)$	(a) 0 (b) 1 (c) 2 (d) -3
8.	$(-4, 1)$	(a) 3 (b) -3 (c) $\frac{1}{2}$ (d) Undefined

In Exercises 9–14, plot the points and find the slope of the line passing through each pair of points.

9. $(-3, -2), (1, 6)$

10. $(2, 4), (4, -4)$

11. $(-6, -1), (-6, 4)$

12. $(0, -10), (-4, 0)$

13. $(1, 2), (-2, -2)$

14. $\left(\frac{7}{8}, \frac{3}{4}\right), \left(\frac{5}{4}, -\frac{1}{4}\right)$

In Exercises 15–20, use the given point on the line and the slope of the line to find three additional points through which the line passes. (The solution is not unique.)

Point	Slope
15. (2, 1)	$m = 0$
16. (−4, 1)	m is undefined.
17. (5, −6)	$m = 1$
18. (10, −6)	$m = -1$
19. (−8, 1)	m is undefined.
20. (−3, −1)	$m = 0$

In Exercises 21–24, determine if the lines L_1 and L_2 passing through the pairs of points are parallel, perpendicular, or neither.

21. L_1: (0, −1), (5, 9)
$\quad L_2$: (0, 3), (4, 1)

22. L_1: (−2, −1), (1, 5)
$\quad L_2$: (1, 3), (5, −5)

23. L_1: (3, 6), (−6, 0)
$\quad L_2$: (0, −1), $\left(5, \frac{7}{3}\right)$

24. L_1: (4, 8), (−4, 2)
$\quad L_2$: (3, −5), $\left(-1, \frac{1}{3}\right)$

25. *Mountain Driving*   When driving down a mountain road, you notice signs warning of a "12% grade." This means that the slope of the road is $-\frac{12}{100}$. Determine the amount of horizontal change in your position if you note from elevation markers that you have descended 2000 feet vertically.

26. *Attic Height*   The "rise to run" in determining the steepness of the roof on a house is 3 to 4. Determine the maximum height in the attic of the house if the house is 30 feet wide (see figure).

Figure for 26

In Exercises 27–32, find the slope and y-intercept (if possible) of the line specified by the equation. Sketch a graph of the line.

27. $5x - y + 3 = 0$
28. $2x + 3y - 9 = 0$
29. $5x - 2 = 0$
30. $3y + 5 = 0$
31. $7x + 6y - 30 = 0$
32. $x - y - 10 = 0$

In Exercises 33–40, find an equation for the line passing through the points.

33. (5, −1), (−5, 5)
34. (4, 3), (−4, −4)
35. $\left(2, \frac{1}{2}\right), \left(\frac{1}{2}, \frac{5}{4}\right)$
36. (−1, 4), (6, 4)
37. (−8, 1), (−8, 7)
38. (1, 1), $\left(6, -\frac{2}{3}\right)$
39. (1, 0.6), (−2, −0.6)
40. (−8, 0.6), (2, −2.4)

In Exercises 41–50, find an equation of the line that passes through the given point and has the indicated slope. Sketch the graph of the line.

Point	Slope
41. (0, −2)	$m = 3$
42. (0, 10)	$m = -1$
43. (−3, 6)	$m = -2$
44. (0, 0)	$m = 4$
45. (4, 0)	$m = -\frac{1}{3}$
46. (−2, −5)	$m = \frac{3}{4}$
47. (6, −1)	m is undefined.
48. (−10, 4)	$m = 0$
49. $\left(4, \frac{5}{2}\right)$	$m = 0$
50. $\left(-\frac{1}{2}, \frac{3}{2}\right)$	m is undefined.

The **intercept** form of the equation of a line with intercepts $(a, 0)$ and $(0, b)$ is

$$\frac{x}{a} + \frac{y}{b} = 1, \qquad a \neq 0, b \neq 0.$$

In Exercises 51–56, use the intercept form to find the equation of the line with the given intercepts.

51. x-intercept: (2, 0)
$\quad y$-intercept: (0, 3)

52. x-intercept: (−3, 0)
$\quad y$-intercept: (0, 4)

53. x-intercept: $\left(-\frac{1}{6}, 0\right)$
$\quad y$-intercept: $\left(0, -\frac{2}{3}\right)$

54. x-intercept: $\left(\frac{2}{3}, 0\right)$
$\quad y$-intercept: (0, −2)

55. Point on line: (1, 2)
$\quad x$-intercept: $(a, 0)$
$\quad y$-intercept: (0, a)
$\quad (a \neq 0)$

56. Point on line: (−3, 4)
$\quad x$-intercept: $(a, 0)$
$\quad y$-intercept: (0, a)
$\quad (a \neq 0)$

In Exercises 57–62, write an equation of the line through the given point (a) parallel to the given line, and (b) perpendicular to the given line.

Point	Line
57. $(2, 1)$	$4x - 2y = 3$
58. $(-3, 2)$	$x + y = 7$
59. $(-6, 4)$	$3x + 4y = 7$
60. $\left(\frac{7}{8}, \frac{3}{4}\right)$	$5x + 3y = 0$
61. $(-1, 0)$	$y = -3$
62. $(2, 5)$	$x = 4$

In Exercises 63–66, you are given the dollar value of a product in 1990 *and* the rate at which the value of the item is expected to change during the next five years. Write a linear equation for the dollar value V of the product in terms of the year t. (Let $t = 0$ represent 1990.)

1990 Value	Rate
63. $2,540	$125 increase per year
64. $156	$4.50 increase per year
65. $20,400	$2,000 increase per year
66. $245,000	$5,600 increase per year

In Exercises 67–70, match the description with a graph. Determine the slope and how it is interpreted in the situation. [The graphs are labeled (a), (b), (c), and (d).]

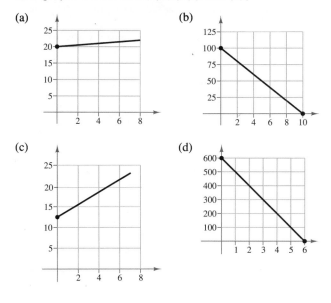

67. A person is paying $10 per week to a friend to repay a $100 loan.

68. An employee is paid $12.50 per hour plus $1.50 for each unit produced per hour.

69. A sales representative receives $20 per day for food plus $0.25 for each mile traveled.

70. A typewriter purchased for $600 depreciates $100 per year.

71. *Temperature* Find the equation of the line giving the relationship between the temperature in degrees Celsius, C, and degrees Fahrenheit, F. Remember that water freezes at 0° Celsius (32° Fahrenheit) and boils at 100° Celsius (212° Fahrenheit).

72. *Temperature* Use the result of Exercise 71 to complete the table.

C		−10°	10°			177°
F	0°			68°	90°	

73. *Annual Salary* Suppose your salary was $28,500 in 1990 and $32,900 in 1992. If your salary follows a linear growth pattern, what will it be in 1995?

74. *College Enrollment* A small college had 2546 students in 1990 and 2702 students in 1992. If the enrollment follows a linear growth pattern, how many students will the college have in 1997?

75. *Straight-Line Depreciation* A small business purchases a piece of equipment for $875. After five years the equipment will be outdated and have no value. Write a linear equation giving the value V of the equipment during the five years it will be used.

76. *Straight-Line Depreciation* A small business purchases a piece of equipment for $25,000. After 10 years the equipment will have to be replaced. Its value at that time is expected to be $2,000. Write a linear equation giving the value V of the equipment during the 10 years it will be used.

77. *Sales Price and List Price* A store is offering a 15% discount on all items in its inventory. Write a linear equation giving the sale price S for an item with a list price L.

78. *Hourly Wages* A manufacturer pays its assembly line workers $11.50 per hour plus $0.75 per unit produced. Write a linear equation for the hourly wages W in terms of the number of units x produced per hour.

79. *Sales Commission*   A salesperson receives a monthly salary of $2500 plus a commission of 7% of her sales. Write a linear equation for the salesperson's monthly wage W in terms of her monthly sales S.

80. *Daily Cost*   A sales representative using his personal car receives $120 per day for lodging and meals plus $0.26 per mile driven. Write a linear equation giving the daily cost C to the company in terms of x, the number of miles driven.

81. *Contracting Purchase*   A contractor purchases a piece of equipment for $36,500. The equipment requires an average expenditure of $5.25 per hour for fuel and maintenance, and the operator is paid $11.50 per hour.

(a) Write a linear equation giving the total cost C of operating this equipment for t hours. (Include the purchase cost for the equipment.)

(b) If customers are charged $27 per hour of machine use, write an equation for the revenue R derived from t hours of use.

(c) Use the formula for profit ($P = R - C$) to write an equation for the profit derived from t hours of use.

(d) *Break-even Point*   Use the result of part (c) to find the number of hours this equipment must be used to yield a profit of 0 dollars.

82. *Real Estate*   A real estate office handles an apartment complex with 50 units. When the rent per unit is $380 per month, all 50 units are occupied. However, when the rent is $425 per month, the average number of occupied units drops to 47. Assume that the relationship between the monthly rent p and the demand x is linear.

(a) Write the equation of the line giving the demand x in terms of the rent p.

(b) Use this equation to predict the number of units occupied if the rent is $455.

(c) Predict the number of units occupied if the rent is $395.

83. *Baseball Salaries*   The average annual salaries of major league baseball players (in 1000s of dollars) from 1979 to 1989 are shown in the scatter plot. Find the equation of the line that you think best fits this data. (Let y represent the average salary and let t represent the year with $t = 0$ corresponding to 1980.)(*Source:* Major League Baseball)

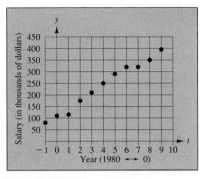

Figure for 83

84. *Quiz and Test Scores*   A mathematics instructor gives regular 20-point quizzes and 100-point exams. Average scores for six students, given as ordered pairs (x, y) where x is the average quiz score and y is the average test score, are $(18, 87), (10, 55), (19, 96), (16, 79), (13, 76)$ and $(15, 82)$.

(a) Plot the points.

(b) Use a straight edge to sketch the "best-fitting" line through the points.

(c) Find an equation for the line sketched in part (b).

(d) Use the equation of part (c) to estimate the average test score for a person with an average quiz score of 17.

[*Note:* The answers are not unique for parts (b)–(d).]

(e) If the instructor added four points to the average test score of everyone in the class, describe the change in the position of the plotted points and the change in the equation of the line.

3.4 Functions

Introduction to Functions / Function Notation / Finding the Domain of a Function /
Applications

Introduction to Functions

Many everyday phenomena involve two quantities that are related to each
other by some rule of correspondence. For example, the simple interest I
earned on \$1000 for one year is related to the annual percentage rate r by
the formula $I = 1000r$, and the distance d traveled on a bicycle in two hours
is related to the speed s of the bicycle by the formula $d = 2s$.

Not all correspondences between two quantities have simple mathematical
formulas. For instance, we commonly match up quantities such as NFL
starting quarterbacks with touchdown passes and days of the year with the
Dow-Jones Industrial Average. In both of these cases, however, there is some
rule of correspondence that matches each item from one set with exactly one
item from a different set. Such a rule of correspondence is a **function.**

DEFINITION OF A FUNCTION

A **function** f from a set A to a set B is a rule of correspondence that
assigns to each element x in the set A exactly one element y in the
set B. The set A is the **domain** (or set of inputs) of the function f,
and the set B contains the **range** (or set of outputs).

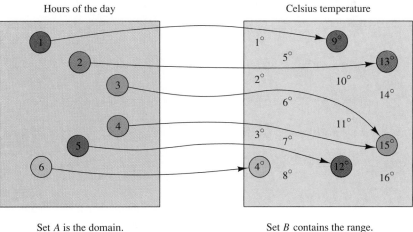

Set A is the domain.
Input: 1, 2, 3, 4, 5, 6

Set B contains the range.
Output: $4°, 9°, 12°, 13°, 15°$

Function from Set A to Set B

FIGURE 3.47

The function in Figure 3.47 can be represented by the following set of ordered pairs.

$$\{(1, 9°), (2, 13°), (3, 15°), (4, 15°), (5, 12°), (6, 4°)\}$$

In each ordered pair, the first coordinate is the input and the second coordinate is the output. From this set and Figure 3.47, note the characteristics of a function.

1. Each element in A must be matched with an element of B.
2. Some elements in B may not be matched with any element in A.
3. Two or more elements of A may be matched with the same element of B.

The converse of the third statement is not true. That is, an element of A (the domain) cannot be matched with two different elements of B.

EXAMPLE 1   Testing for Functions

Let $A = \{a, b, c\}$ and $B = \{1, 2, 3, 4, 5\}$. Does the set of ordered pairs or figures represent a function from set A to set B?

a. $\{(a, 2), (b, 3), (c, 4)\}$ **b.** $\{(a, 4), (b, 5)\}$

c.   **d.**

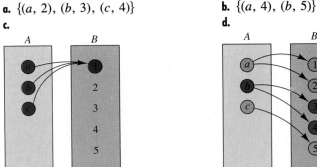

Solution

a. Yes, because each element of A is matched with exactly one element of B.
b. No, because not all elements of A are matched with an element of B.
c. Yes, because it does not matter that each element of A is matched with the same element of B.
d. No, because the element a in A is matched with *two* elements, 1 and 2, in B. This is also true of the element b.

Representing functions by sets of ordered pairs is a common practice in *discrete mathematics*. In algebra, however, it is more common to represent functions by equations or formulas involving two variables. For instance, the equation

$$y = x^2$$

represents the variable y as a function of the variable x. Here, x is the **independent variable** and y is the **dependent variable.** The **domain** of the function is the set of all values taken on by the independent variable x, and the **range** of the function is the set of all values taken on by the dependent variable y.

EXAMPLE 2   Testing for Functions Represented by Equations

Determine whether the equations represent y as a function of x.

a. $x^2 + y = 1$ **b.** $-x + y^2 = 1$

Solution

In each case, to determine whether y is a function of x, it is helpful to solve for y in terms of x.

a. $x^2 + y = 1$ *Given equation*

   $y = 1 - x^2$ *Solve for y*

To each value of x there corresponds just one value for y. Therefore, y is a function of x.

b. $-x + y^2 = 1$ *Given equation*

   $y^2 = 1 + x$ *Add x to both sides*

   $y = \pm\sqrt{1 + x}$ *Solve for y*

The $\pm$ indicates that to a given value of x there correspond two values for y. Therefore, y is *not* a function of x.

Function Notation

When using an equation to represent a function, it is convenient to name the function so that it can be referenced easily. For example, you know that the equation $y = 1 - x^2$, Example 2(a), describes y as a function of x. Suppose you give this function the name "f." Then you can use **function notation.**

Input	*Output*	*Equation*
x	$f(x)$	$f(x) = 1 - x^2$

The symbol $f(x)$ is read as the **value of f at x** or simply "f of x." This corresponds to the y-value for a given x. Thus, you can write $y = f(x)$.

Keep in mind that f is the *name* of the function, while $f(x)$ is the *value* of the function at x. For instance, the function

$$f(x) = 3 - 2x$$

has *function values* denoted by $f(-1)$, $f(0)$, $f(2)$, and so on. To find these values, substitute the specified input values into the equation.

For $x = -1$, $\quad f(-1) = 3 - 2(-1) = 3 + 2 = 5.$

For $x = 0$, $\qquad f(0) = 3 - 2(0) = 3 + 0 = 3.$

For $x = 2$, $\qquad f(2) = 3 - 2(2) = 3 - 4 = -1.$

Although f is generally used as a convenient function name and x as the independent variable, you can use other letters. For instance, $f(x) = x^2 - 4x + 7$, $f(t) = t^2 - 4t + 7$, and $f(s) = s^2 - 4s + 7$ all define the same function. In fact, the role of the independent variable in a function is simply that of a "placeholder." Consequently, the above function could be described by the form

$$f(\blacksquare) = (\blacksquare)^2 - 4(\blacksquare) + 7$$

where the parentheses are used in place of a letter. To evaluate $f(-2)$, simply place -2 in each set of parentheses.

$f(\blacksquare) = (\blacksquare)^2 - 4(\blacksquare) + 7$

$f(-2) = (-2)^2 - 4(-2) + 7$ $\quad$ *Place -2 in each set of parentheses*

$\qquad = 4 + 8 + 7$ $\qquad\qquad$ *Evaluate each term*

$\qquad = 19$ $\qquad\qquad\qquad$ *Simplify*

Similarly, the value of $f(3x)$ is obtained as follows.

$f(\blacksquare) = (\blacksquare)^2 - 4(\blacksquare) + 7$

$f(3x) = (3x)^2 - 4(3x) + 7$ $\quad$ *Place $3x$ in each set of parentheses*

$\qquad = 9x^2 - 12x + 7$ $\qquad$ *Simplify*

EXAMPLE 3 Evaluating a Function

Let $g(x) = -x^2 + 4x + 1$ and find the following.

a. $g(2)$ $\qquad$ **b.** $g(t)$ $\qquad$ **c.** $g(x + 2)$

Solution

a. Replacing x with 2 in $g(x) = -x^2 + 4x + 1$ yields

$$g(2) = -(2)^2 + 4(2) + 1 = -4 + 8 + 1 = 5.$$

b. Replacing x with t yields

$$g(t) = -(t)^2 + 4(t) + 1 = -t^2 + 4t + 1.$$

REMARK Example 3 shows that $g(x + 2) \neq g(x) + g(2)$ because $-x^2 + 5 \neq (-x^2 + 4x + 1) + 5$. In general, $g(u + v) \neq g(u) + g(v)$.

c. Replacing x with $x + 2$ yields

$$g(x + 2) = -(x + 2)^2 + 4(x + 2) + 1$$
$$= -(x^2 + 4x + 4) + 4x + 8 + 1$$
$$= -x^2 - 4x - 4 + 4x + 8 + 1$$
$$= -x^2 + 5.$$

Sometimes a function is defined using more than one equation.

EXAMPLE 4 A Function Defined by Two Equations

Evaluate the function

$$f(x) = \begin{cases} x^2 + 1, & x < 0 \\ x - 1, & x \geq 0 \end{cases}$$

at $x = -1$, 0, and 1.

Solution

Since $x = -1$ is less than 0, use $f(x) = x^2 + 1$ to obtain

$$f(-1) = (-1)^2 + 1 = 2.$$

For $x = 0$, use $f(x) = x - 1$ to obtain

$$f(0) = (0) - 1 = -1.$$

For $x = 1$, use $f(x) = x - 1$ to obtain

$$f(1) = (1) - 1 = 0.$$

Finding the Domain of a Function

The domain of a function may be explicitly described along with the function, or it may be *implied* by the expression used to define the function. The **implied domain** is the set of all real numbers for which the expression is defined. For instance, the function

$$f(x) = \frac{1}{x^2 - 4}$$

has an implied domain that consists of all real x other than $x = \pm 2$. These two values are excluded from the domain because division by zero is undefined. Another common type of implied domain is used to avoid even roots of negative numbers. For example, the function

$$f(x) = \sqrt{x}$$

is defined only for $x \geq 0$. Hence, its implied domain is the interval $[0, \infty)$. In general, the domain of a function *excludes* values that would cause division by zero *or* result in the even root of a negative number.

The *range* of a function is more difficult to find, and can best be obtained from the graph of the function (see Section 3.5).

EXAMPLE 5   Finding the Domain of a Function

Find the domain of each of the functions.

a. f: $\{(-3, 0), (-1, 4), (0, 2), (2, 2), (4, -1)\}$

b. Volume of a sphere: $V = \dfrac{4}{3} \pi r^3$

REMARK   In Example 5(b), note that the domain of a function may be implied by the physical context. For instance, from the equation $V = \frac{4}{3}\pi r^3$, you would have no reason to restrict r to nonnegative values, but the physical context tells you that a sphere cannot have a negative radius.

c. $g(x) = \dfrac{1}{x + 5}$

d. $h(x) = \sqrt{4 - x^2}$

Solution

a. The domain of f consists of all first coordinates in the set of ordered pairs, and is therefore the set

$$\text{Domain} = \{-3, -1, 0, 2, 4\}.$$

b. For the volume of a sphere you must choose nonnegative values for the radius r. Thus, the domain is the set of all real numbers r such that $r \geq 0$.

c. Excluding x-values that yield zero in the denominator, the domain of g is the set of all real numbers $x \neq -5$.

d. Choose x-values for which $4 - x^2 \geq 0$. Using the methods described in Section 2.7, you can conclude that $-2 \leq x \leq 2$. Thus, the domain is the interval $[-2, 2]$.

Applications

EXAMPLE 6   The Dimensions of a Container

A standard soft-drink can has a height of about 4.75 inches and a radius of about 1.3 inches. For this standard can, the ratio of the height to the radius is about 3.65. Suppose you worked in the marketing department of a soft-drink company and were experimenting with a new soft-drink can that was slightly narrower and taller. For your experimental can, the ratio of the height to the radius is 4, as shown in Figure 3.48.

a. Express the volume of the can as a function of the radius r.

b. Express the volume of the can as a function of the height h.

$$\frac{h}{r} = 4$$

FIGURE 3.48

Solution

The volume of a right circular cylinder is

$$V = \pi(\text{radius})^2(\text{height}) = \pi r^2 h.$$

Since the ratio of the height to the radius is 4, you can write $h = 4r$.

a. To write the volume as a function of the radius, use the fact that $h = 4r$.

$$V = \pi r^2 h = \pi r^2(4r) = 4\pi r^3$$

b. To write the volume as a function of the height, use the fact that $r = h/4$.

$$V = \pi\left(\frac{h}{4}\right)^2 h = \frac{\pi h^3}{16}$$

EXAMPLE 7   The Path of a Baseball

A baseball is hit 3 feet above ground at a velocity of 100 feet per second and at an angle of 45° with respect to the ground. The path of the baseball is given by the function

$$y = -0.0032x^2 + x + 3$$

where y and x are measured in feet, as shown in Figure 3.49. (From this equation, note that the height of the baseball is a function of the distance from home plate.) Will the baseball clear a 10-foot fence located 300 feet from home plate?

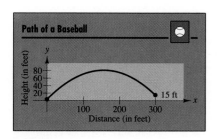

FIGURE 3.49

Solution

When $x = 300$, the height of the baseball is

$$y = -0.0032(300^2) + 300 + 3 = 15 \text{ feet.}$$

Thus, the ball will clear the fence.

One of the basic definitions in calculus employs the ratio

$$\frac{f(x + h) - f(x)}{h}, \quad h \neq 0$$

called a **difference quotient.**

EXAMPLE 8   Evaluating a Difference Quotient

For the function given by $f(x) = x^2 - 4x + 7$, find

$$\frac{f(x + h) - f(x)}{h}.$$

Solution

$$\frac{f(x + h) - f(x)}{h} = \frac{[(x + h)^2 - 4(x + h) + 7] - [x^2 - 4x + 7]}{h}$$

$$= \frac{x^2 + 2xh + h^2 - 4x - 4h + 7 - x^2 + 4x - 7}{h}$$

$$= \frac{2xh + h^2 - 4h}{h}$$

$$= \frac{h(2x + h - 4)}{h}$$

$$= 2x + h - 4, \; h \neq 0$$

Summary of Function Terminology

Function:

A **function** is a relationship between two variables such that to each value of the independent variable there corresponds exactly one value of the dependent variable.

Function Notation: $y = f(x)$

f is the **name** of the function.
y is the **dependent variable.**
x is the **independent variable.**
$f(x)$ is the **value of the function at x.**

Domain:

The **domain** of a function is the set of all values (inputs) of the independent variable for which the function is defined. If x is in the domain of f, f is **defined** at x. If x is not in the domain of f, f is **undefined** at x.

Range:

The **range** of a function is the set of all values (outputs) assumed by the dependent variable (that is, the set of all function values).

Implied Domain:

If f is defined by an algebraic expression and the domain is not specified, then the **implied domain** consists of all real numbers for which the expression is defined.

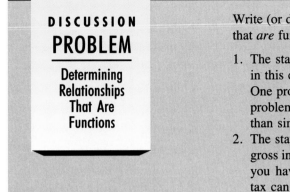

DISCUSSION
PROBLEM

**Determining
Relationships
That Are
Functions**

Write (or discuss) two statements describing relationships in everyday life that *are* functions and two that are *not* functions. Here are two examples.

1. The statement "Your happiness is a function of the grade you receive in this course" is *not* a correct mathematical use of the word function. One problem is that the word "happiness" is ambiguous, and the other problem is that a person's happiness is affected by many more events than simply the grade the person receives in a single course.

2. The statement "Your federal income tax is a function of your adjusted gross income" *is* a correct mathematical use of the word function. Once you have determined your adjusted gross income, then your income tax can be determined.

WARM UP

The following warm-up exercises involve skills that were covered in earlier sections. You will use these skills in the exercise set for this section.

In Exercises 1–4, simplify the expression.

1. $2(-3)^3 + 4(-3) - 7$ **2.** $4(-1)^2 - 5(-1) + 4$

3. $(x + 1)^2 + 3(x + 1) - 4 - (x^2 + 3x - 4)$

4. $(x - 2)^2 - 4(x - 2) - (x^2 - 4)$

In Exercises 5 and 6, solve for y in terms of x.

5. $2x + 5y - 7 = 0$ **6.** $y^2 = x^2$

In Exercises 7–10, solve the inequality.

7. $x^2 - 4 \geq 0$ **8.** $9 - x^2 \geq 0$

9. $x^2 + 2x + 1 \geq 0$ **10.** $x^2 - 3x + 2 \geq 0$

EXERCISES for Section 3.4

In Exercises 1 and 2, determine which of the sets of ordered pairs represents a function from A to B. Give reasons for your answers.

1. $A = \{0, 1, 2, 3\}$ and $B = \{-2, -1, 0, 1, 2\}$
   (a) $\{(0, 1), (1, -2), (2, 0), (3, 2)\}$
   (b) $\{(0, -1), (2, 2), (1, -2), (3, 0), (1, 1)\}$
   (c) $\{(0, 0), (1, 0), (2, 0), (3, 0)\}$
   (d) $\{(0, 2), (3, 0), (1, 1)\}$

2. $A = \{a, b, c\}$ and $B = \{0, 1, 2, 3\}$
   (a) $\{(a, 1), (c, 2), (c, 3), (b, 3)\}$
   (b) $\{(a, 1), (b, 2), (c, 3)\}$
   (c) $\{(1, a), (0, a), (2, c), (3, b)\}$
   (d) $\{(c, 0), (b, 0), (a, 3)\}$

In Exercises 3–10, identify the equations that determine y as a function of x.

3. $x^2 + y^2 = 4$
4. $x = y^2$
5. $x^2 + y = 4$
6. $x + y^2 = 4$
7. $2x + 3y = 4$
8. $x^2 + y^2 - 2x - 4y + 1 = 0$
9. $y^2 = x^2 - 1$
10. $y = \sqrt{x + 5}$

In Exercises 11 and 12, fill in the blanks using the specified function and the value of the independent variable. (The symbol Δx represents a single variable and is read "delta x." This symbol is commonly used in calculus to denote a small change in x.)

11. $f(s) = \dfrac{1}{s + 1}$

    (a) $f(4) = \dfrac{1}{(\ \ ) + 1}$

    (b) $f(0) = \dfrac{1}{(\ \ ) + 1}$

    (c) $f(4x) = \dfrac{1}{(\ \ ) + 1}$

    (d) $f(x + h) = \dfrac{1}{(\ \ ) + 1}$

12. $g(x) = x^2 - 2x$
    (a) $g(2) = (\ \ )^2 - 2(\ \ )$
    (b) $g(-3) = (\ \ )^2 - 2(\ \ )$
    (c) $g(t + 1) = (\ \ )^2 - 2(\ \ )$
    (d) $g(x + \Delta x) = (\ \ )^2 - 2(\ \ )$

In Exercises 13–24, evaluate the function at the specified value of the independent variable and simplify.

13. $f(x) = 2x - 3$
    (a) $f(1)$ (b) $f(-3)$ (c) $f(x - 1)$

14. $g(y) = 7 - 3y$
    (a) $g(0)$ (b) $g\left(\frac{7}{3}\right)$ (c) $g(s + 2)$

15. $h(t) = t^2 - 2t$
    (a) $h(2)$ (b) $h(1.5)$ (c) $h(x + 2)$

16. $V(r) = \frac{4}{3}\pi r^3$
    (a) $V(3)$ (b) $V\left(\frac{3}{2}\right)$ (c) $V(2r)$

17. $f(y) = 3 - \sqrt{y}$
    (a) $f(4)$ (b) $f(0.25)$ (c) $f(4x^2)$

18. $f(x) = \sqrt{x + 8} + 2$
    (a) $f(-8)$ (b) $f(1)$ (c) $f(x - 8)$

19. $q(x) = \dfrac{1}{x^2 - 9}$
    (a) $q(0)$ (b) $q(3)$ (c) $q(y + 3)$

20. $q(t) = \dfrac{2t^2 + 3}{t^2}$
    (a) $q(2)$ (b) $q(0)$ (c) $q(-x)$

21. $f(x) = \dfrac{|x|}{x}$
    (a) $f(2)$ (b) $f(-2)$ (c) $f(x - 1)$

22. $f(x) = |x| + 4$
    (a) $f(2)$ (b) $f(-2)$ (c) $f(x^2)$

23. $f(x) = \begin{cases} 2x + 1, & x < 0 \\ 2x + 2, & x \geq 0 \end{cases}$
    (a) $f(-1)$ (b) $f(0)$ (c) $f(2)$

24. $f(x) = \begin{cases} x^2 + 2, & x \leq 1 \\ 2x^2 + 2, & x > 1 \end{cases}$
    (a) $f(-2)$ (b) $f(1)$ (c) $f(2)$

In Exercises 25–28, find all real values of x such that $f(x) = 0$.

25. $f(x) = 15 - 3x$

26. $f(x) = \dfrac{3x - 4}{5}$

27. $f(x) = x^2 - 9$

28. $f(x) = x^3 - x$

In Exercises 29–38, find the domain of the function.

29. $f(x) = 5x^2 + 2x - 1$

30. $g(x) = 1 - 2x^2$

31. $h(t) = \dfrac{4}{t}$

32. $s(y) = \dfrac{3y}{y + 5}$

33. $g(y) = \sqrt{y - 10}$

34. $f(t) = \sqrt[3]{t + 4}$

35. $f(x) = \sqrt[4]{1 - x^2}$

36. $h(x) = \dfrac{10}{x^2 - 2x}$

37. $g(x) = \dfrac{1}{x} - \dfrac{3}{x + 2}$

38. $f(s) = \dfrac{\sqrt{s - 1}}{s - 4}$

In Exercises 39–42, assume that the domain of f is the set $A = \{-2, -1, 0, 1, 2\}$. Determine the set of ordered pairs representing the function f.

39. $f(x) = x^2$

40. $f(x) = \dfrac{2x}{x^2 + 1}$

41. $f(x) = \sqrt{x + 2}$

42. $f(x) = |x + 1|$

In Exercises 43–46, find the value(s) of x for which $f(x) = g(x)$.

43. $f(x) = x^2$, $g(x) = x + 2$

44. $f(x) = x^2 + 2x + 1$, $g(x) = 3x + 3$

45. $f(x) = \sqrt{3x} + 1$, $g(x) = x + 1$

46. $f(x) = x^4 - 2x^2$, $g(x) = 2x^2$

In Exercises 47–52, find the indicated difference quotient and simplify your answer.

47. $f(x) = x^2 - x + 1$
$$\dfrac{f(2 + h) - f(2)}{h}$$

48. $f(x) = 5x - x^2$
$$\dfrac{f(5 + h) - f(5)}{h}$$

49. $f(x) = x^3$
$$\dfrac{f(x + \Delta x) - f(x)}{\Delta x}$$

50. $f(x) = 2x$
$$\dfrac{f(x + \Delta x) - f(x)}{\Delta x}$$

51. $g(x) = 3x - 1$
$$\dfrac{g(x) - g(3)}{x - 3}$$

52. $f(t) = \dfrac{1}{t}$
$$\dfrac{f(t) - f(1)}{t - 1}$$

53. *Area of a Circle*   Express the area A of a circle as a function of its circumference C.

54. *Area of a Triangle*   Express the area A of an equilateral triangle as a function of the length s of its sides.

55. *Area of a Triangle*   A right triangle is formed in the first quadrant by the x- and y-axes and a line through the point $(1, 2)$ (see figure). Write the area of the triangle as a function of x, and determine the domain of the function.

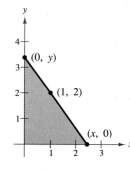

Figure for 55

56. *Area of a Rectangle*   A rectangle is bounded by the *x*-axis and the semicircle $y = \sqrt{25 - x^2}$ (see figure). Write the area of the rectangle as a function of *x*, and determine the domain of the function.

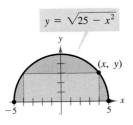

$y = \sqrt{25 - x^2}$

(x, y)

Figure for 56

57. *Volume of a Box*   An open box is made from a square piece of material 12 inches on a side by cutting equal squares from each corner and turning up the sides (see figure). Write the volume *V* of the box as a function of *x*. What is the domain of this function?

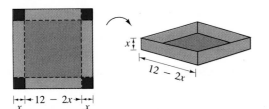

$12 - 2x$

x $12 - 2x$ x

Figure for 57

58. *Volume of a Package*   A rectangular package to be sent by a postal service can have a maximum combined length and girth (perimeter of a cross section) of 108 inches (see figure). Write the volume of such a package as a function of *x*. What is the domain of the function?

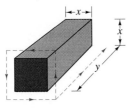

x

x

y

Figure for 58

59. *Height of a Balloon*   A balloon carrying a transmitter ascends vertically from a point 2000 feet from the receiving station (see figure). Let *d* be the distance between the balloon and the receiving station. Express the height of the balloon as a function of *d*. What is the domain of the function?

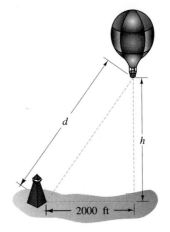

d

h

2000 ft

Figure for 59

60. *Price of Mobile Homes*   The average price of a new mobile home in the United States from 1974 to 1988 can be approximated by the model

$$p(t) = \begin{cases} 19.504 + 1.754t, & -6 \le t \le -1 \\ 19.839 + 0.081t^2, & 0 \le t \le 8 \end{cases}$$

where *p* is the price in thousands of dollars and *t* is the year with $t = 0$ corresponding to 1980 (see figure). Use this model to find the average price of a mobile home in 1978 and 1988. (*Source:* U.S. Bureau of Census, Construction Reports)

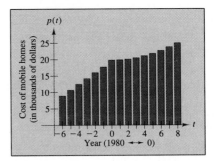

$p(t)$

Cost of mobile homes (in thousands of dollars)

25
20
15
10
5

−6 −4 −2   0   2   4   6   8   *t*

Year (1980 ← 0)

Figure for 60

61. *Cost, Revenue, and Profit* A company produces a product for which the variable cost is $12.30 per unit and the fixed costs are $98,000. The product sells for $17.98. Let x be the number of units produced and sold.
 (a) Write the total cost C as a function of the number of units produced.
 (b) Write the revenue R as a function of the number of units sold.
 (c) Write the profit P as a function of the number of units sold. (*Note: $P = R - C$.*)

62. The inventor of a game believes that the variable cost for producing the game is $0.95 per unit and the fixed costs are $6000. The inventor sells each game for $1.69. Let x be the number of games sold.
 (a) Write the total cost C as a function of the number of games sold.
 (b) Write the average cost per unit $\overline{C} = C/x$ as a function of x.

63. *Charter Bus Fares* For groups of 80 or more people, a charter bus company determines the rate per person according to the formula

Rate $= 8 - 0.05(n - 80)$, $n \geq 80$

where the rate is given in dollars and n is the number of people.
 (a) Express the revenue R for the bus company as a function of n.
 (b) Use the function from part (a) to complete the table.

n	90	100	110	120	130	140	150
$R(n)$							

64. *Fluid Force* The force F (in tons) of water against the face of a dam is the function

$$F(y) = 149.76\sqrt{10}y^{5/2}$$

where y is the depth of the water in feet. Complete the table.

y	5	10	20	30	40
$F(y)$					

SOLVING

A Graphical Approach to Finding a Maximum Revenue and Maximum Profit

In business, the **demand function** gives the price per unit p in terms of the number of units sold x. The demand function whose graph is shown below is

$$p = 40 - 5x^2, \quad 0 \le x \le \sqrt{8} \qquad \text{\textit{Demand function}}$$

where x is measured in millions of units. Note that as the price decreases, the number of units sold increases.

The **revenue** R (in millions of dollars) is determined by multiplying the number of units sold by the price per unit. Thus,

$$R = xp = x(40 - 5x^2), \quad 0 \le x \le \sqrt{8}. \qquad \text{\textit{Revenue function}}$$

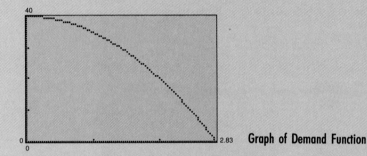

Graph of Demand Function

EXAMPLE 1   Finding the Maximum Revenue

Use a graphing utility to sketch the graph of the revenue function

$$R = 40x - 5x^3, \quad 0 \le x \le \sqrt{8}.$$

How many units should be sold to obtain a maximum revenue? What price per unit should be charged to obtain a maximum revenue?

Solution

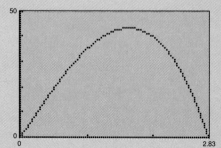

Graph of Revenue Function

To begin, you need to determine a viewing rectangle that will display the part of the graph that is important to this problem. The domain is given, so you can set the x boundaries of the graph between 0 and $\sqrt{8}$. To determine the y-boundaries, however, you need to experiment a little. After calculating several values of R, you could decide to use y boundaries between 0 and 50, as shown in the graph at left. Next, you can use the trace key to find that the maximum revenue of about \$43.5 million occurs when x is approximately 1.64 million units.

To find the price per unit that corresponds to this maximum revenue, you can substitute $x = 1.64$ into the demand function to obtain

$$p = 40 - 5(1.64)^2 \approx \$26.55.$$

EXAMPLE 2   Finding the Maximum Profit

Suppose the cost of producing each unit whose revenue function is discussed in Example 1 is $15. How many units should be sold to obtain a maximum profit? What price per unit should be charged to obtain a maximum profit?

Solution

The total cost C (in millions of dollars) of producing x million units is $C = 15x$. This implies that the profit P (in millions of dollars) obtained by selling x million units is

$$P = R - C$$
$$= (40x - 5x^3) - 15x$$
$$= -5x^3 + 25x.$$

As in Example 1, you can use a graphing utility to sketch the graph of this function. From the graph, you can approximate that the maximum profit of about $21.5 million occurs when x is approximately 1.28 million units. The price per unit that corresponds to this maximum profit is

$$p = 40 - 5(1.28)^2 = \$31.80.$$

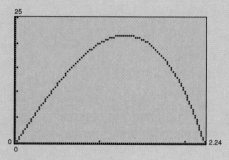

Graph of Profit Function

EXERCISES

(See also: Exercise 61, Section 3.4; Exercise 59, Section 3.5)

1. *Can't Give It Away!*   For the demand function $p = 40 - 5x^2$, match the points (0, 40) and ($\sqrt{8}$, 0) with statement (a) or statement (b). Explain your reasoning.
   (a) At this price, no one is willing to buy the product.
   (b) You can't *give* more than this number of units away.

2. *Exploration*   Use a graphing utility to *zoom in* on the maximum point of the revenue function in Example 1. (Use a setting of $1.62 \leq x \leq 1.65$ and $43.5 \leq y \leq 43.6$.) Use the trace feature of the graphing utility to improve the accuracy of the approximation obtained in Example 1. Do you think this improved accuracy is appropriate to the context of this particular problem? (Did it change the price?)

3. *Exploration*   Find a setting that allows you to graphically improve the accuracy of the solution in Example 2.

4. *Exploration*   In Example 2, suppose that, in addition to the cost of $15 per unit, there is an initial cost of $250,000. How does this change the profit function? Does this affect the *price* that corresponds to a maximum profit? Does it affect the *amount* of the maximum profit? Explain.

5. *Maximum Volume of a Box*   In the Problem Solving with Technology feature on page 39, you were asked to maximize the volume of a box that has a square base and a surface area of 216 square inches. In that problem, x represents the length (in inches) of each side of the base and

   $$V = 54x - \tfrac{1}{2}x^3$$

   represents the volume (in cubic inches). Use a graphing calculator to approximate the dimensions that produce a box of maximum volume. (Use a graph setting that yields an accuracy of 0.0001.)

203

3.5 Graphs of Functions

The Graph of a Function / Increasing and Decreasing Functions / Even and Odd Functions / Summary of Graphs of Common Functions / Shifting, Reflecting, and Stretching Graphs / Step Functions

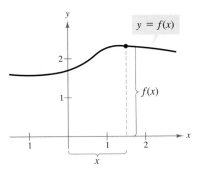

FIGURE 3.50

The Graph of a Function

In Section 3.4 we discussed functions from an algebraic point of view. Here, we look at functions from a geometric perspective. The **graph of a function** f is the collection of ordered pairs $(x, f(x))$ such that x is in the domain of f. As you study this section, remember that

$$x = \text{the directed distance from the } y\text{-axis}$$
$$f(x) = \text{the directed distance from the } x\text{-axis}$$

as shown in Figure 3.50.

You learned in Section 3.4 that the *range* (the set of values assumed by the dependent variable) of a function is often more easily determined from the graph of the function.

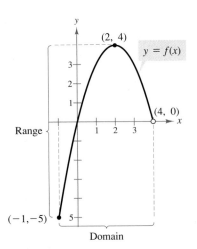

FIGURE 3.51

REMARK The use of dots (open or closed) at the extreme left and right points of a graph indicates that the graph does not extend beyond these points. If no such dots are shown, assume that the graph extends beyond these points.

EXAMPLE 1   **Finding Domain and Range from the Graph of a Function**

Use the graph of the function f, shown in Figure 3.51, to find the following.

a. The domain of f
b. The function values $f(-1)$ and $f(2)$
c. The range of f

Solution

a. The closed dot (on the left) indicates that $x = -1$ is in the domain of f, whereas the open dot (on the right) indicates $x = 4$ is not in the domain. Thus, the domain of f is all x in the interval $[-1, 4)$.
b. Since $(-1, -5)$ is a point on the graph of f, it follows that $f(-1) = -5$. Similarly, since $(2, 4)$ is a point on the graph of f, it follows that $f(2) = 4$.
c. Because the graph does not extend below $f(-1) = -5$ nor above $f(2) = 4$, the range of f is the interval $[-5, 4]$.

By the definition of a function, at most one y-value corresponds to a given x-value. It follows, then, that a vertical line can intersect the graph of a function at most once. This observation provides us with a convenient visual test for functions.

> ### VERTICAL LINE TEST FOR FUNCTIONS
>
> A set of points in a coordinate plane is the graph of y as a function of x if and only if no vertical line intersects the graph at more than one point.

EXAMPLE 2   Vertical Line Test for Functions

Do the graphs in Figure 3.52 represent y as a function of x?

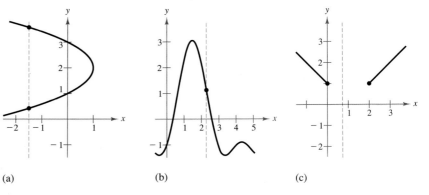

(a)   (b)   (c)

FIGURE 3.52

Solution

a. No, because you can find a vertical line that intersects the graph twice.

b. Yes, because every vertical line intersects the graph at most once.

c. Yes. (Note that if a vertical line does not intersect the graph, it simply means that the function is undefined for this particular value of x.)

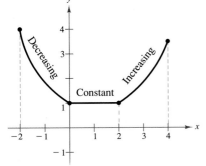

FIGURE 3.53

Increasing and Decreasing Functions

The more you know about the graph of a function, the more you know about the function itself. Consider the graph shown in Figure 3.53. Moving from *left to right*, this graph falls from $x = -2$ to $x = 0$, is constant from $x = 0$ to $x = 2$, and rises from $x = 2$ to $x = 4$. These observations indicate that the function has the following characteristics.

1. The function is **decreasing** on the interval $(-2, 0)$.
2. The function is **constant** on the interval $(0, 2)$.
3. The function is **increasing** on the interval $(2, 4)$.

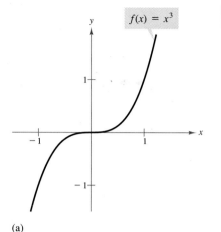

(a)

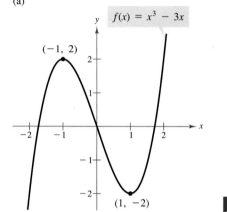

(b)

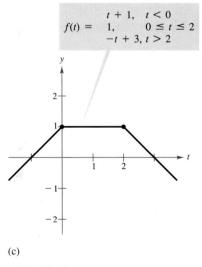

(c)

FIGURE 3.54

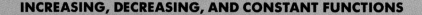

INCREASING, DECREASING, AND CONSTANT FUNCTIONS

A function f is **increasing** on an interval if, for any x_1 and x_2 in the interval,

$$x_1 < x_2 \quad \text{implies} \quad f(x_1) < f(x_2).$$

A function f is **decreasing** on an interval if, for any x_1 and x_2 in the interval,

$$x_1 < x_2 \quad \text{implies} \quad f(x_1) > f(x_2).$$

A function f is **constant** on an interval if, for any x_1 and x_2 in the interval,

$$f(x_1) = f(x_2).$$

EXAMPLE 3 Increasing and Decreasing Functions

In Figure 3.54, determine the open intervals on which each function is increasing, decreasing, or constant.

Solution

a. Although it might appear that there is an interval about zero over which this function is constant, you see that if $x_1 < x_2$, then $f(x_1) = x_1^3 < x_2^3 = f(x_2)$, and conclude that the function is increasing for all real numbers.

b. This function is increasing and decreasing on the following intervals.

Increasing on the interval $(-\infty, -1)$

Decreasing on the interval $(-1, 1)$

Increasing on the interval $(1, \infty)$

c. This function is increasing, constant, and decreasing on the following intervals.

Increasing on the interval $(-\infty, 0)$

Constant on the interval $(0, 2)$

Decreasing on the interval $(2, \infty)$

Even and Odd Functions

In Section 3.2, you learned about the different types of symmetry that a graph can possess. A function is **even** if its graph is symmetric with respect to the y-axis, and a function is **odd** if its graph is symmetric with respect to the origin. Thus, the symmetry tests given in Section 3.2 yield the following tests for even and odd functions.

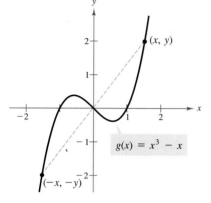

(a)

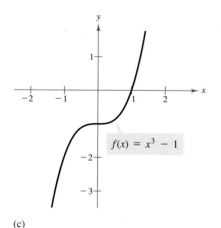

(b)

TESTS FOR EVEN AND ODD FUNCTIONS

A function given by $y = f(x)$ is **even** if, for each x in the domain of f,

$$f(-x) = f(x).$$

A function given by $y = f(x)$ is **odd** if, for each x in the domain of f,

$$f(-x) = -f(x).$$

REMARK   The points at which a function changes its increasing, decreasing, or constant behavior are especially important in producing an accurate graph of the function. These points often identify *maximum* or *minimum* values of the function. Techniques for finding the exact location of these special points are developed in calculus. ◀

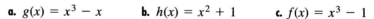

EXAMPLE 4   Even and Odd Functions

Are the functions even, odd, or neither?

a. $g(x) = x^3 - x$ **b.** $h(x) = x^2 + 1$ **c.** $f(x) = x^3 - 1$

Solution

a. Odd, because

$$g(-x) = (-x)^3 - (-x) = -x^3 + x = -(x^3 - x) = -g(x).$$

b. Even, because

$$h(-x) = (-x)^2 + 1 = x^2 + 1 = h(x).$$

c. Substituting $-x$ for x,

$$f(-x) = (-x)^3 - 1 = -x^3 - 1.$$

Since $f(x) = x^3 - 1$ and $-f(x) = -x^3 + 1$, then $f(-x) \neq f(x)$ and $f(-x) \neq -f(x)$. Hence, the function is neither even nor odd.

The graphs of the three functions are shown in Figure 3.55. ◀

FIGURE 3.55

Summary of Graphs of Common Functions

Figure 3.56 shows the graphs of six common functions. You should be familiar with these graphs.

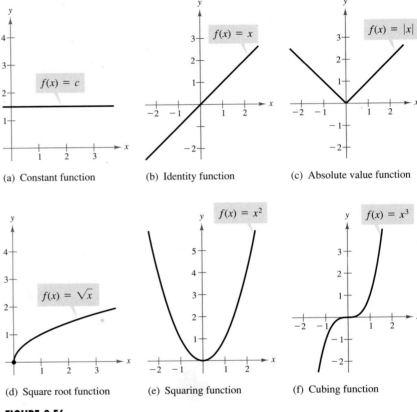

(a) Constant function

(b) Identity function

(c) Absolute value function

(d) Square root function

(e) Squaring function

(f) Cubing function

FIGURE 3.56

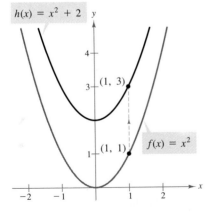

Vertical shift upward: 2 units

FIGURE 3.57

Shifting, Reflecting, and Stretching Graphs

Many functions have graphs that are simple transformations of the common graphs summarized in Figure 3.56. For example, you can obtain the graph of $h(x) = x^2 + 2$ by shifting the graph of $f(x) = x^2$ *up* two units, as shown in Figure 3.57. In function notation, h and f are related as follows.

$$h(x) = x^2 + 2 = f(x) + 2 \qquad \textit{Upward shift of 2}$$

Similarly, you can obtain the graph of $g(x) = (x - 2)^2$ by shifting the graph of $f(x) = x^2$ to the *right* two units, as shown in Figure 3.58. In this case, the functions g and f have the following relationship.

$$g(x) = (x - 2)^2 = f(x - 2) \qquad \textit{Right shift of 2}$$

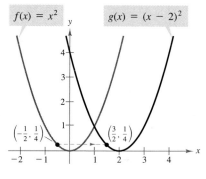

Horizontal shift to the right: 2 units

FIGURE 3.58

REMARK   In part (c) of Figure 3.59, note that the same result is obtained if the vertical shift precedes the horizontal shift.

FIGURE 3.59

VERTICAL AND HORIZONTAL SHIFTS

Let c be a positive real number. **Vertical** and **horizontal shifts** in the graph of $y = f(x)$ are represented as follows.

1. Vertical shift c units **upward:** $\qquad\qquad$ $h(x) = f(x) + c$
2. Vertical shift c units **downward:** $\qquad\;\;$ $h(x) = f(x) - c$
3. Horizontal shift c units to the **right:** $\quad$ $h(x) = f(x - c)$
4. Horizontal shift c units to the **left:** $\qquad$ $h(x) = f(x + c)$

Some graphs can be obtained from a sequence of vertical and horizontal shifts. This is demonstrated in part (c) of Example 5.

EXAMPLE 5   Shifts in the Graph of a Function

Use the graph of $f(x) = x^3$ to sketch the graph of each of the functions.

a. $g(x) = x^3 + 1$ $\qquad$ **b.** $h(x) = (x - 1)^3$ $\qquad$ **c.** $k(x) = (x + 2)^3 + 1$

Solution

See Figure 3.59. Relative to the graph of $f(x) = x^3$, the graph of $g(x) = x^3 + 1$ is an upward shift of one unit, the graph of $h(x) = (x - 1)^3$ is a right shift of one unit, and the graph of $k(x) = (x + 2)^3 + 1$ involves a left shift of two units *and* an upward shift of one unit.

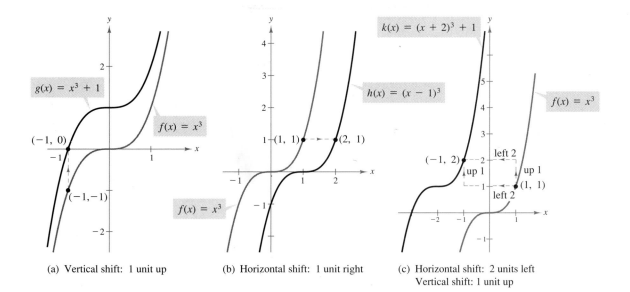

(a) Vertical shift: 1 unit up $\qquad$ (b) Horizontal shift: 1 unit right $\qquad$ (c) Horizontal shift: 2 units left
Vertical shift: 1 unit up

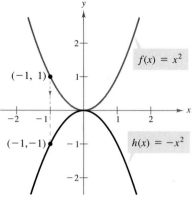

$f(x) = x^2$

$(-1, 1)$

$(-1,-1)$

$h(x) = -x^2$

Reflection

FIGURE 3.60

The second common type of transformation is a **reflection.** For instance, if you assume that the x-axis represents a mirror, then the graph of $h(x) = -x^2 = -(x^2)$ is the mirror image (or reflection) of the graph of $f(x) = x^2$, as shown in Figure 3.60.

REFLECTIONS IN THE COORDINATE AXES

Reflections, in the coordinate axes, of the graph of $y = f(x)$ are represented as follows.

1. Reflection in the x-axis: $\quad h(x) = -f(x)$
2. Reflection in the y-axis: $\quad h(x) = f(-x)$

EXAMPLE 6   Reflections and Shifts

Sketch a graph of each of the functions.

a. $g(x) = -\sqrt{x}$ **b.** $h(x) = \sqrt{-x}$ **c.** $k(x) = -\sqrt{x + 2}$

Solution

The graphs of all three functions are compared to the graph of $f(x) = \sqrt{x}$ in Figure 3.61.

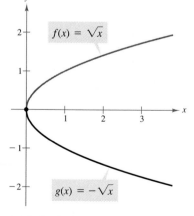

(a) Reflection in x-axis

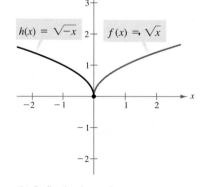

(b) Reflection in y-axis

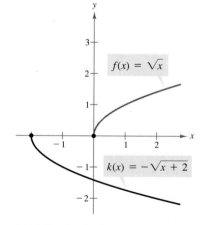

(c) Shift and reflection

FIGURE 3.61

a. The graph of g is a reflection in the x-axis because

$$g(x) = -\sqrt{x} = -f(x).$$

b. The graph of h is a reflection in the y-axis because

$$h(x) = \sqrt{-x} = f(-x).$$

c. From the equation

$$k(x) = -\sqrt{x + 2} = -f(x + 2)$$

the graph of k is a left shift of two units, followed by a reflection in the x-axis, or vice versa. ◢

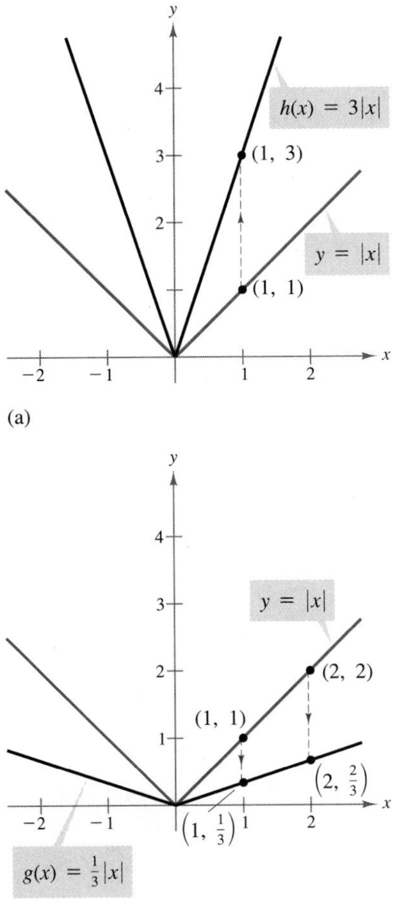

Horizontal shifts, vertical shifts, and reflections are **rigid** transformations because the basic shape of the graph is unchanged. These transformations change only the *position* of the graph in the xy-plane. **Nonrigid** transformations are those that cause a *distortion*—a change in the shape of the original graph. For instance, a nonrigid transformation of the graph of $y = f(x)$ is represented by $y = cf(x)$, where the transformation is a **vertical stretch** if $c > 1$ and a **vertical shrink** if $0 < c < 1$.

EXAMPLE 7 Nonrigid Transformations

Sketch the graph of each of the functions.

a. $h(x) = 3|x|$ b. $g(x) = \dfrac{1}{3}|x|$

Solution

The graphs of both functions and $f(x) = |x|$ are shown in Figure 3.62.

a. The graph of

$$h(x) = 3|x| = 3f(x)$$

is a vertical stretch (multiply each y-value by 3) of the graph of f.

b. Similarly, the equation

$$g(x) = \frac{1}{3}|x| = \frac{1}{3}f(x)$$

indicates that the graph of h is a vertical shrink of the graph of f. ◢

(a)

(b)

FIGURE 3.62

Step Functions

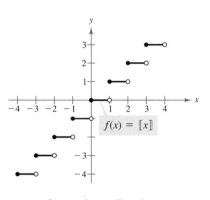

Greatest Integer Function

FIGURE 3.63

EXAMPLE 8   The Greatest Integer Function

The **greatest integer function** is denoted by $[\![x]\!]$ and is defined by

$$f(x) = [\![x]\!] = \text{the greatest integer less than or equal to } x.$$

The graph of this function is shown in Figure 3.63. Note that the graph of the greatest integer function jumps vertically one unit at each integer and is constant (a horizontal line segment) between each pair of consecutive integers. Because of the jumps in its graph, the greatest integer function is an example of a **step function.** Some values of the greatest integer function are as follows.

$$[\![-1]\!] = -1 \qquad [\![-0.5]\!] = -1$$
$$[\![0]\!] = 0 \qquad [\![0.5]\!] = 0$$
$$[\![1]\!] = 1 \qquad [\![1.5]\!] = 1$$

The range of the greatest integer function is the set of all integers.

EXAMPLE 9   The Cost of a Telephone Call

Suppose that the cost of a telephone call between Los Angeles and San Francisco is $0.50 for the first minute and $0.36 for each additional minute. The greatest integer function can be used to create a model for the cost of this call.

$$C = 0.50 + 0.36[\![t]\!]$$

where C is the total cost of the call in dollars and t is the length of the call in minutes. Sketch the graph of this function.

Solution

For calls up to one minute, the cost is $0.50. For calls between one and two minutes, the cost is $0.86, and so on.

Length of Call	$0 \le t < 1$	$1 \le t < 2$	$2 \le t < 3$	$3 \le t < 4$	$4 \le t < 5$
Cost of Call	$0.50	$0.86	$1.22	$1.58	$1.94

Using these values, you can sketch the graph shown in Figure 3.64.

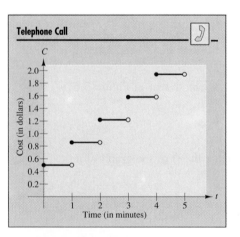

FIGURE 3.64

Use your school's library or some other reference source to find examples of three different functions that represent data between 1980 and 1990. Find one that decreased during the decade, one that increased, and one that was constant. For instance, the value of the dollar decreased, the population of the United States increased, and the land size of the United States remained constant. Can you find three other examples? Present your results graphically.

WARM UP

The following warm-up exercises involve skills that were covered in earlier sections. You will use these skills in the exercise set for this section.

1. Find $f(2)$ for $f(x) = -x^3 + 5x$.

2. Find $f(6)$ for $f(x) = x^2 - 6x$.

3. Find $f(-x)$ for $f(x) = \dfrac{3}{x}$.

4. Find $f(-x)$ for $f(x) = x^2 + 3$.

In Exercises 5 and 6, solve for x.

5. $x^3 - 16x = 0$

6. $2x^2 - 3x + 1 = 0$

In Exercises 7–10, find the domain of the function.

7. $g(x) = \dfrac{4}{x - 4}$

8. $f(x) = \dfrac{2x}{x^2 - 9x + 20}$

9. $h(t) = \sqrt[4]{5 - 3t}$

10. $f(t) = t^3 + 3t - 5$

EXERCISES for Section 3.5

In Exercises 1–6, find the domain and range of the function.

1. $f(x) = \sqrt{x - 1}$

2. $f(x) = 4 - x^2$

3. $f(x) = \sqrt{x^2 - 4}$

4. $f(x) = |x - 2|$

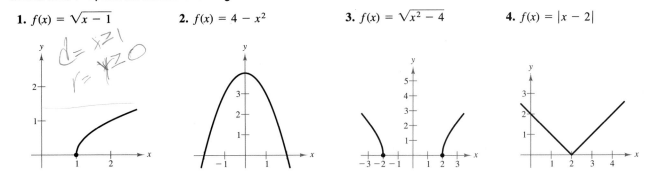

5. $f(x) = \sqrt{25 - x^2}$

6. $f(x) = \dfrac{|x|}{x}$

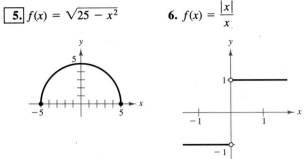

In Exercises 7–12, use the vertical line test to determine if y is a function of x.

7. $y = x^2$

8. $y = x^3 - 1$

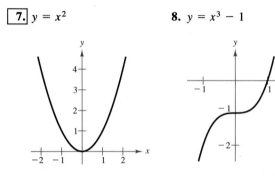

9. $x - y^2 = 0$

10. $x^2 + y^2 = 9$

11. $x^2 = xy - 1$

12. $x = |y|$

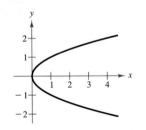

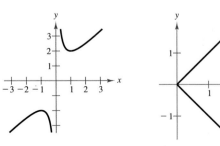

In Exercises 13–20, (a) determine the open intervals over which the function is increasing, decreasing, or constant, and (b) determine if the function is even, odd, or neither.

13. $f(x) = 2x$

14. $f(x) = x^2 - 2x$

15. $f(x) = x^3 - 3x^2$

16. $f(x) = \sqrt{x^2 - 4}$

17. $f(x) = 3x^4 - 6x^2$

18. $f(x) = x^{2/3}$

19. $f(x) = x\sqrt{x + 3}$

20. $f(x) = |x + 1| + |x - 1|$

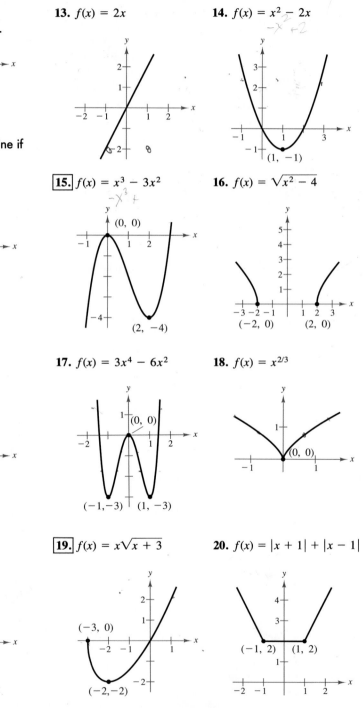
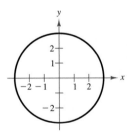

In Exercises 21–26, determine whether the function is even, odd, or neither.

21. $f(x) = x^6 - 2x^2 + 3$ **22.** $h(x) = x^3 - 5$

23. $g(x) = x^3 - 5x$ **24.** $f(x) = x\sqrt{1 - x^2}$

25. $f(t) = t^2 + 2t - 3$ **26.** $g(s) = 4s^{2/3}$

In Exercises 27–40, sketch the graph of the function and determine whether the function is even, odd, or neither.

27. $f(x) = 3$ **28.** $g(x) = x$

29. $f(x) = 5 - 3x$ **30.** $h(x) = x^2 - 4$

31. $g(s) = \dfrac{s^3}{4}$ **32.** $f(t) = -t^4$

33. $f(x) = \sqrt{1 - x}$ **34.** $f(x) = \sqrt{x + 2} - 1$

35. $g(t) = (t - 1)^2 + 2$ **36.** $f(x) = |x + 2|$

37. $f(x) = \begin{cases} x + 3, & x \leq 0 \\ 3, & 0 < x \leq 2 \\ 2x - 1, & x > 2 \end{cases}$

38. $f(x) = \begin{cases} 2x + 1, & x \leq -1 \\ x^2 - 2, & x > -1 \end{cases}$

39. $s(x) = 2[\![x - 1]\!]$ **40.** $g(x) = 6 - [\![x]\!]$

In Exercises 41–50, sketch the graph of the function and determine the interval(s) (if any) on the real axis for which $f(x) \geq 0$.

41. $f(x) = 4 - x$ **42.** $f(x) = 4x + 2$

43. $f(x) = x^2 - 9$ **44.** $f(x) = x^2 - 4x$

45. $f(x) = 1 - x^4$ **46.** $f(x) = \sqrt{x + 2}$

47. $f(x) = x^2 + 1$ **48.** $f(x) = -(1 + |x|)$

49. $f(x) = -5$ **50.** $f(x) = \frac{1}{2}(2 + |x|)$

51. Sketch (on the same set of coordinate axes) a graph of f for $c = -2, 0,$ and 2.
  (a) $f(x) = \frac{1}{2}x + c$
  (b) $f(x) = \frac{1}{2}(x - c)$
  (c) $f(x) = \frac{1}{2}(cx)$

52. Sketch (on the same set of coordinate axes) a graph of f for $c = -2, 0,$ and 2.
  (a) $f(x) = x^3 + c$
  (b) $f(x) = (x - c)^3$
  (c) $f(x) = (x - 2)^3 + c$

53. Use the graph of f to sketch the graphs.
  (a) $y = f(x) + 2$ (b) $y = -f(x)$
  (c) $y = f(x - 2)$ (d) $y = f(x + 3)$
  (e) $y = f(2x)$ (f) $y = f(-x)$

54. Use the graph of f to sketch the graphs.
  (a) $y = f(x) - 1$ (b) $y = f(x + 1)$
  (c) $y = f(x - 1)$ (d) $y = -f(x - 2)$
  (e) $y = f(-x)$ (f) $y = \frac{1}{2}f(x)$

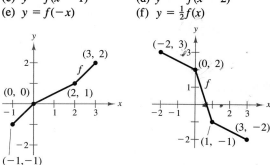

Figure for 53   **Figure for 54**

55. Use the graph of $f(x) = x^2$ (see Figure 3.56(e)) to write formulas for the functions whose graphs are shown in parts (a) and (b) of the figure shown here.

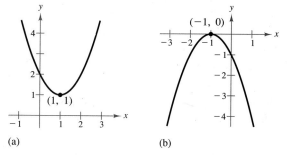

(a)   (b)

Figure for 55

56. Use the graph of $f(x) = x^3$ (see Figure 3.56(f)) to write formulas for the functions whose graphs are shown in parts (a) and (b) of the figure shown here.

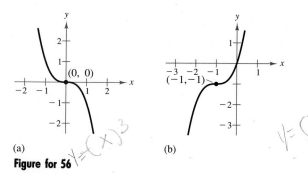

(a)   (b)

Figure for 56

57. *Cost of a Telephone Call* The cost of a telephone call between two cities is \$0.65 for the first minute and \$0.42 for each additional minute (or portion thereof). Use the greatest integer function to create a model for the cost C of a telephone call between the two cities lasting t minutes. Sketch the graph of the function.

58. *Cost of Overnight Delivery* Suppose the cost of sending an overnight package from New York to Atlanta is \$9.80 for the first pound and \$2.50 for each additional pound (or portion thereof). Use the greatest integer function to create a model for the cost C of overnight delivery of a package weighing x pounds. Sketch the graph of the function.

59. *Maximum Profit* The marketing department for a company estimates that the demand for a product is given by $p = 100 - 0.0001x$ where p is the price per unit and x is the number of units. The cost of producing x units is given by $C = 350,000 + 30x$, and the profit for producing and selling x units is given by

$$P = R - C = xp - C.$$

Sketch the graph of the profit function and estimate the number of units that would produce a maximum profit.

60. *Fluorescent Lamp* The number of lumens (time rate of flow of light) L from a fluorescent lamp can be approximated by the model

$$L = -0.294x^2 + 97.744x - 664.875, \quad 20 \le x \le 90$$

where x is the wattage of the lamp. Sketch a graph of the function and estimate the wattage of a bulb necessary to obtain 2000 lumens.

In Exercises 61–64, write the height h of the rectangle as a function of x.

61.
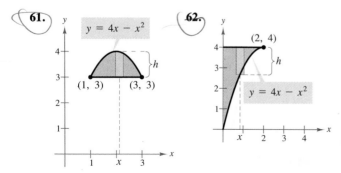

62.

63.
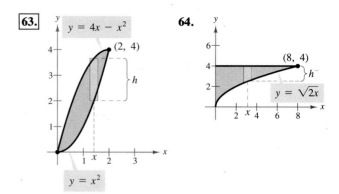

64.

In Exercises 65 and 66, write the length L of the rectangle as a function of y.

65.
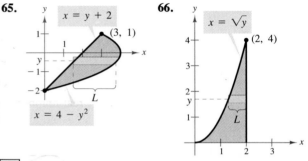

66.

67. Prove that a function of the following form is odd.

$$f(x) = a_{2n+1}x^{2n+1} + a_{2n-1}x^{2n-1} + \ldots + a_3x^3 + a_1x$$

68. Prove that a function of the following form is even.

$$f(x) = a_{2n}x^{2n} + a_{2n-2}x^{2n-2} + \ldots + a_2x^2 + a_0$$

69. The intake pipe of a 100-gallon tank has a flow rate of 10 gallons per minute, and two drain pipes have a flow rate of 5 gallons per minute each. The figure shows the volume V of fluid in the tank as a function of time t. Determine the pipes in which the fluid is flowing in specific subintervals of the 1 hour of time shown on the graph. (There is more than one correct answer.)

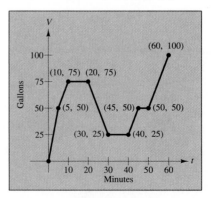

Figure for 69

3.6 Combinations of Functions

Arithmetic Combinations of Functions / Composition of Functions / Applications

Arithmetic Combinations of Functions

Just as two real numbers can be combined by the operations of addition, subtraction, multiplication, and division to form other real numbers, two *functions* can be combined to create new functions. For example, if

$$f(x) = 2x - 3 \quad \text{and} \quad g(x) = x^2 - 1$$

we can form the sum, difference, product, and quotient of f and g as follows.

$$f(x) + g(x) = (2x - 3) + (x^2 - 1) = x^2 + 2x - 4 \qquad \textit{Sum}$$

$$f(x) - g(x) = (2x - 3) - (x^2 - 1) = -x^2 + 2x - 2 \qquad \textit{Difference}$$

$$f(x)g(x) = (2x - 3)(x^2 - 1) = 2x^3 - 3x^2 - 2x + 3 \qquad \textit{Product}$$

$$\frac{f(x)}{g(x)} = \frac{2x - 3}{x^2 - 1}, \quad x \neq \pm 1 \qquad \textit{Quotient}$$

The domain of an arithmetic combination of functions f and g consists of all real numbers that are common to the domains of f and g. In the case of the quotient $f(x)/g(x)$, there is the further restriction that $g(x) \neq 0$.

SUM, DIFFERENCE, PRODUCT, AND QUOTIENT OF FUNCTIONS

Let f and g be two functions with overlapping domains. Then, for all x common to both domains,

1. Sum: $\quad\quad (f + g)(x) = f(x) + g(x)$
2. Difference: $\quad (f - g)(x) = f(x) - g(x)$
3. Product: $\quad (fg)(x) = f(x) \cdot g(x)$
4. Quotient: $\quad \left(\dfrac{f}{g}\right)(x) = \dfrac{f(x)}{g(x)}, \quad g(x) \neq 0.$

EXAMPLE 1   Finding the Sum of Two Functions

Given $f(x) = 2x + 1$ and $g(x) = x^2 + 2x - 1$, find $(f + g)(x)$. Then evaluate this sum when $x = 2$.

Solution

The sum of the functions f and g is

$$
\begin{aligned}
(f + g)(x) &= f(x) + g(x) \\
&= (2x + 1) + (x^2 + 2x - 1) \\
&= x^2 + 4x.
\end{aligned}
$$

When $x = 2$, the value of this sum is

$$(f + g)(2) = 2^2 + 4(2) = 12.$$

EXAMPLE 2 Finding the Difference of Two Functions

Given $f(x) = 2x + 1$ and $g(x) = x^2 + 2x - 1$, find $(f - g)(x)$. Then evaluate this difference when $x = 2$.

Solution

The difference of the functions f and g is

$$\begin{aligned}
(f - g)(x) &= f(x) - g(x) \\
&= (2x + 1) - (x^2 + 2x - 1) \\
&= -x^2 + 2.
\end{aligned}$$

When $x = 2$, the value of this difference is

$$(f - g)(2) = -(2)^2 + 2 = -2.$$

In Examples 1 and 2, both f and g have domains that consist of all real numbers. Thus, the domain of both $(f + g)$ and $(f - g)$ is also the set of all real numbers. Remember that any restrictions on the domains of f or g must be taken into account when forming the sum, difference, product, or quotient of f and g. For instance, the domain of $f(x) = 1/x$ is all $x \neq 0$, and the domain of $g(x) = \sqrt{x}$ is $[0, \infty)$. This implies that the domain of $f + g$ is $(0, \infty)$.

EXAMPLE 3 Finding the Product and Quotient of Two Functions

Find $(fg)(x)$, $(f/g)(x)$, and $(g/f)(x)$ for the functions

$$f(x) = \sqrt{x} \quad \text{and} \quad g(x) = \sqrt{4 - x^2}.$$

Then find the domains of f/g and g/f.

Solution

$$(fg)(x) = f(x)g(x) = \sqrt{x}\sqrt{4 - x^2} = \sqrt{x(4 - x^2)} \quad \textit{Product, } f \cdot g$$

$$\left(\frac{f}{g}\right)(x) = \frac{f(x)}{g(x)} = \frac{\sqrt{x}}{\sqrt{4 - x^2}} \quad \textit{Quotient, } f/g$$

$$\left(\frac{g}{f}\right)(x) = \frac{g(x)}{f(x)} = \frac{\sqrt{4 - x^2}}{\sqrt{x}} \quad \textit{Quotient, } g/f$$

The domain of f is $[0, \infty)$ and the domain of g is $[-2, 2]$. The intersection of these two domains is $[0, 2]$. Thus, we have the following domains for f/g and g/f.

Domain of $\dfrac{f}{g}$: $[0, 2)$ Domain of $\dfrac{g}{f}$: $(0, 2]$

Can you see why these two domains differ slightly?

Composition of Functions

Another way of combining two functions is to form the **composition** of one with the other. For instance, if $f(x) = x^2$ and $g(x) = x + 1$, then the composition of f with g ($f \circ g$) is

$$f(g(x)) = f(x + 1) = (x + 1)^2.$$

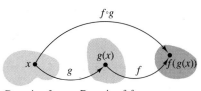

Domain of g Domain of f

FIGURE 3.65

DEFINITION OF COMPOSITION OF TWO FUNCTIONS

The **composition** of the functions f and g is

$$(f \circ g)(x) = f(g(x)).$$

The domain of $f \circ g$ is the set of all x in the domain of g such that $g(x)$ is in the domain of f. (See Figure 3.65.)

EXAMPLE 4   Forming the Composition of f with g

Find $(f \circ g)(x)$ for $f(x) = \sqrt{x}$, $x \geq 0$, and $g(x) = x - 1$, where $x \geq 1$. If possible, find $(f \circ g)(2)$ and $(f \circ g)(0)$.

Solution

$$
\begin{aligned}
(f \circ g)(x) &= f(g(x)) && \textit{Definition of } f \circ g \\
&= f(x - 1) && \textit{Definition of } g(x) \\
&= \sqrt{x - 1}, \quad x \geq 1 && \textit{Definition of } f(x)
\end{aligned}
$$

The domain of $f \circ g$ is $[1, \infty)$. Thus,

$$(f \circ g)(2) = \sqrt{2 - 1} = 1$$

is defined, but $(f \circ g)(0)$ is not defined because 0 is not in the domain of $f \circ g$.

The composition of f with g is generally *not* the same as the composition of g with f.

EXAMPLE 5 Composition of Functions

Given $f(x) = x + 2$ and $g(x) = 4 - x^2$, find the following.

a. $(f \circ g)(x)$

b. $(g \circ f)(x)$

Solution

a.
$$
\begin{aligned}
(f \circ g)(x) &= f(g(x)) && \textit{Definition of } f \circ g \\
&= f(4 - x^2) && \textit{Definition of } g(x) \\
&= (4 - x^2) + 2 && \textit{Definition of } f(x) \\
&= -x^2 + 6
\end{aligned}
$$

b.
$$
\begin{aligned}
(g \circ f)(x) &= g(f(x)) && \textit{Definition of } g \circ f \\
&= g(x + 2) && \textit{Definition of } f(x) \\
&= 4 - (x + 2)^2 && \textit{Definition of } g(x) \\
&= 4 - (x^2 + 4x + 4) \\
&= -x^2 - 4x
\end{aligned}
$$

Note in this case that $(f \circ g)(x) \neq (g \circ f)(x)$.

EXAMPLE 6 A Case in Which $f \circ g = g \circ f$

Given $f(x) = 2x + 3$ and $g(x) = \frac{1}{2}(x - 3)$, find the following.

a. $(f \circ g)(x)$

b. $(g \circ f)(x)$

Solution

REMARK In Example 6, note that the two composite functions $f \circ g$ and $g \circ f$ are equal, and both represent the identity function—$(f \circ g)(x) = (g \circ f)(x) = x$.

a.
$$
(f \circ g)(x) = f(g(x)) = f\left(\frac{1}{2}(x - 3)\right) = 2\left[\frac{1}{2}(x - 3)\right] + 3
$$
$$
= x - 3 + 3 = x
$$

b.
$$
(g \circ f)(x) = g(f(x)) = g(2x + 3) = \frac{1}{2}[(2x + 3) - 3] = \frac{1}{2}(2x) = x
$$

In Examples 4, 5, and 6, we formed the composite of two given functions. In calculus, it is also important to be able to identify two functions that *make up a given* composite function. For instance, the function h given by $h(x) = (3x - 5)^3$ is the composite of f with g, where $f(x) = x^3$ and $g(x) = 3x - 5$. That is,

$$
h(x) = (3x - 5)^3 = [g(x)]^3 = f(g(x)).
$$

To "decompose" a composite function, we look for an "inner" and an "outer" function. In the function h above, $g(x) = 3x - 5$ is the inner function and $f(x) = x^3$ is the outer function.

EXAMPLE 7 Identifying a Composite Function

Express the function $h(x) = 1/(x - 2)^2$ as a composition of two functions f and g.

Solution

Take the inner function to be

$$g(x) = x - 2$$

and the outer function to be

$$f(x) = \frac{1}{x^2} = x^{-2}.$$

Then you can write

$$h(x) = \frac{1}{(x - 2)^2} = (x - 2)^{-2} = f(x - 2) = f(g(x)).$$

Applications

EXAMPLE 8 Bacteria Count

The number of bacteria in a refrigerated food is given by

$$N(T) = 20T^2 - 80T + 500, \qquad 2 \leq T \leq 14$$

where T is the temperature of the food. When the food is removed from refrigeration, the temperature is given by

$$T(t) = 4t + 2, \qquad 0 \leq t \leq 3$$

where t is the time in hours. Find the following.

a. The composite function $N(T(t))$
b. The number of bacteria in the food when $t = 2$ hours
c. The time when the bacteria count reaches 2000

Solution

a. $N(T(t)) = 20(4t + 2)^2 - 80(4t + 2) + 500$

$$= 20(16t^2 + 16t + 4) - 320t - 160 + 500$$
$$= 320t^2 + 320t + 80 - 320t - 160 + 500$$
$$= 320t^2 + 420$$

b. When $t = 2$, the number of bacteria is

$$N = 320(2)^2 + 420 = 1280 + 420 = 1700.$$

c. The bacteria count will reach $N = 2000$ when $320t^2 + 420 = 2000$.

$$320t^2 + 420 = 2000$$

$$320t^2 = 1580$$

$$t^2 = \frac{1580}{320} = \frac{79}{16}$$

$$t = \frac{\sqrt{79}}{4} \approx 2.2 \text{ hours}$$

◢

DISCUSSION

PROBLEM
_____

The Composition of Two Functions

You learned in this section that the composite functions $(f \circ g)(x)$ and $(g \circ f)(x)$ are generally not equal to each other. Mathematically, you can say that in general the operation of forming the composition of two functions is *not commutative*. Discuss some examples of real-life situations in which the order of operations is commutative and some that are not commutative. For instance, the operations of "turning on a calculator" and "pressing the keys 2 $\boxed{+}$ 3 $\boxed{=}$" are not commutative, whereas the operations of "taking off your left shoe" and "taking off your right shoe" would generally be thought of as being commutative (because they produce the same results).

WARM UP
_____

The following warm-up exercises involve skills that were covered in earlier sections. You will use these skills in the exercise set for this section.

In Exercises 1–10, perform the indicated operations and simplify the result.

1. $\dfrac{1}{x} + \dfrac{1}{1 - x}$

2. $\dfrac{2}{x + 3} - \dfrac{2}{x - 3}$

3. $\dfrac{3}{x - 2} - \dfrac{2}{x(x - 2)}$

4. $\dfrac{x}{x - 5} + \dfrac{1}{3}$

5. $(x - 1)\left(\dfrac{1}{\sqrt{x^2 - 1}}\right)$

6. $\left(\dfrac{x}{x^2 - 4}\right)\left(\dfrac{x^2 - x - 2}{x^2}\right)$

7. $(x^2 - 4) \div \left(\dfrac{x + 2}{5}\right)$

8. $\left(\dfrac{x}{x^2 + 3x - 10}\right) \div \left(\dfrac{x^2 + 3x}{x^2 + 6x + 5}\right)$

9. $\dfrac{\left(\dfrac{1}{x}\right) + 5}{3 - \left(\dfrac{1}{x}\right)}$

10. $\dfrac{\left(\dfrac{x}{4}\right) - \left(\dfrac{4}{x}\right)}{x - 4}$

EXERCISES for Section 3.6

In Exercises 1–8, find (a) $(f + g)(x)$, (b) $(f - g)(x)$, (c) $(fg)(x)$, and (d) $(f/g)(x)$. What is the domain of f/g?

1. $f(x) = x + 1$, $g(x) = x - 1$

2. $f(x) = 2x - 5$, $g(x) = 1 - x$

3. $f(x) = x^2$, $g(x) = 1 - x$

4. $f(x) = 2x - 5$, $g(x) = 5$

5. $f(x) = x^2 + 5$, $g(x) = \sqrt{1 - x}$

6. $f(x) = \sqrt{x^2 - 4}$, $g(x) = \dfrac{x^2}{x^2 + 1}$

7. $f(x) = \dfrac{1}{x}$, $g(x) = \dfrac{1}{x^2}$

8. $f(x) = \dfrac{x}{x + 1}$, $g(x) = x^3$

In Exercises 9–20, evaluate the indicated function for $f(x) = x^2 + 1$ and $g(x) = x - 4$.

9. $(f + g)(3)$

10. $(f - g)(-2)$

11. $(f - g)(0)$

12. $(f + g)(1)$

13. $(f - g)(2t)$

14. $(f + g)(t - 1)$

15. $(fg)(4)$

16. $(fg)(-6)$

17. $\left(\dfrac{f}{g}\right)(5)$

18. $\left(\dfrac{f}{g}\right)(0)$

19. $\left(\dfrac{f}{g}\right)(-1) - g(3)$

20. $(2f)(5)$

In Exercises 21–24, find (a) $f \circ g$, (b) $g \circ f$, and (c) $f \circ f$.

21. $f(x) = x^2$, $g(x) = x - 1$

22. $f(x) = \sqrt[3]{x - 1}$, $g(x) = x^3 + 1$

23. $f(x) = 3x + 5$, $g(x) = 5 - x$

24. $f(x) = x^3$, $g(x) = \dfrac{1}{x}$

In Exercises 25–32, find (a) $f \circ g$ and (b) $g \circ f$.

25. $f(x) = \sqrt{x + 4}$, $g(x) = x^2$

26. $f(x) = 3x + 2$, $g(x) = x^2 - 8$

27. $f(x) = \frac{1}{3}x - 3$, $g(x) = 3x + 1$

28. $f(x) = x^4$, $g(x) = x^4$

29. $f(x) = \sqrt{x}$, $g(x) = \sqrt{x}$

30. $f(x) = 2x - 3$, $g(x) = 2x - 3$

31. $f(x) = |x|$, $g(x) = x + 6$

32. $f(x) = x^{2/3}$, $g(x) = x^6$

In Exercises 33–36, use the graphs of f and g (see figure) to evaluate the indicated functions.

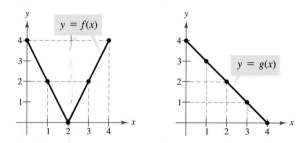

Figures for 33–36

33. (a) $(f + g)(3)$ (b) $\left(\dfrac{f}{g}\right)(2)$

34. (a) $(f - g)(1)$ (b) $(fg)(4)$

35. (a) $(f \circ g)(2)$ (b) $(g \circ f)(2)$

36. (a) $(f \circ g)(1)$ (b) $(g \circ f)(3)$

In Exercises 37–44, find two functions f and g such that $(f \circ g)(x) = h(x)$. (There are many correct answers to these exercises.)

37. $h(x) = (2x + 1)^2$

38. $h(x) = (1 - x)^3$

39. $h(x) = \sqrt[3]{x^2 - 4}$

40. $h(x) = \sqrt{9 - x}$

41. $h(x) = \dfrac{1}{x + 2}$

42. $h(x) = \dfrac{4}{(5x + 2)^2}$

43. $h(x) = (x + 4)^2 + 2(x + 4)$

44. $h(x) = (x + 3)^{3/2}$

In Exercises 45–48, determine the domain of (a) f, (b) g, and (c) $f \circ g$.

45. $f(x) = \sqrt{x}$, $g(x) = x^2 + 1$

46. $f(x) = \dfrac{1}{x}$, $g(x) = x + 3$

47. $f(x) = \dfrac{3}{x^2 - 1}$, $g(x) = x + 1$

48. $f(x) = 2x + 3$, $g(x) = \dfrac{x}{2}$

49. *Stopping Distance* While traveling in a car at x miles per hour, you are required to stop quickly to avoid an accident. The distance the car travels during your reaction time is given by $R(x) = \frac{3}{4}x$. The distance traveled while braking is given by $B(x) = \frac{1}{15}x^2$. Find the function giving total stopping distance T. Graph the functions R, B, and T on the same set of coordinate axes for $0 \le x \le 60$.

50. *Comparing Sales* Suppose you own two fast-food restaurants in town. From 1985 to 1990, the sales for one restaurant have been decreasing according to the function

$$R_1 = 500 - 0.8t^2, \quad t = 5, 6, 7, 8, 9, 10$$

where R_1 represents the sales for the first restaurant (in thousands of dollars) and t represents the calendar year with $t = 5$ corresponding to 1985. During the same six-year period, the sales for the second restaurant have been increasing according to the function

$$R_2 = 250 + 0.78t, \quad t = 5, 6, 7, 8, 9, 10.$$

Write a function that represents the total sales for the two restaurants. Use the *stacked bar graph* in the figure, which represents the total sales during the six-year period, to determine whether the total sales have been increasing or decreasing.

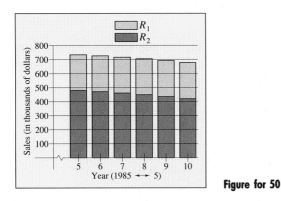

Figure for 50

51. *Ripples* A pebble is dropped into a calm pond, causing ripples in the form of concentric circles (see figure). The radius (in feet) of the outer ripple is given by $r(t) = 0.6t$, where t is the time in seconds after the pebble strikes the water. The area of the circle is given by the function $A(r) = \pi r^2$. Find and interpret $(A \circ r)(t)$.

Figure for 51

52. *Area* A square concrete foundation was prepared as a base for a large cylindrical gasoline tank (see figure).
(a) Express the radius r of the tank as a function of the length x of the sides of the square.
(b) Express the area A of the circular base of the tank as a function of the radius r.
(c) Find and interpret $(A \circ r)(x)$.

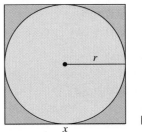

Figure for 52

53. *Cost* The weekly cost of producing x units in a manufacturing process is given by the function $C(x) = 60x + 750$. The number of units produced in t hours is given by $x(t) = 50t$. Find and interpret $(C \circ x)(t)$.

54. *Air Traffic Control* An air traffic controller spots two planes at the same altitude flying toward the same point (see figure). Their flight paths form a right angle at point P. One plane is 150 miles from point P and is moving at 450 miles per hour. The second plane is 200 miles from point P and is moving at 450 miles per hour. Write the distance s between the planes as a function of time t.

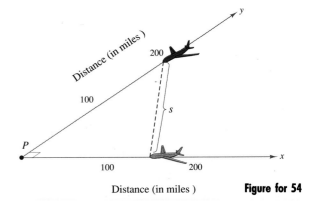

Distance (in miles) **Figure for 54**

55. Prove that the product of two odd functions is an even function.

56. Prove that the product of two even functions is an even function.

57. Prove that the product of an odd function and an even function is odd.

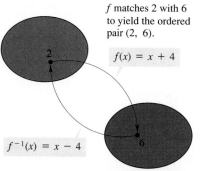

f matches 2 with 6 to yield the ordered pair (2, 6).

$f(x) = x + 4$

$f^{-1}(x) = x - 4$

f^{-1} matches 6 with 2 to yield the ordered pair (6, 2).

FIGURE 3.66

REMARK Don't be confused by the use of -1 to denote the inverse function f^{-1}. In this book f^{-1} will *always* refer to the inverse of the function f and *not* to the reciprocal of $f(x)$.

3.7 Inverse Functions

The Inverse of a Function / The Existence of an Inverse Function / Finding the Inverse of a Function

The Inverse of a Function

You know (from Section 3.4) that one way to represent a function is by a set of ordered pairs. For instance, the function $f(x) = x + 4$ from the set $A = \{1, 2, 3, 4\}$ to the set $B = \{5, 6, 7, 8\}$ can be written as follows.

$$f(x) = x + 4: \{(1, 5), (2, 6), (3, 7), (4, 8)\}$$

By interchanging the first and second coordinates of each of these ordered pairs, you can form the **inverse function** of f, denoted f^{-1}. This is a function from the set B to the set A, and can be written

$$f^{-1}(x) = x - 4: \{(5, 1,), (6, 2), (7, 3), (8, 4)\}.$$

Note that the domain of f is equal to the range of f^{-1}, and vice versa, as shown in Figure 3.66. Also note that the functions f and f^{-1} have the effect of "undoing" each other. In other words, when you form the composition of f with f^{-1} or the composition of f^{-1} with f, you obtain the identity function.

$$f(f^{-1}(x)) = f(x - 4) = (x - 4) + 4 = x$$
$$f^{-1}(f(x)) = f^{-1}(x + 4) = (x + 4) - 4 = x$$

EXAMPLE 1 Finding Inverse Functions Informally

Find the inverse of the following functions.

a. $f(x) = 4x$

b. $f(x) = x - 6$

Verify that both $f(f^{-1}(x))$ and $f^{-1}(f(x))$ are equal to the identity function.

Solution

a. The given function *multiplies* each input by 4. To "undo" this function, *divide* each input by 4. Thus, the inverse function of $f(x) = 4x$ is

$$f^{-1}(x) = \frac{x}{4}.$$

You can verify that both $f(f^{-1}(x))$ and $f^{-1}(f(x))$ are equal to the identity function.

$$f(f^{-1}(x)) = f\left(\frac{x}{4}\right) = 4\left(\frac{x}{4}\right) = x$$

$$f^{-1}(f(x)) = f^{-1}(4x) = \frac{4x}{4} = x$$

b. The given function *subtracts* 6 from each input. To "undo" this function, *add* 6 to each input. Thus, the inverse function of $f(x) = x - 6$ is

$$f^{-1}(x) = x + 6.$$

You can verify that both $f(f^{-1}(x))$ and $f^{-1}(f(x))$ are equal to the identity function.

$$f(f^{-1}(x)) = f(x + 6) = (x + 6) - 6 = x$$
$$f^{-1}(f(x)) = f^{-1}(x - 6) = (x - 6) + 6 = x$$

DEFINITION OF THE INVERSE OF A FUNCTION

Let f and g be two functions such that

$$f(g(x)) = x \qquad \text{for every } x \text{ in the domain of } g$$

and

$$g(f(x)) = x \qquad \text{for every } x \text{ in the domain of } f.$$

Then, the function g is the **inverse** of the function f, denoted f^{-1} (read "*f*-inverse"). Thus, $f(f^{-1}(x)) = x$ and $f^{-1}(f(x)) = x$. The domain of f must be equal to the range of f^{-1}, and vice versa.

Note from this definition that if the function g is the inverse of the function f, then it must also be true that the function f is the inverse of the function g. Thus, the functions f and g are *inverses of each other*.

EXAMPLE 2   Verifying Inverse Functions

Show that the following functions are inverses of each other.

$$f(x) = 2x^3 - 1$$

and

$$g(x) = \sqrt[3]{\frac{x + 1}{2}}$$

Solution

Note that the domain (and the range) of both functions is the entire set of real numbers. To show that f and g are inverses of each other, you need to show that $f(g(x)) = x$ and $g(f(x)) = x$.

$$f(g(x)) = f\left(\sqrt[3]{\frac{x+1}{2}}\right) = 2\left(\sqrt[3]{\frac{x+1}{2}}\right)^3 - 1$$

$$= 2\left(\frac{x+1}{2}\right) - 1$$

$$= x + 1 - 1 = x$$

$$g(f(x)) = g(2x^3 - 1) = \sqrt[3]{\frac{(2x^3 - 1) + 1}{2}}$$

$$= \sqrt[3]{\frac{2x^3}{2}}$$

$$= \sqrt[3]{x^3} = x$$

You can see that the two functions f and g "undo" each other: the function f first cubes the input x, multiplies by 2, and then subtracts 1, whereas the function g first adds 1, then divides by 2, and then takes the cube root of the result. ◢

EXAMPLE 3 Verifying Inverse Functions

Which of the following functions

$$g(x) = \frac{x-2}{5} \quad \text{and} \quad h(x) = \frac{5}{x} + 2$$

is the inverse of the function

$$f(x) = \frac{5}{x-2}?$$

Solution

$$f(g(x)) = f\left(\frac{x-2}{5}\right) = \frac{5}{\left(\dfrac{x-2}{5}\right) - 2} \qquad \textit{Composition of f with g}$$

$$= \frac{25}{(x-2) - 10}$$

$$= \frac{25}{x - 12}$$

$$\neq x$$

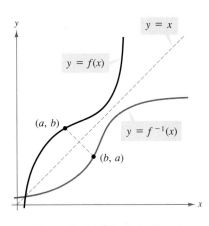

The graph of f^{-1} is a reflection of the graph of f in the line $y = x$.

FIGURE 3.67

Because this composition is not equal to the identity function x, g is *not* the inverse of f.

$$f(h(x)) = f\left(\frac{5}{x} + 2\right) = \frac{5}{\left(\dfrac{5}{x}\right) + 2 - 2} \qquad \text{\textit{Composition of} f \textit{with} h}$$

$$= \frac{5}{\left(\dfrac{5}{x}\right)}$$

$$= x$$

Thus, it appears that h is the inverse of f. Confirm this by showing that the composition of h with f is also equal to the identity function.

The graphs of f and f^{-1} are related to each other in the following way. If the point (a, b) lies on the graph of f, then the point (b, a) lies on the graph of f^{-1} and vice versa. This means that the graph of f^{-1} is a reflection of the graph of f in the line $y = x$, as shown in Figure 3.67.

The Existence of an Inverse Function

A function need not have an inverse function. For instance, the function $f(x) = x^2$ has no inverse [assuming a domain of $(-\infty, \infty)$]. To have an inverse, a function must be **one-to-one,** which means that no two elements in the domain of f correspond to the same element in the range of f.

DEFINITION OF ONE-TO-ONE FUNCTION

A function f is **one-to-one** if, for a and b in its domain,

$$f(a) = f(b) \qquad \text{implies that} \qquad a = b.$$

The function $f(x) = x + 1$ *is* one-to-one because $a + 1 = b + 1$ implies that a and b must be equal. However, the function $f(x) = x^2$ is *not* one-to-one because $a^2 = b^2$ does not imply that $a = b$. For instance, $(-1)^2 = 1^2$ and yet $-1 \neq 1$.

EXISTENCE OF AN INVERSE FUNCTION

A function f has an inverse function f^{-1} if and only if f is one-to-one.

DIRECT VARIATION

The following statements are equivalent.

1. y **varies directly** as x.
2. y is **directly proportional** to x.
3. $y = kx$ for some constant k.

k is the **constant of variation** or the **constant of proportionality.**

In the mathematical model for direct variation, y is a *linear* function of x. That is,

$$y = kx.$$

To set up a mathematical model, you use specific values of x and y to find the value of the constant k.

EXAMPLE 1   Direct Variation

Hooke's Law for a spring states that the distance a spring is stretched (or compressed) varies directly as the force on the spring. A force of 20 pounds stretches the spring 4 inches. (See Figure 3.71.)

a. Write an equation relating the distance stretched to the force applied.
b. How far will a force of 30 pounds stretch the spring?

Solution

a. Let

$d =$ distance spring is stretched (in inches)

$F =$ force (in pounds).

Since distance varies directly as force,

$$d = kF.$$

To find the value of the constant k, use the fact that $d = 4$ when $F = 20$.

$$
\begin{array}{cc}
d & F \\
\downarrow & \downarrow \\
4 & = k(20)
\end{array}
$$

which implies that $k = \frac{4}{20} = \frac{1}{5}$. Thus, the equation relating distance and force is

$$d = \frac{1}{5}F.$$

b. When $F = 30$, the distance is

$$d = \frac{1}{5}F = \frac{1}{5}(30) = 6 \text{ inches.}$$

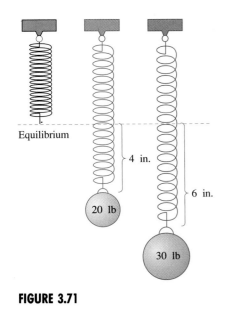

Equilibrium

4 in.

20 lb

6 in.

30 lb

FIGURE 3.71

Direct Variation as nth Power

Another type of direct variation relates one variable to a *power* of another variable. For example, in the formula for the area of a circle

$$A = \pi r^2$$

the area A is directly proportional to the square of the radius r. Note that for this formula, π is the constant of proportionality.

DIRECT VARIATION AS nTH POWER

The following statements are equivalent.

1. y **varies directly as the nth power** of x.
2. y is **directly proportional to the nth power** of x.
3. $y = kx^n$ for some constant k.

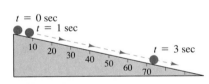

FIGURE 3.72

EXAMPLE 2   Direct Variation as a Power

The distance a ball rolls down an inclined plane is directly proportional to the square of the time it rolls. During the first second the ball rolls 8 feet. (See Figure 3.72.)

a. Write an equation relating the distance traveled to the time.
b. How far will the ball roll during the first 3 seconds?

Solution

a. Letting d be the distance (in feet) the ball rolls and t be the time (in seconds),

$$d = kt^2.$$

Now, since $d = 8$ when $t = 1$, you can see that $k = 8$. Thus, the equation relating distance to time is

$$d = 8t^2.$$

b. When $t = 3$, the distance traveled is

$$d = 8(3^2) = 8(9) = 72 \text{ feet.}$$

In Examples 1 and 2 the direct variations were such that an *increase* in one variable corresponded to an *increase* in the other variable. For example, in the model

$$d = \frac{1}{5}F, \qquad F > 0$$

an increase in F results in an increase in d. You should not, however, assume that this always occurs with direct variation. For example, in the model

$$y = -3x$$

an increase in x results in a *decrease* in y, and yet we say that y varies directly as x.

Inverse Variation

A third type of variation is **inverse variation.**

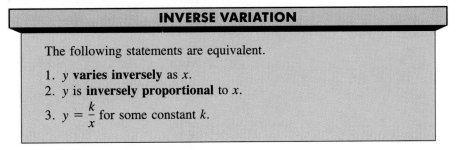

> **INVERSE VARIATION**
>
> The following statements are equivalent.
>
> 1. y **varies inversely** as x.
> 2. y is **inversely proportional** to x.
> 3. $y = \dfrac{k}{x}$ for some constant k.

EXAMPLE 3   Inverse Variation

A gas law states that the volume of an enclosed gas varies directly as the temperature *and* inversely as the pressure. The pressure of a gas is 0.75 kilograms per square centimeter when the temperature is 294° K and the volume is 8000 cubic centimeters.

a. Write an equation relating the pressure, temperature, and volume of this gas.
b. Find the pressure when the temperature is 300° K and the volume is 7000 cubic centimeters.

Solution

a. Let

$$V = \text{volume (in cubic centimeters)}$$
$$P = \text{pressure (in kilograms per square centimeter)}$$
$$T = \text{temperature (in degrees Kelvin)}.$$

Since V varies directly as T *and* inversely as P,

$$V = \frac{kT}{P}.$$

REMARK   If x and y are related by an equation of the form

$$y = \frac{k}{x^n}$$

then y varies inversely as the nth power of x (or y is inversely proportional to the nth power of x).

Note that the same constant of proportionality can be used for the direct variation of T and the inverse variation of P. Now, since $P = 0.75$ when $T = 294$ and $V = 8000$,

$$8000 = \frac{k(294)}{0.75}$$

$$\frac{8000(0.75)}{294} = k$$

$$k = \frac{6000}{294} = \frac{1000}{49}.$$

Thus, the equation relating pressure, temperature, and volume is

$$V = \frac{1000}{49}\left(\frac{T}{P}\right).$$

b. When $T = 300$ and $V = 7000$, the pressure is

$$P = \frac{1000}{49}\left(\frac{300}{7000}\right) = \frac{300}{343} \approx 0.87 \text{ kilogram per square centimeter.}$$

Joint Variation

In Example 3, note that when direct and inverse variation occur in the same statement, we couple them with the word "and." To describe two different *direct* variations in the same statement, you can use the word **jointly.**

JOINT VARIATION

The following statements are equivalent.

1. z **varies jointly** as x and y.
2. z is **jointly proportional** to x and y.
3. $z = kxy$ for some constant k.

REMARK If x, y, and z are related by an equation of the form

$$z = kx^n y^m$$

then z varies jointly as the nth power of x and the mth power of y.

EXAMPLE 4 Joint Variation

The *simple* interest for a certain savings account is jointly proportional to the time and the principal. After one quarter (three months), the interest on a principal of $5000 is $106.25.

a. Write an equation relating the interest, principal, and time.
b. Find the interest after three quarters.

Solution

a. Let I = interest (in dollars), P = principal (in dollars), and t = time (in years). Since I is jointly proportional to P and t,

$$I = kPt.$$

For $I = 106.25$, $P = 5000$, and $t = \frac{1}{4}$,

$$106.25 = k(5000)\left(\frac{1}{4}\right), \text{ which implies that } k = \frac{4(106.25)}{5000} = 0.085.$$

Thus, the equation relating interest, principal, and time is

$$I = 0.085Pt$$

which is the familiar equation for simple interest where the constant of proportionality, 0.085, represents an annual percentage rate of 8.5%.

b. When $P = \$5000$ and $t = \frac{3}{4}$, the interest is

$$I = (0.085)(5000)\left(\frac{3}{4}\right) = \$318.75.$$

DISCUSSION

PROBLEM

Mathematical Models

In this section, you have learned about four basic types of variation: direct variation, direct variation as nth power, inverse variation, and joint variation. Find a model for each of the following situations and classify the model as one of the four basic types of variation. In each case identify the constant of proportionality.

1. Let A represent the area of a rectangle and let l and w represent the length and width of the rectangle. Find a model that gives A in terms of l and w.
2. Let k represent the speed of an automobile in kilometers per hour and let m represent the speed in miles per hour. Find a model that gives k in terms of m.
3. Let V represent the volume of a sphere and let r represent the radius of the sphere. Find a model that gives V in terms of r.
4. Let t represent the time required for an automobile that is traveling a constant rate of r to complete a 50-mile trip. Find a model that gives t in terms of r.

Can you find four other examples like these—one for each of the basic types of variation?

WARM UP

The following warm-up exercises involve skills that were covered in earlier sections. You will use these skills in the exercise set for this section.

In Exercises 1–6, solve for k.

1. $15 = k(45)$

2. $9 = k(4^2)$

3. $20 = \dfrac{k(15)}{32}$

4. $30 = \dfrac{k(0.2)}{0.5}$

5. $110 = k(27)(0.4)$

6. $210 = k(4^2)(16)$

In Exercises 7–10, find the indicated value.

7. Let $d = 2.7r$. Find d when $r = 10$.

8. Let $s = 3tp^3$. Find s when $t = 2$ and $p = \frac{1}{3}$.

9. Let $R = 4t/h$. Find R when $t = 7$ and $h = 13$.

10. Let $M = 14rst$. Find M when $r = 0.01$, $s = 150$, and $t = 7.5$.

EXERCISES for Section 3.8

In Exercises 1–14, find a mathematical model for the verbal statement.

1. A varies directly as the square of r. $A = kr^2$

2. V varies directly as the cube of e.

3. y varies inversely as the square of x. $y = \dfrac{k}{x^2}$

4. h varies inversely as the square root of s. $h = \dfrac{k}{s^{\frac{1}{2}}}$

5. z is proportional to the cube root of u.

6. x is inversely proportional to $t + 1$.

7. z varies jointly as u and v. $z =$

8. V varies jointly as l, w, and h. $V = LWHK$

9. F varies directly as g and inversely as the square of r.

10. z is jointly proportional to the square of x and the cube of y.

11. *Boyle's Law*   For constant temperature, the pressure P of a gas is inversely proportional to the volume V of the gas.

12. *Newton's Law of Cooling*   The rate of change R of the temperature of an object is proportional to the difference between the temperature T of the object and the temperature T_e of the environment in which the object is placed.

13. *Newton's Law of Universal Gravitation*   The gravitational attraction F between two objects of masses m_1 and m_2 is proportional to the product of the masses and inversely proportional to the square of the distance r between the objects.

14. *Logistics Growth*   The rate of growth R of a population is jointly proportional to the size S of the population and the difference between S and the maximum size L that the environment can support.

In Exercises 15–20, write a sentence using variation terminology to describe the formula.

15. Area of a Triangle: $A = \frac{1}{2}bh$

16. Surface Area of a Sphere: $S = 4\pi r^2$

17. Volume of a Sphere: $V = \frac{4}{3}\pi r^3$

18. Volume of a Right Circular Cylinder: $V = \pi r^2 h$

19. Average Speed: $r = \dfrac{d}{t}$

20. Free Vibrations: $\omega = \sqrt{\dfrac{kg}{W}}$

In Exercises 21–36, find a mathematical model representing the statement. (In each case determine the constant of proportionality.)

21. y varies directly as x. ($y = 25$ when $x = 10$.)

22. y is directly proportional to x. ($y = 8$ when $x = 24$.)

23. A varies directly as the square of r. ($A = 9\pi$ when $r = 3$.)

24. s is directly proportional to the square of t. ($s = 64$ when $t = 2$.)

25. y varies inversely as x. ($y = 3$ when $x = 25$.)

26. y is inversely proportional to x. ($y = 7$ when $x = 4$.)

27. h is inversely proportional to the third power of t. ($h = \frac{3}{16}$ when $t = 4$.)

28. R varies inversely as the square of s. ($R = 80$ when $s = \frac{1}{5}$.)

29. z varies jointly as x and y. ($z = 64$ when $x = 4$ and $y = 8$.)

30. z is jointly proportional to x and y. ($z = 32$ when $x = 10$ and $y = 16$.)

31. F is jointly proportional to r and the third power of s. ($F = 4158$ when $r = 11$ and $s = 3$.)

32. P varies directly as x and inversely as the square of y. ($P = \frac{28}{3}$ when $x = 42$ and $y = 9$.)

33. z varies directly as the square of x and inversely as y. ($z = 6$ when $x = 6$ and $y = 4$.)

34. v varies jointly as p and q and inversely as the square of s. ($v = 1.5$ when $p = 4.1$, $q = 6.3$, and $s = 1.2$.)

35. S varies directly as L and inversely as $L - S$. ($S = 4$ when $L = 6$.)

36. P is jointly proportional to S and $L - S$. ($P = 10$ when $S = 4$ and $L = 6$.)

Hooke's Law In Exercises 37–40, use Hooke's Law as stated in Example 1 of this section.

37. A force of 50 pounds stretches a spring 5 inches (see figure).
(a) How far will a force of 20 pounds stretch the spring?
(b) What force is required to stretch the spring 1.5 inches?

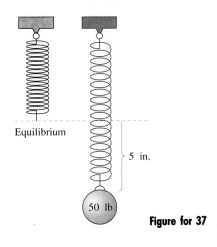

Equilibrium

5 in.

50 lb

Figure for 37

38. A force of 50 pounds stretches a spring 3 inches.
(a) How far will a force of 20 pounds stretch the spring?
(b) What force is required to stretch the spring 1.5 inches?

39. The coiled spring of a toy supports the weight of a child. The spring compresses a distance of 1.9 inches under the weight of a 25-pound child. The toy will not work properly if its spring is compressed more than 3 inches. What is the weight of the heaviest child who should be allowed to use the toy?

40. An overhead garage door has two springs, one on each side of the door (see figure). A force of 15 pounds is required to stretch each spring 1 foot. A pulley system allows the springs to stretch only one-half the distance the door travels. The door moves a total of 8 feet and the springs are at their natural length when the door is open. Find the combined lifting force applied to the door by the springs when the door is in the closed position.

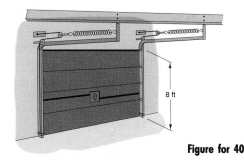

8 ft

Figure for 40

Erosion In Exercises 41 and 42, use the fact that the diameter of a particle moved by a stream varies approximately as the square of the velocity of the stream.

41. A stream with a velocity of $\frac{1}{4}$ mile per hour can move coarse sand particles of about 0.02 inch diameter. What must the velocity be to carry particles with a diameter of 0.12 inch?

42. A stream of velocity v can move particles of diameter d or less. By what factor does d increase when the velocity is doubled?

Electrical Resistance In Exercises 43 and 44, use the fact that the resistance of a wire carrying electrical current is directly proportional to its length and inversely proportional to its cross-sectional area.

43. If #28 copper wire (which has a diameter of 0.0126 inch) has a resistance of 66.17 ohms per thousand feet, what length of #28 copper wire will produce a resistance of 33.5 ohms?

44. A 14-foot piece of copper wire produces a resistance of 0.05 ohms. Use the constant of proportionality of Exercise 43 to find the diameter of the wire.

45. *Free Fall* Neglecting air resistance, the distance s that an object falls varies directly as the square of the time t it has been falling. An object falls a distance of 144 feet in 3 seconds. How far will it fall in 5 seconds?

46. *Stopping Distance* The stopping distance d of an automobile is directly proportional to the square of its speed s. A car required 75 feet to stop when its speed was 30 miles per hour. Estimate the stopping distance if the brakes are applied when the car is traveling at 50 miles per hour.

47. *Comparative Shopping* The prices of 9-inch, 12-inch, and 15-inch diameter pizzas are $6.78, $9.78, and $12.18, respectively. One would expect that the price of a pizza of a certain size would be directly proportional to its surface area. Is that the case for these pizzas? If not, which pizza is the best buy?

48. *Demand for a Product* A company has found that the demand for its product varies inversely as the price of the product. When the price is $3.75, the demand is 500 units. Approximate the demand for a price of $4.25.

49. *Illumination of a Light* The illumination from a light source varies inversely as the square of the distance from the light source. When the distance from a light source is doubled, how does the illumination change?

50. *Safe Load of a Beam* The load that can be safely supported by a horizontal beam varies jointly as the width of the beam and the square of its depth and inversely as the length of the beam. Determine what happens to the safe load under the following conditions.
   (a) The width and length of the beam are doubled.
   (b) The width and depth of the beam are doubled.
   (c) All three of the dimensions are doubled.
   (d) The depth of the beam is halved.

REVIEW EXERCISES for Chapter 3

In Exercises 1–6, find (a) the distance between the two points, (b) the coordinates of the midpoint of the line segment between the two points, (c) an equation of the line through the two points, and (d) an equation of the circle whose diameter is the line segment between the two points.

1. (0, 0), (0, 10)
2. (−1, 4), (2, 0)
3. (2, 1), (14, 6)
4. (−2, 2), (3, −10)
5. (−1, 0), (6, 2)
6. (1, 6), (4, 2)

51. *Fluid Flow* The velocity v of a fluid flowing in a conduit is inversely proportional to the cross-sectional area of the conduit. (Assume the volume of the flow per unit of time is held constant.) Determine the change of velocity of water flowing from a hose when a person places a finger over the end of the hose to decrease its cross-sectional area by 25%.

52. *Fluid Flow* Use the fluid velocity model of Exercise 51 to determine the effect on the velocity of a stream when it is dredged to increase its cross-sectional area by one-third.

In Exercises 53–56, use the given value of k to complete the table for the direct variation model $y = kx^2$. Plot the points on the rectangular coordinate system.

x	2	4	6	8	10
$y = kx^2$					

53. $k = 1$ **54.** $k = 2$
55. $k = \frac{1}{2}$ **56.** $k = \frac{1}{4}$

In Exercises 57–60, use the given value of k to complete the table for the inverse variation model $y = k/x^2$. Plot the points on the rectangular coordinate system.

x	2	4	6	8	10
$y = \dfrac{k}{x^2}$					

57. $k = 2$ **58.** $k = 5$
59. $k = 10$ **60.** $k = 20$

In Exercises 7 and 8, use the Midpoint Formula to estimate the sales of a company for 1991. Assume the sales followed a linear growth pattern.

7.

Year	1989	1993
Sales	640,000	810,000

8.

Year	1989	1993
Sales	3,250,000	5,690,000

In Exercises 9–12, find t so that the three points are collinear.

9. $(-2, 5)$, $(0, t)$, $(1, 1)$ **10.** $(-6, 1)$, $(1, t)$, $(10, 5)$
11. $(1, -4)$, $(t, 3)$, $(5, 10)$ **12.** $(-3, 3)$, $(t, -1)$, $(8, 6)$

In Exercises 13–16, show that the points form the vertices of the indicated polygon.

13. Parallelogram: $(1, 1)$, $(8, 2)$, $(9, 5)$, $(2, 4)$
14. Isosceles triangle: $(4, 5)$, $(1, 0)$, $(-1, 2)$
15. Right triangle: $(-1, -1)$, $(10, 7)$, $(2, 18)$
16. Square: $(-4, 0)$, $(1, -3)$, $(4, 2)$, $(-1, 5)$

In Exercises 17–26, find the intercepts of the graph and check for symmetry with respect to each of the coordinate axes and the origin.

17. $2y^2 = x^3$ **18.** $x^2 + (y + 2)^2 = 4$

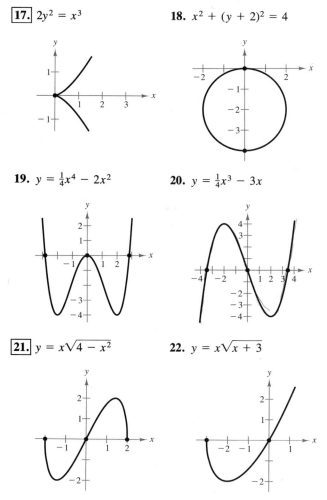

19. $y = \frac{1}{4}x^4 - 2x^2$ **20.** $y = \frac{1}{4}x^3 - 3x$

21. $y = x\sqrt{4 - x^2}$ **22.** $y = x\sqrt{x + 3}$

23. $y = x^3 - 3x^2$ **24.** $x^3 + y^3 - 3xy = 0$

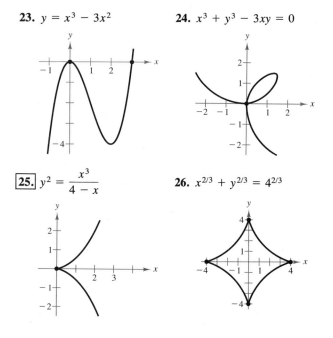

25. $y^2 = \dfrac{x^3}{4 - x}$ **26.** $x^{2/3} + y^{2/3} = 4^{2/3}$

In Exercises 27–30, determine the center and radius of the circle and sketch its graph.

27. $x^2 + y^2 - 12x - 8y + 43 = 0$
28. $x^2 + y^2 - 20x - 10y + 100 = 0$
29. $4x^2 + 4y^2 - 4x - 40y + 92 = 0$
30. $5x^2 + 5y^2 - 14y = 0$

In Exercises 31–44, sketch a graph of the equation.

31. $y - 2x - 3 = 0$ **32.** $3x + 2y + 6 = 0$
33. $x - 5 = 0$ **34.** $y = 8 - |x|$
35. $y = \sqrt{5 - x}$ **36.** $y = \sqrt{x + 2}$
37. $y + 2x^2 = 0$ **38.** $y = x^2 - 4x$
39. $y = \sqrt{25 - x^2}$ **40.** $y = |x^2 - 4x|$
41. $y = |25 - x^2|$ **42.** $y = -(x - 4)^2$
43. $y = \frac{1}{4}(x + 1)^3$ **44.** $y = 4 - (x - 4)^2$

In Exercises 45–48, find an equation of the line that passes through the given point and has the specified slope. Sketch the graph of the line.

Point	Slope
45. $(0, -5)$	$m = \frac{3}{2}$
46. $(-2, 6)$	$m = 0$
47. $(3, 0)$	$m = -\frac{2}{3}$
48. $(5, 4)$	m is undefined.

In Exercises 49 and 50, write an equation of the line through the point (a) parallel to the given line, and (b) perpendicular to the given line.

Point	Line
49. (3, −2)	$5x - 4y = 8$
50. (−8, 3)	$2x + 3y = 5$

51. *Fourth Quarter Sales*   During the second and third quarters of the year, a business had sales of $160,000 and $185,000, respectively. If the growth of sales follows a linear pattern, estimate sales during the fourth quarter.

52. *Dollar Value*   The dollar value of a product in 1990 was $85. The item will increase in value at an expected rate of $3.75 per year. Write a linear equation that gives the dollar value V of the product in terms of the year t. (Let $t = 0$ represent 1990.) Use this model to estimate the dollar value of the product in 1995.

In Exercises 53–56, evaluate the function at the specified values of the independent variable. Simplify your answers.

53. $f(x) = x^2 + 1$
  (a) $f(2)$ (b) $f(-4)$
  (c) $f(t^2)$ (d) $-f(x)$

54. $g(x) = x^{4/3}$
  (a) $g(8)$ (b) $g(t + 1)$
  (c) $\dfrac{g(8) - g(1)}{8 - 1}$ (d) $g(-x)$

55. $h(x) = 6 - 5x^2$
  (a) $h(2)$ (b) $h(x + 3)$
  (c) $\dfrac{h(4) - h(2)}{4 - 2}$ (d) $\dfrac{h(x + \Delta x) - h(x)}{\Delta x}$

56. $f(t) = \sqrt[4]{t}$
  (a) $f(16)$ (b) $f(t + 5)$
  (c) $\dfrac{f(16) - f(0)}{16}$ (d) $f(t + \Delta t)$

In Exercises 57–62, determine the domain of the function.

57. $f(x) = \sqrt{25 - x^2}$

58. $f(x) = 3x + 4$

59. $g(s) = \dfrac{5}{3s - 9}$

60. $f(x) = \sqrt{x^2 + 8x}$

61. $h(x) = \dfrac{x}{x^2 - x - 6}$

62. $h(t) = |t + 1|$

In Exercises 63–68, (a) find f^{-1}, (b) sketch the graphs of f and f^{-1} on the same coordinate plane, and (c) verify that $f^{-1}(f(x)) = x = f(f^{-1}(x))$.

63. $f(x) = \frac{1}{2}x - 3$

64. $f(x) = 5x - 7$

65. $f(x) = \sqrt{x + 1}$

66. $f(x) = x^3 + 2$

67. $f(x) = x^2 - 5, \quad x \geq 0$

68. $f(x) = \sqrt[3]{x + 1}$

In Exercises 69–72, restrict the domain of the function f to an interval where the function is increasing and determine f^{-1} over that interval.

69. $f(x) = 2(x - 4)^2$

70. $f(x) = |x - 2|$

71. $f(x) = \sqrt{x^2 - 4}$

72. $f(x) = x^{4/3}$

In Exercises 73–80, let $f(x) = 3 - 2x$, $g(x) = \sqrt{x}$ and $h(x) = 3x^2 + 2$, and find the indicated value.

73. $(f - g)(4)$

74. $(f + h)(5)$

75. $(fh)(1)$

76. $\left(\dfrac{g}{h}\right)(1)$

77. $(h \circ g)(7)$

78. $(g \circ f)(-2)$

79. $g^{-1}(3)$

80. $(h \circ f^{-1})(1)$

81. *Vertical Motion*   The velocity of a ball thrown vertically upward from ground level is

$$v(t) = -32t + 48$$

where t is the time in seconds and v is the velocity in feet per second.
(a) Find the velocity when $t = 1$.
(b) Find the time when the ball reaches its maximum height. (*Hint:* Find the time when $v(t) = 0$.)
(c) Find the velocity when $t = 2$.

82. *Cost and Profit*   A company produces a product for which the variable cost is \$5.35 per unit and the fixed costs are \$16,000. The company sells the product for \$8.20, and can sell all that it produces.
(a) Find the total cost as a function of the number of units produced x.
(b) Find the profit as a function of x.

83. *Dimensions of a Rectangle*   A wire 24 inches long is cut into four pieces to form a rectangle whose shortest side has length x. Express the area A of the rectangle as a function of x. Determine the domain of the function and sketch its graph over that domain.

84. *Cost of a Phone Call*   Suppose the cost of a telephone call between Dallas and Philadelphia is \$0.70 for the first minute and \$0.38 for each additional minute (or portion thereof). A model for the total cost of the phone call is

$$C = 0.70 + 0.38[\![x]\!]$$

where C is the total cost of the call in dollars and x is the length of the call in minutes. Sketch the graph of this function.

In Exercises 85–88, find a mathematical model representing the statement. (In each case determine the constant of proportionality.)

85. F is jointly proportional to x and the square root of y. ($F = 6$ when $x = 9$ and $y = 4$.)

86. R varies inversely as the cube of x. ($R = 128$ when $x = 2$.)

87. z varies directly as the square of x and inversely as y. ($z = 16$ when $x = 5$ and $y = 2$.)

88. w varies jointly as x and y and inversely as the cube of z. ($w = \frac{44}{9}$ when $x = 12$, $y = 11$, and $z = 6$.)

89. *Wind Power*   The power P produced by a wind turbine is proportional to the cube of the wind speed S. A wind speed of 27 miles per hour produces a power output of 750 kilowatts. Find the output for a wind speed of 40 miles per hour.

90. *Frictional Force*   The frictional force F between the tires and the road required to keep a car on a curved section of a highway is directly proportional to the square of the speed s of the car. If the speed of the car is doubled, the force will change by what factor?

CUMULATIVE TEST for Chapters 1–3

Take this test as you would take a test in class. After you are done, check your work with the answers given in the back of the book.

1. Evaluate: $\left(\frac{12}{5} \div 9\right) - \left(10 \times \frac{1}{45}\right)$

2. Simplify: $(3z^{-4})(6^2 z^2)$

3. Simplify: $\dfrac{8x^2 y^{-3}}{30x^{-1} y^2}$

4. Simplify: $\sqrt{24x^4 y^3}$

5. Expand and simplify: $(x - 2)(x^2 + x - 3)$

6. Factor completely: $x - 5x^2 - 6x^3$

7. Subtract and simplify: $\dfrac{2}{s + 3} - \dfrac{1}{s + 1}$

8. Simplify: $\dfrac{\dfrac{2x + 1}{2\sqrt{x}} - \sqrt{x}}{2x + 1}$

9. Solve: $-5(x + 3) + 2x = 3(2x - 7)$

10. Solve: $2\left(x + \dfrac{10}{x}\right) = 13$

11. Solve: $x^2 + 4x + 5 = 0$

12. Solve: $\sqrt{9x + 7} - x = 3$

13. Solve and sketch the solution on the real number line: $\left|\dfrac{x - 2}{3}\right| < 1$

14. Given the points $(-1, 3)$ and $(5, 0)$, find:
(a) the distance between them.
(b) an equation of a line through them.

15. Sketch a graph of the equation $2x - 3y - 6 = 0$.

16. Sketch a graph of the equation $x^2 + y^2 - 6y = 0$.

17. Determine the domain of the function $g(x) = \dfrac{1}{x - 5}$.

18. Given the function $f(x) = 3 - x^2$ evaluate and/or simplify each of the following.
(a) $f(3)$ (b) $\dfrac{f(1 + h) - f(1)}{h}$

19. Find $f \circ g$ if $f(x) = \sqrt{x}$ and $g(x) = x^2 + 3$.

20. Find the inverse of the function $F(x) = 2x + 5$.

21. An inheritance of $12,000 is divided between two investments earning 7.5% and 9% simple interest. How much is in each investment if the total interest for one year is $960?

22. Three gallons of a mixture is 60% water by volume. Determine the number of gallons of water that must be added to bring the mixture to 75% water.

23. A group of n people decides to buy a $36,000 minibus for a charitable organization. Each person will pay an equal share of the cost. If 3 additional people were to join the group, the cost per person would decrease by $1,000. Find n.

OVERVIEW

The material in this chapter is both classic and current. It is classic because it discusses the classic problem in algebra—how to find the solutions of a polynomial equation. It is current because it shows how such solutions can be used to answer questions about real life. For instance, Exercise 58 on page 280 uses the solution of a polynomial equation to find the air-fuel ratio in a car engine that produces a given emission of nitric oxide.

As you study, remember that many of the results can be viewed in four different ways, as summarized on page 261. That is, if $x = a$ is the *zero* of a polynomial function *f*, then *a* is a *solution* of the polynomial equation $f(x) = 0$, $(x - a)$ is a *factor* of the polynomial $f(x)$, and $(a, 0)$ is an *x-intercept* of the graph of *f*.

Polynomial Functions: Graphs and Zeros

4.1 Quadratic Functions

The Graph of a Quadratic Function / The Standard Form of a Quadratic Function / Applications

The Graph of a Quadratic Function

In this chapter you will study polynomial functions, the most widely used functions in algebra.

DEFINITION OF POLYNOMIAL FUNCTION

Let n be a nonnegative integer and let $a_n, a_{n-1}, \ldots, a_2, a_1, a_0$ be real numbers with $a_n \neq 0$. The function

$$f(x) = a_n x^n + a_{n-1} x^{n-1} + \cdots + a_2 x^2 + a_1 x + a_0$$

is a **polynomial function of x with degree n.**

The polynomial function $f(x) = a$, $a \neq 0$, has degree 0 and is a **constant function.** The polynomial function $f(x) = ax + b$, $a \neq 0$, has degree 1 and is a **linear function.** In Chapter 3, you saw that the graph of the linear

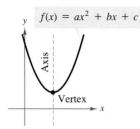

$f(x) = ax^2 + bx + c$

$a > 0$: Parabola opens upward.

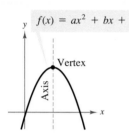

$f(x) = ax^2 + bx + c$

$a < 0$: Parabola opens downward.

FIGURE 4.1

function $f(x) = ax + b$ is a line whose slope is a and whose y-intercept is $(0, b)$. In this section we look at second-degree polynomial functions (**quadratic functions**).

DEFINITION OF QUADRATIC FUNCTION

Let a, b, and c be real numbers with $a \neq 0$. The function of x given by

$$f(x) = ax^2 + bx + c$$

is a **quadratic function.** The graph of a quadratic function is a **parabola.**

All parabolas are symmetric with respect to a line called the **axis of symmetry,** or simply the **axis** of the parabola. The point where the axis intersects the parabola is the **vertex** of the parabola, as shown in Figure 4.1. If the leading coefficient is positive, then the graph of $f(x) = ax^2 + bx + c$ is a parabola that opens upward, and if the leading coefficient is negative, then the graph is a parabola that opens downward.

The simplest type of quadratic function is $f(x) = ax^2$. Its graph is a parabola whose vertex is $(0, 0)$. If $a > 0$, then the vertex is the *minimum* point on the graph, and if $a < 0$, then the vertex is the *maximum* point on the graph, as shown in Figure 4.2. When sketching the graph of $f(x) = ax^2$, it is helpful to use the graph of $y = x^2$ as a reference, as discussed in Section 3.5.

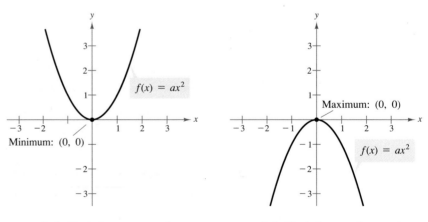

$a > 0$: Parabola opens upward. $a < 0$: Parabola opens downward.

FIGURE 4.2

EXAMPLE 1 Sketching the Graphs of Simple Quadratic Functions

Sketch the graphs of the functions.

a. $f(x) = \dfrac{1}{3}x^2$

b. $g(x) = 2x^2$

REMARK In Example 1, note that the coefficient a determines how widely the parabola given by $f(x) = ax^2$ opens. If $|a|$ is small, the parabola opens more widely than if $|a|$ is large.

Solution

a. Compared with $y = x^2$, each output of f "shrinks" by a factor of $\frac{1}{3}$. The result is a parabola that opens upward and is broader than the parabola represented by $y = x^2$, as shown in Figure 4.3.

b. Each output of g "stretches" by a factor of 2, creating the more narrow parabola shown in Figure 4.4.

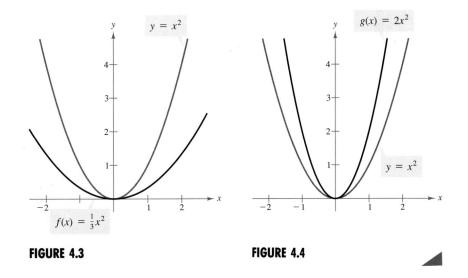

FIGURE 4.3   **FIGURE 4.4**

Recall from Section 3.5 that the graphs of $y = f(x \pm c)$, $y = f(x) \pm c$, and $y = -f(x)$ are rigid transformations of the graph of $y = f(x)$.

$y = f(x \pm c)$	*Horizontal shift*
$y = f(x) \pm c$	*Vertical shift*
$y = -f(x)$	*Reflection*

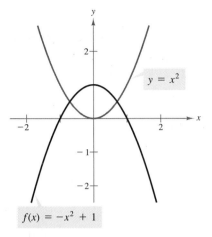

$f(x) = -x^2 + 1$

FIGURE 4.5

EXAMPLE 2 Sketching a Parabola

Sketch the graphs of the quadratic functions.

a. $f(x) = -x^2 + 1$ **b.** $g(x) = (x + 2)^2 - 3$

Solution

a. With respect to the graph of $y = x^2$, the negative coefficient in $f(x) = -x^2 + 1$ reflects the graph *downward* and the positive constant term shifts the vertex *up one unit*. The graph of f is shown in Figure 4.5. Note that the axis of the parabola is the y-axis and the vertex is $(0, 1)$.

b. With respect to the graph of $y = x^2$, the graph of $g(x) = (x + 2)^2 - 3$ is obtained by a horizontal shift two units *to the left* and a vertical shift three units *down,* as shown in Figure 4.6. Note that the axis of the parabola is the vertical line $x = -2$ and the vertex is $(-2, -3)$.

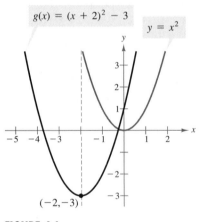
$g(x) = (x + 2)^2 - 3$
$y = x^2$
$(-2,-3)$

FIGURE 4.6

The Standard Form of a Quadratic Function

The equation in Example 2(b) is written in **standard form:**

$$f(x) = a(x - h)^2 + k.$$

This form is especially convenient for sketching a parabola because it identifies the vertex of the parabola.

STANDARD FORM OF A QUADRATIC FUNCTION

The quadratic function

$$f(x) = a(x - h)^2 + k, \qquad a \neq 0$$

is said to be in **standard form.** The graph of f is a parabola whose axis is the vertical line $x = h$ and whose vertex is the point (h, k). If $a > 0$, the parabola opens upward and if $a < 0$, the parabola opens downward.

EXAMPLE 3 Writing a Quadratic Function in Standard Form

Sketch the graph of $f(x) = 2x^2 + 8x + 7$ and identify the vertex and x-intercepts.

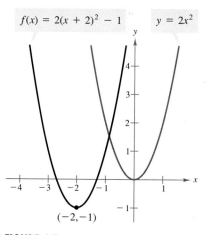

$f(x) = 2(x + 2)^2 - 1$ $y = 2x^2$

$(-2, -1)$

FIGURE 4.7

REMARK When completing the square to write a function, $f(x)$, in standard form, bear in mind that you are rewriting an *expression*, not an equation (as was the case with circles in Section 3.2). Thus, you must do all algebraic steps on *one* side of the equal sign and no factors can be divided out.

Solution

Write the quadratic function in standard form by completing the square. Notice that the first step is to factor out any coefficient of x^2 that is different from 1.

$$
\begin{aligned}
f(x) &= 2x^2 + 8x + 7 && \text{Given form} \\
&= 2(x^2 + 4x) + 7 && \text{Factor 2 out of x terms} \\
&= 2(x^2 + 4x + 4 - 4) + 7 && \text{Add and subtract 4 within} \\
& && \text{parentheses} \\
&= 2(x^2 + 4x + 4) - 2(4) + 7 && \text{Regroup terms} \\
&= 2(x^2 + 4x + 4) - 8 + 7 && \text{Simplify} \\
&= 2(x + 2)^2 - 1 && \text{Standard form}
\end{aligned}
$$

2^2

From the standard form, you can see that the graph of f is a parabola that opens upward with vertex $(-2, -1)$. This corresponds to a left shift of two units and a downward shift of one unit relative to the graph of $y = 2x^2$, as shown in Figure 4.7. The x-intercepts can be found by using the quadratic formula to solve $2x^2 + 8x + 7 = 0$ or by extracting square roots for the equation $2(x + 2)^2 - 1 = 0$. In either case, $x = -2 \pm (\sqrt{2}/2)$.

EXAMPLE 4 Writing a Quadratic Function in Standard Form

Sketch the graph of $f(x) = -x^2 + 6x - 8$ and identify the vertex.

Solution

As in Example 3, we begin by writing the quadratic function in standard form.

$$
\begin{aligned}
f(x) &= -x^2 + 6x - 8 && \text{Given form} \\
&= -(x^2 - 6x) - 8 && \text{Factor } -1 \text{ out of x terms} \\
&= -(x^2 - 6x + 9 - 9) - 8 && \text{Add and subtract 9 within} \\
& && \text{parentheses} \\
&= -(x^2 - 6x + 9) - (-9) - 8 && \text{Regroup terms} \\
&= -(x - 3)^2 + 1 && \text{Standard form}
\end{aligned}
$$

3^2

Thus, the graph of f is a parabola that opens downward with vertex at $(3, 1)$, as shown in Figure 4.8.

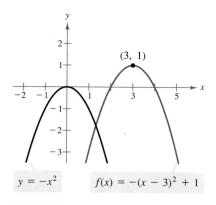

$(3, 1)$

$y = -x^2$ $f(x) = -(x - 3)^2 + 1$

FIGURE 4.8

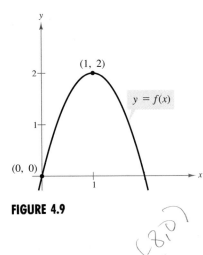

FIGURE 4.9

EXAMPLE 5   Finding the Equation of a Parabola

Find an equation for the parabola whose vertex is (1, 2) and passes through the point (0, 0), as shown in Figure 4.9.

Solution

Since the parabola has a vertex at $(h, k) = (1, 2)$, the equation must have the form

$$f(x) = a(x - 1)^2 + 2. \qquad \textit{Standard form}$$

Because the parabola passes through the point (0, 0), it follows that $f(0) = 0$. Thus,

$$0 = a(0 - 1)^2 + 2 \quad \rightarrow \quad a = -2$$

which implies that the equation is

$$f(x) = -2(x - 1)^2 + 2 = -2x^2 + 4x.$$

Applications

We know that the vertex (h, k) of a parabola is either its lowest point (Example 3) or its highest point (Example 4) and that $k = f(h)$ is the corresponding minimum or maximum value of f. Many applications involve finding the maximum or minimum value of a quadratic function. By writing the quadratic function $f(x) = ax^2 + bx + c$ in standard form

$$f(x) = a\left(x + \frac{b}{2a}\right)^2 + \left(c - \frac{b^2}{4a}\right)$$

you can see that the vertex occurs at $x = -b/2a$, and you can determine the following.

1. If $a > 0$, then the quadratic function $f(x) = ax^2 + bx + c$ has a *minimum* that occurs at $x = -b/2a$.
2. If $a < 0$, then the quadratic function $f(x) = ax^2 + bx + c$ has a *maximum* that occurs at $x = -b/2a$.

In either case, you can find the minimum or maximum by evaluating the function at $x = -b/2a$.

EXAMPLE 6   The Maximum Height of a Baseball

A baseball is hit 3 feet above ground at a velocity of 100 feet per second and at an angle of 45 degrees with respect to the ground. The path of the baseball is given by the function

$$f(x) = -0.0032x^2 + x + 3$$

where $f(x)$ is the height of the baseball (in feet) and x is the distance from home plate (in feet). What is the maximum height reached by the baseball?

Solution

For this quadratic function,

$$f(x) = ax^2 + bx + c$$
$$= -0.0032x^2 + x + 3.$$

Thus, $a = -0.0032$ and $b = 1$. Since the function has a maximum when $x = -b/2a$, you can conclude that the baseball reaches its maximum height when it is

$$x = -\frac{b}{2a} = -\frac{1}{2(-0.0032)} = 156.25 \text{ feet}$$

from home plate. At this distance, the maximum height is

$$f(156.25) = -0.0032(156.25)^2 + 156.25 + 3 = 81.125 \text{ feet.}$$

The path of the baseball is shown in Figure 4.10.

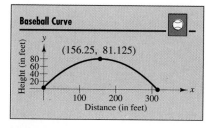

Baseball Curve

(156.25, 81.125)

Height (in feet) — 80, 60, 40, 20

Distance (in feet) — 100, 200, 300

FIGURE 4.10

EXAMPLE 7   Charitable Contributions

According to a survey conducted in 1990 by *Independent Sector*, the percent of their income that Americans give to charities is related to their household income. For families with an annual income of $100,000 or less, the percent is approximately

$$P = 0.0014x^2 - 0.1529x + 5.855, \qquad 5 \le x \le 100$$

where P is the percentage of annual income given and x is the annual income in 1000s of dollars. According to this model, what income level corresponds to the least percentage of charitable contributions?

Solution

There are two ways to answer this question. One is to sketch the graph of the quadratic function, as shown in Figure 4.11. From this graph, it appears that the minimum percentage corresponds to an income level of about $55,000. The other way to answer the question is to use the fact that the minimum point of the parabola occurs when $x = -b/2a$. For this function, you have $a = 0.0014$ and $b = -0.1529$. Thus,

$$x = -\frac{b}{2a} = -\frac{-0.1529}{2(0.0014)} \approx 54.6.$$

From this x-value, you can conclude that the minimum percentage corresponds to an income level of about $54,600.

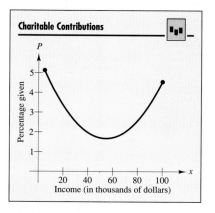

Charitable Contributions

P

Percentage given — 5, 4, 3, 2, 1

Income (in thousands of dollars) — 20, 40, 60, 80, 100

FIGURE 4.11

DISCUSSION

PROBLEM

Finding an
Equation for a
Curve

The parabola in Figure 4.12 has an equation of the form $y = x^2 + bx + c$. Try to find the equation for this parabola, and write a short paragraph about (or discuss) the method you used.

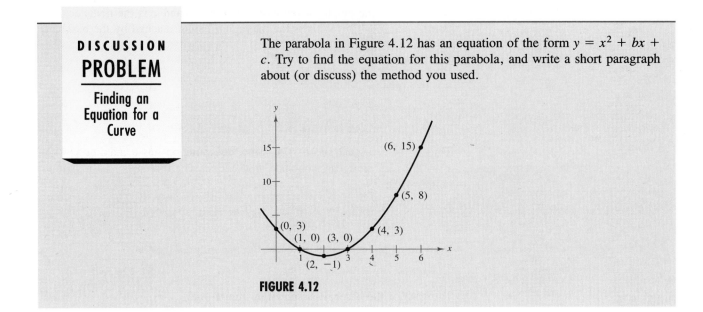

FIGURE 4.12

WARM UP

The following warm-up exercises involve skills that were covered in earlier sections. You will use these skills in the exercise set for this section.

In Exercises 1–4, solve the quadratic equations by factoring.

1. $2x^2 + 11x - 6 = 0$ **2.** $5x^2 - 12x - 9 = 0$

3. $3 + x - 2x^2 = 0$ **4.** $x^2 + 20x + 100 = 0$

In Exercises 5–8, solve the quadratic equations by completing the square.

5. $x^2 - 6x + 4 = 0$ **6.** $x^2 + 4x + 1 = 0$

7. $2x^2 - 16x + 25 = 0$ **8.** $3x^2 + 30x + 74 = 0$

In Exercises 9 and 10, use the Quadratic Formula to solve the quadratic equations.

9. $x^2 + 3x + 3 = 0$ **10.** $x^2 + 3x - 3 = 0$

EXERCISES for Section 4.1

In Exercises 1–6, match the given quadratic function with the correct graph. [The graphs are labeled (a), (b), (c), (d), (e), and (f).]

1. $f(x) = (x - 3)^2$
2. $f(x) = (x + 5)^2$
3. $f(x) = x^2 - 4$
4. $f(x) = 5 - x^2$
5. $f(x) = 4 - (x - 1)^2$
6. $f(x) = (x + 2)^2 - 2$

(a)
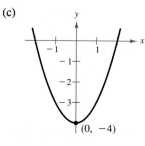

(b)
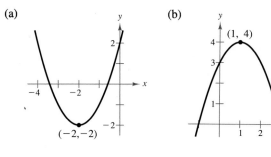

(c)
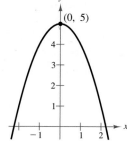

(d)
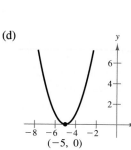

(e)
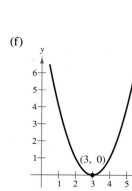

(f)

In Exercises 7–12, find an equation for the parabola.

7.
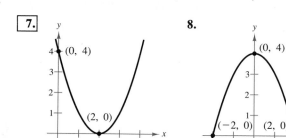

8.
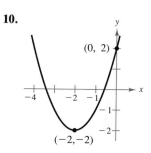

9.
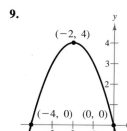

10.
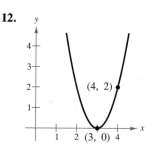

11.
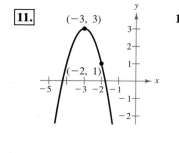

12.

In Exercises 13–30, sketch the graph of the quadratic function. Identify the vertex and intercepts.

13. $f(x) = x^2 - 5$
14. $f(x) = \frac{1}{2}x^2 - 4$
15. $f(x) = 16 - x^2$
16. $h(x) = 25 - x^2$
17. $f(x) = (x + 5)^2 - 6$
18. $f(x) = (x - 6)^2 + 3$
19. $h(x) = x^2 - 8x + 16$
20. $g(x) = x^2 + 2x + 1$
21. $f(x) = -(x^2 + 2x - 3)$
22. $g(x) = x^2 + 8x + 11$
23. $f(x) = x^2 - x + \frac{5}{4}$
24. $f(x) = x^2 + 3x + \frac{1}{4}$
25. $f(x) = -x^2 + 2x + 5$
26. $f(x) = -x^2 - 4x + 1$
27. $h(x) = 4x^2 - 4x + 21$
28. $f(x) = 2x^2 - x + 1$
29. $f(x) = 2x^2 - 16x + 31$
30. $g(x) = \frac{1}{2}(x^2 + 4x - 2)$

In Exercises 31–34, find the quadratic function that has the indicated vertex and whose graph passes through the given point.

31. Vertex: (3, 4); point: (1, 2)
32. Vertex: (2, 3); point: (0, 2)
33. Vertex: (5, 12); point: (7, 15)
34. Vertex: (−2, −2); point: (−1, 0)

In Exercises 35–40, find two quadratic functions whose graphs have the given x-intercepts. (One function has a graph that opens upward and the other has a graph that opens downward.)

35. (−1, 0), (3, 0)
36. $\left(-\frac{5}{2}, 0\right)$, (2, 0)
37. (0, 0), (10, 0)
38. (4, 0), (8, 0)
39. (−3, 0), $\left(-\frac{1}{2}, 0\right)$
40. (−5, 0), (5, 0)

In Exercises 41–44, find two positive real numbers that satisfy the requirements.

41. The sum is 110 and the product is a maximum.
42. The sum is S and the product is a maximum.
43. The sum of the first and twice the second is 24 and the product is a maximum.
44. The sum of two numbers is 50 and the product is a maximum.

Maximum Area In Exercises 45 and 46, consider a rectangle of length x and perimeter P (see figure). (a) Express the area A as a function of x and determine the domain of the function. (b) Sketch the graph of the area function. (c) Find the length and width of the rectangle of maximum area.

Figure for 45 and 46

45. $P = 100$ feet
46. $P = 36$ meters

47. *Maximum Area* A rancher has 200 feet of fencing to enclose two adjacent rectangular corrals (see figure). What dimensions will produce a maximum enclosed area?

Figure for 47

48. *Maximum Area* An indoor physical fitness room consists of a rectangular region with a semicircle on each end (see figure). The perimeter of the room is to be a 200-meter running track. What dimensions will produce a maximum area of the rectangle?

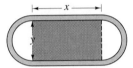

Figure for 48

49. *Maximum Revenue* Find the number of units that produce a maximum revenue

$$R = 900x - 0.1x^2$$

where R is the total revenue in dollars and x is the number of units sold.

50. *Maximum Revenue* Find the number of units that produce a maximum revenue

$$R = 100x - 0.0002x^2.$$

where R is the total revenue in dollars and x is the number of units sold.

51. *Minimum Cost* A manufacturer of lighting fixtures has daily production costs of

$$C = 800 - 10x + 0.25x^2$$

where C is the total cost in dollars and x is the number of units produced. How many fixtures should be produced each day to yield a minimum cost?

52. *Maximum Profit* Let x be the amount (in hundreds of dollars) a company spends on advertising, and let P be the profit, where

$$P = 230 + 20x - 0.5x^2.$$

What expenditure for advertising gives the maximum profit?

53. *Trajectory of a Ball*   The height y (in feet) of a ball thrown by a child is

$$y = -\tfrac{1}{12}x^2 + 2x + 4$$

where x is the horizontal distance (in feet) from where the ball is thrown (see figure).
(a) Sketch the path of the ball.
(b) How high was the ball when it left the child's hand? (*Note:* Find y when $x = 0$.)
(c) How high was the ball when it was at its maximum height?
(d) How far from the child did the ball strike the ground?

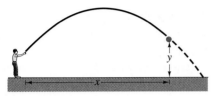

Figure for 53

54. *Maximum Height of a Dive*   The path of a dive is

$$y = -\tfrac{4}{9}x^2 + \tfrac{24}{9}x + 10$$

where y is the height in feet and x is the horizontal distance from the end of the diving board in feet (see figure). What is the maximum height of the dive?

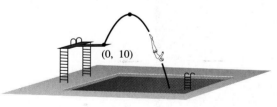

(0, 10)

Figure for 54

55. *Forestry*   The number of board feet in a 16-foot log is approximated by the model

$$V = 0.77x^2 - 1.32x - 9.31, \qquad 5 \le x \le 40$$

where V is the number of board feet and x is the diameter of the log at the small end in inches. (One board foot is a measure of volume equivalent to a board that is 12 inches wide, 12 inches long, and 1 inch thick.)
(a) Sketch a graph of the function.
(b) Estimate the number of board feet in a 16-foot log with a diameter of 16 inches.
(c) Estimate the diameter of a 16-foot log that scaled 500 board feet when the lumber was sold.

56. *Automobile Aerodynamics*   The amount of horsepower y required to overcome wind drag on a certain automobile is approximated by

$$y = 0.002x^2 + 0.005x - 0.029, \qquad 0 \le x \le 100$$

where x is the speed of the car in miles per hour.
(a) Sketch a graph of the function.
(b) Estimate the maximum speed of the car if the power required to overcome wind drag is not to exceed 10 horsepower.

57. Complete the square for the quadratic function $f(x) = ax^2 + bx + c$ ($a \ne 0$) and show that the vertex is at

$$\left(-\frac{b}{2a},\ -\frac{b^2 - 4ac}{4a}\right).$$

58. Use Exercise 57 to verify the vertices found in Exercises 19 and 20.

59. Assume that the function $f(x) = ax^2 + bx + c$ ($a \ne 0$) has two real zeros. Show that the x-coordinate of the vertex of the graph is the average of the zeros of f. (*Hint:* Use the Quadratic Formula.)

60. Create a quadratic function with zeros at $x = -4$ and $x = 2$. (The answer is not unique.)

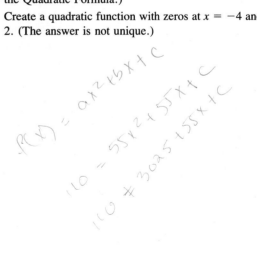

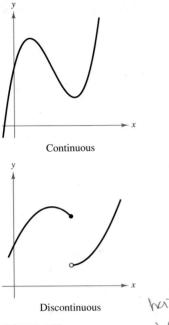

Continuous

Discontinuous

FIGURE 4.13

has to be a function (handwritten annotation)

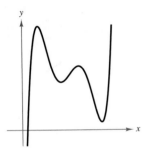

Polynomial functions have smooth, rounded graphs.

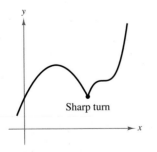

Sharp turn

The graph of a polynomial function *cannot* have a sharp pointed turn.

FIGURE 4.14

4.2 Polynomial Functions of Higher Degree

Graphs of Polynomial Functions / The Leading Coefficient Test /
Zeros of Polynomial Functions / The Intermediate Value Theorem

Graphs of Polynomial Functions

At this point you should be able to sketch an accurate graph of polynomial functions of degrees 0, 1, and 2.

Function	Graph
$f(x) = a$	Horizontal line
$f(x) = ax + b$	Line of slope a
$f(x) = ax^2 + bx + c$	Parabola

The graphs of polynomial functions of degree greater than 2 are more difficult to sketch. However, in this section you will learn how to recognize some of the basic features of the graphs of polynomial functions. Using these features and point-plotting, intercepts, and symmetry, you should be able to make reasonably accurate sketches *by hand*. Of course, if you have a graphing facility such as a graphing calculator or graphing software for a computer, then the task is easier.

The graph of a polynomial function is **continuous.** Essentially, this means that the graph of a polynomial function has no breaks, as shown in Figure 4.13.

Another feature of the graph of a polynomial function is that it has only smooth, rounded turns, as shown in Figure 4.14. Notice that the graph of a polynomial function cannot have a sharp turn. (For instance, the graph of $f(x) = |x|$ has a sharp turn at the point $(0, 0)$.)

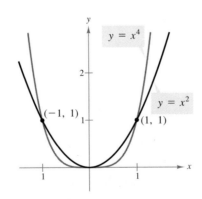

If n is even, the graph of $y = x^n$ *touches* axis at x-intercept.

FIGURE 4.15

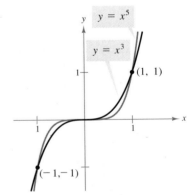

If n is odd, the graph of $y = x^n$ *crosses* axis at x-intercept.

The polynomial functions that have the simplest graphs are monomials of the form

$$f(x) = x^n$$

where n is an integer greater than zero. From Figure 4.15, we see that when n is *even* the graph is similar to the graph of $f(x) = x^2$, and when n is *odd* the graph is similar to the graph of $f(x) = x^3$. Moreover, the greater the value of n, the flatter the graph is on the interval $[-1, 1]$. You can check this by plotting a few points in this interval.

EXAMPLE 1   Sketching Transformations of Monomial Functions

Sketch the graphs of the polynomial functions.

a. $f(x) = -x^5$ **b.** $g(x) = x^4 + 1$ **c.** $h(x) = (x + 1)^4$

Solution

a. Since the degree of f is odd, the graph is similar to the graph of $y = x^3$. Moreover, the negative coefficient reflects the graph in the x-axis. Plotting the intercept $(0, 0)$ and the points $(1, -1)$ and $(-1, 1)$, we obtain the graph shown in Figure 4.16.

b. In this case, the graph of g is an upward shift, by one unit, of the graph of $y = x^4$ (see Figure 4.15). Thus, we obtain the graph shown in Figure 4.17.

c. The graph of h is a left shift, by one unit, of the graph of $y = x^4$ and it is shown in Figure 4.18.

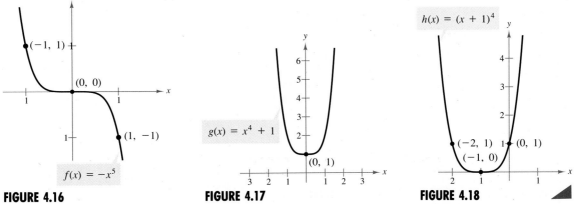

FIGURE 4.16   **FIGURE 4.17**   **FIGURE 4.18**

The Leading Coefficient Test

In Example 1, note that the three graphs eventually rise or fall without bound as x moves to the right or left. Symbolically, we write

$$f(x) \rightarrow \infty \qquad \text{as} \qquad x \rightarrow \infty$$

to mean that $f(x)$ increases without bound as x moves to the right without bound. (The infinity symbol ∞ indicates unboundedness.) Whether the graph of a polynomial eventually rises or falls can be determined by the function's degree (even or odd) and by its leading coefficient, as indicated by the **Leading Coefficient Test.**

Leading Coefficient Test

As x moves without bound to the left or to the right, the graph of the polynomial function $f(x) = a_n x^n + \cdots + a_1 x + a_0$ eventually rises or falls in the following manner. (*Note:* The dashed portions of the graphs indicate that the test determines *only* the right and left behavior of the graph.)

1. **When n is *odd*:**

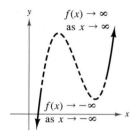

If the leading coefficient is positive ($a_n > 0$), then the graph falls to the left and rises to the right.

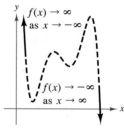

If the leading coefficient is negative ($a_n < 0$), then the graph rises to the left and falls to the right.

2. **When n is *even*:**

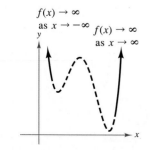

If the leading coefficient is positive ($a_n > 0$), then the graph rises to the left and right.

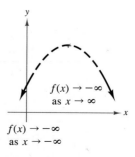

If the leading coefficient is negative ($a_n < 0$), then the graph falls to the left and right.

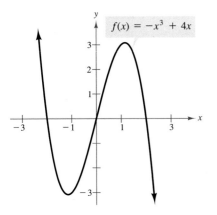

FIGURE 4.19

EXAMPLE 2   Applying the Leading Coefficient Test

Use the Leading Coefficient Test to determine the right and left behavior of the graphs of the polynomial functions.

a. $f(x) = -x^3 + 4x$ **b.** $f(x) = x^4 - 5x^2 + 4$ **c.** $f(x) = x^5 - x$

Solution

a. Because the degree is odd and the leading coefficient is negative, the graph rises to the left and falls to the right, as shown in Figure 4.19.

b. Because the degree is even and the leading coefficient is positive, the graph rises to the left and right, as shown in Figure 4.20.

c. Because the degree is odd and the leading coefficient is positive, the graph falls to the left and rises to the right, as shown in Figure 4.21.

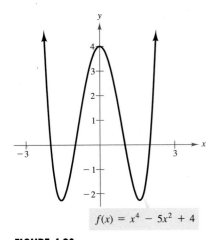

FIGURE 4.20

Zeros of Polynomial Functions

A **zero** of a function f is a number x for which $f(x) = 0$. For instance, 2 is a zero for the function $f(x) = x - 2$ because $f(2) = 2 - 2 = 0$. Similarly, 0 and -3 are zeros of the function $f(x) = x^2 + 3x$ because $f(0) = 0^2 + 3(0) = 0$ and $f(-3) = (-3)^3 + 3(-3) = 0$.

It can be shown that for a polynomial function f of degree n, the following statements are true.

1. The graph of f has, at most, $n - 1$ turning points. (Turning points are points at which the graph changes from increasing to decreasing or vice versa.)
2. The function f has, at most, n real zeros. (We will discuss this result in detail in Section 4.5 when we present the Fundamental Theorem of Algebra.)

Finding the zeros of polynomial functions is one of the most important problems in algebra. There is a strong interplay between graphical and algebraic approaches to this problem. Sometimes you can use information about the graph of a function to help find its zeros, and in other cases you can use information about the zeros of a function to help sketch its graph.

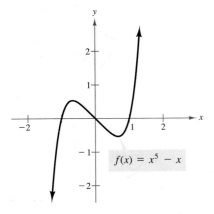

FIGURE 4.21

REAL ZEROS OF POLYNOMIAL FUNCTIONS

If f is a polynomial function and a is a real zero of f, then the following statements are equivalent.

1. $x = a$ is a *zero* of the function f.
2. $x = a$ is a *solution* of the polynomial equation $f(x) = 0$.
3. $(x - a)$ is a *factor* of the polynomial $f(x)$.
4. $(a, 0)$ is an *x-intercept* of the graph of f.

Finding zeros of polynomial functions is closely related to factoring and finding x-intercepts, as demonstrated in Examples 3, 4, and 5.

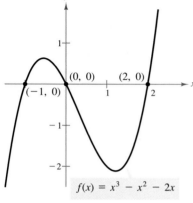

FIGURE 4.22

EXAMPLE 3   Finding Zeros of a Third-Degree Polynomial Function

Find all real zeros of $f(x) = x^3 - x^2 - 2x$.

Solution

$$\begin{aligned}
f(x) &= x^3 - x^2 - 2x && \textit{Given function} \\
&= x(x^2 - x - 2) && \textit{Remove common monomial factor} \\
&= x(x - 2)(x + 1) && \textit{Factor completely}
\end{aligned}$$

Thus, the real zeros are $x = 0$, $x = 2$, and $x = -1$, and the corresponding x-intercepts are $(0, 0)$, $(2, 0)$, and $(-1, 0)$, as shown in Figure 4.22. In the figure, note that the graph has two turning points. This is consistent with the fact that a third-degree polynomial can have *at most* two turning points.

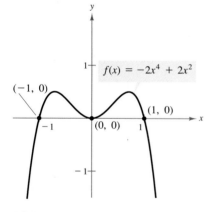

FIGURE 4.23

EXAMPLE 4   Finding Zeros of a Fourth-Degree Polynomial Function

Find all real zeros of $f(x) = -2x^4 + 2x^2$.

Solution

$$\begin{aligned}
f(x) &= -2x^4 + 2x^2 && \textit{Given function} \\
&= -2x^2(x^2 - 1) && \textit{Remove common monomial factor} \\
&= -2x^2(x - 1)(x + 1) && \textit{Factor completely}
\end{aligned}$$

Thus, the real zeros are $x = 0$, $x = 1$, and $x = -1$, and the corresponding x-intercepts are $(0, 0)$, $(1, 0)$, and $(-1, 0)$, as shown in Figure 4.23. Note in the figure that the graph has three turning points, which is consistent with the fact that a fourth-degree polynomial can have at most three turning points.

In Example 4, the real zero arising from $-2x^2 = 0$ is a **repeated zero.** In general, we say that a factor $(x - a)^k$ yields a repeated zero $x = a$ of **multiplicity** k. If k is odd, then the graph *crosses* the x-axis at $x = a$. If k is even, then the graph *touches* (but does not cross) the x-axis at $x = a$, as shown in Figure 4.23.

EXAMPLE 5   Finding Zeros of a Fifth-Degree Polynomial Function

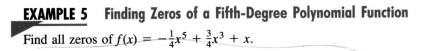

Find all zeros of $f(x) = -\frac{1}{4}x^5 + \frac{3}{4}x^3 + x$.

Solution

$$
\begin{aligned}
f(x) &= -\frac{1}{4}x^5 + \frac{3}{4}x^3 + x \\
&= \frac{1}{4}(-x^5 + 3x^3 + 4x) \\
&= -\frac{1}{4}x(x^4 - 3x^2 - 4) \\
&= -\frac{1}{4}x(x^2 - 4)(x^2 + 1) = -\frac{1}{4}x(x - 2)(x + 2)(x^2 + 1)
\end{aligned}
$$

Thus, the real zeros are $x = 0$, $x = 2$, and $x = -2$, and the corresponding intercepts are $(0, 0)$, $(2, 0)$, and $(-2, 0)$. Note that $x^2 + 1 = 0$ has no real solutions and so produces no real zeros of the function. (See Figure 4.24.)

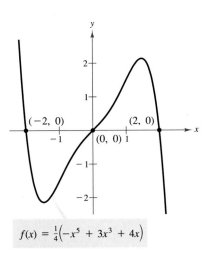

$f(x) = \frac{1}{4}\left(-x^5 + 3x^3 + 4x\right)$

FIGURE 4.24

To find mathematical models, you often need to create a polynomial function that has a particular set of zeros.

EXAMPLE 6   Finding a Polynomial Function with Given Zeros

Find polynomial functions with the following zeros.

a. $-2, -1, 1, 2$ **b.** $-\frac{1}{2}, 3, 3$

Solution

a. For each of the given zeros, we form a corresponding factor. For instance, the zero given by $x = -2$ corresponds to the factor $(x + 2)$. Thus, we can write the function as

$$
\begin{aligned}
f(x) &= (x + 2)(x + 1)(x - 1)(x - 2) \\
&= (x^2 - 4)(x^2 - 1) \\
&= x^4 - 5x^2 + 4.
\end{aligned}
$$

Note that other polynomials could be found if you consider *repeated* roots arising from factors like $(x + 2)^2$ or $(x - 1)^3$.

b. Note that the zero $x = -\frac{1}{2}$ corresponds to either $\left(x + \frac{1}{2}\right)$ or $(2x + 1)$. To avoid fractions, choose the second factor.

$$
\begin{aligned}
f(x) &= (2x + 1)(x - 3)^2 \\
&= (2x + 1)(x^2 - 6x + 9) \\
&= 2x^3 - 11x^2 + 12x + 9
\end{aligned}
$$

EXAMPLE 7   Sketching the Graph of a Polynomial Function

Sketch the graph of $f(x) = 3x^4 - 4x^3$.

Solution

Because the leading coefficient is positive and the degree is even, the graph eventually rises to the left and to the right, as shown in Figure 4.25. By factoring

$$f(x) = 3x^4 - 4x^3 = x^3(3x - 4)$$

we see that the zeros of f are $x = 0$ and $x = \frac{4}{3}$ (both of odd multiplicity). Thus, the x-intercepts occur at $(0, 0)$ and $\left(\frac{4}{3}, 0\right)$. Finally, we plot a few additional points as indicated in the accompanying table and obtain the graph shown in Figure 4.26.

x	-1	0.5	1	1.5
$f(x)$	7	-0.3125	-1	1.6875

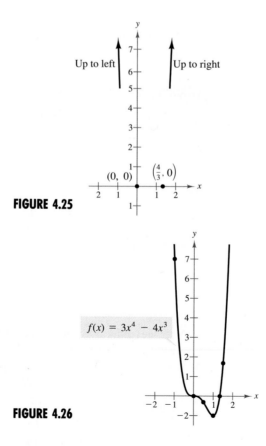

FIGURE 4.25

$f(x) = 3x^4 - 4x^3$

FIGURE 4.26

The Intermediate Value Theorem

The **Intermediate Value Theorem** verifies the existence of real zeros of polynomial functions. It implies that if $(a, f(a))$ and $(b, f(b))$ are two points on the graph of a polynomial such that $f(a) \neq f(b)$, then for any number d between $f(a)$ and $f(b)$ there must be a number c between a and b such that $f(c) = d$. (See Figure 4.27.)

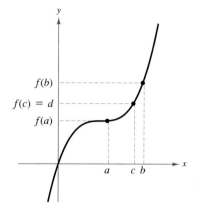

If d lies between $f(a)$ and $f(b)$, then there exists c between a and b such that $f(c) = d$.

FIGURE 4.27

INTERMEDIATE VALUE THEOREM

Let a and b be real numbers such that $a < b$. If f is a polynomial function such that $f(a) \neq f(b)$, then in the interval $[a, b]$, f takes on every value between $f(a)$ and $f(b)$.

This theorem will help you locate the real zeros of a polynomial function in the following way. If you can find a value $x = a$ where a polynomial function is positive, and another value $x = b$ where it is negative, then you can conclude that the function has at least one real zero between these two values. For example, the function $f(x) = x^3 + x^2 + 1$ is negative when $x = -2$ and positive when $x = -1$. Therefore, it follows from the Intermediate Value Theorem that f must have a real zero somewhere between -2 and -1, as shown in Figure 4.28.

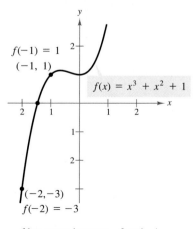

f has a zero between -2 and -1.

FIGURE 4.28

EXAMPLE 8 Approximating a Zero of a Polynomial Function

Use the Intermediate Value Theorem to approximate the real zero of $f(x) = x^3 - x^2 + 1$.

Solution

Begin by computing a few function values as follows.

x	-2	-1	0	1
$f(x)$	-11	-1	1	1

Since $f(-1)$ is negative and $f(0)$ is positive, you can conclude from the Intermediate Value Theorem that the function has a zero between -1 and 0. To pinpoint this zero more closely, divide the interval $[-1, 0]$ into tenths and evaluate the function at each point. After doing this, we find that

$$f(-0.8) = -0.152 \quad \text{and} \quad f(-0.7) = 0.167.$$

Thus, f must have a zero between -0.8 and -0.7, as shown in Figure 4.29. By continuing the process, you can approximate this zero to any desired accuracy.

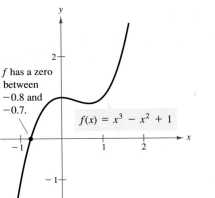

f has a zero between -0.8 and -0.7.

$f(x) = x^3 - x^2 + 1$

FIGURE 4.29

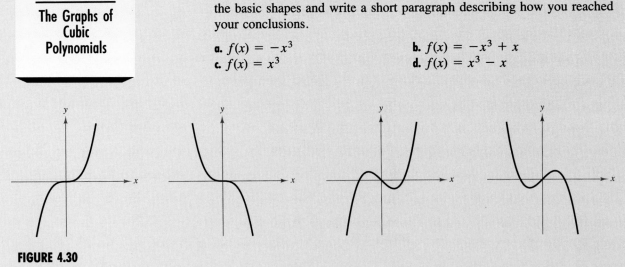

DISCUSSION PROBLEM

The Graphs of Cubic Polynomials

The graphs of cubic polynomials can be categorized according to the four basic shapes shown in Figure 4.30.

Match the graph of each of the following four functions with one of the basic shapes and write a short paragraph describing how you reached your conclusions.

a. $f(x) = -x^3$ **b.** $f(x) = -x^3 + x$
c. $f(x) = x^3$ **d.** $f(x) = x^3 - x$

FIGURE 4.30

WARM UP

The following warm-up exercises involve skills that were covered in earlier sections. You will use these skills in the exercise set for this section.

In Exercises 1–6, factor the expressions completely.

1. $12x^2 + 7x - 10$

2. $25x^3 - 60x^2 + 36x$

3. $12z^4 + 17z^3 + 5z^2$

4. $y^3 + 125$

5. $x^3 + 3x^2 - 4x - 12$

6. $x^3 + 2x^2 + 3x + 6$

In Exercises 7–10, find all real solutions to the equation.

7. $5x^2 + 8 = 0$

8. $x^2 - 6x + 4 = 0$

9. $4x^2 + 4x - 11 = 0$

10. $x^4 - 18x^2 + 81 = 0$

EXERCISES for Section 4.2

In Exercises 1 and 2, use the graph of the given equation to sketch the graphs of the specified transformations.

1. $y = x^3$

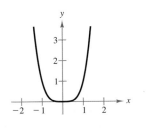

(a) $f(x) = (x - 2)^3$

(b) $f(x) = x^3 - 2$

(c) $f(x) = (x - 2)^3 - 2$

(d) $f(x) = -\frac{1}{2}x^3$

2. $y = x^4$

(a) $f(x) = (x + 3)^4$

(b) $f(x) = x^4 - 3$

(c) $f(x) = 4 - x^4$

(d) $f(x) = \frac{1}{2}(x - 1)^4$

In Exercises 3–10, match the polynomial functions with the correct graph. [The graphs are labeled (a), (b), (c), (d), (e), (f), (g), and (h).]

3. $f(x) = -3x + 5$

4. $f(x) = x^2 - 2x$

5. $f(x) = -2x^2 - 8x - 9$

6. $f(x) = 3x^3 - 9x + 1$

7. $f(x) = -\frac{1}{3}x^3 + x - \frac{2}{3}$

8. $f(x) = -\frac{1}{4}x^4 + 2x^2$

9. $f(x) = 3x^4 + 4x^3$

10. $f(x) = x^5 - 5x^3 + 4x$

(a)

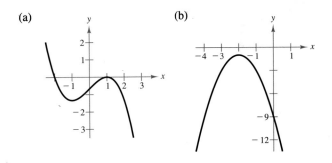

(b)

(c)

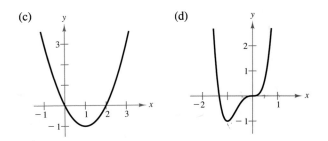

(d)

(e)

(f)

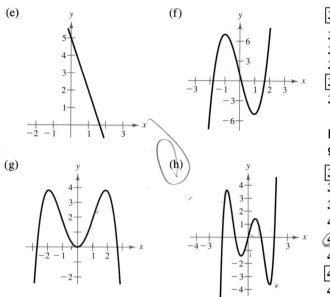

(g)

(h)

31. $g(t) = \frac{1}{2}t^4 - \frac{1}{2}$

32. $f(x) = x^5 + x^3 - 6x$

33. $f(x) = 2x^4 - 2x^2 - 40$

34. $g(t) = t^5 - 6t^3 + 9t$

35. $f(x) = 5x^4 + 15x^2 + 10$

36. $f(x) = x^3 - 4x^2 - 25x + 100$

In Exercises 37–46, find a polynomial function that has the given zeros.

37. 0, 10

38. 0, -3

39. 2, -6

40. -4, 5

41. 0, -2, -3

42. 0, 2, 5

43. 4, -3, 3, 0

44. -2, -1, 0, 1, 2

45. $1 + \sqrt{3}$, $1 - \sqrt{3}$

46. 2, $4 + \sqrt{5}$, $4 - \sqrt{5}$

In Exercises 11–20, determine the right-hand and left-hand behavior of the graph of the polynomial function.

11. $f(x) = 2x^2 - 3x + 1$

12. $f(x) = \frac{1}{3}x^3 + 5x$

13. $g(x) = 5 - \frac{7}{2}x - 3x^2$

14. $f(x) = -2.1x^5 + 4x^3 - 2$

15. $f(x) = 2x^5 - 5x + 7.5$

16. $h(x) = 1 - x^6$

17. $f(x) = 6 - 2x + 4x^2 - 5x^3$

18. $f(x) = \dfrac{3x^4 - 2x + 5}{4}$

19. $h(t) = -\frac{2}{3}(t^2 - 5t + 3)$

20. $f(s) = -\frac{7}{8}(s^3 + 5s^2 - 7s + 1)$

In Exercises 47–62, sketch the graph of the function.

47. $f(x) = -\frac{3}{2}$

48. $h(x) = \frac{1}{3}x - 3$

49. $f(t) = \frac{1}{4}(t^2 - 2t + 15)$

50. $g(x) = -x^2 + 10x - 16$

51. $f(x) = x^3 - 3x^2$

52. $f(x) = 1 - x^3$

53. $f(x) = x^3 - 4x$

54. $f(x) = \frac{1}{4}x^4 - 2x^2$

55. $g(t) = -\frac{1}{4}(t - 2)^2(t + 2)^2$

56. $f(x) = x^2(x - 4)$

57. $f(x) = \frac{1}{5}(x + 1)^2(x - 3)(2x - 9)$

58. $f(x) = \frac{1}{5}(x + 2)^2(3x - 5)^2$

59. $h(x) = \frac{1}{3}x^3(x - 4)^2$

60. $g(x) = \frac{1}{10}(x + 1)^2(x - 3)^3$

61. $f(x) = 1 - x^6$

62. $g(x) = 1 - (x + 1)^6$

In Exercises 21–36, find all the real zeros of the polynomial function.

21. $f(x) = x^2 - 25$

22. $f(x) = 49 - x^2$

23. $h(t) = t^2 - 6t + 9$

24. $f(x) = x^2 + 10x + 25$

25. $f(x) = x^2 + x - 2$

26. $f(x) = \frac{1}{2}x^2 + \frac{5}{2}x - \frac{3}{2}$

27. $f(x) = 3x^2 - 12x + 3$

28. $g(x) = 5(x^2 - 2x - 1)$

29. $f(t) = t^3 - 4t^2 + 4t$

30. $f(x) = x^4 - x^3 - 20x^2$

In Exercises 63–66, follow the procedure given in Example 8 to estimate the zero of $f(x)$ in the given interval $[a, b]$. Give your approximation to the nearest tenth. (If you have a graphics calculator or graphing software, use it to help approximate the zero.)

Function	Interval
63. $f(x) = x^3 + x - 1$	$[0, 1]$
64. $f(x) = x^5 + x + 1$	$[-1, 0]$
65. $f(x) = x^4 - 10x^2 - 11$	$[3, 4]$
66. $f(x) = -x^3 + 3x^2 + 9x - 2$	$[4, 5]$

67. *Volume of a Box*   An open box is made from a 12-inch-square piece of material by cutting equal squares from each corner and turning up the sides (see figure).

(a) Verify that the volume of the box is given by $V(x) = 4x(6 - x)^2$.

(b) Determine the domain of the function V.

(c) Sketch the graph of the function and use the graph to estimate the value of x for which $V(x)$ is maximum.

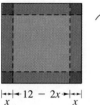
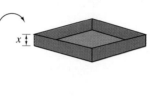

$|{\leftarrow}{\rightarrow}|{\leftarrow}\,12 - 2x\,{\rightarrow}|{\leftarrow}{\rightarrow}|$
$x \qquad\qquad\qquad x$

Figure for 67

68. *Volume of a Box*   An open box with locking tabs is made from a 12-inch-square piece of material. This is done by cutting equal squares from each corner and folding along the dashed lines, as shown in the figure.

(a) Verify that the volume of the box is given by $V(x) = 8x(3 - x)(6 - x)$.

(b) Determine the domain of the function V.

(c) Sketch the graph of the function and use the graph to estimate the value of x for which $V(x)$ is maximum.

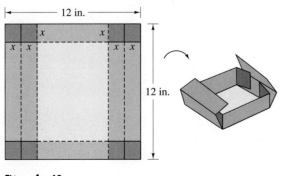

Figure for 68

69. *Advertising Expenses*   The total revenue for a soft drink company is related to its advertising expense by the function

$$R = \frac{1}{50,000}(-x^3 + 600x^2), \qquad 0 \le x \le 400$$

where R is the total revenue in millions of dollars and x is the amount spent on advertising (in 10,000s of dollars). Use the graph of this function, shown in the figure, to estimate the point on the graph at which the function is increasing most rapidly. This is the **point of diminishing returns** because any expense above this amount will yield less return per dollar invested in advertising.

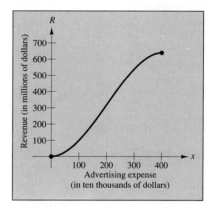

Figure for 69

SOLVING

A Graphical Approach to Finding the Zeros of a Function

You can use the *zoom* feature of a graphing utility to approximate a function's real zeros (the x-intercepts of its graph) to any desired accuracy. The four graphs below show four steps in approximating the positive zero of $f(x) = x^2 - 5$ to be $x \approx 2.236$. (The actual zero is $x = \sqrt{5} \approx 2.236068$.) By repeated zooming, you can obtain whatever accuracy you need.

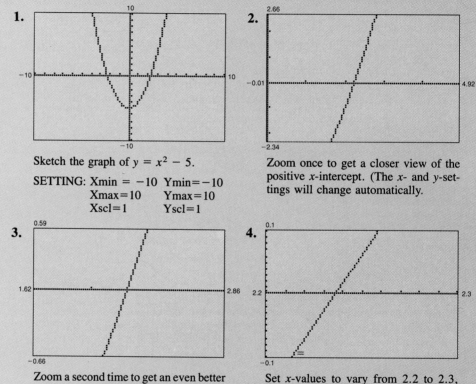

1. Sketch the graph of $y = x^2 - 5$.

SETTING: Xmin = −10   Ymin=−10
              Xmax=10   Ymax=10
              Xscl=1   Yscl=1

2. Zoom once to get a closer view of the positive x-intercept. (The x- and y-settings will change automatically.

3. Zoom a second time to get an even better view. Use the cursor keys to determine that the x-intercept is about 2.2.

4. Set x-values to vary from 2.2 to 2.3, with an x-scale of 0.01. Using the trace key, you can approximate the x-intercept to be 2.236.

SETTING: Xmin = 2.2   Ymin=−.1
              Xmax=2.3   Ymax=.1
              Xscl=.01   Yscl=0.01

The equation $x^2 - 5 = 0$ that is solved graphically above is, of course, easy to solve algebraically. Example 1 uses a graphic approach to solving an equation that would be difficult to solve algebraically.

EXAMPLE 1   Finding Points of Intersection of Two Graphs

Find the points of intersection of the circle and parabola given by

$$x^2 + y^2 - 3x + 5y - 11 = 0 \quad \text{and} \quad y = x^2 - 4x + 5.$$

Solution

To begin, you can sketch the graph of both equations on the same graphing utility screen. If you are using a graphing utility that graphs *functions*, rather than *relations*, you should begin by writing the circle as the union of two functions.

$$y = \tfrac{1}{2}(-5 + \sqrt{69 + 12x - 4x^2})$$ *Top half of circle*

$$y = \tfrac{1}{2}(-5 - \sqrt{69 + 12x - 4x^2})$$ *Bottom half of circle*

$$y = x^2 - 4x + 5$$ *Parabola*

From the graph, you can see that the parabola intersects the circle twice. The coordinates of the points of intersection are roughly (1, 1.9) and (2.8, 1.7). To obtain a more accurate approximation of the points of the intersection, you could use the zoom feature of the graphing utility. (See Exercise 1 below.)

Another approach is to substitute $x^2 - 4x + 5$ for y in the equation of the circle. This produces a fourth-degree polynomial equation that can be solved for x.

$$x^2 + y^2 - 3x + 5y - 11 = 0$$

$$x^2 + (x^2 - 4x + 5)^2 - 3x + 5(x^2 - 4x + 5) - 11 = 0$$

$$x^4 - 8x^3 + 32x^2 - 63x + 39 = 0$$

Using a graphing utility, you can approximate the solutions of this equation to be $x \approx 1.055$ and $x \approx 2.841$. ◣

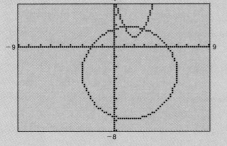

EXERCISES

(See also: Exercises 63–66, Section 4.2; Exercise set for Section 4.6)

1. *Exploration* Using a setting of $1.05 \le x \le 1.06$ and $1.89 \le y \le 1.90$, sketch the graphs of the top half of the circle and the parabola on the same screen. Then use the trace feature to approximate (accurate to 3 decimal places) the y-coordinate of the point of the intersection that is shown on the screen.

2. *Exploration*
   (a) Find a graphing utility setting that will allow you to approximate the solution $x = 1.055$ to two more decimal places of accuracy.
   (b) Find a graphing utility setting that will allow you to approximate the solution $x = 2.841$ to two more decimal places of accuracy.

3. *It Doesn't Look Like a Circle* The graphs of
   $$y = \sqrt{36 - x^2} \quad \text{and} \quad y = -\sqrt{36 - x^2}$$
   form the top and bottom halves of a circle. Use a graphing utility to sketch both graphs on the same display screen. Which of the following produces a result that "looks" like a circle? Describe a general rule that can be used with your graphing utility to make the circles look like circles.

   (a) SETTING   Xmin=−10   Ymin=−10
                     Xmax=10   Ymax=10
                     Xscl=1   Yscl=1

   (b) SETTING   Xmin=−9   Ymin=−6
                     Xmax=9   Ymax=6
                     Xscl=1   Yscl=1

Long Division of Polynomials

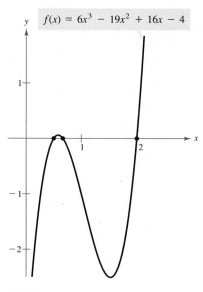

FIGURE 4.31

Up to this point in the text we have added, subtracted, and multiplied polynomials. In this section, we look at a procedure for *dividing* polynomials. This procedure has many important applications and is especially valuable in factoring and finding the zeros of polynomial functions.

To begin, consider the graph of

$$f(x) = 6x^3 - 19x^2 + 16x - 4,$$

shown in Figure 4.31. Notice that a zero of f occurs at $x = 2$. [Try verifying this by evaluating $f(x)$ at $x = 2$.] Since $x = 2$ is a zero of the polynomial function f, $(x - 2)$ is a factor of $f(x)$. This means that there exists a second-degree polynomial $q(x)$ such that

$$f(x) = (x - 2) \cdot q(x).$$

To find $q(x)$, we use **long division of polynomials.** Study the following example to see how this process works.

EXAMPLE 1   Long Division of Polynomials

Divide the polynomial

$$f(x) = 6x^3 - 19x^2 + 16x - 4$$

by $x - 2$, and use the result to factor $f(x)$ completely.

Solution

$$
\begin{array}{r}
\text{Partial quotients} \\
\downarrow \quad \downarrow \quad \downarrow \\
6x^2 - 7x + 2 \\
x - 2 \overline{) 6x^3 - 19x^2 + 16x - 4} \\
\underline{6x^3 - 12x^2} \qquad\qquad\qquad \textit{Multiply: } 6x^2(x-2) \\
-7x^2 + 16x \qquad\qquad \textit{Subtract} \\
\underline{-7x^2 + 14x} \qquad\qquad \textit{Multiply: } -7x(x-2) \\
2x - 4 \qquad \textit{Subtract} \\
\underline{2x - 4} \qquad \textit{Multiply: } 2(x-2) \\
0 \qquad \textit{Subtract}
\end{array}
$$

We see that

$$6x^3 - 19x^2 + 16x - 4 = (x - 2)(6x^2 - 7x + 2)$$

and, factoring the quadratic $6x^2 - 7x + 2$,

$$6x^3 - 19x^2 + 16x - 4 = (x - 2)(2x - 1)(3x - 2).$$

Note that this factorization agrees with the graph of f (Figure 4.31) in that the three x-intercepts occur at

$$x = 2, \qquad x = \frac{1}{2}, \qquad \text{and} \qquad x = \frac{2}{3}.$$

In Example 1, $x - 2$ is a factor of the polynomial $6x^3 - 19x^2 + 16x - 4$, and the long division process produced a remainder of zero. Often, long division will produce a nonzero remainder. For instance, if we divide $x^2 + 3x + 5$ by $x + 1$, we obtain the following.

$$
\begin{array}{r}
x + 2 \qquad \leftarrow \text{Quotient} \\
\text{Divisor} \rightarrow x + 1 \overline{)\, x^2 + 3x + 5} \qquad \leftarrow \text{Dividend} \\
\underline{x^2 + x} \\
2x + 5 \\
\underline{2x + 2} \\
3 \qquad \leftarrow \text{Remainder}
\end{array}
$$

We write this result as

$$
\underbrace{\frac{\overbrace{x^2 + 3x + 5}^{\text{Dividend}}}{\underbrace{x + 1}_{\text{Divisor}}}}_{} = \overbrace{x + 2}^{\text{Quotient}} + \underbrace{\frac{\overbrace{3}^{\text{Remainder}}}{\underbrace{x + 1}_{\text{Divisor}}}}_{}
$$

or

$$x^2 + 3x + 5 = (x + 2)(x + 1) + 3.$$

This example illustrates the well-known theorem, the **Division Algorithm.**

THE DIVISION ALGORITHM

If $f(x)$ and $d(x)$ are polynomials such that $d(x) \neq 0$, and the degree of $d(x)$ is less than or equal to the degree of $f(x)$, then there exist unique polynomials $q(x)$ and $r(x)$ such that

$$f(x) = d(x)q(x) + r(x)$$

$$\underset{\text{Dividend}}{\uparrow} \quad \underset{\text{Divisor}}{\uparrow} \quad \underset{\text{Quotient}}{\nwarrow} \quad \underset{\text{Remainder}}{\nwarrow}$$

where $r(x) = 0$ or the degree of $r(x)$ is less than the degree of $d(x)$. If the remainder $r(x)$ is zero, then $d(x)$ **divides evenly** into $f(x)$.

REMARK The Division Algorithm can also be written as

$$\frac{f(x)}{d(x)} = q(x) + \frac{r(x)}{d(x)}.$$

In the Division Algorithm the rational expression $f(x)/d(x)$ is **improper** because the degree of $f(x)$ is greater than or equal to the degree of $d(x)$. On the other hand, the rational expression $r(x)/d(x)$ is **proper** because the degree of $r(x)$ is less than the degree of $d(x)$.

EXAMPLE 2   Long Division of Polynomials

Divide $x^3 - 1$ by $x - 1$.

Solution

Because there is no x^2-term or x-term in the dividend, we line up the subtraction by using zero coefficients (or leaving a space) for the missing terms.

$$
\begin{array}{r}
x^2 + x + 1 \\
x - 1 \overline{)x^3 + 0x^2 + 0x - 1} \\
\underline{x^3 - x^2} \\
x^2 \\
\underline{x^2 - x} \\
x - 1 \\
\underline{x - 1} \\
0
\end{array}
$$

Thus, $x - 1$ divides evenly into $x^3 - 1$ and we can write

$$\frac{x^3 - 1}{x - 1} = x^2 + x + 1.$$

EXAMPLE 3   Long Division of Polynomials

$$
\begin{array}{r}
2x^2 \qquad\ + 1 \\
x^2 + 2x - 3 \overline{)2x^4 + 4x^3 - 5x^2 + 3x - 2} \\
\underline{2x^4 + 4x^3 - 6x^2} \\
x^2 + 3x - 2 \\
\underline{x^2 + 2x - 3} \\
x + 1
\end{array}
$$

Note that the first subtraction eliminated two terms from the dividend. When this happens, the quotient skips a term. Thus, we can write

$$\frac{2x^4 + 4x^3 - 5x^2 + 3x - 2}{x^2 + 2x - 3} = 2x^2 + 1 + \frac{x + 1}{x^2 + 2x - 3}.$$

Synthetic Division

Synthetic division is a shortcut for long division by polynomials of the form $x - k$. To see how it works, take another look at Example 1.

$$
\begin{array}{r}
6x^2 - 7x + 2 \\
x - 2\overline{)6x^3 - 19x^2 + 16x - 4} \\
\underline{6x^3 - 12x^2} \\
-7x^2 + 16x \\
\underline{-7x^2 + 14x} \\
2x - 4 \\
\underline{2x - 4} \\
0
\end{array}
$$

You can retain the essential steps of this division tableau by using only the coefficients, as follows.

$$
\begin{array}{r}
6 \quad -7 \quad 2 \\
-2\overline{)6 \quad -19 \quad 16 \quad -4} \\
\underline{6 \quad -12} \\
-7 \quad 16 \\
\underline{-7 \quad 14} \\
2 \quad -4 \\
\underline{2 \quad -4} \\
0
\end{array}
$$

Since the coefficients shown in color are duplicates of those in the quotient or the dividend, you can omit them and condense vertically.

$$
\begin{array}{r}
6 \quad -7 \quad 2 \\
-2\overline{)6 \quad -19 \quad 16 \quad -4} \\
\underline{-12 \quad 14 \quad -4} \\
0
\end{array}
$$

Now, move the quotient to the bottom row.

$$
\begin{array}{r}
-2\overline{)6 \quad -19 \quad 16 \quad -4} \\
\underline{-12 \quad 14 \quad -4} \\
6 \quad -7 \quad 2 \quad 0
\end{array}
$$

Finally, change from subtraction to addition (and reduce the likelihood of errors) by changing the sign of the divisor and of row two. This produces the following synthetic division array.

$$
\begin{array}{r|rrrr}
2 & 6 & -19 & 16 & -4 \\
& & 12 & -14 & 4 \\
\hline
& 6 & -7 & 2 & 0
\end{array}
$$

We summarize the pattern for synthetic division of a cubic polynomial as follows. (The pattern for higher-degree polynomials is similar.)

SYNTHETIC DIVISION (FOR A CUBIC POLYNOMIAL)

To divide $ax^3 + bx^2 + cx + d$ by $x - k$, use the following pattern.

Coefficients of quotient

Vertical Pattern: Add terms.
Diagonal Pattern: Multiply by k.

REMARK Synthetic division works *only* for divisors of the form $x - k$. You cannot use synthetic division to divide a polynomial by a quadratic such as $x^2 - 3$.

EXAMPLE 4 **Using Synthetic Division**

Use synthetic division to divide $x^4 - 10x^2 - 2x + 4$ by $x + 3$.

Solution

You can set up the array as follows. (Note that you must include a zero for each missing term in the dividend.)

Divisor: $x - (-3)$ Dividend: $x^4 - 10x^2 - 2x + 4$

$$
\begin{array}{r|rrrrr}
-3 & 1 & 0 & -10 & -2 & 4 \\
   &   & -3 & 9 & 3 & -3 \\
\hline
   & 1 & -3 & -1 & 1 & 1 \\
\end{array}
$$
$\leftarrow$ Remainder: 1

Quotient: $x^3 - 3x^2 - x + 1$

Thus, you have

$$\frac{x^4 - 10x^2 - 2x + 4}{x + 3} = x^3 - 3x^2 - x + 1 + \frac{1}{x + 3}.$$

The Remainder and Factor Theorems

The remainder obtained in the synthetic division process has an important interpretation, as given in the **Remainder Theorem.**

THE REMAINDER THEOREM

If a polynomial $f(x)$ is divided by $x - k$, then the remainder is

$r = f(k)$.

Proof

From the Division Algorithm, we have

$$f(x) = (x - k)q(x) + r(x)$$

and since either $r(x) = 0$ or the degree of $r(x)$ is less than the degree of $x - k$, we know that $r(x)$ must be a constant. That is, $r(x) = r$. Now, by evaluating $f(x)$ at $x = k$, we have

$$f(k) = (k - k)q(k) + r = (0)q(k) + r = r. \qquad \blacktriangle$$

EXAMPLE 5   Evaluating a Polynomial by the Remainder Theorem

Use the Remainder Theorem to evaluate the following function at $x = -2$.

$$f(x) = 3x^3 + 8x^2 + 5x - 7$$

Solution

Using synthetic division, we obtain the following.

$$
\begin{array}{r|rrrr}
-2 & 3 & 8 & 5 & -7 \\
   &   & -6 & -4 & -2 \\
\hline
   & 3 & 2 & 1 & -9
\end{array}
$$

Since the remainder is $r = -9$, we conclude that

$$f(-2) = -9.$$

This means that $(-2, -9)$ is a point on the graph of f. Check this by substituting $x = -2$ in the original function. $\qquad \blacktriangle$

FACTOR THEOREM

A polynomial $f(x)$ has a factor $(x - k)$ if and only if $f(k) = 0$.

Proof

Using the Division Algorithm with the factor $(x - k)$, we have

$$f(x) = (x - k)q(x) + r(x).$$

By the Remainder Theorem, $r(x) = r = f(k)$, and we have

$$f(x) = (x - k)q(x) + f(k)$$

where $q(x)$ is a polynomial of lesser degree than $f(x)$. If $f(k) = 0$, then

$$f(x) = (x - k)q(x)$$

and we see that $(x - k)$ is a factor of $f(x)$. Conversely, if $(x - k)$ is a factor of $f(x)$, then division of $f(x)$ by $(x - k)$ yields a remainder of 0. Hence, by the Remainder Theorem, we have $f(k) = 0$. $\qquad \blacktriangle$

EXAMPLE 6   Using Synthetic Division to Find Factors of a Polynomial

Show that $(x - 2)$ and $(x + 3)$ are factors of the polynomial

$$f(x) = 2x^4 + 7x^3 - 4x^2 - 27x - 18.$$

Then find the remaining factors of $f(x)$.

Solution

Use synthetic division with 2 and -3 *successively* to obtain the following.

$$
\begin{array}{r|rrrrr}
2 & 2 & 7 & -4 & -27 & -18 \\
  &   & 4 & 22 & 36 & 18 \\
\hline
  & 2 & 11 & 18 & 9 & 0
\end{array}
$$
0 remainder

$\rightarrow (x - 2)$ *is a factor*

$$
\begin{array}{r|rrrr}
-3 & 2 & 11 & 18 & 9 \\
   &   & -6 & -15 & -9 \\
\hline
   & 2 & 5 & 3 & 0
\end{array}
$$
Use new coefficients
0 remainder

$\rightarrow (x + 3)$ *is a factor*

Since the resulting quadratic factors as

$$2x^2 + 5x + 3 = (2x + 3)(x + 1)$$

the complete factorization of $f(x)$ is

$$f(x) = (x - 2)(x + 3)(2x + 3)(x + 1).$$

This factorization implies that f has four real zeros: 2, -3, $-\frac{3}{2}$, and -1.

In summary, the remainder r, obtained in the synthetic division of $f(x)$ by $x - k$, provides the following information.

1. The remainder r gives the value of f at $x = k$. That is, $r = f(k)$.
2. If $r = 0$, then $(x - k)$ is a factor of $f(x)$.
3. If $r = 0$, then $(k, 0)$ is an x-intercept of the graph of f.

DISCUSSION

PROBLEM

Finding Patterns in Polynomial Division

Complete the polynomial divisions.

1. $\dfrac{x^2 - 1}{x - 1} = $ ▬▬

2. $\dfrac{x^3 - 1}{x - 1} = $ ▬▬

3. $\dfrac{x^4 - 1}{x - 1} = $ ▬▬

Describe the pattern that you obtain and use your result to find a formula for the polynomial division

$$\frac{x^n - 1}{x - 1}.$$

WARM UP

The following warm-up exercises involve skills that were covered in earlier sections. You will use these skills in the exercise set for this section.

In Exercises 1–4, write the expression in standard polynomial form.

1. $(x - 1)(x^2 + 2) + 5$ **2.** $(x^2 - 3)(2x + 4) + 8$

3. $(x^2 + 1)(x^2 - 2x + 3) - 10$ **4.** $(x + 6)(2x^3 - 3x) - 5$

In Exercises 5 and 6, factor the polynomials.

5. $x^2 - 4x + 3$ **6.** $4x^3 - 10x^2 + 6x$

In Exercises 7–10, find a polynomial function that has the given zeros.

7. $0, 3, 4$ **8.** $-6, 1$

9. $-3, 1 + \sqrt{2}, 1 - \sqrt{2}$ **10.** $1, -2, 2 + \sqrt{3}, 2 - \sqrt{3}$

EXERCISES for Section 4.3

In Exercises 1–14, divide by long division.

	Dividend	Divisor		Dividend	Divisor
1.	$2x^2 + 10x + 12$	$x + 3$	**20.**	$3x^3 - 16x^2 - 72$	$x - 6$
2.	$5x^2 - 17x - 12$	$x - 4$	**21.**	$5x^3 - 6x^2 + 8$	$x - 4$
3.	$4x^3 - 7x^2 - 11x + 5$	$4x + 5$	**22.**	$5x^3 + 6x + 8$	$x + 2$
4.	$6x^3 - 16x^2 + 17x - 6$	$3x - 2$	**23.**	$10x^4 - 50x^3 - 800$	$x - 6$
5.	$x^4 + 5x^3 + 6x^2 - x - 2$	$x + 2$	**24.**	$x^5 - 13x^4 - 120x + 80$	$x + 3$
6.	$x^3 + 4x^2 - 3x - 12$	$x^2 - 3$	**25.**	$x^3 + 512$	$x + 8$
7.	$7x + 3$	$x + 2$	**26.**	$5x^3$	$x + 3$
8.	$8x - 5$	$2x + 1$	**27.**	$-3x^4$	$x - 2$
9.	$6x^3 + 10x^2 + x + 8$	$2x^2 + 1$	**28.**	$-3x^4$	$x + 2$
10.	$x^3 - 9$	$x^2 + 1$	**29.**	$5 - 3x + 2x^2 - x^3$	$x + 1$
11.	$x^4 + 3x^2 + 1$	$x^2 - 2x + 3$	**30.**	$180x - x^4$	$x - 6$
12.	$x^5 + 7$	$x^3 - 1$	**31.**	$4x^3 + 16x^2 - 23x - 15$	$x + \frac{1}{2}$
13.	$2x^3 - 4x^2 - 15x + 5$	$(x - 1)^2$	**32.**	$3x^3 - 4x^2 + 5$	$x - \frac{3}{2}$
14.	x^4	$(x - 1)^3$			

In Exercises 15–32, divide by synthetic division.

	Dividend	Divisor
15.	$3x^3 - 17x^2 + 15x - 25$	$x - 5$
16.	$5x^3 + 18x^2 + 7x - 6$	$x + 3$
17.	$4x^3 - 9x + 8x^2 - 18$	$x + 2$
18.	$9x^3 - 16x - 18x^2 + 32$	$x - 2$
19.	$-x^3 + 75x - 250$	$x + 10$

In Exercises 33–40, use synthetic division to show that x is a solution of the third-degree polynomial equation, and use the result to factor the polynomial completely.

Polynomial Equation	Value of x
33. $x^3 - 7x + 6 = 0$	$x = 2$
34. $x^3 - 28x - 48 = 0$	$x = -4$
35. $2x^3 - 15x^2 + 27x - 10 = 0$	$x = \frac{1}{2}$
36. $48x^3 - 80x^2 + 41x - 6 = 0$	$x = \frac{2}{3}$

Polynomial Equation	Value of x
37. $x^3 + 2x^2 - 3x - 6 = 0$	$x = \sqrt{3}$
38. $x^3 + 2x^2 - 2x - 4$	$x = \sqrt{2}$
39. $x^3 - 3x^2 + 2 = 0$	$x = 1 + \sqrt{3}$
40. $x^3 - x^2 - 13x - 3 = 0$	$x = 2 - \sqrt{5}$

In Exercises 41–44, express the function in the form $f(x) = (x - k)q(x) + r$ for the given value of k, and demonstrate that $f(k) = r$.

Function	Value of k
41. $f(x) = x^3 - x^2 - 14x + 11$	$k = 4$
42. $f(x) = \frac{1}{3}(15x^4 + 10x^3 - 6x^2 + 17x + 14)$	$k = -\frac{2}{3}$
43. $f(x) = x^3 + 3x^2 - 2x - 14$	$k = \sqrt{2}$
44. $f(x) = 4x^3 - 6x^2 - 12x - 4$	$k = 1 - \sqrt{3}$

In Exercises 45–50, use synthetic division to find the required function values.

45. $f(x) = 4x^3 - 13x + 10$
(a) $f(1)$ (b) $f(-2)$ (c) $f\left(\frac{1}{2}\right)$ (d) $f(8)$

46. $g(x) = x^6 - 4x^4 + 3x^2 + 2$
(a) $g(2)$ (b) $g(-4)$ (c) $g(3)$ (d) $g(-1)$

47. $h(x) = 3x^3 + 5x^2 - 10x + 1$
(a) $h(3)$ (b) $h\left(\frac{1}{3}\right)$ (c) $h(-2)$ (d) $h(-5)$

48. $f(x) = 0.4x^4 - 1.6x^3 + 0.7x^2 - 2$
(a) $f(1)$ (b) $f(-2)$ (c) $f(5)$ (d) $f(-10)$

49. $f(x) = x^3 - 2x^2 - 11x + 52$
(a) $f(5)$ (b) $f(-4)$ (c) $f(1.2)$ (d) $f(2)$

50. $g(x) = x^3 - x^2 + 25x - 25$
(a) $g(5)$ (b) $g\left(\frac{1}{5}\right)$ (c) $g(-1.5)$ (d) $g(-1)$

In Exercises 51–56, simplify the rational expression.

51. $\dfrac{4x^3 - 8x^2 + x + 3}{2x - 3}$

52. $\dfrac{x^3 + x^2 - 64x - 64}{x + 8}$

53. $\dfrac{x^3 + 3x^2 - x - 3}{x + 1}$

54. $\dfrac{2x^3 + 3x^2 - 3x - 2}{x - 1}$

55. $\dfrac{x^4 + 6x^3 + 11x^2 + 6x}{x^2 + 3x + 2}$

56. $\dfrac{x^4 + 9x^3 - 5x^2 - 36x + 4}{x^2 - 4}$

57. *Power of an Engine* The horsepower y developed by a compact car engine is approximated by the model

$$y = -1.42x^3 + 5.04x^2 + 32.45x - 0.75, \quad 1 \le x \le 5$$

where x is the engine speed in thousands of revolutions

per minute. Note on the graph that there are two engine speeds that develop 110 horsepower, one of which is 5000 rpm. Approximate the other engine speed.

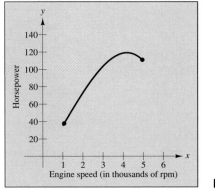

Horsepower vs. Engine speed (in thousands of rpm)

Figure for 57

58. *Automobile Emissions* The number of parts per million of nitric oxide emissions y from a certain car engine is approximated by the model

$$y = -5.05x^3 + 3857x - 38,411.25, \quad 13 \le x \le 18$$

where x is the air-fuel ratio. Note on the graph that there are two air-fuel ratios that produce 2400 parts per million of nitric oxide, one of which is 15. Approximate the other air-fuel mixture.

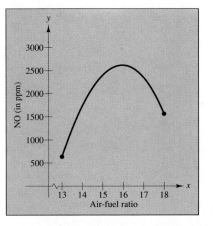

NO (in ppm) vs. Air-fuel ratio

Figure for 58

59. Use the form $f(x) = (x - k)q(x) + r$ to create a cubic function that (a) passes through the point $(2, 5)$ and rises to the right, and (b) passes through the point $(-3, 1)$ and falls to the right. (The answers are not unique.)

4.4 Real Zeros of Polynomial Functions

Descartes's Rule of Signs / The Rational Zero Test /
Bounds for Real Zeros of Polynomial Functions

Descartes's Rule of Signs

In this and the following section we will present some additional aids for finding zeros of polynomial functions.

In Section 4.2, we noted that an nth degree polynomial function can have *at most n* real zeros. Of course, many nth degree polynomials do not have that many real zeros. For instance, $f(x) = x^2 + 1$ has no real zeros, and $f(x) = x^3 + 1$ has only one real zero. **Descartes's Rule of Signs** sheds more light on the number of real zeros that a polynomial can have.

REMARK   When there is only one variation in sign, Descartes's Rule of Signs guarantees the existence of exactly one positive (or negative) real zero.

DESCARTES'S RULE OF SIGNS

Let $f(x) = a_n x^n + a_{n-1} x^{n-1} + \cdots + a_2 x^2 + a_1 x + a_0$ be a polynomial with real coefficients and $a_0 \neq 0$.

1. The number of *positive real zeros* of f is either equal to the number of variations in sign of $f(x)$ or is less than that number by an even integer.
2. The number of *negative real zeros* of f is either equal to the number of variations in sign of $f(-x)$ or is less than that number by an even integer.

Variation in sign means that two consecutive coefficients have opposite signs. For example, the polynomial

$$\overset{+\ \text{to}\ -}{} \quad \overset{+\ \text{to}\ -}{}$$
$$f(x) = 3x^3 - 5x^2 + 6x - 4$$
$$\underset{-\ \text{to}\ +}{}$$

has *three* variations in sign, whereas

$$f(-x) = 3(-x)^3 - 5(-x)^2 + 6(-x) - 4$$
$$= -3x^3 - 5x^2 - 6x - 4$$

has no variations in sign. Thus, from Descartes's Rule of Signs, the polynomial $f(x) = 3x^3 - 5x^2 + 6x - 4$ has either three positive real zeros or one positive real zero, and has no negative real zeros. From the graph of f shown in Figure 4.32, you can see that the function has only one positive real zero.

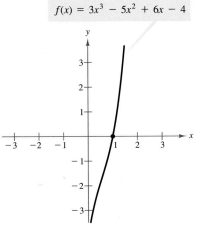

$f(x) = 3x^3 - 5x^2 + 6x - 4$

FIGURE 4.32

EXAMPLE 1   Using Descartes's Rule of Signs

Apply Descartes's Rule of Signs to the polynomial function

$$f(x) = 2x^4 + 7x^3 - 4x^2 - 27x - 18.$$

Solution

Because $f(x)$ has only *one* variation in sign, f must have *exactly one* positive real zero. Moreover, because

$$f(-x) = 2(-x)^4 + 7(-x)^3 - 4(-x)^2 - 27(-x) - 18$$
$$= 2x^4 - 7x^3 - 4x^2 + 27x - 18$$

has *three* variations in sign, f has either three negative zeros or one negative zero. This result agrees with Example 6 in Section 4.3, where the zeros of f were determined to be 2, -3, $-\frac{3}{2}$, and -1.

EXAMPLE 2   Using Descartes's Rule of Signs

Apply Descartes's Rule of Signs to the polynomial function

$$f(x) = x^3 - x + 1.$$

Solution

Because $f(x)$ has two variations in sign, it follows that f can have either two or no positive real zeros. Moreover, because

$$f(-x) = -x^3 + x + 1$$

has one variation in sign, f has exactly one negative real zero. The graph of f in Figure 4.33 shows that f actually has only one real zero (it is a negative number between -2 and -1).

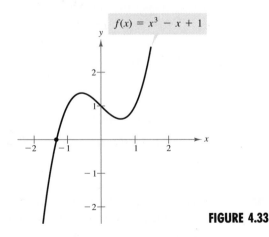

$$f(x) = x^3 - x + 1$$

FIGURE 4.33

The Rational Zero Test

The **Rational Zero Test** relates the possible rational zeros of a polynomial (having integer coefficients) to the leading coefficient and to the constant term of the polynomial.

THE RATIONAL ZERO TEST

If the polynomial $f(x) = a_n x^n + a_{n-1} x^{n-1} + \cdots + a_2 x^2 + a_1 x + a_0$ has *integer* coefficients, then every rational zero of f has the form

$$\text{Rational zero} = \frac{p}{q}$$

where p and q have no common factors other than 1, and

p = a factor of the constant term a_0

q = a factor of the leading coefficient a_n.

The feasibility of this theorem can be seen by exploring a special case. Consider a third-degree polynomial $P(x)$ having $\frac{2}{3}$ as a zero.

$$
\begin{aligned}
P(x) &= (3x - 2)(ax^2 + bx + c) && \textit{Factored form}\\
&= 3ax^3 - 2ax^2 + 3bx^2 - 2bx + 3cx - 2c\\
&= 3ax^3 + (3b - 2a)x^2 + (3c - 2b)x - 2c && \textit{Standard form}
\end{aligned}
$$

Note that the numerator of $\frac{2}{3}$ is a factor of the constant term, $-2c$, and that the denominator is a factor of the leading coefficient, $3a$, as specified by the theorem.

To use the Rational Zero Test, we first list all rational numbers whose numerators are factors of the constant term and whose denominators are factors of the leading coefficient.

$$\text{Possible rational zeros} = \frac{\text{factors of constant term}}{\text{factors of leading coefficient}}$$

Having formed this list of *possible rational zeros*, we use a trial-and-error method to determine which, if any, are actual zeros of the polynomial. Note that when the leading coefficient is 1, then the possible rational zeros are simply the factors of the constant term.

EXAMPLE 3   Rational Zero Test with Leading Coefficient of 1

Find the rational zeros of $f(x) = x^3 + x + 1$.

Solution

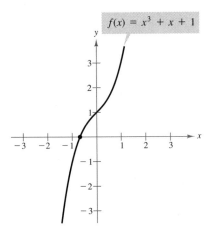

FIGURE 4.34

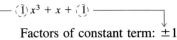

Factors of constant term: ± 1

Factors of leading coefficient: ± 1

Since 1 and -1 are the only factors of the leading coefficient, the possible rational zeros are simply the factors of the constant term, ± 1.

By testing these possible zeros, you can see that neither works.

$$f(1) = (1)^3 + 1 + 1 = 3$$
$$f(-1) = (-1)^3 + (-1) + 1 = -1$$

Thus, the given polynomial has *no* rational zeros. Note from the graph of f in Figure 4.34 that f does have one real zero (between -1 and 0). However, by the Rational Zero Test, you know that this real zero is *not* a rational number.

EXAMPLE 4   Rational Zero Test with Leading Coefficient of 1

Find the rational zeros of $f(x) = x^4 - x^3 + x^2 - 3x - 6$.

Solution

For $f(x) = x^4 - x^3 + x^2 - 3x - 6$, the leading coefficient is 1. Hence, the possible rational zeros are the factors of the constant term.

$$\pm 1, \ \pm 2, \ \pm 3, \ \pm 6$$

A test of these possible zeros would show that $x = -1$ and $x = 2$ are the only two that work. Check the others to be sure.

If the leading coefficient of a polynomial is not 1, the list of possible rational zeros can increase dramatically. In such cases the search can be shortened in several ways. (1) A programmable calculator can be used to speed up the calculations. (2) A rough sketch, possibly with a graphics calculator or graphing software, may give a good estimate of the location of the zeros. (3) Synthetic division can be used to test the possible rational zeros.

To see how to use synthetic division to test the possible rational zeros, let's take another look at the function

$$f(x) = x^4 - x^3 + x^2 - 3x - 6$$

given in Example 4. To test that $x = -1$ and $x = 2$ are zeros of f, you can apply synthetic division.

$$
\begin{array}{r|rrrr}
-1 & 1 & -1 & 1 & -3 & -6 \\
   &   & -1 & 2 & -3 & 6 \\
\hline
   & 1 & -2 & 3 & -6 & 0 \\
\end{array}
$$

$$
\begin{array}{r|rrrr}
2 & 1 & -2 & 3 & -6 \\
  &   & 2 & 0 & 6 \\
\hline
  & 1 & 0 & 3 & 0 \\
\end{array}
$$

Thus,

$$f(x) = (x + 1)(x - 2)(x^2 + 3).$$

Since the factor $(x^2 + 3)$ produces no real zeros, $x = -1$ and $x = 2$ are the *only* real zeros of f.

Finding the first zero is often the most difficult part. After that, the search is simplified by working with the lower-degree polynomial obtained in synthetic division.

EXAMPLE 5   Using the Rational Zero Test

Find the rational zeros of $f(x) = 2x^3 + 3x^2 - 8x + 3$.

Solution

Since the leading coefficient is 2 and the constant term is 3, the possible rational zeros are as follows.

$$\frac{\text{Factors of } 3}{\text{Factors of } 2} = \frac{\pm 1, \ \pm 3}{\pm 1, \ \pm 2} = \pm 1, \ \pm 3, \ \pm\frac{1}{2}, \ \pm\frac{3}{2}$$

By synthetic division, we determine that $x = 1$ is a zero.

$$
\begin{array}{r|rrrr}
1 & 2 & 3 & -8 & 3 \\
  &   & 2 & 5 & -3 \\
\hline
  & 2 & 5 & -3 & 0 \\
\end{array}
$$

Thus, $f(x)$ factors as

$$
\begin{aligned}
f(x) &= (x - 1)(2x^2 + 5x - 3) \\
     &= (x - 1)(2x - 1)(x + 3)
\end{aligned}
$$

and we conclude that the zeros of f are $x = 1$, $x = \frac{1}{2}$ and $x = -3$.   ◢

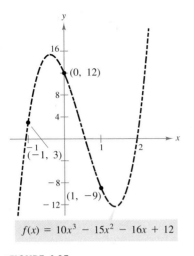

$f(x) = 10x^3 - 15x^2 - 16x + 12$

FIGURE 4.35

EXAMPLE 6   Using the Rational Zero Test

Find all the real zeros of $f(x) = 10x^3 - 15x^2 - 16x + 12$.

Solution

Since the leading coefficient is 10 and the constant term is 12, we have a long list of possible rational zeros.

$$\frac{\text{Factors of } 12}{\text{Factors of } 10} = \frac{\pm 1, \pm 2, \pm 3, \pm 4, \pm 6, \pm 12}{\pm 1, \pm 2, \pm 5, \pm 10}$$

With so many possibilities (32, in fact), it is worth our time to stop and make a rough sketch of this function. From Figure 4.35, it looks like three reasonable choices would be $x = -\frac{6}{5}$, $x = \frac{1}{2}$, and $x = 2$. Testing these by synthetic division shows that only $x = 2$ works. Thus, we have

$$f(x) = (x - 2)(10x^2 + 5x - 6).$$

Using the Quadratic Formula, we find that the two additional zeros are irrational numbers:

$$x = \frac{-5 + \sqrt{265}}{20} \approx 0.5639 \quad \text{and} \quad x = \frac{-5 - \sqrt{265}}{20} \approx -1.0639.$$

Bounds for Real Zeros of Polynomial Functions

The final test for zeros of a polynomial function is related to the sign pattern in the last row of the synthetic division tableau. This test can give us an upper or a lower bound of the real zeros of f. A real number b is an **upper bound** for the real zeros of f if no zeros are greater than b. Similarly, b is a **lower bound** if no real zeros of f are less than b.

LOWER AND UPPER BOUND RULE

Let $f(x)$ be a polynomial with real coefficients and a positive leading coefficient. Suppose $f(x)$ is divided by $x - c$, using synthetic division.

1. If $c > 0$ and each number in the last row is either positive or zero, then c is an *upper bound* for the real zeros of f.
2. If $c < 0$ and the numbers in the last row are alternatively positive and negative (zero entries count as positive or negative), then c is a *lower bound* for the real zeros of f.

Note in Example 7 that we use all three tests presented in this section to search for the real zeros of a polynomial function. In addition, we show how to handle rational coefficients by factoring out the reciprocal of their least common denominator.

EXAMPLE 7   A Polynomial Function with Rational Coefficients

Find the real zeros of

$$f(x) = x^3 - \frac{2}{3}x^2 + \frac{1}{2}x - \frac{1}{3}.$$

Solution

To find the rational zeros, rewrite $f(x)$ by factoring out the reciprocal of the least common denominator of the coefficients.

$$f(x) = \frac{6}{6}x^3 - \frac{4}{6}x^2 + \frac{3}{6}x - \frac{2}{6} = \frac{1}{6}(6x^3 - 4x^2 + 3x - 2)$$

Now, for the purpose of finding the zeros of f, you can drop the factor $\frac{1}{6}$. This is legitimate because the zeros of f are the same as the zeros of

$$g(x) = 6x^3 - 4x^2 + 3x - 2$$

which has the following possible rational zeros.

$$\frac{\text{Factors of 2}}{\text{Factors of 6}} = \frac{\pm 1, \pm 2}{\pm 1, \pm 2, \pm 3, \pm 6} = \pm 1, \pm \frac{1}{2}, \pm \frac{1}{3}, \pm \frac{1}{6}, \pm \frac{2}{3}, \pm 2$$

Since $f(x)$ has three variations in sign and $f(-x)$ has none, we conclude by Descartes's Rule of Signs that there are three positive real zeros or one positive real zero, and no negative zeros. Trying $x = 1$, we obtain the following.

$$
\begin{array}{r|rrrr}
1 & 6 & -4 & 3 & -2 \\
  &   & 6 & 2 & 5 \\
\hline
  & 6 & 2 & 5 & 3 \leftarrow \text{All positive entries}
\end{array}
$$

Thus, $x = 1$ is not a zero, but because the last row has all positive entries, $x = 1$ is an upper bound for the real zeros. Thus, we restrict our search to zeros between 0 and 1. Choosing $x = \frac{2}{3}$, we obtain the following.

$$
\begin{array}{r|rrrr}
\frac{2}{3} & 6 & -4 & 3 & -2 \\
  &   & 4 & 0 & 2 \\
\hline
  & 6 & 0 & 3 & 0
\end{array}
$$

Thus, $f(x)$ factors as

$$f(x) = \frac{1}{6}\left(x - \frac{2}{3}\right)(6x^2 + 3)$$

$$= \frac{1}{6}\left(\frac{1}{3}\right)(3x - 2)(3)(2x^2 + 1)$$

$$= \frac{1}{6}(3x - 2)(2x^2 + 1).$$

Since $2x^2 + 1$ has no real zeros, we conclude that $x = \frac{2}{3}$ is the only real zero of f.

Additional Hints for Finding Zeros of Polynomials

1. If the terms of $f(x)$ have a common monomial factor, it should be factored out before applying the tests in this section. For instance, by writing

$$f(x) = x^4 - 5x^3 + 3x^2 + x$$
$$= x(x^3 - 5x^2 + 3x + 1)$$

   you can see that $x = 0$ is a zero of f and the remaining zeros can be obtained by analyzing the cubic factor.

2. If you are able to find all but two zeros of $f(x)$, then you are home free because you can always use the Quadratic Formula on the remaining quadratic factor. For instance, if you succeeded in writing

$$f(x) = x^4 - 5x^3 + 3x^2 + x$$
$$= x(x - 1)(x^2 - 4x - 1)$$

   then you can apply the Quadratic Formula to $x^2 - 4x - 1$ to find the two remaining zeros.

DISCUSSION

PROBLEM

Comparing Real Zeros and Rational Zeros

Compare the *real* zeros of a polynomial function with the *rational* zeros of a polynomial function. Then answer the following questions.

1. Is it possible for a polynomial function to have no rational zeros but to have real zeros? If so, give an example.
2. If a polynomial function has three real zeros, and only one of them is a rational number, then must the other two zeros be irrational numbers?
3. Consider a cubic polynomial function, $f(x) = ax^3 + bx^2 + cx + d$, where $a \neq 0$. Is it possible that f has no real zeros? If so, give an example. Is it possible that f has no rational zeros? If so, give an example.

WARM UP

The following warm-up exercises involve skills that were covered in earlier sections. You will use these skills in the exercise set for this section.

In Exercises 1 and 2, find a polynomial function with integer coefficients having the given zeros.

1. $-1, \frac{2}{3}, 3$

2. $-2, 0, \frac{3}{4}, 2$

In Exercises 3 and 4, divide by synthetic division.

3. $\dfrac{x^5 - 9x^3 + 5x + 18}{x + 3}$

4. $\dfrac{3x^4 + 17x^3 + 10x^2 - 9x - 8}{x + (2/3)}$

In Exercises 5–8, use the given zero to find all the real zeros of f.

Polynomial Equation	Value of x
5. $f(x) = 2x^3 + 11x^2 + 2x - 4$	$x = \frac{1}{2}$
6. $f(x) = 6x^3 - 47x^2 - 124x - 60$	$x = 10$
7. $f(x) = 4x^3 - 13x^2 - 4x + 6$	$x = -\frac{3}{4}$
8. $f(x) = 10x^3 + 51x^2 + 48x - 28$	$x = \frac{2}{5}$

In Exercises 9 and 10, find all real solutions of the polynomial equation.

9. $x^4 - 3x^2 + 2 = 0$

10. $x^4 - 7x^2 + 12 = 0$

EXERCISES for Section 4.4

In Exercises 1–10, use Descartes's Rule of Signs to determine the possible number of positive and negative zeros of the function.

1. $f(x) = x^3 + 3$

2. $g(x) = x^3 + 3x^2$

3. $h(x) = 3x^4 + 2x^2 + 1$

4. $h(x) = 2x^4 - 3x + 2$

5. $g(x) = 2x^3 - 3x^2 - 3$

6. $f(x) = 4x^3 - 3x^2 + 2x - 1$

7. $f(x) = -5x^3 + x^2 - x + 5$

8. $g(x) = 5x^5 + 10x$

9. $h(x) = 4x^2 - 8x + 3$

10. $f(x) = 3x^3 + 2x^2 + x + 3$

In Exercises 11–16, use the Rational Zero Test to list all possible rational zeros of f. Verify that the zeros of f on the graph are contained in the list.

11. $f(x) = x^3 + x^2 - 4x - 4$

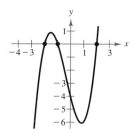

12. $f(x) = -3x^3 + 20x^2 - 36x + 16$

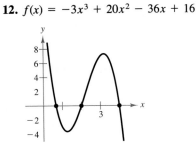

13. $f(x) = -4x^3 + 15x^2 - 8x - 3$

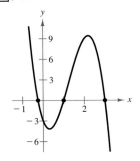

14. $f(x) = 4x^3 - 12x^2 - x + 15$

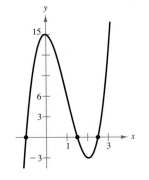

15. $f(x) = -2x^4 + 13x^3 - 21x^2 + 2x + 8$

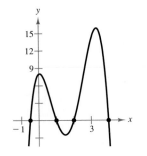

16. $f(x) = 4x^4 - 17x^2 + 4$

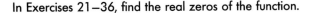

In Exercises 17–20, use synthetic division to determine if the x-value is an upper bound of the zeros of f, a lower bound of the zeros of f, or neither.

17. $f(x) = x^4 - 4x^3 + 15$
    (a) $x = 4$ (b) $x = -1$ (c) $x = 3$

18. $f(x) = 2x^3 - 3x^2 - 12x + 8$
    (a) $x = 2$ (b) $x = 4$ (c) $x = -1$

19. $f(x) = x^4 - 4x^3 + 16x - 16$
    (a) $x = -1$ (b) $x = -3$ (c) $x = 5$

20. $f(x) = 2x^4 - 8x + 3$
    (a) $x = 1$ (b) $x = 3$ (c) $x = -4$

In Exercises 21–36, find the real zeros of the function.

21. $f(x) = x^3 - 6x^2 + 11x - 6$
22. $f(x) = x^3 - 7x - 6$
23. $g(x) = x^3 - 4x^2 - x + 4$
24. $h(x) = x^3 - 9x^2 + 20x - 12$
25. $h(t) = t^3 + 12t^2 + 21t + 10$
26. $f(x) = x^3 + 6x^2 + 12x + 8$
27. $f(x) = x^3 - 4x^2 + 5x - 2$
28. $p(x) = x^3 - 9x^2 + 27x - 27$
29. $C(x) = 2x^3 + 3x^2 - 1$
30. $f(x) = 3x^3 - 19x^2 + 33x - 9$
31. $f(x) = 4x^3 - 3x - 1$
32. $f(z) = 12z^3 - 4z^2 - 27z + 9$
33. $f(y) = 4y^3 + 3y^2 + 8y + 6$
34. $g(x) = 3x^3 - 2x^2 + 15x - 10$
35. $f(x) = x^4 - 3x^2 + 2$
36. $P(t) = t^4 - 7t^2 + 12$

In Exercises 37–44, find all real solutions of the polynomial equation.

37. $z^4 - z^3 - 2z - 4 = 0$

38. $x^4 - x^3 - 29x^2 - x - 30 = 0$

39. $x^4 - 13x^2 - 12x = 0$

40. $2y^4 + 7y^3 - 26y^2 + 23y - 6 = 0$

41. $2x^4 - 11x^3 - 6x^2 + 64x + 32 = 0$

42. $x^5 - x^4 - 3x^3 + 5x^2 - 2x = 0$

43. $x^5 - 7x^4 + 10x^3 + 14x^2 - 24x = 0$

44. $6x^4 - 11x^3 - 51x^2 + 99x - 27 = 0$

In Exercises 45–48, (a) list the possible rational zeros of f, (b) sketch the graph of f so that some of the possible zeros in part (a) can be disregarded, and then (c) determine all real zeros of f.

45. $f(x) = 32x^3 - 52x^2 + 17x + 3$

46. $f(x) = 6x^3 - x^2 - 13x + 8$

47. $f(x) = 4x^3 + 7x^2 - 11x - 18$

48. $f(x) = 2x^3 + 5x^2 - 21x - 10$

In Exercises 49–52, find the rational zeros of the polynomial function.

49. $P(x) = x^4 - \frac{25}{4}x^2 + 9$

50. $f(x) = x^3 - \frac{3}{2}x^2 - \frac{23}{2}x + 6$

51. $f(x) = x^3 - \frac{1}{4}x^2 - x + \frac{1}{4}$

52. $f(z) = z^3 + \frac{11}{6}z^2 - \frac{1}{2}z - \frac{1}{3}$

In Exercises 53–56, match the cubic equation with the number of rational and irrational zeros (a), (b), (c), or (d).

53. $f(x) = x^3 - 1$ **54.** $f(x) = x^3 - 2$

55. $f(x) = x^3 - x$ **56.** $f(x) = x^3 - 2x$

    (a) Rational zeros: 0   (b) Rational zeros: 3
        Irrational zeros: 1   Irrational zeros: 0

    (c) Rational zeros: 1   (d) Rational zeros: 1
        Irrational zeros: 2   Irrational zeros: 0

57. *Dimensions of a Box*   An open box is made from a rectangular piece of material, 9 inches by 5 inches, by cutting equal squares from each corner and turning up the sides (see figure). Find the dimensions of the box, given that the volume is to be 18 cubic inches.

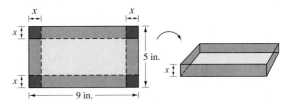

Figure for 57

58. *Dimensions of a Box*   An open box is made from a rectangular piece of material, 12 inches by 10 inches, by cutting equal squares from each corner and turning up the sides. Find the dimensions of the box, given that the volume is to be 96 cubic inches.

59. *Dimensions of a Package*   A rectangular package to be sent by a postal service can have a maximum combined length and girth (perimeter of a cross section) of 108 inches (see figure). Find the dimensions of the package, given that the volume is to be 11,664 cubic inches.

60. *Dimensions of a Package*   A rectangular package to be sent by a postal service can have a maximum combined length and girth (perimeter of a cross section) of 120 inches (see figure). Find the dimensions of the package, given that the volume is to be 16,000 cubic inches.

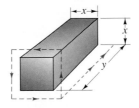

Figure for 59 and 60

4.5 The Fundamental Theorem of Algebra

The Fundamental Theorem of Algebra / Conjugate Pairs / Factoring a Polynomial

The Fundamental Theorem of Algebra

We have been using the fact that an nth degree polynomial can have at most n real zeros. In this section we show that, in the complex number system, every nth degree polynomial function has *precisely* n zeros. This important result is derived from the **Fundamental Theorem of Algebra,** first proved by the famous German mathematician Carl Friedrich Gauss (1777–1855).

THE FUNDAMENTAL THEOREM OF ALGEBRA

If $f(x)$ is a polynomial of degree n, where $n > 0$, then f has at least one zero in the complex number system.

Using the Fundamental Theorem of Algebra and the equivalence of zeros and factors, we can state the following theorem.

LINEAR FACTORIZATION THEOREM

If $f(x)$ is a polynomial of degree n, where $n > 0$, then f has precisely n linear factors

$$f(x) = a(x - c_1)(x - c_2) \ldots (x - c_n)$$

where $c_1, c_2, \ldots, c_n$ are complex numbers and a is the leading coefficient of $f(x)$.

Proof

Using the Fundamental Theorem, we know that f must have at least one zero, c_1. Consequently, $(x - c_1)$ is a factor of $f(x)$, and we have

$$f(x) = (x - c_1)f_1(x).$$

If the degree of $f_1(x)$ is greater than zero, we again apply the Fundamental Theorem to conclude that f_1 must have a zero c_2, which implies that

$$f(x) = (x - c_1)(x - c_2)f_2(x).$$

It is clear that the degree of $f_1(x)$ is $n - 1$, that the degree of $f_2(x)$ is $n - 2$, and that we can repeatedly apply the Fundamental Theorem n times until we obtain

$$f(x) = a(x - c_1)(x - c_2) \cdots (x - c_n)$$

where a is the leading coefficient of the polynomial $f(x)$. ◢

Note that neither the Fundamental Theorem of Algebra nor the Linear Factorization Theorem explains *how* to find the zeros or factors of a polynomial. We call such theorems **existence theorems.** To find the zeros of a polynomial function, we still rely on the techniques developed in earlier parts of the text.

The Linear Factorization Theorem states that an nth degree polynomial has precisely n linear factors. Therefore, it follows that an nth degree polynomial *function* has precisely n zeros. Remember, however, that these n zeros can be real or complex, and that they may be repeated.

EXAMPLE 1   Zeros of Polynomial Functions

a. The first-degree polynomial function

$$f(x) = x - 2$$

has exactly *one* zero: $x = 2$.

b. Counting multiplicity, the second-degree polynomial function

$$f(x) = x^2 - 6x + 9 = (x - 3)(x - 3)$$

has exactly *two* zeros: $x = 3$ and $x = 3$.

c. The third-degree polynomial function

$$f(x) = x^3 + 4x = x(x - 2i)(x + 2i)$$

has exactly *three* zeros: $x = 0$, $x = 2i$, and $x = -2i$.

d. The fourth-degree polynomial function

$$f(x) = x^4 - 1 = (x - 1)(x + 1)(x - i)(x + i)$$

has exactly *four* zeros: $x = 1$, $x = -1$, $x = i$, and $x = -i$. ◢

Example 2 shows how we can use the methods described in previous sections—Descartes's Rule of Signs, Rational Zero Test, synthetic division, and factoring—to find all the zeros of a polynomial function, including the complex zeros.

EXAMPLE 2 Finding the Zeros of a Polynomial Function

Write the polynomial function

$$f(x) = x^5 + x^3 + 2x^2 - 12x + 8$$

as the product of linear factors and list all of its zeros.

Solution

Descartes's Rule of Signs indicates two or no positive real zeros and one negative real zero. Moreover, the possible rational zeros are ± 1, ± 2, ± 4, and ± 8. Synthetic division produces the following.

$$
\begin{array}{r|rrrrrr}
1 & 1 & 0 & 1 & 2 & -12 & 8 \\
  &   & 1 & 1 & 2 & 4 & -8 \\
\hline
  & 1 & 1 & 2 & 4 & -8 & 0
\end{array}
$$

$\longrightarrow$ *1 is a zero*

$$
\begin{array}{r|rrrrr}
1 & 1 & 1 & 2 & 4 & -8 \\
  &   & 1 & 2 & 4 & 8 \\
\hline
  & 1 & 2 & 4 & 8 & 0
\end{array}
$$

$\longrightarrow$ *1 is a repeated zero*

$$
\begin{array}{r|rrrr}
-2 & 1 & 2 & 4 & 8 \\
   &   & -2 & 0 & -8 \\
\hline
   & 1 & 0 & 4 & 0
\end{array}
$$

$\longrightarrow$ *-2 is a zero*

Thus, we have

$$f(x) = x^5 + x^3 + 2x^2 - 12x + 8$$
$$= (x - 1)(x - 1)(x + 2)(x^2 + 4).$$

By factoring $x^2 + 4$ as

$$x^2 - (-4) = (x - \sqrt{-4})(x + \sqrt{-4}) = (x - 2i)(x + 2i)$$

we have

$$f(x) = (x - 1)(x - 1)(x + 2)(x - 2i)(x + 2i)$$

which gives the following five zeros of f.

$$1, \quad 1, \quad -2, \quad 2i, \quad \text{and} \quad -2i$$

Note from the graph of f shown in Figure 4.36 that the *real* zeros are the only ones that appear as x-intercepts.

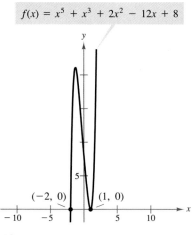

$f(x) = x^5 + x^3 + 2x^2 - 12x + 8$

FIGURE 4.36

Conjugate Pairs

In Example 2, note that the two complex zeros are **conjugates** because they are of the form

$$a + bi \qquad \text{and} \qquad a - bi.$$

REMARK Be sure you see that this result is true only if the polynomial function has *real coefficients*. For instance, the result applies to the function $f(x) = x^2 + 1$, but not to the function $g(x) = x - i$.

COMPLEX ZEROS OCCUR IN CONJUGATE PAIRS

Let $f(x)$ be a polynomial function that has real coefficients. If $a + bi$, where $b \neq 0$ is a zero of the function, then the conjugate $a - bi$ is also a zero of the function.

EXAMPLE 3 Finding a Polynomial with Given Zeros

Find a *fourth-degree* polynomial function, with real coefficients, that has -1, -1, and $3i$ as zeros.

Solution

Since $3i$ is a zero *and* the polynomial is stated to have real coefficients, the conjugate $-3i$ must also be a zero. Thus, from the Linear Factorization Theorem, $f(x)$ can be written as

$$f(x) = a(x + 1)(x + 1)(x - 3i)(x + 3i).$$

For simplicity, let $a = 1$, and obtain

$$\begin{aligned} f(x) &= (x^2 + 2x + 1)(x^2 + 9) \\ &= x^4 + 2x^3 + 10x^2 + 18x + 9. \end{aligned}$$

EXAMPLE 4 Finding a Polynomial with Given Zeros

Find a *cubic* polynomial function f, with real coefficients, that has 2 and $1 - i$ as zeros, and such that $f(1) = 3$.

Solution

Because $1 - i$ is a zero of f, so is $1 + i$. Therefore, we have

$$\begin{aligned} f(x) &= a(x - 2)[x - (1 - i)][x - (1 + i)] \\ &= a(x - 2)[x^2 - x(1 + i) - x(1 - i) + 1 - i^2] \\ &= a(x - 2)(x^2 - 2x + 2) \\ &= a(x^3 - 4x^2 + 6x - 4). \end{aligned}$$

To find the value of a, we use the fact that $f(1) = 3$ and obtain $f(1) = a(1 - 4 + 6 - 4) = 3$. Thus, $a = -3$ and we conclude that

$$\begin{aligned} f(x) &= -3(x^3 - 4x^2 + 6x - 4) \\ &= -3x^3 + 12x^2 - 18x + 12. \end{aligned}$$

Factoring a Polynomial

The Linear Factorization Theorem states that any nth degree polynomial can be written as the product of n linear factors.

$$f(x) = a(x - c_1)(x - c_2)(x - c_3) \cdots (x - c_n)$$

However, this result includes the possibility that some of the values of c_i are complex. The following result tells us that even if you do not want to get involved with "complex factors," you can still write $f(x)$ as the product of linear and/or quadratic factors.

FACTORS OF A POLYNOMIAL

Every polynomial of degree $n > 0$ with real coefficients can be written as the product of linear and quadratic factors with real coefficients, where the quadratic factors have no real zeros.

Proof

To begin, we use the Linear Factorization Theorem to conclude that $f(x)$ can be *completely* factored in the form

$$f(x) = a(x - c_1)(x - c_2)(x - c_3) \cdots (x - c_n).$$

If each c_i is real, there is nothing more to prove. If any c_i is complex ($c_i = a + bi$, $b \neq 0$), then because the coefficients of $f(x)$ are real, we know that the conjugate $c_j = a - bi$ is also a zero. By multiplying the corresponding factors, we obtain

$$(x - c_i)(x - c_j) = [x - (a + bi)][x - (a - bi)]$$
$$= x^2 - 2ax + (a^2 + b^2)$$

where each coefficient is real. ◢

A quadratic factor with no real zeros is **irreducible over the reals.** Be sure you see that this is not the same as being *irreducible over the rationals*. For example, the quadratic

$$x^2 + 1 = (x - i)(x + i)$$

is irreducible over the reals (and therefore over the rationals). On the other hand, the quadratic

$$x^2 - 2 = \left(x - \sqrt{2}\right)\left(x + \sqrt{2}\right)$$

is irreducible over the rationals, but it is *reducible* over the reals.

EXAMPLE 5   Factoring a Polynomial

Write the polynomial $f(x) = x^4 - x^2 - 20$

a. as the product of factors that are irreducible over the *rationals*,
b. as the product of linear factors and quadratic factors that are irreducible over the *reals*, and
c. in completely factored form.

Solution

a. We begin by factoring the polynomial into the product of two quadratic polynomials.

$$x^4 - x^2 - 20 = (x^2 - 5)(x^2 + 4)$$

Both of these factors are irreducible over the rationals.

b. By factoring over the reals, we have

$$x^4 - x^2 - 20 = (x + \sqrt{5})(x - \sqrt{5})(x^2 + 4)$$

where the quadratic factor is irreducible over the reals.

c. In completely factored form, we have

$$x^4 - x^2 - 20 = (x + \sqrt{5})(x - \sqrt{5})(x - 2i)(x + 2i).$$

EXAMPLE 6   Finding the Zeros of a Polynomial Function

Find all the zeros of

$$f(x) = x^4 - 3x^3 + 6x^2 + 2x - 60$$

given that $1 + 3i$ is a zero of f.

Solution

Since complex zeros occur in conjugate pairs, $1 - 3i$ is also a zero of f. This means that both

$$[x - (1 + 3i)] \qquad \text{and} \qquad [x - (1 - 3i)]$$

are factors of $f(x)$. Multiplying these two factors produces

$$[x - (1 + 3i)][x - (1 - 3i)] = [(x - 1) - 3i][(x - 1) + 3i]$$
$$= (x - 1)^2 - 9i^2$$
$$= x^2 - 2x + 10.$$

Using long division, we can divide $x^2 - 2x + 10$ into $f(x)$ to obtain the following.

$$
\require{enclose}
\begin{array}{r}
x^2 - x - 6 \\[-3pt]
x^2 - 2x + 10 \enclose{longdiv}{x^4 - 3x^3 + 6x^2 + 2x - 60} \\
\end{array}
$$

$$
\begin{array}{r}
x^4 - 2x^3 + 10x^2 \\ \hline
-x^3 - 4x^2 + 2x \\
-x^3 + 2x^2 - 10x \\ \hline
-6x^2 + 12x - 60 \\
-6x^2 + 12x - 60 \\ \hline
0
\end{array}
$$

Therefore, we have

$$f(x) = (x^2 - 2x + 10)(x^2 - x - 6)$$
$$= (x^2 - 2x + 10)(x - 3)(x + 2)$$

and we conclude that the zeros of f are

$$1 + 3i, \quad 1 - 3i, \quad 3, \quad \text{and} \quad -2.$$

Throughout this chapter, we have basically stated the results and examples in terms of *zeros of polynomial functions*. Be sure you see that the same results could have been stated in terms of *solutions of polynomial equations*. This is true because the zeros of the polynomial function

$$f(x) = a_n x^n + a_{n-1} x^{n-1} + \cdots + a_2 x^2 + a_1 x + a_0$$

are precisely the solutions of the polynomial equation

$$a_n x^n + a_{n-1} x^{n-1} + \cdots + a_2 x^2 + a_1 x + a_0 = 0.$$

DISCUSSION

PROBLEM

Factoring a
Polynomial

In a short paper or discussion, summarize the various techniques used to factor a polynomial. Include all of the techniques covered up to this point in the text.

WARM UP

The following warm-up exercises involve skills that were covered in earlier sections. You will use these skills in the exercise set for this section.

In Exercises 1–4, write each complex number in standard form and give its complex conjugate.

1. $4 - \sqrt{-29}$

2. $-5 - \sqrt{-144}$

3. $-1 + \sqrt{-32}$

4. $6 + \sqrt{-1/4}$

In Exercises 5–10, perform the indicated operations and write the answers in standard form.

5. $(-3 + 6i) - (10 - 3i)$

6. $(12 - 4i) + 20i$

7. $(4 - 2i)(3 + 7i)$

8. $(2 - 5i)(2 + 5i)$

9. $\dfrac{1 + i}{1 - i}$

10. $(3 + 2i)^3$

EXERCISES for Section 4.5

In Exercises 1–26, find all the zeros of the function and write the polynomial as a product of linear factors.

1. $f(x) = x^2 + 25$

2. $f(x) = x^2 - x + 56$

3. $h(x) = x^2 - 4x + 1$

4. $g(x) = x^2 + 10x + 23$

5. $f(x) = x^4 - 81$

6. $f(y) = y^4 - 625$

7. $f(z) = z^2 - 2z + 2$

8. $h(x) = x^3 - 3x^2 + 4x - 2$

9. $g(x) = x^3 - 6x^2 + 13x - 10$

10. $f(x) = x^3 - 2x^2 - 11x + 52$

11. $f(t) = t^3 - 3t^2 - 15t + 125$

12. $f(x) = x^3 + 11x^2 + 39x + 29$

13. $f(x) = x^3 + 24x^2 + 214x + 740$

14. $f(s) = 2s^3 - 5s^2 + 12s - 5$

15. $f(x) = 16x^3 - 20x^2 - 4x + 15$

16. $f(x) = 9x^3 - 15x^2 + 11x - 5$

17. $h(x) = x^3 - x + 6$

18. $h(x) = x^3 + 9x^2 + 27x + 35$

19. $f(x) = 5x^3 - 9x^2 + 28x + 6$

20. $g(x) = 3x^3 - 4x^2 + 8x + 8$

21. $g(x) = x^4 - 4x^3 + 8x^2 - 16x + 16$

22. $h(x) = x^4 + 6x^3 + 10x^2 + 6x + 9$

23. $f(x) = x^4 + 10x^2 + 9$

24. $f(x) = x^4 + 29x^2 + 100$

25. $f(x) = 2x^4 + 5x^3 + 4x^2 + 5x + 2$

26. $g(x) = x^5 - 8x^4 + 28x^3 - 56x^2 + 64x - 32$

In Exercises 27–36, find a polynomial with integer coefficients that has the given zeros.

27. $1, 5i, -5i$

28. $4, 3i, -3i$

29. $2, 4 + i, 4 - i$

30. $6, -5 + 2i, -5 - 2i$

31. $i, -i, 6i, -6i$

32. $2, 2, 2, 4i, -4i$

33. $-5, -5, 1 + \sqrt{3}i$

34. $\frac{2}{3}, -1, 3 + \sqrt{2}i$

35. $\frac{3}{4}, -2, -\frac{1}{2} + i$

36. $0, 0, 4, 1 + i$

In Exercises 37–40, write the polynomial (a) as the product of factors that are irreducible over the *rationals*, (b) as the product of linear and quadratic factors that are irreducible over the *reals*, and (c) in completely factored form.

37. $f(x) = x^4 + 6x^2 - 27$

38. $f(x) = x^4 - 2x^3 - 3x^2 + 12x - 18$
    (*Hint:* One factor is $x^2 - 6$.)

39. $f(x) = x^4 - 4x^3 + 5x^2 - 2x - 6$
    (*Hint:* One factor is $x^2 - 2x - 2$.)

40. $f(x) = x^4 - 3x^3 - x^2 - 12x - 20$
    (*Hint:* One factor is $x^2 + 4$.)

In Exercises 41–50, use the given zero of f to find all the zeros of f.

Function	Zero of f
41. $f(x) = 2x^3 + 3x^2 + 50x + 75$	$r = 5i$
42. $f(x) = x^3 + x^2 + 9x + 9$	$r = 3i$
43. $f(x) = 2x^4 - x^3 + 7x^2 - 4x - 4$	$r = 2i$
44. $g(x) = x^3 - 7x^2 - x + 87$	$r = 5 + 2i$
45. $g(x) = 4x^3 + 23x^2 + 34x - 10$	$r = -3 + i$
46. $h(x) = 3x^3 - 4x^2 + 8x + 8$	$r = 1 - \sqrt{3}i$
47. $f(x) = x^4 + 2x^3 - 9x^2 - 20x +$ 44	$r = -3 + \sqrt{2}i$
48. $f(x) = 4x^4 + 4x^3 + 33x^2 -$ $38x + 10$	$r = -1 - 3i$
49. $h(x) = 8x^3 - 14x^2 + 18x - 9$	$r = (1 - \sqrt{5}i)/2$
50. $f(x) = 25x^3 - 55x^2 - 54x - 18$	$r = (-2 + \sqrt{2}i)/5$

51. *Maximum Height* A baseball is thrown upward from ground level with an initial velocity of 48 feet per second and its height h in feet is given by

$$h = -16t^2 + 48t, \quad 0 \le t \le 3$$

where t is the time in seconds. Suppose you are told the ball reaches a height of 64 feet. Explain why this is not possible.

52. *Profit* The demand equation for a certain product is given by $p = 140 - 0.0001x$, where p is the unit price (in dollars) of the product and x is the number of units produced and sold. The cost equation for the product is $C = 80x + 150{,}000$, where C is the total cost (in dollars) and x is the number of units produced. The total profit obtained by producing and selling x units is given by

$$P = R - C = xp - C.$$

Suppose you are the marketing manager for this product, and you are asked to determine a price p that would yield a profit of 9 million dollars. Explain why this is not possible.

53. Find a quadratic function f (with integer coefficients) that has $\pm\sqrt{b}i$ as zeros. Assume that b is a positive integer.

54. Find a quadratic function f (with integer coefficients) that has a and bi as zeros. Assume that b is a positive integer.

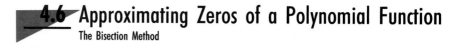

4.6 Approximating Zeros of a Polynomial Function
The Bisection Method

The Bisection Method

Throughout history, mathematicians have devoted a great deal of time and effort to developing methods for finding the zeros of polynomial functions. We have not looked at all of these methods in this chapter, but we have looked at the basic ones: graphical methods, factorization methods, the Rational Zero Test, and Descartes's Rule of Signs.

Generally, the higher the degree of a polynomial function, the more difficult it is to find its zeros. In practical applications involving polynomials of degree 3 or greater, we often must be content with an approximation technique for finding zeros. Most approximation methods involve an **iterative process,** meaning that the method is applied repeatedly to obtain better and better approximations.

The **Bisection Method** is one of the simpler approximation methods. To apply this method, we must find two x-values—one at which the function is

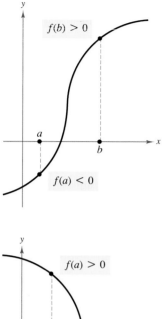

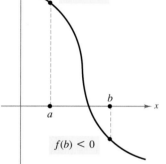

FIGURE 4.37

positive and one at which it is negative. By the Intermediate Value Theorem, we know that the function has at least one zero between the two *x*-values, as shown in Figure 4.37.

In Figure 4.37, we can see that a zero must occur somewhere in the interval (a, b). To apply the Bisection Method, we cut this interval in half and consider the two intervals

$$\left(a, \frac{a+b}{2}\right) \quad \text{and} \quad \left(\frac{a+b}{2}, b\right).$$

Depending upon the value of $f(x)$ at the midpoint $(a + b)/2$, we apply the Bisection Method again to one of these intervals.

EXAMPLE 1   Using the Bisection Method

Use the Bisection Method to approximate the real zero of

$$f(x) = x^3 - x^2 + 1$$

to within 0.001 unit.

Solution

We begin by making a sketch of f, as shown in Figure 4.38. Recall from Example 8 of Section 4.2 that

$$f(-0.8) = -0.152 < 0 \quad \text{and} \quad f(-0.7) = 0.167 > 0$$

which implies that f has a zero between -0.8 and -0.7. Using the midpoint of this interval, we approximate the zero to be

$$c = \frac{-0.8 + (-0.7)}{2} = -0.75.$$

The maximum error of this approximation is one-half the length of the interval.

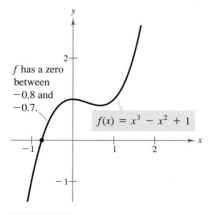

FIGURE 4.38

That is,

$$\text{Maximum error} \leq \frac{-0.7 - (-0.8)}{2} = 0.05.$$

Now calculate the value of $f(-0.75)$ and make one of the following deductions.

1. If $f(-0.75) = 0$, then $= -0.75$ is a zero of f.
2. If $f(-0.75) > 0$, then a zero occurs between -0.8 and -0.75.
3. If $f(-0.75) < 0$, then a zero occurs between -0.75 and -0.7.

Because $f(-0.75) = 0.015625$ is positive, we choose $(-0.8, -0.75)$ as our new interval. The results of several iterations are shown in Table 4.1. After the seventh iteration, the maximum error is less than 0.001, and we approximate the zero of f to be

$$c = -0.75546875 \approx -0.755.$$

TABLE 4.1

	a	c	b	$f(a)$	$f(c)$	$f(b)$	Maximum Error
1	-0.8	-0.75	-0.7	-0.1520	0.0156	0.1670	0.05
2	-0.8	-0.775	-0.75	-0.1520	-0.0661	0.0156	0.025
3	-0.775	-0.7625	-0.75	-0.0661	-0.0247	0.0156	0.0125
4	-0.7625	-0.7563	-0.75	-0.0247	-0.0044	0.0156	0.0063
5	-0.7563	-0.7531	-0.75	-0.0044	0.0056	0.0156	0.0031
6	-0.7563	-0.7547	-0.7531	-0.0044	0.0006	0.0056	0.0016
7	-0.7563	-0.7555	-0.7547	-0.0044	-0.0019	0.0006	0.0008

By continuing the process in Table 4.1, we could approximate the zero to *any* desired accuracy, and we say that the sequence of successively better approximations *converges* to the zero of the function. The convergence of the Bisection Method is relatively slow and several iterations are usually necessary to obtain a very fine accuracy.

In Example 1, the function had only one real zero. However, the Bisection Method can be applied just as well to functions that have several real zeros.

EXAMPLE 2 Using the Bisection Method

Use the Bisection Method to approximate each of the real zeros of

$$f(x) = x^3 - 2x^2 - x + 1$$

to within 0.001 unit.

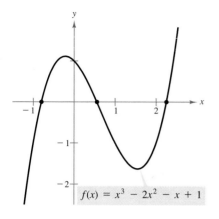

$f(x) = x^3 - 2x^2 - x + 1$

FIGURE 4.39

Solution

From Figure 4.39, we see that this function has three real zeros: one in the interval $[-1, 0]$, one in the interval $[0, 1]$, and one in the interval $[2, 3]$.

The following BASIC program can be used to approximate the first of these three zeros.

BASIC PROGRAM FOR BISECTION METHOD

```
10 REM DEFINE THE FUNCTION
20      DEF FNY(X) = X^3-2*X^2-X+1
30 REM A = LEFT ENDPOINT, B = RIGHT ENDPOINT
40      A = -1
50      B = 0
60      C = (A+B)/2
70 REM ME = MAXIMUM ERROR
80      ME = (B-A)/2
90      PRINT "A = ";A,
           "C = ";C,
           "B = ";B,
           "MAX ERR = ";ME
100 REM TEST THE SIZE OF THE ERROR
110      IF ME < .001 GOTO 150
120 REM TEST THE SIGN OF F(C)
130      IF FNY(A)*FNY(C) < 0 THEN B = C: GOTO
           60
140      IF FNY(B)*FNY(C) < 0 THEN A = C: GOTO
           60
150      END
```

PRINTOUT FROM BASIC PROGRAM

```
A = -1.0000 C = -0.5000 B =  0.0000 MAX ERR = 0.5000
A = -1.0000 C = -0.7500 B = -0.5000 MAX ERR = 0.2500
A = -1.0000 C = -0.8750 B = -0.7500 MAX ERR = 0.1250
A = -0.8750 C = -0.8125 B = -0.7500 MAX ERR = 0.0625
A = -0.8125 C = -0.7813 B = -0.7500 MAX ERR = 0.0313
A = -0.8125 C = -0.7969 B = -0.7813 MAX ERR = 0.0156
A = -0.8125 C = -0.8047 B = -0.7969 MAX ERR = 0.0078
A = -0.8047 C = -0.8008 B = -0.7969 MAX ERR = 0.0039
A = -0.8047 C = -0.8027 B = -0.8008 MAX ERR = 0.0020
A = -0.8027 C = -0.8018 B = -0.8008 MAX ERR = 0.0010
```

From this printout, we approximate one of the zeros of f to be

$x \approx -0.802$.

By rerunning the program (with different interval endpoints) we approximate the other two zeros to be $x \approx 0.556$ and $x \approx 2.247$.

DISCUSSION
PROBLEM
Linear
Interpolation

The Bisection Method is only one of several techniques that can be used to approximate the real zeros of a polynomial function. Another technique is *linear interpolation*, as described in the *Discussion Problem* in Section 3.3. In that problem we considered the graph of

$$f(x) = x^3 + x + 1$$

as shown in Figure 4.40. When $x = -0.69$ the value of y is negative and when $x = -0.68$ the value of y is positive, which implies that f has a real zero between -0.69 and -0.68. Approximate this real zero using both linear interpolation and the Bisection Method. Then write a short paragraph telling which method you prefer and why.

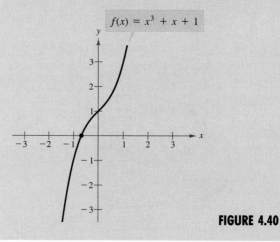

FIGURE 4.40

WARM UP

The following warm-up exercises involve skills that were covered in earlier sections. You will use these skills in the exercise set for this section.

In Exercises 1–4, evaluate the function at the given value of x.

1. $f(x) = x^3 + 2x^2 + 3$, $x = 1.5$

2. $f(x) = 2x^3 - 6x + 10$, $x = -2.6$

3. $f(x) = -x^3 + 7x^2 - 3x - 12$, $x = 3.4$

4. $f(x) = -2x^3 - 3x^2 + x - 2$, $x = -3.1$

In Exercises 5–10, sketch a graph of the function.

5. $f(x) = x^2 - 4x + 1$

6. $f(x) = -x^2 - 2x + 4$

7. $g(x) = -x^3 + 2$

8. $g(x) = x^3 - 4$

9. $h(x) = x^3 - 2x^2 + 1$

10. $h(x) = -x^4 + 2x^2 + 3$

EXERCISES for Section 4.6

In Exercises 1–4, the value of the function is negative for one of the x-values and positive for the other. Determine which is positive and which is negative.

Function	x-values
1. $f(x) = 2x^3 - 4x^2 + 5x - 5$	$x = 1.4, x = 1.5$
2. $f(x) = -x^3 + 5x - 3$	$x = -2.6, x = -2.5$
3. $f(x) = -x^3 + 4x^2 - 5x + 3$	$x = 2.4, x = 2.5$
4. $f(x) = 3x^3 - 5x - 4$	$x = 1.5, x = 1.6$

In Exercises 5–14, use the Bisection Method to approximate the real zero(s) of the function. List your approximation(s) correct to two decimal places.

5. $f(x) = x^3 + x - 1$ **6.** $f(x) = x^5 + x - 1$

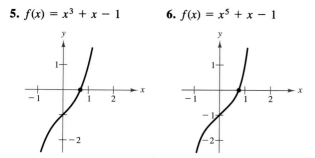

7. $f(x) = 2x^3 - 6x^2 + 6x - 1$

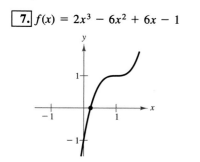

8. $f(x) = 4x^3 - 12x^2 + 12x - 3$

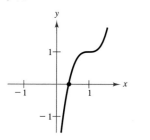

9. $f(x) = -x^3 + 3x^2 - x + 1$

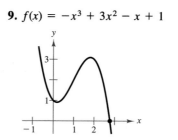

10. $f(x) = x^3 - 3x - 1$ **11.** $f(x) = x^4 - x - 3$

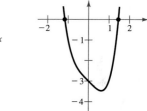

12. $f(x) = x^3 + 2x + 1$ **13.** $f(x) = x^3 - 27x - 27$

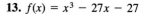
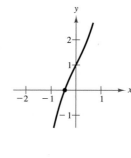
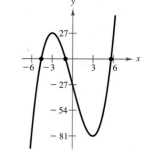

14. $f(x) = x^3 - 3.9x^2 + 4.79x - 1.881$

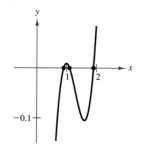

In Exercises 15–20, approximate the zero of the function in the given interval. Give your approximation correct to two decimal places.

Function	Interval
15. $f(x) = 4x^3 + 14x - 8$	$[0, 1]$
16. $f(x) = 4x^3 - 14x^2 - 2$	$[3, 4]$
17. $f(x) = x^3 - 4x^2 - 3x - 9$	$[4, 5]$
18. $f(x) = x^4 + x - 3$	$[1, 2]$
19. $f(x) = 7x^4 - 42x^3 + 43x^2 + 216x - 324$	$[1, 2]$
20. $f(x) = 3x^4 - 12x^3 + 27x^2 + 4x - 4$	$[0, 1]$

21. *Advertising Costs*   A company that produces portable cassette players estimates that the profit from a particular model is given by

$$P = -76x^3 + 4830x^2 - 320,000, \qquad 0 \le x \le 60$$

where P is the profit in dollars and x is the advertising expense in 10,000s of dollars (see figure). Using this model, find the smaller of two advertising amounts that yields a profit of $2,500,000.

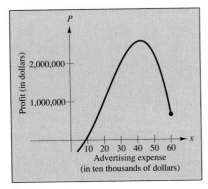

Figure for 21

22. *Engine Power*   The torque produced by a compact automobile engine is approximated by the model

$$T = 0.808x^3 - 17.974x^2 + 71.248x + 110.843,$$
$$1 \le x \le 5$$

where T is the torque in foot-pounds and x is the engine speed in thousands of revolutions per minute (see figure). Approximate the two engine speeds that yield a torque of 170 foot-pounds.

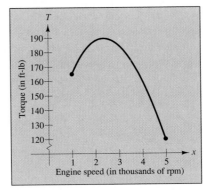

Figure for 22

23. *Medicine*   The concentration of a certain chemical in the bloodstream t hours after injection into muscle tissue is given by

$$C = \frac{3t^2 + t}{t^3 + 50}.$$

The concentration is greater when $3t^4 + 2t^3 - 300t - 50 = 0$. Approximate this time to the nearest tenth of an hour.

24. *Transportation Cost*   The ordering and transportation cost C of the components used in manufacturing a certain product is given by

$$C = 100\left(\frac{200}{x^2} + \frac{x}{x + 30}\right), \qquad 1 \le x$$

where C is measured in thousands of dollars, and x is the order size in hundreds. The cost is at a minimum when $3x^3 - 40x^2 - 2400x - 36,000 = 0$. Approximate the optimal order size to the nearest hundred units.

25. Create a fourth degree polynomial function with integer coefficients and irrational zeros $2 \pm \sqrt{2}$ and $6 \pm \sqrt{3}$.

REVIEW EXERCISES for Chapter 4

In Exercises 1–4, sketch the graph of the quadratic function. Identify the vertex and the intercepts.

1. $f(x) = \left(x + \frac{3}{2}\right)^2 + 1$

2. $f(x) = (x - 4)^2 - 4$

3. $f(x) = \frac{1}{3}(x^2 + 5x - 4)$

4. $f(x) = 3x^2 - 12x + 11$

In Exercises 5 and 6, find the quadratic function that has the indicated vertex and whose graph passes through the given point.

5. Vertex: $(1, -4)$; point: $(2, -3)$

6. Vertex: $(2, 3)$; point: $(-1, 6)$

In Exercises 7–14, find the maximum or minimum value of the quadratic function.

7. $g(x) = x^2 - 2x$

8. $f(x) = x^2 + 8x + 10$

9. $f(x) = 6x - x^2$

10. $h(x) = 3 + 4x - x^2$

11. $f(t) = -2t^2 + 4t + 1$

12. $h(x) = 4x^2 + 4x + 13$

13. $h(x) = x^2 + 5x - 4$

14. $f(x) = 4x^2 + 4x + 5$

15. *Maximum Area*   A rectangle is inscribed in the region bounded by the x-axis, the y-axis, and the graph of $x + 2y - 6 = 0$ (see figure). Find the coordinates (x, y) that yield a maximum area for the rectangle.

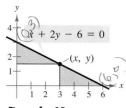

Figure for 15

16. *Maximum Area*   The perimeter of a rectangle is 200 feet. Let x represent the width of the rectangle and write a quadratic function that expresses the area of the rectangle in terms of x. Of all possible rectangles with a perimeter of 200 feet, find the dimensions of the one that has the greatest area.

17. *Minimum Cost*   A textile manufacturer has daily production costs of

$$C = 10,000 - 10x + 0.045x^2$$

where C is the total cost in dollars and x is the number of units produced. How many fixtures should be produced each day to yield a minimum cost?

18. *Maximum Profit*   The profit for a company is

$$P = -0.0002x^2 + 140x - 250,000$$

where x is the number of units sold. What sales level will yield a maximum profit?

19. *Maximum Profit*   A real estate office handles 50 apartment units. When the rent is $540 per month, all units are occupied. However, for each $30 increase in rent, one unit becomes vacant. Each occupied unit requires an average of $18 per month for service and repairs. What rent should be charged to realize the most profit?

20. *Minimum Cost*   A manufacturer has daily production costs of

$$C = 20,000 - 120x + 0.055x^2$$

where C is the total cost in dollars and x is the number of units produced. How many units should be produced each day to yield a minimum cost?

In Exercises 21–24, determine the right-hand and left-hand behavior of the graph of the polynomial function.

21. $f(x) = -x^2 + 6x + 9$

22. $f(x) = \frac{1}{2}x^3 + 2x$

23. $g(x) = \frac{3}{4}(x^4 + 3x^2 + 2)$

24. $h(x) = -x^5 - 7x^2 + 10x$

In Exercises 25–32, sketch the graph of the function.

25. $f(x) = -(x - 2)^3$

26. $f(x) = (x + 1)^3$

27. $g(x) = x^4 - x^3 - 2x^2$

28. $h(x) = -2x^3 - x^2 + x$

29. $f(t) = t^3 - 3t$

30. $f(x) = -x^3 + 3x - 2$

31. $f(x) = x(x + 3)^2$

32. $f(t) = t^4 - 4t^2$

In Exercises 33–40, perform the division.

33. $\dfrac{24x^2 - x - 8}{3x - 2}$

34. $\dfrac{4x + 7}{3x - 2}$

35. $\dfrac{x^4 + x^3 - x^2 + 2x}{x^2 + 2x}$

36. $\dfrac{5x^3 - 13x^2 - x + 2}{x^2 - 3x + 1}$

37. $\dfrac{x^4 - 3x^2 + 2}{x^2 - 1}$

38. $\dfrac{3x^4}{x^2 - 1}$

39. $\dfrac{x^4 - 3x^3 + 4x^2 - 6x + 3}{x^2 + 2}$

40. $\dfrac{6x^4 + 10x^3 + 13x^2 - 5x + 2}{2x^2 - 1}$

In Exercises 41–46, use synthetic division to perform the division.

41. $\dfrac{0.25x^4 - 4x^3}{x - 2}$

42. $\dfrac{2x^3 + 2x^2 - x + 2}{x - (1/2)}$

43. $\dfrac{6x^4 - 4x^3 - 27x^2 + 18x}{x - (2/3)}$

44. $\dfrac{0.1x^3 + 0.3x^2 - 0.5}{x - 5}$

45. $\dfrac{2x^3 - 5x^2 + 12x - 5}{x - (1 + 2i)}$

46. $\dfrac{9x^3 - 15x^2 + 11x - 5}{x - [(1/3) + (2/3)i]}$

In Exercises 47–50, use synthetic division to determine whether the values of x are zeros of the function.

47. $f(x) = 2x^3 + 3x^2 - 20x - 21$
  (a) $x = 4$ (b) $x = -1$ (c) $x = -\frac{7}{2}$ (d) $x = 0$

48. $f(x) = 20x^4 + 9x^3 - 14x^2 - 3x$
  (a) $x = -1$ (b) $x = \frac{3}{4}$ (c) $x = 0$ (d) $x = 1$

49. $f(x) = 2x^3 + 7x^2 - 18x - 30$
  (a) $x = 1$ (b) $x = \frac{5}{2}$ (c) $x = -3 + \sqrt{3}$ (d) $x = 0$

50. $f(x) = 3x^3 - 26x^2 + 364x - 232$
  (a) $x = 4 - 10i$ (b) $x = 4$ (c) $x = \frac{2}{3}$ (d) $x = -1$

In Exercises 51–54, use synthetic division to find the specified value of the function.

51. $g(x) = 2x^4 - 17x^3 + 58x^2 - 77x + 26$
  (a) $g(-2)$ (b) $g\left(\frac{1}{2}\right)$

52. $h(x) = 5x^5 - 2x^4 - 45x + 18$
  (a) $h(2)$ (b) $h(\sqrt{3})$

53. $f(x) = x^4 + 10x^3 - 24x^2 + 20x + 44$
  (a) $f(-3)$ (b) $f(\sqrt{2}i)$

54. $g(t) = 2t^5 - 5t^4 - 8t + 20$
  (a) $g(-4)$ (b) $g(\sqrt{2})$

In Exercises 55–58, find a polynomial with integer coefficients that has the given zeros.

55. $-1, -1, \frac{1}{3}, -\frac{1}{2}$

56. $5, 1 - \sqrt{2}, 1 + \sqrt{2}$

57. $\frac{2}{3}, 4, \sqrt{3}i, -\sqrt{3}i$

58. $2, -3, 1 - 2i, 1 + 2i$

In Exercises 59 and 60, use Descartes's Rule of Signs to determine the possible number of positive and negative zeros of the function.

59. $g(x) = 5x^3 + 3x^2 - 6x + 9$

60. $h(x) = -2x^5 + 4x^3 - 2x^2 + 5$

In Exercises 61 and 62, use the Rational Zero Test to list all possible rational zeros of f.

61. $f(x) = -4x^3 + 8x^2 - 3x + 15$

62. $f(x) = 3x^4 + 4x^3 - 5x^2 - 8$

In Exercises 63–68, find all the zeros of the function.

63. $f(x) = 4x^3 - 11x^2 + 10x - 3$

64. $f(x) = 10x^3 + 21x^2 - x - 6$

65. $f(x) = 6x^3 - 5x^2 + 24x - 20$

66. $f(x) = x^3 - 1.3x^2 - 1.7x + 0.6$

67. $f(x) = 6x^4 - 25x^3 + 14x^2 + 27x - 18$

68. $f(x) = 5x^4 + 126x^2 + 25$

In Exercises 69–72, use the Bisection Method to find the zero of the function (in the given interval) to the nearest hundredth.

Function	Interval
69. $f(x) = x^4 + 2x - 1$	$[0, 1]$
70. $g(x) = x^3 - 3x^2 + 3x + 2$	$[-1, 0]$
71. $h(x) = x^3 - 6x^2 + 12x - 10$	$[3, 4]$
72. $f(x) = x^5 + 2x^3 - 3x - 20$	$[1, 2]$

73. *Volume Marker*   A spherical tank of radius 50 feet (see figure) will be two-thirds full when the depth of the fluid is $x + 50$ feet, where

$$3x^3 - 22,500x + 250,000 = 0.$$

Use the Bisection Method to approximate x to within 0.01 unit.

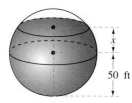

50 ft

Figure for 73

74. *Age of the Groom*   The average age of the groom in a marriage for the given age of the bride can be approximated by the model

$$y = -0.00428x^2 + 1.442x - 3.136, \qquad 20 \le x \le 55$$

where y is the age of the groom and x is the age of the bride. For what age of the bride is the average age of the groom 30? (*Source:* U.S. National Center for Health Statistics)

CHAPTER 5

OVERVIEW

The first two sections of this chapter are about rational functions. The graphs of such functions differ from graphs of polynomial functions in that they can have horizontal and vertical *asymptotes*.

There are many real-life examples of asymptotic behavior. For instance, Example 8 on page 320 shows the asymptotic behavior of the cost of removing smokestack emissions. In the example, notice that the cost becomes prohibitive as the percentage of pollutants removed approaches 100%.

The last two sections of the chapter are about conics. There are four types of conics: circles, parabolas, ellipses, and hyperbolas. Conics have reflective properties that make them valuable in constructing such objects as mirrors, satellite dishes, and bridges.

Rational Functions and Conic Sections

5.1 Rational Functions

Introduction / Horizontal and Vertical Asymptotes /
Sketching the Graph of a Rational Function / Slant Asymptotes / Applications

Introduction

A **rational function** can be written in the form

$$f(x) = \frac{p(x)}{q(x)}$$

where $p(x)$ and $q(x)$ are polynomials and $q(x)$ is not the zero polynomial. In this section we assume that $p(x)$ and $q(x)$ have no common factors. Some examples of rational functions are

$$f(x) = \frac{1}{x + 2}, \qquad g(x) = \frac{x - 1}{(x + 1)(x - 2)}, \qquad \text{and} \qquad h(x) = \frac{x}{x^2 + 1}.$$

Unlike polynomial functions, whose domains consist of all real numbers, rational functions often have restricted domains to avoid division by zero. In the examples above, the domain of f excludes $x = -2$, and the domain of g excludes $x = 2$ and $x = -1$. On the other hand, the domain of h is all real numbers because there are no real values of x that make the denominator $x^2 + 1$ equal to zero.

In general, the *domain* of a rational function of x includes all real numbers except x-values that make the denominator zero. Much of our discussion of rational functions will focus on their graphical behavior near these x-values.

EXAMPLE 1   Finding the Domain of a Rational Function

Find the domain of

$$f(x) = \frac{1}{x}$$

and discuss the behavior of f near any excluded x-values.

Solution

Because the denominator is zero when $x = 0$, the domain of f is all real numbers except $x = 0$. To determine the behavior of f near this excluded value, we evaluate $f(x)$ to the left and right of $x = 0$, as indicated in the tables.

x approaches 0 from the left

x	-1	-0.5	-0.1	-0.01	-0.001	$\rightarrow 0$
$f(x)$	-1	-2	-10	-100	-1000	$\rightarrow -\infty$

x approaches 0 from the right

x	$0 \leftarrow$	0.001	0.01	0.1	0.5	1
$f(x)$	$\infty \leftarrow$	1000	100	10	2	1

Note that as x approaches 0 *from the left*, $f(x)$ decreases without bound, whereas as x approaches 0 *from the right*, $f(x)$ increases without bound. The graph of f is shown in Figure 5.1.

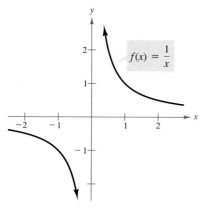

$$f(x) = \frac{1}{x}$$

FIGURE 5.1

Horizontal and Vertical Asymptotes

In Example 1, the behavior of $f(x) = 1/x$ near $x = 0$ is denoted by

$$f(x) \rightarrow -\infty \text{ as } x \rightarrow 0^-$$
  f(x) decreases without bound as x approaches 0 from the left.

$$f(x) \rightarrow \infty \text{ as } x \rightarrow 0^+.$$
  f(x) increases without bound as x approaches 0 from the right.

The line $x = 0$ is a **vertical asymptote** of the graph of f, as shown in Figure 5.2.

Figure 5.2 shows that the graph of f also has a **horizontal asymptote**— the line $y = 0$. This means that the values of $f(x) = 1/x$ approach zero as x increases or decreases without bound.

$f(x) \to 0$ as $x \to -\infty$ *f(x) approaches 0 as x decreases without bound.*

$f(x) \to 0$ as $x \to \infty$ *f(x) approaches 0 as x increases without bound.*

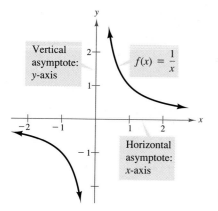

Vertical asymptote: y-axis

$f(x) = \dfrac{1}{x}$

Horizontal asymptote: x-axis

FIGURE 5.2

REMARK The graphs of

$f(x) = 1/x$ (Figure 5.2) and
$f(x) = (2x + 1)/(x + 1)$ (Figure 5.3)

are **hyperbolas.** We will say more about this type of curve in Sections 5.3 and 5.4.

DEFINITION OF VERTICAL AND HORIZONTAL ASYMPTOTES

1. The line $x = a$ is a **vertical asymptote** of the graph of f if

$$f(x) \to \infty \quad \text{or} \quad f(x) \to -\infty$$

as $x \to a$, either from the right or from the left.
2. The line $y = b$ is a **horizontal asymptote** of the graph of f if

$$f(x) \to b$$

as $x \to \infty$ or $x \to -\infty$.

Though the graph of a rational function will never intersect its vertical asymptote, it may intersect its horizontal asymptote. Eventually, however, the distance between the horizontal asymptote and the points on the graph must approach zero (as $x \to \infty$ or $x \to -\infty$). Figure 5.3 shows the horizontal and vertical asymptotes of the graphs of three rational functions.

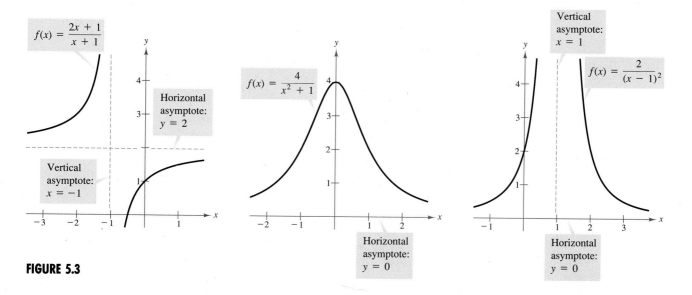

$f(x) = \dfrac{2x + 1}{x + 1}$

Horizontal asymptote: $y = 2$

Vertical asymptote: $x = -1$

$f(x) = \dfrac{4}{x^2 + 1}$

Horizontal asymptote: $y = 0$

Vertical asymptote: $x = 1$

$f(x) = \dfrac{2}{(x - 1)^2}$

Horizontal asymptote: $y = 0$

FIGURE 5.3

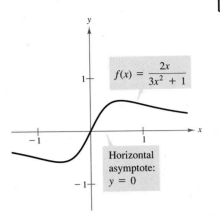

(a)

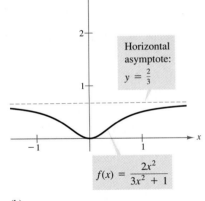

(b)

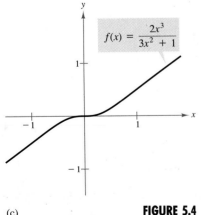

(c)

FIGURE 5.4

ASYMPTOTES OF A RATIONAL FUNCTION

Let f be the rational function given by

$$f(x) = \frac{p(x)}{q(x)} = \frac{a_n x^n + a_{n-1} x^{n-1} + \cdots + a_1 x + a_0}{b_m x^m + b_{m-1} x^{m-1} + \cdots + b_1 x + b_0}$$

where $p(x)$ and $q(x)$ have no common factors. Then the following are true.

1. The graph of f has *vertical* asymptotes at the zeros of $q(x)$.
2. The graph of f has one or no *horizontal* asymptotes determined as follows.
   (a) If $n < m$, then the graph of f has the x-axis ($y = 0$) as a horizontal asymptote.
   (b) If $n = m$, then the graph of f has the line $y = a_n / b_m$ as a horizontal asymptote.
   (c) If $n > m$, then the graph of f has no horizontal asymptote.

We can apply this theorem by comparing the degrees of the numerator and denominator.

EXAMPLE 2 Finding Horizontal Asymptotes of Rational Functions

a. The graph of

$$f(x) = \frac{2x}{3x^2 + 1}$$

has the x-axis as a horizontal asymptote, as shown in Figure 5.4(a). Note that the degree of the numerator is *less than* the degree of the denominator.

b. The graph of

$$f(x) = \frac{2x^2}{3x^2 + 1}$$

has the line $y = \frac{2}{3}$ as a horizontal asymptote, as shown in Figure 5.4(b). Note that the degree of the numerator is *equal to* the degree of the denominator, and the horizontal asymptote is given by the ratio of the leading coefficients of the numerator and denominator.

c. The graph of

$$f(x) = \frac{2x^3}{3x^2 + 1}$$

has no horizontal asymptote because the degree of the numerator is *greater than* the degree of the denominator. (See Figure 5.4(c).)

In Example 2(b), note that if the degree of the numerator and denominator of a rational function are equal ($m = n$), then the horizontal asymptote is given by the ratio of the leading coefficients. For instance, for the function

$$f(x) = \frac{3x^3 - 7x^2 + 2}{-4x^3 + 5x}$$

the horizontal asymptote is given by

$$y = \frac{\text{leading coefficient of numerator}}{\text{leading coefficient of denominator}} = \frac{3}{-4}.$$

Sketching the Graph of a Rational Function

Guidelines for Graphing Rational Functions

Let $f(x) = p(x)/q(x)$, where $p(x)$ and $q(x)$ are polynomials with no common factors.

1. Find and plot the y-intercept (if any) by evaluating $f(0)$.
2. Find the zeros of the numerator (if any) by solving the equation $p(x) = 0$. Then plot the corresponding x-intercepts.
3. Find the zeros of the denominator (if any) by solving the equation $q(x) = 0$. Then sketch the corresponding vertical asymptotes.
4. Find and sketch the horizontal asymptote (if any) by using the rule for finding the horizontal asymptote of a rational function.
5. Plot at least one point both *between and beyond* each x-intercept and vertical asymptote.
6. Use smooth curves to complete the graph between and beyond the vertical asymptotes.

REMARK   Testing for symmetry can be useful, especially for simple rational functions. For example, the graph of $f(x) = 1/x$ is symmetrical with respect to the origin, and the graph of $g(x) = 1/x^2$ is symmetrical with respect to the y-axis.

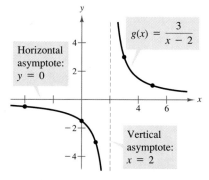

FIGURE 5.5

EXAMPLE 3   Sketching the Graph of a Rational Function

Sketch the graph of

$$g(x) = \frac{3}{x - 2}.$$

Solution

Note that the numerator and denominator have no common factors.

y-intercept	$\left(0, -\frac{3}{2}\right)$, from $g(0) = -\frac{3}{2}$
x-intercept	None, since $3 \neq 0$
Vertical Asymptote	$x = 2$, zero of denominator
Horizontal Asymptote	$y = 0$, degree of $p(x) <$ degree of $q(x)$

Additional Points

x	-4	1	3	5
$g(x)$	-0.5	-3	3	1

By plotting the intercepts, asymptotes, and a few additional points, we obtain the graph shown in Figure 5.5.

REMARK   The graph of g is a vertical stretch and a right shift of the graph of $f(x) = 1/x$ because

$$g(x) = \frac{3}{x - 2} = 3\left(\frac{1}{x - 2}\right) = 3f(x - 2).$$

◢

EXAMPLE 4   Sketching the Graph of a Rational Function

Sketch the graph of

$$f(x) = \frac{2x - 1}{x}.$$

Solution

Note that the numerator and denominator have no common factors.

y-intercept	None, because $x = 0$ is not in the domain.
x-intercept	$\left(\frac{1}{2}, 0\right)$, from $2x - 1 = 0$
Vertical Asymptote	$x = 0$, zero of denominator
Horizontal Asymptote	$y = 2$, degree of $p(x) =$ degree of $q(x)$

Additional Points

x	-4	-1	$\frac{1}{4}$	4
$f(x)$	2.25	3	-2	1.75

By plotting the intercepts, asymptotes, and a few additional points, we obtain the graph shown in Figure 5.6.

◢

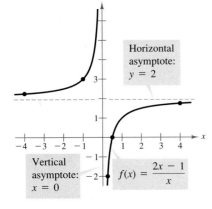

Horizontal asymptote: $y = 2$

Vertical asymptote: $x = 0$

    $f(x) = \dfrac{2x - 1}{x}$

FIGURE 5.6

EXAMPLE 5   Sketching the Graph of a Rational Function

Sketch the graph of

$$f(x) = \frac{x}{x^2 - x - 2}.$$

Solution

By factoring the denominator, we have

$$f(x) = \frac{x}{x^2 - x - 2} = \frac{x}{(x + 1)(x - 2)}.$$

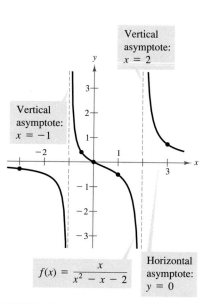

Vertical asymptote: x = 2

Vertical asymptote: x = -1

$$f(x) = \frac{x}{x^2 - x - 2}$$

Horizontal asymptote: y = 0

FIGURE 5.7

Thus, the numerator and denominator have no common factors.

y-intercept	(0, 0), because $f(0) = 0$
x-intercept	(0, 0)
Vertical Asymptote	$x = -1$, $x = 2$, zeros of denominator
Horizontal Asymptote	$y = 0$, degree of $p(x) <$ degree of $q(x)$
Additional Points	

x	-3	-0.5	1	3
$f(x)$	-0.3	0.4	-0.5	0.75

By plotting the intercepts, asymptotes, and a few additional points, we obtain the graph shown in Figure 5.7. Note that the graph crosses its horizontal asymptote at (0, 0).

EXAMPLE 6   Sketching the Graph of a Rational Function

Sketch the graph of

$$f(x) = \frac{2(x^2 - 9)}{x^2 - 4}.$$

Solution

By factoring the numerator and denominator, we have

$$f(x) = \frac{2(x^2 - 9)}{x^2 - 4} = \frac{2(x - 3)(x + 3)}{(x - 2)(x + 2)}.$$

Thus, the numerator and denominator have no common factors.

y-intercept	$\left(0, \frac{9}{2}\right)$, from $f(0) = \frac{9}{2}$
x-intercept	$(-3, 0)$ and $(3, 0)$
Vertical Asymptote	$x = -2$, $x = 2$, zeros of denominator
Horizontal Asymptote	$y = 2$, degree of $p(x) =$ degree of $q(x)$
Symmetry	With respect to y-axis, since $f(-x) = f(x)$
Additional Points	

x	0.5	2.5	6
$f(x)$	4.67	-2.44	1.69

By plotting the intercepts, asymptotes, and a few additional points, we obtain the graph shown in Figure 5.8.

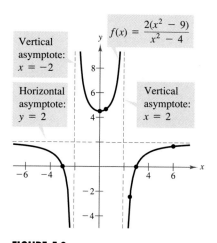

Vertical asymptote: x = -2

Horizontal asymptote: y = 2

Vertical asymptote: x = 2

$$f(x) = \frac{2(x^2 - 9)}{x^2 - 4}$$

FIGURE 5.8

Slant Asymptotes

If the degree of the numerator of a rational function is exactly *one more* than the degree of its denominator, then the graph of the function has a **slant asymptote.** For example, the graphs of

$$f(x) = \frac{x^2 - x}{x + 1} \quad \text{and} \quad g(x) = \frac{-x^3 + x^2 + 4}{x^2}$$

have slant asymptotes as shown in Figure 5.9.

To find the equation of a slant asymptote, use long division. For instance, by dividing $x + 1$ into $x^2 - x$, we have

$$f(x) = \frac{x^2 - x}{x + 1} = \underbrace{x - 2}_{\substack{\text{Slant asymptote} \\ (y = x - 2)}} + \frac{2}{x + 1}.$$

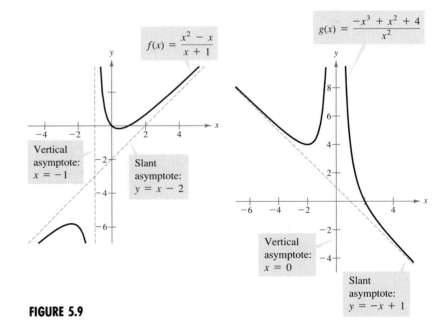

FIGURE 5.9

EXAMPLE 7   A Rational Function with a Slant Asymptote

Sketch the graph of

$$f(x) = \frac{x^2 - x - 2}{x - 1}.$$

Solution

As a preliminary step, we write $f(x)$ in two different ways. Factoring the numerator

$$f(x) = \frac{x^2 - x - 2}{x - 1} = \frac{(x - 2)(x + 1)}{x - 1}$$

allows us to recognize the x-intercepts, and long division

$$f(x) = \frac{x^2 - x - 2}{x - 1} = x - \frac{2}{x - 1}$$

allows us to recognize that the line $y = x$ is a slant asymptote of the graph of f.

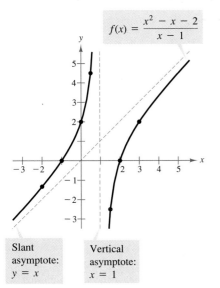

$f(x) = \dfrac{x^2 - x - 2}{x - 1}$

Slant
asymptote:
$y = x$

Vertical
asymptote:
$x = 1$

FIGURE 5.10

y-intercept	$(0, 2)$ because $f(0) = 2$
x-intercepts	$(-1, 0)$ and $(2, 0)$
Vertical Asymptote	$x = 1$
Slant Asymptote	$y = x$

Additional Points

x	-2	0.5	1.5	3
$f(x)$	-1.33	4.5	-2.5	2

The graph of f is shown in Figure 5.10.

Finally, note that it is possible for the graph of a rational function to cross its horizontal or its slant asymptote, as shown in Figure 5.11.

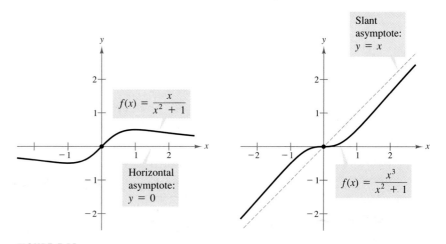

$f(x) = \dfrac{x}{x^2 + 1}$

Horizontal
asymptote:
$y = 0$

Slant
asymptote:
$y = x$

$f(x) = \dfrac{x^3}{x^2 + 1}$

FIGURE 5.11

Applications

There are many examples of asymptotic behavior in real-life problems.

EXAMPLE 8   Cost-Benefit Model for Smokestack Emission

A utility company burns coal to generate electricity. The cost of removing a certain *percentage* of the pollutants from the stack emission is typically not a linear function. That is, if it costs C dollars to remove 25% of the pollutants, it would cost more than $2C$ dollars to remove 50% of the pollutants. As the percentage of removed pollutants approaches 100%, the cost tends to become prohibitive. Suppose the cost C of removing $p\%$ of the smokestack pollutants is given by

$$C = \frac{80,000p}{100 - p}.$$

Sketch the graph of this function. Suppose you are a member of a state legislature that is considering a law that will require utility companies to remove 90% of the pollutants from their smokestack emissions. If the current law requires 85% removal, how much additional expense is the new law asking the utility company to spend?

Solution

The graph of this function is shown in Figure 5.12. Note that the graph has a vertical asymptote at $p = 100$. Since the current law requires 85% removal, the current cost to the utility company is

$$C = \frac{80,000(85)}{100 - 85} \approx \$453,333.$$

If the new law increases the percentage removal to 90%, the cost to the utility company will be

$$C = \frac{80,000(90)}{100 - 90} = \$720,000.$$

Therefore, the new law requires the utility company to spend an additional

$$720,000 - 453,333 = \$266,667.$$

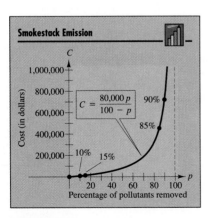

FIGURE 5.12

DISCUSSION

PROBLEM

Common Factors
in the
Numerator
and
Denominator

In the guidelines for sketching the graph of a rational function, we noted that the rational function should have no factor that is common to its numerator and denominator. To see why we required this, consider the function given by

$$f(x) = \frac{x(x-1)}{x}$$

which has a common factor of x in the numerator and denominator. Sketch the graph of this function. Does it have a vertical asymptote at $x = 0$?

WARM UP

The following warm-up exercises involve skills that were covered in earlier sections. You will use these skills in the exercise set for this section.

In Exercises 1–4, factor the polynomials.

1. $x^2 - 3x - 10$ **2.** $x^2 - 7x + 10$

3. $x^3 + 4x^2 + 3x$ **4.** $x^3 - 4x^2 - 2x + 8$

In Exercises 5–8, sketch the graph of the equation.

5. $y = 2$ **6.** $x = -1$

7. $y = x - 2$ **8.** $y = -x + 1$

In Exercises 9 and 10, use long division to write the rational expression as the sum of a polynomial and a rational expression.

9. $\dfrac{x^2 + 5x + 6}{x - 4}$ **10.** $\dfrac{x^2 + 5x + 6}{x + 4}$

EXERCISES for Section 5.1

In Exercises 1–8, match the rational function with its graph. [The graphs are labeled (a), (b), (c), (d), (e), (f), (g), and (h).]

1. $f(x) = \dfrac{2}{x + 1}$ **2.** $f(x) = \dfrac{1}{x - 4}$

3. $f(x) = \dfrac{x + 1}{x}$ **4.** $f(x) = \dfrac{1 - 2x}{x}$

5. $f(x) = \dfrac{x - 2}{x - 1}$ **6.** $f(x) = -\dfrac{x + 2}{x + 1}$

7. $f(x) = \dfrac{x^2 + 1}{x}$ **8.** $f(x) = \dfrac{x^2 - 2x}{x - 1}$

(a)   (b)

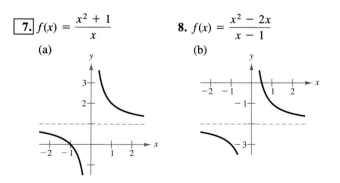

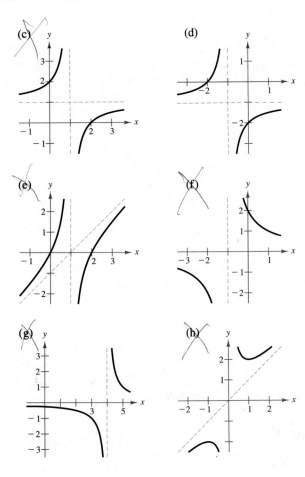

(c)

(d)

(e)

(f)

(g)

(h)

In Exercises 19–22, use the graph of $f(x) = 1/x$ to sketch the graph of the equation.

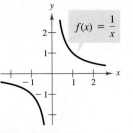

$f(x) = \dfrac{1}{x}$

Figure for 19–22

19. $y = f(x) + 1 = \dfrac{1}{x} + 1$

20. $y = f(x - 1) = \dfrac{1}{x - 1}$

21. $y = -f(x) = -\dfrac{1}{x}$

22. $y = f(x + 2) = \dfrac{1}{x + 2}$

In Exercises 23–26, use the graph of $f(x) = 4/x^2$ to sketch the graph of the equation.

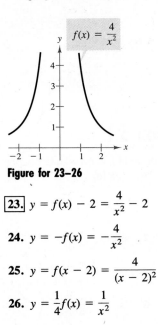

$f(x) = \dfrac{4}{x^2}$

Figure for 23–26

In Exercises 9–18, find the domain of the function and identify any horizontal, vertical, or slant asymptotes.

9. $f(x) = \dfrac{1}{x^2}$

10. $f(x) = \dfrac{4}{(x - 2)^3}$

11. $f(x) = \dfrac{2 + x}{2 - x}$

12. $f(x) = \dfrac{1 - 5x}{1 + 2x}$

13. $f(x) = \dfrac{x^3}{x^2 - 1}$

14. $f(x) = \dfrac{2x^2}{x + 1}$

15. $f(x) = \dfrac{3x^2 + 1}{x^2 + 9}$

16. $f(x) = \dfrac{3x^2 + x - 5}{x^2 + 1}$

17. $f(x) = \dfrac{5x^4}{x^2 + 1}$

18. $f(x) = \dfrac{x^2}{x + 1}$

23. $y = f(x) - 2 = \dfrac{4}{x^2} - 2$

24. $y = -f(x) = -\dfrac{4}{x^2}$

25. $y = f(x - 2) = \dfrac{4}{(x - 2)^2}$

26. $y = \dfrac{1}{4}f(x) = \dfrac{1}{x^2}$

In Exercises 27–30, use the graph of $g(x) = 8/x^3$ to sketch the graph of the equation.

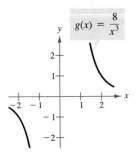

$g(x) = \dfrac{8}{x^3}$

Figure for 27–30

27. $y = g(x + 2) = \dfrac{8}{(x + 2)^3}$

28. $y = g(x) + 1 = \dfrac{8}{x^3} + 1$

29. $y = -g(x) = -\dfrac{8}{x^3}$

30. $y = \dfrac{1}{8}g(x) = \dfrac{1}{x^3}$

In Exercises 31–54, sketch the graph of the rational function. As sketching aids, check for intercept, symmetry, vertical asymptotes, and horizontal asymptotes.

31. $f(x) = \dfrac{1}{x + 2}$

32. $f(x) = \dfrac{1}{x - 3}$

33. $h(x) = \dfrac{-1}{x + 2}$

34. $g(x) = \dfrac{1}{3 - x}$

35. $f(x) = \dfrac{x + 1}{x + 2}$

36. $f(x) = \dfrac{x - 2}{x - 3}$

37. $f(x) = \dfrac{2 + x}{1 - x}$

38. $f(x) = \dfrac{3 - x}{2 - x}$

39. $f(t) = \dfrac{3t + 1}{t}$

40. $f(t) = \dfrac{1 - 2t}{t}$

41. $g(x) = \dfrac{1}{x + 2} + 2$

42. $h(x) = \dfrac{1}{x - 3} + 1$

43. $C(x) = \dfrac{5 + 2x}{1 + x}$

44. $P(x) = \dfrac{1 - 3x}{1 - x}$

45. $f(x) = \dfrac{x^2}{x^2 + 9}$

46. $f(x) = 2 - \dfrac{3}{x^2}$

47. $h(x) = \dfrac{x^2}{x^2 - 9}$

48. $g(x) = \dfrac{x}{x^2 - 9}$

49. $f(x) = -\dfrac{1}{(x - 2)^2}$

50. $g(x) = -\dfrac{x}{(x - 2)^2}$

51. $f(x) = -\dfrac{1}{(x - 2)^2} + 3$

52. $g(x) = -\dfrac{x}{(x - 2)^2} + 2$

53. $f(x) = \dfrac{3x}{x^2 - x - 2}$

54. $f(x) = \dfrac{2x}{x^2 + x - 2}$

In Exercises 55–64, sketch the graph of the rational function. As sketching aids, check for intercepts, symmetry, vertical asymptotes, and slant asymptotes.

55. $f(x) = \dfrac{2x^2 + 1}{x}$

56. $f(x) = \dfrac{1 - x^2}{x}$

57. $g(x) = \dfrac{x^2 + 1}{x}$

58. $h(x) = \dfrac{x^2}{x - 1}$

59. $f(x) = \dfrac{x^3}{x^2 - 1}$

60. $g(x) = \dfrac{x^3}{2x^2 - 8}$

61. $f(x) = \dfrac{x^2 - x + 1}{x - 1}$

62. $f(x) = \dfrac{2x^2 - 5x + 5}{x - 2}$

63. $f(x) = \dfrac{x^2 + 5x + 8}{x + 3}$

64. $f(x) = \dfrac{2x^2 + x}{x + 1}$

65. *Cost of Clean Water*   The cost in millions of dollars for removing $p\%$ of the industrial and municipal pollutants discharged into a river is

$$C = \frac{255p}{100 - p}, \qquad 0 \le p < 100.$$

(a) Find the cost of removing 10% of the pollutants.
(b) Find the cost of removing 40%.
(c) Find the cost of removing 75%.
(d) According to this model, would it be possible to remove 100% of the pollutants?

66. *Recycling Costs*   In a pilot project, a rural township was given recycling bins for separating and storing recyclable products. The cost in dollars for giving bins to $p\%$ of the population is

$$C = \frac{25{,}000p}{100 - p}, \qquad 0 \le p < 100.$$

(a) Find the cost of 15% of the population receiving the bins.
(b) Find the cost of 50% receiving the bins.
(c) Find the cost of 90% receiving the bins.
(d) According to this model, would it be possible to give bins to 100% of the residents?

67. *Deer Population*   The game commission introduces 50 deer into newly acquired state game lands. The population of the herd is

$$N = \frac{10(5 + 3t)}{1 + 0.04t}, \qquad 0 \le t$$

where t is the time in years (see figure).
(a) Find the population when t is 5, 10, and 25.
(b) What is the limiting size of the herd as time increases?

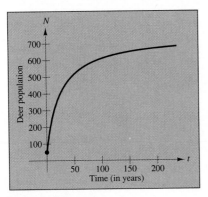

Figure for 67

68. *Food Consumption*   A biology class performs an experiment comparing the quantity of food consumed by a certain kind of moth with the quantity supplied (see figure). The model for their experimental data is given by

$$y = \frac{1.568x - 0.001}{6.360x + 1}$$

where x is the quantity (mg) of food supplied and y is the quantity (mg) eaten. At what level of consumption will the moth become satiated?

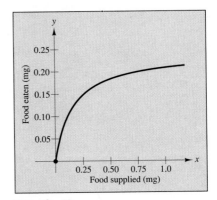

Figure for 68

69. *Average Cost*   The cost of producing x units of a product is $C = 150{,}000 + 0.25x$, and therefore the average cost per unit is

$$\overline{C} = \frac{C}{x} = \frac{150{,}000 + 0.25x}{x}, \qquad 0 < x.$$

Sketch the graph of this average cost function and find the average cost for producing $x = 1000$, $x = 10{,}000$, and $x = 100{,}000$ units.

70. *Average Cost*   The cost of producing x units of a product is $C = 0.2x^2 + 10x + 5$, and therefore the average cost per unit is

$$\overline{C} = \frac{C}{x} = \frac{0.2x^2 + 10x + 5}{x}, \qquad 0 < x.$$

Sketch the graph of the average cost function, and estimate the number of units that should be produced to minimize the average cost per unit.

71. *Minimum Area*   A right triangle is formed in the first quadrant by the *x*-axis, the *y*-axis, and a line segment through the point (2, 3) (see figure).

(a) Show that an equation of the line segment is

$$y = \frac{3(x - a)}{2 - a}, \qquad 0 \le x \le a.$$

(b) Show that the area of the triangle is

$$A = \frac{-3a^2}{2(2 - a)}.$$

(c) Sketch the graph of the area function of part (b), and from the graph estimate the value of *a* that yields a minimum area.

Figure for 71

72. *Human Memory Model*   Psychologists have developed mathematical models to predict performance as a function of the number of trials *n* for a certain task. Consider the learning curve given by

$$P = \frac{0.5 + 0.9(n - 1)}{1 + 0.9(n - 1)}, \qquad 0 < n$$

where *P* is the percentage of correct responses after *n* trials.

(a) Complete the following table for this model.

n	1	2	3	4	5	6	7	8	9	10
P										

(b) According to this model, what is the limiting percentage of correct responses as *n* increases?

SOLVING

A Graphical Approach to Finding the Average Cost

A graphing utility can be used to investigate asymptotic behavior. For instance, the following example shows how to use a graphing utility to visualize how the average cost of producing a product changes as the number of units produced increases.

EXAMPLE 1   Finding the Average Cost of a Product

You have started a small business to produce copies of compact discs. Your initial investment is $250,000 and the cost of producing each disc is $0.32. Describe the *average cost per unit* of producing each disc as a function of x (the number of units produced).

Solution

The total cost C of producing x units is

$$C = \boxed{\begin{array}{c} Initial \\ cost \end{array}} + \boxed{\begin{array}{c} Cost\ per \\ unit \end{array}} \cdot \boxed{\begin{array}{c} Number \\ of\ units \end{array}}$$

$$= 250,000 + 0.32x.$$

To obtain the average cost per unit $\overline{C}$ of producing x units, divide the total cost by x.

$$\overline{C} = \frac{C}{x} = \frac{250,000 + 0.32x}{x} = \frac{250,000}{x} + 0.32$$

The following table shows the average cost per unit for several different levels of production.

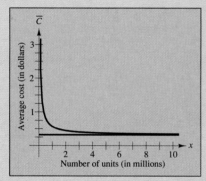

Number of units, x	100	1000	10,000	100,000	1,000,000	10,000,000
Total cost, C	$250,032	$250,320	$253,200	$282,000	$570,000	$3,450,000
Average cost, $\overline{C}$	$2500.32	$250.32	$25.32	$2.82	$0.57	$0.345

The graph of the average cost function is shown at the left together with the graph of the horizontal line $y = 0.32$. Notice that as the number of units produced increases, the average cost per unit gets closer and closer to the unit cost of $0.32.

◢

In Example 1, it is assumed that the production level can continue to increase without having to invest additional capital. Example 2 shows how the problem changes when additional capital investments are required.

EXAMPLE 2 Finding the Minimum Average Cost

As the production level in Example 1 increases and the business grows, suppose that you must invest additional capital to buy more equipment, hire more employees, advertise, build new offices, and so on. You have determined that the additional investments are proportional to the square of the number of units produced, and you have derived the following model for the total cost of producing x units.

$$C = 250,000 + 0.32x + 0.00001x^2.$$

Describe the average cost per unit as a function of the number of units produced.

Solution

With this model for the total cost, the average cost function is

$$\overline{C} = \frac{250,000 + 0.32x + 0.00001x^2}{x} = \frac{250,000}{x} + 0.32 + 0.00001x.$$

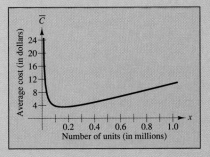

The graph of the average cost function is shown at the left. With this model, notice that the average cost decreases until the production level reaches about 150,000 or 160,000 units. Then, with increased production, the average cost per unit increases. ◤

EXERCISES

(*See also: Exercises 69–70, Section 5.1*)

1. *Exploration* In Example 1, would it be possible to produce enough units so that the average cost is exactly $0.32? Would it be possible to produce enough units so that the average cost is arbitrarily close to $0.32? Explain.

2. *Exploration* In Example 1, how many units should be produced to obtain an average cost per unit of $0.33? To answer this question, did you use a numerical approach, an algebraic approach, or a graphical approach? Explain why you chose the approach you used.

3. *Exploration* In Example 2, use the zoom feature of a graphing utility to approximate the *minimum* average cost. At what production level does this minimum average cost occur?

4. *Slant Asymptote* The graph shown in Example 2 has a slant asymptote. Find its equation, sketch its graph, and interpret it in the context of the problem.

5. *Economy of Scale* In business, the expression *economy of scale* means that the average cost per unit tends to decrease as the production level increases. Judging from Examples 1 and 2 *and* your own knowledge of business, what factors could temper an economy of scale in a business?

6. *Testing a Hypothesis* Which graph has the x-axis as an asymptote? Which has a horizontal asymptote that is not the x-axis? Which has a slant asymptote?

(a) $y = \dfrac{3}{2x + 1}, \quad 0 \le x$

(b) $y = \dfrac{3x}{2x + 1}, \quad 0 \le x$

(c) $y = \dfrac{3x^2}{2x + 1}, \quad 0 \le x$

Form a hypothesis about the horizontal or slant asymptotes of the graph of

$$y = \frac{ax^n}{bx + 1}.$$

Test your hypothesis with examples.

327

5.2 Partial Fractions

Introduction / Partial Fraction Decomposition

Introduction

In Section 1.6, you learned how to combine two or more rational expressions using a least common denominator (LCD). In this section, we reverse the problem (a useful procedure in calculus) and write a given rational expression as the sum of two or more simpler rational expressions. For example, the rational expression $(x + 7)/(x^2 - x - 6)$ can be written as the sum of two fractions with first-degree denominators. That is,

$$\frac{x + 7}{x^2 - x - 6} = \frac{2}{x - 3} + \frac{-1}{x + 2}.$$

Each fraction on the right side of the equation is a **partial fraction,** and together they make up the **partial fraction decomposition** of the left side.

In Chapter 4 we noted that it is theoretically possible to write any polynomial as the product of linear and irreducible quadratic factors. For instance,

$$x^5 + x^4 - x - 1 = (x - 1)(x + 1)^2(x^2 + 1)$$

where $(x - 1)$ is a linear factor, $(x + 1)$ is a repeated linear factor, and $(x^2 + 1)$ is an irreducible quadratic factor.

We can use this factorization to find the partial fraction decomposition of any rational expression having $x^5 + x^4 - x - 1$ as its denominator. Specifically, if $N(x)$ is a polynomial of degree 4 or less, then the partial fraction decomposition of $N(x)/(x^5 + x^4 - x - 1)$ has the form

$$\frac{N(x)}{x^5 + x^4 - x - 1} = \frac{N(x)}{(x - 1)(x + 1)^2(x^2 + 1)}$$
$$= \frac{A}{x - 1} + \frac{B}{x + 1} + \frac{C}{(x + 1)^2} + \frac{Dx + F}{x^2 + 1}.$$

Note that the factor $(x + 1)^2$ results in *two* fractions: one for $(x + 1)$ and one for $(x + 1)^2$. If $(x + 1)^3$ were a factor, then we would use three fractions: one for $(x + 1)$, one for $(x + 1)^2$, and one for $(x + 1)^3$. In general, the number of fractions resulting from a repeated factor is equal to the exponent of the factor. Note also that an irreducible quadratic factor like $x^2 + 1$ must have a linear numerator.

Decomposition of $N(x)/D(x)$ into Partial Fractions

1. *Divide if improper:* If $N(x)/D(x)$ is an improper fraction, then divide the denominator into the numerator to obtain

$$\frac{N(x)}{D(x)} = (\text{polynomial}) + \frac{N_1(x)}{D(x)}$$

and apply Steps 2, 3, and 4 (below) to the proper rational expression $N_1(x)/D(x)$.

2. *Factor denominator:* Completely factor the denominator into factors of the form

$$(px + q)^m \quad \text{and} \quad (ax^2 + bx + c)^n$$

where $(ax^2 + bx + c)$ is irreducible.

3. *Linear factors:* For *each* factor of the form $(px + q)^m$, the partial fraction decomposition must include the following sum of m fractions.

$$\frac{A_1}{(px + q)} + \frac{A_2}{(px + q)^2} + \cdots + \frac{A_m}{(px + q)^m}$$

4. *Quadratic factors:* For *each* factor of the form $(ax^2 + bx + c)^n$, the partial fraction decomposition must include the following sum of n fractions.

$$\frac{B_1x + C_1}{ax^2 + bx + c} + \frac{B_2x + C_2}{(ax^2 + bx + c)^2} + \cdots + \frac{B_nx + C_n}{(ax^2 + bx + c)^n}$$

Partial Fraction Decomposition

Algebraic techniques for determining the constants in the numerators of the partial fractions are demonstrated in the examples that follow. Note that the techniques vary slightly, depending on the type of factors of the denominator: linear or quadratic, distinct or repeated.

EXAMPLE 1   Distinct Linear Factors

Write the partial fraction decomposition for

$$\frac{x + 7}{x^2 - x - 6}.$$

Solution

Since $x^2 - x - 6 = (x - 3)(x + 2)$, we include one partial fraction with a constant numerator for each linear factor of the denominator and write

$$\frac{x + 7}{x^2 - x - 6} = \frac{A}{x - 3} + \frac{B}{x + 2}.$$

Multiplying both sides of this equation by the least common denominator, $(x - 3)(x + 2)$, leads to the **basic equation**

$$x + 7 = A(x + 2) + B(x - 3). \qquad \textit{Basic equation}$$

Because this equation is true for all x, you can substitute any *convenient* values of x which will help determine the constants A and B. Values of x that are especially convenient are ones that make the factors $(x + 2)$ and $(x - 3)$ equal to zero. For instance, let $x = -2$. Then

$$-2 + 7 = A(0) + B(-5) \qquad \textit{Substitute convenient value of x}$$
$$5 = -5B$$
$$-1 = B.$$

To solve for A, let $x = 3$ to obtain

$$3 + 7 = A(5) + B(0) \qquad \textit{Substitute convenient value of x}$$
$$10 = 5A$$
$$2 = A.$$

Therefore, the decomposition is

$$\frac{x + 7}{x^2 - x - 6} = \frac{2}{x - 3} - \frac{1}{x + 2}$$

as indicated at the beginning of this section. Check this result by combining the two partial fractions on the right side of the equation. ◢

EXAMPLE 2 Repeated Linear Factors

Write the partial fraction decomposition for

$$\frac{5x^2 + 20x + 6}{x^3 + 2x^2 + x}.$$

Solution

Since the denominator factors as

$$x^3 + 2x^2 + x = x(x^2 + 2x + 1) = x(x + 1)^2$$

we include one fraction with a constant numerator for each power of x and $(x + 1)$ and write

$$\frac{5x^2 + 20x + 6}{x(x + 1)^2} = \frac{A}{x} + \frac{B}{x + 1} + \frac{C}{(x + 1)^2}.$$

Multiplying by the LCD, $x(x + 1)^2$, leads to the basic equation

$$5x^2 + 20x + 6 = A(x + 1)^2 + Bx(x + 1) + Cx. \qquad \textit{Basic equation}$$

Letting $x = -1$ eliminates the A and B terms and yields

$$5 - 20 + 6 = 0 + 0 - C \qquad \textit{Substitute convenient value of } x$$
$$C = 9.$$

Letting $x = 0$ eliminates the B and C terms and yields

$$6 = A(1) + 0 + 0 \qquad \textit{Substitute convenient value of } x$$
$$6 = A.$$

At this point, we have exhausted the most convenient choices for x, so to find the value of B, use *any other value* for x along with the known values of A and C. Thus, using $x = 1$, $A = 6$, and $C = 9$,

$$5 + 20 + 6 = A(4) + B(2) + C$$
$$31 = 6(4) + 2B + 9$$
$$-2 = 2B$$
$$-1 = B.$$

Therefore, the partial fraction decomposition is

$$\frac{5x^2 + 20x + 6}{x(x + 1)^2} = \frac{6}{x} - \frac{1}{x + 1} + \frac{9}{(x + 1)^2}.$$

◢

The procedure used to solve for the constants $A, B, C, \ldots$ in Examples 1 and 2 works well when the factors of the denominator are linear. However, when the denominator contains irreducible quadratic factors, we use a different procedure, which involves writing the right side of the basic equation in polynomial form and *equating the coefficients* of like terms.

EXAMPLE 3 Distinct Linear and Quadratic Factors

Write the partial fraction decomposition for

$$\frac{3x^2 + 4x + 4}{x^3 + 4x}.$$

Solution

Since the denominator factors as

$$x^3 + 4x = x(x^2 + 4)$$

we include one partial fraction with a constant numerator and one partial fraction with a linear numerator and write

$$\frac{3x^2 + 4x + 4}{x^3 + 4x} = \frac{A}{x} + \frac{Bx + C}{x^2 + 4}.$$

Multiplying by the LCD, $x(x^2 + 4)$, yields the basic equation

$$3x^2 + 4x + 4 = A(x^2 + 4) + (Bx + C)x. \qquad \textit{Basic equation}$$

Expanding this basic equation and collecting like terms produces

$$3x^2 + 4x + 4 = Ax^2 + 4A + Bx^2 + Cx$$
$$= (A + B)x^2 + Cx + 4A. \qquad \textit{Polynomial form}$$

Finally, because two polynomials are equal if and only if the coefficients of like terms are equal,

$$3x^2 + 4x + 4 = (A + B)x^2 + Cx + 4A \qquad \textit{Equate coefficients of like terms}$$

we obtain the following equations.

$$3 = A + B, \qquad 4 = C, \qquad \text{and} \qquad 4 = 4A$$

Thus, $A = 1$ and $C = 4$. Moreover, substituting $A = 1$ in the equation $3 = A + B$ yields

$$3 = 1 + B$$
$$2 = B.$$

Therefore, the partial fraction decomposition is

$$\frac{3x^2 + 4x + 4}{x^3 + 4x} = \frac{1}{x} + \frac{2x + 4}{x^2 + 4}.$$

EXAMPLE 4   Repeated Quadratic Factors

Write the partial fraction decomposition for

$$\frac{8x^3 + 13x}{(x^2 + 2)^2}.$$

Solution

We include one partial fraction with a linear numerator for each power of $(x^2 + 2)$, and write

$$\frac{8x^3 + 13x}{(x^2 + 2)^2} = \frac{Ax + B}{x^2 + 2} + \frac{Cx + D}{(x^2 + 2)^2}.$$

Multiplying by the LCD, $(x^2 + 2)^2$, yields the basic equation

$$8x^3 + 13x = (Ax + B)(x^2 + 2) + Cx + D \qquad \textit{Basic equation}$$
$$= Ax^3 + 2Ax + Bx^2 + 2B + Cx + D$$
$$= Ax^3 + Bx^2 + (2A + C)x + (2B + D). \qquad \textit{Polynomial form}$$

REMARK   By equating coefficients of like terms in Examples 3 and 4, you obtained several equations involving A, B, C, and D, which were solved by *substitution*. In a later chapter we will discuss a more general method for solving systems of equations.

Equating coefficients of like terms,

$$8x^3 + 0x^2 + 13x + 0 = Ax^3 + Bx^2 + (2A + C)x + (2B + D)$$

produces

$$8 = A, \quad 0 = B, \quad 13 = 2A + C, \quad \text{and} \quad 0 = 2B + D. \quad \textit{Equate coefficients}$$

Finally, use the values $A = 8$ and $B = 0$ to obtain the following.

$$13 = 2A + C = 2(8) + C \qquad 0 = 2B + D = 2(0) + D$$
$$-3 = C \qquad\qquad\qquad\qquad 0 = D$$

Therefore,

$$\frac{8x^3 + 13x}{(x^2 + 2)^2} = \frac{8x}{x^2 + 2} + \frac{-3x}{(x^2 + 2)^2}.$$

Guidelines for Solving the Basic Equation

Linear Factors

1. Substitute the *zeros* of the distinct linear factors into the basic equation.
2. For repeated linear factors, use the coefficients determined above to rewrite the basic equation. Then substitute *other* convenient values for x and solve for the remaining coefficients.

Quadratic Factors

1. Expand the basic equation.
2. Collect terms according to powers of x.
3. Equate the coefficients of like terms to obtain equations involving A, B, C, and so on.
4. Use substitution to solve for A, B, C, ...

Keep in mind that for *improper* rational expressions like

$$\frac{N(x)}{D(x)} = \frac{2x^3 + x^2 - 7x + 7}{x^2 + x - 2}$$

you must first divide to obtain the form

$$\frac{N(x)}{D(x)} = (\text{polynomial}) + \frac{N_1(x)}{D(x)}.$$

The proper rational expression $N_1(x)/D(x)$ is then decomposed into its partial fractions by the usual methods.

DISCUSSION

PROBLEM

You Be
the
Instructor

Suppose you were tutoring a student in algebra. In trying to find a partial fraction decomposition, your student wrote the following.

$$\frac{x^2 + 1}{x(x - 1)} = \frac{A}{x} + \frac{B}{x - 1}$$

$$\frac{x^2 + 1}{x(x - 1)} = \frac{A(x - 1)}{x(x - 1)} + \frac{Bx}{x(x - 1)}$$

$$x^2 + 1 = A(x - 1) + Bx \qquad\qquad \textit{Basic equation}$$

By substituting $x = 0$ and $x = 1$ into the basic equation, your student concluded that $A = -1$ and $B = 2$. However, in checking this solution, your student obtained

$$\frac{-1}{x} + \frac{2}{x - 1} = \frac{(-1)(x - 1) + 2(x)}{x(x - 1)} = \frac{x + 1}{x(x - 1)} \neq \frac{x^2 + 1}{x(x - 1)}.$$

What went wrong?

WARM UP

The following warm-up exercises involve skills that were covered in earlier sections. You will use these skills in the exercise set for this section.

In Exercises 1–10, find the sum and simplify.

1. $\dfrac{2}{x} + \dfrac{3}{x + 1}$

2. $\dfrac{5}{x + 2} + \dfrac{3}{x}$

3. $\dfrac{7}{x - 2} - \dfrac{3}{2x - 1}$

4. $\dfrac{2}{x + 5} - \dfrac{5}{x + 12}$

5. $\dfrac{1}{x - 3} + \dfrac{3}{(x - 3)^2} - \dfrac{5}{(x - 3)^3}$

6. $\dfrac{-5}{x + 2} + \dfrac{4}{(x + 2)^2}$

7. $\dfrac{-3}{x} + \dfrac{3x - 1}{x^2 + 3}$

8. $\dfrac{5}{x + 1} - \dfrac{x - 6}{x^2 + 5}$

9. $\dfrac{3}{x^2 + 1} + \dfrac{x - 3}{(x^2 + 1)^2}$

10. $\dfrac{x}{x^2 + x + 1} - \dfrac{x - 1}{(x^2 + x + 1)^2}$

EXERCISES for Section 5.2

In Exercises 1–36, write the partial fraction decomposition for the rational expression.

1. $\dfrac{1}{x^2 - 1}$

2. $\dfrac{1}{4x^2 - 9}$

3. $\dfrac{1}{x^2 + x}$

4. $\dfrac{3}{x^2 - 3x}$

5. $\dfrac{1}{2x^2 + x}$

6. $\dfrac{5}{x^2 + x - 6}$

7. $\dfrac{3}{x^2 + x - 2}$

8. $\dfrac{x + 1}{x^2 + 4x + 3}$

9. $\dfrac{5 - x}{2x^2 + x - 1}$

10. $\dfrac{3x^2 - 7x - 2}{x^3 - x}$

11. $\dfrac{x^2 + 12x + 12}{x^3 - 4x}$

12. $\dfrac{x + 2}{x(x - 4)}$

13. $\dfrac{4x^2 + 2x - 1}{x^2(x + 1)}$

14. $\dfrac{2x - 3}{(x - 1)^2}$

15. $\dfrac{x - 1}{x^3 + x^2}$

16. $\dfrac{4x^2 - 1}{2x(x + 1)^2}$

17. $\dfrac{3x}{(x - 3)^2}$

18. $\dfrac{6x^2 + 1}{x^2(x - 1)^3}$

19. $\dfrac{x^2 - 1}{x(x^2 + 1)}$

20. $\dfrac{x}{(x - 1)(x^2 + x + 1)}$

21. $\dfrac{x^2}{x^4 - 2x^2 - 8}$

22. $\dfrac{2x^2 + x + 8}{(x^2 + 4)^2}$

23. $\dfrac{x}{16x^4 - 1}$

24. $\dfrac{x^2 - 4x + 7}{(x + 1)(x^2 - 2x + 3)}$

25. $\dfrac{x^2 + x + 2}{(x^2 + 2)^2}$

26. $\dfrac{x^3}{(x + 2)^2(x - 2)^2}$

27. $\dfrac{x^2 + 5}{(x + 1)(x^2 - 2x + 3)}$

28. $\dfrac{x + 1}{x^3 + x}$

29. $\dfrac{2x^3 - 4x^2 - 15x + 5}{x^2 - 2x - 8}$

30. $\dfrac{x^3 - x + 3}{x^2 + x - 2}$

31. $\dfrac{x^4}{(x - 1)^3}$

32. $\dfrac{x^2 - x}{x^2 + x + 1}$

33. $\dfrac{1}{a^2 - x^2}$, a is a constant

34. $\dfrac{1}{x(x + a)}$, a is a constant

35. $\dfrac{1}{y(L - y)}$, L is a constant

36. $\dfrac{1}{(x + 1)(n - x)}$, n is a positive integer

5.3 Conic Sections

Introduction / Parabolas / Ellipses / Hyperbolas

Introduction

Conic sections were discovered during the classical Greek period, 600 to 300 B.C. This early Greek study was largely concerned with the geometrical properties of conics. It was not until the early 17th century that the broad applicability of conics became apparent and played a prominent role in the early development of calculus.

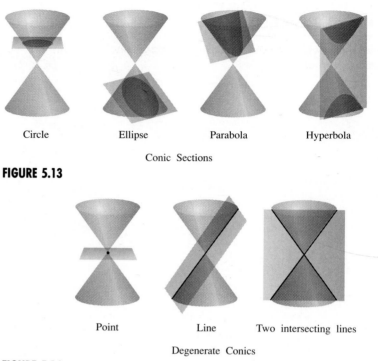

Conic Sections

FIGURE 5.13

Degenerate Conics

FIGURE 5.14

A **conic section** (or simply **conic**) is the intersection of a plane and a double-napped cone. Notice from Figure 5.13 that in the formation of the four basic conics, the intersecting plane does not pass through the vertex of the cone. When the plane does pass through the vertex, the resulting figure is a **degenerate conic,** as shown in Figure 5.14.

There are several ways to approach the study of conics. We could begin by defining conics in terms of the intersections of planes and cones, as the Greeks did, or we could define them algebraically, in terms of the general second-degree equation

$$Ax^2 + Bxy + Cy^2 + Dx + Ey + F = 0.$$

However, we will use a third approach, in which each of the conics is defined as a *locus* (collection) of points satisfying a certain geometric property. For example, in Section 3.2, we saw how the definition of a circle as *the collection of all points (x, y) that are equidistant from a fixed point (h, k)* led easily to the standard equation of a circle, $(x - h)^2 + (y - k)^2 = r^2$.

Parabolas

In Section 4.1, we determined that the graph of the quadratic function $f(x) = ax^2 + bx + c$ is a parabola that opens upward or downward. The following definition of a parabola is more general in the sense that it is independent of the orientation of the parabola.

DEFINITION OF A PARABOLA

A **parabola** is the set of all points (x, y) that are equidistant from a fixed line (the **directrix**) and a fixed point (the **focus**) (not on the line).

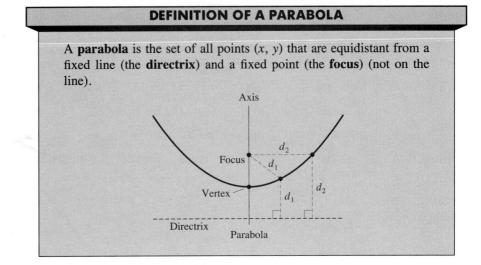

The midpoint between the focus and the directrix is the **vertex,** and the line passing through the focus and the vertex is the **axis** of the parabola.

STANDARD EQUATION OF A PARABOLA (VERTEX AT ORIGIN)

The **standard form of the equation of a parabola** with a vertex at $(0, 0)$ and directrix $y = -p$ is

$$x^2 = 4py, \quad p \neq 0. \qquad \text{\textit{Vertical axis}}$$

For directrix $x = -p$, the equation is

$$y^2 = 4px, \quad p \neq 0. \qquad \text{\textit{Horizontal axis}}$$

The focus is on the axis p units (directed distance) from the vertex.

Proof

Since the two cases are similar, we give a proof for the first case only. Suppose the directrix $(y = -p)$ is parallel to the x-axis. In Figure 5.15 we assume $p > 0$, and since p is the directed distance from the vertex to the focus, the focus must lie above the vertex. Since the point (x, y) is equidistant from $(0, p)$ and $y = -p$, we can apply the distance formula to obtain

$$\sqrt{(x - 0)^2 + (y - p)^2} = y + p$$
$$x^2 + (y - p)^2 = (y + p)^2$$
$$x^2 + y^2 - 2py + p^2 = y^2 + 2py + p^2$$
$$x^2 = 4py.$$

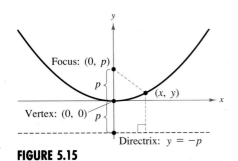

FIGURE 5.15

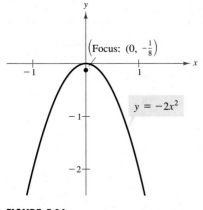

FIGURE 5.16

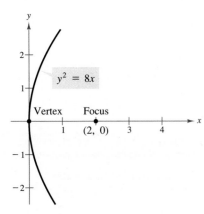

FIGURE 5.17

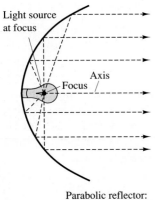

Parabolic reflector:
Light is reflected
in parallel rays.

FIGURE 5.18

EXAMPLE 1   Finding the Focus of a Parabola

Find the focus of the parabola whose equation is $y = -2x^2$.

Solution

Since the squared term in the equation involves x, we know that the axis is vertical, and we have

$$x^2 = 4py. \qquad\qquad \textit{Standard form}$$

Writing the given equation in this form,

$$x^2 = -\frac{1}{2}y$$

$$x^2 = 4\left(-\frac{1}{8}\right)y. \qquad\qquad \textit{Standard form}$$

Thus, $p = -\frac{1}{8}$. Since p is negative, the parabola opens downward (see Figure 5.16), and the focus of the parabola is $(0, p) = \left(0, -\frac{1}{8}\right)$. ◢

EXAMPLE 2   A Parabola with a Horizontal Axis

Write the standard form of the equation of the parabola with vertex at the origin and focus at $(2, 0)$.

Solution

The axis of the parabola is horizontal, passing through $(0, 0)$ and $(2, 0)$, as shown in Figure 5.17. Thus, the standard form is

$$y^2 = 4px.$$

Since the focus is $p = 2$ units from the vertex, the equation is

$$y^2 = 4(2)x$$

$$y^2 = 8x.$$ ◢

Parabolas occur in a wide variety of applications. For instance, a parabolic reflector can be formed by revolving a parabola about its axis. The resulting surface has the property that all incoming rays parallel to the axis are reflected through the focus of the parabola—this is the principle behind the construction of the parabolic mirrors used in reflecting telescopes. Conversely, the light rays emanating from the focus of a parabolic reflector used in a flashlight are all parallel to one another, as shown in Figure 5.18.

REMARK Be sure you understand that the term *parabola* is a technical term used in mathematics and does not simply refer to *any* U-shaped curve. ◢

Ellipses

DEFINITION OF AN ELLIPSE

An **ellipse** is the set of all points (x, y), the sum of whose distances from two distinct fixed points (**foci**) is constant.

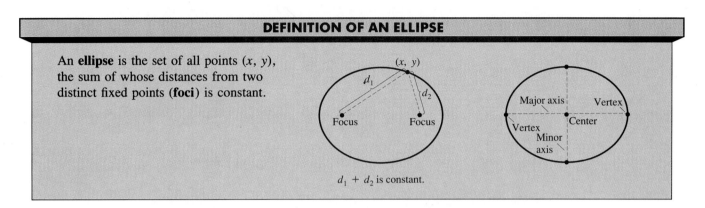

$d_1 + d_2$ is constant.

The line through the foci intersects the ellipse at two points (**vertices**). The chord joining the vertices is the **major axis,** and its midpoint is the **center** of the ellipse. The chord perpendicular to the major axis at the center is the **minor axis** of the ellipse.

You can visualize the definition of an ellipse by imagining two thumbtacks placed at the foci, as shown in Figure 5.19. If the ends of a fixed length of string are fastened to the thumbtacks and the string is drawn taut with a pencil, the path traced by the pencil will be an ellipse.

The standard form of the equation of an ellipse takes one of two forms, depending upon whether the major axis is horizontal or vertical.

FIGURE 5.19

STANDARD EQUATION OF AN ELLIPSE (CENTER AT ORIGIN)

The **standard form of the equation of an ellipse** with center at the origin and major and minor axes of lengths $2a$ and $2b$ (where $0 < b < a$) is

$$\frac{x^2}{a^2} + \frac{y^2}{b^2} = 1 \qquad \text{or}$$

$$\frac{x^2}{b^2} + \frac{y^2}{a^2} = 1, \qquad 0 < b < a.$$

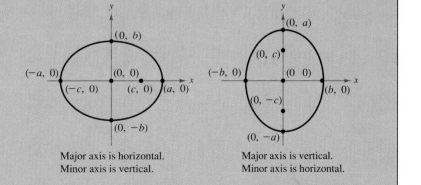

Major axis is horizontal.
Minor axis is vertical.

Major axis is vertical.
Minor axis is horizontal.

The vertices and foci lie on the major axis, a and c units, respectively, from the center. Moreover, a, b, and c are related by the equation $c^2 = a^2 - b^2$.

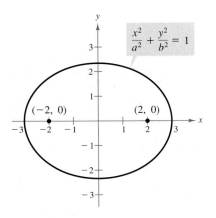

FIGURE 5.20

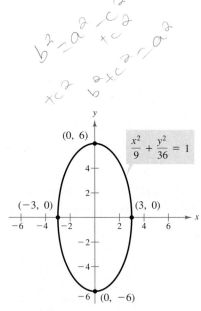

FIGURE 5.21

EXAMPLE 3   Finding the Standard Equation of an Ellipse

Find the standard form of the equation of the ellipse that has a major axis of length 6 and foci at $(-2, 0)$ and $(2, 0)$, as shown in Figure 5.20.

Solution

Since the foci occur at $(-2, 0)$ and $(2, 0)$, the center of the ellipse is $(0, 0)$, and the major axis is horizontal. Thus, the ellipse has an equation of the form

$$\frac{x^2}{a^2} + \frac{y^2}{b^2} = 1. \qquad\qquad \textit{Standard form}$$

Since the length of the major axis is 6,

$$2a = 6 \qquad\qquad \textit{Length of major axis}$$

which implies that $a = 3$. Moreover, the distance from the center to either focus is $c = 2$. Finally,

$$b^2 = a^2 - c^2 = 3^2 - 2^2 = 9 - 4 = 5$$

which yields the equation

$$\frac{x^2}{9} + \frac{y^2}{5} = 1.$$

EXAMPLE 4   Sketching an Ellipse

Sketch the ellipse given by $4x^2 + y^2 = 36$ and identify the vertices.

Solution

$$4x^2 + y^2 = 36 \qquad\qquad \textit{Given equation}$$

$$\frac{4x^2}{36} + \frac{y^2}{36} = \frac{36}{36}$$

$$\frac{x^2}{3^2} + \frac{y^2}{6^2} = 1 \qquad\qquad \textit{Standard form}$$

Since the denominator of the y^2 term is larger than the denominator of the x^2 term, we conclude that the major axis is vertical. Moreover, since $a = 6$, the vertices are $(0, -6)$ and $(0, 6)$. Finally, since $b = 3$, the endpoints of the minor axis are $(-3, 0)$ and $(3, 0)$, as shown in Figure 5.21.

In Example 4 note that from the standard form of the equation we can sketch the ellipse by locating the endpoints of the two axes. Since 3^2 is the denominator of the x^2 term, we move three units to the *right and left* of the center to locate the endpoints of the horizontal axis. Similarly, since 6^2 is the denominator of the y^2 term, we move six units *up and down* from the center to locate the endpoints of the vertical axis.

Hyperbolas

The definition of a **hyperbola** is similar to that of an ellipse. The difference is that, for an ellipse, the *sum* of the distances between the foci and a point on the ellipse is constant, while, for a hyperbola, the *difference* of these distances is constant.

DEFINITION OF A HYPERBOLA

A **hyperbola** is the set of all points (x, y), the difference of whose distances from two distinct fixed points (**foci**) is constant.

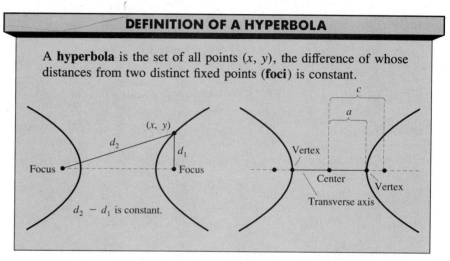

The graph of a hyperbola has two disconnected parts (**branches**). The line through the two foci intersects the hyperbola at two points (**vertices**). The line segment connecting the vertices is the **transverse axis,** and the midpoint of the transverse axis is the **center** of the hyperbola.

STANDARD EQUATION OF A HYPERBOLA (CENTER AT ORIGIN)

The **standard form of the equation of a hyperbola** with center at $(0, 0)$ is

$$\frac{x^2}{a^2} - \frac{y^2}{b^2} = 1 \qquad \text{or} \qquad \frac{y^2}{a^2} - \frac{x^2}{b^2} = 1.$$

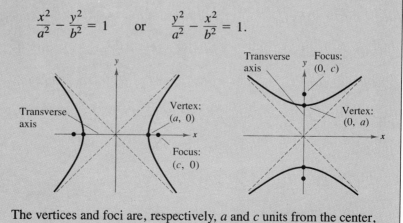

The vertices and foci are, respectively, a and c units from the center, and $b^2 = c^2 - a^2$.

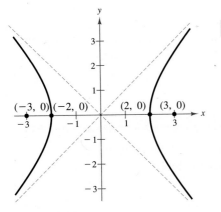

FIGURE 5.22

EXAMPLE 5   Finding the Standard Equation of a Hyperbola

Find the standard form of the equation of the hyperbola with foci at $(-3, 0)$ and $(3, 0)$ and vertices at $(-2, 0)$ and $(2, 0)$, as shown in Figure 5.22.

Solution

It can be seen that $c = 3$ because the foci are three units from the center. Moreover, $a = 2$ because the vertices are two units from the center. Thus, it follows that

$$b^2 = c^2 - a^2 = 3^2 - 2^2 = 9 - 4 = 5.$$

Since the transverse axis is horizontal, the standard form of the equation is

$$\frac{x^2}{a^2} - \frac{y^2}{b^2} = 1.$$

Finally, substitute $a^2 = 4$ and $b^2 = 5$ to obtain

$$\frac{x^2}{4} - \frac{y^2}{5} = 1.$$ *Standard form*

An important aid in sketching the graph of a hyperbola is the determination of its **asymptotes,** as shown in Figure 5.23. Each hyperbola has two asymptotes that intersect at the center of the hyperbola. Furthermore, the asymptotes pass through the corners of a rectangle of dimensions $2a$ by $2b$. The line segment of length $2b$, joining $(0, b)$ and $(0, -b)$ [or $(-b, 0)$ and $(b, 0)$], is the **conjugate axis** of the hyperbola.

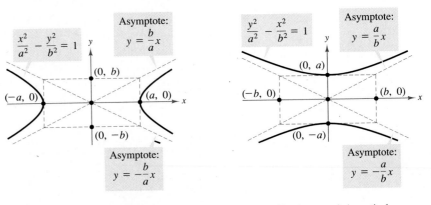

Transverse axis is horizontal.   Transverse axis is vertical.

FIGURE 5.23

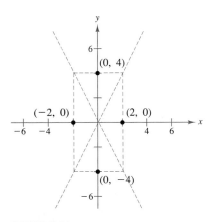

FIGURE 5.24

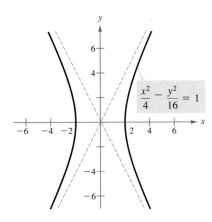

FIGURE 5.25

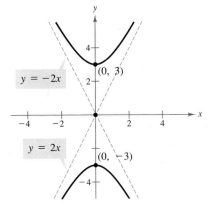

FIGURE 5.26

ASYMPTOTES OF A HYPERBOLA (CENTER AT ORIGIN)

The **asymptotes of a hyperbola** with center at $(0, 0)$ are

$$y = \frac{b}{a}x \quad \text{and} \quad y = -\frac{b}{a}x \quad \textit{Transverse axis is horizontal}$$

or

$$y = \frac{a}{b}x \quad \text{and} \quad y = -\frac{a}{b}x. \quad \textit{Transverse axis is vertical}$$

EXAMPLE 6   Sketching the Graph of a Hyperbola

Sketch the graph of the hyperbola whose equation is $4x^2 - y^2 = 16$.

Solution

$$4x^2 - y^2 = 16 \qquad\qquad\qquad \textit{Given equation}$$

$$\frac{4x^2}{16} - \frac{y^2}{16} = \frac{16}{16}$$

$$\frac{x^2}{4} - \frac{y^2}{16} = 1 \qquad\qquad\qquad \textit{Standard form}$$

Because the x^2-term is positive, we conclude that the transverse axis is horizontal and the vertices occur at $(-2, 0)$ and $(2, 0)$. Moreover, the endpoints of the conjugate axis occur at $(0, -4)$ and $(0, 4)$, and we are able to sketch the rectangle shown in Figure 5.24. Finally, by drawing the asymptotes through the corners of this rectangle, we complete the sketch shown in Figure 5.25.

EXAMPLE 7   Finding the Standard Equation of a Hyperbola

Find the standard form of the equation of the hyperbola having vertices at $(0, -3)$ and $(0, 3)$ and with asymptotes $y = -2x$ and $y = 2x$, as shown in Figure 5.26.

Solution

Since the transverse axis is vertical, we have asymptotes of the form

$$y = \frac{a}{b}x \quad \text{and} \quad y = -\frac{a}{b}x.$$

Thus,

$$\frac{a}{b} = 2$$

and since $a = 3$, we can determine that $b = \frac{3}{2}$. Finally, we see that the hyperbola has the following equation.

$$\frac{y^2}{3^2} - \frac{x^2}{(3/2)^2} = 1 \qquad\qquad \textit{Standard form}$$

DISCUSSION

PROBLEM

Hyperbolas in Applications

At the beginning of this section, we mentioned that each type of conic section can be formed by the intersection of a plane and a double-napped cone. Figure 5.27 shows three examples of how such an intersection can occur in physical situations.

FIGURE 5.27

Identify the cone and hyperbola (or portion of a hyperbola) in each of the three situations. Can you think of other examples of physical situations in which hyperbolas are formed?

WARM UP

The following warm-up exercises involve skills that were covered in earlier sections. You will use these skills in the exercise set for this section.

In Exercises 1–4, rewrite the equations so that they have no fractions.

1. $\dfrac{x^2}{16} + \dfrac{y^2}{9} = 1$ **2.** $\dfrac{x^2}{32} + \dfrac{4y^2}{32} = \dfrac{32}{32}$

3. $\dfrac{x^2}{1/4} - \dfrac{y^2}{4} = 1$ **4.** $\dfrac{3x^2}{1/9} + \dfrac{4y^2}{9} = 1$

In Exercises 5–8, solve for c. (Assume $c > 0$.)

5. $c^2 = 3^2 - 1^2$ **6.** $c^2 = 2^2 + 3^2$

7. $c^2 + 2^2 = 4^2$ **8.** $c^2 - 1^2 = 2^2$

In Exercises 9 and 10, find the distance between the point and the origin.

9. $(0, -4)$ **10.** $(-2, 0)$

EXERCISES for Section 5.3

In Exercises 1–8, match the equation with its graph. (The graphs are labeled (a), (b), (c), (d), (e), (f), (g), and (h).)

1. $x^2 = 4y$

2. $x^2 = -4y$

3. $y^2 = 4x$

4. $y^2 = -4x$

5. $\dfrac{x^2}{1} + \dfrac{y^2}{4} = 1$

6. $\dfrac{x^2}{4} + \dfrac{y^2}{1} = 1$

7. $\dfrac{x^2}{1} - \dfrac{y^2}{4} = 1$

8. $\dfrac{y^2}{4} - \dfrac{x^2}{1} = 1$

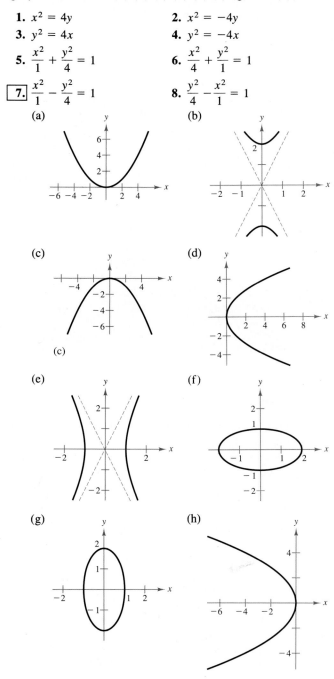

(a)

(b)

(c)

(d)

(c)

(e)

(f)

(g)

(h)

In Exercises 9–16, find the vertex and focus of the parabola and sketch its graph.

9. $y = 4x^2$

10. $y = 2x^2$

11. $y^2 = -6x$

12. $y^2 = 3x$

13. $x^2 + 8y = 0$

14. $x + y^2 = 0$

15. $y^2 - 8x = 0$

16. $x^2 + 12y = 0$

In Exercises 17–26, find an equation of the specified parabola with vertex at the origin.

17. Focus: $\left(0, -\frac{3}{2}\right)$

18. Focus: $(2, 0)$

19. Focus: $(-2, 0)$

20. Focus: $(0, -2)$

21. Directrix: $y = -1$

22. Directrix: $x = 3$

23. Directrix: $y = 2$

24. Directrix: $x = -2$

25. Horizontal axis and passes through the point $(4, 6)$

26. Vertical axis and passes through the point $(-2, -2)$

In Exercises 27–34, find the center and vertices of the ellipse and sketch its graph.

27. $\dfrac{x^2}{25} + \dfrac{y^2}{16} = 1$

28. $\dfrac{x^2}{144} + \dfrac{y^2}{169} = 1$

29. $\dfrac{x^2}{16} + \dfrac{y^2}{25} = 1$

30. $\dfrac{x^2}{169} + \dfrac{y^2}{144} = 1$

31. $\dfrac{x^2}{9} + \dfrac{y^2}{5} = 1$

32. $\dfrac{x^2}{28} + \dfrac{y^2}{64} = 1$

33. $5x^2 + 3y^2 = 15$

34. $x^2 + 4y^2 = 4$

In Exercises 35–42, find an equation of the specified ellipse with center at the origin.

35. Vertices: $(0, \pm 2)$; minor axis of length 2

36. Vertices: $(\pm 2, 0)$; minor axis of length 3

37. Vertices: $(\pm 5, 0)$; foci: $(\pm 2, 0)$

38. Vertices: $(0, \pm 8)$; foci: $(0, \pm 4)$

39. Foci: $(\pm 5, 0)$; major axis of length 12

40. Foci: $(\pm 2, 0)$; major axis of length 8

41. Vertices: $(0, \pm 5)$; passes through the point $(4, 2)$

42. Major axis vertical; passes through points $(0, 4)$ and $(2, 0)$

In Exercises 43–50, find the center and vertices of the hyperbola and sketch its graph, using asymptotes as an aid.

43. $x^2 - y^2 = 1$

44. $\dfrac{x^2}{9} - \dfrac{y^2}{16} = 1$

45. $\dfrac{y^2}{1} - \dfrac{x^2}{4} = 1$

46. $\dfrac{y^2}{9} - \dfrac{x^2}{1} = 1$

47. $\dfrac{y^2}{25} - \dfrac{x^2}{144} = 1$

48. $\dfrac{x^2}{36} - \dfrac{y^2}{4} = 1$

49. $2x^2 - 3y^2 = 6$

50. $3y^2 - 5x^2 = 15$

In Exercises 51–58, find an equation of the specified hyperbola with center at the origin.

51. Vertices: $(0, \pm2)$; foci: $(0, \pm4)$

52. Vertices: $(\pm3, 0)$; foci: $(\pm5, 0)$

53. Vertices: $(\pm1, 0)$; asymptotes: $y = \pm3x$

54. Vertices: $(0, \pm3)$; asymptotes: $y = \pm3x$

55. Foci: $(0, \pm8)$; asymptotes: $y = \pm4x$

56. Foci: $(\pm10, 0)$; asymptotes: $y = \pm\frac{3}{4}x$

57. Vertices: $(0, \pm3)$; passes through the point $(-2, 5)$

58. Vertices: $(\pm2, 0)$; passes through the point $(3, \sqrt{3})$

59. *Satellite Antenna* The receiver in a parabolic television dish antenna is three feet from the vertex and is located at the focus (see figure). Find an equation of a cross section of the reflector. (Assume that the dish is directed upward and the vertex is at the origin.)

60. *Suspension Bridge* Each cable of a suspension bridge is suspended (in the shape of a parabola) between two towers that are 400 feet apart and 50 feet above the roadway (see figure). The cables touch the roadway midway between the towers. Find an equation for the parabolic shape of each cable.

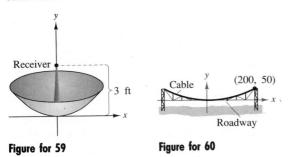

Figure for 59

Figure for 60

61. *Fireplace Arch* A fireplace arch is to be constructed in the shape of a semi-ellipse. The opening is to have a height of two feet at the center and a width of five feet along the base (see figure). The contractor draws the outline of the ellipse by the method shown in Figure 5.19. Where should the tacks be placed and what should the length of the piece of string be?

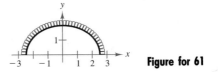

Figure for 61

62. *Mountain Tunnel* A semi-elliptical arch over a tunnel for a road through a mountain has a major axis of 100 feet and its height at the center is 30 feet (see figure). Determine the height of the arch five feet from the edge of the tunnel.

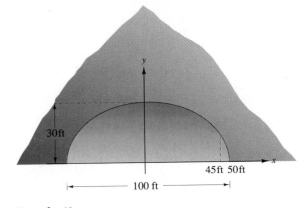

Figure for 62

63. Sketch a graph of the ellipse consisting of all points (x, y) such that the sum of the distances between (x, y) and two fixed points is 16 units and the foci are located at the centers of the two sets of concentric circles in the figure.

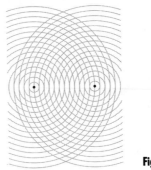

Figure for 63

64. A line segment through a focus with endpoints on the ellipse and perpendicular to the major axis is a **latus rectum** of the ellipse. Therefore, an ellipse has two latera recta. Knowing the length of the latera recta is helpful in sketching an ellipse because it yields other points on the curve (see figure). Show that the length of each latus rectum is $2b^2/a$.

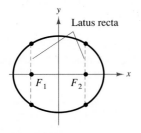

Figure for 64

In Exercises 65–68, sketch the graph of the ellipse, making use of the latus recta (see Exercise 64).

65. $\dfrac{x^2}{4} + \dfrac{y^2}{1} = 1$ **66.** $\dfrac{x^2}{9} + \dfrac{y^2}{16} = 1$

67. $9x^2 + 4y^2 = 36$ **68.** $5x^2 + 3y^2 = 15$

69. *LORAN* Long-distance radio navigation for aircraft and ships uses synchronized pulses transmitted by widely separated transmitting stations. These pulses travel at the speed of light (186,000 miles per second). The difference in the times of arrival of these pulses at an aircraft or ship is constant on a hyperbola having the transmitting stations as foci. Assume that two stations, 300 miles apart, are positioned on the rectangular coordinate system at points with coordinates $(-150, 0)$ and $(150, 0)$ and that a ship is traveling on a path with coordinates $(x, 75)$ (see figure). Find the x-coordinate of the position of the ship if the time difference between the pulses from the transmitting stations is 1000 microseconds (0.001 second).

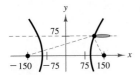

Figure for 69

70. *Hyperbolic Mirror* A hyperbolic mirror (used in some telescopes) has the property that a light ray directed at the focus will be reflected to the other focus (see figure). The focus of a hyperbolic mirror has coordinates $(12, 0)$. Find the vertex of the mirror if its mount has coordinates $(12, 12)$.

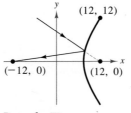

Figure for 70

71. Use the definition of an ellipse to derive the standard form of the equation of an ellipse.

72. Use the definition of a hyperbola to derive the standard form of the equation of a hyperbola.

73. Discuss the change in shape and orientation of the graph of the ellipse

$$\frac{x^2}{a^2} + \frac{y^2}{16} = 1$$

as a increases continuously from 1 to 8.

5.4 Conic Sections and Translations

Vertical and Horizontal Shifts of Conics / Writing Equations of Conics in Standard Form

Vertical and Horizontal Shifts of Conics

In Section 5.3 you looked at conic sections whose graphs were in *standard position*. In this section you will study the equations of conic sections that have been shifted vertically or horizontally in the plane.

STANDARD FORMS OF EQUATIONS OF CONICS

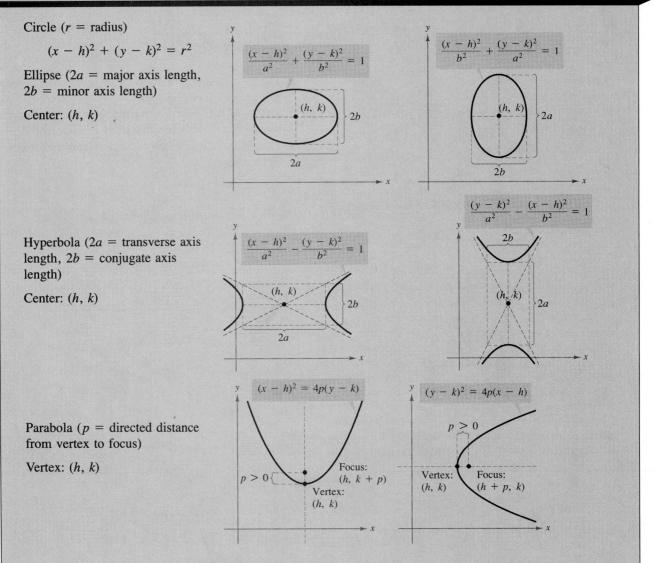

Circle (r = radius)

$$(x - h)^2 + (y - k)^2 = r^2$$

Ellipse ($2a$ = major axis length, $2b$ = minor axis length)

Center: (h, k)

$$\frac{(x - h)^2}{a^2} + \frac{(y - k)^2}{b^2} = 1$$

$$\frac{(x - h)^2}{b^2} + \frac{(y - k)^2}{a^2} = 1$$

Hyperbola ($2a$ = transverse axis length, $2b$ = conjugate axis length)

Center: (h, k)

$$\frac{(x - h)^2}{a^2} - \frac{(y - k)^2}{b^2} = 1$$

$$\frac{(y - k)^2}{a^2} - \frac{(x - h)^2}{b^2} = 1$$

Parabola (p = directed distance from vertex to focus)

Vertex: (h, k)

$$(x - h)^2 = 4p(y - k)$$

Focus: $(h, k + p)$
Vertex: (h, k)

$$(y - k)^2 = 4p(x - h)$$

Vertex: (h, k)
Focus: $(h + p, k)$

EXAMPLE 1 **Equations of Conic Sections**

a. The graph of

$$(x - 1)^2 + (y + 2)^2 = 3^2$$

is a circle whose center is the point $(1, -2)$ and whose radius is 3, as shown in Figure 5.28(a).

b. The graph of

$$\frac{(x - 2)^2}{3^2} + \frac{(y - 1)^2}{2^2} = 1$$

is an ellipse whose center is the point $(2, 1)$. The major axis of the ellipse is horizontal and of length $2(3) = 6$. The minor axis of the ellipse is vertical and of length $2(2) = 4$, as shown in Figure 5.28(b).

c. The graph of

$$\frac{(x - 3)^2}{1^2} - \frac{(y - 2)^2}{3^2} = 1$$

is a hyperbola whose center is the point $(3, 2)$. The transverse axis is horizontal and of length $2(1) = 2$. The conjugate axis is vertical and of length $2(3) = 6$, as shown in Figure 5.28(c).

d. The graph of

$$(x - 2)^2 = 4(-1)(y - 3)$$

is a parabola whose vertex is the point $(2, 3)$. The axis of the parabola is vertical. The focus of the parabola is one unit above or below the vertex. Moreover, since $p = -1$, it follows that the focus lies *below* the vertex, as shown in Figure 5.28(d).

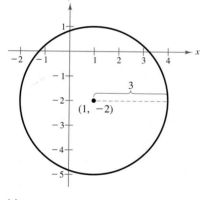

(a)

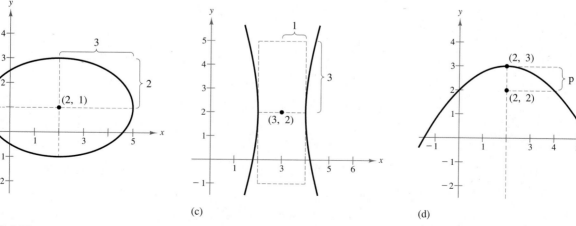

(b)

(c)

(d)

FIGURE 5.28

Writing Equations of Conics in Standard Form

To write the equation of a conic in standard form, we complete the square.

REMARK Note in Example 2 that p is the *directed distance* from the vertex to the focus. Because the axis of the parabola is vertical and $p = -1$, the focus is one unit *below* the vertex, and the parabola opens downward.

EXAMPLE 2 Finding the Standard Form of a Parabola

Find the vertex and focus of the parabola given by

$$x^2 - 2x + 4y - 3 = 0.$$

Solution

$x^2 - 2x + 4y - 3 = 0$	*Given equation*
$x^2 - 2x = -4y + 3$	*Group terms*
$x^2 - 2x + 1 = -4y + 3 + 1$	*Add 1 to both sides*
$(x - 1)^2 = -4y + 4$	*Completed square form*
$(x - 1)^2 = 4(-1)(y - 1)$	*Standard form*
$(x - h)^2 = 4p(y - k)$	

From this standard form, we see that $h = 1$, $k = 1$, and $p = -1$. Since the axis is vertical and p is negative, the parabola opens downward. The vertex and focus are

Vertex $(h, k) = (1, 1)$

Focus $(h, k + p) = (1, 0)$.

The graph of this parabola is shown in Figure 5.29.

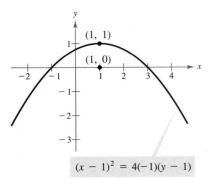

$(x - 1)^2 = 4(-1)(y - 1)$

FIGURE 5.29

EXAMPLE 3 Sketching an Ellipse

Sketch the graph of the ellipse whose equation is

$$x^2 + 4y^2 + 6x - 8y + 9 = 0.$$

Solution

$x^2 + 4y^2 + 6x - 8y + 9 = 0$	*Given equation*
$(x^2 + 6x + \blacksquare) + (4y^2 - 8y + \blacksquare) = -9$	*Group terms*
$(x^2 + 6x + \blacksquare) + 4(y^2 - 2y + \blacksquare) = -9$	*Factor 4 out of y-terms*
$(x^2 + 6x + 9) + 4(y^2 - 2y + 1) = -9 + 9 + 4(1)$	*Add 9 and 4 to both sides*
$(x + 3)^2 + 4(y - 1)^2 = 4$	*Completed square form*
$\dfrac{(x + 3)^2}{4} + \dfrac{(y - 1)^2}{1} = 1$	*Standard form*
$\dfrac{(x - h)^2}{a^2} + \dfrac{(y - k)^2}{b^2} = 1$	

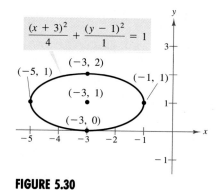

$$\frac{(x + 3)^2}{4} + \frac{(y - 1)^2}{1} = 1$$

FIGURE 5.30

From this standard form, we see that the center is $(h, k) = (-3, 1)$. Since the denominator of the x-term is $4 = a^2 = 2^2$, the endpoints of the major axis lie two units to the right and left of the center. Similarly, since the denominator of the y-term is $1 = b^2 = 1^2$, the endpoints of the minor axis lie one unit up and down from the center. The graph of this ellipse is shown in Figure 5.30.

EXAMPLE 4   Sketching a Hyperbola

Sketch the graph of the hyperbola given by the equation

$$y^2 - 4x^2 + 4y + 24x - 41 = 0.$$

Solution

$y^2 - 4x^2 + 4y + 24x - 41 = 0$	*Given equation*
$(y^2 + 4y + \blacksquare) - (4x^2 - 24x + \blacksquare) = 41$	*Group terms*
$(y^2 + 4y + \blacksquare) - 4(x^2 - 6x + \blacksquare) = 41$	*Factor 4 out of x-terms*
$(y^2 + 4y + 4) - 4(x^2 - 6x + 9) = 41 + 4 - 4(9)$	*Add 4, subtract 36*
$(y + 2)^2 - 4(x - 3)^2 = 9$	*Completed square form*
$\dfrac{(y + 2)^2}{9} - \dfrac{4(x - 3)^2}{9} = 1$	*Divide both sides by 9*
$\dfrac{(y + 2)^2}{9} - \dfrac{(x - 3)^2}{9/4} = 1$	*Change 4 to $\dfrac{1}{1/4}$*
$\dfrac{(y + 2)^2}{3^2} - \dfrac{(x - 3)^2}{(3/2)^2} = 1$	*Standard form*
$\dfrac{(y - h)^2}{a^2} - \dfrac{(x - h)^2}{b^2} = 1$	

From the standard form, we see that the transverse axis is vertical and the center lies at $(h, k) = (3, -2)$. Since the denominator of the y-term is $a^2 = 3^2$, we know that the vertices occur three units above and below the center.

$$(3, -5) \quad \text{and} \quad (3, 1) \qquad \textit{Vertices}$$

To sketch the hyperbola, we draw a rectangle whose top and bottom pass through the vertices. Since the denominator of the x-term is $b^2 = \left(\frac{3}{2}\right)^2$, we locate the sides of the rectangle $\frac{3}{2}$ units to the right and left of center, as shown in Figure 5.31. Finally, we sketch the asymptotes by drawing lines through the opposite corners of the rectangle. Using these asymptotes, we complete the graph of the hyperbola, as shown in Figure 5.31.

$$\frac{(y + 2)^2}{3^2} - \frac{(x - 3)^2}{(3/2)^2} = 1$$

FIGURE 5.31

To find the foci in Example 4 we first find c.

$$c^2 = a^2 + b^2 = 9 + \frac{9}{4} = \frac{45}{4} \rightarrow c = \frac{3\sqrt{5}}{2}$$

Because the transverse axis is vertical, the foci lie c units above and below the center.

$$\left(3, -2 + \frac{3\sqrt{5}}{2}\right) \quad \text{and} \quad \left(3, -2 - \frac{3\sqrt{5}}{2}\right) \qquad \textit{Foci}$$

EXAMPLE 5 Writing the Equation of an Ellipse

Write the standard form of the equation of the ellipse whose vertices are $(2, -2)$ and $(2, 4)$. The length of the minor axis of the ellipse is 4, as shown in Figure 5.32.

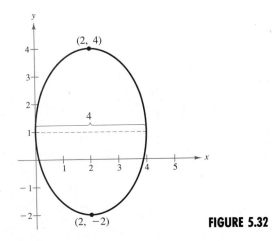

FIGURE 5.32

Solution

The center of the ellipse lies at the midpoint of its vertices. Thus, the center is

$$(h, k) = (2, 1). \qquad \textit{Center}$$

Since the vertices lie on a vertical line and are six units apart, it follows that the major axis is vertical and has a length of $2a = 6$. Thus, $a = 3$. Moreover, since the minor axis has a length of 4, it follows that $2b = 4$, which implies that $b = 2$. Therefore, the standard form of the equation of the ellipse is as follows.

$$\frac{(x - h)^2}{a^2} + \frac{(y - k)^2}{b^2} = 1 \qquad \textit{Major axis is vertical}$$

$$\frac{(x - 2)^2}{3^2} + \frac{(y - 1)^2}{2^2} = 1 \qquad \textit{Standard form}$$

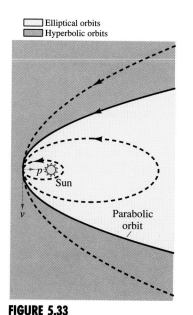

Elliptical orbits
Hyperbolic orbits

FIGURE 5.33

An interesting application of conic sections involves the orbits of comets in our solar system. Of the 610 comets identified prior to 1970, 245 have elliptical orbits, 295 have parabolic orbits, and 70 have hyperbolic orbits. For example, Halley's comet has an elliptical orbit, and we can predict the reappearance of this comet every 75 years. The center of the sun is a focus of each of these orbits, and each orbit has a vertex at the point where the comet is closest to the sun, as shown in Figure 5.33.

If p is the distance between the vertex and the focus, and v is the speed of the comet at the vertex, then the orbit is:

$$\text{an } \textit{ellipse} \text{ if } \quad v < \sqrt{\frac{2GM}{p}}$$

$$\text{a } \textit{parabola} \text{ if } \quad v = \sqrt{\frac{2GM}{p}}$$

$$\text{a } \textit{hyperbola} \text{ if } \quad v > \sqrt{\frac{2GM}{p}}$$

where M is the mass of the sun and $G \approx 6.67(10^{-8})$ cm^3/(gm · sec^2).

DISCUSSION

PROBLEM

Asymptotes of a Hyperbola

In Section 5.3 you learned how to find equations for the asymptotes of a hyperbola whose center is the origin. Describe how to find equations for the asymptotes of a hyperbola whose center is the point (h, k). Then use your procedure to find the asymptotes of the hyperbola given by

$$\frac{(x - 1)^2}{3^2} - \frac{(y - 2)^2}{2^2} = 1.$$

WARM UP

The following warm-up exercises involve skills that were covered in earlier sections. You will use these skills in the exercise set for this section.

In Exercises 1–10, identify the conic represented by each equation.

1. $\dfrac{x^2}{4} - \dfrac{y^2}{4} = 1$ 2. $\dfrac{x^2}{9} + \dfrac{y^2}{1} = 1$ 3. $2x + y^2 = 0$

4. $\dfrac{x^2}{9} - \dfrac{y^2}{4} = 1$ 5. $\dfrac{x^2}{4} + \dfrac{y^2}{16} = 1$ 6. $4x^2 + 4y^2 = 25$

7. $\dfrac{y^2}{4} - \dfrac{x^2}{2} = 1$ 8. $x^2 - 6y = 0$ 9. $3x - y^2 = 0$

10. $\dfrac{x^2}{9/4} + \dfrac{y^2}{4} = 1$

EXERCISES for Section 5.4

In Exercises 1–12, find the vertex, focus, and directrix of the parabola, and sketch its graph.

1. $(x - 1)^2 + 8(y + 2) = 0$

2. $(x + 3) + (y - 2)^2 = 0$

3. $\left(y + \frac{1}{2}\right)^2 = 2(x - 5)$ **4.** $\left(x + \frac{1}{2}\right)^2 = 4(y - 3)$

5. $y = \frac{1}{4}(x^2 - 2x + 5)$ **6.** $y = -\frac{1}{6}(x^2 + 4x - 2)$

7. $4x - y^2 - 2y - 33 = 0$

8. $y^2 + x + y = 0$

9. $y^2 + 6y + 8x + 25 = 0$

10. $x^2 - 2x + 8y + 9 = 0$

11. $y^2 - 4y - 4x = 0$ **12.** $y^2 - 4x - 4 = 0$

In Exercises 13–20, find an equation of the specified parabola.

13. Vertex: $(3, 2)$; focus: $(1, 2)$

14. Vertex: $(-1, 2)$; focus: $(-1, 0)$

15. Vertex: $(0, 4)$; directrix: $y = 2$

16. Vertex: $(-2, 1)$; directrix: $x = 1$

17. Focus: $(2, 2)$; directrix: $x = -2$

18. Focus: $(0, 0)$; directrix: $y = 4$

19.

20.

In Exercises 21–28, find the center, foci, and vertices of the ellipse and sketch the graph.

21. $\dfrac{(x - 1)^2}{9} + \dfrac{(y - 5)^2}{25} = 1$

22. $(x + 2)^2 + \dfrac{(y + 4)^2}{1/4} = 1$

23. $9x^2 + 4y^2 + 36x - 24y + 36 = 0$

24. $9x^2 + 4y^2 - 36x + 8y + 31 = 0$

25. $16x^2 + 25y^2 - 32x + 50y + 16 = 0$

26. $9x^2 + 25y^2 - 36x - 50y + 61 = 0$

27. $12x^2 + 20y^2 - 12x + 40y - 37 = 0$

28. $36x^2 + 9y^2 + 48x - 36y + 43 = 0$

In Exercises 29–36, find an equation for the specified ellipse.

29. Vertices: $(0, 2)$, $(4, 2)$; minor axis of length 2

30. Foci: $(0, 0)$, $(4, 0)$; major axis of length 8

31. Foci: $(0, 0)$, $(0, 8)$; major axis of length 16

32. Center: $(2, -1)$; vertex: $\left(2, \frac{1}{2}\right)$; minor axis of length 2

33. Vertices: $(3, 1)$, $(3, 9)$; minor axis of length 6

34. Center: $(3, 2)$; $a = 3c$; foci: $(1, 2)$, $(5, 2)$

35. Center: $(0, 4)$; $a = 2c$; vertices: $(-4, 4)$, $(4, 4)$

36. Vertices: $(5, 0)$, $(5, 12)$; endpoints of the minor axis: $(0, 6)$, $(10, 6)$

In Exercises 37–46, find the center, vertices, and foci of the hyperbola and sketch the graph, using asymptotes as an aid.

37. $\dfrac{(x - 1)^2}{4} - \dfrac{(y + 2)^2}{1} = 1$

38. $\dfrac{(x + 1)^2}{144} - \dfrac{(y - 4)^2}{25} = 1$

39. $(y + 6)^2 - (x - 2)^2 = 1$

40. $\dfrac{(y - 1)^2}{1/4} - \dfrac{(x + 3)^2}{1/9} = 1$

41. $9x^2 - y^2 - 36x - 6y + 18 = 0$

42. $x^2 - 9y^2 + 36y - 72 = 0$

43. $9y^2 - x^2 + 2x + 54y + 62 = 0$

44. $16y^2 - x^2 + 2x + 64y + 63 = 0$

45. $x^2 - 9y^2 + 2x - 54y - 80 = 0$

46. $9x^2 - y^2 + 54x + 10y + 55 = 0$

In Exercises 47–54, find an equation for the specified hyperbola.

47. Vertices: $(2, 0)$, $(6, 0)$; foci: $(0, 0)$, $(8, 0)$

48. Vertices: $(2, 3)$, $(2, -3)$; foci: $(2, 5)$, $(2, -5)$

49. Vertices: $(4, 1)$, $(4, 9)$; foci: $(4, 0)$, $(4, 10)$

50. Vertices: $(-2, 1)$, $(2, 1)$; foci: $(-3, 1)$, $(3, 1)$

51. Vertices: $(2, 3)$, $(2, -3)$; passes through the point $(0, 5)$

52. Vertices: $(-2, 1)$, $(2, 1)$; passes through the point $(4, 3)$

53. Vertices: $(0, 2)$, $(6, 2)$; asymptotes: $y = \frac{2}{3}x$, $y = 4 - \frac{2}{3}x$

54. Vertices: $(3, 0)$, $(3, 4)$; asymptotes: $y = \frac{2}{3}x$, $y = 4 - \frac{2}{3}x$

In Exercises 55–62, classify the graph of each equation as a circle, a parabola, an ellipse, or a hyperbola.

55. $x^2 + y^2 - 6x + 4y + 9 = 0$

56. $x^2 + 4y^2 - 6x + 16y + 21 = 0$

57. $4x^2 - y^2 - 4x - 3 = 0$

58. $y^2 - 4y - 4x = 0$

59. $4x^2 + 3y^2 + 8x - 24y + 51 = 0$

60. $4y^2 - 2x^2 - 4y - 8x - 15 = 0$

61. $25x^2 - 10x - 200y - 119 = 0$

62. $4x^2 + 4y^2 - 16y + 15 = 0$

63. *Satellite Orbit* A satellite in a 100-mile-high circular orbit around the earth has a velocity of approximately 17,500 miles per hour. If this velocity is multiplied by $\sqrt{2}$, then the satellite will have the minimum velocity necessary to escape the earth's gravity and it will follow a parabolic path with the center of the earth as the focus (see figure).
(a) Find the escape velocity of the satellite.
(b) Find an equation of its path (assume the radius of the earth is 4,000 miles).

64. *Fluid Flow* Water is flowing from a horizontal pipe 48 feet above the ground. The falling stream of water has the shape of a parabola whose vertex $(0, 48)$ is at the end of the pipe (see figure). The stream of water strikes the ground at the point $(10\sqrt{3}, 0)$. Find the equation of the path taken by the water.

In Exercises 65–71, e is the eccentricity of the ellipse and is defined by $e = c/a$. It measures the flatness of the ellipse.

65. Find an equation of the ellipse with vertices $(\pm 5, 0)$ and eccentricity $e = \frac{3}{5}$.

66. Find an equation of the ellipse with vertices $(0, \pm 8)$ and eccentricity $e = \frac{1}{2}$.

67. *Orbit of the Earth* The earth moves in an elliptical orbit with the sun at one of the foci (see figure). The length of half of the major axis is 92.957×10^6 miles and the eccentricity is 0.017. Find the smallest distance and the greatest distance of the earth from the sun.

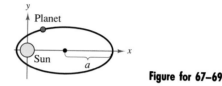

Figure for 67–69

68. *Orbit of Pluto* The planet Pluto moves in an elliptical orbit with the sun at one of the foci (see figure). The length of half of the major axis is 3.666×10^9 miles and the eccentricity is 0.248. Find the smallest distance and the greatest distance of Pluto from the sun.

69. *Orbit of Saturn* The planet Saturn moves in an elliptical orbit with the sun at one of the foci (see figure). The smallest distance and the greatest distance of the planet from the sun are 1.3495×10^9 kilometers and 1.5045×10^9, respectively. Find the eccentricity of the orbit.

70. *Satellite Orbit* The first artificial satellite to orbit the earth was Sputnik I (launched by the Soviet Union in 1957). Its highest point above the earth's surface was 583 miles, and its lowest point was 132 miles (see figure). Assume that the center of the earth is the focus of the elliptical orbit and the radius of the earth is 4000 miles. Find the eccentricity of the orbit.

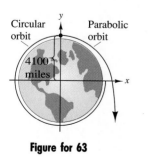

Figure for 63

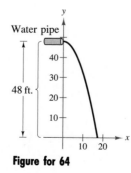

Figure for 64

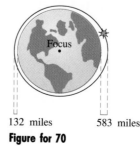

Figure for 70

71. Show that the equation of an ellipse can be written as

$$\frac{(x - h)^2}{a^2} + \frac{(y - k)^2}{a^2(1 - e^2)} = 1.$$

Note that as e approaches zero, the ellipse approaches a circle of radius a.

REVIEW EXERCISES for Chapter 5

In Exercises 1–4, find the domain of the function and identify any horizontal or vertical asymptotes.

1. $f(x) = \dfrac{4}{x + 3}$

2. $f(x) = \dfrac{2x^2 + 5x - 3}{x^2 + 2}$

3. $g(x) = \dfrac{x^2}{x^2 - 4}$

4. $g(x) = \dfrac{1}{(x - 3)^2}$

In Exercises 5–20, sketch the graph of the rational function. As sketching aids, check for intercept, symmetry, vertical asymptotes, horizontal asymptotes, and slant asymptotes.

5. $g(x) = \dfrac{2 + x}{1 - x}$

6. $h(x) = \dfrac{x - 3}{x - 2}$

7. $f(x) = \dfrac{x}{x^2 + 1}$

8. $f(x) = \dfrac{2x}{x^2 + 4}$

9. $p(x) = \dfrac{x^2}{x^2 + 1}$

10. $s(x) = \dfrac{2x^2}{x^2 + 4}$

11. $h(x) = \dfrac{4}{(x - 1)^2}$

12. $g(x) = \dfrac{-2}{(x + 3)^2}$

13. $f(x) = \dfrac{-5}{x^2}$

14. $f(x) = \dfrac{4}{x}$

15. $y = \dfrac{x}{x^2 - 1}$

16. $y = \dfrac{2x}{x^2 - 4}$

17. $y = \dfrac{2x^2}{x^2 - 4}$

18. $y = \dfrac{1}{x + 3} + 2$

19. $f(x) = \dfrac{2x^3}{x^2 + 1}$

20. $g(x) = \dfrac{x^2 + 1}{x + 1}$

21. *Average Cost* A business has a cost of $C = 0.5x + 500$ for producing x units. The average cost per unit is

$$\overline{C} = \frac{C}{x} = \frac{0.5x + 500}{x}, \qquad 0 < x.$$

Determine the average cost per unit as x increases without bound. (Find the horizontal asymptote.)

22. *Average Cost* The cost of producing x units of a product is C and therefore the average cost per unit is

$$\overline{C} = \frac{C}{x} = \frac{100,000 + 0.9x}{x}, \qquad 0 < x.$$

Sketch a graph of the average cost function and find the average cost of producing $x = 1000$, $x = 10,000$, and $x = 100,000$ units.

23. *Seizure of Illegal Drugs* The cost in millions of dollars for the federal government to seize $p\%$ of an illegal drug as it enters the country is given by

$$C = \frac{528p}{100 - p}, \qquad 0 \le p < 100.$$

(a) Find the cost of seizing 25%.
(b) Find the cost of seizing 50%.
(c) Find the cost of seizing 75%.
(d) According to this model, would it be possible to seize 100% of the drug?

24. *Capillary Attraction* The rise of distilled water in tubes x inches in diameter is approximated by the model

$$y = \frac{0.80 - 0.54x}{(1 + 2.72x)^2}, \qquad 0 < x$$

where y is measured in inches (see figure). Approximate the diameter of the tube that will cause the water to rise 0.1 inch.

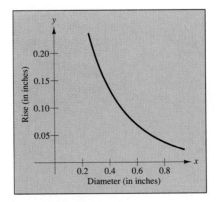

Figure for 24

25. *Photosynthesis* The amount y of CO_2 uptake in milligrams per square decimeter per hour at optimal temperatures and with the natural supply of CO_2 is approximated by the model

$$y = \frac{18.47x - 2.96}{0.23x + 1}, \qquad 0 < x$$

where x is the light intensity in watts per square meter. Sketch a graph of the function and determine the limiting amount of CO_2 uptake.

26. *Population of Fish* The Parks and Wildlife Commission introduces 80,000 fish into a large manmade lake. The population of the fish in thousands is given by

$$N = \frac{20(4 + 3t)}{1 + 0.05t}, \qquad 0 \le t$$

where t is the time in years.
(a) Find the population when t is 5, 10, and 25.
(b) What is the limiting number of fish in the lake as time increases?

In Exercises 27–34, write the partial fraction decomposition for the rational expression.

27. $\dfrac{4 - x}{x^2 + 6x + 8}$

28. $\dfrac{-x}{x^2 + 3x + 2}$

29. $\dfrac{x^2}{x^2 + 2x - 15}$

30. $\dfrac{9}{x^2 - 9}$

31. $\dfrac{x^2 + 2x}{x^3 - x^2 + x - 1}$

32. $\dfrac{4x - 2}{3(x - 1)^2}$

33. $\dfrac{3x^3 + 4x}{(x^2 + 1)^2}$

34. $\dfrac{4x^2}{(x - 1)(x^2 + 1)}$

In Exercises 35–46, identify the conic and sketch its graph.

35. $4x - y^2 = 0$
36. $8y + x^2 = 0$
37. $x^2 - 6x + 2y + 9 = 0$
38. $y^2 - 12y - 8x + 20 = 0$
39. $x^2 + y^2 - 2x - 4y + 5 = 0$ *Coefficient are the same*
40. $16x^2 + 16y^2 - 16x + 24y - 3 = 0$
41. $4x^2 + y^2 = 16$
42. $2x^2 + 6y^2 = 18$
43. $x^2 + 9y^2 + 10x - 18y + 25 = 0$
44. $4x^2 + y^2 - 16x + 15 = 0$
45. $5y^2 - 4x^2 = 20$
46. $x^2 - 9y^2 + 10x + 18y + 7 = 0$

In Exercises 47–50, find an equation of the specified parabola.

47. Vertex: $(4, 2)$; focus: $(4, 0)$
48. Vertex: $(2, 0)$; focus: $(0, 0)$
49. Vertex: $(0, 2)$; passes through point $(-1, 0)$; horizontal axis
50. Vertex: $(2, 2)$; directrix: $y = 0$

In Exercises 51–54, find an equation of the specified ellipse.

51. Vertices: $(-3, 0)$, $(7, 0)$; foci: $(0, 0)$, $(4, 0)$
52. Vertices: $(2, 0)$, $(2, 4)$; foci: $(2, 1)$, $(2, 3)$
53. Vertices: $(0, \pm6)$; passes through the point $(2, 2)$
54. Vertices: $(0, 1)$, $(4, 1)$; endpoints of the minor axis: $(2, 0)$, $(2, 2)$

In Exercises 55–58, find an equation of the specified hyperbola.

55. Vertices: $(0, \pm1)$; foci: $(0, \pm3)$
56. Vertices: $(2, 2)$, $(-2, 2)$; foci: $(4, 2)$, $(-4, 2)$
57. Foci: $(0, 0)$, $(8, 0)$; asymptotes: $y = \pm2(x - 4)$
58. Foci: $(3, \pm2)$; asymptotes: $y = \pm2(x - 3)$

59. *Satellite Antenna*   A cross section of a large parabolic antenna (see figure) is given by

$$y = \frac{x^2}{200}, \quad -100 \le x \le 100.$$

The receiving and transmitting equipment is positioned at the focus. Find the coordinates of the focus.

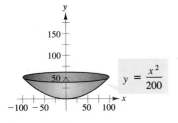

Figure for 59

60. *Semi-elliptical Archway*   A semi-elliptical archway is to be formed over the entrance to an estate. The arch is to be set on pillars that are 10 feet apart and is to have a height (atop the pillars) of 4 feet (see figure). Where should the foci be placed in order to sketch the elliptic arch?

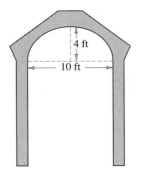

Figure for 60

C H A P T E R 6

O V E R V I E W

This chapter discusses two types of functions—exponential and logarithmic functions.

A quantity is said to grow or decay *exponentially* if its rate of growth is proportional to its size. Many real-life quantities grow or decay exponentially. For instance, the balance in a savings account *grows* exponentially (see Example 9 on page 366), and the amount of a radioactive material *decays* exponentially (see Example 10 on page 367).

In the chapter, you will study two uses of logarithmic functions. First, you will study real-life situations that can be modeled with logarithmic functions. Second, you will learn how to use the inverse relationship between exponential and logarithmic functions to solve exponential and logarithmic equations.

Exponential and Logarithmic Functions

6.1 Exponential Functions

Introduction / Graphs of Exponential Functions / The Natural Base *e* / Compound Interest / Other Applications

Introduction

Thus far in the text, we have dealt only with **algebraic functions,** which include polynomial functions and rational functions. In this chapter you will study two types of nonalgebraic functions—*exponential* functions and *logarithmic* functions. These functions are **transcendental functions.**

Exponential functions are widely used in describing economic and physical phenomena such as compound interest, population growth, memory retention, and decay of radioactive material. Exponential functions involve a *constant base* and a *variable exponent* such as

$$f(x) = 2^x \quad \text{or} \quad g(x) = 3^{-x}.$$

DEFINITION OF EXPONENTIAL FUNCTION

The **exponential function** f with base a is denoted by

$$f(x) = a^x$$

where $a > 0$, $a \neq 1$, and x is any real number.

359

REMARK We exclude the base $a = 1$ because it yields $f(x) = 1^x = 1$, which is a constant function, not an exponential function.

In Sections 1.2 and 1.3 you learned to evaluate a^x for integer and rational values of x. For example, we know that

$$8^3 = 8 \cdot 8 \cdot 8 = 512 \qquad \text{and} \qquad 8^{2/3} = (\sqrt[3]{8})^2 = (2)^2 = 4.$$

However, to evaluate 8^x for any real number x, we need to interpret forms with *irrational* exponents, such as $8^{\sqrt{2}}$ and 8^{π}. A technical definition of such forms is beyond the scope of this text. For our purposes, it is sufficient to think of

$$a^{\sqrt{2}} \qquad (\text{where } \sqrt{2} \approx 1.414214)$$

as that value which has the successively closer approximations

$$a^{1.4},\ a^{1.41},\ a^{1.414},\ a^{1.4142},\ a^{1.41421},\ a^{1.414214},\ \dots .$$

Consequently, we assume in this text that a^x exists for all real x and that the properties of exponents (Section 1.3) can be extended to cover exponential functions. For instance,

$$a^{-x} = \frac{1}{a^x} = \left(\frac{1}{a}\right)^x.$$

EXAMPLE 1 **Using a Calculator to Evaluate Exponential Expressions**

Number	*Keystrokes*	*Display*
$(1.085)^3$	1.085 $\boxed{y^x}$ $\boxed{3}$ $\boxed{=}$	1.277289
$12^{5/7}$	12 $\boxed{y^x}$ $\boxed{(}$ $\boxed{5}$ $\boxed{\div}$ $\boxed{7}$ $\boxed{)}$ $\boxed{=}$	5.899888
$2^{-\pi}$	2 $\boxed{y^x}$ π $\boxed{+/-}$ $\boxed{=}$	0.1133147

Graphs of Exponential Functions

EXAMPLE 2 **Graphs of $y = a^x$**

On the same coordinate plane, sketch the graphs of the following functions.

a. $f(x) = 2^x$ **b.** $g(x) = 4^x$

Solution

Table 6.1 lists some values for each function, and Figure 6.1 shows their graphs. Note that both graphs are increasing. Moreover, the graph of $g(x) = 4^x$ is increasing more rapidly than the graph of $f(x) = 2^x$.

TABLE 6.1

x	-2	-1	0	1	2	3
a. $f(x) = 2^x$	$\frac{1}{4}$	$\frac{1}{2}$	1	2	4	8
b. $g(x) = 4^x$	$\frac{1}{16}$	$\frac{1}{4}$	1	4	16	64

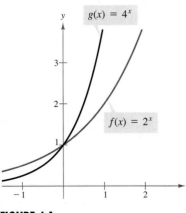

FIGURE 6.1

EXAMPLE 3 Graphs of $y = a^{-x}$

On the same coordinate plane, sketch the graphs of the following functions.

a. $F(x) = 2^{-x}$ **b.** $G(x) = 4^{-x}$

Solution

Table 6.2 lists some values for each function, and Figure 6.2 shows their graphs. Note that both graphs are decreasing. Moreover, the graph of $G(x) = 4^{-x}$ is decreasing more rapidly than the graph of $F(x) = 2^{-x}$.

TABLE 6.2

x	-3	-2	-1	0	1	2
a. $F(x) = 2^{-x}$	8	4	2	1	$\frac{1}{2}$	$\frac{1}{4}$
b. $G(x) = 4^{-x}$	64	16	4	1	$\frac{1}{4}$	$\frac{1}{16}$

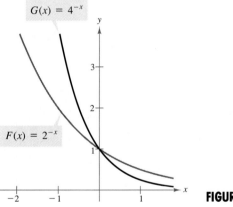

FIGURE 6.2

Comparing the functions in Examples 2 and 3, observe that

$$F(x) = 2^{-x} = f(-x) \qquad \text{and} \qquad G(x) = 4^{-x} = g(-x).$$

Consequently, the graph of F is a reflection (in the y-axis) of the graph of f. The graphs of G and g have the same relationship. This is verified by a comparison of the graphs in Figures 6.1 and 6.2.

     The graphs in Figures 6.1 and 6.2 are typical of the exponential functions a^x and a^{-x}. They have one y-intercept and one horizontal asymptote (the x-axis), and they are continuous. The basic characteristics of these exponential functions are summarized in Figure 6.3.

REMARK Since $2^{-x} = \left(\frac{1}{2}\right)^x$, Examples 2 and 3 show that the function $f(x) = a^x$ *increases* if $a > 1$ and *decreases* if $0 < a < 1$.

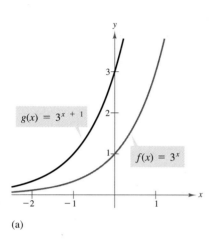

(a)

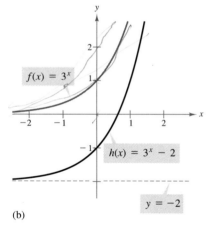

(b)

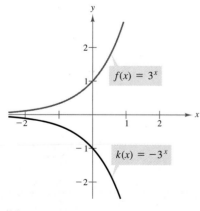

(c)

FIGURE 6.4

FIGURE 6.3

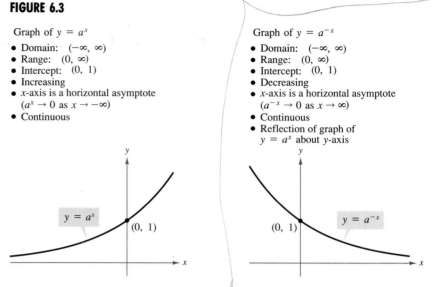

Graph of $y = a^x$
- Domain: $(-\infty, \infty)$
- Range: $(0, \infty)$
- Intercept: $(0, 1)$
- Increasing
- x-axis is a horizontal asymptote
  ($a^x \to 0$ as $x \to -\infty$)
- Continuous

Graph of $y = a^{-x}$
- Domain: $(-\infty, \infty)$
- Range: $(0, \infty)$
- Intercept: $(0, 1)$
- Decreasing
- x-axis is a horizontal asymptote
  ($a^{-x} \to 0$ as $x \to \infty$)
- Continuous
- Reflection of graph of
  $y = a^x$ about y-axis

Characteristics of the Exponential Functions a^x and $a^{-x} (a > 1)$

In the following example, we use the graph of a^x to sketch the graphs of functions of the form $f(x) = b \pm a^{x+c}$.

EXAMPLE 4   Sketching Graphs of Exponential Functions

Sketch the graph of each of the following.

a. $g(x) = 3^{x+1}$ **b.** $h(x) = 3^x - 2$ **c.** $k(x) = -3^x$

Solution

The graph of each of these three functions is similar to the graph of $f(x) = 3^x$, as shown in Figure 6.4.

a. Because $g(x) = 3^{x+1} = f(x + 1)$, the graph of g can be obtained by shifting the graph of f one unit to the left.

b. Because $h(x) = 3^x - 2 = f(x) - 2$, the graph of h can be obtained by shifting the graph of f down two units. The horizontal asymptote is $y = -2$.

c. Because $k(x) = -3^x = -f(x)$, the graph of k can be obtained by reflecting the graph of f in the x-axis. ◢

The Natural Base e

We used an unspecified base a to introduce exponential functions. It happens that in many applications the convenient choice for a base is the irrational number

$$e \approx 2.71828 \ldots$$

called the **natural base.** The function $f(x) = e^x$ is the **natural exponential function.** Its graph is shown in Figure 6.5. Be sure you see that for the exponential function $f(x) = e^x$, e is the constant $2.71828\ldots$, whereas x is the variable.

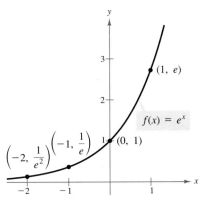

FIGURE 6.5

EXAMPLE 5 Evaluating the Natural Exponential Function

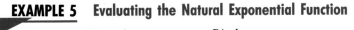

Number	Keystrokes	Display
$3e^2$	3 ⊠ 2 e^x =	22.1671683
e^{-1}	1 +/− e^x	0.3678794
$e^{0.12(4)}$	.12 ⊠ 4 = e^x	1.6160744

Some calculators have a natural exponential key e^x. On such a calculator you can evaluate e^2 by entering the sequence 2 e^x. Other calculators require the two-key sequence **INV** **ln x** to evaluate exponential functions. On such a calculator you can evaluate e^2 using the following sequence.

2 **INV** **ln x**   *Display: 7.3890561*

Similarly, to evaluate e^{-1}, enter the sequence

1 +/− **INV** **ln x** .   *Display: 0.3678794*

EXAMPLE 6 Sketching the Graph of a Natural Exponential Function

Sketch the graphs of the following natural exponential functions.

a. $f(x) = 2e^{0.24x}$ **b.** $g(x) = \dfrac{1}{2}e^{-0.58x}$

Solution

To sketch these two graphs, we use a calculator to plot several points on each graph, as shown in Table 6.3. Then we connect the points with a smooth curve, as shown in Figure 6.6. Note that the graph in part (a) is increasing whereas the graph in part (b) is decreasing.

TABLE 6.3

x	−3	−2	−1	0	1	2	3
a. $f(x) = 2e^{0.24x}$	0.974	1.238	1.573	2.000	2.542	3.232	4.109
b. $g(x) = \frac{1}{2}e^{-0.58x}$	2.849	1.595	0.893	0.500	0.280	0.157	0.088

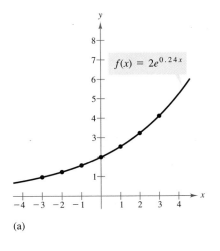

(a)

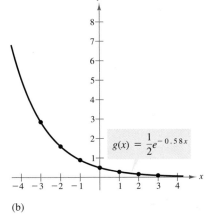

(b)

FIGURE 6.6

Example 7 gives an indication of how the irrational number e arises in applications. We will refer to this example when we develop the formula for continuous compounding of interest.

EXAMPLE 7 Approximation of the Number e

Evaluate the expression

$$\left(1 + \frac{1}{n}\right)^n$$

for several large values of n to see that the values approach $e \approx 2.71828$ as n increases without bound.

Solution

Using the keystroke sequence

n $\boxed{1/x}$ $\boxed{+}$ 1 $\boxed{=}$ $\boxed{y^x}$ n $\boxed{=}$

we obtain the values shown in the table.

n	10	100	1,000	10,000	100,000	1,000,000
$\left(1 + \frac{1}{n}\right)^n$	2.59374	2.70481	2.71692	2.71815	2.71827	2.71828

From this table, it seems reasonable to conclude that

$$\left(1 + \frac{1}{n}\right)^n \to e \quad \text{as} \quad n \to \infty.$$

Compound Interest

One of the most familiar examples of exponential growth is an investment earning *continuously compounded interest*. Suppose a principal P is invested at an annual percentage rate r, compounded once a year. If the interest is added to the principal at the end of the year, then the balance is

$$P_1 = P + Pr = P(1 + r).$$

This pattern of multiplying the previous principal by $1 + r$ is then repeated each successive year, as shown in Table 6.4.

TABLE 6.4

Time in years	Balance after each compounding
0	$P = P$
1	$P_1 = P(1 + r)$
2	$P_2 = P_1(1 + r) = P(1 + r)(1 + r) = P(1 + r)^2$
3	$P_3 = P_2(1 + r) = P(1 + r)^2(1 + r) = P(1 + r)^3$
$\vdots$	$\vdots$
n	$P_n = P(1 + r)^n$

To accommodate more frequent (quarterly, monthly, or daily) compounding of interest, we let n be the number of compoundings per year and t be the number of years. Then the rate per compounding is r/n and the account balance after t years is

$$A = P\left(1 + \frac{r}{n}\right)^{nt}. \qquad \textit{Amount with n compoundings per year}$$

If we let the number of compoundings, n, increase without bound, we approach **continuous compounding.** In the formula for n compoundings per year, let $m = n/r$. This produces

$$A = P\left(1 + \frac{r}{n}\right)^{nt} = P\left(1 + \frac{1}{m}\right)^{mrt} = P\left[\left(1 + \frac{1}{m}\right)^m\right]^{rt}.$$

As m increases without bound, we know from Example 7 that

$$\left(1 + \frac{1}{m}\right)^m \rightarrow e.$$

Hence, for continuous compounding, it follows that

$$P\left[\left(1 + \frac{1}{m}\right)^m\right]^{rt} \rightarrow P[e]^{rt}$$

and we write $A = Pe^{rt}$.

FORMULAS FOR COMPOUND INTEREST

After t years, the balance A in an account with principal P and annual percentage rate r (expressed as a decimal) is given by the following formulas.

1. For n compoundings per year: $A = P\left(1 + \dfrac{r}{n}\right)^{nt}$

2. For continuous compounding: $A = Pe^{rt}$

EXAMPLE 8   Finding the Balance for Compound Interest

A sum of $9,000 is invested at an annual percentage rate of 8.5%, compounded annually. Find the balance in the account after three years.

Solution

In this case, $P = 9,000$, $r = 8.5\% = 0.085$, $n = 1$, and $t = 3$. Using the formula

$$A = P\left(1 + \frac{r}{n}\right)^{nt}$$

we have

$$A = 9,000(1 + 0.085)^3 = 9,000(1.085)^3 \approx \$11,495.60.$$

EXAMPLE 9   Compounding n Times and Compounding Continuously

A total of $12,000 is invested at an annual percentage rate of 9%. Find the balance after five years if it is compounded

a. quarterly   **b.** continuously.

Solution

a. For quarterly compoundings, we have $n = 4$. Thus, in five years at 9%, the balance is

$$A = P\left(1 + \frac{r}{n}\right)^{nt} = 12,000\left(1 + \frac{0.09}{4}\right)^{4(5)} = \$18,726.11.$$

b. Compounding continuously, the balance is

$$A = Pe^{rt} = 12,000e^{0.09(5)} = \$18,819.75.$$

Note that continuous compounding yields

$$\$18,819.75 - \$18,726.11 = \$93.64$$

more than quarterly compounding.

Other Applications

You have already seen that exponential functions can be used as models for continuously compounded interest. Throughout this chapter you will also encounter several other types of applications that have exponential models.

EXAMPLE 10   An Application Involving Radioactive Decay

Let y represent the mass of a particular radioactive element whose half-life is 25 years. After t years, the mass (in grams) is given by

$$y = 10\left(\frac{1}{2}\right)^{t/25}.$$

a. What is the initial mass (when $t = 0$)?
b. How much of the initial mass is present after 80 years?

Solution

a. When $t = 0$, the mass is

$$y = 10\left(\frac{1}{2}\right)^0 = 10(1) = 10 \text{ grams}.$$

b. When $t = 80$, the mass is

$$y = 10\left(\frac{1}{2}\right)^{80/25} = 10(0.5)^{3.2} \approx 1.088 \text{ grams}.$$

The graph of this function is shown in Figure 6.7.

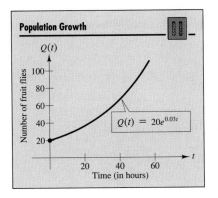

Radioactive Decay

$y = 10\left(\frac{1}{2}\right)^{t/25}$

FIGURE 6.7

EXAMPLE 11   Population Growth

The number of fruit flies in an experimental population after t hours is given by

$$Q(t) = 20e^{0.03t}, \qquad t \geq 0.$$

a. Find the initial number of fruit flies in the population.
b. How large is the population of fruit flies after 72 hours?
c. Sketch the graph of Q.

Solution

a. To find the initial population, we evaluate $Q(t)$ at $t = 0$.

$$Q(0) = 20e^{0.03(0)} = 20e^0 = 20(1) = 20 \text{ flies}$$

b. After 72 hours, the population size is

$$Q(72) = 20e^{(0.03)(72)} = 20e^{2.16} \approx 173 \text{ flies}.$$

c. To sketch the graph of Q, we evaluate $Q(t)$ for several values of t (rounded to the nearest integer) and plot the corresponding points, as shown in Figure 6.8.

Population Growth

$Q(t) = 20e^{0.03t}$

FIGURE 6.8

REMARK Many animal populations have a growth pattern described by the function $Q(t) = ce^{kt}$ where c is the original population, $Q(t)$ is the population at time t, and k is a constant determined by the rate of growth. Informally, we can write this as $(Then) = (Now)(e^{kt})$.

t	0	5	10	20	40	60
$20e^{0.03t}$	20	23	27	36	66	121

DISCUSSION
PROBLEM
Exponential Growth

Consider the following sequences of numbers.

Sequence 1   2, 4, 6, 8, 10, 12, . . .

Sequence 2   2, 4, 8, 16, 32, 64, . . .

The first sequence is given by $f(n) = 2n$, $n = 1, 2, 3, 4, \ldots$. This type of growth is **linear growth.** The second sequence is given by $f(n) = 2^n$, $n = 1, 2, 3, 4, \ldots$. This type of growth is **exponential growth.** Which of the following sequences represents linear growth and which represents exponential growth? Can you find a linear function and an exponential function that represent the sequences?

1. 3, 6, 9, 12, 15, . . .
2. 3, 9, 27, 81, 243, . . .

We will say more about these two types of growth later. At that time, we will see that sequences that represent linear growth are called *arithmetic sequences* and sequences that represent exponential growth are called *geometric sequences.*

WARM UP

The following warm-up exercises involve skills that were covered in earlier sections. You will use these skills in the exercise set for this section.

In Exercises 1–10, use the properties of exponents to simplify the expressions.

1. $5^{2x}(5^{-x})$

2. $3^{-x}(3^{3x})$

3. $\dfrac{4^{5x}}{4^{2x}}$

4. $\dfrac{10^{2x}}{10^x}$

5. $(4^x)^2$

6. $(4^{2x})^5$

7. $\left(\dfrac{2^x}{3^x}\right)^{-1}$

8. $(4^{6x})^{1/2}$

9. $(2^{3x})^{-1/3}$

10. $(16^x)^{1/4}$

EXERCISES for Section 6.1

In Exercises 1–14, use a calculator to evaluate the given quantity. Round your answers to three decimal places.

1. $(3.4)^{5.6}$

2. $(1.005)^{400}$

3. $1000(1.06)^{-5}$

4. $5000(2^{-1.5})$

5. $\sqrt[4]{763}$

6. $\sqrt[3]{4395}$

7. $8^{2\pi}$

8. $5^{-\pi}$

9. $100^{\sqrt{2}}$

10. $0.6^{\sqrt{3}}$

11. e^2

12. $e^{1/2}$

13. $e^{-3/4}$

14. $e^{3.2}$

In Exercises 15–22, match the exponential function with its graph. [The graphs are labeled (a), (b), (c), (d), (e), (f), (g), and (h).]

15. $f(x) = 3^x$

16. $f(x) = -3^x$

17. $f(x) = 3^{-x}$

18. $f(x) = -3^{-x}$

19. $f(x) = 3^x - 4$

20. $f(x) = 3^x + 1$

21. $f(x) = -3^{x-2}$

22. $f(x) = 3^{x-2}$

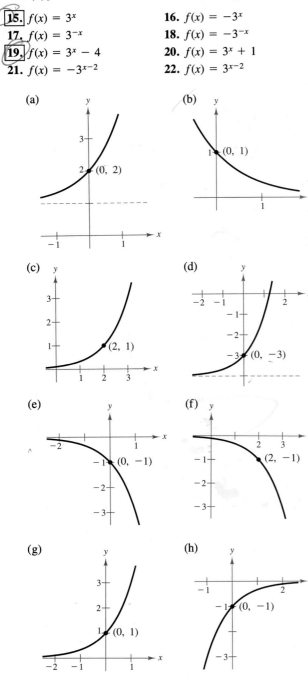

In Exercises 23–40, sketch the graph of the given exponential function.

23. $g(x) = 5^x$

24. $f(x) = \left(\frac{3}{2}\right)^x$

25. $f(x) = \left(\frac{1}{5}\right)^x = 5^{-x}$

26. $h(x) = \left(\frac{3}{2}\right)^{-x}$

27. $h(x) = 5^{x-2}$

28. $g(x) = \left(\frac{3}{2}\right)^{x+2}$

29. $g(x) = 5^{-x} - 3$

30. $f(x) = \left(\frac{3}{2}\right)^{-x} + 2$

31. $f(x) = 3^{x-2} + 1$

32. $f(x) = 4^{x+1} - 2$

33. $y = -e^{-x}$

34. $y = e^{x/2}$

35. $s(t) = 2e^{0.12t}$

36. $s(t) = 3e^{-0.2t}$

37. $f(x) = e^{2x}$

38. $h(x) = e^{x-2}$

39. $g(x) = 1 + e^{-x}$

40. $N(t) = 1000e^{-0.2t}$

In Exercises 41–44, complete the table to determine the balance A for P dollars invested at rate r for t years and compounded n times per year.

n	1	2	4	12	365	Continuous compounding
A						

41. $P = \$2500,\ r = 12\%,\ t = 10$ years

42. $P = \$1000,\ r = 10\%,\ t = 10$ years

43. $P = \$2500,\ r = 12\%,\ t = 20$ years

44. $P = \$1000,\ r = 10\%,\ t = 40$ years

In Exercises 45–48, complete the following table to determine the amount of money P that should be invested at rate r to produce a final balance of $\$100,000$ in t years.

t	1	10	20	30	40	50
P						

45. $r = 9\%$, compounded continuously

46. $r = 12\%$, compounded continuously

47. $r = 10\%$, compounded monthly

48. $r = 7\%$, compounded daily

49. *Trust Fund* On the day of your grandchild's birth, you deposited $\$25,000$ in a trust fund that pays 8.75% interest, compounded continuously. Determine the balance in this account on your grandchild's 25th birthday.

50. *Trust Fund*   Suppose you deposited $5000 in a trust fund that pays 7.5% interest, compounded continuously. In the trust fund, you specify that the balance will be given to the college from which you graduated after the money has earned interest for 50 years. How much will your college receive after 50 years?

51. *Demand Function*   The demand equation for a certain product is given by

$$p = 500 - 0.5e^{0.004x}.$$

Find the price p for a demand of (a) $x = 1000$ units and (b) $x = 1500$ units.

52. *Demand Function*   The demand equation for a certain product is given by

$$p = 5000\left(1 - \frac{4}{4 + e^{-0.002x}}\right).$$

Find the price p for a demand of (a) $x = 100$ units and (b) $x = 500$ units.

53. *Bacteria Growth*   A certain type of bacteria increases according to the model

$$P(t) = 100e^{0.2197t}$$

where t is the time in hours. Find (a) $P(0)$, (b) $P(5)$, and (c) $P(10)$.

54. *Population Growth*   The population of a town increases according to the model

$$P(t) = 2500e^{0.0293t}$$

where t is the time in years, with $t = 0$ corresponding to 1990. Use the model to approximate the population in (a) 1995, (b) 2000, and (c) 2010.

55. *Radioactive Decay*   Let Q represent the mass of radium (Ra^{226}) whose half-life is 1620 years. The quantity of radium present after t years is given by

$$Q = 25\left(\frac{1}{2}\right)^{t/1620}.$$

(a) Determine the initial quantity (when $t = 0$).
(b) Determine the quantity present after 1000 years.
(c) Sketch the graph of this function over the interval $t = 0$ to $t = 5000$.

56. *Radioactive Decay*   Let Q represent the mass of carbon 14 (C^{14}) whose half-life is 5,730 years. The quantity of carbon 14 present after t years is given by

$$Q = 10\left(\frac{1}{2}\right)^{t/5730}.$$

(a) Determine the initial quantity (when $t = 0$).
(b) Determine the quantity present after 2,000 years.
(c) Sketch the graph of this function over the interval $t = 0$ to $t = 10,000$.

57. *Forest Defoliation*   To estimate the amount of defoliation caused by the gypsy moth during a given year, a forester counts the number of egg masses on $\frac{1}{40}$ of an acre the preceding fall. The percentage of defoliation y is approximated by

$$y = \frac{300}{3 + 17e^{-1.57x}}$$

where x is the number of egg masses in thousands. Estimate the percentage of defoliation if 2000 egg masses are counted. (*Source:* Department of Environmental Resources)

58. *Inflation*   If the annual rate of inflation averages 5% over the next 10 years, then the approximate cost C of goods or services during any year in that decade will be given by

$$C(t) = P(1.05)^t$$

where t is the time in years and P is the present cost. If the price of an oil change for your car is presently $19.95, estimate the price 10 years from now.

59. *Depreciation*   After t years, the value of a car that cost you $20,000 is given by

$$V(t) = 20,000\left(\frac{3}{4}\right)^t.$$

Sketch a graph of the function and determine the value of the car two years after it was purchased.

60. Create two exponential functions of the form $f(x) = C2^{kx}$ with y-intercepts $(0, 10)$. (The answer is not unique.)

61. Given the exponential function $f(x) = a^x$, show that
(a) $f(u + v) = f(u) \cdot f(v)$, and (b) $f(2x) = [f(x)]^2$.

A Graphical Approach to Compound Interest

A graphing utility can be used to investigate the rate of growth of different types of compound interest.

EXAMPLE 1   Comparing Balances

You are depositing $1000 into a savings account. Which of the following will produce a larger balance?

a. 6% annual percentage rate, compounded annually
b. 6% annual percentage rate, compounded continuously
c. 6.25% annual percentage rate, compounded quarterly

Solution

Option (b) is better than option (a) because, for a given annual percentage rate, continuous compounding yields a larger balance than compounding n times per year. Distinguishing between the second and third options is not as straightforward—the higher percentage rate favors option (c), but the "more frequent" compounding favors option (b). One way to compare all three options is to sketch their graphs on the same display screen.

Option (a)	Option (b)	Option (c)
$A = 1000(1 + 0.06)^t$	$A = 1000e^{0.06t}$	$A = 1000\left(1 + \frac{0.0625}{4}\right)^{4t}$

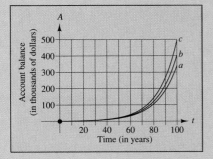

The graphs of all three functions are shown at the left. On the graph, the t-values vary from 0 years through 100 years. Note that for the first 50 years, there is little difference in the graphs. Between 50 and 100 years, however, the balances obtained with the three options begin to differ significantly. At the end of 100 years, the balances are (a) $339,302, (b) $403,429, and (c) $493,575.

Thus, your conclusion could be that for any length of time, option (c) is better than option (b), and option (b) is better than option (a). Moreover, as the time increases, the difference between the three options increases. ◢

To help distinguish between different interest rates and different types of compounding, banks use the concept of *effective yield*. The **effective yield** of a savings plan is the percentage increase in the balance at the end of *one* year. For instance, in Example 1 the balances at the end of one year are (a) $1060.00, (b) $1061.84, and (c) $1063.98. The effective yields are below.

Effective Yield (a)	*Effective Yield* (b)	*Effective Yield* (c)
6.000%	6.184%	6.398%

Because option (c) has the largest effective yield, it is the best option and will yield the highest balance.

371

If you were to create a retirement plan with a regular savings account, the income tax on the interest would be due each year. With a *tax deferred* retirement plan, the interest is allowed to build without being taxed until the account reaches its maturity.

EXAMPLE 2   To Defer or Not to Defer

You deposit $25,000 in an account to accrue interest for 40 years. The account pays 8% compounded annually. Assume that the income tax on the earned interest is 30%. Which of the following plans produces a larger balance after all income tax is paid?

a. *Deferred*   The income tax on the interest that is earned is paid in one lump sum at the end of 40 years.

b. *Not Deferred*   The income tax on the interest that is earned each year is paid at the end of that year.

Solution

a. The untaxed balance at the end of 40 years is

$$A = 25,000(1 + 0.08)^{40} = \$543,113.04.$$

The income tax due is $0.3(518,113.04) = \$155,433.91$, so you are left with a balance of $387,679.13.

b. You can reason that only 70% of the earned interest will remain in the account each year. The taxed balance at the end of t years is $A = 25,000 [1 + 0.08(0.7)]^t$, which implies that the balance at the end of 40 years is

$$A = 25,000[1 + 0.08(0.7)]^{40} = \$221,053.16.$$

Thus, the tax deferred plan will produce a significantly larger balance at the end of 40 years. The balances of the two plans are compared at left.

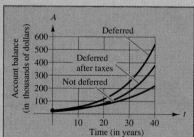
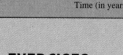

EXERCISES

(*See also: Exercises 41–50, Section 6.1*)

1. *Comparing Savings Plans*   Which would produce a larger balance: an annual percentage rate of 8.05% compounded monthly, or an annual percentage rate of 8% compounded continuously? Explain.

2. *Exploration*   You deposit $1000 into each of two savings accounts. The interest for the accounts is paid according to the two options described in Exercise 1. How long would it take for the balance in one of the accounts to exceed the balance in the other account by $100? By $100,000?

3. No income tax is due on the interest earned in some types of investments. You deposit $25,000 into an account. Which plan produces a larger balance in 40 years (after taxes)?

(a) *Tax-free*   The account pays 5% compounded annually. There is no income tax due on the earned interest.

(b) *Tax-deferred*   The account pays 7% compounded annually. At maturity, the earned interest is taxable at a rate of 40%.

372

✳ inverseof an exponetal function

6.2 Logarithmic Functions

Introduction / The Common Logarithmic Function / Graphs of Logarithmic Functions / The Natural Logarithmic Function / Applications

Introduction

In Section 3.7 we discussed the concept of the inverse of a function. If a function has the property that no horizontal line intersects the graph of a function more than once, then the function must have an inverse. In Section 6.1, every function of the form $f(x) = a^x$ passes the "horizontal line test," and therefore must have an inverse. This inverse function is the **logarithmic function with base a** (see Figure 6.9).

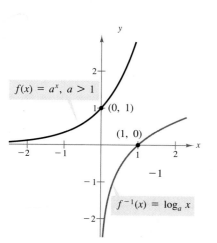

$f(x) = a^x, a > 1$

$(0, 1)$

$(1, 0)$

$f^{-1}(x) = \log_a x$

Inverse Functions
Domain of $\log_a x$ is Range of a^x.

FIGURE 6.9

DEFINITION OF LOGARITHMIC FUNCTION

For $x > 0$ and $0 < a \neq 1$,

$$y = \log_a x \text{ if and only if } a^y = x.$$

The function given by

$$f(x) = \log_a x$$

is the **logarithmic function with base a.**

REMARK The equations $y = \log_a x$ and $a^y = x$ are equivalent. The first equation is in logarithmic form and the second is in exponential form.

When evaluating logarithms, remember that *a logarithm is an exponent*. This means that $\log_a x$ is the exponent to which a must be raised to obtain x. For instance, $\log_2 8 = 3$ because 2 must be raised to the third power to obtain 8. That is,

$$3 = \log_2 8 \quad \text{if and only if} \quad 2^3 = 8.$$

Base

Logarithm is an exponent

EXAMPLE 1 Evaluating Logarithms

a. $\log_2 32 = 5$ because $2^5 = 32$.

b. $\log_3 27 = 3$ because $3^3 = 27$.

c. $\log_4 2 = \dfrac{1}{2}$ because $4^{1/2} = \sqrt{4} = 2$.

d. $\log_{10} \dfrac{1}{100} = -2$ because $10^{-2} = \dfrac{1}{10^2} = \dfrac{1}{100}$.

e. $\log_3 1 = 0$ because $3^0 = 1$.

f. $\log_2 2 = 1$ because $2^1 = 2$.

The Common Logarithmic Function

The logarithmic function with base 10 is the **common logarithmic function.** On most calculators, this function is denoted by $\boxed{\text{log}}$. You can tell whether this key denotes base 10 by entering 10 $\boxed{\text{log}}$. The display should be 1. The common logarithmic function is often written as $\log x$, without denoting the base 10.

EXAMPLE 2 Evaluating Logarithms on a Calculator

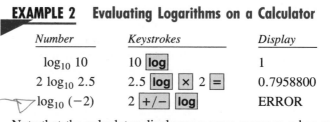

Number	Keystrokes	Display
$\log_{10} 10$	10 $\boxed{\text{log}}$	1
$2 \log_{10} 2.5$	2.5 $\boxed{\text{log}}$ $\boxed{\times}$ 2 $\boxed{=}$	0.7958800
$\log_{10}(-2)$	2 $\boxed{+/-}$ $\boxed{\text{log}}$	ERROR

Note that the calculator displays an error message when you try to evaluate $\log_{10}(-2)$. The reason for this is that the domain of every logarithmic function is the set of *positive real numbers*.

The following properties follow directly from the definition of the logarithmic function with base a.

PROPERTIES OF LOGARITHMS

1. $\log_a 1 = 0$ because 0 is the power to which a must be raised to obtain 1.
2. $\log_a a = 1$ because 1 is the power to which a must be raised to obtain a.
3. $\log_a a^x = x$ because x is the power to which a must be raised to obtain a^x.

Graphs of Logarithmic Functions

To sketch the graph of $y = \log_a x$, we can use the fact that the graphs of inverse functions are reflections of each other in the line $y = x$.

EXAMPLE 3 Graphs of Exponential and Logarithmic Functions

On the same coordinate plane, sketch the graphs of the following functions.

a. $f(x) = 2^x$ **b.** $g(x) = \log_2 x$

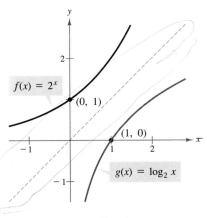

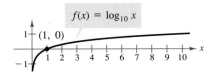

Inverse Functions

FIGURE 6.10

FIGURE 6.11

Solution

a. For $f(x) = 2^x$, we make a table of values.

x	-2	-1	0	1	2	3
$f(x) = 2^x$	$\frac{1}{4}$	$\frac{1}{2}$	1	2	4	8

By plotting these points and connecting them with a smooth curve, we have the graph shown in Figure 6.10.

b. Since $g(x) = \log_2 x$ is the inverse of $f(x) = 2^x$, the graph of g is obtained by reflecting the graph of f in the line $y = x$, as shown in Figure 6.10.

EXAMPLE 4   Sketching the Graph of a Logarithmic Function

Sketch the graph of the logarithmic function $f(x) = \log_{10} x$.

Solution

We begin by making a table of values. Note that some of the values can be obtained without a calculator, while others require a calculator. We plot the corresponding points and sketch the graph in Figure 6.11.

x	Without a calculator				With a calculator		
	$\frac{1}{100}$	$\frac{1}{10}$	1	10	2	5	8
$\log_{10} x$	-2	-1	0	1	0.301	0.699	0.903

The nature of the graph in Figure 6.11 is typical of functions of the form $f(x) = \log_a x$, $a > 1$. They have one x-intercept and one vertical asymptote, and their domains are all positive numbers, $(0, \infty)$. We summarize the basic characteristics of logarithmic graphs in Figure 6.12.

REMARK   In Figure 6.12, note that the vertical asymptote occurs at $x = 0$, where $\log_a x$ is *undefined*.

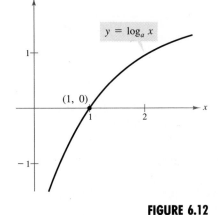

Graph of $y = \log_a x$, $a > 1$
- Domain: $(0, \infty)$
- Range: $(-\infty, \infty)$
- Intercept: $(1, 0)$
- Increasing

- y-axis is a vertical asymptote
  ($\log_a x \to -\infty$ as $x \to 0^+$)
- Continuous
- Reflection of graph of $y = a^x$
  about the line $y = x$

FIGURE 6.12

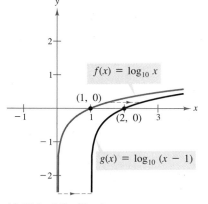

(a) Right shift of 1 unit
   Vertical asymptote is $x = 1$.

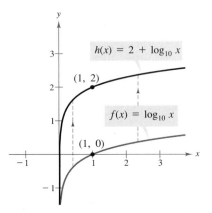

(b) Upward shift of 2 units
   Vertical asymptote remains $x = 0$.

FIGURE 6.13

In the following example we use the graph of $\log_a x$ to sketch the graphs of functions of the form $y = b \pm \log_a(x + c)$. The function $f(x) = \log_a(bx + c)$ has a domain which consists of all x such that $bx + c > 0$. The vertical asymptote occurs when $bx + c = 0$, and the x-intercept occurs when $bx + c = 1$.

EXAMPLE 5   Sketching the Graphs of Logarithmic Functions

Sketch the graphs of the following functions.

a. $g(x) = \log_{10}(x - 1)$ **b.** $h(x) = 2 + \log_{10} x$

Solution

The graph of each of these functions is similar to the graph of $f(x) = \log_{10} x$, as shown in Figure 6.13.

a. Because $g(x) = \log_{10}(x - 1) = f(x - 1)$, the graph of g can be obtained by shifting the graph of f one unit to the right.
b. Because $h(x) = 2 + \log_{10} x = 2 + f(x)$, the graph of h can be obtained by shifting the graph of f two units up.

The Natural Logarithmic Function

As with exponential functions, the most widely used base for logarithmic functions is the number e. The logarithmic function with base e is the **natural logarithmic function,** denoted by the special symbol ln x, read as "el en of x."

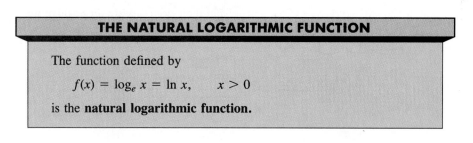

THE NATURAL LOGARITHMIC FUNCTION

The function defined by
$$f(x) = \log_e x = \ln x, \qquad x > 0$$
is the **natural logarithmic function.**

The three properties of logarithms listed at the beginning of this section are also valid for natural logarithms.

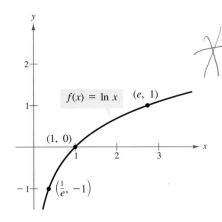

FIGURE 6.14

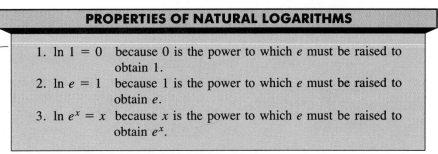

PROPERTIES OF NATURAL LOGARITHMS

1. $\ln 1 = 0$ because 0 is the power to which e must be raised to obtain 1.
2. $\ln e = 1$ because 1 is the power to which e must be raised to obtain e.
3. $\ln e^x = x$ because x is the power to which e must be raised to obtain e^x.

The graph of the natural logarithmic function is shown in Figure 6.14.

EXAMPLE 6   Evaluating the Natural Logarithmic Function

a. $\ln \dfrac{1}{e} = \ln e^{-1} = -1$ *Property 3*

b. $\ln e^2 = 2$ *Property 3*

On most calculators, the natural logarithm is denoted by $\boxed{\ln x}$.

EXAMPLE 7   Evaluating the Natural Logarithmic Function

Number	Calculator Steps	Display
$\ln 2$	2 $\boxed{\ln x}$	0.6931472
$\ln 0.3$	.3 $\boxed{\ln x}$	-1.2039728
$\ln(-1)$	1 $\boxed{+/-}$ $\boxed{\ln x}$	ERROR

Be sure you see that $\ln(-1)$ gives an error. This occurs because the domain of $\ln x$ is the set of positive real numbers. (See Figure 6.14.) Hence, $\ln(-1)$ is undefined.

EXAMPLE 8   Finding the Domain of Logarithmic Functions

Find the domain of the following functions.

a. $f(x) = \log_3(x - 2)$ **b.** $g(x) = \ln(2 - x)$
c. $h(x) = \log_{10}(x^2 - 1)$

Solution

a. Because $\log_3(x - 2)$ is defined only if $x - 2 > 0$, it follows that the domain of f is $(2, \infty)$.
b. Because $\ln(2 - x)$ is defined only if $2 - x > 0$, it follows that the domain of g is $(-\infty, 2)$. The graph of g is shown in Figure 6.15.
c. Because $\log_{10}(x^2 - 1)$ is defined only if $x^2 - 1 > 0$, it follows by the methods of Section 2.8 that the domain of h is all real numbers in the interval $(-\infty, -1)$ or the interval $(1, \infty)$.

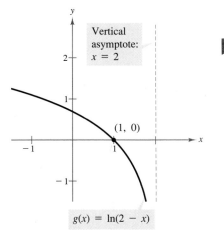

FIGURE 6.15

Applications

EXAMPLE 9   Human Memory Model

Students participating in a psychological experiment attended several lectures on a subject. Every month for a year after that, the students were tested to see how much of the material they remembered. The average scores for the group were given by the *human memory model*

$$f(t) = 75 - 6 \ln(t + 1), \qquad 0 \le t \le 12$$

where t is the time in months.

a. What was the average score on the original ($t = 0$) exam?
b. What was the average score at the end of $t = 2$ months?
c. What was the average score at the end of $t = 6$ months?
d. Sketch the graph of f.

Solution

a. The original average score was

$$f(0) = 75 - 6 \ln(0 + 1) = 75 - 6(0) = 75.$$

b. After two months, the average score was

$$f(2) = 75 - 6 \ln 3 \approx 75 - 6(1.0986) \approx 68.4.$$

c. After six months, the average score was

$$f(6) = 75 - 6 \ln 7 \approx 75 - 6(1.9459) \approx 63.3.$$

d. Several points are shown in the following table, and the graph of f is shown in Figure 6.16.

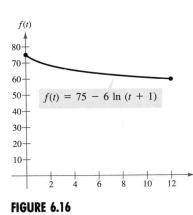

$f(t)$

$f(t) = 75 - 6 \ln (t + 1)$

FIGURE 6.16

t	0	1	2	6	8	12
$f(t)$	75	70.8	68.4	63.3	61.8	59.6

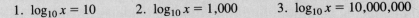

DISCUSSION

PROBLEM

The Graph of a Logarithmic Function

In the summary of the characteristics of the logarithmic function $f(x) = \log_a x$, we stated that the range of the function is $(-\infty, \infty)$. From this, we can conclude that the value of $\log_a x$ can be made as large as we want. Can you find values of x that satisfy the following equations?

1. $\log_{10} x = 10$ 2. $\log_{10} x = 1,000$ 3. $\log_{10} x = 10,000,000$

WARM UP

The following warm-up exercises involve skills that were covered in earlier sections. You will use these skills in the exercise set for this section.

In Exercises 1–4, solve for x.

1. $2^x = 8$ **2.** $4^x = 1$
3. $10^x = 0.1$ **4.** $e^x = e$

In Exercises 5 and 6, evaluate the given expressions. Round to three decimal places.

5. e^2 **6.** e^{-1}

In Exercises 7–10, describe how the graph of g is related to the graph of f.

7. $g(x) = f(x + 2)$ **8.** $g(x) = -f(x)$
9. $g(x) = -1 + f(x)$ **10.** $g(x) = f(-x)$

EXERCISES for Section 6.2

In Exercises 1–16, evaluate the given expression without using a calculator.

1. $\log_2 16$ **2.** $\log_4 64$
3. $\log_5\left(\frac{1}{25}\right)$ **4.** $\log_2\left(\frac{1}{8}\right)$
5. $\log_{16} 4$ **6.** $\log_{27} 9$
7. $\log_7 1$ **8.** $\log_{10} 1000$
9. $\log_{10} 0.01$ **10.** $\log_{10} 10$
11. $\ln e^3$ **12.** $\ln \frac{1}{e}$
13. $\ln e^{-2}$ **14.** $\ln 1$
15. $\log_a a^2$ **16.** $\log_a \frac{1}{a}$

In Exercises 17–26, use the definition of a logarithm to write the given equation in logarithmic form. For instance, the logarithmic form of $2^3 = 8$ is $\log_2 8 = 3$.

17. $5^3 = 125$ **18.** $8^2 = 64$
19. $81^{1/4} = 3$ **20.** $9^{3/2} = 27$
21. $6^{-2} = \frac{1}{36}$ **22.** $10^{-3} = 0.001$
23. $e^3 = 20.0855 \ldots$ **24.** $e^0 = 1$
25. $e^x = 4$ **26.** $u^v = w$

In Exercises 27–34, use a calculator to evaluate the logarithm. Round to three decimal places.

27. $\log_{10} 345$
28. $\log_{10}\left(\frac{4}{5}\right)$
29. $\log_{10} (0.48)$
30. $\log_{10} 12.5$
31. $\ln 18.42$
32. $\ln \sqrt{42}$
33. $\ln(1 + \sqrt{3})$
34. $\ln(\sqrt{5} - 2)$

In Exercises 35–38, sketch the graphs of f and g on the same coordinate plane to demonstrate that one is the inverse of the other.

35. $f(x) = 3^x$, $g(x) = \log_3 x$
36. $f(x) = 5^x$, $g(x) = \log_5 x$
37. $f(x) = e^x$, $g(x) = \ln x$
38. $f(x) = 10^x$, $g(x) = \log_{10} x$

In Exercises 39–44, use the graph of $y = \ln x$ to match the given function to its graph. [The graphs are labeled (a), (b), (c), (d), (e), and (f).]

39. $f(x) = \ln x + 2$

40. $f(x) = -\ln x$

41. $f(x) = -\ln(x + 2)$

42. $f(x) = \ln(x - 1)$

43. $f(x) = \ln(1 - x)$

44. $f(x) = -\ln(-x)$

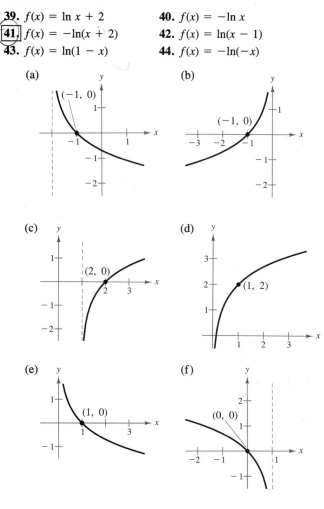

(a)

(b)

(c)

(d)

(e)

(f)

In Exercises 45–56, find the domain, vertical asymptote, and x-intercept of the logarithmic function and sketch its graph.

45. $f(x) = \log_4 x$

46. $g(x) = \log_6 x$

47. $h(x) = \log_4(x - 3)$

48. $f(x) = -\log_6(x + 2)$

49. $y = -\log_3 x + 2$

50. $y = \log_5(x - 1) + 4$

51. $y = \log_{10}\left(\dfrac{x}{5}\right)$

52. $y = \log_{10}(-x)$

53. $f(x) = \ln(x - 2)$

54. $h(x) = \ln(x + 1)$

55. $g(x) = \ln(-x)$

56. $f(x) = \ln(3 - x)$

57. *Human Memory Model* Students in a mathematics class were given an exam and then tested monthly with an equivalent exam. The average score for the class was given by the human memory model

$$f(t) = 80 - 17 \log_{10}(t + 1), \qquad 0 \le t \le 12$$

where t is the time in months.
(a) What was the average score on the original exam ($t = 0$)?
(b) What was the average score after 4 months?
(c) What was the average score after 10 months?

58. *Population Growth* The population of a town will double in

$$t = \frac{10 \ln 2}{\ln 67 - \ln 50}$$

years. Find t.

59. *World Population Growth* The time in years for the world population to double if it is increasing at a continuous rate of r is given by

$$t = \frac{\ln 2}{r}.$$

Complete the table.

r	0.005	0.010	0.015	0.020	0.025	0.030
t						

60. *Investment Time* A principal P invested at $9\frac{1}{2}\%$ and compounded continuously increases to an amount K times the original principal after t years, where t is given by

$$t = \frac{\ln K}{0.095}.$$

(a) Complete the table.

K	1	2	4	6	8	10	12
t							

(b) Use the table in part (a) to graph this function.

Ventilation Rates In Exercises 61 and 62, use the model

$$y = 80.4 - 11 \ln x$$

which approximates the minimum required ventilation rate in terms of the air space per child in a public school classroom. In the model, x is the air space per child in cubic feet and y is the ventilation rate in cubic feet per minute (see figure).

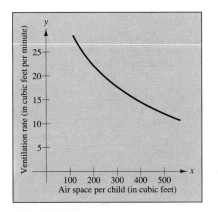

Figure for 61 and 62

61. Use the model to approximate the required ventilation rate if there are 300 cubic feet of air space per child.

62. A classroom is designed for 30 students, and the air-conditioning system in the room has the capacity to move 450 cubic feet of air per minute.
(a) Determine the ventilation rate per child assuming the room is filled to capacity.
(b) Use the figure to estimate the air space required per child.
(c) Determine the minimum number of square feet of floor space required for the room if the ceiling height is 30 feet.

Monthly Payment In Exercises 63–66, use the model

$$t = \frac{5.315}{-6.7968 + \ln x}, \qquad 1000 < x$$

which approximates the length of a home mortgage of $120,000 at 10% in terms of the monthly payment. In the model, t is the length of the mortgage in years and x is the monthly payment in dollars (see figure).

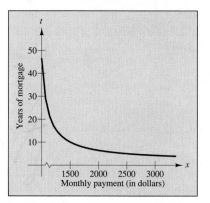

Figure for 63–66

63. Use the model to approximate the length of a home mortgage (for $120,000 at 10%) that has a monthly payment of $1167.41.

64. Use the model to approximate the length of a home mortgage (for $120,000 at 10%) that has a monthly payment of $1068.45.

65. Approximate the total amount paid over the term of a mortgage with a monthly payment of $1167.41.

66. Approximate the total amount paid over the term of a mortgage with a monthly payment of $1068.45.

67. *Work* The work (in foot-pounds) done in compressing an initial volume of 9 cubic feet at a pressure of 15 pounds per square inch to a volume of 3 cubic feet is

$$W = 19,440(\ln 9 - \ln 3).$$

Find W.

68. *Sound Intensity* The relationship between the number of decibels β and the intensity of a sound I in watts per centimeter squared is given by

$$\beta = 10 \log_{10}\left(\frac{I}{10^{-16}}\right).$$

Determine the number of decibels of a sound with an intensity of 10^{-4} watts per centimeter squared.

69. (a) Use a calculator to complete the table for the function

$$f(x) = \frac{\ln x}{x}.$$

x	1	5	10	10^2	10^4	10^6
$f(x)$						

(b) Use the table in part (a) to determine what $f(x)$ approaches as x increases without bound.

70. Answer the following for the function $f(x) = \log_{10} x$. Do not use a calculator.
(a) What is the domain of f?
(b) Find f^{-1}.
(c) If x is a real number between 1000 and 10,000, then determine the interval in which $f(x)$ will be found.
(d) Determine the interval in which x will be found if $f(x)$ is negative.
(e) If $f(x)$ is increased by one unit, then x must have been increased by what factor?
(f) If $f(x_1) = 3n$ and $f(x_2) = n$, then find the ratio of x_1 to x_2.

6.3 Properties of Logarithms

Change of Base / Properties of Logarithms / Rewriting Logarithmic Expressions

Change of Base

Most calculators have only two types of "log keys," one for common logarithms (base 10) and one for natural logarithms (base e). Although common logs and natural logs are the most frequently used, we occasionally need to evaluate logarithms to other bases. To do this, the *change of base formula* is useful. (This formula is derived in Example 10 in Section 6.4.)

> ### CHANGE OF BASE FORMULA
>
> Let a, b, and x be positive real numbers such that $a \neq 1$ and $b \neq 1$. Then $\log_a x$ is given by
>
> $$\log_a x = \frac{\log_b x}{\log_b a}.$$

REMARK   One way to look at the change of base formula is that logarithms to base a are simply *constant multiples* of logarithms to base b. The constant multiplier is $1/(\log_b a)$.

EXAMPLE 1   Changing Bases

Use *common* logarithms to evaluate the following.

a. $\log_4 30$ **b.** $\log_2 14$

Solution

a. Using the change of base formula with $a = 4$, $b = 10$, and $x = 30$, we convert to common logarithms and obtain

$$\log_4 30 = \frac{\log_{10} 30}{\log_{10} 4} \approx \frac{1.47712}{0.60206} \approx 2.4534.$$

b. Using the change of base formula with $a = 2$, $b = 10$, and $x = 14$, we convert to common logarithms and obtain

$$\log_2 14 = \frac{\log_{10} 14}{\log_{10} 2} \approx \frac{1.14613}{0.30103} \approx 3.8074.$$

EXAMPLE 2   Changing Bases

Use *natural* logarithms to evaluate the following.

a. $\log_4 30$ **b.** $\log_2 14$

Solution

a. Using the change of base formula with $a = 4$, $b = e$, and $x = 30$, we convert to natural logarithms and obtain

$$\log_4 30 = \frac{\ln 30}{\ln 4} \approx \frac{3.40120}{1.38629} \approx 2.4534.$$

b. Using the change of base formula with $a = 2$, $b = e$, and $x = 14$, we convert to natural logarithms and obtain

$$\log_2 14 = \frac{\ln 14}{\ln 2} \approx \frac{2.63906}{0.693147} \approx 3.8074.$$

Note that the results agree with those obtained in Example 1 using common logarithms.

Properties of Logarithms

We know from the previous section that the logarithmic function with base a is the *inverse* of the exponential function with base a. Thus, it makes sense that the properties of exponents should have corresponding properties involving logarithms. For instance, the exponential property

$$a^0 = 1$$

has the corresponding logarithmic property

$$\log_a 1 = 0.$$

In this section we will show how to use the logarithmic properties that correspond to the following three exponential properties.

1. $a^n a^m = a^{n+m}$ 2. $\dfrac{a^n}{a^m} = a^{n-m}$ 3. $(a^n)^m = a^{nm}$

PROPERTIES OF LOGARITHMS

Let a be a positive number such that $a \neq 1$, and let n be a real number. If u and v are positive real numbers, then the following properties are true.

Base a Logarithm	*Natural Logarithm*
1. $\log_a(uv) = \log_a u + \log_a v$	1. $\ln(uv) = \ln u + \ln v$
2. $\log_a \dfrac{u}{v} = \log_a u - \log_a v$	2. $\ln \dfrac{u}{v} = \ln u - \ln v$
3. $\log_a u^n = n \log_a u$	3. $\ln u^n = n \ln u$

Proof

We give a proof of Property 1 and leave the other two proofs for you. To prove Property 1, let

$$x = \log_a u \quad \text{and} \quad y = \log_a v.$$

The corresponding exponential forms of these two equations are

$$a^x = u \quad \text{and} \quad a^y = v.$$

Multiplying u and v produces $uv = a^x a^y = a^{x+y}$. The corresponding logarithmic form of $uv = a^{x+y}$ is

$$\log_a(uv) = x + y.$$

Hence, $\log_a(uv) = \log_a u + \log_a v.$ ◢

REMARK There is no general property that can be used to rewrite $\log_a(u \pm v)$. Specifically

$$\log_a(x + y) \quad \text{DOES NOT EQUAL} \quad \log_a x + \log_a y.$$

("The log of a sum does *not* equal the sum of the logs.") ◢

EXAMPLE 3 Using Properties of Logarithms

Given $\log_b 2 \approx 0.693$, $\log_b 3 \approx 1.099$, and $\log_b 7 \approx 1.946$, use the properties of logarithms to approximate the following.

a. $\log_b 6$ **b.** $\log_b \dfrac{7}{27}$

Solution

a. $\log_b 6 = \log_b(2 \cdot 3)$

$\quad\quad = \log_b 2 + \log_b 3$ *Property 1*

$\quad\quad \approx 0.693 + 1.099$

$\quad\quad = 1.792$

b. $\log_b \dfrac{7}{27} = \log_b 7 - \log_b 27$ *Property 2*

$\quad\quad\quad = \log_b 7 - \log_b 3^3$

$\quad\quad\quad = \log_b 7 - 3 \log_b 3$ *Property 3*

$\quad\quad\quad \approx 1.946 - 3(1.099)$

$\quad\quad\quad = -1.351$

EXAMPLE 4 Using Properties of Logarithms

Use the properties of logarithms to verify that

$$-\ln \frac{1}{a} = \ln a.$$

Solution

$$-\ln \frac{1}{a} = -\ln(a^{-1}) = -(-1)\ln a = \ln a$$

Try verifying this result on your calculator using $a = 2$.

Rewriting Logarithmic Expressions

The properties of logarithms are useful for rewriting logarithmic expressions in forms that simplify the operations of algebra. This is true because they convert complicated products, quotients, and exponential forms into simpler sums, differences, and products, respectively.

EXAMPLE 5 Rewriting the Logarithm of a Product

Use the properties of logarithms to rewrite

$$\log_{10} 5x^3y$$

as the sum of logarithms.

Solution

$$\log_{10} 5x^3y = \log_{10} 5 + \log_{10} x^3y$$
$$= \log_{10} 5 + \log_{10} x^3 + \log_{10} y$$
$$= \log_{10} 5 + 3 \log_{10} x + \log_{10} y$$

EXAMPLE 6 Rewriting the Logarithm of a Quotient

Use the properties of logarithms to rewrite

$$\ln \frac{\sqrt{3x - 5}}{7}$$

as the sum and/or difference of logarithms.

Solution

$$\ln \frac{\sqrt{3x - 5}}{7} = \ln(3x - 5)^{1/2} - \ln 7$$
$$= \frac{1}{2}\ln(3x - 5) - \ln 7$$

In Examples 5 and 6, we used the properties of logarithms to *expand* logarithmic expressions. In Examples 7 and 8, we reverse the procedure and use the properties of logarithms to *condense* logarithmic expressions.

EXAMPLE 7 Condensing a Logarithmic Expression

Rewrite as the logarithm of a single quantity.

$$\frac{1}{2} \log_{10} x - 3 \log_{10}(x + 1)$$

Solution

$$\frac{1}{2} \log_{10} x - 3 \log_{10}(x + 1) = \log_{10} x^{1/2} - \log_{10}(x + 1)^3 \qquad \textit{Property 3}$$

$$= \log_{10} \frac{\sqrt{x}}{(x + 1)^3} \qquad \textit{Property 2}$$

EXAMPLE 8 Condensing a Logarithmic Expression

Rewrite as the logarithm of a single quantity.

$$2 \ln(x + 2) + \ln x$$

Solution

$$2 \ln(x + 2) + \ln x = \ln(x + 2)^2 + \ln x \qquad \textit{Property 3}$$

$$= \ln x(x + 2)^2 \qquad \textit{Property 1}$$

When applying the properties of logarithms to a logarithmic function, you should take note of the domain of the function. For example, the domain of $f(x) = \ln x^2$ is all real $x \neq 0$, whereas the domain of $g(x) = 2 \ln x$ is all real $x > 0$.

DISCUSSION

PROBLEM

Demonstrating Properties of Logarithms

For each of the following statements: (a) If it is *true*, state a property of logarithms to support your answer. (b) If it is *false*, use your calculator to support your answer.

1. $\log(3 + 5) = \log 3 + \log 5$ 2. $\log 9 = \log 36 - \log 4$
3. $\dfrac{\log 3}{\log 2} = \log \dfrac{3}{2}$ 4. $\log 28 = \log 4 \cdot \log 7$
5. $\log 3^5 = 3 \log 5$ 6. $\log 81 = 4 \log 3$

WARM UP

The following warm-up exercises involve skills that were covered in earlier sections. You will use these skills in the exercise set for this section.

In Exercises 1–4, evaluate the expressions without using a calculator.

1. $\log_7 49$ **2.** $\log_2\left(\frac{1}{32}\right)$

3. $\ln \frac{1}{e^2}$ **4.** $\log_{10} 0.001$

In Exercises 5–8, simplify the expressions.

5. $e^2 e^3$ **6.** $\frac{e^2}{e^3}$

7. $(e^2)^3$ **8.** $(e^2)^0$

In Exercises 9 and 10, rewrite the expressions in exponential form.

9. $\frac{1}{x^2}$ **10.** $\sqrt{x}$

EXERCISES for Section 6.3

In Exercises 1–4, use the change of base formula to write the given logarithm as a multiple of a common logarithm. For instance, $\log_2 3 = (1/\log_{10} 2)\log_{10} 3$.

1. $\log_3 5$ **2.** $\log_4 10$

3. $\log_2 x$ **4.** $\ln 5$

In Exercises 5–8, use the change of base formula to write the given logarithm as a multiple of a natural logarithm. For instance, $\log_2 3 = (1/\ln 2)\ln 3$.

5. $\log_3 5$ **6.** $\log_4 10$

7. $\log_2 x$ **8.** $\log_{10} 5$

In Exercises 9–16, evaluate the logarithm using the change of base formula. Do the problem twice, once with common logarithms and once with natural logarithms. Round to three decimal places.

9. $\log_3 7$ **10.** $\log_7 4$

11. $\log_{1/2} 4$ **12.** $\log_4(0.55)$

13. $\log_9(0.4)$ **14.** $\log_{20} 125$

15. $\log_{15} 1250$ **16.** $\log_{1/3}(0.015)$

In Exercises 17–36, use the properties of logarithms to write the expression as a sum, difference, and/or multiple of logarithms.

17. $\log_{10} 5x$ **18.** $\log_{10} 10z$

19. $\log_{10} \frac{5}{x}$ **20.** $\log_{10} \frac{y}{2}$

21. $\log_8 x^4$ **22.** $\log_6 z^{-3}$

23. $\ln \sqrt{z}$ **24.** $\ln \sqrt[3]{t}$

25. $\ln xyz$ **26.** $\ln \frac{xy}{z}$

27. $\ln \sqrt{a - 1}$ **28.** $\ln\left(\frac{x^2 - 1}{x^3}\right)$

29. $\ln z(z - 1)^2$ **30.** $\ln \sqrt{\frac{x^2}{y^3}}$

31. $\ln \sqrt[3]{\frac{x}{y}}$ **32.** $\ln \frac{x}{\sqrt{x^2 + 1}}$

33. $\ln \frac{x^4\sqrt{y}}{z^5}$ **34.** $\ln \sqrt{x^2(x + 2)}$

35. $\log_b \frac{x^2}{y^2 z^3}$ **36.** $\log_b \frac{\sqrt{x}y^4}{z^4}$

In Exercises 37–56, write the expression as the logarithm of a single quantity.

37. $\ln x + \ln 2$

38. $\ln y + \ln z$

39. $\log_4 z - \log_4 y$

40. $\log_5 8 - \log_5 t$

41. $2 \log_2(x + 4)$

42. $-4 \log_6 2x$

43. $\frac{1}{3} \log_3 5x$

44. $\frac{3}{2} \log_7(z - 2)$

45. $\ln x - 3 \ln(x + 1)$

46. $2 \ln 8 + 5 \ln z$

47. $\ln(x - 2) - \ln(x + 2)$

48. $3 \ln x + 2 \ln y - 4 \ln z$

49. $\ln x - 2[\ln(x + 2) + \ln(x - 2)]$

50. $4[\ln z + \ln(z + 5)] - 2 \ln(z - 5)$

51. $\frac{1}{3}[2 \ln(x + 3) + \ln x - \ln(x^2 - 1)]$

52. $2[\ln x - \ln(x + 1) - \ln(x - 1)]$

53. $\frac{1}{3}[\ln y + 2 \ln(y + 4)] - \ln(y - 1)$

54. $\frac{1}{2}[\ln(x + 1) + 2 \ln(x - 1)] + 3 \ln x$

55. $2 \ln 3 - \frac{1}{2} \ln(x^2 + 1)$ **56.** $\frac{3}{2} \ln 5t^6 - \frac{3}{4} \ln t^4$

In Exercises 57–70, approximate the logarithm using the properties of logarithms, given $\log_b 2 \approx 0.3562$, $\log_b 3 \approx 0.5646$, and $\log_b 5 \approx 0.8271$.

57. $\log_b 6$

58. $\log_b 15$

59. $\log_b\left(\frac{2}{3}\right)$

60. $\log_b\left(\frac{5}{3}\right)$

61. $\log_b 25$

62. $\log_b 18$ $(18 = 2 \cdot 3^2)$

63. $\log_b \sqrt{2}$

64. $\log_b\left(\frac{9}{2}\right)$

65. $\log_b \frac{1}{4}$

66. $\log_b \sqrt[3]{75}$

67. $\log_b \sqrt{5b}$

68. $\log_b(3b^2)$

69. $\log_b \dfrac{(4.5)^3}{\sqrt{3}}$

70. $\log_b 1$

In Exercises 71–76, find the exact value of the logarithm.

71. $\log_3 9$

72. $\log_6 \sqrt[3]{6}$

73. $\log_4 16^{1.2}$

74. $\log_5\left(\frac{1}{125}\right)$

75. $\ln e^{4.5}$

76. $\ln \sqrt[4]{e^3}$

In Exercises 77–84, use the properties of logarithms to simplify the given logarithmic expression.

77. $\log_4 8$

78. $\log_5\left(\frac{1}{15}\right)$

79. $\log_7 \sqrt{70}$

80. $\log_2(4^2 \cdot 3^4)$

81. $\log_5\left(\frac{1}{250}\right)$

82. $\log_{10}\left(\frac{9}{300}\right)$

83. $\ln(5e^6)$

84. $\ln \dfrac{6}{e^2}$

85. *Sound Intensity*   The relationship between the number of decibels β and the intensity of a sound I in watts per meter squared is given by

$$\beta = 10 \log_{10}\left(\frac{I}{10^{-16}}\right).$$

Use properties of logarithms to write the formula in simpler form, and determine the number of decibels of a sound with an intensity of 10^{-10} watts per meter squared.

86. Approximate the natural logarithm of as many integers as possible between 1 and 20 given that $\ln 2 \approx 0.6931$, $\ln 3 \approx 1.0986$, and $\ln 5 \approx 1.6094$.

87. Prove that $\log_b \dfrac{u}{v} = \log_b u - \log_b v$.

88. Prove that $\log_b u^n = n \log_b u$.

6.4 Solving Exponential and Logarithmic Equations

Introduction / Solving Exponential Equations / Solving Logarithmic Equations / Applications

Introduction

So far in this chapter, you have studied the definitions, graphs, and properties of exponential and logarithmic functions. Now we will concentrate on procedures for *solving equations* involving these exponential and logarithmic functions. The solution procedures are based on the fact that the exponential and logarithmic functions are inverses of each other.

PROPERTIES OF EXPONENTIAL AND LOGARITHMIC FUNCTIONS

Let $f(x) = a^x$ and $g(x) = \log_a x$, with $a > 1$.

Inverse Properties	*Reason*
1. $\log_a a^x = x$	$(g \circ f)(x) = x$
$\ln e^x = x$	Replace base a by base e
2. $a^{\log_a x} = x$	$(f \circ g)(x) = x$
$e^{\ln x} = x$	Replace base a by base e

One-to-One Properties	*Reason*
3. $x = y$ if and only if $\log_a x = \log_a y$	g is one-to-one
4. $x = y$ if and only if $a^x = a^y$	f is one-to-one

To solve a simple equation like $2^x = 32$, we can rewrite it as $2^x = 2^5$ and use Property 4 to conclude that $x = 5$. However, to solve for x in the equation $e^x = 7$, we use two different properties.

$e^x = 7$	*Given equation*
$\ln e^x = \ln 7$	*Take logarithm of both sides*
$x = \ln 7$	*Property 1*

This example suggests that to solve an exponential equation, you take the logarithms of both sides. On the other hand, to solve a logarithmic equation you rewrite it in exponential form.

Guidelines for Solving Exponential and Logarithmic Equations

1. *To solve an exponential equation*, first isolate the exponential expression, then take the logarithm of both sides and solve for the variable.
2. *To solve a logarithmic equation*, rewrite the equation in exponential form and solve for the variable.

Solving Exponential Equations

EXAMPLE 1   Solving an Exponential Equation

$e^x = 72$	*Given equation*
$\ln e^x = \ln 72$	*Take log of both sides*
$x = \ln 72$	*Inverse property of logs and exponents*
$x \approx 4.277$	

A check, using your calculator, will show that $e^{4.277} \approx 72$.

EXAMPLE 2   Solving an Exponential Equation

$$4e^{2x} = 5 \qquad \qquad \text{Given equation}$$

$$e^{2x} = \frac{5}{4} \qquad \qquad \text{Divide both sides by 4}$$

$$\ln e^{2x} = \ln \frac{5}{4} \qquad \qquad \text{Take log of both sides}$$

$$2x = \ln \frac{5}{4} \qquad \qquad \text{Inverse property of logs and exponents}$$

$$x = \frac{1}{2} \ln \frac{5}{4} \qquad \qquad \text{Divide both sides by 2}$$

$$x \approx 0.112$$

Thus, the solution is $x = \frac{1}{2} \ln \frac{5}{4}$. Check this solution in the original equation.

When an equation involves two or more exponential expressions, we can still use a procedure similar to that demonstrated in the first two examples. However, the algebra is a bit more complicated. Study the next example carefully.

EXAMPLE 3   Solving an Exponential Equation

$$e^{2x} - 3e^x + 2 = 0 \qquad \qquad \text{Given equation}$$

$$(e^x)^2 - 3e^x + 2 = 0 \qquad \qquad \text{Quadratic form}$$

$$(e^x - 2)(e^x - 1) = 0 \qquad \qquad \text{Factor}$$

$$e^x - 2 = 0 \qquad e^x - 1 = 0 \qquad \text{Set factors to zero}$$

$$e^x = 2 \qquad \qquad e^x = 1$$

$$x = \ln 2 \qquad \qquad x = 0 \qquad \text{Solutions}$$

Thus, the equation has two solutions: $x = \ln 2$ and $x = 0$. Check each solution in the original equation.

Examples 1 through 3 all deal with exponential equations in which the base is e. The same approach can be used to solve exponential equations involving other bases.

EXAMPLE 4   A Base Other Than e

$$2^x = 10 \qquad\qquad \text{Given equation}$$

$$\ln 2^x = \ln 10 \qquad\qquad \text{Take log of both sides}$$

$$x \ln 2 = \ln 10 \qquad\qquad \text{Property of logarithms}$$

$$x = \frac{\ln 10}{\ln 2} \qquad\qquad \text{Divide both sides by } \ln 2$$

Thus, the equation has one solution: $x = \ln 10 / \ln 2 \approx 3.32$. Try checking this solution in the original equation. (*Note:* Using the change of base formula, this solution could be written as $x = \log_2 10$.)

EXAMPLE 5   A Base Other Than e

$$4^{x+3} = 7^x \qquad\qquad \text{Given equation}$$

$$\ln 4^{x+3} = \ln 7^x \qquad\qquad \text{Take ln of both sides}$$

$$(x + 3)\ln 4 = x \ln 7 \qquad\qquad \text{Property 3}$$

$$x \ln 4 + 3 \ln 4 = x \ln 7 \qquad\qquad \text{Distributive Property}$$

$$x \ln 4 - x \ln 7 = -3 \ln 4 \qquad\qquad \text{Collect like terms}$$

$$x(\ln 4 - \ln 7) = -3 \ln 4 \qquad\qquad \text{Factor out } x$$

$$x = \frac{-3 \ln 4}{\ln 4 - \ln 7} \qquad\qquad \text{Divide}$$

$$x \approx 7.432$$

Solving Logarithmic Equations

To solve a logarithmic equation, convert it to an equivalent exponential equation by using the definition $\log_a x = y$ if and only if $x = a^y$. For example,

$$\ln x = 3 \qquad\qquad \text{Logarithmic form}$$

$$x = e^3. \qquad\qquad \text{Exponential form (by definition)}$$

Such a conversion to exponential form encompasses two properties of exponential functions.

$$\ln x = 3 \qquad\qquad \text{Given equation}$$

$$e^{\ln x} = e^3 \qquad\qquad \text{Exponentiate both sides (Property 4)}$$

$$x = e^3 \qquad\qquad \text{Exponential form (Property 2)}$$

This latter procedure is sometimes called *exponentiating* both sides of an equation.

EXAMPLE 6 Solving a Logarithmic Equation

$$\ln x = 2 \qquad\qquad\qquad \textit{Given equation}$$

$$x = e^2 \qquad\qquad\qquad \textit{Exponential form}$$

$$x \approx 7.389$$

Thus, the solution is $x = e^2$. Check this solution in the original equation.

EXAMPLE 7 Solving a Logarithmic Equation

$$5 + 2 \ln x = 4 \qquad\qquad \textit{Given equation}$$

$$2 \ln x = -1 \qquad\qquad \textit{Subtract 5 from both sides}$$

$$\ln x = -\frac{1}{2} \qquad\qquad \textit{Divide both sides by 2}$$

$$x = e^{-1/2} \qquad\qquad \textit{Exponential form}$$

$$x \approx 0.607$$

Thus, the equation has one solution: $x = e^{-1/2}$. Check this solution in the original equation.

EXAMPLE 8 Solving a Logarithmic Equation

$$2 \ln 3x = 4 \qquad\qquad \textit{Given equation}$$

$$\ln 3x = 2 \qquad\qquad \textit{Divide both sides by 2}$$

$$3x = e^2 \qquad\qquad \textit{Exponential form}$$

$$x = \frac{1}{3}e^2 \qquad\qquad \textit{Divide both sides by 3}$$

$$x \approx 2.463$$

Thus, the equation has one solution: $x = \frac{1}{3}e^2$. Check this solution in the original equation.

The techniques used to solve the equations involving logarithmic expressions can produce extraneous solutions.

EXAMPLE 9 Solving a Logarithmic Equation

$$\ln(x - 2) + \ln(2x - 3) = 2 \ln x \qquad \text{\textit{Given equation}}$$

$$\ln(x - 2)(2x - 3) = \ln x^2 \qquad \text{\textit{Properties of logarithms}}$$

$$\ln(2x^2 - 7x + 6) = \ln x^2$$

$$2x^2 - 7x + 6 = x^2 \qquad \text{\textit{One-to-one property of exponentials}}$$

$$x^2 - 7x + 6 = 0 \qquad \text{\textit{Quadratic form}}$$

$$(x - 6)(x - 1) = 0 \qquad \text{\textit{Factor}}$$

$$x - 6 = 0 \quad \rightarrow \quad x = 6 \qquad \text{\textit{Set 1st factor equal to 0}}$$

$$x - 1 = 0 \quad \rightarrow \quad x = 1 \qquad \text{\textit{Set 2nd factor equal to 0}}$$

Finally, by checking these two "solutions" in the original equation, we find that $x = 1$ is not valid. Do you see why? Thus, the only solution is $x = 6$.

EXAMPLE 10 The Change of Base Formula

Prove the change of base formula given in Section 6.3.

$$\log_a x = \frac{\log_b x}{\log_b a}$$

Solution

We begin by letting

$$y = \log_a x$$

and writing the equivalent exponential form

$$a^y = x.$$

Now, taking the logarithm *with base b* of both sides, we have

$$\log_b a^y = \log_b x$$

$$y \log_b a = \log_b x$$

$$y = \frac{\log_b x}{\log_b a}$$

$$\log_a x = \frac{\log_b x}{\log_b a}.$$

Applications

EXAMPLE 11 Waste Processed for Energy Recovery

The amount of municipal waste processed for energy recovery in the United States from 1960 to 1986 can be approximated by the equation

$$y = 0.00643e^{0.00533t^2}$$

where y is the amount of waste (in pounds per person) that was processed for energy recovery and t represents the calendar year with $t = 0$ corresponding to 1960 (see Figure 6.17). According to this model, during which year did the amount of waste reach 0.2 pounds? (*Source:* Franklin Associates, *Characterization of Municipal Solid Waste in U.S.*)

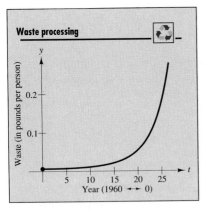

Waste processing

y axis: Waste (in pounds per person): 0.1, 0.2

t axis: 5, 10, 15, 20, 25 — Year (1960 ⟶ 0)

FIGURE 6.17

Solution

$0.00643e^{0.00533t^2} = y$	*Given model*
$0.00643e^{0.00533t^2} = 0.2$	*Let y equal 0.2*
$e^{0.00533t^2} \approx 31.104$	*Divide both sides by 0.00643*
$\ln e^{0.00533t^2} \approx \ln 31.104$	*Take log of both sides*
$0.00533t^2 \approx 3.437$	*Inverse property of logs and exponents*
$t^2 \approx 644.905$	*Divide both sides by 0.00533*
$t \approx 25.4$	*Extract positive square root*

Thus, the solution is $t \approx 25.4$ years. Since $t = 0$ represents 1960, it follows that the amount of waste would have reached 0.2 pounds per person in 1985.

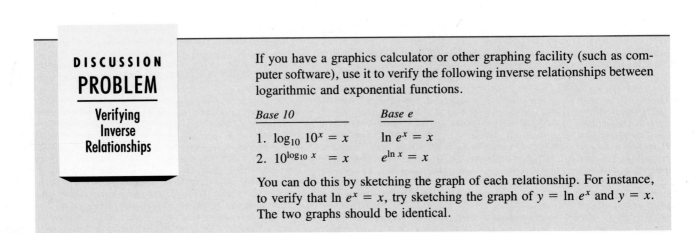

DISCUSSION

PROBLEM

Verifying Inverse Relationships

If you have a graphics calculator or other graphing facility (such as computer software), use it to verify the following inverse relationships between logarithmic and exponential functions.

Base 10

1. $\log_{10} 10^x = x$
2. $10^{\log_{10} x} = x$

Base e

$\ln e^x = x$

$e^{\ln x} = x$

You can do this by sketching the graph of each relationship. For instance, to verify that $\ln e^x = x$, try sketching the graph of $y = \ln e^x$ and $y = x$. The two graphs should be identical.

WARM UP

The following warm-up exercises involve skills that were covered in earlier sections. You will use these skills in the exercise set for this section.

In Exercises 1–6, solve for x.

1. $x \ln 2 = \ln 3$

2. $(x - 1)\ln 4 = 2$

3. $2xe^2 = e^3$

4. $4xe^{-1} = 8$

5. $x^2 - 4x + 5 = 0$

6. $2x^2 - 3x + 1 = 0$

In Exercises 7–10, simplify the expressions.

7. $\log_{10} 100^x$

8. $\log_4 64^x$

9. $\ln e^{2x}$

10. $\ln e^{-x^2}$

EXERCISES for Section 6.4

In Exercises 1–10, solve for x without the aid of a calculator.

1. $4^x = 16$

2. $3^x = 243$

3. $7^x = \frac{1}{49}$

4. $8^x = 4$

5. $\left(\frac{3}{4}\right)^x = \frac{27}{64}$

6. $3^{x-1} = 27$

7. $\log_4 x = 3$

8. $\log_5 5x = 2$

9. $\log_{10} x = -1$

10. $\ln(2x - 1) = 0$

In Exercises 11–16, apply the inverse properties of $\ln x$ and e^x to simplify the given expression.

11. $\ln e^{x^2}$

12. $\ln e^{2x-1}$

13. $e^{\ln(5x+2)}$

14. $-1 + \ln e^{2x}$

15. $e^{\ln x^2}$

16. $-8 + e^{\ln x^3}$

In Exercises 17–50, solve the given exponential equation. Round to three decimal places.

17. $e^x = 10$

18. $e^x = 6500$

19. $2e^x = 39$

20. $4e^x = 91$

21. $e^x - 5 = 10$

22. $e^x + 6 = 38$

23. $7 - 2e^x = 5$

24. $-14 + 3e^x = 11$

25. $e^{3x} = 12$

26. $e^{2x} = 50$

27. $500e^{-x} = 300$

28. $1000e^{-4x} = 75$

29. $3e^{3x/2} = 962$

30. $6e^{1-x} = 25$

31. $e^{2x} - 4e^x - 5 = 0$

32. $e^{2x} - 5e^x + 6 = 0$

33. $3(1 + e^{2x}) = 4$

34. $20(100 - e^{x/2}) = 500$

35. $\dfrac{400}{1 + e^{-x}} = 200$

36. $\dfrac{3000}{2 + e^{-2x}} = 1200$

37. $10^x = 42$

38. $10^x = 570$

39. $3^{2x} = 80$

40. $6^{5x} = 3000$

41. $5^{-t/2} = 0.20$

42. $4^{-3t} = 0.10$

43. $\frac{1}{3}(10^{2x}) = 12$

44. $8(10^{3x}) = 12$

45. $3(5^{x-1}) = 21$

46. $2^{3-x} = 565$

47. $5^{x+2} = 3^{2x-1}$

48. $6^{3x-5} = 2^{7x}$

49. $\left(1 + \dfrac{0.10}{12}\right)^{12t} = 2$

50. $\left(1 + \dfrac{0.065}{365}\right)^{365t} = 4$

In Exercises 51–70, solve the given logarithmic equation. Round to three decimal places.

51. $\ln x = 5$

52. $\ln x = -4.5$

53. $\ln 2x = 2.4$

54. $\ln 4x = 1$

55. $2 \log_6 4x = 0$

56. $3 \log_2 5x = 10$

57. $\ln \sqrt{x + 2} = 1$

58. $\ln(x + 1)^2 = 2$

59. $\ln x + \ln(x - 2) = 1$

60. $\ln x + \ln(x + 3) = 1$

61. $\log_{10}(z - 3) = 2$

62. $\log_{10} x^2 = 6$

63. $\log_{10}(x + 4) - \log_{10} x = \log_{10}(x + 2)$

64. $\log_2 x + \log_2(x + 2) = \log_2(x + 6)$

65. $\log_3 x + \log_3(x^2 - 8) = \log_3 8x$

66. $\ln(x + 1) - \ln(x - 2) = \ln x^2$

67. $\ln(x + 5) = \ln(x - 1) - \ln(x + 1)$

68. $\log_{10} x - \log_{10}(2x - 1) = 0$

69. $\log_2(x + 5) - \log_2(x - 2) = 3$

70. $\log_4 x - \log_4(x - 1) = \frac{1}{2}$

Compound Interest   In Exercises 71 and 72, find the time required for a $1000 investment to double at interest rate *r,* compounded continuously.

71. *r* = 0.085   **72.** *r* = 0.12

Compound Interest   In Exercises 73 and 74, find the time required for a $1000 investment to triple at interest rate *r,* compounded continuously.

73. *r* = 0.085   **74.** *r* = 0.12

75. *Demand Function*   The demand equation for a certain product is given by

$$p = 500 - 0.5(e^{0.004x}).$$

Find the demand *x* for a price of (a) *p* = $350 and (b) *p* = $300.

76. *Demand Function*   The demand equation for a certain product is given by

$$p = 5000\left(1 - \frac{4}{4 + e^{-0.002x}}\right).$$

Find the demand *x* for a price of (a) *p* = $600 and (b) *p* = $400.

77. *Forest Yield*   The yield *V* (in millions of cubic feet per acre) for a forest at age *t* years is given by

$$V = 6.7e^{-48.1/t}.$$

Find the time necessary to have a yield of (a) 1.3 million cubic feet and (b) 2 million cubic feet.

78. *Human Memory Model*   In a group project in learning theory, a mathematical model for the proportion *P* of correct responses after *n* trials was found to be

$$P = \frac{0.83}{1 + e^{-0.2n}}.$$

After how many trials will 60% of the responses be correct?

79. *Average Heights*   The percentage of American males between the ages of 18 and 24 who are no more than *x* inches tall is given by

$$m(x) = \frac{100}{1 + e^{-0.6114(x-69.71)}}$$

where *m* is the percentage and *x* is the height in inches (see figure). (*Source:* U.S. National Center for Health Statistics) The function giving the percentages *f* for females for the same ages is given by

$$f(x) = \frac{100}{1 + e^{-0.66607(x-64.51)}}.$$

What is the median height of each sex?

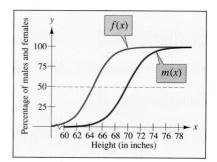

Figure for 79

80. *Trees per Acre*   The number of trees per acre *N* of a certain species is approximated by the model

$$N = 68 \cdot 10^{-0.04x}, \qquad 5 \le x \le 40$$

where *x* is the average diameter (in inches) of the trees three feet above the ground. Use the model to approximate the average diameter of the trees in a test plot when *N* = 21.

6.5 Exponential and Logarithmic Applications
Compound Interest / Growth and Decay / Logistics Growth Models / Logarithmic Models

Compound Interest

The behavior of many physical, economic, and social phenomena can be described by exponential and logarithmic functions. In this section, we look at four basic types of applications: (1) compound interest, (2) growth and decay, (3) logistics models, and (4) intensity models. The problems presented in this section require the full range of solution techniques studied in this chapter.

From Section 6.1, recall the following two compound interest formulas, where A is the account balance, P is the principal, r is the annual percentage rate, and t is the time in years.

n Compoundings per Year	*Continuous Compounding*
$A = P\left(1 + \dfrac{r}{n}\right)^{nt}$	$A = Pe^{rt}$

EXAMPLE 1   Doubling Time for an Investment

An investment is made in a trust fund at an annual percentage rate of 9.5%, compounded quarterly. How long will it take for the investment to double in value?

Solution

For quarterly compounding, we use the formula

$$A = P\left(1 + \frac{r}{4}\right)^{4t}.$$

Using $r = 0.095$, the time required for the investment to double is given by solving for t in the equation $2P = A$.

$$2P = P\left(1 + \frac{0.095}{4}\right)^{4t} \qquad \text{\textit{2P = A}}$$

$$2 = (1.02375)^{4t} \qquad \text{\textit{Divide both sides by P}}$$

$$\ln 2 = \ln(1.02375)^{4t} \qquad \text{\textit{Take ln of both sides}}$$

$$\ln 2 = 4t \ln(1.02375)$$

$$t = \frac{\ln 2}{4 \ln(1.02375)} \approx 7.4$$

Therefore, it will take approximately 7.4 years for the investment to double in value with quarterly compounding.

Try reworking Example 1 using continuous compounding. To do this you will need to solve the equation

$$2P = Pe^{0.095t}.$$

The solution is $t \approx 7.3$ years, which makes sense because the principal should double more quickly with continuous compounding than with quarterly compounding.

From Example 1, we see that the time required for an investment to double in value is independent of the amount invested. In general, the **doubling time** is as follows.

n Compoundings per Year	*Continuous Compounding*
$t = \dfrac{\ln 2}{n \ln[1 + (r/n)]}$	$t = \dfrac{\ln 2}{r}$

EXAMPLE 2   Finding an Annual Percentage Rate

An investment of $10,000 is compounded continuously. What annual percentage rate will produce a balance of $25,000 in 10 years?

Solution

We use the formula

$$A = Pe^{rt}$$

with $P = 10,000$, $A = 25,000$, and $t = 10$, and solve the following equation for r.

$$10,000e^{10r} = 25,000$$
$$e^{10r} = 2.5$$
$$10r = \ln 2.5$$
$$r = \frac{1}{10} \ln 2.5 \approx 0.0916$$

Thus, the annual percentage rate must be approximately 9.16%.

EXAMPLE 3   The Effective Yield for an Investment

A deposit is compounded continuously at an annual percentage rate of 7.5%. Find the simple interest rate that would yield the same balance at the end of one year (**effective yield**).

Solution

Using the formula $A = Pe^{rt}$ with $r = 0.075$ and $t = 1$, the balance at the end of one year is

$$A = Pe^{0.075(1)}$$
$$\approx P(1.0779)$$
$$= P(1 + 0.0779). \qquad\qquad A = P(1 + r)$$

Since the formula for simple interest after one year is

$$A = P(1 + r)$$

we conclude that the effective yield is approximately 7.79%.   ◢

Growth and Decay

The balance in an account earning *continuously compounded* interest is one example of a quantity that increases over time according to the **exponential growth model**

$$Q(t) = Ce^{kt}.$$

In this model, $Q(t)$ is the size of the population (balance, weight, and so forth) at any time t, C is the original population (when $t = 0$), and k is a constant determined by the rate of growth. If $k > 0$, the population *grows* (increases) over time, and if $k < 0$ it *decays* (decreases) over time. Example 11 of Section 6.1 is an example of population growth. Recall from Section 6.1 that we can remember this growth model as $(Then) = (Now)(e^{kt})$.

EXAMPLE 4   Exponential Decay

◤ Radioactive iodine is a by-product of some types of nuclear reactors. Its half-life is 60 days. That is, after 60 days, a given amount of radioactive iodine will have decayed to half the original amount. Suppose a contained nuclear accident occurs and gives off an initial amount C of radioactive iodine.

a. Write an equation for the amount of radioactive iodine present at any time t following the accident.
b. How long will it take for the radioactive iodine to decay to a level of 20% of the original amount?

Solution

a. We first need to find the rate k, in the exponential model $Q(t) = Ce^{kt}$. Knowing that half the original amount remains after $t = 60$ days, we obtain

$$Q(60) = Ce^{k(60)} = \frac{1}{2}C$$

$$e^{60k} = \frac{1}{2}$$

$$60k = -\ln 2$$

$$k = \frac{-\ln 2}{60} \approx -0.0116.$$

Thus, the exponential model is

$$Q(t) = Ce^{-0.0116t}.$$

b. The time required to decay to 20% of the original amount is given by

$$Q(t) = Ce^{-0.0116t} = (0.2)C$$

$$e^{-0.0116t} = 0.2$$

$$-0.0116t = \ln 0.2$$

$$t = \frac{\ln 0.2}{-0.0116} \approx 139 \text{ days.}$$

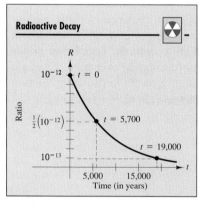

FIGURE 6.18

In living organic material the ratio of radioactive carbon isotopes (Carbon 14) to the number of nonradioactive carbon isotopes (Carbon 12) is about 1 to 10^{12}. When organic material dies, its Carbon 12 content remains fixed, whereas its radioactive Carbon 14 begins to decay with a half-life of about 5700 years. To estimate the age of dead organic material, scientists use the following formula, which denotes the ratio of Carbon 14 to Carbon 12 present at any time t (in years).

$$R = \frac{1}{10^{12}}e^{-t/8223}$$

The graph of R is shown in Figure 6.18. Note that R decreases as the time t increases.

EXAMPLE 5   Carbon Dating

Suppose the Carbon 14/Carbon 12 ratio of a newly discovered fossil is

$$R = \frac{1}{10^{13}}.$$

Estimate the age of the fossil.

Solution

In the carbon dating model, we substitute the given value of R to obtain

$$\frac{1}{10^{12}}e^{-t/8223} = R \qquad \text{\textit{Given model}}$$

$$\frac{e^{-t/8223}}{10^{12}} = \frac{1}{10^{13}} \qquad \text{\textit{Let R equal 1/10}}^{13}$$

$$e^{-t/8223} = \frac{1}{10} \qquad \text{\textit{Multiply both sides by 10}}^{12}$$

$$\ln e^{-t/8223} = \ln \frac{1}{10} \qquad \text{\textit{Take log of both sides}}$$

$$-\frac{t}{8223} \approx -2.3026 \qquad \text{\textit{Inverse property of logs and exponents}}$$

$$t \approx 18{,}934. \qquad \text{\textit{Multiply both sides by 8223}}$$

Thus, to the nearest thousand years, we estimate the age of the fossil to be 19,000 years.

REMARK The carbon dating model in Example 5 assumed that the Carbon 14/Carbon 12 ratio was one part in 10,000,000,000,000. Suppose an error in measurement occurred and the actual ratio was only one part in 8,000,000,000,000. The fossil age corresponding to the actual ratio would then be approximately 17,000 years. Check this result.

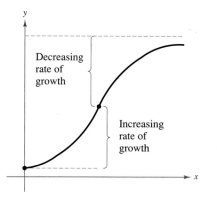

Logistic Curve

FIGURE 6.19

Logistics Growth Models

Some populations initially have rapid growth, followed by a declining rate of growth, as indicated by the graph in Figure 6.19. One model for describing this type of growth pattern is the **logistics curve** given by the function

$$y = \frac{a}{1 + be^{(-(x-c)/d)}}$$

where y is the population size and x is the time. An example would be a bacteria culture allowed to grow initially under ideal conditions, followed by less favorable conditions that inhibit growth. A logistics growth curve is also called a **sigmoidal curve**.

EXAMPLE 6 Spread of a Virus

On a college campus of 5000 students, one student returned from vacation with a contagious flu virus. The spread of the virus through the student body is given by

$$y = \frac{5000}{1 + 4999e^{-0.8t}}$$

where y is the total number infected after t days. The college will cancel classes when 40% or more of the students are ill.

a. How many students are infected after five days?
b. After how many days will the college cancel classes?

Solution

a. After five days, the number of students infected is

$$y = \frac{5000}{1 + 4999e^{-0.8(5)}} = \frac{5000}{1 + 4999e^{-4}} \approx 54.$$

b. In this case, the number of students infected is $(0.40)(5000) = 2000$. Therefore, we solve for t in the following equation.

$$2000 = \frac{5000}{1 + 4999e^{-0.8t}}$$

$$1 + 4999e^{-0.8t} = 2.5$$

$$e^{-0.8t} \approx 0.0003$$

$$\ln(e^{-0.8t}) \approx \ln 0.0003$$

$$-0.8t \approx -8.1115$$

$$t \approx 10.1$$

Hence, after 10 days, at least 40% of the students will be infected, and the college will cancel classes. The graph of the function is shown in Figure 6.20.

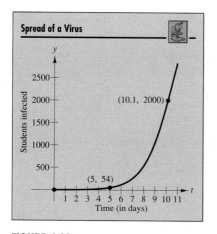

Spread of a Virus

(10.1, 2000)

(5, 54)

Students infected

Time (in days)

FIGURE 6.20

Logarithmic Models

Sound and shock waves can be measured by the **intensity model**

$$S = K \log_{10} \frac{I}{I_0}$$

where I is the intensity of the stimulus wave, I_0 is the **threshold intensity** (the smallest value of I that can be detected by the listening device), and K determines the units in which S is measured. Sound heard by the human ear is measured in decibels. One **decibel** is considered to be the smallest detectable difference in the loudness of two sounds.

EXAMPLE 7   Magnitude of Earthquakes

On the Richter Scale, the magnitude R of an earthquake of intensity I is given by

$$R = \log_{10} \frac{I}{I_0}$$

where $I_0 = 1$ is the minimum intensity used for comparison. Find the intensity per unit of area for the following earthquakes. (Intensity is a measure of the wave energy of an earthquake.)

a. San Francisco in 1906, $R = 8.6$
b. Mexico City in 1978, $R = 7.85$
c. San Francisco Bay Area in 1989, $R = 7.1$

Solution

a. Since $I_0 = 1$ and $R = 8.6$, we have

$$8.6 = \log_{10} I$$
$$I = 10^{8.6} \approx 398,107,171.$$

b. For Mexico City, we have $7.85 = \log_{10} I$, and

$$I = 10^{7.85} \approx 70,795,000.$$

c. For $R = 7.1$, we have $7.1 = \log_{10} I$, and

$$I = 10^{7.1} \approx 12,589,254.$$

Note that an increase of 1.5 units on the Richter Scale (from 7.1 to 8.6) represents an intensity change by a factor of

$$\frac{398,107,171}{12,589,254} \approx 31.6.$$

In other words, the "great San Francisco earthquake" in 1906 had a magnitude that was about 32 times more than the one in 1989. ◢

DISCUSSION
PROBLEM

Comparing Population Models

The population of the United States from 1800 to 1990 is given (in millions) in the table.

t	0	1	2	3	4	5	6	7	8	9
Year	1800	1810	1820	1830	1840	1850	1860	1870	1880	1890
Population	5.31	7.23	9.64	12.87	17.07	23.19	31.44	39.82	50.16	62.95

t	10	11	12	13	14	15	16	17	18	19
Year	1900	1910	1920	1930	1940	1950	1960	1970	1980	1990
Population	75.99	91.97	105.71	122.78	131.67	151.33	179.32	203.30	226.55	250.00

Using the statistical procedure *least squares regression analysis*, we found the best quadratic and exponential models for this data. Which of the following two equations is a better model for the population of the United States between 1800 and 1990? Describe the method you used to reach your conclusion.

Quadratic Model

$$P = 0.662t^2 + 0.211t + 6.165$$

Exponential Model

$$P = 7.7899e^{0.2013t}$$

WARM UP

The following warm-up exercises involve skills that were covered in earlier sections. You will use these skills in the exercise set for this section.

In Exercises 1–6, sketch the graph of the equation.

1. $y = 2^{0.25x}$

2. $y = 2^{-0.25x}$

3. $y = 4 \log_2 x$

4. $y = \ln(x - 3)$

5. $y = e^{-x^2/5}$

6. $y = \dfrac{2}{1 + e^{-x}}$

In Exercises 7–10, solve the equation for x. Round to three decimal places.

7. $3e^{2x} = 7$

8. $2e^{-0.2x} = 0.002$

9. $4 \ln 5x = 14$

10. $6 \ln 2x = 12$

EXERCISES for Section 6.5

Compound Interest In Exercises 1–10, complete the table for a savings account in which interest is compounded continuously.

Initial investment	Annual % rate	Effective yield	Time to double	Amount after 10 years
1. $1,000	12%			
2. $20,000	$10\frac{1}{2}$%			
3. $750			$7\frac{3}{4}$ yr	
4. $10,000			5 yr	
5. $500				$1,292.85
6. $2,000		4.5%		
7.	11%			$19,205.00
8.	8%			$20,000.00
9. $5,000		8.33%		
10. $250		12.19%		

Compound Interest In Exercises 11 and 12, determine the principal P which must be invested at rate r, compounded monthly, so that $500,000 will be available for retirement in t years.

11. $r = 7\frac{1}{2}$%, $t = 20$

12. $r = 12$%, $t = 40$

Compound Interest In Exercises 13 and 14, determine the time necessary for $1000 to double if it is invested at interest rate r compounded (a) annually, (b) monthly, (c) daily, and (d) continuously.

13. $r = 11$%

14. $r = 10\frac{1}{2}$%

15. *Compound Interest* Complete the following table for the time t necessary for P dollars to triple if interest is compounded continuously at rate r.

r	2%	4%	6%	8%	10%	12%
t						

16. *Compound Interest* Complete the following table for the time t necessary for P dollars to triple if interest is compounded annually at rate r.

r	2%	4%	6%	8%	10%	12%
t						

In Exercises 17–22, complete the table for the given radioactive isotope.

Isotope	Half-life (years)	Initial quantity	Amount after 1000 years	Amount after 10,000 years
17. Ra^{226}	1,620	10 g		
18. Ra^{226}	1,620		1.5 g	
19. C^{14}	5,730			2 g
20. C^{14}	5,730	3 g		
21. Pu^{230}	24,360		2.1 g	
22. Pu^{230}	24,360			0.4 g

In Exercises 23–26, find the constant k such that the exponential function $y = Ce^{kt}$ passes through the given points on the graph.

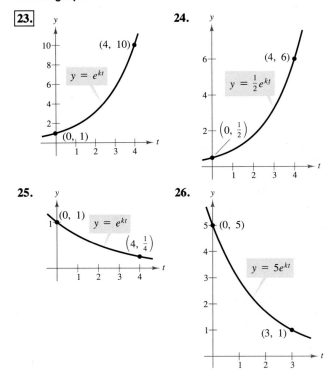

23.

$y = e^{kt}$

(4, 10)

(0, 1)

24.

$y = \frac{1}{2}e^{kt}$

(4, 6)

$\left(0, \frac{1}{2}\right)$

25.

(0, 1)

$y = e^{kt}$

$\left(4, \frac{1}{4}\right)$

26.

(0, 5)

$y = 5e^{kt}$

(3, 1)

27. *Population*   The population P of a city is given by

$$P = 105,300e^{0.015t}$$

where t is the time in years, with $t = 0$ corresponding to 1990. According to this model, in what year will the city have a population of 150,000?

28. *Population*   The population P of a city is given by

$$P = 240,360e^{0.012t}$$

where t is the time in years, with $t = 0$ corresponding to 1990. According to this model, in what year will the city have a population of 250,000?

29. *Population*   The population P of a city is given by

$$P = 2500e^{kt}$$

where t is the time in years, with $t = 0$ corresponding to the year 1990. In 1945, the population was 1350. Find the value of k and use this result to predict the population in the year 2010.

30. *Population*   The population P of a city is given by

$$P = 140,500e^{kt}$$

where t is the time in years, with $t = 0$ corresponding to the year 1990. In 1960, the population was 100,250. Find the value of k and use this result to predict the population in the year 2000.

31. *Population*   The population of Dhaka, Bangladesh was 4.22 million in 1990, and its projected population for the year 2000 is 6.49 million. (*Source:* U.S. Bureau of the Census) Find the exponential growth model $y = Ce^{kt}$ for the population growth of Dhaka by letting $t = 0$ correspond to 1990. Use the model to predict the population of the city in 2010.

32. *Population*   The population of Houston, Texas was 2.30 million in 1990, and its projected population for the year 2000 is 2.65 million. (*Source:* U.S. Bureau of the Census) Find the exponential growth model $y = Ce^{kt}$ for the population growth of Houston by letting $t = 0$ correspond to 1990. Use the model to predict the population of the city in 2010.

33. *Bacteria Growth*   The number of bacteria N in a culture is given by the model

$$N = 100e^{kt}$$

where t is the time in hours, with $t = 0$ corresponding to the time when $N = 100$. When $t = 5$, $N = 300$. How long will it take the population to double in size?

34. *Bacteria Growth*   The number of bacteria N in a culture is given by the model

$$N = 250e^{kt}$$

where t is the time in hours, with $t = 0$ corresponding to the time when $N = 250$. When $t = 10$, $N = 280$. How long will it take the population to double in size?

35. *Radioactive Decay*   The half-life of radioactive radium (Ra^{226}) is 1620 years. What percentage of a present amount of radioactive radium will remain after 100 years?

36. *Radioactive Decay* C^{14} dating assumes that the carbon dioxide on earth today has the same radioactive content as it did centuries ago. If this is true, then the amount of C^{14} absorbed by a tree that grew several centuries ago should be the same as the amount of C^{14} absorbed by a tree growing today. A piece of ancient charcoal contains only 15% as much of the radioactive carbon as a piece of modern charcoal. How long ago was the tree burned to make the ancient charcoal if the half-life of C^{14} is 5730 years?

37. *Depreciation*   A certain car that cost $22,000 new has a depreciated value of $16,500 after one year. Find the value of the car when it is three years old by using the exponential model $y = Ce^{kt}$.

38. *Depreciation*   A computer that cost $4600 new has a depreciated value of $3000 after two years. Find the value of the computer after three years by using the exponential model $y = Ce^{kt}$.

39. *Sales*   The sales S (in thousands of units) of a new product after it has been on the market t years are given by

$$S(t) = 100(1 - e^{kt}).$$

(a) Find S as a function of t if 15,000 units have been sold after one year.

(b) How many units will be sold after five years?

40. *Learning Curve*   The management at a factory has found that the maximum number of units a worker can produce in a day is 30. The learning curve for the number of units N produced per day after a new employee has worked t days is given by

$$N = 30(1 - e^{kt}).$$

After 20 days on the job, a worker produced 19 units per day.

(a) Find the learning curve for this worker (first, find the value of k).

(b) How many days should pass before this worker is producing 25 units per day?

41. *Stocking a Lake with Fish*   A certain lake was stocked with 500 fish and the fish population increased according to the logistics curve

$$p(t) = \frac{10,000}{1 + 19e^{-t/5}}$$

where t is measured in months (see figure).

(a) Estimate the fish population after five months.

(b) After how many months will the fish population be 2000?

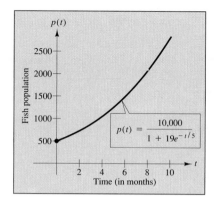

$$p(t) = \frac{10,000}{1 + 19e^{-t/5}}$$

Figure for 41

42. *Endangered Species*   A conservation organization releases 100 animals of an endangered species into a game preserve. The organization believes that the preserve has a carrying capacity of 1000 animals and that the growth of the herd will be modeled by the logistics curve

$$p(t) = \frac{1000}{1 + 9e^{-0.1656t}}$$

where t is measured in years (see figure).

(a) Estimate the population after five years.

(b) After how many years will the population be 500?

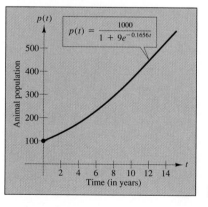

$$p(t) = \frac{1000}{1 + 9e^{-0.1656t}}$$

Figure for 42

43. *Sales and Advertising*   The sales S (in thousands of units) of a product after spending x hundred dollars in advertising is given by

$$S = 10(1 - e^{kx}).$$

(a) Find S as a function of x if 2500 units are sold when $500 is spent on advertising.

(b) Estimate the number of units that will be sold if advertising expenditures are raised to $700.

44. *Sales and Advertising*   After discontinuing all advertising for a certain product in 1988, the manufacturer noted that sales began to drop according to the model

$$S = \frac{500,000}{1 + 0.6e^{kt}}$$

where S represents the number of units sold and t represents the calendar year with $t = 0$ corresponding to 1988.
(a) Find k if the company sold 300,000 units in 1990.
(b) According to this model, what will sales be in 1993?

Earthquake Magnitudes   In Exercises 45 and 46, use the Richter Scale (Example 7) for measuring the magnitude of earthquakes.

45. Find the magnitude R of an earthquake of intensity I (let $I_0 = 1$).
(a) $I = 80,500,000$ (b) $I = 48,275,000$
46. Find the intensity I of an earthquake measuring R on the Richter Scale (let $I_0 = 1$).
(a) Colombia in 1906, $R = 8.6$
(b) Los Angeles in 1971, $R = 6.7$

Intensity of Sound   In Exercises 47–50 use the following information to determine the level of sound (in decibels) for the given sound intensity. The level of sound β, in decibels, with an intensity of I is given by

$$\beta(I) = 10 \log_{10} \frac{I}{I_0}$$

where I_0 is an intensity of 10^{-16} watts per square centimeter, corresponding roughly to the faintest sound that can be heard.

47. (a) $I = 10^{-14}$ watts per square centimeter (faint whisper)
(b) $I = 10^{-9}$ watts per square centimeter (busy street corner)
(c) $I = 10^{-6.5}$ watts per square centimeter (air hammer)
(d) $I = 10^{-4}$ watts per square centimeter (threshold of pain)
48. (a) $I = 10^{-13}$ watts per square centimeter (whisper)
(b) $I = 10^{-7.5}$ watts per square centimeter (DC-8 four miles from takeoff)
(c) $I = 10^{-7}$ watts per square centimeter (diesel truck at 25 feet)
(d) $I = 10^{-4.5}$ watts per square centimeter (auto horn at three feet)
49. *Noise Level*   Due to the installation of noise suppression materials, the noise level in an auditorium was reduced from 93 to 80 decibels. Find the percentage decrease in the intensity level of the noise because of the installation of these materials.

50. *Noise Level*   Due to the installation of a muffler, the noise level in an engine was reduced from 88 to 72 decibels. Find the percentage decrease in the intensity level of the noise because of the installation of the muffler.

Acidity   In Exercises 51–56, use the acidity model given by

$$pH = -\log_{10}[H^+]$$

where acidity (pH) is a measure of the hydrogen ion concentration $[H^+]$ (measured in moles of hydrogen per liter) of a solution.

51. Find the pH if $[H^+] = 2.3 \times 10^{-5}$.
52. Find the pH if $[H^+] = 11.3 \times 10^{-6}$.
53. Compute $[H^+]$ for a solution in which pH $= 5.8$.
54. Compute $[H^+]$ for a solution in which pH $= 3.2$.
55. A certain fruit has a pH of 2.5 and an antacid tablet has a pH of 9.5. The hydrogen ion concentration of the fruit is how many times the concentration of the tablet?
56. If the pH of a solution is decreased by one unit, the hydrogen ion concentration is increased by what factor?

57. *Estimating the Time of Death*   At 8:30 A.M., a coroner was called to the home of a person who had died during the night. In order to estimate the time of death, the coroner took the person's temperature twice. At 9:00 A.M. the temperature was 85.7°, and at 9:30 A.M. the temperature was 82.8°. From these two temperatures the coroner was able to determine that the time elapsed since death and the body temperature were related by the formula

$$t = -2.5 \ln \frac{T - 70}{98.6 - 70}$$

where t is the time in hours that has elapsed since the person died and T is the temperature (in degrees Fahrenheit) of the person's body at 9:00 A.M. Assume that the person had a normal body temperature of 98.6° at death, and that the room temperature was a constant 70°. (This formula is derived from a general cooling principle called Newton's Law of Cooling.) Use this formula to estimate the time of death of the person.

58. *Population Growth*   From Exercises 31 and 32, it is obvious that the populations of the two different cities are growing at different rates. What constant in the equation $y = Ce^{kt}$ is affected by those different growth rates? Discuss the relationship between the different growth rates and the magnitude of the constant.

REVIEW EXERCISES for Chapter 6

In Exercises 1–6, match the function with the sketch of its graph. [The graphs are labeled (a), (b), (c), (d), (e), and (f).]

1. $f(x) = 2^x$

2. $f(x) = 2^{-x}$

3. $f(x) = -2^x$

4. $f(x) = 2^x + 1$

5. $f(x) = \log_2 x$

6. $f(x) = \log_2(x - 1)$

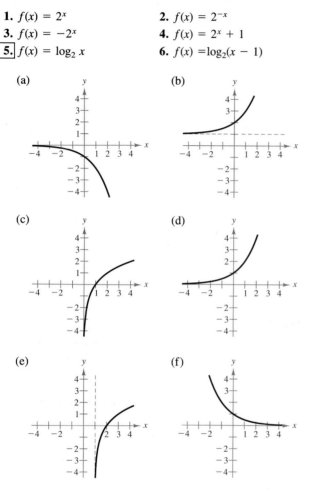

(a)

(b)

(c)

(d)

(e)

(f)

In Exercises 7–14, sketch the graph of the function.

7. $f(x) = 6^x$

8. $f(x) = 0.3^x$

9. $g(x) = 6^{-x}$

10. $g(x) = 0.3^{-x}$

11. $h(x) = e^{-x/2}$

12. $h(x) = 2 - e^{-x/2}$

13. $f(x) = e^{x+2}$

14. $s(t) = 4e^{-2/t}, \quad t > 0$

In Exercises 15 and 16, complete the table to determine the balance A for P dollars invested at rate r for t years and compounded n times per year.

n	1	2	4	12	365	Continuous compounding
A						

15. $P = \$3500$, $r = 10.5\%$, $t = 10$ years

16. $P = \$2000$, $r = 12\%$, $t = 30$ years

In Exercises 17 and 18, complete the table to determine the amount of money P that should be invested at rate r to produce a final balance of \$200,000 in t years.

t	1	10	20	30	40	50
P						

17. $r = 8\%$, compounded continuously

18. $r = 10\%$, compounded monthly

19. *Trust Fund*   On the day your child was born you deposited \$50,000 into a trust fund that pays 8.75% interest, compounded continuously. Determine the balance in the account at the time of your child's 35th birthday.

20. *Depreciation*   After t years, the value of a car that cost \$14,000 is given by

$$V(t) = 14{,}000\left(\frac{3}{4}\right)^t.$$

Sketch a graph of the function and determine the value of the car two years after it was purchased.

21. *Drug Decomposition*   A solution of a certain drug contained 500 units per milliliter when prepared. It was analyzed after 40 days and found to contain 300 units per milliliter. Assuming that the rate of decomposition is proportional to the amount present, the equation giving the amount A after t days is

$$A = 500e^{-0.013t}.$$

Use this model to find A when $t = 60$.

22. *Waiting Times* The average time between incoming calls at a switchboard is three minutes. The probability of waiting less than t minutes until the next incoming call is approximated by the model

$$F(t) = 1 - e^{-t/3}.$$

If a call has just come in, find the probability that the next call will be within (a) $\frac{1}{2}$ minute, (b) two minutes, (c) five minutes.

23. *Fuel Efficiency* A certain automobile gets 28 miles per gallon of gasoline for speeds up to 50 miles per hour. Over 50 miles per hour, the number of miles per gallon drops at the rate of 12% for each 10 miles per hour. If s is the speed and y is the number of miles per gallon, then

$$y = 28e^{0.6-0.012s}, \qquad s \geq 50.$$

Use this function to complete the table.

Speed	50	55	60	65	70
Miles per gallon					

24. *Inflation* If the annual rate of inflation averages 5% over the next 10 years, then the approximate cost C of goods or services during any year in that decade will be given by

$$C(t) = P(1.05)^t$$

where t is the time in years and P is the present cost. If the price of a tire for your car is presently $69.95, estimate the price 10 years from now.

In Exercises 25–30, sketch the graph of the function.

25. $g(x) = \log_3 x$

26. $g(x) = \log_5 x$

27. $f(x) = \ln x + 3$

28. $f(x) = \ln (x - 3)$

29. $h(x) = \ln(e^{x-1})$

30. $f(x) = \frac{1}{4} \ln x$

In Exercises 31 and 32, use the definition of a logarithm to write the given equation in logarithmic form.

31. $4^3 = 64$

32. $25^{3/2} = 125$

In Exercises 33–40, evaluate the given expression without using a calculator.

33. $\log_{10} 1000$

34. $\log_9 3$

35. $\log_3 \dfrac{1}{9}$

36. $\log_4 \dfrac{1}{16}$

37. $\ln e^7$

38. $\log_a \dfrac{1}{a}$

39. $\ln 1$

40. $\ln e^{-3}$

In Exercises 41–44, evaluate the logarithm using the change of base formula. Do each problem twice, once with common logarithms and once with natural logarithms. Round to three decimal places.

41. $\log_4 9$

42. $\log_{1/2} 5$

43. $\log_{12} 200$

44. $\log_3 0.28$

In Exercises 45–50, use the properties of logarithms to write the expression as a sum, difference, and/or multiple of logarithms.

45. $\log_5 5x^2$

46. $\log_7 \dfrac{\sqrt{x}}{4}$

47. $\log_{10} \dfrac{5\sqrt{y}}{x^2}$

48. $\ln \left| \dfrac{x-1}{x+1} \right|$

49. $\ln[(x^2 + 1)(x - 1)]$

50. $\ln \sqrt[5]{\dfrac{4x^2 - 1}{4x^2 + 1}}$

In Exercises 51–56, write the expression as the logarithm of a single quantity.

51. $\log_2 5 + \log_2 x$

52. $\log_6 y - 2 \log_6 z$

53. $\frac{1}{2} \ln |2x - 1| - 2 \ln |x + 1|$

54. $5 \ln |x - 2| - \ln |x + 2| - 3 \ln |x|$

55. $\ln 3 + \frac{1}{3} \ln(4 - x^2) - \ln x$

56. $3[\ln x - 2 \ln(x^2 + 1)] + 2 \ln 5$

In Exercises 57–60, determine whether the statement or equation is true or false.

57. The domain of the function $f(x) = \ln x$ is the set of all real numbers.

58. $\log_b b^{2x} = 2x$

59. $\ln(x + y) = \ln x + \ln y$

60. $e^{x-1} = \dfrac{e^x}{e}$

In Exercises 61–64, approximate the logarithm using the properties of logarithms given $\log_b 2 \approx 0.3562$, $\log_b 3 \approx 0.5646$, and $\log_b 5 \approx 0.8271$.

61. $\log_b 25$

62. $\log_b\left(\frac{25}{9}\right)$

63. $\log_b \sqrt{3}$

64. $\log_b 30$

65. *Snow Removal* The number of miles s of roads cleared of snow is approximated by the model

$$s = 25 - \frac{13 \ln(h/12)}{\ln 3}, \quad 2 \le h \le 15$$

where h is the depth of the snow in inches. Use this model to find s when $h = 10$ inches.

66. *Climb Rate* The time t, in minutes, for a small plane to climb to an altitude of h feet is given by

$$t = 50 \log_{10} \frac{18,000}{18,000 - h}$$

where 18,000 feet is its absolute ceiling. Find the time for the plane to climb to an altitude of 4000 feet.

In Exercises 67–72, solve the exponential equation. Round to three decimal places.

67. $e^x = 12$

68. $e^{3x} = 25$

69. $3e^{-5x} = 132$

70. $14e^{3x+2} = 560$

71. $e^{2x} - 7e^x + 10 = 0$

72. $e^{2x} - 6e^x + 8 = 0$

In Exercises 73–76, solve the logarithmic equation. Round to three decimal places.

73. $\ln 3x = 8.2$

74. $2 \ln 4x = 15$

75. $\ln x - \ln 3 = 2$

76. $\ln \sqrt{x + 1} = 2$

In Exercises 77–80, find the exponential function $y = Ce^{kt}$ that passes through the two points.

77. $(0, 2)$, $(4, 3)$

78. $\left(0, \frac{1}{2}\right)$, $(5, 5)$

79. $(0, 4)$, $\left(5, \frac{1}{2}\right)$

80. $(0, 2)$, $(5, 1)$

81. *Demand Function* The demand equation for a certain product is given by

$$p = 500 - 0.5e^{0.004x}.$$

Find the demand x for a price of (a) $p = \$450$ and (b) $p = \$400$.

82. *Typing Speed* In a typing class, the average number of words per minute typed after t weeks of lessons was found to be

$$N = \frac{157}{1 + 5.4e^{-0.12t}}.$$

Find the time necessary to type (a) 50 words per minute and (b) 75 words per minute.

83. *Compound Interest* A deposit of $750 is made in a savings account for which the interest is compounded continuously. The balance will double in $7\frac{3}{4}$ years.
(a) What is the annual percentage rate for this account?
(b) Find the balance in the account after 10 years.
(c) Find the effective yield.

84. *Compound Interest* A deposit of $10,000 is made in a savings account for which the interest is compounded continuously. The balance will double in five years.
(a) What is the annual percentage rate for this account?
(b) Find the balance after one year.
(c) Find the effective yield.

85. *Sound Intensity* The relationship between the number of decibels β and the intensity of a sound I in watts per centimeter squared is given by

$$\beta = 10 \log_{10}\left(\frac{I}{10^{-16}}\right).$$

Determine the intensity of a sound in watts per centimeter squared if the decibel level is 125.

86. *Earthquake Magnitudes* On the Richter Scale, the magnitude R of an earthquake of intensity I is given by

$$R = \log_{10} \frac{I}{I_0}$$

where $I_0 = 1$ is the minimum intensity used for comparison. Find the intensity per unit of area for the following R.
(a) $R = 8.4$
(b) $R = 6.85$
(c) $R = 9.1$

CUMULATIVE TEST for Chapters 4–6

Take this test as you would take a test in class. After you are done, check your work with the answers given in the back of the book.

1. Sketch a graph of the function $h(x) = -(x^2 + 4x)$.

2. Sketch a graph of the function $f(t) = \frac{1}{4}t(t - 2)^2$.

3. Perform the division: $\dfrac{6x^3 - 4x^2}{2x^2 + 1}$

4. Use synthetic division to perform the division: $\dfrac{3x^3 - 5x + 4}{x - 2}$

5. Find all the zeros of the function $f(x) = x^3 + 2x^2 + 4x + 8$.

6. Use the Bisection Method to approximate the real zero of the function $g(x) = x^3 + 3x^2 - 6$ to the nearest hundredth.

7. Sketch a graph of the function $g(s) = \dfrac{2s}{s - 3}$.

8. Sketch a graph of the function $g(s) = \dfrac{2s^2}{s - 3}$.

9. Sketch a graph of the equation $6x - y^2 = 0$.

10. Sketch a graph of the equation $\dfrac{(x - 2)^2}{4} + \dfrac{(y + 1)^2}{9} = 1$.

11. Find an equation of the parabola with vertex $(3, -2)$, vertical axis, and passing through the point $(0, 4)$.

12. Find an equation of the hyperbola with foci $(0, 0)$ and $(0, 4)$ and asymptotes $y = \pm\frac{1}{2}x + 2$.

13. Sketch a graph of the function $f(x) = 6(2^{-x})$.

14. Sketch a graph of the function $g(x) = \log_3 x$.

15. Evaluate without the aid of a calculator: $\log_5 125$

16. Use the properties of logarithms to write the expression $2 \ln x - \frac{1}{2} \ln(x + 5)$ as the logarithm of a single quantity.

17. Solve: $6e^{2x} = 72$

18. Solve: $\log_2 x + \log_2 5 = 6$

19. Let x be the amount (in hundreds of dollars) that a company spends on advertising, and let P be the profit, where

$$P = 230 + 20x - \tfrac{1}{2}x^2.$$

What amount of advertising will yield maximum profit?

20. On the day a grandchild is born, a grandparent deposits $2500 into a fund earning 7.5%, compounded continuously. Determine the balance in the account at the time of the grandchild's 25th birthday.

CHAPTER 7

OVERVIEW

In this chapter, you will study ways to solve systems of equations. A system of equations can be used to model many different types of real-life situations. For instance, Example 7 on page 419 uses a system of equations to compare the sales of compact disc players and turntables from 1983 through 1987. During these years, the sale of turntables decreased at a linear rate and the sale of compact disc players increased at a linear rate. The solution of the system represents the time at which the sale of compact disc players overtook the sale of turntables.

Solving a system of equations can involve many arithmetic steps. Because it is easy to make errors, be sure to check your solution(s) in the original system or problem statement.

Systems of Equations and Inequalities

7.1 Systems of Equations

The Method of Substitution / Graphical Approach to Finding Solutions / Applications

The Method of Substitution

Up to this point in the text most problems have involved either a function of one variable or a single equation in two variables. However, many problems in science, business, and engineering involve two or more equations in two or more variables. To solve such problems, you need to find solutions of a **system of equations.**

Here is an example of a system of two equations in x and y.

$$2x + y = 5 \qquad \text{\textit{Equation 1}}$$
$$3x - 2y = 4 \qquad \text{\textit{Equation 2}}$$

A **solution** of this system is an ordered pair that satisfies each equation in the system, and when you find the set of all solutions you are **solving the system of equations.** For instance, the ordered pair (2, 1) is a solution of this system. To check this you can substitute 2 for x and 1 for y into *each* equation, as follows.

$$2(2) + 1 = 5 \qquad \text{\textit{Equation 1 checks}}$$
$$3(2) - 2(1) = 4 \qquad \text{\textit{Equation 2 checks}}$$

413

There are several different ways to solve systems of equations. In this chapter we will consider three of the most common techniques. We begin with the **method of substitution.** This method has five basic steps, which can be labeled *solve*, *substitute*, *solve*, *back-substitute*, and *check*. We illustrate these five steps in Example 1.

EXAMPLE 1   Solving a System of Two Equations in Two Variables

Solve the following system of equations.

$$x + y = 4 \qquad\qquad \text{\textit{Equation 1}}$$
$$x - y = 2 \qquad\qquad \text{\textit{Equation 2}}$$

Solution

Solving for y in Equation 1, you get

$$y = 4 - x.$$

Substituting $(4 - x)$ for y in Equation 2, you obtain a single-variable equation, which you then *solve* for x.

$$x - (4 - x) = 2$$
$$x - 4 + x = 2$$
$$2x = 6$$
$$x = 3$$

REMARK The term *back-substitution* implies that you work *backwards*. First solve for one of the variables, and then substitute that value *back* into one of the equations in the system to find the value of the other variable.

Finally, by *back-substituting* $x = 3$ into the equation $y = 4 - x$, you obtain

$$y = 4 - x = 4 - 3 = 1.$$

Thus, the solution is the ordered pair $(3, 1)$.

Check

Equation 1: $3 + 1 = 4$ *Check solution in each equation in the original system.*
Equation 2: $3 - 1 = 2$

Because many steps are required to solve a system of equations, it is very easy to make errors in arithmetic. Thus, we *strongly* suggest that you always *check your solution by substituting it into each equation in the original system.*

Method of Substitution

To solve a system of two equations in two variables, use the following process.

1. *Solve* one of the equations for one variable in terms of the other.
2. *Substitute* the expression found in Step 1 into the other equation to obtain an equation of one variable.
3. *Solve* the equation obtained in Step 2.
4. *Back-substitute* the solution in Step 3 into the expression obtained in Step 1 to find the value of the other variable.
5. *Check* the solution to see that it satisfies *each* of the original equations.

EXAMPLE 2   Solving a System by Substitution: One-Solution Case

A total of $12,000 is invested in two funds paying 9% and 11% simple interest. If the yearly interest is $1,180, how much of the $12,000 is invested at each rate?

Solution

Verbal Model

$$\frac{9\%}{\text{fund}} + \frac{11\%}{\text{fund}} = \frac{\text{Total}}{\text{investment}}$$

$$\frac{9\%}{\text{interest}} + \frac{11\%}{\text{interest}} = \frac{\text{Total}}{\text{interest}}$$

Labels

9% fund = x,   9% interest = $0.09x$
11% fund = y,   11% interest = $0.11y$
Total investment = $12,000,   Total interest = $1,180

System of Equations

$$x + \quad y = 12{,}000 \quad \textit{Equation 1}$$
$$0.09x + 0.11y = \quad 1{,}180 \quad \textit{Equation 2}$$

To begin, it is convenient to multiply both sides of the second equation by 100 to obtain $9x + 11y = 118{,}000$. This eliminates the need to work with decimals. Then the following steps are performed.

1. Solve for x in Equation 1.

$$x = 12{,}000 - y$$

2. Substitute this expression for x into the new Equation 2.

$$9(12{,}000 - y) + 11y = 118{,}000$$

3. Solve for y.

$$108{,}000 - 9y + 11y = 118{,}000$$
$$2y = 10{,}000$$
$$y = 5{,}000 \text{ (dollars)}$$

4. Back-substitute the value $y = 5,000$ to solve for x.

$$x = 12,000 - 5,000 = 7,000 \text{ (dollars)}$$

Therefore, the solution is the ordered pair (7000, 5000). Check to see that $x = 7,000$ and $y = 5,000$ satisfy each of the original equations. ◢

Note that the equations in Examples 1 and 2 are linear. That is, the variables x and y occurred to the first power only. The method of substitution can also be used to solve systems in which one or both of the equations are nonlinear.

EXAMPLE 3   Solving a System by Substitution: Two-Solution Case

Solve the following system of equations.

$$x^2 - x - y = 1 \qquad\qquad \textit{Equation 1}$$
$$-x + y = -1 \qquad\qquad \textit{Equation 2}$$

Solution

1. Solve for y in Equation 2.

$$-x + y = -1$$
$$y = x - 1$$

2. Substitute this expression for y into Equation 1.

$$x^2 - x - (x - 1) = 1$$

3. Solve for x.

$$x^2 - 2x + 1 = 1$$
$$x^2 - 2x = 0$$
$$x(x - 2) = 0$$
$$x = 0, x = 2$$

4. Back-substitute these values of x to solve for the corresponding values of y.

$$\text{For } x = 0: y = 0 - 1 = -1$$
$$\text{For } x = 2: y = 2 - 1 = 1$$

Thus, there are two solutions: $(0, -1)$ and $(2, 1)$. Check these solutions in the original system to see that both work. ◢

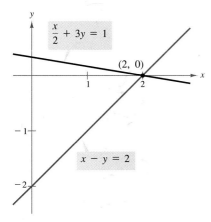

(a) One Point of Intersection

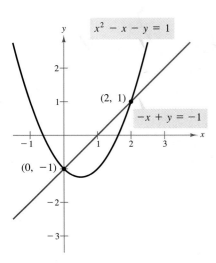

(b) Two Points of Intersection

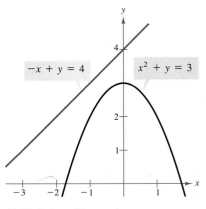

(c) No Points of Intersection

FIGURE 7.1

EXAMPLE 4 **Solving a System by Substitution: No-Solution Case**

Solve the following system of equations.

$$-x + y = 4 \qquad \qquad \textit{Equation 1}$$
$$x^2 + y = 3 \qquad \qquad \textit{Equation 2}$$

Solution

1. In this case, you solve for y in Equation 1.

$$y = x + 4$$

2. Substitute this expression for y into Equation 2.

$$x^2 + (x + 4) = 3$$

3. Solve for x.

$$x^2 + x + 4 = 3$$
$$x^2 + x + 1 = 0$$
$$x = \frac{-1 \pm \sqrt{1^2 - 4(1)(1)}}{2} \qquad \textit{Quadratic Formula}$$

Since the discriminant is negative, the equation $x^2 + x + 1 = 0$ has no (real) solution. Hence, this system has no (real) solution. ◢

Graphical Approach to Finding Solutions

From Examples 2, 3, and 4 you can see that a system of two equations in two unknowns can have exactly one solution, more than one solution, or no solution. In practice, you can gain insight about the location and number of solutions of a system of equations by graphing each of the equations on the same coordinate plane. The solutions of the system correspond to the **points of intersection** of the graphs. For instance, in Figure 7.1(a) the two equations graph as two lines with a *single point* of intersection. The two equations in Example 3 graph as a parabola and a line with *two points* of intersection, as shown in Figure 7.1(b). Moreover, the two equations in Example 4 graph as a line and a parabola that happen to have *no points* of intersection, as shown in Figure 7.1(c).

Occasionally, the graphical approach to solving a system of equations is easier than the method of substitution.

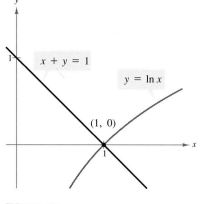

FIGURE 7.2

EXAMPLE 5   Solving a System of Equations

Solve the following system of equations.

$$y = \ln x \qquad \qquad \textit{Equation 1}$$
$$x + y = 1 \qquad \qquad \textit{Equation 2}$$

Solution

The graph of each equation is shown in Figure 7.2. From this sketch it is clear that there is only one point of intersection. Also, it appears that $(1, 0)$ is the solution point, and you can confirm this by checking these coordinates in *both* equations.

Check   Let $x = 1$ and $y = 0$.

$$0 = \ln 1 \qquad \qquad \textit{Equation 1 checks}$$
$$1 + 0 = 1 \qquad \qquad \textit{Equation 2 checks}$$

REMARK   Example 5 shows you the value of a graphical approach to solving systems of equations in two variables. Notice what would have happened if you had tried only the substitution method in Example 5. By substituting $y = \ln x$ into $x + y = 1$, you obtain $x + \ln x = 1$. It would be difficult to solve this equation for x using standard algebraic techniques.

Applications

The total cost C of producing x units of a product typically has two components—the initial cost and the cost per unit. When enough units have been sold so that the total revenue R equals the total cost, sales have reached the **break-even point.** The break-even point corresponds to the point of intersection of the cost and revenue curves.

EXAMPLE 6   An Application: Break-Even Analysis

A small business invests \$10,000 in equipment to produce a product. Each unit of the product costs \$0.65 to produce and is sold for \$1.20. How many items must be sold before the business breaks even?

Solution

The total cost of producing x units is

$$C = \underbrace{0.65x}_{\substack{\text{Cost per} \\ \text{unit}}} + \underbrace{10{,}000}_{\substack{\text{Initial} \\ \text{cost}}} \qquad \qquad \textit{Equation 1}$$

and the revenue obtained by selling x units is

$$R = \underbrace{1.2x}_{\substack{\text{Price} \\ \text{per unit}}}. \qquad \qquad \textit{Equation 2}$$

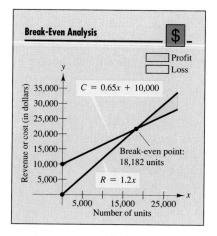

FIGURE 7.3

Since the break-even point occurs when $R = C$, we have

$$1.2x = 0.65x + 10,000 \qquad \textit{Solve for R and C}$$
$$0.55x = 10,000$$
$$x = \frac{10,000}{0.55} \approx 18,182 \text{ units.}$$

Note in Figure 7.3 that sales less than the break-even point correspond to an overall loss, while sales greater than the break-even point correspond to a profit.

EXAMPLE 7 Compact Disc Players and Turntables

Between 1983 and 1987 the number of compact disc players sold each year in the United States was *increasing* and the number of turntables was *decreasing*. Two models that approximate the sales are

$$S = -1700 + 496t \qquad \textit{Compact disc players}$$
$$S = 1972 - 82t \qquad \textit{Turntables}$$

where S represents the annual sales in thousands of units and t represents the calendar year with $t = 3$ corresponding to 1983 (see Figure 7.4). According to these two models, when would you expect the sales of compact disc players to have exceeded the sales of turntables? (*Source:* Dealerscope Merchandising)

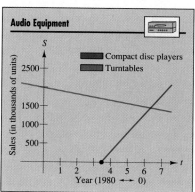

FIGURE 7.4

Solution

Since the first equation has already been solved for S in terms of t, substitute this value into the second equation and solve for t as follows.

$$-1700 + 496t = 1972 - 82t$$
$$496t + 82t = 1972 + 1700$$
$$578t = 3672$$
$$t \approx 6.4$$

Thus, from the given models, you would expect that the sales of compact disc players exceeded the sales of turntables sometime during 1986.

DISCUSSION PROBLEM

Points of Intersection of Two Graphs

In this section, you learned that the graphs of two equations can intersect at zero, one, or more than one point. Sketch the graphs of the following systems. Which represents a system with no solution? Which represents a system with one solution? Which represents a system with two solutions?

1. $x^2 + y^2 = 4$
   $x + y = 2$

2. $x^2 + y^2 = 8$
   $x + y = 4$

3. $x^2 + y^2 = 4$
   $x + y = 6$

Find an example of three other systems of equations, one with no solution, one with one solution, and one with two solutions.

WARM UP

The following warm-up exercises involve skills that were covered in earlier sections. You will use these skills in the exercise set for this section.

In Exercises 1–4, sketch the graph of the equation.

1. $y = -\frac{1}{3}x + 6$

2. $y = 2(x - 3)$

3. $x^2 + y^2 = 4$

4. $y = 5 - (x - 3)^2$

In Exercises 5–8, perform the indicated operations and simplify.

5. $(3x + 2y) - 2(x + y)$

6. $(-10u + 3v) + 5(2u - 8v)$

7. $x^2 + (x - 3)^2 + 6x$

8. $y^2 - (y + 1)^2 + 2y$

In Exercises 9 and 10, solve the equation.

9. $3x + (x - 5) = 15 + 4$

10. $y^2 + (y - 2)^2 = 2$

EXERCISES for Section 7.1

In Exercises 1–10, solve the system by the method of substitution.

1. $2x + y = 4$
   $-x + y = 1$

2. $x - y = -5$
   $x + 2y = 4$

3. $x - y = -3$
   $x^2 - y = -1$

4. $3x - y = -2$
   $x^3 - y = 0$

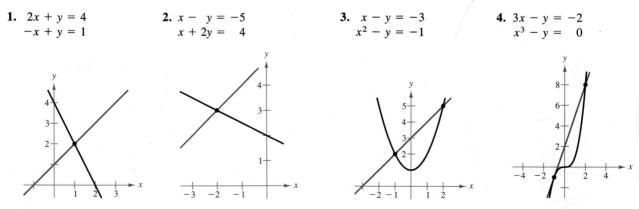

5. $x + 3y = 15$
$x^2 + y^2 = 25$

6. $x \quad\quad - y = 0$
$x^3 - 5x + y = 0$

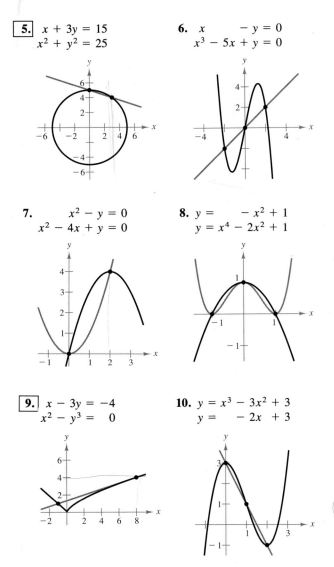

7. $x^2 - y = 0$
$x^2 - 4x + y = 0$

8. $y = \quad - x^2 + 1$
$y = x^4 - 2x^2 + 1$

9. $x - 3y = -4$
$x^2 - y^3 = \quad 0$

10. $y = x^3 - 3x^2 + 3$
$y = \quad - 2x + 3$

19. $x - y = 0$
$2x + y = 0$

20. $x - 2y = 0$
$3x - y = 0$

21. $y = 2x$
$y = x^2 + 1$

22. $x + y = 4$
$x^2 + y = 2$

23. $3x - 7y + 6 = 0$
$x^2 - y^2 = 4$

24. $x^2 + y^2 = 25$
$2x + y = 10$

25. $x^2 + y^2 = 5$
$x - y = 1$

26. $y = \quad x^3 - 2x^2 + x - 1$
$y = -x^2 + 3x \quad\quad - 1$

27. $y = x^4 - 2x^2 + 1$
$y = 1 - x^2$

28. $x^2 + y = 4$
$2x - y = 1$

29. $xy - 1 = 0$
$2x - 4y + 7 = 0$

30. $x - 2y = 1$
$y = \sqrt{x - 1}$

In Exercises 31–42, find all points of intersection of the graphs of the given pair of equations. [*Hint:* A graphical approach, as demonstrated in Example 5, may be helpful.]

31. $x + y \quad = 4$
$x^2 + y^2 - 4x = 0$

32. $x - y + 3 = 0$
$x^2 - 4x + 7 = y$

33. $2x - y + 3 = 0$
$x^2 + y^2 - 4x = 0$

34. $3x - 2y = 0$
$x^2 - y^2 = 4$

35. $x^2 + y^2 = 25$
$(x - 8)^2 + y^2 = 41$

36. $x^2 + y^2 = 8$
$y = x^2$

37. $y \quad = e^x$
$x - y + 1 = 0$

38. $x + 2y = 8$
$y = \log_2 x$

39. $y = \sqrt{x}$
$y = x$

40. $x - y = 3$
$x - y^2 = 1$

41. $x^2 + y^2 = 169$
$x^2 - 8y = 104$

42. $x^2 + y^2 = 4$
$2x^2 - y = 2$

In Exercises 11–30, solve the system by the method of substitution.

11. $x - y = 0$
$5x - 3y = 10$

12. $x + 2y = 1$
$5x - 4y = -23$

13. $2x - y + 2 = 0$
$4x + y - 5 = 0$

14. $6x - 3y - 4 = 0$
$x + 2y - 4 = 0$

15. $30x - 40y - 33 = 0$
$10x + 20y - 21 = 0$

16. $1.5x + 0.8y = 2.3$
$0.3x - 0.2y = 0.1$

17. $\frac{1}{5}x + \frac{1}{2}y = 8$
$x + y = 20$

18. $\frac{1}{2}x + \frac{3}{4}y = 10$
$\frac{3}{2}x - y = 4$

Break-Even Analysis In Exercises 43–46, find the sales necessary to break even $(R = C)$ for the given cost C of x units, and the given revenue R obtained by selling x units. (Round your answer to the nearest whole unit.)

43. $C = 8650x + 250,000,$ $\quad R = 9950x$

44. $C = 5.5\sqrt{x} + 10,000,$ $\quad R = 3.29x$

45. $C = 2.65x + 350,000,$ $\quad R = 4.15x$

46. $C = 0.08x + 50,000,$ $\quad R = 0.25x$

47. *Break-Even Point* Suppose you are setting up a small business and have invested $16,000 to produce an item that will sell for $5.95. If each unit can be produced for $3.45, how many units must be sold to break even?

48. *Break-Even Point* Suppose you are setting up a small business and have an initial investment of $5000. The unit cost of the product is $21.60, and the selling price is $34.10. How many units must be sold to break even?

49. *Investment Portfolio* A total of $25,000 is invested in two funds paying 8% and 8.5% simple interest. If the yearly interest is $2,060, how much of the $25,000 is invested at each rate?

50. *Investment Portfolio* A total of $18,000 is invested in two funds paying 7.75% and 8.25% simple interest. If the yearly interest is $1,455, how much of the $18,000 is invested at each rate?

51. *Choice of Two Jobs* Suppose you are offered two different jobs selling dental supplies. One company offers a straight commission of 6% of sales. The other company offers a salary of $250 per week *plus* 3% of the sales. How much would you have to sell in a week in order to make the straight commission offer better?

52. *Choice of Two Jobs* Suppose you are offered two different jobs selling college textbooks. One company offers an annual salary of $20,000 *plus* a year-end bonus of 1% of your total sales. The other company offers a salary of $15,000 *plus* a year-end bonus of 2% of your total sales. How much would you have to sell in a year in order to make the second offer better than the first?

53. *Log Volume* You are offered two different rules for estimating the number of board feet in a log that is 16 feet long. One is the *Doyle Log Rule* and is modeled by

$$V = (D - 4)^2, \quad 5 \le D \le 40$$

and the other is the *Scribner Log Rule* and is modeled by

$$V = 0.79D^2 - 2D - 4, \quad 5 \le D \le 40$$

where D is the diameter of the log and V is its volume in board feet (see figure).
(a) For what diameter do the two scales agree?
(b) If you were selling large logs, which scale would you want to be used?

54. *Market Equilibrium* The supply and demand curves for a business dealing with wheat are given by

Supply: $p = 1.45 + 0.00014x^2$
Demand: $p = (2.388 - 0.007x)^2$

where p is the price in dollars per bushel and x is the quantity in bushels per day (see figure). Find the market equilibrium.

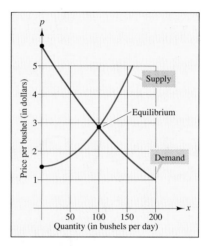

Figure for 54

55. *Area* What are the dimensions of a rectangular tract of land if its perimeter is 40 miles and its area is 96 square miles?

56. *Area* What are the dimensions of an isosceles right triangle with a 2-inch-long hypotenuse and an area of 1 square inch?

57. Find an equation of a line whose graph intersects the graph of the parabola $y = x^2$ at (a) two points, (b) one point, and (c) no points. (The answers are not unique.)

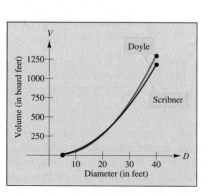

Figure for 53

7.2 Systems of Linear Equations in Two Variables

The Method of Elimination / Graphical Interpretation of Solutions / Applications

The Method of Elimination

In Section 7.1, we discussed two methods of solving a system of equations (by substitution and by graphing). In this section, we discuss the **method of elimination.** The key step in the method of elimination is to obtain, for one of the variables, coefficients that differ only in sign so that by *adding* the two equations this variable will be eliminated. This is true for the following system.

$$
\begin{array}{ll}
3x + 5y = 7 & \textit{Equation 1} \\
\underline{-3x - 2y = -1} & \textit{Equation 2} \\
3y = 6 & \textit{Add equations}
\end{array}
$$

Note that by adding the two equations, you eliminated the variable x and obtained a single equation in y. Solving this equation for y produces $y = 2$, which you back-substitute into one of the original equations to solve for x.

EXAMPLE 1   The Method of Elimination

Solve the following system of linear equations.

$$
\begin{array}{ll}
3x + 2y = 4 & \textit{Equation 1} \\
5x - 2y = 8 & \textit{Equation 2}
\end{array}
$$

Solution

Begin by noting that the coefficients for y differ only in sign. Therefore, by adding the two equations, you can eliminate y.

$$
\begin{array}{ll}
3x + 2y = 4 & \textit{Equation 1} \\
\underline{5x - 2y = 8} & \textit{Equation 2} \\
8x = 12 & \textit{Add equations}
\end{array}
$$

Therefore, $x = \frac{3}{2}$. By back-substituting this value into the first equation, you can solve for y, as follows.

$$
\begin{array}{ll}
3x + 2y = 4 & \textit{Equation 1} \\
3\left(\dfrac{3}{2}\right) + 2y = 4 & \textit{Replace x by } \frac{3}{2} \\
\phantom{3\left(\dfrac{3}{2}\right) + 2}y = -\dfrac{1}{4} & \textit{Solve for y}
\end{array}
$$

Therefore, the solution is $\left(\frac{3}{2}, -\frac{1}{4}\right)$. Check this solution in the original system of linear equations.

Try using the method of substitution to solve the system given in Example 1. Which method do you think is easier? Many people find that the method of elimination is more efficient.

The Method of Elimination

To use the **method of elimination** to solve a system of two linear equations in x and y, use the following steps.

1. Obtain coefficients for x (or y) that differ only in sign by multiplying all terms of one or both equations by suitably chosen constants.
2. Add the equations to eliminate one variable and solve the resulting equation.
3. Back-substitute the value obtained in Step 2 into either of the original equations and solve for the other variable.
4. Check your solution in both of the original equations.

To obtain coefficients (for one of the variables) that differ only in sign we often need to multiply one or both of the equations by a suitable constant.

EXAMPLE 2   The Method of Elimination

Solve the following system of linear equations.

$$2x - 3y = -7 \qquad \text{\textit{Equation 1}}$$
$$3x + y = -5 \qquad \text{\textit{Equation 2}}$$

Solution

For this system, you can obtain coefficients that differ only in sign by multiplying the second equation by 3.

$$
\begin{array}{llll}
2x - 3y = -7 & \rightarrow & 2x - 3y = -7 & \textit{Equation 1} \\
3x + y = -5 & \rightarrow & \underline{9x + 3y = -15} & \textit{Multiply Equation 2 by 3} \\
& & 11x = -22 & \textit{Add equations}
\end{array}
$$

Thus, you see that $x = -2$. By back-substituting this value of x into the first equation, you can solve for y.

$$
\begin{array}{ll}
2x - 3y = -7 & \textit{Equation 1} \\
2(-2) - 3y = -7 & \textit{Replace x by } -2 \\
-3y = -3 & \textit{Add 4 to both sides} \\
y = 1 & \textit{Solve for y}
\end{array}
$$

Therefore, the solution is $(-2, 1)$. Check to see that it satisfies both original equations.

In Example 2, the two systems of linear equations

$$2x - 3y = -7 \qquad \text{and} \qquad 2x - 3y = -7$$
$$3x + y = -5 \qquad \qquad \qquad 9x + 3y = -15$$

are **equivalent** because they have precisely the same solution set. The operations that can be performed on a system of linear equations to produce an equivalent system are (1) interchange two equations, (2) multiply an equation by a nonzero constant, and (3) add a multiple of an equation to another equation.

EXAMPLE 3   The Method of Elimination: One-Solution Case

Solve the following system of linear equations.

$$5x + 3y = 9 \qquad \qquad Equation \ 1$$
$$2x - 4y = 14 \qquad \qquad Equation \ 2$$

Solution

You can obtain coefficients that differ only in sign by multiplying the first equation by 4 and the second equation by 3.

$$5x + 3y = \ \ 9 \quad \rightarrow \quad 20x + 12y = 36 \qquad Multiply \ Equation \ 1 \ by \ 4$$
$$2x - 4y = 14 \quad \rightarrow \quad \underline{\ \ 6x - 12y = 42} \qquad Multiply \ Equation \ 2 \ by \ 3$$
$$26x \qquad = 78 \qquad Add \ equations$$

From this equation, you can see that $x = 3$. By back-substituting this value of x into the second equation, you can solve for y, as follows.

$$2x - 4y = 14 \qquad \qquad Equation \ 2$$
$$2(3) - 4y = 14 \qquad \qquad Replace \ x \ by \ 3$$
$$-4y = 8$$
$$y = -2 \qquad \qquad Solve \ for \ y$$

Therefore, the solution is $(3, -2)$. Check the solution in each of the original equations.

Graphical Interpretation of Solutions

As you observed in Section 7.1, it is possible for a *general* system of equations to have exactly one solution, two or more solutions, or no solution. For a system of *linear* equations you can strengthen this result somewhat. Specifically, if a system of linear equations has two different solutions, then it must have an infinite number of solutions! To see why this is true, consider the following graphical interpretations of a system of two linear equations in two variables. (Remember that the graph of a linear equation in two variables is a straight line.)

> ## Graphical Interpretation of Solutions
>
> For a system of two linear equations in two variables, the number of solutions is given by one of the following.
>
> | *Number of Solutions* | *Graphical Interpretation* |
> |---|---|
> | 1. Exactly one solution | The two lines intersect at one point. |
> | 2. Infinitely many solutions | The two lines are identical. |
> | 3. No solution | The two lines are parallel. |
>
> These three possibilities are shown in Figure 7.5.

A system of linear equations is **consistent** if it has at least one solution, and it is **inconsistent** if it has no solution.

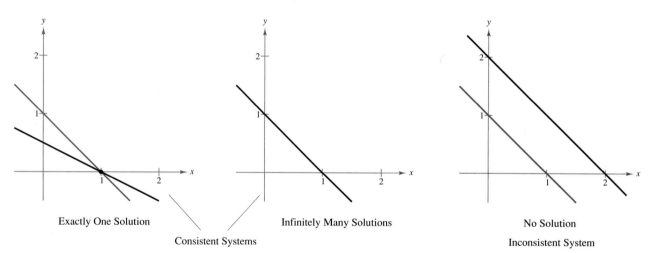

Exactly One Solution   Infinitely Many Solutions   No Solution

Consistent Systems   Inconsistent System

FIGURE 7.5

EXAMPLE 4   The Method of Elimination: No-Solution Case

Solve the following system of linear equations.

$$x - 2y = 3 \qquad \text{\textit{Equation 1}}$$
$$-2x + 4y = 1 \qquad \text{\textit{Equation 2}}$$

Solution

To obtain coefficients that differ only in sign, multiply the first equation by 2.

$$
\begin{array}{rcll}
x - 2y = 3 & \rightarrow & 2x - 4y = 6 & \textit{Multiply Equation 1 by 2} \\
-2x + 4y = 1 & \rightarrow & \underline{-2x + 4y = 1} & \textit{Equation 2} \\
& & 0 = 7 & \textit{False statement}
\end{array}
$$

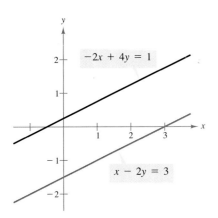

$-2x + 4y = 1$

$x - 2y = 3$

FIGURE 7.6

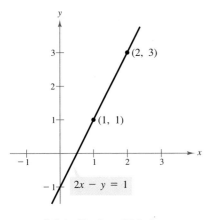

(2, 3)

(1, 1)

$2x - y = 1$

Infinite Number of Solutions

FIGURE 7.7

Since there are no values of x and y for which $0 = 7$, the system is inconsistent and has no solution. The lines corresponding to the two equations given in this system are shown in Figure 7.6. Note that the two lines are parallel, and therefore have no point of intersection.

EXAMPLE 5 The Method of Elimination: Many-Solutions Case

Solve the following system of linear equations.

$$2x - y = 1 \qquad \text{Equation 1}$$
$$4x - 2y = 2 \qquad \text{Equation 2}$$

Solution

To obtain coefficients that differ only in sign, multiply the second equation by $-\frac{1}{2}$.

$$2x - y = 1 \quad \rightarrow \quad 2x - y = 1 \qquad \text{Equation 1}$$
$$\underline{4x - 2y = 2} \quad \rightarrow \quad \underline{-2x + y = -1} \qquad \text{Multiply Equation 2 by } -\frac{1}{2}$$
$$\qquad\qquad\qquad\qquad\qquad\quad 0 = 0 \qquad \text{Add equations}$$

Since the two equations turn out to be equivalent (have the same solution set), the system has infinitely many solutions. The solution set consists of all points (x, y) lying on the line $2x - y = 1$, as shown in Figure 7.7.

In Example 5, you could have reached the same conclusion by multiplying the first equation by 2 to obtain

$$4x - 2y = 2 \qquad \text{New Equation 1}$$
$$4x - 2y = 2. \qquad \text{Equation 2}$$

EXAMPLE 6 Solving a Linear System Having Decimal Coefficients

Solve the following system of linear equations.

$$0.02x - 0.05y = -0.38 \qquad \text{Equation 1}$$
$$0.03x + 0.04y = 1.04 \qquad \text{Equation 2}$$

Solution

Because the coefficients in this system have two decimal places, we begin by multiplying each equation by 100. (This produces an equivalent system in which the coefficients are all integers.)

$$2x - 5y = -38 \qquad \text{Revised Equation 1}$$
$$3x + 4y = 104 \qquad \text{Revised Equation 2}$$

Now, to obtain coefficients that differ only in sign, we multiply the first equation by 3 and the second equation by -2.

$$2x - 5y = -38 \quad \rightarrow \quad 6x - 15y = -114 \qquad \textit{Multiply Equation 1 by 3}$$
$$\underline{3x + 4y = 104} \quad \rightarrow \quad \underline{-6x - 8y = -208} \qquad \textit{Multiply Equation 2 by } -2$$
$$-23y = -322 \qquad \textit{Add equations}$$

Thus, we find that

$$y = \frac{-322}{-23} = 14.$$

Back-substituting this value into Equation 2 produces the following.

$$3x + 4y = 104 \qquad\qquad \textit{Equation 2}$$
$$3x + 4(14) = 104 \qquad\qquad \textit{Replace y by 14}$$
$$3x = 48$$
$$x = 16 \qquad\qquad \textit{Solve for x}$$

Therefore, the solution is (16, 14). Check this solution in each of the original equations in the system.

Applications

We stated at the beginning of this chapter that systems of linear equations have many applications in science, business, health services, and government. The question that may come to mind is, How can I tell which application problems can be solved using a system of linear equations? The answer comes from the following considerations.

1. Does the problem involve more than one unknown quantity?
2. Are there two (or more) equations or conditions to be satisfied?

If one or both of these conditions occur, then the appropriate mathematical model for the problem may be a system of linear equations. Example 7 shows how to construct such a model.

EXAMPLE 7   An Application of a Linear System

An airplane flying into a headwind travels the 2000-mile flying distance between two cities in 4 hours and 24 minutes. On the return flight, the same distance is traveled in 4 hours. Find the ground speed of the plane and the speed of the wind, assuming that both remain constant.

Solution

The two unknown quantities are the speeds of the wind and of the plane. If r_1 is the speed of the plane and r_2 is the speed of the wind, then

$$r_1 - r_2 = \text{speed of the plane } \textit{against} \text{ the wind}$$
$$r_1 + r_2 = \text{speed of the plane } \textit{with} \text{ the wind}$$

as shown in Figure 7.8. Using the formula

$$\text{Distance} = (\text{Rate})(\text{Time})$$

for these two speeds, you obtain the following equations.

$$2000 = (r_1 - r_2)\left(4 + \frac{24}{60}\right)$$

$$2000 = (r_1 + r_2)(4)$$

These two equations simplify as follows.

$$5000 = 11r_1 - 11r_2 \qquad \textit{Equation 1}$$

$$500 = r_1 + r_2 \qquad \textit{Equation 2}$$

By elimination, the solution is

$$r_1 = \frac{5250}{11} \approx 477.27 \text{ miles per hour}$$

$$r_2 = \frac{250}{11} \approx 22.73 \text{ miles per hour.}$$

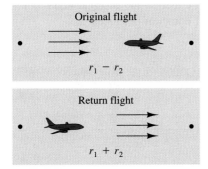

Original flight

$r_1 - r_2$

Return flight

$r_1 + r_2$

FIGURE 7.8

In a free market, the demand for many products is related to the price of the product. As the price of a product decreases, the demand by *consumers* increases. For *producers* the opposite is true. In other words, as the price of a product decreases, the supply tends to decrease.

EXAMPLE 8   Finding the Point of Equilibrium

Suppose the demand and supply functions for a certain type of calculator are given by

$$p = 150 - 10x \qquad \textit{Demand equation}$$

$$p = 60 + 20x \qquad \textit{Supply equation}$$

where p is the price in dollars and x represents the number of units in millions. Find the **point of equilibrium** for this market by solving this system of equations. (The point of equilibrium is the price p and the number of units x that satisfy both the demand and the supply equations.)

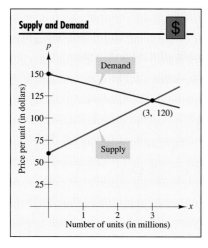

Supply and Demand $

p

Price per unit (in dollars)

150
125
100
75
50
25

Demand

(3, 120)

Supply

1   2   3   x
Number of units (in millions)

FIGURE 7.9

Solution

For this system, you use the method of substitution (because the two equations are given in a form in which p is written in terms of x). By substituting the value of p given in the second equation into the first equation, you obtain the following.

$$p = 150 - 10x \quad \textit{First equation}$$
$$60 + 20x = 150 - 10x \quad \textit{Replace p by 60 + 20x}$$
$$30x = 90 \quad \textit{Add 10x and subtract 60}$$
$$x = 3 \quad \textit{Divide both sides by 30}$$

Thus, the point of equilibrium occurs when the demand and supply are each 3 million units. (See Figure 7.9.) The price that corresponds to this x-value is obtained by back-substituting $x = 3$ into either of the original equations. For instance, back-substituting into the first equation produces

$$p = 150 - 10(3) = 150 - 30 = \$120.$$

Try back-substituting $x = 3$ into the second equation to see that you obtain the same price. ◢

DISCUSSION

PROBLEM

Creating
Consistent
and
Inconsistent
Systems

Consider the following system of linear equations.

$$x - y = 4$$
$$-2x + 2y = k$$

1. Find the value of k so that the system has an infinite number of solutions.
2. Find one value of k so that the system has no solutions.
3. Can the system have a unique solution? Why or why not?

WARM UP

The following warm-up exercises involve skills that were covered in earlier sections. You will use these skills in the exercise set for this section.

In Exercises 1 and 2, sketch the graph of the equation.

1. $2x + y = 4$ **2.** $5x - 2y = 3$

In Exercises 3 and 4, find an equation of the line passing through the two points.

3. $(-1, 3)$, $(4, 8)$ **4.** $(2, 6)$, $(5, 1)$

In Exercises 5 and 6, determine the slope of the line.

5. $3x + 6y = 4$ **6.** $7x - 4y = 10$

In Exercises 7–10, determine whether the lines represented by the pair of equations are parallel, perpendicular, or neither.

7. $2x - 3y = -10$
$3x + 2y = 11$

8. $4x - 12y = 5$
$-2x + 6y = 3$

9. $5x + y = 2$
$3x + 2y = 1$

10. $x - 3y = 2$
$6x + 2y = 4$

EXERCISES for Section 7.2

In Exercises 1–10, solve the linear system by elimination. Label each line with the appropriate equation.

1. $2x + y = 4$
$x - y = 2$

2. $x + 3y = 2$
$-x + 2y = 3$

3. $x - y = 0$
$3x - 2y = -1$

4. $2x - y = 2$
$4x + 3y = 24$

5. $x - y = 1$
$-2x + 2y = 5$

6. $3x + 2y = 2$
$6x + 4y = 14$

7. $3x - 2y = 6$
$-6x + 4y = -12$

8. $x - 2y = 5$
$6x + 2y = 7$

9. $9x - 3y = -1$
$3x + 6y = -5$

10. $5x + 3y = 18$
$2x - 7y = -1$

In Exercises 11–30, solve the system by elimination.

11. $x + 2y = 4$
    $x - 2y = 1$

12. $3x - 5y = 2$
    $2x + 5y = 13$

13. $2x + 3y = 18$
    $5x - y = 11$

14. $x + 7y = 12$
    $3x - 5y = 10$

15. $3x + 2y = 10$
    $2x + 5y = 3$

16. $8r + 16s = 20$
    $16r + 50s = 55$

17. $2u + v = 120$
    $u + 2v = 120$

18. $5u + 6v = 24$
    $3u + 5v = 18$

19. $6r - 5s = 3$
    $10s - 12r = 5$

20. $1.8x + 1.2y = 4$
    $9x + 6y = 3$

21. $\dfrac{x}{4} + \dfrac{y}{6} = 1$
    $x - y = 3$

22. $\dfrac{2}{3}x + \dfrac{1}{6}y = \dfrac{2}{3}$
    $4x + y = 4$

23. $\dfrac{x + 3}{4} + \dfrac{y - 1}{3} = 1$
    $2x - y = 12$

24. $\dfrac{x - 1}{2} + \dfrac{y + 2}{3} = 4$
    $x - 2y = 5$

25. $2.5x - 3y = 1.5$
    $10x - 12y = 6$

26. $0.02x - 0.05y = -0.19$
    $0.03x + 0.04y = 0.52$

27. $0.05x - 0.03y = 0.21$
    $0.07x + 0.02y = 0.16$

28. $0.2x - 0.5y = -27.8$
    $0.3x + 0.4y = 68.7$

29. $4b + 3m = 3$
    $3b + 11m = 13$

30. $3b + 3m = 7$
    $3b + 5m = 3$

In Exercises 31 and 32, the graphs of the two equations appear to be parallel. Yet, when the system is solved algebraically, we find that the system does have a solution. Find the solution and explain why it does not appear on the portion of the graph that is shown.

31. $200y - x = 200$
    $199y - x = -198$

32. $25x - 24y = 0$
    $13x - 12y = 120$

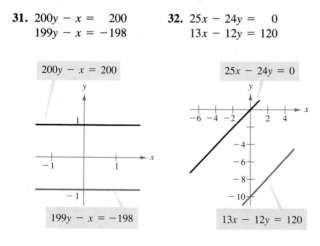

33. *Airplane Speed* An airplane flying into a headwind travels the 1800-mile flying distance between two cities in 3 hours and 36 minutes. On the return flight, the distance is traveled in 3 hours. Find the ground speed of the plane and the speed of the wind, assuming that both remain constant.

34. *Airplane Speed* Two planes start from the same airport and fly in opposite directions. The second plane starts one-half hour after the first plane, but its speed is 50 miles per hour faster. Find the ground speed of each plane if 2 hours after the first plane starts, the planes are 2000 miles apart.

35. *Acid Mixture* Ten gallons of a 30% acid solution are obtained by mixing a 20% solution with a 50% solution. How much of each must be used?

36. *Fuel Mixture* Five hundred gallons of 82-octane gasoline are obtained by mixing 80-octane gasoline with 86-octane gasoline. How much of each must be used?

37. *Investment Portfolio* A total of $12,000 is invested in two corporate bonds that pay 10.5% and 12% simple interest. The annual interest is $1,380. How much is invested in each bond?

38. *Investment Portfolio* A total of $32,000 is invested in two municipal bonds that pay 5.75% and 6.25% simple interest. The annual interest is $1,930. How much is invested in each bond?

39. *Ticket Sales* Five hundred tickets were sold for a certain performance of a play. The tickets for adults and children sold for $7.50 and $4.00, respectively, and the receipts for the performance were $3312.50. How many of each kind of ticket were sold?

40. *Shoe Sales* Suppose you are the manager of a shoe store. On Saturday night you are going over the receipts of the previous week's sales. Two hundred and forty pairs of tennis shoes were sold. One style sold for $66.95 and the other sold for $84.95. The total receipts were $17,652. The cash register that was supposed to record the number of each type of shoe sold malfunctioned. Can you recover the information? If so, how many shoes of each type were sold?

Supply and Demand In Exercises 41–46, find the point of equilibrium for each pair of supply and demand equations.

Demand	Supply
41. $p = 50 - 0.5x$	$p = 0.125x$
42. $p = 60 - x$	$p = 10 + \frac{7}{3}x$

Demand	Supply
43. $p = 300 - x$	$p = 100 + x$
44. $p = 100 - 0.05x$	$p = 25 + 0.1x$
45. $p = 140 - 0.00002x$	$p = 80 + 0.00001x$
46. $p = 400 - 0.0002x$	$p = 225 + 0.0005x$

47. *Driving Distances* On a 300-mile trip two people do the driving. One person drives three times as far as the other. Find the distance that each person drives.

48. *Truck Scheduling* A contractor is hiring two trucking companies to haul 1600 tons of crushed stone for a highway construction project. The contracts state that one company is to haul four times as much as the other. Find the amount hauled by each.

Fitting a Line to Data In Exercises 49–56, find the *least squares regression line* $y = ax + b$ for the points

$(x_1, y_1), (x_2, y_2), \ldots, (x_n, y_n)$.

To find the line, solve the following system for a and b. (If you are unfamiliar with summation notation, look at the discussion in Section 9.1.)

$$nb + \left(\sum_{i=1}^{n} x_i\right)a = \sum_{i=1}^{n} y_i$$

$$\left(\sum_{i=1}^{n} x_i\right)b + \left(\sum_{i=1}^{n} x_i^2\right)a = \sum_{i=1}^{n} x_i y_i$$

49. $5b + 10a = 20.2$
$10b + 30a = 50.1$

50. $5b + 10a = 11.7$
$10b + 30a = 25.6$

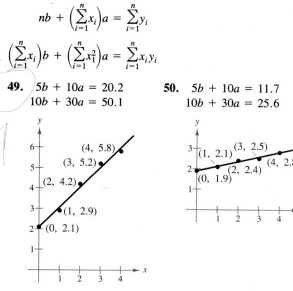

51. $7b + 21a = 35.1$
$21b + 91a = 114.2$

52. $6b + 15a = 23.6$
$15b + 55a = 48.8$

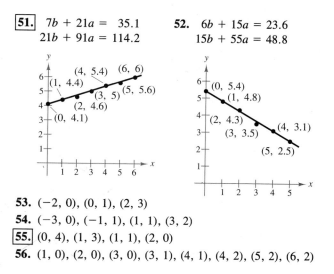

53. $(-2, 0), (0, 1), (2, 3)$

54. $(-3, 0), (-1, 1), (1, 1), (3, 2)$

55. $(0, 4), (1, 3), (1, 1), (2, 0)$

56. $(1, 0), (2, 0), (3, 0), (3, 1), (4, 1), (4, 2), (5, 2), (6, 2)$

57. *Demand Function* A store manager wants to know the demand of a certain product as a function of the price. The daily sales for the different prices of the product are given in the table.

Price (x)	$1.00	$1.25	$1.50
Demand (y)	450	375	330

Use the technique demonstrated in Exercises 49–56 to find the line that best fits the given data. Then use the line to predict the demand when the price is $1.40.

58. *Yield* A farmer used four test plots to determine the relationship between wheat yield in bushels per acre and the amount of fertilizer in hundreds of pounds per acre. The results are given in the table.

Fertilizer (x)	1.0	1.5	2.0	2.5
Yield (y)	32	41	48	53

Use the technique demonstrated in Exercises 49–56 to find the line that best fits the given data. Then use the line to estimate the yield for a fertilizer application of 160 pounds per acre.

In Exercises 59 and 60, find a system of linear equations having the given solution. (The answers are not unique.)

59. $\left(3, \frac{5}{2}\right)$

60. $(8, -2)$

USING TECHNOLOGY

SOLVING

Using Technology to Fit Models to Data

Many of the models in this text were created with a statistical method called *least squares regression analysis*. This procedure is tedious to perform by hand, but can be performed quite efficiently with a computer or graphing calculator.

EXAMPLE 1 Fitting a Line to Data

For 1978 through 1989, the numbers (in millions) of morning and evening newspapers sold each day in the United States are as shown in the table. Use the data to project the number of morning and evening newspapers that will be sold each day in 1995. In the table, $t = 0$ represents 1980. (*Source:* Editor and Publisher Company)

Year, t	−2	−1	0	1	2	3
Morning	27.7	28.6	29.4	30.6	33.2	33.8
Evening	34.3	33.6	32.8	30.9	29.3	28.8

Year, t	4	5	6	7	8	9
Morning	35.4	36.4	37.4	39.1	40.4	40.7
Evening	27.7	26.4	25.1	23.7	22.2	21.8

Solution

Begin by finding a computer or graphing calculator program that will perform linear regression analysis. (Such a program is a built-in feature of some calculators.) After entering the data and running the program, you should obtain the following models. (Both models have a correlation coefficient of $r^2 > 0.99$, which means that the models are very good fits for the data.)

$$y = 29.956 + 1.267t \qquad \text{\textit{Morning paper circulation}}$$
$$y = 32.243 - 1.198t \qquad \text{\textit{Evening paper circulation}}$$

With these models, you can project the 1995 newspaper sales. *If* the sales through 1995 continue to follow the pattern from 1978 through 1989, then the 1995 sales of morning papers should be about

$$y = 29.956 + 1.267(15) \approx 49.0 \text{ million} \qquad \textit{Morning}$$

and the sales of evening papers should be about

$$y = 32.243 - 1.198(15) \approx 14.3 \text{ million.} \qquad \textit{Evening}$$

The graphs of the data points and their models are shown at the left.

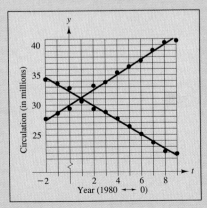

EXERCISES

(See also: Exercises 49–58, Section 7.2)

1. *Sunday Paper Circulation* The numbers (in millions) of Sunday newspapers sold each week in the United States from 1978 through 1989 are shown in the table below. Find a linear model that represents this data. Use your model to project the number of Sunday newspapers to be sold each week in 1995.

2. *Newspaper Companies* The numbers of morning, evening, and Sunday newspaper companies in the United States from 1978 through 1989 are shown in the table below. Find a linear model for each of these three sets of data.

3. *Average Sales* From 1978 through 1989, the circulation of morning newspapers increased. However, because the number of morning newspaper companies also increased, the competition for morning newspaper readers became keener. Did the average circulation per morning newspaper company increase or decrease? Explain.

4. *Average Sales* From 1978 through 1989, the circulation of evening newspapers decreased. However, because the number of evening newspaper companies also decreased, the competition for evening newspaper readers became less keen. Did the average circulation per evening newspaper company increase or decrease? Explain.

5. *Average Sales* From 1978 through 1989, the circulation of Sunday newspapers increased. However, because the number of Sunday newspaper companies also increased, the competition for Sunday newspaper readers became keener. Did the average circulation per Sunday newspaper company increase or decrease? Explain.

6. *Which Would You Choose?* If you had the opportunity to invest in a company that produced only one type of newspaper (morning, evening, or Sunday), which would you choose? Explain your reasoning.

7. *Households and Population* For 1978 through 1989, the population and number of households in the United States (both in millions) are given in the table below. From this information, would you say that the percent of Americans who read newspapers was increasing or decreasing from 1978 through 1989? Explain your reasoning.

Year, t	−2	−1	0	1	2	3	4	5	6	7	8	9
Sunday	54.0	54.4	54.7	55.2	56.3	56.7	57.5	58.8	58.9	60.1	61.5	62.0

Table for Exercise 1

Year, t	−2	−1	0	1	2	3	4	5	6	7	8	9
Morning	355	382	387	408	434	446	458	482	499	511	529	530
Evening	1419	1405	1388	1352	1310	1284	1257	1220	1188	1166	1141	1125
Sunday	696	720	736	755	768	772	783	798	802	820	840	847

Table for Exercise 2

Year, t	−2	−1	0	1	2	3	4	5	6	7	8	9
Households	77.6	78.9	80.8	81.6	83.0	83.9	85.4	86.8	88.5	89.5	91.1	92.8
Population	222.6	225.1	227.7	229.9	232.2	234.3	236.3	238.5	240.7	242.8	245.1	247.4

Table for Exercise 7

7.3 Linear Systems in More Than Two Variables

Row-Echelon Form and Back-Substitution / Gaussian Elimination / Nonsquare Systems / Applications

Row-Echelon Form and Back-Substitution

The method of elimination can be applied to a system of linear equations in more than two variables. In fact, this method easily adapts to computer use for solving linear systems with dozens of variables.

When using elimination to solve a system of linear equations, the goal is to rewrite the system in a form to which back-substitution can be applied. To see how this works, consider the following two systems of linear equations.

$$\begin{aligned} x - 2y + 3z &= 9 \\ -x + 3y &= -4 \\ 2x - 5y + 5z &= 17 \end{aligned} \qquad \begin{aligned} x - 2y + 3z &= 9 \\ y + 3z &= 5 \\ z &= 2 \end{aligned}$$

Clearly, the system on the right is easier to solve. This system is in **row-echelon form,** which means that it follows a stair-step pattern and has leading coefficients of 1. To solve such a system, use back-substitution, working from the bottom equation to the top equation, as demonstrated in Example 1.

EXAMPLE 1   Back-Substitution to Solve a System in Row-Echelon Form

Solve the following system of linear equations.

$$\begin{aligned} x - 2y + 3z &= 9 && \text{\textit{Equation 1}} \\ y + 3z &= 5 && \text{\textit{Equation 2}} \\ z &= 2 && \text{\textit{Equation 3}} \end{aligned}$$

Solution

From Equation 3, you already know the value of z. To solve for y, substitute $z = 2$ into Equation 2 to obtain

$$\begin{aligned} y + 3(2) &= 5 && \text{\textit{Substitute } z = 2} \\ y &= -1. && \text{\textit{Solve for } y} \end{aligned}$$

Finally, substitute $y = -1$ and $z = 2$ into Equation 1 to obtain

$$\begin{aligned} x - 2(-1) + 3(2) &= 9 && \text{\textit{Substitute } y = -1, z = 2} \\ x &= 1. && \text{\textit{Solve for } x} \end{aligned}$$

Thus, the solution is $x = 1$, $y = -1$, and $z = 2$.

Gaussian Elimination

Two systems of equations are **equivalent** if they have precisely the same solution set. To solve a system that is not in row-echelon form, first change it to an *equivalent* system that is in row-echelon form by using the following operations.

Operations that Lead to Equivalent Systems of Equations

Each of the following operations on a system of linear equations produce an *equivalent* system.

1. Interchange two equations.
2. Multiply one of the equations by a nonzero constant.
3. Add a multiple of an equation to another equation.

Rewriting a system of linear equations in row-echelon form usually involves a *chain* of equivalent systems, each of which is obtained by using one of the three basic operations. This process is called **Gaussian elimination,** after the German mathematician Carl Friedrich Gauss (1777–1855).

EXAMPLE 2   Using Elimination to Solve a Linear System

Solve the following system of linear equations.

$$\begin{aligned} x - 2y + 3z &= 9 \\ -x + 3y &= -4 \\ 2x - 5y + 5z &= 17 \end{aligned}$$

Solution

Although there are several ways to begin, the goal is to develop a systematic procedure that can be applied to large systems. Work from the upper left corner, saving the x in the upper left position and eliminating the other x's from the first column.

$$\begin{aligned} x - 2y + 3z &= 9 \\ y + 3z &= 5 \\ 2x - 5y + 5z &= 17 \end{aligned}$$

> Adding the first equation to the second equation produces a new second equation.

$$\begin{aligned} x - 2y + 3z &= 9 \\ y + 3z &= 5 \\ -y - z &= -1 \end{aligned}$$

> Adding -2 times the first equation to the third equation produces a new third equation.

Now that all but the first x has been eliminated from the first column, go to work on the second column. (You need to eliminate y from the third equation.)

$$\begin{aligned} x - 2y + 3z &= 9 \\ y + 3z &= 5 \\ 2z &= 4 \end{aligned}$$

Adding the second equation to the third equation produces a a new third equation.

Finally, you need a coefficient of 1 for z in the third equation.

$$\begin{aligned} x - 2y + 3z &= 9 \\ y + 3z &= 5 \\ z &= 2 \end{aligned}$$

Multiplying the third equation by $\frac{1}{2}$ produces a new third equation.

This is the same system you solved in Example 1, and, as in that example, the solution is

$$x = 1, \qquad y = -1, \qquad \text{and} \qquad z = 2.$$

The solution can also be written as the **ordered triple** $(1, -1, 2)$.

In Example 2, you can check the solution $x = 1$, $y = -1$, and $z = 2$ as follows.

Equation 1: $(1) - 2(-1) + 3(2) = 9$ *Check solution in each equation of original system.*

Equation 2: $-(1) + 3(-1) = -4$

Equation 3: $2(1) - 5(-1) + 5(2) = 17$

We now look at an inconsistent system—one that has no solution. The key to recognizing an inconsistent system is that at some stage in the elimination process, you obtain an absurdity such as $0 = 7$. This is demonstrated in Example 3.

EXAMPLE 3   An Inconsistent System

Solve the following system of linear equations.

$$\begin{aligned} x - 3y + z &= 1 \\ 2x - y - 2z &= 2 \\ x + 2y - 3z &= -1 \end{aligned}$$

Solution

$$x - 3y + z = 1$$
$$5y - 4z = 0$$
$$x + 2y - 3z = -1$$

> Adding -2 times the first equation to the second equation produces a new second equation.

$$x - 3y + z = 1$$
$$5y - 4z = 0$$
$$5y - 4z = -2$$

> Adding -1 times the first equation to the third equation produces a new third equation.

$$x - 3y + z = 1$$
$$5y - 4z = 0$$
$$0 = -2$$

> Adding -1 times the second equation to the third equation produces a new third equation.

Because the third "equation" is absurd, this system is inconsistent and has no solution. Moreover, because this system is equivalent to the original system, the original system also has no solution. ◣

As with a system of linear equations in two variables, the solutions of a system of linear equations in more than two variables must fall into one and only one of the following categories.

1. There is exactly one solution.
2. There are infinitely many solutions.
3. There is no solution.

When a system of equations has no solution, you simply state that it is *inconsistent*. If a system has exactly one solution, you list the value of each variable. However, for systems that have infinitely many solutions, you encounter a certain awkwardness in listing the solutions. For example, you might give the solutions to a system in three variables as

$(a, a + 1, 2a)$,   where a is any real number.

This means that for each real number a, you have a valid solution to the system. A few of the infinitely many possible solutions are found by letting $a = -1$, 0, 1, and 2 to obtain $(-1, 0, -2)$, $(0, 1, 0)$, $(1, 2, 2)$, and $(2, 3, 4)$, respectively. Now consider the solutions represented by

$(b - 1, b, 2b - 2)$,   where b is any real number.

Here again a few possible solutions are $(-1, 0, -2)$, $(0, 1, 0)$, $(1, 2, 2)$, and $(2, 3, 4)$, found by letting $b = 0$, 1, 2, and 3, respectively. Note that both descriptions result in the same collection of solutions. Thus, when comparing descriptions of an infinite solution set, keep in mind that there is more than one way to describe the set.

EXAMPLE 4 A System with Infinitely Many Solutions

Solve the following system of linear equations.

$$
\begin{aligned}
x + y - 3z &= -1 && \text{\textit{Equation 1}} \\
y - z &= 0 && \text{\textit{Equation 2}} \\
-x + 2y \phantom{{}-3z} &= 1 && \text{\textit{Equation 3}}
\end{aligned}
$$

Solution

Begin by rewriting the system in row-echelon form, as follows.

$$
\begin{aligned}
x + y - 3z &= -1 \\
y - z &= 0 \\
3y - 3z &= 0
\end{aligned}
$$

Adding the first equation to the third equation produces a new third equation.

$$
\begin{aligned}
x + y - 3z &= -1 \\
y - z &= 0 \\
0 &= 0
\end{aligned}
$$

Adding -3 times the second equation to the third equation produces a new third equation.

This means that Equation 3 is *dependent* on Equations 1 and 2 in the sense that it gives no additional information about the variables. Thus, the original system is equivalent to the system

$$
\begin{aligned}
x + y - 3z &= -1 \\
y - z &= 0.
\end{aligned}
$$

In this last equation, you solve for y in terms of z to obtain $y = z$. Back-substituting for y into the previous equation, you find x in terms of z, as follows.

$$
\begin{aligned}
x + z - 3z &= -1 \\
x - 2z &= -1 \\
x &= 2z - 1
\end{aligned}
$$

Finally, letting $z = a$, the solutions to the given system are all of the form

$$
x = 2a - 1, \quad y = a, \quad z = a
$$

where a is a real number. Thus, every ordered triple of the form

$$
(2a - 1, a, a), \quad a \text{ is a real number}
$$

is a solution of the system.

Nonsquare Systems

So far we have only considered **square** systems, for which the number of equations is equal to the number of variables. In a **nonsquare** system, the number of equations differs from the number of variables. It can be shown that a system of linear equations cannot have a unique solution unless there are at least as many equations as there are variables in the system.

EXAMPLE 5   A System with Fewer Equations than Variables

Solve the following system of linear equations.

$$x - 2y + z = 2 \qquad \text{\textit{Equation 1}}$$
$$2x - y - z = 1 \qquad \text{\textit{Equation 2}}$$

Solution

$$x - 2y + z = 2$$
$$3y - 3z = -3 \leftarrow$$

Adding -2 times the first equation to the second equation produces a new second equation.

$$x - 2y + z = 2$$
$$y - z = -1 \leftarrow$$

Multiplying the second equation by $\frac{1}{3}$ produces a new second equation.

Solving for y in terms of z, you get $y = z - 1$, and back-substitution into Equation 1 yields

$$x - 2(z - 1) + z = 2$$
$$x - 2z + 2 + z = 2$$
$$x = z.$$

Finally, by letting $z = a$, you have the solution

$$x = a, \qquad y = a - 1, \qquad \text{and} \qquad z = a$$

where a is a real number. Thus, every ordered triple of the form

$$(a, a - 1, a), \qquad a \text{ is a real number}$$

is a solution of the system.

Applications

We conclude this section with three applications involving systems of linear equations in three variables.

In Example 6 we show how to fit a parabola through three given points in the plane. This procedure can be generalized to fit an nth degree polynomial function to $n + 1$ points in the plane. The only restriction to the procedure is that (since we are trying to fit a *function* to the points) every point must have a distinct x-coordinate.

EXAMPLE 6   An Application: Moving Object

The height at time t of an object that is moving in a (vertical) line with constant acceleration a is given by the **position equation**

$$s = \frac{1}{2}at^2 + v_0 t + s_0.$$

The height s is measured in feet, t is measured in seconds, v_0 is the initial velocity (at time $t = 0$), and s_0 is the initial height. Find the values of a, v_0, and s_0, if $s = 52$ feet at 1 second, $s = 52$ feet at 2 seconds, and $s = 20$ feet at 3 seconds.

Solution

By substituting the three values of t and s into the position equation, you obtain three linear equations in a, v_0, and s_0.

When $t = 1$: $\frac{1}{2}a(1^2) + v_0(1) + s_0 = 52$

When $t = 2$: $\frac{1}{2}a(2^2) + v_0(2) + s_0 = 52$

When $t = 3$: $\frac{1}{2}a(3^2) + v_0(3) + s_0 = 20$

By multiplying the first and third equations by 2, this system can be rewritten

$$a + 2v_0 + 2s_0 = 104$$
$$2a + 2v_0 + s_0 = 52$$
$$9a + 6v_0 + 2s_0 = 40$$

and you apply elimination as follows.

$$\begin{aligned} a + 2v_0 + 2s_0 &= 104 \\ -2v_0 - 3s_0 &= -156 \\ 9a + 6v_0 + 2s_0 &= 40 \end{aligned}$$

Adding -2 times the first equation to the second equation produces a new second equation.

$$\begin{aligned} a + 2v_0 + 2s_0 &= 104 \\ -2v_0 - 3s_0 &= -156 \\ -12v_0 - 16s_0 &= -896 \end{aligned}$$

Adding -9 times the first equation to the third equation produces a new third equation.

$$\begin{aligned} a + 2v_0 + 2s_0 &= 104 \\ -2v_0 - 3s_0 &= -156 \\ 2s_0 &= 40 \end{aligned}$$

Adding -6 times the second equation to the third equation produces a new third equation.

By back-substituting $s_0 = 20$ (obtained from the third equation), we find that $v_0 = 48$ and ultimately $a = -32$. Thus, the position equation for this object is

$$s = -16t^2 + 48t + 20.$$

EXAMPLE 7   An Application: Curve-Fitting

Find a quadratic function

$$f(x) = ax^2 + bx + c$$

whose graph passes through the points $(-1, 3)$, $(1, 1)$, and $(2, 6)$.

Solution

Since the graph of f passes through the points $(-1, 3)$, $(1, 1)$, and $(2, 6)$, you have

$$f(-1) = a(-1)^2 + b(-1) + c = 3$$
$$f(1) = a(1)^2 + b(1) + c = 1$$
$$f(2) = a(2)^2 + b(2) + c = 6.$$

This produces the following system of linear equations in the variables a, b, and c.

$$\begin{aligned} a - b + c &= 3 \\ a + b + c &= 1 \\ 4a + 2b + c &= 6 \end{aligned}$$

The solution to this system turns out to be

$$a = 2, \quad b = -1, \quad \text{and} \quad c = 0.$$

Thus, the equation of the parabola passing through the three given points is

$$f(x) = 2x^2 - x$$

as shown in Figure 7.10.

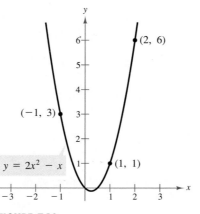

FIGURE 7.10

EXAMPLE 8   An Investment Portfolio

Suppose an investor has a portfolio totaling $450,000, and wishes to allocate this amount to the following types of investments: (1) certificates of deposit, (2) municipal bonds, (3) blue-chip stocks, and (4) growth or speculative stocks. The certificates of deposit pay 9% annually, and the municipal bonds pay 6% annually. Over a five-year period the investor expects the blue-chip stocks to return 10% annually, and expects the growth stocks to return 15% annually. The investor wants a combined annual return of 8%, and also wants to have only one-third of the portfolio invested in stocks. How much should be allocated to each type of investment?

Solution

To solve this problem, let C, M, B, and G represent the amounts in the four types of investments. Because the total investment is $450,000, you can write the following equation.

$$C + M + B + G = 450{,}000$$

A second equation can be derived from the fact that the combined annual return should be 8%.

$$0.09C + 0.06M + 0.10B + 0.15G = 0.08(450{,}000)$$

Finally, because only one-third of the investment should be allocated to stocks, you can write

$$B + G = \frac{1}{3}(450{,}000).$$

These three equations make up the following system.

$$
\begin{aligned}
C + \quad M + \quad B + \quad G &= 450{,}000 \\
0.09C + 0.06M + 0.1B + 0.15G &= \quad 36{,}000 \\
B + \quad G &= 150{,}000
\end{aligned}
$$

Using elimination, you find that the system has infinitely many solutions, which can be written as follows.

$$C = -\frac{5}{3}a + 100{,}000, \quad M = \frac{5}{3}a + 200{,}000, \quad B = -a + 150{,}000, \quad G = a$$

Thus, the investor has many different options. One possible solution is to choose $a = 30{,}000$, which yields the following portfolio.

1. Certificates of deposit: $50,000
2. Municipal bonds: $250,000
3. Blue-chip stocks: $120,000
4. Growth or speculative stocks: $30,000

EXAMPLE 9   An Application: Partial Fractions

Write the partial fraction decomposition for

$$\frac{3x + 4}{x^3 - 2x - 4}.$$

Solution

Since

$$x^3 - 2x - 4 = (x - 2)(x^2 + 2x + 2)$$

you write

$$\frac{3x + 4}{x^3 - 2x - 4} = \frac{A}{x - 2} + \frac{Bx + C}{x^2 + 2x + 2}$$

$$3x + 4 = A(x^2 + 2x + 2) + (Bx + C)(x - 2) \qquad \textit{Basic equation}$$

$$3x + 4 = (A + B)x^2 + (2A - 2B + C)x + (2A - 2C).$$

By equating coefficients of like powers on opposite sides of the expanded equation, you obtain the following system of linear equations in A, B, and C.

$$
\begin{aligned}
A + B \quad\quad &= 0 \\
2A - 2B + C &= 3 \\
2A \quad\quad - 2C &= 4
\end{aligned}
$$

The solution of this system is $A = 1$, $B = -1$, and $C = -1$. Therefore, the partial fraction decomposition is

$$\frac{3x + 4}{x^3 - 2x - 4} = \frac{1}{x - 2} + \frac{-x - 1}{x^2 + 2x + 2}$$

$$= \frac{1}{x - 2} - \frac{x + 1}{x^2 + 2x + 2}.$$

DISCUSSION

PROBLEM

Computer
Software

In addition to a graphing calculator, there are many computer software programs that will solve a system of linear equations. One is called MATRIXPAD and is available from D.C. Heath and Company. Use this software (or some other program) to solve the examples given in this section. For example, a screen from MATRIXPAD showing the solution to Example 7 is given below.

T register	3	3	|	0
Z register	3	3	|	0
Y register	3	4	|	0

1	−1	1	3
1	1	1	1
4	2	1	6

X register	3	4	|	0

1	0	0	2
0	1	0	−1
0	0	1	0

MATRIXPAD   © Copyright D.C. Heath and Company

Press ⟨H⟩ for Help   ⟨Q⟩ to Quit program

Try using this, or some other computer software, to solve some of the systems of linear equations that are given in this section.

WARM UP

The following warm-up exercises involve skills that were covered in earlier sections. You will use these skills in the exercise set for this section.

In Exercises 1–4, solve the system of linear equations.

1. $x + y = 25$
$\quad\quad\ y = 10$

2. $2x - 3y = 4$
$\quad 6x\quad\quad = -12$

3. $x + y = 32$
$\ x - y = 24$

4. $2r - s = 5$
$\quad r + 2s = 10$

In Exercises 5–8, determine whether the ordered triple is a solution of the equation.

5. $5x - 3y + 4z = 2$
$(-1, -2, 1)$

6. $x - 2y + 12z = 9$
$(6, 3, 2)$

7. $2x - 5y + 3z = -9$
$(a - 2, a + 1, a)$

8. $-5x + y + z = 21$
$(a - 4, 4a + 1, a)$

In Exercises 9 and 10, solve for x in terms of a.

9. $x + 2y - 3z = 4$
$y = 1 - a, z = a$

10. $x - 3y + 5z = 4$
$y = 2a + 3, z = a$

EXERCISES for Section 7.3

In Exercises 1–26, solve the system of linear equations.

1. $x + y + z = 6$
$2x - y + z = 3$
$3x \quad - z = 0$

2. $x + y + z = 2$
$-x + 3y + 2z = 8$
$4x + y \quad = 4$

3. $4x + y - 3z = 11$
$2x - 3y + 2z = 9$
$x + y + z = -3$

4. $2x \quad + 2z = 2$
$5x + 3y \quad = 4$
$3y - 4z = 4$

5. $\quad 6y + 4z = -12$
$3x + 3y \quad = 9$
$2x \quad - 3z = 10$

6. $2x + 4y + z = -4$
$2x - 4y + 6z = 13$
$4x - 2y + z = 6$

7. $3x - 2y + 4z = 1$
$x + y - 2z = 3$
$2x - 3y + 6z = 8$

8. $5x - 3y + 2z = 3$
$2x + 4y - z = 7$
$x - 11y + 4z = 3$

9. $3x + 3y + 5z = 1$
$3x + 5y + 9z = 0$
$5x + 9y + 17z = 0$

10. $2x + y + 3z = 1$
$2x + 6y + 8z = 3$
$6x + 8y + 18z = 5$

11. $x + 2y - 7z = -4$
$2x + y + z = 13$
$3x + 9y - 36z = -33$

12. $2x + y - 3z = 4$
$4x \quad + 2z = 10$
$-2x + 3y - 13z = -8$

13. $x \quad + 4z = 13$
$4x - 2y + z = 7$
$2x - 2y - 7z = -19$

14. $4x - y + 5z = 11$
$x + 2y - z = 5$
$5x - 8y + 13z = 7$

15. $x - 2y + 5z = 2$
$3x + 2y - z = -2$

16. $x - 3y + 2z = 18$
$5x - 13y + 12z = 80$

17. $2x - 3y + z = -2$
$-4x + 9y \quad = 7$

18. $2x + 3y + 3z = 7$
$4x + 18y + 15z = 44$

19. $x \quad + 3w = 4$
$2y - z - w = 0$
$3y \quad - 2w = 1$
$2x - y + 4z \quad = 5$

20. $x + y + z + w = 6$
$2x + 3y \quad - w = 0$
$-3x + 4y + z + 2w = 4$
$x + 2y - z + w = 0$

21. $x \quad + 4z = 1$
$x + y + 10z = 10$
$2x - y + 2z = -5$

22. $3x - 2y - 6z = -4$
$-3x + 2y + 6z = 1$
$x - y - 5z = -3$

23. $4x + 3y + 17z = 0$
$5x + 4y + 22z = 0$
$4x + 2y + 19z = 0$

24. $2x + 3y \quad = 0$
$4x + 3y - z = 0$
$8x + 3y + 3z = 0$

25. $5x + 5y - z = 0$
$10x + 5y + 2z = 0$
$5x + 15y - 9z = 0$

26. $12x + 5y + z = 0$
$12x + 4y - z = 0$

In Exercises 27–30, find the equation of the parabola
$y = ax^2 + bx + c$
that passes through the given points.

27. $(0, -4), (1, 1), (2, 10)$

28. $(0, 5)$, $(1, 6)$, $(2, 5)$

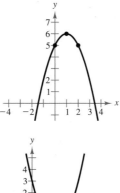

29. $(1, 0)$, $(2, -1)$, $(3, 0)$

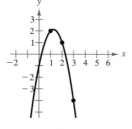

30. $(1, 2)$, $(2, 1)$, $(3, -4)$

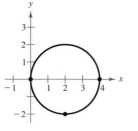

In Exercises 31–34, find the equation of the circle

$$x^2 + y^2 + Dx + Ey + F = 0$$

that passes through the given points.

31. $(0, 0)$, $(2, -2)$, $(4, 0)$

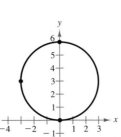

32. $(0, 0)$, $(0, 6)$, $(-3, 3)$

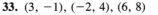

33. $(3, -1)$, $(-2, 4)$, $(6, 8)$

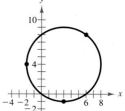

34. $(0, 0)$, $(0, 2)$, $(3, 0)$

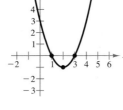

Vertical Motion   In Exercises 35–38, find the position equation

$$s = \tfrac{1}{2}at^2 + v_0 t + s_0$$

for an object moving vertically and at the given heights at the specified times.

35. At $t = 1$ second, $s = 128$ feet
At $t = 2$ seconds, $s = 80$ feet
At $t = 3$ seconds, $s = 0$ feet

36. At $t = 1$ second, $s = 48$ feet
At $t = 2$ seconds, $s = 64$ feet
At $t = 3$ seconds, $s = 48$ feet

37. At $t = 1$ second, $s = 452$ feet
At $t = 3$ seconds, $s = 260$ feet
At $t = 4$ seconds, $s = 116$ feet

38. At $t = 2$ seconds, $s = 132$ feet
At $t = 3$ seconds, $s = 100$ feet
At $t = 4$ seconds, $s = 36$ feet

39. *Investments*   An inheritance of $16,000 was divided among three investments yielding a total of $990 in interest per year. The interest rates for the three investments were 5%, 6%, and 7%. Find the amount placed in each investment if the 5% and 6% investments were $3000 and $2000 less than the 7% investment, respectively.

40. *Investments*   Suppose you receive a total of $1520 a year in interest from three investments. The interest rates for the three investments are 5%, 7%, and 8%. The 5% investment is half of the 7% investment, and the 7% investment is $1500 less than the 8% investment. What is the amount of each investment?

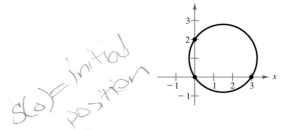

41. *Borrowing*   A small corporation borrowed $775,000 to expand its product line. Some of the money was borrowed at 8%, some at 9%, and some at 10%. How much was borrowed at each rate if the annual interest was $67,500 and the amount borrowed at 8% was four times the amount borrowed at 10%?

42. *Borrowing*   A small corporation borrowed $800,000 to expand its product line. Some of the money was borrowed at 8%, some at 9%, and some at 10%. How much was borrowed at each rate if the annual interest was $67,000 and the amount borrowed at 8% was five times the amount borrowed at 10%?

Investment Portfolio   In Exercises 43 and 44, consider an investor with a portfolio totaling $500,000 that is to be allocated among the following types of investments: (1) certificates of deposit, (2) municipal bonds, (3) blue-chip stocks, and (4) growth or speculative stocks. How much should be allocated to each type of investment?

43. The certificates of deposit pay 10% annually, and the municipal bonds pay 8% annually. Over a five-year period, the investor expects the blue-chip stocks to return 12% annually, and expects the growth stocks to return 13% annually. The investor wants a combined annual return of 10% and also wants to have only one-fourth of the portfolio invested in stocks.

44. The certificates of deposit pay 9% annually, and the municipal bonds pay 5% annually. Over a five-year period, the investor expects the blue-chip stocks to return 12% annually, and expects the growth stocks to return 14% annually. The investor wants a combined annual return of 10% and also wants to have only one-fourth of the portfolio invested in stocks.

45. *Crop Spraying*   A mixture of 12 gallons of chemical A, 16 gallons of chemical B, and 26 gallons of chemical C is required to kill a certain destructive crop insect. Commercial spray X contains 1, 2, and 2 parts, respectively, of these chemicals. Commercial spray Y contains only chemical C. Commercial spray Z contains only chemicals A and B in equal amounts. How much of each type of commercial spray is needed to get the desired mixture?

46. *Chemistry*   A chemist needs 10 liters of a 25% acid solution. The solution is to be mixed from three solutions whose concentrations are 10%, 20%, and 50%, respectively. How many liters of each solution should the chemist use to satisfy the following?

(a) Use as little as possible of the 50% solution.

(b) Use as much as possible of the 50% solution.

(c) Use two liters of the 50% solution.

47. *Truck Scheduling*   A small company that manufactures products A and B has an order for 15 units of product A and 16 units of product B. The company has trucks of three different sizes that can haul the products, as shown in the table.

	Product	
Truck	A	B
Large	6	3
Medium	4	4
Small	0	3

How many trucks of each size are needed to deliver the order? (Give *two* possible solutions.)

48. *Electrical Networks*   Applying Kirchhoff's Laws to the electrical network in the accompanying figure, the currents I_1, I_2, and I_3 must be the solution to the system

$$\begin{array}{rcrcrcl} I_1 & - & I_2 & + & I_3 & = & 0 \\ 3I_1 & + & 2I_2 & & & = & 7 \\ & & 2I_2 & + & 4I_3 & = & 8 \end{array}$$

where the current is measured in amperes. Find the currents.

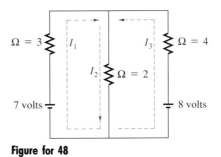

Figure for 48

49. *Pulley System*   A system of pulleys that are assumed fric-tionless and without mass is loaded with 128-pound and 32-pound weights (see figure). The tensions t_1 and t_2 in the ropes and the acceleration a of the 32-pound weight are found by solving the system

$$
\begin{aligned}
t_1 - 2t_2 &= 0 \\
t_1 \quad\; - 2a &= 128 \\
t_2 + a &= 32
\end{aligned}
$$

where t_1 and t_2 are measured in pounds and a is in feet per second squared. Solve the system.

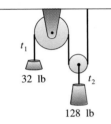

32 lb t_2

128 lb

Figure for 49

50. *Pulley System*   If the 32-pound weight is replaced by a 64-pound weight in the pulley system of Exercise 49, it is modeled by the following system of equations.

$$
\begin{aligned}
t_1 - 2t_2 &= 0 \\
t_1 \quad\; - 2a &= 128 \\
t_2 + 2a &= 64
\end{aligned}
$$

Solve the system and use your answer for the acceleration to describe what (if anything) is happening in the system.

Partial Fractions   In Exercises 51–54, write the partial fraction decomposition for the rational fraction (see Example 9).

51. $\dfrac{1}{x^3 - x} = \dfrac{A}{x} + \dfrac{B}{x - 1} + \dfrac{C}{x + 1}$

52. $\dfrac{3}{x^2 + x - 2} = \dfrac{A}{x - 1} + \dfrac{B}{x + 2}$

53. $\dfrac{x^2 - 3x - 3}{x(x - 2)(x + 3)} = \dfrac{A}{x} + \dfrac{B}{x - 2} + \dfrac{C}{x + 3}$

54. $\dfrac{12}{x(x - 2)(x + 3)} = \dfrac{A}{x} + \dfrac{B}{x - 2} + \dfrac{C}{x + 3}$

Fitting a Parabola   In Exercises 55–58, find the **least squares regression parabola** $y = ax^2 + bx + c$ for the points

$(x_1, y_1), (x_2, y_2), \ldots, (x_n, y_n).$

To find the parabola, solve the following system of linear equations for a, b, and c.

$$nc + \left(\sum_{i=1}^{n} x_i\right)b + \left(\sum_{i=1}^{n} x_i^2\right)a = \sum_{i=1}^{n} y_i$$

$$\left(\sum_{i=1}^{n} x_i\right)c + \left(\sum_{i=1}^{n} x_i^2\right)b + \left(\sum_{i=1}^{n} x_i^3\right)a = \sum_{i=1}^{n} x_i y_i$$

$$\left(\sum_{i=1}^{n} x_i^2\right)c + \left(\sum_{i=1}^{n} x_i^3\right)b + \left(\sum_{i=1}^{n} x_i^4\right)a = \sum_{i=1}^{n} x_i^2 y_i$$

55.
$$
\begin{aligned}
4c \quad\;\; + 40a &= 19 \\
40b \quad\quad &= -12 \\
40c \quad\;\; + 544a &= 160
\end{aligned}
$$

56.
$$
\begin{aligned}
5c \quad\;\; + 10a &= 8 \\
10b \quad\quad &= 12 \\
10c \quad\;\; + 34a &= 22
\end{aligned}
$$

57.
$$
\begin{aligned}
4c + 9b + 29a &= 20 \\
9c + 29b + 99a &= 70 \\
29c + 99b + 353a &= 254
\end{aligned}
$$

58.
$$
\begin{aligned}
4c + 6b + 14a &= 25 \\
6c + 14b + 36a &= 21 \\
14c + 36b + 98a &= 33
\end{aligned}
$$

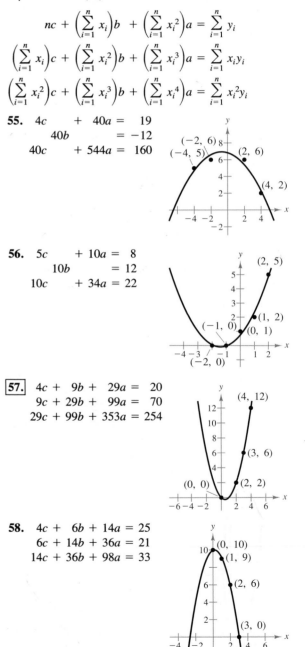

59. *Stopping Distances*   In testing the new braking system on an automobile, the speed in miles per hour and the stopping distance in feet were recorded (see table).

Speed (x)	20	30	40	50	60
Stopping distance (y)	25	55	105	188	300

(a) Plot the data on the rectangular coordinate system.
(b) Fit a least squares regression parabola to the data and sketch the graph of the parabola on the coordinate system of part (a).

60. *Reproduction*   A wildlife management team studied the reproduction rates of deer in five tracts of a wildlife preserve. Each tract contained five acres. In each tract the number of females and the percentage of females that had offspring the following year were counted. The results are given in the table.

Number (x)	80	100	120	140	160
Percentage (y)	80	75	68	55	30

(a) Plot the data on the rectangular coordinate system.
(b) Fit a least squares regression parabola to the data and sketch the graph of the parabola on the coordinate system of part (a).

In Exercises 61 and 62, find a system of linear equations having the given solution. (The answers are not unique.)

61. $(4, -1, 2)$ **62.** $\left(-\frac{3}{2}, 4, -7\right)$

7.4 Systems of Inequalities

The Graph of an Inequality / Systems of Inequalities / Applications

The Graph of an Inequality

The following statements are inequalities in two variables:

$$3x - 2y < 6 \quad \text{and} \quad 2x^2 + 3y^2 \geq 6.$$

An ordered pair (a, b) is a **solution of an inequality** in x and y if the inequality is true when a and b are substituted for x and y, respectively. The **graph** of an inequality is the collection of all solutions of the inequality. To sketch the graph of an inequality such as $3x - 2y < 6$, begin by sketching the graph of the *corresponding equation* $3x - 2y = 6$. This graph is made with a dashed line for the strict inequalities $<$ or $>$ and a solid line for the inequalities $\leq$ or $\geq$. The graph of the equation will normally separate the plane into two or more regions. In each such region, one of the following must be true.

1. *All* points in the region are solutions of the inequality.
2. *No* points in the region are solutions of the inequality.

Thus, you can determine whether the points in an entire region satisfy the inequality by simply testing *one* point in the region.

Sketching the Graph of an Inequality in Two Variables

1. Replace the inequality sign by an equal sign, and sketch the graph of the resulting equation. (Use a dashed line for $<$ or $>$ and a solid line for $\leq$ or $\geq$.)
2. Test one point in each of the regions formed by the graph in Step 1. If the point satisfies the inequality, then shade the entire region to denote that every point in the region satisfies the inequality.

EXAMPLE 1 Sketching the Graph of an Inequality

Sketch the graph of the inequality $y \geq x^2 - 1$.

Solution

The graph of the corresponding *equation* $y = x^2 - 1$ is a parabola, as shown in Figure 7.11. By testing a point *above* the parabola (0, 0), and a point *below* the parabola (0, −2), you see that the points that satisfy the inequality are those lying above (or on) the parabola.

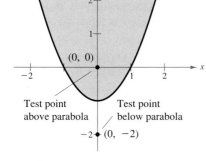

Test point
above parabola

Test point
below parabola

$(0, -2)$

FIGURE 7.11

The inequality given in Example 1 is a nonlinear inequality in two variables. In this section, however, you will work primarily with **linear inequalities** of the form

$$ax + by < c \qquad ax + by \leq c$$
$$ax + by > c \qquad ax + by \geq c.$$

The graph of each of these linear inequalities is a half-plane lying on one side of the line $ax + by = c$.

The simplest linear inequalities are those corresponding to horizontal or vertical lines.

EXAMPLE 2 Sketching the Graph of a Linear Inequality

Sketch the graphs of the following linear inequalities.

a. $x > -2$
b. $y \leq 3$

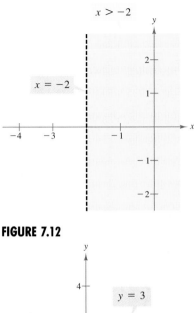

FIGURE 7.12

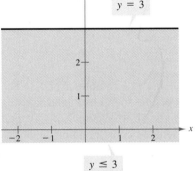

FIGURE 7.13

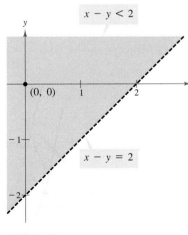

FIGURE 7.14

Solution

a. The graph of the corresponding equation $x = -2$ is a vertical line. The points that satisfy the inequality $x > -2$ are those lying to the right of this line, as shown in Figure 7.12.

b. The graph of the corresponding equation $y = 3$ is a horizontal line. The points that satisfy the inequality $y \leq 3$ are those lying below (or on) this line, as shown in Figure 7.13.

EXAMPLE 3 Sketching the Graph of a Linear Inequality

Sketch the graph of $x - y < 2$.

Solution

The graph of the corresponding equation $x - y = 2$ is a line, as shown in Figure 7.14. Since the origin $(0, 0)$ satisfies the inequality, the graph consists of the half-plane lying above the line. (Try checking a point below the line. Regardless of which point you choose, you will see that it does not satisfy the inequality.)

For a linear inequality in two variables, you can sometimes simplify the graphing procedure by writing the inequality in *slope-intercept* form. For instance, by writing $x - y < 2$ in the form

$$y > x - 2$$

you can see that the solution points lie *above* the line $y = x - 2$, as shown in Figure 7.14. Similarly, by writing the inequality $3x - 2y > 5$ in the form

$$y < \frac{3}{2}x - \frac{5}{2}$$

you see that the solutions lie *below* the line $y = \frac{3}{2}x - \frac{5}{2}$.

Systems of Inequalities

Many practical problems in business, science, and engineering involve systems of linear inequalities. Here are two examples of such systems.

$$
\begin{aligned}
2x - y &\leq 5 \\
x + 2y &> 2
\end{aligned}
\qquad
\begin{aligned}
x + y &\leq 12 \\
3x - 4y &\leq 15 \\
x &\geq 0 \\
y &\geq 0
\end{aligned}
$$

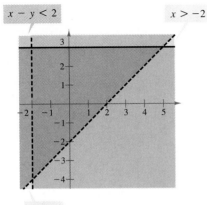

$x - y < 2$

$x > -2$

$y \le 3$

A **solution** of a system of inequalities in x and y is an ordered pair (x, y) that satisfies each inequality in the system. For instance, $(2, 4)$ is a solution of the system on the right because $x = 2$ and $y = 4$ satisfy each of the four inequalities in the system.

To sketch the graph of a system of inequalities in two variables, first sketch the graph of each individual inequality (on the same coordinate system) and then find the region that is *common* to every graph in the system. For systems of linear inequalities, it is helpful to find the *vertices* of the solution region, as shown in the following example.

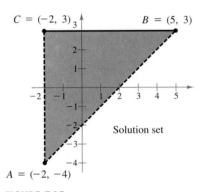

$C = (-2, 3)$ $B = (5, 3)$

Solution set

$A = (-2, -4)$

FIGURE 7.15

EXAMPLE 4   Solving a System of Inequalities

Sketch the graph (and label the vertices) of the solution set of the following system.

$$x - y < 2$$
$$x > -2$$
$$y \le 3$$

Solution

You already sketched the graph of each inequality in Examples 2 and 3. The triangular region common to all three graphs can be found by superimposing the graphs on the same coordinate plane, as shown in Figure 7.15. To find the vertices of the region, solve the three systems of corresponding equations obtained by taking *pairs* of equations representing the boundaries of the individual regions.

Vertex A: $(-2, -4)$	Vertex B: $(5, 3)$	Vertex C: $(-2, 3)$
Obtained by solving the system	Obtained by solving the system	Obtained by solving the system
$x - y = 2$	$x - y = 2$	$x = -2$
$x = -2.$	$y = 3.$	$y = 3.$

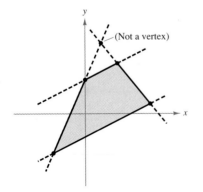

(Not a vertex)

Boundary lines can intersect at a point that is not a vertex.

FIGURE 7.16

For the triangular region shown in Figure 7.15, each point of intersection of a pair of boundary lines corresponds to a vertex. With more complicated regions, two border lines can sometimes intersect at a point that is not a vertex of the region, as shown in Figure 7.16. In order to keep track of which points of intersection are actually vertices of the region, we suggest that you make a careful sketch of the region and refer to your sketch as you find each point of intersection.

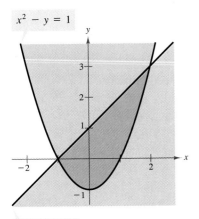

$x^2 - y = 1$

$-x + y = 1$

FIGURE 7.17

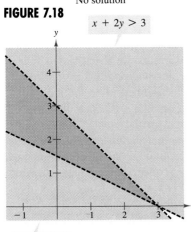

$x + y > 3$

$x + y < -1$

No solution

FIGURE 7.18

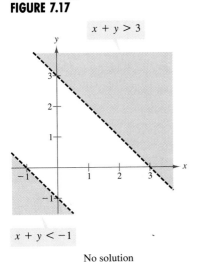

$x + 2y > 3$

$x + y < 3$

Unbounded region

EXAMPLE 5   Solving a System of Inequalities

Sketch the region containing all points that satisfy the following system.

$$x^2 - y \le 1$$
$$-x + y \le 1$$

Solution

As shown in Figure 7.17, the points that satisfy the inequality $x^2 - y \le 1$ are the points lying above (or on) the parabola given by $y = x^2 - 1$. The points satisfying the inequality $-x + y \le 1$ are the points lying on or below the line given by $y = x + 1$. To find the points of intersection of the parabola and the line, solve the following system of corresponding equations.

$$x^2 - y = 1$$
$$-x + y = 1$$

Using the method of substitution, you find the solutions to be $(-1, 0)$ and $(2, 3)$, as shown in Figure 7.17.   ◢

When solving a system of inequalities, you should be aware that the system might have no solution. For instance, the system

$$x + y > 3$$
$$x + y < -1$$

has no solution points, because the quantity $(x + y)$ cannot be both less than -1 and greater than 3, as shown in Figure 7.18.

Another possibility is that the solution set of a system of inequalities can be unbounded. For instance, the solution set of

$$x + y < 3$$
$$x + 2y > 3$$

forms an *infinite wedge*, as shown in Figure 7.19.

FIGURE 7.19

Applications

EXAMPLE 6   An Application of a System of Inequalities

The liquid portion of a diet is to provide at least 300 calories, 36 units of vitamin A, and 90 units of vitamin C daily. A cup of dietary drink X provides 60 calories, 12 units of vitamin A, and 10 units of vitamin C. A cup of dietary drink Y provides 60 calories, 6 units of vitamin A, and 30 units of vitamin C. Set up a system of linear inequalities that describes the minimum daily requirements for calories and vitamins.

Solution

Let

$$x = \text{number of cups of dietary drink X}$$
$$y = \text{number of cups of dietary drink Y.}$$

Then, to meet the minimum daily requirements, the following inequalities must be satisfied.

For Calories: $60x + 60y \geq 300$
For Vitamin A: $12x + 6y \geq 36$
For Vitamin C: $10x + 30y \geq 90$
$$x \geq 0$$
$$y \geq 0$$

The last two inequalities are included because x and y cannot be negative. The graph of this system of inequalities is shown in Figure 7.20. (More is said about this application in Section 7.5, Example 6.)

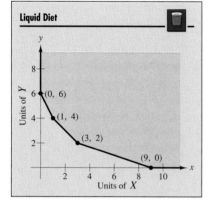

FIGURE 7.20

In Example 8 in Section 7.2 we discussed the *point of equilibrium* for a demand and supply function. In the next example, we discuss two related concepts that economists call **consumer surplus** and **producer surplus.** As shown in Figure 7.21, the consumer surplus is defined to be the area of the region that lies *below* the demand curve, *above* the horizontal line passing through the equilibrium point, and to the right of the y-axis. Similarly, the producer surplus is defined to be the area of the region that lies *above* the supply curve, *below* the horizontal line passing through the equilibrium point, and to the right of the y-axis. In general terms, consumer surplus is a measure of the amount of money that consumers would have been willing to pay *above what they actually paid*. Similarly, producer surplus is a measure of the amount of money that producers would have been willing to receive *below what they actually received.*

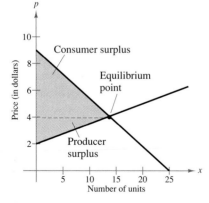

FIGURE 7.21

EXAMPLE 7 **Consumer and Producer Surplus**

Suppose the demand and supply functions for a certain type of calculator are given by

$$p = 150 - 10x \qquad \textit{Demand equation}$$
$$p = 60 + 20x \qquad \textit{Supply equation}$$

where p is the price in dollars and x represents the number of units in millions. Find the consumer and producer surplus for these two equations.

Solution

To begin, find the point of equilibrium by solving the equation

$$60 + 20x = 150 - 10x.$$

In Example 8 of Section 7.2, you found that the solution was $x = 3$, which corresponded to an equilibrium price of $p = \$120$. Thus, the consumer surplus and producer surplus are the areas of the triangular regions given by the following sets of inequalities.

Consumer Surplus	*Producer Surplus*
$p \leq 150 - 10x$	$p \geq 60 + 20x$
$p \geq 120$	$p \leq 120$
$x \geq 0$	$x \geq 0$

Using Figure 7.22 and the formula for the area of a triangle, you find that the consumer surplus is

$$\text{Consumer surplus} = \frac{1}{2}(\text{base})(\text{height}) = \frac{1}{2}(30)(3) = \$45 \text{ million}$$

and the producer surplus is

$$\text{Producer surplus} = \frac{1}{2}(\text{base})(\text{height}) = \frac{1}{2}(60)(3) = \$90 \text{ million}.$$

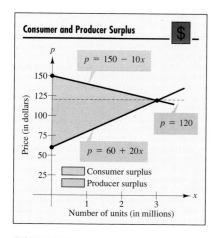

Consumer and Producer Surplus

$p = 150 - 10x$

$p = 60 + 20x$

$p = 120$

Consumer surplus

Producer surplus

Number of units (in millions)

FIGURE 7.22

DISCUSSION

PROBLEM

You Be
the
Instructor

Suppose you are tutoring a student in algebra and want to construct some practice problems for your student. Write a paragraph describing how you could write a system of linear inequalities that had a given region as its solution. Then apply your procedure to find two systems of linear inequalities that have the following regions as solutions (see Figures 7.23 and 7.24).

1. Region with (0, 0), (0, 2), (3, 0), and (2, 1) as vertices.

2. Region with (0, 0), (0, 3), (4, 0), and (3, 2) as vertices.

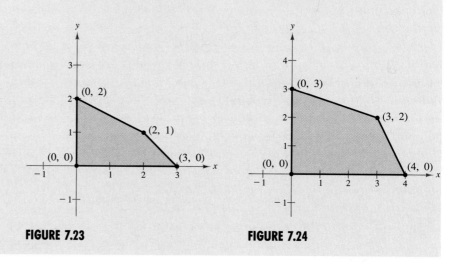

FIGURE 7.23   **FIGURE 7.24**

WARM UP

The following warm-up exercises involve skills that were covered in earlier sections. You will use these skills in the exercise set for this section.

In Exercises 1–6, identify the graph of each equation.

1. $x + y = 3$

2. $4x - y = 8$

3. $y = x^2 - 4$

4. $y = -x^2 + 1$

5. $x^2 + y^2 = 9$

6. $\dfrac{x^2}{4} + \dfrac{y^2}{9} = 1$

In Exercises 7–10, solve the system of equations.

7. $\begin{aligned} x + 2y &= 3 \\ 4x - 7y &= -3 \end{aligned}$

8. $\begin{aligned} 2x - 3y &= 4 \\ x + 5y &= 2 \end{aligned}$

9. $\begin{aligned} x^2 + y &= 5 \\ 2x - 4y &= 0 \end{aligned}$

10. $\begin{aligned} x^2 + y^2 &= 13 \\ x + y &= 5 \end{aligned}$

EXERCISES for Section 7.4

In Exercises 1–8, match the inequality with its graph. [The graphs are labeled (a), (b), (c), (d), (e), (f), (g), and (h).]

1. $x > 3$

2. $y \le 2$

3. $2x + 3y \le 6$

4. $2x - y \ge -2$

5. $x^2 + y^2 < 4$ *a*

6. $(x - 2)^2 + (y - 3)^2 > 4$

7. $xy > 2$ *B*

8. $y \le 4 - x^2$

In Exercises 9–20, sketch the graph of the inequality.

9. $x \ge 2$

10. $x \le 4$

11. $y \ge -1$

12. $y \le 3$

13. $y < 2 - x$

14. $y > 2x - 4$

15. $2y - x \ge 4$

16. $5x + 3y \ge -15$

17. $(x + 1)^2 + (y - 2)^2 < 9$

18. $y^2 - x < 0$

19. $y \le \dfrac{1}{1 + x^2}$

20. $y < \ln x$

In Exercises 21–40, sketch the graph of the solution of the system of inequalities.

21. $\begin{aligned} x + y &\le 1 \\ -x + y &\le 1 \\ y &\ge 0 \end{aligned}$

22. $\begin{aligned} 3x + 2y &< 6 \\ x &> 0 \\ y &> 0 \end{aligned}$

23. $\begin{aligned} x + y &\le 5 \\ x &\ge 2 \\ y &\ge 0 \end{aligned}$

24. $\begin{aligned} 2x + y &\ge 2 \\ x &\le 2 \\ y &\le 1 \end{aligned}$

25. $\begin{aligned} -3x + 2y &< 6 \\ x + 4y &> -2 \\ 2x + y &< 3 \end{aligned}$

26. $\begin{aligned} x - 7y &> -36 \\ 5x + 2y &> 5 \\ 6x - 5y &> 6 \end{aligned}$

27. $\begin{aligned} 2x + y &> 2 \\ 6x + 3y &< 2 \end{aligned}$

28. $\begin{aligned} x - 2y &< -6 \\ 5x - 3y &> -9 \end{aligned}$

29. $\begin{aligned} x &\ge 1 \\ x - 2y &\le 3 \\ 3x + 2y &\ge 9 \\ x + y &\le 6 \end{aligned}$

30. $\begin{aligned} x - y^2 &> 0 \\ x - y &< 2 \end{aligned}$

31. $\begin{aligned} x^2 + y^2 &\le 9 \\ x^2 + y^2 &\ge 1 \end{aligned}$

32. $\begin{aligned} x^2 + y^2 &\le 25 \\ 4x - 3y &\le 0 \end{aligned}$

33. $\begin{aligned} x &> y^2 \\ x &< y + 2 \end{aligned}$

34. $\begin{aligned} x &< 2y - y^2 \\ 0 &< x + y \end{aligned}$

35. $\begin{aligned} y &\le \sqrt{3x} + 1 \\ y &\ge x + 1 \end{aligned}$

36. $\begin{aligned} y &< -x^2 + 2x + 3 \\ y &> x^2 - 4x + 3 \end{aligned}$

37. $\begin{aligned} y &< x^3 - 2x + 1 \\ y &> -2x \\ x &\le 1 \end{aligned}$

38. $\begin{aligned} y &\ge x^4 - 2x^2 + 1 \\ y &\le 1 - x^2 \end{aligned}$

39. $\begin{aligned} x^2y &\ge 1 \\ 0 &< x \le 4 \\ y &\le 4 \end{aligned}$

40. $\begin{aligned} y &\le e^{-x^2/2} \\ y &\ge 0 \\ -2 &\le x \le 2 \end{aligned}$

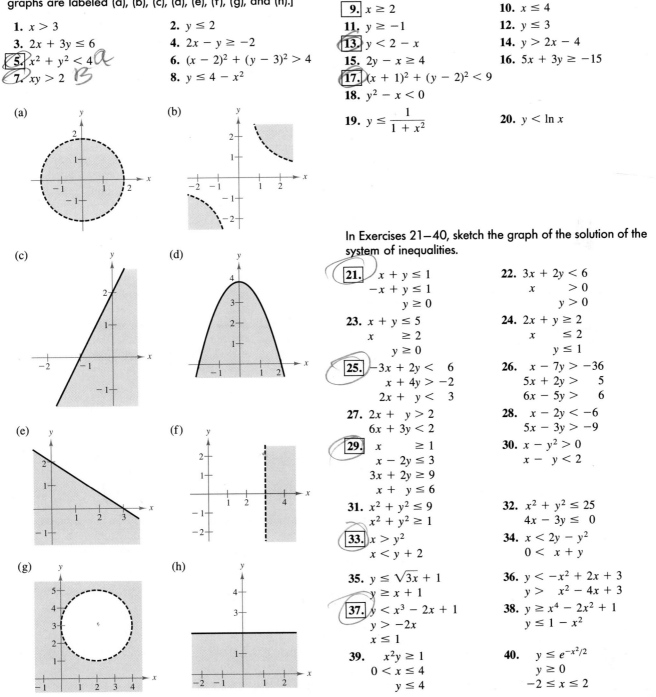

(a)

(b)

(c)

(d)

(e)

(f)

(g)

(h)

In Exercises 41–46, derive a set of inequalities to describe the region.

41. Rectangular region with vertices at (2, 1), (5, 1), (5, 7), and (2, 7)

42. Parallelogram region with vertices at (0, 0), (4, 0), (1, 4), and (5, 4)

43. Triangular region with vertices at (0, 0), (5, 0), and (2, 3)

44. Triangular region with vertices at (−1, 0), (1, 0), and (0, 1)

45. Sector of a circle   **46.** Sector of a circle

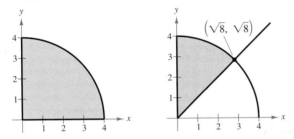

47. *Furniture Production*   A furniture company can sell all the tables and chairs it produces. Each table requires 1 hour in the assembly center and $1\frac{1}{3}$ hours in the finishing center. Each chair requires $1\frac{1}{2}$ hours in the assembly center and $1\frac{1}{2}$ hours in the finishing center. The company's assembly center is available 12 hours per day, and its finishing center is available 15 hours per day. If x is the number of tables produced per day and y is the number of chairs produced per day, find a system of inequalities describing all possible production levels. Sketch the graph of the system.

48. *Computer Inventory*   A store sells two models of a certain brand of computer. Because of the demand, it is necessary to stock twice as many units of model A as units of model B. The cost to the store for the two models is $800 and $1,200, respectively. The management does not want more than $20,000 in computer inventory at any one time, and it wants at least four model A computers and two model B computers in inventory at all times. Devise a system of inequalities describing all possible inventory levels, and sketch the graph of the system.

49. *Investment*   A person plans to invest $20,000 in two different interest-bearing accounts. Each account is to contain at least $5,000. Moreover, one account should have at least twice the amount that is in the other account. Find a system of inequalities to describe the various amounts that can be deposited in each account, and sketch the graph of the system.

50. *Concert Ticket Sales*   Two types of tickets are to be sold for a concert. One type costs $15 per ticket and the other type costs $25 per ticket. The promoter of the concert must sell at least 15,000 tickets, including 8,000 of the $15 tickets and 4,000 of the $25 tickets. Moreover, the gross receipts must total at least $275,000 in order for the concert to be held. Find a system of inequalities describing the different numbers of tickets that can be sold and sketch the graph of the system.

51. *Diet Supplement*   A dietitian is asked to design a special diet supplement using two different foods. Each ounce of food X contains 20 units of calcium, 15 units of iron, and 10 units of vitamin B. Each ounce of food Y contains 10 units of calcium, 10 units of iron, and 20 units of vitamin B. The minimum daily requirements in the diet are 280 units of calcium, 160 units of iron, and 180 units of vitamin B. Find a system of inequalities describing the different amounts of food X and food Y that can be used in the diet and sketch the graph of the system.

52. *Diet Supplement*   A dietitian is asked to design a special diet supplement using two different foods. Each ounce of food X contains 20 units of calcium, 15 units of iron, and 10 units of vitamin B. Each ounce of food Y contains 10 units of calcium, 10 units of iron, and 20 units of vitamin B. The minimum daily requirements in the diet are 300 units of calcium, 150 units of iron, and 200 units of vitamin B. Find a system of inequalities describing the different amounts of food X and food Y that can be used in the diet and sketch the graph of the system.

Consumer and Producer Surplus   In Exercises 53–58, find the consumer surplus and producer surplus for the given pair of supply and demand equations.

Demand	Supply
53. $p = 50 - 0.5x$	$p = 0.125x$
54. $p = 60 - x$	$p = 10 + \frac{7}{3}x$
55. $p = 300 - x$	$p = 100 + x$
56. $p = 100 - 0.05x$	$p = 25 + 0.1x$
57. $p = 140 - 0.00002x$	$p = 80 + 0.00001x$
58. $p = 400 - 0.0002x$	$p = 225 + 0.0005x$

59. *Physical Fitness Facility* An indoor running track is to be constructed with a space for body-building equipment inside the track (see figure). The inside track must be at least 125 meters long, and the body-building space must have an area of at least 500 square meters. Find a system of inequalities describing the various sizes of the track, and sketch the graph of the system.

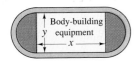

Figure for 59

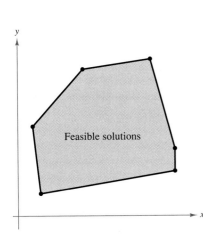

The objective function has its optimal value at one of the vertices of the region determined by the constraints.

FIGURE 7.25

![7.5] **Linear Programming**
The Graphical Approach to Linear Programming / Applications

The Graphical Approach to Linear Programming

Many applications in business and economics involve a process called **optimization,** in which we are required to find the minimum cost, the maximum profit, or the minimum use of resources. In this section we discuss one type of optimization problem called **linear programming.**

A two-dimensional linear programming problem consists of a linear **objective function** and a system of linear inequalities (**constraints**). The objective function gives the quantity that is to be maximized (or minimized), and the constraints determine the set of **feasible solutions.**

For example, consider a linear programming problem in which you are asked to maximize the value of

$$z = ax + by \qquad \text{*Objective function*}$$

subject to a set of constraints that determine the region indicated in Figure 7.25. Because every point in the region satisfies each constraint, it is not clear how you should go about finding the point that yields a maximum value of z. Fortunately, it can be shown that if there is an optimal solution, it must occur at one of the vertices of the region. In other words, *you can find the maximum value by testing z at each of the vertices*, as shown in Example 1.

Optimal Solution of Linear Programming Problem

If a linear programming problem has a solution, it must occur at a vertex of the set of feasible solutions. If the problem has more than one solution, then at least one of them must occur at a vertex of the set of feasible solutions. In either case, the value of the objective function is unique.

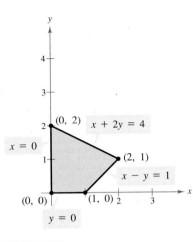

FIGURE 7.26

REMARK   In Example 1, try testing some of the *interior* points in the region. You will see that the corresponding values of z are less than 8.

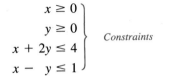

EXAMPLE 1   Solving a Linear Programming Problem

Find the maximum value of

$$z = 3x + 2y \qquad \textit{Objective function}$$

subject to the following constraints.

$$\left.\begin{array}{r} x \geq 0 \\ y \geq 0 \\ x + 2y \leq 4 \\ x - y \leq 1 \end{array}\right\} \textit{Constraints}$$

Solution

The constraints form the region shown in Figure 7.26. At the four vertices of this region, the objective function has the following values.

At $(0, 0)$: $z = 3(0) + 2(0) = 0$

At $(1, 0)$: $z = 3(1) + 2(0) = 3$

At $(2, 1)$: $z = 3(2) + 2(1) = 8$ *Maximum value of z*

At $(0, 2)$: $z = 3(0) + 2(2) = 4$

Thus, the maximum value of z is 8, and this occurs when $x = 2$ and $y = 1$.

To see why the maximum value of the objective function in Example 1 must occur at a vertex, consider writing the objective function in the form

$$y = -\frac{3}{2}x + \frac{z}{2}$$

where $z/2$ is the y-intercept of the objective function. This equation represents a family of lines, each of slope $-\frac{3}{2}$. Of these infinitely many lines, you want the one that has the largest z-value, while still intersecting the region determined by the constraints. In other words, of all the lines whose slope is $-\frac{3}{2}$, you want the one that has the largest y-intercept *and* intersects the given region, as shown in Figure 7.27. It should be clear that such a line will pass through one (or more) of the vertices of the region.

The steps used in Example 1 can be outlined as follows.

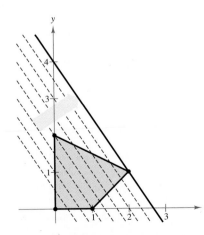

FIGURE 7.27

Graphical Method of Solving a Linear Programming Problem

To solve a linear programming problem involving two variables by the graphical method, use the following steps.

1. Sketch the region corresponding to the system of constraints. (The points inside or on the boundary of the region are called feasible solutions.)
2. Find the vertices of the region.
3. Test the objective function at each of the vertices and select the values of the variables that optimize the objective function. For a bounded region, both a minimum and maximum value will exist. (For an unbounded region, *if* an optimal solution exists, then it will occur at a vertex.)

These guidelines will work whether the objective function is to be maximized *or* minimized. For instance, in Example 1 the same test used to find the maximum value of z can be used to conclude that the minimum value of z is 0, and this occurs at the vertex $(0, 0)$.

EXAMPLE 2 Solving a Linear Programming Problem

Find the maximum value of the objective function

$$z = 4x + 6y \qquad \text{\textit{Objective function}}$$

where $x \geq 0$ and $y \geq 0$, subject to the following constraints.

$$\left. \begin{array}{r} -x + y \leq 11 \\ x + y \leq 27 \\ 2x + 5y \leq 90 \end{array} \right\} \quad \textit{Constraints}$$

Solution

The region bounded by the constraints is shown in Figure 7.28. By testing the objective function at each vertex, you obtain the following.

At $(0, 0)$: $z = 4(0) + 6(0) = 0$

At $(0, 11)$: $z = 4(0) + 6(11) = 66$

At $(5, 16)$: $z = 4(5) + 6(16) = 116$

At $(15, 12)$: $z = 4(15) + 6(12) = 132$ *Maximum value of z*

At $(27, 0)$: $z = 4(27) + 6(0) = 108$

Thus, the maximum value of z is 132 when $x = 15$ and $y = 12$.

$2x + 5y = 90$
$(5, 16)$
$-x + y = 11$
$(0, 11)$
$(15, 12)$
$x + y = 27$
$x = 0$
$(27, 0)$
$(0, 0)$
$y = 0$

FIGURE 7.28

REMARK In Example 3, note that the steps used to find the minimum value are precisely the same ones you would use to find the maximum value. In other words, once you have evaluated the objective function at the vertices of the feasible region, you simply choose the largest value as the maximum and the smallest value as the minimum.

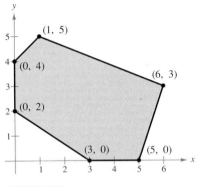

FIGURE 7.29

EXAMPLE 3 Minimizing an Objective Function

Find the minimum value of the objective function

$$z = 5x + 7y \qquad \textit{Objective function}$$

where $x \geq 0$ and $y \geq 0$, subject to the following constraints.

$$\left.\begin{array}{r} 2x + 3y \geq 6 \\ 3x - y \leq 15 \\ -x + y \leq 4 \\ 2x + 5y \leq 27 \end{array}\right\} \quad \textit{Constraints}$$

Solution

The region bounded by the constraints is shown in Figure 7.29. By testing the objective function at each vertex, we obtain the following.

At $(0, 2)$: $z = 5(0) + 7(2) = 14$ *Minimum value of z*

At $(0, 4)$: $z = 5(0) + 7(4) = 28$

At $(1, 5)$: $z = 5(1) + 7(5) = 40$

At $(6, 3)$: $z = 5(6) + 7(3) = 51$

At $(5, 0)$: $z = 5(5) + 7(0) = 25$

At $(3, 0)$: $z = 5(3) + 7(0) = 15$

Thus, the minimum value of z is 14 when $x = 0$ and $y = 2$.

When solving a linear programming problem, it is possible that the maximum (or minimum) value occurs at *two* different vertices. For instance, at the vertices of the region shown in Figure 7.30, the objective function

$$z = 2x + 2y \qquad \textit{Objective function}$$

has the following values.

At $(0, 0)$: $z = 2(0) + 2(0) = 0$

At $(0, 4)$: $z = 2(0) + 2(4) = 8$

At $(2, 4)$: $z = 2(2) + 2(4) = 12$ *Maximum value of z*

At $(5, 1)$: $z = 2(5) + 2(1) = 12$ *Maximum value of z*

At $(5, 0)$: $z = 2(5) + 2(0) = 10$

In this case, the objective function has a maximum value not only at the vertices $(2, 4)$ and $(5, 1)$; it also has a maximum value (of 12) at *any point on the line segment connecting these two vertices*. Note that the objective function, $y = -x + \frac{1}{2}z$, has the same slope as the line through the vertices $(2, 4)$ and $(5, 1)$.

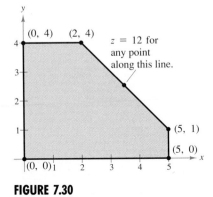

FIGURE 7.30

Some linear programming problems have no optimal solution. This can occur if the region determined by the constraint is *unbounded*. Example 4 illustrates such a problem.

EXAMPLE 4   An Unbounded Region

Find the maximum value of

$$z = 4x + 2y \qquad \textit{Objective function}$$

where $x \geq 0$ and $y \geq 0$, subject to the following constraints.

$$\left.\begin{array}{r} x + 2y \geq 4 \\ 3x + y \geq 7 \\ -x + 2y \leq 7 \end{array}\right\} \quad \textit{Constraints}$$

Solution

The region determined by the constraints is shown in Figure 7.31. For this unbounded region, there is no maximum value of z. Note that the point $(x, 0)$ lies in the region for all values of $x \geq 4$. By choosing x to be large, you can obtain values of $z = 4(x) + 2(0) = 4x$ to be as large as you want. Thus, there is no maximum value of z. For this problem, there *is* a minimum value of $z = 10$, which occurs at the vertex $(2, 1)$.

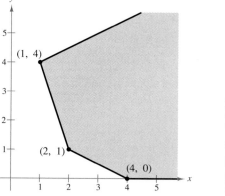

FIGURE 7.31

Applications

EXAMPLE 5   A Business Application: Maximum Profit

A manufacturer wants to maximize the profit for two products. The first product yields a profit of $1.50 per unit, and the second product yields a profit of $2.00 per unit. Market tests and available resources have indicated the following constraints.

1. The combined production level should not exceed 1200 units per month.
2. The demand for product II is less than or equal to half of the demand for product I.
3. The production level of product I is less than or equal to 600 units plus three times the production level of product II.

Solution

If x is the number of units of product I and y is the number of units of product II, then the objective function (for the combined profit) is given by

$$P = 1.5x + 2y. \qquad \textit{Objective function}$$

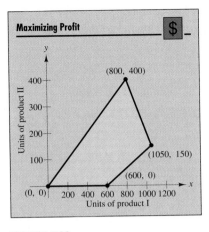

Maximizing Profit

FIGURE 7.32

The three constraints translate into the following linear inequalities.

1. $x + y \leq 1200$ $\rightarrow$ $x + y \leq 1200$
2. $y \leq \frac{1}{2}x$ $\rightarrow$ $x - 2y \geq 0$
3. $x \leq 3y + 600$ $\rightarrow$ $x - 3y \leq 600$

Since neither x nor y can be negative, you also have the two additional constraints of $x \geq 0$ and $y \geq 0$. Figure 7.32 shows the region determined by the constraints. To find the maximum profit, test the value of P at the vertices of the region.

At $(0, 0)$: $P = 1.5(0) \quad + 2(0) \quad = 0$
At $(800, 400)$: $P = 1.5(800) \quad + 2(400) = 2000$ *Maximum profit*
At $(1050, 150)$: $P = 1.5(1050) + 2(150) = 1875$
At $(600, 0)$: $P = 1.5(600) \quad + 2(0) \quad = 900$

Thus, the maximum profit is $2000, and it occurs when the monthly production consists of 800 units of product I and 400 units of product II.

EXAMPLE 6   An Application: Minimum Cost

In Example 6 in Section 7.4, we set up a system of linear equations for the following problem. The liquid portion of a diet is to provide at least 300 calories, 36 units of vitamin A, and 90 units of vitamin C daily. A cup of dietary drink X provides 60 calories, 12 units of vitamin A, and 10 units of vitamin C. A cup of dietary drink Y provides 60 calories, 6 units of vitamin A, and 30 units of vitamin C. Now, suppose that dietary drink X costs $0.12 per cup and drink Y costs $0.15 per cup. How many cups of each drink should be consumed each day to minimize the cost and still meet the stated daily requirements?

Solution

Begin by letting x be the number of cups of dietary drink X and y be the number of cups of dietary drink Y. Moreover, to meet the minimum daily requirements, the following inequalities must be satisfied.

$$\left. \begin{array}{lrcl} \textit{For Calories:} & 60x + 60y & \geq & 300 \\ \textit{For Vitamin A:} & 12x + 6y & \geq & 36 \\ \textit{For Vitamin C:} & 10x + 30y & \geq & 90 \\ & x & \geq & 0 \\ & y & \geq & 0 \end{array} \right\} \textit{Constraints}$$

The cost C is given by

$$C = 0.12x + 0.15y. \qquad \textit{Objective function}$$

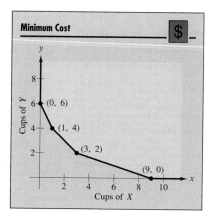

FIGURE 7.33

The graph of the region corresponding to the constraints is shown in Figure 7.33. To determine the minimum cost, test C at each vertex of the region as follows.

At $(0, 6)$:　$C = 0.12(0) + 0.15(6) = 0.90$
At $(1, 4)$:　$C = 0.12(1) + 0.15(4) = 0.72$
At $(3, 2)$:　$C = 0.12(3) + 0.15(2) = 0.66$　*Minimum value of C*
At $(9, 0)$:　$C = 0.12(9) + 0.15(0) = 1.08$

Thus, the minimum cost is $0.66 per day, and this occurs when three cups of drink X and two cups of drink Y are consumed each day.　◀

EXAMPLE 7　Maximum Profit

A small computer keyboard company makes two popular models, for which the demand is much greater than the current supply (both models have substantial back orders). Both models take 1 hour to assemble. However, Model TT1 requires only 7.5 minutes to test, whereas Model TT2 requires 30 minutes to test. With the company's current facilities, there are 45,000 hours per month available for assembly, and 15,000 hours per month available for testing. The profit for Model TT1 is $50 per unit, and the profit for Model TT2 is $80 per unit. What is the greatest monthly profit the company can make without increasing its current facilities?

Solution

To begin to solve this problem, it is helpful to organize the information in table form, as shown in Table 7.1.

TABLE 7.1

	Model TT1 (x units)	Model TT2 (y units)	Maximum Hours
Assembly time per unit	1 hour	1 hour	45,000
Test time per unit	$\frac{1}{8}$ hour	$\frac{1}{2}$ hour	15,000
Profit per unit	$50	$80	—

From the third row in the table, you see that the total profit for selling x units of Model TT1 and y units of Model TT2 is

$$P = 50x + 80y.$$　*Objective function*

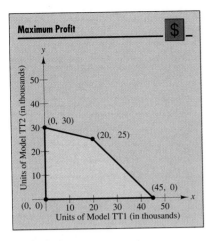

Maximum Profit

FIGURE 7.34

This equation represents the objective function that you need to maximize. The constraints under which you are allowed to work are given by the following two inequalities.

$$x + y \le 45{,}000$$
$$\frac{1}{8}x + \frac{1}{2}y \le 15{,}000$$

In addition, you further require that $x \ge 0$ and $y \ge 0$. Thus, the possible values of x and y are those that lie in (or on the boundary of) the region shown in Figure 7.34. At the four vertices, you obtain the following profits.

$$P = 50(0) + 80(0) = \$0$$
$$P = 50(0) + 80(30{,}000) = \$2{,}400{,}000$$
$$P = 50(20{,}000) + 80(25{,}000) = \$3{,}000{,}000 \quad \textit{Maximum value of P}$$
$$P = 50(45{,}000) + 80(0) = \$2{,}250{,}000$$

Therefore, the maximum profit can be obtained by producing 20,000 units of Model TT1 and 25,000 units of Model TT2.

DISCUSSION

PROBLEM

Creating
a
Linear
Programming
Problem

Consider the following linear programming problem.

Objective Function: $\qquad z = ax + by$

Constraints: $\qquad\qquad x \ge 0$
$$y \ge 0$$
$$x + 2y \le 8$$
$$x + y \le 5$$

The region determined by these constraints is shown in Figure 7.35. Find, if possible, an objective function that has a *maximum* at the indicated vertex of the region.

1. Maximum at $(0, 4)$
2. Maximum at $(2, 3)$
3. Maximum at $(5, 0)$
4. Maximum at $(0, 0)$

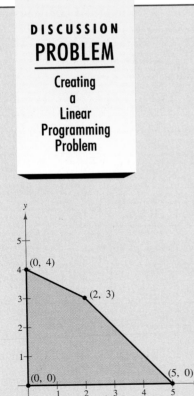

FIGURE 7.35

WARM UP

The following warm-up exercises involve skills that were covered in earlier sections. You will use these skills in the exercise set for this section.

In Exercises 1–4, sketch the graph of the linear equation.

1. $y + x = 3$

2. $y - x = 12$

3. $x = 0$

4. $y = 4$

In Exercises 5–8, solve the system of equations.

5. $x + y = 4$
   $\quad x \quad\;\; = 0$

6. $x + 2y = 12$
   $\qquad\quad y = 0$

7. $\;\; x + \;\; y = 4$
   $\; 2x + 3y = 9$

8. $\;\; x + 2y = 12$
   $\; 2x + \;\; y = 9$

In Exercises 9 and 10, sketch the graph of the inequality.

9. $2x + 3y \geq 18$

10. $4x + 3y \geq 12$

EXERCISES for Section 7.5

In Exercises 1–12, find the minimum and maximum values of the given objective function, subject to the indicated constraints. (For each exercise, the graph of the region determined by the constraints is provided.)

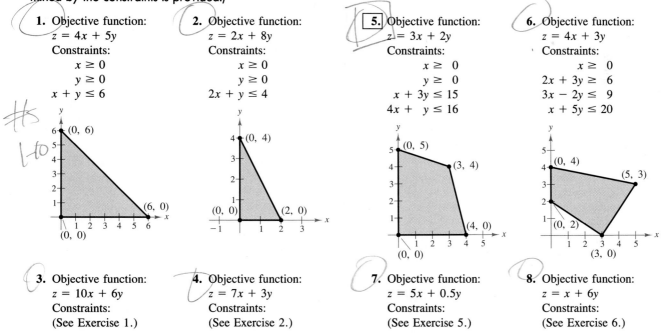

1. Objective function:
$z = 4x + 5y$
Constraints:
$\quad x \geq 0$
$\quad y \geq 0$
$\quad x + y \leq 6$

2. Objective function:
$z = 2x + 8y$
Constraints:
$\quad x \geq 0$
$\quad y \geq 0$
$\quad 2x + y \leq 4$

5. Objective function:
$z = 3x + 2y$
Constraints:
$\quad x \geq 0$
$\quad y \geq 0$
$\quad x + 3y \leq 15$
$\quad 4x + \;\; y \leq 16$

6. Objective function:
$z = 4x + 3y$
Constraints:
$\quad x \geq 0$
$\quad 2x + 3y \geq 6$
$\quad 3x - 2y \leq 9$
$\quad x + 5y \leq 20$

3. Objective function:
$z = 10x + 6y$
Constraints:
(See Exercise 1.)

4. Objective function:
$z = 7x + 3y$
Constraints:
(See Exercise 2.)

7. Objective function:
$z = 5x + 0.5y$
Constraints:
(See Exercise 5.)

8. Objective function:
$z = x + 6y$
Constraints:
(See Exercise 6.)

9. Objective function:
$z = 10x + 7y$
Constraints:
$$0 \le x \le 60$$
$$0 \le y \le 45$$
$$5x + 6y \le 420$$

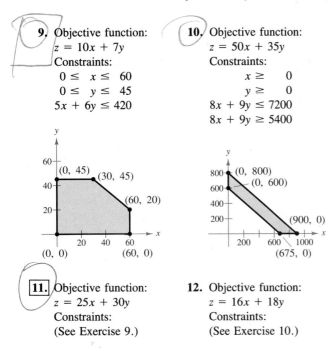

(0, 45) (30, 45)
(60, 20)
(0, 0) (60, 0)

10. Objective function:
$z = 50x + 35y$
Constraints:
$$x \ge 0$$
$$y \ge 0$$
$$8x + 9y \le 7200$$
$$8x + 9y \ge 5400$$

(0, 800)
(0, 600)
(900, 0)
(675, 0)

11. Objective function:
$z = 25x + 30y$
Constraints:
(See Exercise 9.)

12. Objective function:
$z = 16x + 18y$
Constraints:
(See Exercise 10.)

In Exercises 13–24, sketch the region determined by the constraints. Then find the minimum and maximum values of the objective function, subject to the constraints.

13. Objective function:
$z = 6x + 10y$
Constraints:
$$x \ge 0$$
$$y \ge 0$$
$$2x + 5y \le 10$$

14. Objective function:
$z = 7x + 8y$
Constraints:
$$x \ge 0$$
$$y \ge 0$$
$$x + \frac{1}{2}y \le 4$$

15. Objective function:
$z = 9x + 24y$
Constraints:
(See Exercise 13.)

16. Objective function:
$z = 7x + 2y$
Constraints:
(See Exercise 14.)

17. Objective function:
$z = 4x + 5y$
Constraints:
$$x \ge 0$$
$$y \ge 0$$
$$4x + 3y \ge 27$$
$$x + y \ge 8$$
$$3x + 5y \ge 30$$

18. Objective function:
$z = 4x + 5y$
Constraints:
$$x \ge 0$$
$$y \ge 0$$
$$2x + 2y \le 10$$
$$x + 2y \le 6$$

19. Objective function:
$z = 2x + 7y$
Constraints:
(See Exercise 17.)

20. Objective function:
$z = 2x - y$
Constraints:
(See Exercise 18.)

21. Objective function:
$z = 4x + y$
Constraints:
$$x \ge 0$$
$$y \ge 0$$
$$x + 2y \le 40$$
$$x + y \ge 30$$
$$2x + 3y \ge 72$$

22. Objective function:
$z = x$
Constraints:
$$x \ge 0$$
$$y \ge 0$$
$$2x + 3y \le 60$$
$$2x + y \le 28$$
$$4x + y \le 48$$

23. Objective function:
$z = x + 4y$
Constraints:
(See Exercise 21.)

24. Objective function:
$z = y$
Constraints:
(See Exercise 22.)

In Exercises 25–28, maximize the objective function subject to the constraints $3x + y \le 15$ and $4x + 3y \le 30$, where $x \ge 0$ and $y \ge 0$.

25. $z = 2x + y$

26. $z = 5x + y$

27. $z = x + y$

28. $z = 3x + y$

In Exercises 29–32, maximize the objective function subject to the constraints $x + 4y \le 20$, $x + y \le 8$, and $3x + 2y \le 21$, where $x \ge 0$ and $y \ge 0$.

29. $z = x + 5y$

30. $z = 2x + 4y$

31. $z = 4x + 5y$

32. $z = 4x + y$

33. *Maximum Profit* A merchant plans to sell two models of home computers at costs of $250 and $400, respectively. The $250 model yields a profit of $45 and the $400 model yields a profit of $50. The merchant estimates that the total monthly demand will not exceed 250 units. Find the number of units of each model that should be stocked in order to maximize profit. Assume that the merchant does not want to invest more than $70,000 in computer inventory.

34. *Maximum Profit* A fruit grower has 150 acres of land available to raise two crops, A and B. It takes one day to trim an acre of crop A and two days to trim an acre of crop B, and there are 240 days per year available for trimming. It takes 0.3 day to pick an acre of crop A and 0.1 day to pick an acre of crop B, and there are 30 days per year available for picking. Find the number of acres of each fruit that should be planted to maximize profit, assuming that the profit is $140 per acre for crop A and $235 per acre for crop B.

35. *Minimum Cost* A farming cooperative mixes two brands of cattle feed. Brand X costs $25 per bag and contains 2 units of nutritional element A, 2 units of element B, and 2 units of element C. Brand Y costs $20 per bag and contains 1 unit of nutritional element A, 9 units of element

B, and 3 units of element C. Find the number of bags of each brand that should be mixed to produce a mixture having a minimum cost per bag. The minimum requirements of nutrients A, B, and C are 12 units, 36 units, and 24 units, respectively.

36. *Minimum Cost*   Two gasolines, type A and type B, have octane ratings of 80 and 92, respectively. Type A costs $1.13 per gallon and type B costs $1.28 per gallon. Determine the blend of minimum cost with an octane rating of at least 90. (*Hint:* Let x be the fraction of each gallon that is type A and y be the fraction that is type B.)

37. *Maximum Profit*   A manufacturer produces two models of bicycles. The time (in hours) required for assembling, painting, and packaging each model is as follows.

	Model A	Model B
Assembling	2	2.5
Painting	4	1
Packaging	1	0.75

The total time available for assembling, painting, and packaging is 4000 hours, 4800 hours, and 1500 hours, respectively. The profit per unit for each model is $45 (model A) and $50 (model B). How many of each type should be produced to obtain a maximum profit?

38. *Maximum Profit*   A manufacturer produces two models of bicycles. The time (in hours) required for assembling, painting, and packaging each model is as follows.

	Model A	Model B
Assembling	2.5	3
Painting	2	1
Packaging	0.75	1.25

The total time available for assembling, painting, and packaging is 4000 hours, 2500 hours, and 1500 hours, respectively. The profit per unit for each model is $50 (model A) and $52 (model B). How many of each type should be produced to obtain a maximum profit?

39. *Maximum Revenue*   An accounting firm has 900 hours of staff time and 100 hours of reviewing time available each week. It charges $2000 for an audit and $300 for a tax return. Each audit takes 100 hours of staff time and 10 hours of review time. Each tax return takes 12.5 hours of staff time and 2.5 hours of review time. What number of audits and tax returns will yield the maximum revenue?

40. *Maximum Revenue*   The accounting firm in Exercise 39 lowers its charge for an audit to $1000. What number of audits and tax returns will yield the maximum revenue?

In Exercises 41–46, the given linear programming problem has an unusual characteristic. Sketch a graph of the solution region for the problem and describe the unusual characteristic. In each problem, the objective function is to be maximized.

41. Objective function:
$z = 2.5x + y$
Constraints:
$x \geq 0$
$y \geq 0$
$3x + 5y \leq 15$
$5x + 2y \leq 10$

42. Objective function:
$z = x + y$
Constraints:
$x \geq 0$
$y \geq 0$
$-x + y \leq 1$
$-x + 2y \leq 4$

43. Objective function:
$z = -x + 2y$
Constraints:
$x \geq 0$
$y \geq 0$
$x \leq 10$
$x + y \leq 7$

44. Objective function:
$z = x + y$
Constraints:
$x \geq 0$
$y \geq 0$
$-x + y \leq 0$
$-3x + y \geq 3$

45. Objective function:
$z = 3x + 4y$
Constraints:
$x \geq 0$
$y \geq 0$
$x + y \leq 1$
$2x + y \leq 4$

46. Objective function:
$z = x + 2y$
Constraints:
$x \geq 0$
$y \geq 0$
$x + 2y \leq 4$
$2x + y \leq 4$

In Exercises 47 and 48, determine the values of t such that the objective function has a maximum value at the indicated vertex.

47. Objective function:
$z = 3x + ty$
Constraints:
$x \geq 0$
$y \geq 0$
$x + 3y \leq 15$
$4x + y \leq 16$

(a) $(0, 5)$
(b) $(3, 4)$

48. Objective function:
$z = 3x + ty$
Constraints:
$x \geq 0$
$y \geq 0$
$x + 2y \leq 4$
$x - y \leq 1$

(a) $(2, 1)$
(b) $(0, 2)$

REVIEW EXERCISES for Chapter 7

In Exercises 1–8, solve the system by the method of substitution.

1. $x + y = 2$
$\quad x - y = 0$

2. $2x = 3(y - 1)$
$\qquad\quad y = x$

3. $x^2 - y^2 = 9$
$\quad\;\; x - y = 1$

4. $x^2 + y^2 = 169$
$\quad 3x + 2y = 39$

5. $y = 2x^2$
$\quad y = x^4 - 2x^2$

6. $x = y + 3$
$\quad x = y^2 + 1$

7. $y^2 - 2y + x = 0$
$\qquad\quad x + y = 0$

8. $y = 2x^2 - 4x + 1$
$\quad y = x^2 - 4x + 3$

9. *Break-Even Point* Suppose you are setting up a small business and have made an initial investment of $10,000. The unit cost of the product is $2.85 and the selling price is $4.95. How many units must you sell to break even? (Round your answer to the nearest whole unit.)

10. *Choice of Two Jobs* You are offered two different jobs selling personal computers. One company offers an annual salary of $22,500 plus a year-end bonus of 1.5% of your total sales. The other company offers a salary of $20,000 plus a year-end bonus of 2% of your total sales. How much would you have to sell in order to make the second offer better than the first?

In Exercises 11–18, solve the system by elimination.

11. $2x - y = 2$
$\quad 6x + 8y = 39$

12. $40x + 30y = 24$
$\quad 20x - 50y = -14$

13. $0.2x + 0.3y = 0.14$
$\quad 0.4x + 0.5y = 0.20$

14. $12x + 42y = -17$
$\quad 30x - 18y = 19$

15. $\quad\;\; 3x - 2y = 0$
$\quad 3x + 2(y + 5) = 10$

16. $7x + 12y = 63$
$\quad 2x + 3y = 15$

17. $1.25x - 2y = 3.5$
$\qquad 5x - 8y = 14$

18. $1.5x + 2.5y = 8.5$
$\quad 6x + 10y = 24$

19. *Acid Mixture* One hundred gallons of a 60% acid solution are obtained by mixing a 75% solution with a 50% solution. How many gallons of each must be used to obtain the desired mixture?

20. *Cassette Tape Sales* Suppose you are the manager of a music store. At the end of the week you are going over receipts for the previous week's sales. Six hundred and fifty cassette tapes were sold. One type of cassette sold for $9.95 and the other sold for $14.95. The total cassette receipts were $7717.50. The cash register that was supposed to record the number of each type of cassette sold malfunctioned. Can you recover the information? If so, how many of each type of cassette were sold?

21. *Flying Speeds* Two planes leave Pittsburgh and Philadelphia at the same time, each going to the other city. Because of the wind, one plane flies 25 miles per hour faster than the other. Find the ground speed of each plane if the cities are 275 miles apart and the planes pass one another (at different altitudes) after 40 minutes of flying time.

22. *Dimensions of a Rectangle* The perimeter of a rectangle is 480 meters, and its length is 150% of its width. Find the dimensions of the rectangle.

Supply and Demand In Exercises 23 and 24, find the point of equilibrium for each pair of supply and demand equations.

Demand	Supply
23. $p = 37 - 0.0002x$	$p = 22 + 0.00001x$
24. $p = 120 - 0.0001x$	$p = 45 + 0.0002x$

In Exercises 25–30, solve the system of equations.

25. $\quad x + 2y + 6z = 4$
$\quad -3x + 2y - z = -4$
$\quad 4x \quad\;\; + 2z = 16$

26. $x + 3y - z = 13$
$\quad 2x \qquad - 5z = 23$
$\quad 4x - y - 2z = 14$

27. $\quad x - 2y + z = -6$
$\quad 2x - 3y \qquad = -7$
$\quad -x + 3y - 3z = 11$

28. $2x + \qquad 6z = -9$
$\quad 3x - 2y + 11z = -16$
$\quad 3x - y + 7z = -11$

29. $2x + 5y - 19z = 34$
$\quad 3x + 8y - 31z = 54$

30. $2x + y + z + 2w = -1$
$\quad 5x - 2y + z - 3w = 0$
$\quad -x + 3y + 2z + 2w = 1$
$\quad 3x + 2y + 3z - 5w = 12$

In Exercises 31 and 32, find the equation of the parabola

$$y = ax^2 + bx + c$$

that passes through the given points.

31. $(0, -6), (1, -3), (2, 4)$ **32.** $(-5, 0), (1, -6), (2, 14)$

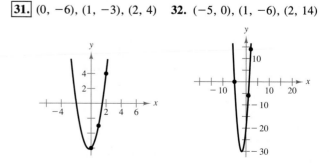

In Exercises 33 and 34, find the equation of the circle

$$x^2 + y^2 + Dx + Ey + F = 0$$

that passes through the given points.

33. $(2, 2), (5, -1), (-1, -1)$

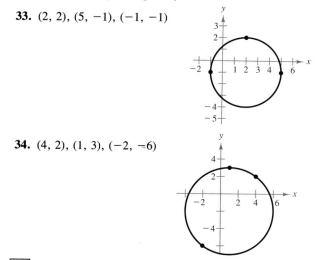

34. $(4, 2), (1, 3), (-2, -6)$

35. *Crop Spraying*   A mixture of 6 gallons of chemical A, 8 gallons of chemical B, and 13 gallons of chemical C is required to kill a certain destructive crop insect. Commercial spray X contains 1, 2, and 2 parts, respectively, of these chemicals. Commercial spray Y contains only chemical C. Commercial spray Z contains chemicals A, B, and C in equal amounts. How much of each type of commercial spray is needed to get the desired mixture?

36. *Investments*   An inheritance of $20,000 was divided among three investments yielding $1,780 in interest per year. The interest rates for the three investments were 7%, 9%, and 11%. Find the amount placed in each investment if the second and third were $3,000 and $1,000 less than the first, respectively.

37. *Fitting a Line to Data*   Find the least squares regression line $y = ax + b$ for the points

$$(x_1, y_1), (x_2, y_2), \ldots , (x_n, y_n).$$

To find the line, solve the following system of linear equations for a and b.

$$5b + 10a = 17.8$$
$$10b + 30a = 45.7$$

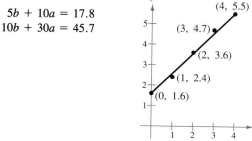

38. *Fitting a Parabola to Data*   Find the least squares regression parabola $y = ax^2 + bx + c$ for the points

$$(x_1, y_1), (x_2, y_2), \ldots , (x_n, y_n).$$

To find the parabola, solve the following system of linear equations for a, b, and c.

$$5c \quad + 10a = 9.1$$
$$10b \quad\quad = 8.0$$
$$10c + 34a = 19.8$$

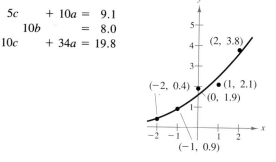

In Exercises 39–46, sketch a graph of the solution set of the system of inequalities.

39. $x + 2y \le 160$
    $3x + y \le 180$
       $x \ge 0$
       $y \ge 0$

40. $2x + 3y \le 24$
    $2x + y \le 16$
       $x \ge 0$
       $y \ge 0$

41. $3x + 2y \ge 24$
    $x + 2y \ge 12$
    $2 \le x \le 15$
       $y \le 15$

42. $2x + y \ge 16$
    $x + 3y \ge 18$
    $0 \le x \le 25$
    $0 \le y \le 25$

43. $y < x + 1$
    $y > x^2 - 1$

44. $y \le 6 - 2x - x^2$
    $y \ge x + 6$

45. $2x - 3y \ge 0$
    $2x - y \le 8$
       $y \ge 0$

46. $x^2 + y^2 \le 9$
    $(x - 3)^2 + y^2 \le 9$

In Exercises 47 and 48, derive a set of inequalities to describe the region.

47. Parallelogram with vertices at $(1, 5), (3, 1), (6, 10), (8, 6)$

48. Triangle with vertices at $(1, 2), (6, 7), (8, 1)$

In Exercises 49 and 50, determine a system of inequalities that models the description and sketch a graph of the solution of the system.

49. *Fruit Distribution*   A Pennsylvania fruit grower has 1500 bushels of apples that are to be divided between markets in Harrisburg and Philadelphia. These two markets need at least 400 bushels and 600 bushels, respectively.

50. *Inventory Costs* A warehouse operator has 24,000 square feet of floor space in which to store two products. Each unit of product I requires 20 square feet of floor space and costs $12 per day to store. Each unit of product II requires 30 square feet of floor space and costs $8 per day to store. The total storage cost per day cannot exceed $12,400.

In Exercises 51 and 52, find the consumer surplus and producer surplus for the pair of supply and demand equations.

Demand	Supply
51. $p = 160 - 0.0001x$	$p = 70 + 0.0002x$
52. $p = 130 - 0.0002x$	$p = 30 + 0.0003x$

In Exercises 53–56, find the required optimum value of the objective function subject to the indicated constraints.

53. Maximize the objective function:
$z = 3x + 4y$
Constraints:
$$x \geq 0$$
$$y \geq 0$$
$$2x + 5y \leq 50$$
$$4x + y \leq 28$$

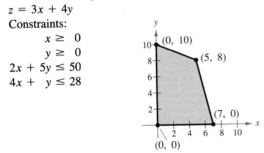

54. Minimize the objective function:
$z = 10x + 7y$
Constraints:
$$x \geq 0$$
$$y \geq 0$$
$$2x + y \geq 100$$
$$x + y \geq 75$$

55. Minimize the objective function:
$z = 1.75x + 2.25y$
Constraints:
$$x \geq 0$$
$$y \geq 0$$
$$2x + y \geq 25$$
$$3x + 2y \geq 45$$

56. Maximize the objective function:
$z = 50x + 70y$
Constraints:
$$x \geq 0$$
$$y \geq 0$$
$$x + 2y \leq 1500$$
$$5x + 2y \leq 3500$$

57. *Maximum Profit* A manufacturer produces products A and B yielding profits of $18 and $24, respectively. Each product must go through three processes with the required times per unit as shown in the following table.

Process	Hours for product A	Hours for product B	Hours available per day
I	4	2	24
II	1	2	9
III	1	1	8

Find the daily production level for each unit to maximize the profit.

58. *Maximum Revenue* A student is working part-time as a cosmetologist to pay college expenses. The student may work no more than 21 hours per week. Haircuts cost $17 and require an average of 20 minutes; permanents cost $60 and require an average of 1 hour and 10 minutes. What combination of haircuts and/or perms will yield maximum revenue?

59. *Minimum Cost* A pet supply company mixes two brands of dry dog food. Brand X costs $15 per bag and contains 8 units of nutritional element A, 1 unit of nutritional element B, and 2 units of nutritional element C. Brand Y costs $30 per bag and contains 2 units of nutritional element A, 1 unit of nutritional element B, and 7 units of nutritional element C. Each bag of dog food must contain at least 16 units, 5 units, and 20 units of nutritional elements A, B, and C, respectively. Find the number of bags of brands X and Y that should be mixed to produce a mixture meeting the minimum nutritional requirements and having a minimum cost per bag.

60. *Minimum Cost* Two gasolines, type A and type B, have octane ratings of 80 and 92, respectively. Type A costs $1.25 per gallon and type B costs $1.55 per gallon. Determine the blend of minimum cost with an octane rating of at least 88. (*Hint:* Let x be the fraction of each gallon that is type A, and let y be the fraction that is type B.)

OVERVIEW

This chapter shows how matrices can be used to model a variety of problems, beginning with their use in solving systems of linear equations.

The examples in the chapter tend to use matrices that have only a few rows or columns. You should realize, however, that matrices that model real life problems are often much larger. For instance, it is easy to imagine the need for very large matrices to model inventory levels, such as that described in Exercise 43 on page 500.

Adding, subtracting, and multiplying matrices can involve many arithmetic steps. Technology can minimize the tedium and the likelihood of calculation errors. If you have access to a calculator or computer software that performs matrix operations, you may want to use it.

Matrices and Determinants

8.1 Matrices and Systems of Linear Equations

Matrices / Elementary Row Operations / Gaussian Elimination with Back-Substitution / Gauss-Jordan Elimination

Matrices

In mathematics we always look for valid shortcuts to solving problems. In this section we look at a streamlined technique for solving systems of linear equations. This technique involves the use of a rectangular array of real numbers—a **matrix.**

DEFINITION OF A MATRIX

If m and n are positive integers, then an $m \times n$ **matrix** (read "m by n") is a rectangular array

$$\begin{bmatrix} a_{11} & a_{12} & a_{13} & \cdots & a_{1n} \\ a_{21} & a_{22} & a_{23} & \cdots & a_{2n} \\ a_{31} & a_{32} & a_{33} & \cdots & a_{3n} \\ \vdots & \vdots & \vdots & & \vdots \\ a_{m1} & a_{m2} & a_{m3} & \cdots & a_{mn} \end{bmatrix} \Bigg\} \; m \text{ rows}$$

$$\underbrace{\hphantom{a_{11} \quad a_{12} \quad a_{13}}}_{n \text{ columns}}$$

in which each **entry,** a_{ij}, of the matrix is a number. An $m \times n$ matrix has m **rows** (horizontal lines) and n **columns** (vertical lines).

REMARK The plural of matrix is *matrices.*

The entry in the ith row and jth column is denoted by the *double subscript* notation a_{ij}. We call i the **row subscript** because it gives the position in the horizontal lines, and j the **column subscript** because it gives the position in the vertical lines.

A matrix having m rows and n columns is of **order $m \times n$.** If $m = n$, the matrix is **square** of order n. For a square matrix, the entries a_{11}, a_{22}, a_{33}, . . . are the **main diagonal** entries.

EXAMPLE 1   Examples of Matrices

The following matrices have the indicated orders.

a. Order: 1×4

$$\begin{bmatrix} 1 & -3 & 0 & \frac{1}{2} \end{bmatrix}$$

b. Order: 2×2

$$\begin{bmatrix} 0 & 0 \\ 0 & 0 \end{bmatrix}$$

c. Order: 2×3

$$\begin{bmatrix} -1 & 0 & 5 \\ 2 & 1 & -4 \end{bmatrix}$$

d. Order: 3×2

$$\begin{bmatrix} 5 & 0 \\ 2 & -2 \\ -7 & 4 \end{bmatrix}$$

A matrix that has only one row is a **row matrix** and a matrix that has only one column is a **column matrix.**

A matrix derived from a system of linear equations (each written in standard form with the constant term on the right) is the **augmented matrix** of the system. Moreover, the matrix derived from the coefficients of the system (but which does not include the constant terms) is the **coefficient matrix** of the system. Here is an example.

System	*Augmented Matrix*	*Coefficient Matrix*
$\begin{aligned} x - 4y + 3z &= 5 \\ -x + 3y - z &= -3 \\ 2x \quad\;\; - 4z &= 6 \end{aligned}$	$\left[\begin{array}{ccc:c} 1 & -4 & 3 & 5 \\ -1 & 3 & -1 & -3 \\ 2 & 0 & -4 & 6 \end{array}\right]$	$\begin{bmatrix} 1 & -4 & 3 \\ -1 & 3 & -1 \\ 2 & 0 & -4 \end{bmatrix}$

REMARK Note the use of 0 for the missing y-variable in the third equation, and also note the fourth column of constant terms in the augmented matrix.

When forming either the coefficient matrix or the augmented matrix of a system, you should begin by vertically aligning the variables in the equations and using 0's for the missing variables.

Given System	*Line Up Variables*	*Form Augmented Matrix*
$\begin{aligned} x + 3y &= 9 \\ -y + 4z &= -2 \\ x - 5z &= 0 \end{aligned}$	$\begin{aligned} x + 3y \quad\quad &= 9 \\ -y + 4z &= -2 \\ x \quad\quad - 5z &= 0 \end{aligned}$	$\left[\begin{array}{ccc:c} 1 & 3 & 0 & 9 \\ 0 & -1 & 4 & -2 \\ 1 & 0 & -5 & 0 \end{array}\right]$

Elementary Row Operations

In Section 7.3 you studied three operations that can be used on a system of linear equations to produce an equivalent system.

1. Interchange two equations.
2. Multiply an equation by a nonzero constant.
3. Add a multiple of an equation to another equation.

In matrix terminology these three operations correspond to **elementary row operations.** An elementary row operation on an augmented matrix of a given system of linear equations produces a new augmented matrix corresponding to a new (but equivalent) system of linear equations. Two matrices are **row-equivalent** if one can be obtained from the other by a sequence of elementary row operations.

ELEMENTARY ROW OPERATIONS

1. Interchange two rows.
2. Multiply a row by a nonzero constant.
3. Add a multiple of a row to another row.

Although elementary row operations are simple to perform, they involve a lot of arithmetic. Because it is easy to make a mistake, we suggest that you get into the habit of noting the elementary row operations performed in each step so that you can go back and check your work.

EXAMPLE 2   Elementary Row Operations

a. Interchange the first and second rows.

Original Matrix

$$\begin{bmatrix} 0 & 1 & 3 & 4 \\ -1 & 2 & 0 & 3 \\ 2 & -3 & 4 & 1 \end{bmatrix}$$

New Row Equivalent Matrix

$$\begin{matrix} \curvearrowright R_2 \\ \curvearrowright R_1 \end{matrix} \begin{bmatrix} -1 & 2 & 0 & 3 \\ 0 & 1 & 3 & 4 \\ 2 & -3 & 4 & 1 \end{bmatrix}$$

b. Multiply the first row by $\frac{1}{2}$.

Original Matrix

$$\begin{bmatrix} 2 & -4 & 6 & -2 \\ 1 & 3 & -3 & 0 \\ 5 & -2 & 1 & 2 \end{bmatrix}$$

New Row Equivalent Matrix

$$\frac{1}{2}R_1 \rightarrow \begin{bmatrix} 1 & -2 & 3 & -1 \\ 1 & 3 & -3 & 0 \\ 5 & -2 & 1 & 2 \end{bmatrix}$$

c. Add -2 times the first row to the third row.

Original Matrix

$$\begin{bmatrix} 1 & 2 & -4 & 3 \\ 0 & 3 & -2 & -1 \\ 2 & 1 & 5 & -2 \end{bmatrix}$$

New Row Equivalent Matrix

$$-2R_1 + R_3 \rightarrow \begin{bmatrix} 1 & 2 & -4 & 3 \\ 0 & 3 & -2 & -1 \\ 0 & -3 & 13 & -8 \end{bmatrix}$$

Note that you write the elementary row operation beside the row that you are *changing*.

◣

In Section 7.3 you used Gaussian elimination with back-substitution to solve a system of linear equations. We now demonstrate the matrix version of Gaussian elimination. The two methods are essentially the same. The basic difference is that with matrices you do not need to keep writing the variables.

◣ EXAMPLE 3   Using Elementary Row Operations to Solve a System

Linear System		*Associated Augmented Matrix*

$$\begin{aligned} x - 2y + 3z &= 9 \\ -x + 3y \phantom{{}+ 3z} &= -4 \\ 2x - 5y + 5z &= 17 \end{aligned}$$

$$\begin{bmatrix} 1 & -2 & 3 & \vdots & 9 \\ -1 & 3 & 0 & \vdots & -4 \\ 2 & -5 & 5 & \vdots & 17 \end{bmatrix}$$

Add the first equation to the second equation.

Add the first row to the second row $(R_1 + R_2)$.

$$\begin{aligned} x - 2y + 3z &= 9 \\ y + 3z &= 5 \\ 2x - 5y + 5z &= 17 \end{aligned}$$

$R_1 + R_2 \rightarrow$
$$\begin{bmatrix} 1 & -2 & 3 & \vdots & 9 \\ 0 & 1 & 3 & \vdots & 5 \\ 2 & -5 & 5 & \vdots & 17 \end{bmatrix}$$

Add -2 times the first equation to the third equation.

Add -2 times the first row to the third row $(-2R_1 + R_3)$.

$$\begin{aligned} x - 2y + 3z &= 9 \\ y + 3z &= 5 \\ -y - z &= -1 \end{aligned}$$

$-2R_1 + R_3 \rightarrow$
$$\begin{bmatrix} 1 & -2 & 3 & \vdots & 9 \\ 0 & 1 & 3 & \vdots & 5 \\ 0 & -1 & -1 & \vdots & -1 \end{bmatrix}$$

Add the second equation to the third equation.

Add the second row to the third row $(R_2 + R_3)$.

$$\begin{aligned} x - 2y + 3z &= 9 \\ y + 3z &= 5 \\ 2z &= 4 \end{aligned}$$

$R_2 + R_3 \rightarrow$
$$\begin{bmatrix} 1 & -2 & 3 & \vdots & 9 \\ 0 & 1 & 3 & \vdots & 5 \\ 0 & 0 & 2 & \vdots & 4 \end{bmatrix}$$

Multiply the third equation by $\frac{1}{2}$.

Multiply the third row by $\frac{1}{2}$.

$$\begin{aligned} x - 2y + 3z &= 9 \\ y + 3z &= 5 \\ z &= 2 \end{aligned}$$

$\frac{1}{2}R_3 \rightarrow$
$$\begin{bmatrix} 1 & -2 & 3 & \vdots & 9 \\ 0 & 1 & 3 & \vdots & 5 \\ 0 & 0 & 1 & \vdots & 2 \end{bmatrix}$$

At this point, you can use back-substitution to find that the solution is $x = 1$, $y = -1$, and $z = 2$, as you did in Section 7.3.

◣

The last matrix in Example 3 is in **row-echelon form.** The term *echelon* refers to the stair-step pattern formed by the nonzero elements of the matrix. To be in this form, a matrix must have the following properties.

ROW-ECHELON FORM AND REDUCED ROW-ECHELON FORM

A matrix in **row-echelon form** has the following properties.

1. All rows consisting entirely of zeros occur at the bottom of the matrix.
2. For each row that does not consist entirely of zeros, the first nonzero entry is 1 (a **leading 1**).
3. For two successive (nonzero) rows, the leading 1 in the higher row is farther to the left than the leading 1 in the lower row.

A matrix in *row-echelon form* is in **reduced row-echelon form** if every column that has a leading 1 has zeros in every position above and below its leading 1.

EXAMPLE 4   Row-Echelon Form

The following matrices are in row-echelon form.

a. $\begin{bmatrix} 1 & 2 & -1 & 4 \\ 0 & 1 & 0 & 3 \\ 0 & 0 & 1 & -2 \end{bmatrix}$
   b. $\begin{bmatrix} 0 & 1 & 0 & 5 \\ 0 & 0 & 1 & 3 \\ 0 & 0 & 0 & 0 \end{bmatrix}$

c. $\begin{bmatrix} 1 & -5 & 2 & -1 & 3 \\ 0 & 0 & 1 & 3 & -2 \\ 0 & 0 & 0 & 1 & 4 \\ 0 & 0 & 0 & 0 & 1 \end{bmatrix}$
   d. $\begin{bmatrix} 1 & 0 & 0 & -1 \\ 0 & 1 & 0 & 2 \\ 0 & 0 & 1 & 3 \\ 0 & 0 & 0 & 0 \end{bmatrix}$

The matrices in (b) and (d) also happen to be in *reduced* row-echelon form. The following matrices are not in row-echelon form.

e. $\begin{bmatrix} 1 & 2 & -3 & 4 \\ 0 & 2 & 1 & -1 \\ 0 & 0 & 1 & -3 \end{bmatrix}$
   f. $\begin{bmatrix} 1 & 2 & -1 & 2 \\ 0 & 0 & 0 & 0 \\ 0 & 1 & 2 & -4 \end{bmatrix}$

Every matrix is row equivalent to a matrix in row-echelon form. For instance, in Example 4, you can change the matrix in part (e) to row-echelon form by multiplying its second row by $\frac{1}{2}$.

Gaussian Elimination with Back-Substitution

Guidelines for using Gaussian elimination with back-substitution to solve a system of linear equations are summarized as follows.

Gaussian Elimination with Back-Substitution

1. Write the augmented matrix of the system of linear equations.
2. Use elementary row operations to rewrite the augmented matrix in row-echelon form.
3. Write the system of linear equations corresponding to the matrix in row-echelon form, and use back-substitution to find the solution.

Gaussian elimination with back-substitution works well for solving systems of linear equations with a computer. For this algorithm, the order in which the elementary row operations are performed is important. Operate from *left to right by columns*, using elementary row operations to obtain zeros in all entries directly below the leading 1's.

EXAMPLE 5   Gaussian Elimination with Back-Substitution

Solve the following system.

$$
\begin{aligned}
y + z - 2w &= -3 \\
x + 2y - z \phantom{{}- 3w} &= 2 \\
2x + 4y + z - 3w &= -2 \\
x - 4y - 7z - w &= -19
\end{aligned}
$$

Solution

The augmented matrix for this system is

$$
\left[
\begin{array}{cccc:c}
0 & 1 & 1 & -2 & -3 \\
1 & 2 & -1 & 0 & 2 \\
2 & 4 & 1 & -3 & -2 \\
1 & -4 & -7 & -1 & -19
\end{array}
\right].
$$

Begin by obtaining a leading 1 in the upper left corner by interchanging the first and second rows, and then proceed to obtain zeros elsewhere in the first column.

$$
\begin{array}{c}
{\small R_2} \\
{\small R_1}
\end{array}
\left[
\begin{array}{cccc:c}
1 & 2 & -1 & 0 & 2 \\
0 & 1 & 1 & -2 & -3 \\
2 & 4 & 1 & -3 & -2 \\
1 & -4 & -7 & -1 & -19
\end{array}
\right]
\quad
\begin{array}{l}
\textit{First column has leading 1} \\
\textit{in upper left corner.}
\end{array}
$$

$$
\begin{array}{c}
{} \\
{} \\
-2R_1 + R_3 \rightarrow \\
-R_1 + R_4 \rightarrow
\end{array}
\left[
\begin{array}{cccc:c}
1 & 2 & -1 & 0 & 2 \\
0 & 1 & 1 & -2 & -3 \\
0 & 0 & 3 & -3 & -6 \\
0 & -6 & -6 & -1 & -21
\end{array}
\right]
\quad
\begin{array}{l}
\textit{First column has zeros} \\
\textit{below its leading 1.}
\end{array}
$$

Now that the first column is in the desired form, change the second, third, and fourth columns as follows.

$$6R_2 + R_4 \rightarrow \begin{bmatrix} 1 & 2 & -1 & 0 & \vdots & 2 \\ 0 & 1 & 1 & -2 & \vdots & -3 \\ 0 & 0 & 3 & -3 & \vdots & -6 \\ 0 & 0 & 0 & -13 & \vdots & -39 \end{bmatrix}$$ *Second column has zeros below its leading 1.*

$$\tfrac{1}{3}R_3 \rightarrow \begin{bmatrix} 1 & 2 & -1 & 0 & \vdots & 2 \\ 0 & 1 & 1 & -2 & \vdots & -3 \\ 0 & 0 & 1 & -1 & \vdots & -2 \\ 0 & 0 & 0 & -13 & \vdots & -39 \end{bmatrix}$$ *Third column has zeros below its leading 1.*

$$-\tfrac{1}{13}R_4 \rightarrow \begin{bmatrix} 1 & 2 & -1 & 0 & \vdots & 2 \\ 0 & 1 & 1 & -2 & \vdots & -3 \\ 0 & 0 & 1 & -1 & \vdots & -2 \\ 0 & 0 & 0 & 1 & \vdots & 3 \end{bmatrix}$$ *Fourth column has a leading 1.*

The matrix is now in row-echelon form, and the corresponding system of linear equations is

$$\begin{aligned} x + 2y - z & = 2 \\ y + z - 2w &= -3 \\ z - w &= -2 \\ w &= 3. \end{aligned}$$

Using back-substitution, you can determine that the solution is

$$x = -1, \quad y = 2, \quad z = 1, \quad \text{and} \quad w = 3.$$

You can now check for errors in your elementary row operations by substituting these values in each equation in the *original* system. (If they don't check, then you know that you made an error in the back-substitution or one of the elementary row operations.) ◢

When solving a system of linear equations, remember that it is possible for the system to have no solution. If, in the elimination process, you obtain a row with zeros except for the last entry, it is unnecessary to continue the elimination process. You can simply conclude that the system is inconsistent. For instance, Gaussian elimination applied to the system

$$\begin{aligned} x - y + 2z &= 4 \\ x + z &= 6 \\ 2x - 3y + 5z &= 4 \\ 3x + 2y - z &= 1 \end{aligned}$$ yields the matrix $$\begin{bmatrix} 1 & -1 & 2 & \vdots & 4 \\ 0 & 1 & -1 & \vdots & 2 \\ 0 & 0 & 0 & \vdots & -2 \\ 0 & 5 & -7 & \vdots & -11 \end{bmatrix}.$$

Note that the third row of this matrix consists of zeros except for the last entry. This means that the original system of linear equations is *inconsistent*.

Gauss-Jordan Elimination

With Gaussian elimination, you apply elementary row operations to a matrix to obtain a (row-equivalent) row-echelon form. A second method of elimination, called **Gauss-Jordan elimination,** after Carl Friedrich Gauss and Wilhelm Jordan (1842–1899), continues the reduction process until a *reduced* row-echelon form is obtained. We demonstrate this procedure in the following example.

EXAMPLE 6 Gauss-Jordan Elimination

Use Gauss-Jordan elimination to solve the following system.

$$\begin{aligned} x - 2y + 3z &= 9 \\ -x + 3y &= -4 \\ 2x - 5y + 5z &= 17 \end{aligned}$$

Solution

In Example 3 you used Gaussian elimination to obtain the following row-echelon form.

$$\begin{bmatrix} 1 & -2 & 3 & \vdots & 9 \\ 0 & 1 & 3 & \vdots & 5 \\ 0 & 0 & 1 & \vdots & 2 \end{bmatrix}$$

Now, rather than using back-substitution, you will apply elementary row operations until you obtain a matrix in reduced row-echelon form. To do this, you must produce zeros above each of the leading 1's, as follows.

$$\begin{array}{c} 2R_2 + R_1 \to \\ \\ \\ \end{array} \begin{bmatrix} 1 & 0 & 9 & \vdots & 19 \\ 0 & 1 & 3 & \vdots & 5 \\ 0 & 0 & 1 & \vdots & 2 \end{bmatrix}$$

Second column has zeros above its leading 1.

$$\begin{array}{c} -9R_3 + R_1 \to \\ -3R_3 + R_2 \to \\ \\ \end{array} \begin{bmatrix} 1 & 0 & 0 & \vdots & 1 \\ 0 & 1 & 0 & \vdots & -1 \\ 0 & 0 & 1 & \vdots & 2 \end{bmatrix}$$

Third column has zeros above its leading 1.

Now, converting back to a system of linear equations, you have

$$\begin{aligned} x &= 1 \\ y &= -1 \\ z &= 2. \end{aligned}$$

The beauty of Gauss-Jordan elimination is that, from the reduced row-echelon form, you can simply read the solution.

It is worth noting that the row-echelon form for a matrix is not unique. That is, two different sequences of elementary row operations may yield different row-echelon forms. For instance, the following sequence of elementary row operations on the matrix in Example 3 produces a slightly different row-echelon form.

$$\begin{bmatrix} 1 & -2 & 3 & \vdots & 9 \\ -1 & 3 & 0 & \vdots & -4 \\ 2 & -5 & 5 & \vdots & 17 \end{bmatrix}$$

$$\begin{matrix} R_2 \\ R_1 \end{matrix} \begin{bmatrix} -1 & 3 & 0 & \vdots & -4 \\ 1 & -2 & 3 & \vdots & 9 \\ 2 & -5 & 5 & \vdots & 17 \end{bmatrix}$$

$$-R_1 \rightarrow \begin{bmatrix} 1 & -3 & 0 & \vdots & 4 \\ 1 & -2 & 3 & \vdots & 9 \\ 2 & -5 & 5 & \vdots & 17 \end{bmatrix}$$

$$\begin{matrix} -R_1 + R_2 \rightarrow \\ -2R_1 + R_3 \rightarrow \end{matrix} \begin{bmatrix} 1 & -3 & 0 & \vdots & 4 \\ 0 & 1 & 3 & \vdots & 5 \\ 0 & 1 & 5 & \vdots & 9 \end{bmatrix}$$

$$-R_2 + R_3 \rightarrow \begin{bmatrix} 1 & -3 & 0 & \vdots & 4 \\ 0 & 1 & 3 & \vdots & 5 \\ 0 & 0 & 2 & \vdots & 4 \end{bmatrix}$$

$$\tfrac{1}{2}R_3 \rightarrow \begin{bmatrix} 1 & -3 & 0 & \vdots & 4 \\ 0 & 1 & 3 & \vdots & 5 \\ 0 & 0 & 1 & \vdots & 2 \end{bmatrix}$$

However, the *reduced* row-echelon form for a given matrix *is* unique. You should try applying Gauss-Jordan elimination to this matrix to see that you obtain the same reduced row-echelon form as in Example 6.

The elimination procedures described in this section employ an algorithmic approach that is easily adapted to computer use. However, the procedure makes no effort to avoid fractional coefficients. For instance, if the system given in Example 6 had been listed as

$$\begin{aligned} 2x - 5y + 5z &= 17 \\ x - 2y + 3z &= 9 \\ -x + 3y &= -4 \end{aligned}$$

the procedure would then have required multiplying the first row by $\frac{1}{2}$, which would have introduced fractions in the first row. For hand computations, fractions can sometimes be avoided by judiciously choosing the order in which the elementary row operations are applied.

The next example demonstrates how Gauss-Jordan elimination can be used to solve a system with an infinite number of solutions.

EXAMPLE 7   A System with an Infinite Number of Solutions

Solve the following system of linear equations.

$$2x + 4y - 2z = 0$$
$$3x + 5y \quad\quad = 1$$

Solution

Using Gauss-Jordan elimination, the augmented matrix reduces as follows.

$$\begin{bmatrix} 2 & 4 & -2 & \vdots & 0 \\ 3 & 5 & 0 & \vdots & 1 \end{bmatrix} \qquad \tfrac{1}{2}R_1 \rightarrow \begin{bmatrix} 1 & 2 & -1 & \vdots & 0 \\ 3 & 5 & 0 & \vdots & 1 \end{bmatrix}$$

$$-3R_1 + R_2 \rightarrow \begin{bmatrix} 1 & 2 & -1 & \vdots & 0 \\ 0 & -1 & 3 & \vdots & 1 \end{bmatrix}$$

$$-R_2 \rightarrow \begin{bmatrix} 1 & 2 & -1 & \vdots & 0 \\ 0 & 1 & -3 & \vdots & -1 \end{bmatrix}$$

$$-2R_2 + R_1 \rightarrow \begin{bmatrix} 1 & 0 & 5 & \vdots & 2 \\ 0 & 1 & -3 & \vdots & -1 \end{bmatrix}$$

The corresponding system of equations is

$$x + 5z = 2$$
$$y - 3z = -1.$$

Solving for x and y in terms of z, you have $x = -5z + 2$ and $y = 3z - 1$. Then, letting $z = a$, the solution set has the form

$$(-5a + 2, 3a - 1, a), \qquad \text{where } a \text{ is a real number.}$$

You have looked at two elimination methods for solving a system of linear equations. Which is better? To some degree, it depends on personal preference. For hand computations, Gaussian elimination with back-substitution is often preferred. However, you will encounter other applications in which Gauss-Jordan elimination is better. Thus, you should know both methods.

DISCUSSION
PROBLEM

Comparing Gaussian Elimination with Gauss-Jordan Elimination

Solve the following system of linear equations in two ways: once with Gaussian elimination with back-substitution and once with Gauss-Jordan elimination. Then discuss, or write a short paragraph describing, the benefits of one method over the other.

$$3x - 2y + z = -6$$
$$-x + y - 2z = 1$$
$$2x + 2y - 3z = -1$$

WARM UP

The following warm-up exercises involve skills that were covered in earlier sections. You will use these skills in the exercise set for this section.

In Exercises 1–4, evaluate the given expression.

1. $2(-1) - 3(5) + 7(2)$

2. $-4(-3) + 6(7) + 8(-3)$

3. $11\left(\frac{1}{2}\right) - 7\left(-\frac{3}{2}\right) - 5(2)$

4. $\frac{2}{3}\left(\frac{1}{2}\right) + \frac{4}{3}\left(-\frac{1}{3}\right)$

In Exercises 5 and 6, determine whether $x = 1$, $y = 3$, and $z = -1$ is a solution of the system of linear equations.

5.
$$\begin{array}{rcrcrcr} 4x & - & 2y & + & 3z & = & -5 \\ x & + & 3y & - & z & = & 11 \\ -x & + & 2y & & & = & 5 \end{array}$$

6.
$$\begin{array}{rcrcrcr} -x & + & 2y & + & z & = & 4 \\ 2x & & & - & 3z & = & 5 \\ 3x & + & 5y & - & 2z & = & 21 \end{array}$$

In Exercises 7–10, use back-substitution to solve the system of linear equations.

7.
$$\begin{array}{rcl} 2x - 3y &=& 4 \\ y &=& 2 \end{array}$$

8.
$$\begin{array}{rcl} 5x + 4y &=& 0 \\ y &=& -3 \end{array}$$

9.
$$\begin{array}{rcl} x - 3y + z &=& 0 \\ y - 3z &=& 8 \\ z &=& 2 \end{array}$$

10.
$$\begin{array}{rcl} 2x - 5y + 3z &=& -2 \\ y - 4z &=& 0 \\ z &=& 1 \end{array}$$

EXERCISES for Section 8.1

In Exercises 1–6, determine the order of the matrix.

1. $\begin{bmatrix} 4 & -2 \\ 7 & 0 \\ 0 & 8 \end{bmatrix}$

2. $[5 \quad -3 \quad 8 \quad 7]$

3. $\begin{bmatrix} -9 \\ 2 \\ 36 \\ 11 \\ 3 \end{bmatrix}$

4. $\begin{bmatrix} 11 & 0 & 8 & 5 & 5 \\ -3 & 7 & 15 & 0 & 10 \\ 0 & 6 & 3 & 3 & 9 \\ 12 & 4 & 16 & 9 & 0 \\ 1 & 1 & 6 & 7 & 8 \end{bmatrix}$

5. $\begin{bmatrix} 33 & 45 \\ -9 & 20 \end{bmatrix}$

6. $[4]$

In Exercises 7–10, determine whether the matrix is in row-echelon form. If it is, determine if it is also in reduced row-echelon form.

7. $\begin{bmatrix} 1 & 0 & 0 & 0 \\ 0 & 1 & 1 & 5 \\ 0 & 0 & 0 & 0 \end{bmatrix}$

8. $\begin{bmatrix} 1 & 0 & 2 & 1 \\ 0 & 1 & -3 & 10 \\ 0 & 0 & 1 & 0 \end{bmatrix}$

9. $\begin{bmatrix} 2 & 0 & 4 & 0 \\ 0 & -1 & 3 & 6 \\ 0 & 0 & 1 & 5 \end{bmatrix}$

10. $\begin{bmatrix} 1 & 3 & 0 & 0 \\ 0 & 0 & 1 & 8 \\ 0 & 0 & 0 & 0 \end{bmatrix}$

In Exercises 11–14, fill in the blanks using the elementary row operations to form a row equivalent matrix.

11. $\begin{bmatrix} 1 & 4 & 3 \\ 2 & 10 & 5 \end{bmatrix}$ $\overset{-6}{}$
$\begin{bmatrix} 1 & 4 & -3 \\ 0 & \blacksquare & -1 \end{bmatrix}$

12. $\begin{bmatrix} 3 & 6 & 8 \\ 4 & -3 & 6 \end{bmatrix}$
$\begin{bmatrix} 1 & \blacksquare & \frac{8}{3} \\ 4 & -3 & 6 \end{bmatrix}$

13. $\begin{bmatrix} 1 & 1 & 4 & -1 \\ 3 & 8 & 10 & 3 \\ -2 & 1 & 12 & 6 \end{bmatrix}$
$\begin{bmatrix} 1 & 1 & 4 & -1 \\ 0 & 5 & \blacksquare & \blacksquare \\ 0 & 3 & \blacksquare & \blacksquare \end{bmatrix}$
$\begin{bmatrix} 1 & 1 & 4 & -1 \\ 0 & 1 & \blacksquare & \blacksquare \\ 0 & 3 & 20 & 4 \end{bmatrix}$

14. $\begin{bmatrix} 2 & 4 & 8 & 3 \\ 1 & -1 & -3 & 2 \\ 2 & 6 & 4 & 9 \end{bmatrix}$
$\begin{bmatrix} 1 & \blacksquare & \blacksquare & \blacksquare \\ 1 & -1 & -3 & 2 \\ 2 & 6 & 4 & 9 \end{bmatrix}$
$\begin{bmatrix} 1 & 2 & 4 & \frac{3}{2} \\ 0 & \blacksquare & -7 & \frac{1}{2} \\ 0 & 2 & \blacksquare & \blacksquare \end{bmatrix}$

15. Perform the indicated *sequence* of elementary row operations to write the given matrix in reduced row-echelon form.

$$\begin{bmatrix} 1 & 2 & 3 \\ 2 & -1 & -4 \\ 3 & 1 & -1 \end{bmatrix}$$

(a) Add (-2) times Row 1 to Row 2. (Only Row 2 should change.)

(b) Add (-3) times Row 1 to Row 3. (Only Row 3 should change.)

(c) Add (-1) times Row 2 to Row 3.

(d) Multiply Row 2 by $\left(-\frac{1}{5}\right)$.

(e) Add (-2) times Row 2 to Row 1.

16. Perform the indicated *sequence* of elementary row operations to write the given matrix in reduced row-echelon form.

$$\begin{bmatrix} 7 & 1 \\ 0 & 2 \\ -3 & 4 \\ 4 & 1 \end{bmatrix}$$

(a) Add Row 3 to Row 4. (Only Row 4 should change.)

(b) Interchange Rows 1 and 4. (Note that the first element in the matrix is now 1, and it was obtained without introducing fractions.)

(c) Add (3) times Row 1 to Row 3.

(d) Add (-7) times Row 1 to Row 4.

(e) Multiply Row 2 by $\frac{1}{2}$.

(f) Add the appropriate multiple of Row 2 to Rows 1, 3, and 4.

In Exercises 17–20, write the matrix in row-echelon form. Remember that the row-echelon form for a given matrix is not unique.

17. $\begin{bmatrix} 1 & 1 & 0 & 5 \\ -2 & -1 & 2 & -10 \\ 3 & 6 & 7 & 14 \end{bmatrix}$

18. $\begin{bmatrix} 1 & 2 & -1 & 3 \\ 3 & 7 & -5 & 14 \\ -2 & -1 & -3 & 8 \end{bmatrix}$

19. $\begin{bmatrix} 1 & -1 & -1 & 1 \\ 5 & -4 & 1 & 8 \\ -6 & 8 & 18 & 0 \end{bmatrix}$

20. $\begin{bmatrix} 1 & -3 & 0 & -7 \\ -3 & 10 & 1 & 23 \\ 4 & -10 & 2 & -24 \end{bmatrix}$

In Exercises 21–24, write the matrix in *reduced* row-echelon form.

21. $\begin{bmatrix} 3 & 3 & 3 \\ -1 & 0 & -4 \\ 2 & 4 & -2 \end{bmatrix}$

22. $\begin{bmatrix} 1 & 3 & 2 \\ 5 & 15 & 9 \\ 2 & 6 & 10 \end{bmatrix}$

23. $\begin{bmatrix} 1 & 2 & 3 & -5 \\ 1 & 2 & 4 & -9 \\ -2 & -4 & -4 & 3 \\ 4 & 8 & 11 & -14 \end{bmatrix}$

24. $\begin{bmatrix} 1 & -3 \\ -1 & 8 \\ 0 & 4 \\ -2 & 10 \end{bmatrix}$

In Exercises 25–28, write the system of linear equations represented by the augmented matrix.

25. $\begin{bmatrix} 4 & 3 & \vdots & 8 \\ 1 & -2 & \vdots & 3 \end{bmatrix}$

26. $\begin{bmatrix} 9 & -4 & \vdots & 0 \\ 6 & 1 & \vdots & -4 \end{bmatrix}$

27. $\begin{bmatrix} 1 & 0 & 2 & \vdots & -10 \\ 0 & 3 & -1 & \vdots & 5 \\ 4 & 2 & 0 & \vdots & 3 \end{bmatrix}$

28. $\begin{bmatrix} 5 & 8 & 2 & 0 & \vdots & -1 \\ -2 & 15 & 5 & 1 & \vdots & 9 \\ 1 & 6 & -7 & 0 & \vdots & -3 \end{bmatrix}$

In Exercises 29–32, write the system of linear equations represented by the augmented matrix. Then use back-substitution to find the solution. (Use variables x, y, and z.)

29. $\begin{bmatrix} 1 & -2 & \vdots & 4 \\ 0 & 1 & \vdots & -3 \end{bmatrix}$

30. $\begin{bmatrix} 1 & 5 & \vdots & 0 \\ 0 & 1 & \vdots & -1 \end{bmatrix}$

31. $\begin{bmatrix} 1 & -1 & 2 & \vdots & 4 \\ 0 & 1 & -1 & \vdots & 2 \\ 0 & 0 & 1 & \vdots & -2 \end{bmatrix}$

32. $\begin{bmatrix} 1 & 2 & -2 & \vdots & -1 \\ 0 & 1 & 1 & \vdots & 9 \\ 0 & 0 & 1 & \vdots & -3 \end{bmatrix}$

In Exercises 33–36, an augmented matrix that represents a system of linear equations (in variables x, y, and z) has been reduced using Gauss-Jordan elimination. Write the solution represented by the augmented matrix.

33. $\begin{bmatrix} 1 & 0 & \vdots & 7 \\ 0 & 1 & \vdots & -5 \end{bmatrix}$

34. $\begin{bmatrix} 1 & 0 & \vdots & -2 \\ 0 & 1 & \vdots & 4 \end{bmatrix}$

35. $\begin{bmatrix} 1 & 0 & 0 & \vdots & -4 \\ 0 & 1 & 0 & \vdots & -8 \\ 0 & 0 & 1 & \vdots & 2 \end{bmatrix}$

36. $\begin{bmatrix} 1 & 0 & 0 & \vdots & 3 \\ 0 & 1 & 0 & \vdots & -1 \\ 0 & 0 & 1 & \vdots & 0 \end{bmatrix}$

In Exercises 37–58, solve the system of equations. Use Gaussian elimination with back-substitution or Gauss-Jordan elimination.

37. $x + 2y = 7$
$\qquad 2x + y = 8$

38. $2x + 6y = 16$
$\qquad 2x + 3y = 7$

39. $-3x + 5y = -22$
$\qquad 3x + 4y = 4$
$\qquad 4x - 8y = 32$

40. $x + 2y = 0$
$\qquad x + y = 6$
$\qquad 3x - 2y = 8$

41. $8x - 4y = 7$
$\qquad 5x + 2y = 1$

42. $2x - y = -0.1$
$\qquad 3x + 2y = 1.6$

43. $-x + 2y = 1.5$
$\qquad 2x - 4y = 3$

44. $x - 3y = 5$
$\qquad -2x + 6y = -10$

45. $x \qquad - 3z = -2$
$\qquad 3x + y - 2z = 5$
$\qquad 2x + 2y + z = 4$

46. $2x - y + 3z = 24$
$\qquad 2y - z = 14$
$\qquad 7x - 5y = 6$

47. $x + y - 5z = 3$
$\qquad x \qquad - 2z = 1$
$\qquad 2x - y - z = 0$

48. $2x \qquad + 3z = 3$
$\qquad 4x - 3y + 7z = 5$
$\qquad 8x - 9y + 15z = 9$

49. $x + 2y + z = 8$
$\qquad 3x + 7y + 6z = 26$

50. $4x + 12y - 7z - 20w = 22$
$\qquad 3x + 9y - 5z - 28w = 30$

51. $3x + 3y + 12z = 6$
$\qquad x + y + 4z = 2$
$\qquad 2x + 5y + 20z = 10$
$\qquad -x + 2y + 8z = 4$

52. $2x + 10y + 2z = 6$
$\qquad x + 5y + 2z = 6$
$\qquad x + 5y + z = 3$
$\qquad -3x - 15y - 3z = -9$

53. $2x + y - z + 2w = -6$
$\qquad 3x + 4y + w = 1$
$\qquad x + 5y + 2z + 6w = -3$
$\qquad 5x + 2y - z - w = 3$

54. $x + 2y + 2z + 4w = 11$
$\qquad 3x + 6y + 5z + 12w = 30$

55. $x + 2y = 0$
$\qquad -x - y = 0$

56. $x + 2y = 0$
$\qquad 2x + 4y = 0$

57. $x + y + z = 0$
$\qquad 2x + 3y + z = 0$
$\qquad 3x + 5y + z = 0$

58. $x + 2y + z + 3w = 0$
$\qquad x - y + w = 0$
$\qquad y - z + 2w = 0$

59. *Borrowing Money* A small corporation borrowed $1,500,000—some of it at 8%, some at 9%, and some at 12%. How much was borrowed at each rate if the annual interest was $133,000 and the amount borrowed at 8% was four times the amount borrowed at 12%?

60. *Borrowing Money* A small corporation borrowed $500,000—some of it at 9%, some at 10%, and some at 12%. How much was borrowed at each rate if the annual interest was $52,000 and the amount borrowed at 10% was $2\frac{1}{2}$ times the amount borrowed at 9%?

In Exercises 61–64, find the specified equation that passes through the given points.

61. Parabola: $y = ax^2 + bx + c$

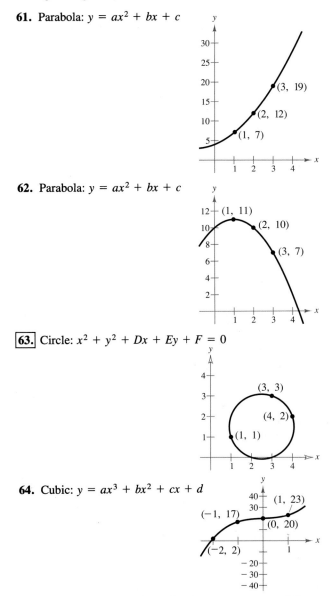

62. Parabola: $y = ax^2 + bx + c$

63. Circle: $x^2 + y^2 + Dx + Ey + F = 0$

64. Cubic: $y = ax^3 + bx^2 + cx + d$

65. The given augmented matrix represents a system of linear equations (in variables x, y, and z) that has been reduced using Gauss-Jordan elimination. Write a system of equations with nonzero coefficients that is represented by the reduced matrix. (The answer is not unique.)

$$\begin{bmatrix} 1 & 0 & 3 & : & -2 \\ 0 & 1 & 4 & : & 1 \\ 0 & 0 & 0 & : & 0 \end{bmatrix}$$

8.2 Operations with Matrices

Equality of Matrices / Matrix Addition and Scalar Multiplication / Matrix Multiplication / Applications

Equality of Matrices

In Section 8.1 you used matrices to solve systems of linear equations. Matrices, however, can do much more than that. There is a rich mathematical theory of matrices, and its applications are numerous. This section and the next introduce some fundamentals of matrix theory. It is standard mathematical convention to represent matrices in any of the following three ways.

1. A matrix can be denoted by an uppercase letter such as A, B, and C.
2. A matrix can be denoted by a representative element enclosed in brackets, such as $[a_{ij}]$, $[b_{ij}]$, and $[c_{ij}]$.
3. A matrix can be denoted by a rectangular array of numbers such as

$$A = [a_{ij}] = \begin{bmatrix} a_{11} & a_{12} & a_{13} & \cdots & a_{1n} \\ a_{21} & a_{22} & a_{23} & \cdots & a_{2n} \\ a_{31} & a_{32} & a_{33} & \cdots & a_{3n} \\ \vdots & \vdots & \vdots & & \vdots \\ a_{m1} & a_{m2} & a_{m3} & \cdots & a_{mn} \end{bmatrix}.$$

Two matrices are **equal** if their corresponding entries are equal.

DEFINITION OF EQUALITY OF MATRICES

Two matrices $A = [a_{ij}]$ and $B = [b_{ij}]$ are **equal** if they have the same order ($m \times n$) and

$$a_{ij} = b_{ij}$$

for $1 \leq i \leq m$ and $1 \leq j \leq n$.

EXAMPLE 1 Equality of Matrices

Solve for a_{11}, a_{12}, a_{21}, and a_{22} in the following matrix equation.

$$\begin{bmatrix} a_{11} & a_{12} \\ a_{21} & a_{22} \end{bmatrix} = \begin{bmatrix} 2 & -1 \\ -3 & 0 \end{bmatrix}$$

Solution

Because two matrices are equal only if corresponding entries are equal, you can conclude that

$$a_{11} = 2, \quad a_{12} = -1, \quad a_{21} = -3, \quad \text{and} \quad a_{22} = 0.$$

Matrix Addition and Scalar Multiplication

Add two matrices (of the same order) by adding their corresponding entries.

DEFINITION OF MATRIX ADDITION

If $A = [a_{ij}]$ and $B = [b_{ij}]$ are matrices of order $m \times n$, then their **sum** is the $m \times n$ matrix given by

$$A + B = [a_{ij} + b_{ij}].$$

The sum of two matrices of different orders is undefined.

EXAMPLE 2 Addition of Matrices

a. $\begin{bmatrix} -1 & 2 \\ 0 & 1 \end{bmatrix} + \begin{bmatrix} 1 & 3 \\ -1 & 2 \end{bmatrix} = \begin{bmatrix} -1+1 & 2+3 \\ 0-1 & 1+2 \end{bmatrix} = \begin{bmatrix} 0 & 5 \\ -1 & 3 \end{bmatrix}$

b. $\begin{bmatrix} 1 \\ -3 \\ -2 \end{bmatrix} + \begin{bmatrix} -1 \\ 3 \\ 2 \end{bmatrix} = \begin{bmatrix} 0 \\ 0 \\ 0 \end{bmatrix}$

c. The sum of

$$A = \begin{bmatrix} 2 & 1 & 0 \\ 4 & 0 & -1 \\ 3 & -2 & 2 \end{bmatrix} \quad \text{and} \quad B = \begin{bmatrix} 0 & 1 \\ -1 & 3 \\ 2 & 4 \end{bmatrix}$$

is undefined.

When working with matrices, we usually refer to numbers as **scalars.** In this text, scalars will always be real numbers. We multiply a matrix A by a scalar c by multiplying each entry in A by c.

DEFINITION OF SCALAR MULTIPLICATION

If $A = [a_{ij}]$ is an $m \times n$ matrix and c is a scalar, then the **scalar multiple** of A by c is the $m \times n$ matrix given by

$$cA = [ca_{ij}].$$

Use $-A$ to represent the scalar product $(-1)A$. Moreover, if A and B are of the same order, then $A - B$ represents the sum of A and $(-1)B$. That is,

$$A - B = A + (-1)B. \qquad \textit{Subtraction of matrices}$$

EXAMPLE 3   Scalar Multiplication and Matrix Subtraction

For the matrices

$$A = \begin{bmatrix} 1 & 2 & 4 \\ -3 & 0 & -1 \\ 2 & 1 & 2 \end{bmatrix} \quad \text{and} \quad B = \begin{bmatrix} 2 & 0 & 0 \\ 1 & -4 & 3 \\ -1 & 3 & 2 \end{bmatrix}$$

find the following.

a. $3A$ **b.** $3A - B$

Solution

a. $3A = 3\begin{bmatrix} 1 & 2 & 4 \\ -3 & 0 & -1 \\ 2 & 1 & 2 \end{bmatrix} = \begin{bmatrix} 3(1) & 3(2) & 3(4) \\ 3(-3) & 3(0) & 3(-1) \\ 3(2) & 3(1) & 3(2) \end{bmatrix} = \begin{bmatrix} 3 & 6 & 12 \\ -9 & 0 & -3 \\ 6 & 3 & 6 \end{bmatrix}$

b. $3A - B = \begin{bmatrix} 3 & 6 & 12 \\ -9 & 0 & -3 \\ 6 & 3 & 6 \end{bmatrix} - \begin{bmatrix} 2 & 0 & 0 \\ 1 & -4 & 3 \\ -1 & 3 & 2 \end{bmatrix} = \begin{bmatrix} 1 & 6 & 12 \\ -10 & 4 & -6 \\ 7 & 0 & 4 \end{bmatrix}$

It is often convenient to rewrite the scalar multiple cA by factoring c out of every entry in the matrix. For instance, in the following example, the scalar $\frac{1}{2}$ has been factored out of the matrix.

$$\begin{bmatrix} \frac{1}{2} & -\frac{3}{2} \\ \frac{5}{2} & \frac{1}{2} \end{bmatrix} = \frac{1}{2}\begin{bmatrix} 1 & -3 \\ 5 & 1 \end{bmatrix}$$

The properties of matrix addition and scalar multiplication are similar to those of addition and multiplication of real numbers, and we summarize them in the following list.

MATRIX ADDITION AND SCALAR MULTIPLICATION

If A, B, and C are $m \times n$ matrices and c and d are scalars, then the following properties are true.

1. $A + B = B + A$ *Commutative Property of Addition*
2. $A + (B + C) = (A + B) + C$ *Associative Property of Addition*
3. $(cd)A = c(dA)$ *Associative Property of Scalar Multiplication*
4. $1A = A$ *Scalar Identity*
5. $c(A + B) = cA + cB$ *Distributive Property*
6. $(c + d)A = cA + dA$ *Distributive Property*

Note that the Associative Property of matrix addition allows you to write expressions like $A + B + C$ without ambiguity because the same sum occurs no matter how the matrices are grouped. In other words, you obtain the same sum whether you group $A + B + C$ as $(A + B) + C$ or as $A + (B + C)$. This same reasoning applies to sums of four or more matrices.

One important property of addition of real numbers is that the number 0 is the additive identity. That is, $c + 0 = c$ for any real number c. For matrices, a similar property holds. That is, if A is an $m \times n$ matrix and O is the $m \times n$ **zero matrix** consisting entirely of zeros, then

$$A + O = A.$$

In other words, O is the **additive identity** for the set of all $m \times n$ matrices. For example, the following matrix is the additive identity for the set of all 2×3 matrices.

$$O = \begin{bmatrix} 0 & 0 & 0 \\ 0 & 0 & 0 \end{bmatrix} \qquad \textit{Zero 2 × 3 matrix}$$

Similarly, the additive identity for the set of all 3×4 matrices is

$$O = \begin{bmatrix} 0 & 0 & 0 & 0 \\ 0 & 0 & 0 & 0 \\ 0 & 0 & 0 & 0 \end{bmatrix}. \qquad \textit{Zero 3 × 4 matrix}$$

The algebra of real numbers and the algebra of matrices have many similarities.* For example, compare the following solutions.

Real Numbers *(Solve for x)*	*m × n Matrices* *(Solve for X)*
$x + a = b$	$X + A = B$
$x + a + (-a) = b + (-a)$	$X + A + (-A) = B + (-A)$
$x + 0 = b - a$	$X + O = B - A$
$x = b - a$	$X = B - A$

The process of solving a matrix equation is demonstrated in Example 4.

*There are also some important differences, which will be discussed later.

EXAMPLE 4   Solving a Matrix Equation

Solve for X in the equation $3X + A = B$, where

$$A = \begin{bmatrix} 1 & -2 \\ 0 & 3 \end{bmatrix} \quad \text{and} \quad B = \begin{bmatrix} -3 & 4 \\ 2 & 1 \end{bmatrix}.$$

Solution

Begin by solving the given equation for X to obtain

$$3X = B - A \quad \rightarrow \quad X = \frac{1}{3}(B - A).$$

Now, using the given matrices A and B, you have

$$X = \frac{1}{3}\left(\begin{bmatrix} -3 & 4 \\ 2 & 1 \end{bmatrix} - \begin{bmatrix} 1 & -2 \\ 0 & 3 \end{bmatrix} \right) = \frac{1}{3}\begin{bmatrix} -4 & 6 \\ 2 & -2 \end{bmatrix} = \begin{bmatrix} -\frac{4}{3} & 2 \\ \frac{2}{3} & -\frac{2}{3} \end{bmatrix}.$$

Matrix Multiplication

The third basic matrix operation is **matrix multiplication.** At first glance, the definition may seem unusual. You will see later, however, that this definition of the product of two matrices has many practical applications.

DEFINITION OF MATRIX MULTIPLICATION

If $A = [a_{ij}]$ is an $m \times n$ matrix and $B = [b_{ij}]$ is an $n \times p$ matrix, then the **product** AB is the $m \times p$ matrix

$$AB = [c_{ij}]$$

where $c_{ij} = a_{i1}b_{1j} + a_{i2}b_{2j} + a_{i3}b_{3j} + \cdots + a_{in}b_{nj}.$

This definition indicates a *row-by-column* multiplication, where the entry c_{ij} in the ith row and the jth column of the product AB is obtained by multiplying the entries in the ith row of A by the corresponding entries in the jth column of B and then adding the results. The following example illustrates the process.

EXAMPLE 5   Finding the Product of Two Matrices

Find the product AB where

$$A = \begin{bmatrix} -1 & 3 \\ 4 & -2 \\ 5 & 0 \end{bmatrix} \quad \text{and} \quad B = \begin{bmatrix} -3 & 2 \\ -4 & 1 \end{bmatrix}.$$

Solution

First note that the product AB is defined because the number of columns of A is equal to the number of rows of B. Moreover, the product AB has order 3×2 and will take the form

$$\begin{bmatrix} -1 & 3 \\ 4 & -2 \\ 5 & 0 \end{bmatrix} \begin{bmatrix} -3 & 2 \\ -4 & 1 \end{bmatrix} = \begin{bmatrix} c_{11} & c_{12} \\ c_{21} & c_{22} \\ c_{31} & c_{32} \end{bmatrix}.$$

To find c_{11} (the entry in the first row and first column of the product), multiply corresponding entries in the first row of A and the first column of B. That is,

$$c_{11} = (-1)(-3) + (3)(-4) = -9$$

$$\begin{bmatrix} -1 & 3 \\ 4 & -2 \\ 5 & 0 \end{bmatrix} \begin{bmatrix} -3 & 2 \\ -4 & 1 \end{bmatrix} = \begin{bmatrix} -9 & c_{12} \\ c_{21} & c_{22} \\ c_{31} & c_{32} \end{bmatrix}.$$

Similarly, to find c_{12}, multiply corresponding entries in the first row of A and the second column of B to obtain

$$c_{12} = (-1)(2) + (3)(1) = 1$$

$$\begin{bmatrix} -1 & 3 \\ 4 & -2 \\ 5 & 0 \end{bmatrix} \begin{bmatrix} -3 & 2 \\ -4 & 1 \end{bmatrix} = \begin{bmatrix} -9 & 1 \\ c_{21} & c_{22} \\ c_{31} & c_{32} \end{bmatrix}.$$

Continuing this pattern produces the following results.

$$\begin{aligned}
c_{21} &= (4)(-3) + (-2)(-4) = -4 \\
c_{22} &= (4)(2) + (-2)(1) = 6 \\
c_{31} &= (5)(-3) + (0)(-4) = -15 \\
c_{32} &= (5)(2) + (0)(1) = 10
\end{aligned}$$

Thus, the product is

$$\begin{aligned}
AB &= \begin{bmatrix} -1 & 3 \\ 4 & -2 \\ 5 & 0 \end{bmatrix} \begin{bmatrix} -3 & 2 \\ -4 & 1 \end{bmatrix} \\
&= \begin{bmatrix} (-1)(-3) + (3)(-4) & (-1)(2) + (3)(1) \\ (4)(-3) + (-2)(-4) & (4)(2) + (-2)(1) \\ (5)(-3) + (0)(-4) & (5)(2) + (0)(1) \end{bmatrix} \\
&= \begin{bmatrix} -9 & 1 \\ -4 & 6 \\ -15 & 10 \end{bmatrix}.
\end{aligned}$$

Be sure you understand that for the product of two matrices to be defined, the number of columns of the first matrix must equal the number of rows of the second matrix. That is, the middle two indices must be the same and the outside two indices give the order of the product, as shown in the following diagram.

$$\begin{array}{ccc} A & B & = & AB \\ m \times n & n \times p & & m \times p \end{array}$$

equal

order of AB

The general pattern for matrix multiplication is as follows. To obtain the entry in the ith row and the jth column of the product AB, use the ith row of A and the jth column of B.

$$\begin{bmatrix} a_{11} & a_{12} & a_{13} & \cdots & a_{1n} \\ a_{21} & a_{22} & a_{23} & \cdots & a_{2n} \\ a_{31} & a_{32} & a_{33} & \cdots & a_{3n} \\ \vdots & \vdots & \vdots & & \vdots \\ a_{i1} & a_{i2} & a_{i3} & \cdots & a_{in} \\ \vdots & \vdots & \vdots & & \vdots \\ a_{m1} & a_{m2} & a_{m3} & \cdots & a_{mn} \end{bmatrix} \begin{bmatrix} b_{11} & b_{12} & \cdots & \cdots & b_{1p} \\ b_{21} & b_{22} & \cdots & \cdots & b_{2p} \\ b_{31} & b_{32} & \cdots & \cdots & b_{3p} \\ \vdots & \vdots & & & \vdots \\ b_{n1} & b_{n2} & \cdots & \cdots & b_{np} \end{bmatrix} = \begin{bmatrix} c_{11} & c_{12} & \cdots & c_{1j} & \cdots & c_{1p} \\ c_{21} & c_{22} & \cdots & c_{2j} & \cdots & c_{2p} \\ \vdots & \vdots & & \vdots & & \vdots \\ c_{i1} & c_{i2} & \cdots & c_{ij} & \cdots & c_{ip} \\ \vdots & \vdots & & \vdots & & \vdots \\ c_{m1} & c_{m2} & \cdots & c_{mj} & \cdots & c_{mp} \end{bmatrix}$$

$$a_{i1}b_{1j} + a_{i2}b_{2j} + a_{i3}b_{3j} + \cdots + a_{in}b_{nj} = c_{ij}$$

EXAMPLE 6 Matrix Multiplication

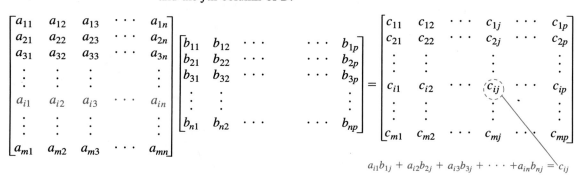

a. $\begin{bmatrix} 1 & 0 & 3 \\ 2 & -1 & -2 \end{bmatrix} \begin{bmatrix} -2 & 4 & 2 \\ 1 & 0 & 0 \\ -1 & 1 & -1 \end{bmatrix} = \begin{bmatrix} -5 & 7 & -1 \\ -3 & 6 & 6 \end{bmatrix}$

2×3 3×3 2×3

b. $[1 \quad -2 \quad -3] \begin{bmatrix} 2 \\ -1 \\ 1 \end{bmatrix} = [1]$

1×3 3×1 1×1

REMARK In parts (b) and (c) of Example 6, note that the two products are different. Matrix multiplication is not, in general, commutative. That is, for most matrices, $AB \neq BA$.

c. $\begin{bmatrix} 2 \\ -1 \\ 1 \end{bmatrix} [1 \quad -2 \quad -3] = \begin{bmatrix} 2 & -4 & -6 \\ -1 & 2 & 3 \\ 1 & -2 & -3 \end{bmatrix}$

3×1 1×3 3×3

d. The product AB for

$$A = \begin{bmatrix} -2 & 1 \\ 1 & -3 \\ 1 & 4 \end{bmatrix} \quad \text{and} \quad B = \begin{bmatrix} -2 & 3 & 1 & 4 \\ 0 & 1 & -1 & 2 \\ 2 & -1 & 0 & 1 \end{bmatrix}$$

$$3 \times 2 \qquad\qquad\qquad 3 \times 4$$

is not defined (nor is the product BA).

◢

PROPERTIES OF MATRIX MULTIPLICATION

If A, B, and C are matrices and c is a scalar, then the following properties are true.

1. $A(BC) = (AB)C$ *Associative Property of Multiplication*
2. $A(B + C) = AB + AC$ *Left Distributive Property*
3. $(A + B)C = AC + BC$ *Right Distributive Property*
4. $c(AB) = (cA)B = A(cB)$

The $n \times n$ matrix that consists of 1's on its main diagonal and 0's elsewhere is the **identity matrix of order** n and is denoted by

$$I_n = \begin{bmatrix} 1 & 0 & 0 & \cdots & 0 \\ 0 & 1 & 0 & \cdots & 0 \\ 0 & 0 & 1 & \cdots & 0 \\ \vdots & \vdots & \vdots & & \vdots \\ 0 & 0 & 0 & \cdots & 1 \end{bmatrix}. \quad \textit{Identity matrix}$$

Note that an identity matrix must be *square*. When the order is understood to be n, I_n is often denoted simply by I. If A is an $n \times n$ matrix, then the identity matrix has the property that

$$AI_n = A \quad \text{and} \quad I_n A = A.$$

For example,

$$\begin{bmatrix} 3 & -2 & 5 \\ 1 & 0 & 4 \\ -1 & 2 & -3 \end{bmatrix} \begin{bmatrix} 1 & 0 & 0 \\ 0 & 1 & 0 \\ 0 & 0 & 1 \end{bmatrix} = \begin{bmatrix} 3 & -2 & 5 \\ 1 & 0 & 4 \\ -1 & 2 & -3 \end{bmatrix}$$

and

$$\begin{bmatrix} 1 & 0 & 0 \\ 0 & 1 & 0 \\ 0 & 0 & 1 \end{bmatrix} \begin{bmatrix} 3 & -2 & 5 \\ 1 & 0 & 4 \\ -1 & 2 & -3 \end{bmatrix} = \begin{bmatrix} 3 & -2 & 5 \\ 1 & 0 & 4 \\ -1 & 2 & -3 \end{bmatrix}.$$

Applications

EXAMPLE 7 An Application of Matrix Multiplication

Two softball teams submit equipment lists to their sponsors.

	Women's Team	Men's Team
Bats	12	15
Balls	45	38
Gloves	15	17

Each bat costs $21, each ball costs $4, and each glove costs $30. Use matrices to find the total cost of equipment for each team.

Solution

The equipment lists can be written in matrix form as

$$E = \begin{bmatrix} 12 & 15 \\ 45 & 38 \\ 15 & 17 \end{bmatrix}$$

and the cost per item can be written in matrix form as

$$C = [21 \quad 4 \quad 30].$$

The total cost of equipment for each team is given by the product

$$21(12) + 4(45) + 30(15) = 882 \quad \textit{(Women's team)}$$

$$CE = [21 \quad 4 \quad 30] \begin{bmatrix} 12 & 15 \\ 45 & 38 \\ 15 & 17 \end{bmatrix} = [882 \quad 977].$$

$$21(15) + 4(38) + 30(17) = 977 \quad \textit{(Men's team)}$$

Thus, the total cost of equipment for the women's team is $882, and the total cost of equipment for the men's team is $977.

Another useful application of matrix multiplication is in representing a system of linear equations. Note how the system

$$a_{11}x_1 + a_{12}x_2 + a_{13}x_3 = b_1$$
$$a_{21}x_1 + a_{22}x_2 + a_{23}x_3 = b_2$$
$$a_{31}x_1 + a_{32}x_2 + a_{33}x_3 = b_3$$

can be written as the matrix equation $AX = B$, where A is the *coefficient matrix* of the system, and X and B are column matrices.

$$\begin{bmatrix} a_{11} & a_{12} & a_{13} \\ a_{21} & a_{22} & a_{23} \\ a_{31} & a_{32} & a_{33} \end{bmatrix} \begin{bmatrix} x_1 \\ x_2 \\ x_3 \end{bmatrix} = \begin{bmatrix} b_1 \\ b_2 \\ b_3 \end{bmatrix}$$

$$A \qquad\quad X \;=\; B$$

EXAMPLE 8   Solving a System of Linear Equations

Solve the matrix equation $AX = B$ for X, where

Coefficient matrix   Constant matrix

$$A = \begin{bmatrix} 1 & -2 & 1 \\ 0 & 1 & 2 \\ 2 & 3 & -2 \end{bmatrix} \quad \text{and} \quad B = \begin{bmatrix} -4 \\ 4 \\ 2 \end{bmatrix}.$$

Solution

As a system of linear equations, $AX = B$ is as follows.

$$
\begin{aligned}
x_1 - 2x_2 + x_3 &= -4 \\
x_2 + 2x_3 &= 4 \\
2x_1 + 3x_2 - 2x_3 &= 2
\end{aligned}
$$

Using Gauss-Jordan elimination on the augmented matrix of this system, you obtain

$$\left[\begin{array}{ccc:c} 1 & 0 & 0 & -1 \\ 0 & 1 & 0 & 2 \\ 0 & 0 & 1 & 1 \end{array}\right].$$

Thus, the solution of the system of linear equations is $x_1 = -1$, $x_2 = 2$, and $x_3 = 1$, and the solution of the matrix equation is

$$X = \begin{bmatrix} x_1 \\ x_2 \\ x_3 \end{bmatrix} = \begin{bmatrix} -1 \\ 2 \\ 1 \end{bmatrix}.$$

DISCUSSION

PROBLEM

Diagonal
Matrices

A square matrix is a **diagonal matrix** if each entry that is not on the main diagonal is zero. For instance,

$$A = \begin{bmatrix} -1 & 0 & 0 \\ 0 & 2 & 0 \\ 0 & 0 & 0 \end{bmatrix} \quad \text{and} \quad B = \begin{bmatrix} 2 & 0 & 0 & 0 \\ 0 & -1 & 0 & 0 \\ 0 & 0 & 3 & 0 \\ 0 & 0 & 0 & -2 \end{bmatrix}$$

are diagonal matrices. Describe a quick rule for multiplying a diagonal matrix by itself. Then illustrate your rule to find the matrices $A^2 = AA$ and $B^2 = BB$ for the given matrices A and B.

WARM UP

The following warm-up exercises involve skills that were covered in earlier sections. You will use these skills in the exercise set for this section.

In Exercises 1 and 2, evaluate the expressions.

1. $-3\left(-\frac{5}{6}\right) + 10\left(-\frac{3}{4}\right)$

2. $-22\left(\frac{5}{2}\right) + 6(8)$

In Exercises 3 and 4, determine whether the matrices are in *reduced* row-echelon form.

3. $\begin{bmatrix} 0 & 1 & 0 & -5 \\ 1 & 0 & 3 & 2 \\ 0 & 0 & 1 & 0 \end{bmatrix}$

4. $\begin{bmatrix} 1 & 0 & 0 & 2 & 3 \\ 0 & 0 & 0 & 0 & 0 \\ 0 & 1 & 1 & 3 & 10 \end{bmatrix}$

In Exercises 5 and 6, write the augmented matrix for each system of linear equations.

5. $-5x + 10y = 12$
$7x - 3y = 0$
$-x + 7y = 25$

6. $10x + 15y - 9z = 42$
$6x - 5y = 0$

In Exercises 7–10, solve the systems of linear equations represented by the augmented matrices.

7. $\left[\begin{array}{cc:c} 1 & 0 & 0 \\ 0 & 1 & 2 \end{array}\right]$

8. $\left[\begin{array}{ccc:c} 1 & 0 & -1 & 2 \\ 0 & 1 & 1 & 3 \end{array}\right]$

9. $\left[\begin{array}{ccc:c} 1 & 2 & 1 & 0 \\ 0 & 0 & 1 & -1 \\ 0 & 0 & 0 & 0 \end{array}\right]$

10. $\left[\begin{array}{ccc:c} 1 & -1 & 0 & 3 \\ 0 & 1 & -2 & 1 \\ 0 & 0 & 1 & -1 \end{array}\right]$

EXERCISES for Section 8.2

In Exercises 1–4, find x and y.

1. $\begin{bmatrix} x & -2 \\ 7 & y \end{bmatrix} = \begin{bmatrix} -4 & -2 \\ 7 & 22 \end{bmatrix}$

2. $\begin{bmatrix} -5 & x \\ y & 8 \end{bmatrix} = \begin{bmatrix} -5 & 13 \\ 12 & 8 \end{bmatrix}$

3. $\begin{bmatrix} 16 & 4 & 5 & 4 \\ -3 & 13 & 15 & 6 \\ 0 & 2 & 4 & 0 \end{bmatrix} = \begin{bmatrix} 16 & 4 & 2x+1 & 4 \\ -3 & 13 & 15 & 3x \\ 0 & 2 & 3y-5 & 0 \end{bmatrix}$

4. $\begin{bmatrix} x+2 & 8 & -3 \\ 1 & 2y & 2x \\ 7 & -2 & y+2 \end{bmatrix} = \begin{bmatrix} 2x+6 & 8 & -3 \\ 1 & 18 & -8 \\ 7 & -2 & 11 \end{bmatrix}$

In Exercises 5–10, find (a) $A + B$, (b) $A - B$, (c) $3A$, and (d) $3A - 2B$.

5. $A = \begin{bmatrix} 1 & -1 \\ 2 & -1 \end{bmatrix}$, $B = \begin{bmatrix} 2 & -1 \\ -1 & 8 \end{bmatrix}$

6. $A = \begin{bmatrix} 1 & 2 \\ 2 & 1 \end{bmatrix}$, $B = \begin{bmatrix} -3 & -2 \\ 4 & 2 \end{bmatrix}$

7. $A = \begin{bmatrix} 6 & -1 \\ 2 & 4 \\ -3 & 5 \end{bmatrix}$, $B = \begin{bmatrix} 1 & 4 \\ -1 & 5 \\ 1 & 10 \end{bmatrix}$

8. $A = \begin{bmatrix} 2 & 1 & 1 \\ -1 & -1 & 4 \end{bmatrix}$, $B = \begin{bmatrix} 2 & -3 & 4 \\ -3 & 1 & -2 \end{bmatrix}$

9. $A = \begin{bmatrix} 2 & 2 & -1 & 0 & 1 \\ 1 & 1 & -2 & 0 & -1 \end{bmatrix}$,

$B = \begin{bmatrix} 1 & 1 & -1 & 1 & 0 \\ -3 & 4 & 9 & -6 & -7 \end{bmatrix}$

10. $A = \begin{bmatrix} 3 \\ 2 \\ -1 \end{bmatrix}$, $B = \begin{bmatrix} -4 \\ 6 \\ 2 \end{bmatrix}$

In Exercises 11–16, find (a) AB, (b) BA, and if possible (c) A^2. (Note: $A^2 = AA$.)

11. $A = \begin{bmatrix} 1 & 2 \\ 4 & 2 \end{bmatrix}$, $B = \begin{bmatrix} 2 & -1 \\ -1 & 8 \end{bmatrix}$

12. $A = \begin{bmatrix} 2 & -1 \\ 1 & 4 \end{bmatrix}$, $B = \begin{bmatrix} 0 & 0 \\ 3 & -3 \end{bmatrix}$

13. $A = \begin{bmatrix} 3 & -1 \\ 1 & 3 \end{bmatrix}$, $B = \begin{bmatrix} 1 & -3 \\ 3 & 1 \end{bmatrix}$

14. $A = \begin{bmatrix} 1 & -1 \\ 1 & 1 \end{bmatrix}$, $B = \begin{bmatrix} 1 & 3 \\ -3 & 1 \end{bmatrix}$

15. $A = \begin{bmatrix} 1 & -1 & 7 \\ 2 & -1 & 8 \\ 3 & 1 & -1 \end{bmatrix}$, $B = \begin{bmatrix} 1 & 1 & 2 \\ 2 & 1 & 1 \\ 1 & -3 & 2 \end{bmatrix}$

16. $A = [3 \quad 2 \quad 1]$, $B = \begin{bmatrix} 2 \\ 3 \\ 0 \end{bmatrix}$

In Exercises 17–24, find AB, if possible.

17. $A = \begin{bmatrix} 2 & 1 \\ -3 & 4 \\ 1 & 6 \end{bmatrix}$, $B = \begin{bmatrix} 0 & -1 & 0 \\ 4 & 0 & 2 \\ 8 & -1 & 7 \end{bmatrix}$

18. $A = \begin{bmatrix} 0 & -1 & 0 \\ 4 & 0 & 2 \\ 8 & -1 & 7 \end{bmatrix}$, $B = \begin{bmatrix} 2 & 1 \\ -3 & 4 \\ 1 & 6 \end{bmatrix}$

19. $A = \begin{bmatrix} -1 & 3 \\ 4 & -5 \\ 0 & 2 \end{bmatrix}$, $B = \begin{bmatrix} 1 & 2 \\ 0 & 7 \end{bmatrix}$

20. $A = \begin{bmatrix} 1 & 0 & 0 \\ 0 & 4 & 0 \\ 0 & 0 & -2 \end{bmatrix}$, $B = \begin{bmatrix} 3 & 0 & 0 \\ 0 & -1 & 0 \\ 0 & 0 & 5 \end{bmatrix}$

21. $A = \begin{bmatrix} 5 & 0 & 0 \\ 0 & -8 & 0 \\ 0 & 0 & 7 \end{bmatrix}$, $B = \begin{bmatrix} \frac{1}{5} & 0 & 0 \\ 0 & -\frac{1}{8} & 0 \\ 0 & 0 & \frac{1}{2} \end{bmatrix}$

22. $A = \begin{bmatrix} 0 & 0 & 5 \\ 0 & 0 & -3 \\ 0 & 0 & 4 \end{bmatrix}$, $B = \begin{bmatrix} 6 & -11 & 4 \\ 8 & 16 & 4 \\ 0 & 0 & 0 \end{bmatrix}$

23. $A = \begin{bmatrix} 6 \\ -2 \\ 1 \\ 6 \end{bmatrix}$, $B = [10 \quad 12]$

24. $A = \begin{bmatrix} 1 & 0 & 3 & -2 & 4 \\ 6 & 13 & 8 & -17 & 10 \end{bmatrix}$,

$B = \begin{bmatrix} 1 & 6 \\ 4 & 2 \end{bmatrix}$

In Exercises 25–28, solve for X given

$A = \begin{bmatrix} -2 & -1 \\ 1 & 0 \\ 3 & -4 \end{bmatrix}$ and $B = \begin{bmatrix} 0 & 3 \\ 2 & 0 \\ -4 & -1 \end{bmatrix}$.

25. $X = 3A - 2B$ **26.** $2X = 2A - B$

27. $2X + 3A = B$ **28.** $2A + 4B = -2X$

In Exercises 29–32, find matrices A, X, and B such that the given system of linear equations can be written as the matrix equation $AX = B$. Solve the system of equations.

29. $-x + y = 4$ **30.** $2x + 3y = 5$
      $-2x + y = 0$ $x + 4y = 10$

31. $x - 2y + 3z = 9$ **32.** $x + y - 3z = -1$
      $-x + 3y - z = -6$ $-x + 2y = 1$
      $2x - 5y + 5z = 17$ $-y + z = 0$

In Exercises 33–36, find $f(A) = a_0 I_n + a_1 A + a_2 A^2 + \cdots + a_n A^n$.

33. $f(x) = x^2 - 5x + 2$, $A = \begin{bmatrix} 2 & 0 \\ 4 & 5 \end{bmatrix}$

34. $f(x) = x^2 - 7x + 6$, $A = \begin{bmatrix} 5 & 4 \\ 1 & 2 \end{bmatrix}$

35. $f(x) = x^3 - 10x^2 + 31x - 30$,

$A = \begin{bmatrix} 3 & 1 & 4 \\ 0 & 2 & 6 \\ 0 & 0 & 5 \end{bmatrix}$

36. $f(x) = x^2 - 10x + 24$, $A = \begin{bmatrix} 8 & -4 \\ 2 & 2 \end{bmatrix}$

37. If a, b, and c are real numbers such that $c \neq 0$, and $ac = bc$, then $a = b$. However, if A, B, and C are matrices such that $AC = BC$, then A is *not* necessarily equal to B. Illustrate this using the following matrices.

$$A = \begin{bmatrix} 1 & 2 & 3 \\ 0 & 5 & 4 \\ 3 & -2 & 1 \end{bmatrix}, B = \begin{bmatrix} 4 & -6 & 3 \\ 5 & 4 & 4 \\ -1 & 0 & 1 \end{bmatrix}, \text{ and}$$

$$C = \begin{bmatrix} 0 & 0 & 0 \\ 0 & 0 & 0 \\ 4 & -2 & 3 \end{bmatrix}.$$

38. If a and b are real numbers such that $ab = 0$, then $a = 0$ or $b = 0$. However, if A and B are matrices such that $AB = 0$, then it is *not* necessarily true that $A = 0$ or $B = 0$. Illustrate this using the following matrices.

$$A = \begin{bmatrix} 3 & 3 \\ 4 & 4 \end{bmatrix}, \quad B = \begin{bmatrix} 1 & -1 \\ -1 & 1 \end{bmatrix}$$

39. *Factory Production*   A certain corporation has three factories, each of which manufactures two products. The number of units of product i produced at factory j in one day is represented by a_{ij} in the matrix

$$A = \begin{bmatrix} 60 & 40 & 20 \\ 30 & 90 & 60 \end{bmatrix}.$$

Find the production levels if production is increased by 20%. (*Hint:* Since an increase of 20% corresponds to 100% + 20%, multiply the given matrix by 1.20.)

40. *Factory Production*   A certain corporation has four factories, each of which manufactures two products. The number of units of product i produced at factory j in one day is represented by a_{ij} in the matrix

$$A = \begin{bmatrix} 100 & 90 & 70 & 30 \\ 40 & 20 & 60 & 60 \end{bmatrix}.$$

Find the production levels if production is increased by 10%. (*Hint:* Since an increase of 10% corresponds to 100% + 10%, multiply the given matrix by 1.10.)

41. *Crop Production*   A fruit grower raises two crops that are shipped to three outlets. The number of units of product i that are shipped to outlet j is represented by a_{ij} in the matrix

$$A = \begin{bmatrix} 100 & 75 & 75 \\ 125 & 150 & 100 \end{bmatrix}.$$

The profit per unit is represented by the matrix

$B = [\$3.75 \quad \$7.00].$

Find the product BA, and state what each entry of the product represents.

42. *Total Revenue*   A manufacturer produces three different models of a given product that are shipped to two different warehouses. The number of units of model i that are shipped to warehouse j is represented by a_{ij} in the matrix

$$A = \begin{bmatrix} 5000 & 4000 \\ 6000 & 10000 \\ 8000 & 5000 \end{bmatrix}.$$

The price per unit is represented by the matrix

$B = [\$20.50 \quad \$26.50 \quad \$29.50].$

Find the product BA, and state what each entry of the product represents.

43. *Inventory Levels*   A company sells five different models of computers through three retail outlets. The inventory of each model in the three outlets is given by the matrix S

$$\begin{array}{c} \text{Model} \\ \overbrace{\begin{array}{ccccc} A & B & C & D & E \end{array}} \\ S = \begin{bmatrix} 3 & 2 & 2 & 3 & 0 \\ 0 & 2 & 3 & 4 & 3 \\ 4 & 2 & 1 & 3 & 2 \end{bmatrix} \begin{array}{l} 1 \\ 2 \\ 3 \end{array} \left.\begin{array}{l} \\ \\ \end{array}\right\} \text{Outlet} \end{array}$$

and the wholesale and retail price for each model is given by matrix T.

$$\begin{array}{c} \text{Price} \\ \overbrace{\begin{array}{cc} \text{Wholesale} & \text{Retail} \end{array}} \\ T = \begin{bmatrix} \$840 & \$1100 \\ \$1200 & \$1350 \\ \$1450 & \$1650 \\ \$2650 & \$3000 \\ \$3050 & \$3200 \end{bmatrix} \begin{array}{l} A \\ B \\ C \\ D \\ E \end{array} \left.\begin{array}{l} \\ \\ \\ \\ \end{array}\right\} \text{Model} \end{array}$$

(a) What is the retail price of the inventory at Outlet 1?
(b) What is the wholesale price of the inventory at Outlet 3?
(c) Compute ST and interpret the result.

44. *Labor/Wage Requirements*   A company that makes boats has the following labor-hour and wage requirements.

Labor-hour requirements (per boat)

	Department		
	Cutting	Assembly	Packaging
$S = \begin{bmatrix}$ 1.0 hour	0.5 hour	0.2 hour $\end{bmatrix}$ Small		
1.6 hours	1.0 hour	0.2 hour Medium		
2.5 hours	2.0 hours	0.4 hour Large		

$S = \begin{bmatrix} 1.0 \text{ hour} & 0.5 \text{ hour} & 0.2 \text{ hour} \\ 1.6 \text{ hours} & 1.0 \text{ hour} & 0.2 \text{ hour} \\ 2.5 \text{ hours} & 2.0 \text{ hours} & 0.4 \text{ hour} \end{bmatrix} \begin{array}{l} \text{Small} \\ \text{Medium} \\ \text{Large} \end{array} \left.\begin{array}{l} \\ \\ \end{array}\right\} \text{Boat size}$

Wage requirements (per hour)

Plant

$$T = \begin{bmatrix} \$12 & \$10 \\ \$9 & \$8 \\ \$6 & \$5 \end{bmatrix} \begin{matrix} \text{Cutting} \\ \text{Assembly} \\ \text{Packaging} \end{matrix} \Big\} \text{Department}$$

(a) What is the labor cost for a medium boat at Plant B?
(b) What is the labor cost for a large boat at Plant A?
(c) Compute ST and interpret the result.

45. *Voting Preference*   The matrix

From

$$P = \begin{bmatrix} 0.6 & 0.1 & 0.1 \\ 0.2 & 0.7 & 0.1 \\ 0.2 & 0.2 & 0.8 \end{bmatrix} \begin{matrix} R \\ D \\ I \end{matrix} \Big\} \text{To}$$

is called a *stochastic matrix*. Each entry p_{ij} ($i \neq j$) represents the proportion of the voting population that changes from party i to party j, and p_{ii} represents the proportion that remains loyal to the party from one election to the next. Find P^2. (This matrix gives the transition probabilities from the first election to the third.)

46. *Voting Preference*   Use a computer or calculator to find P^3, P^4, P^5, P^6, P^7, and P^8 for the matrix given in Exercise 45. Can you detect a pattern as P is raised to higher and higher powers?

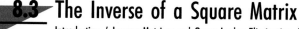

8.3   The Inverse of a Square Matrix

Introduction / Inverse Matrices and Gauss-Jordan Elimination / Systems of Linear Equations

Introduction

This section further develops the algebra of matrices to include the solution of matrix equations involving matrix multiplication. To begin, consider the real number equation $ax = b$. To solve this equation for x, multiply both sides of the equation by a^{-1} (provided $a \neq 0$).

$$ax = b$$
$$(a^{-1}a)x = a^{-1}b$$
$$(1)x = a^{-1}b$$
$$x = a^{-1}b$$

The number a^{-1} is the *multiplicative inverse* of a because it has the property that $a^{-1}a = 1$. The definition of a multiplicative inverse of a matrix is similar.

DEFINITION OF AN INVERSE OF A MATRIX

Let A be a square matrix of order n. If there exists a matrix A^{-1} such that

$$AA^{-1} = I_n = A^{-1}A$$

then A^{-1} is called the **inverse** of A.

REMARK   The symbol A^{-1} is read "A inverse."

If a matrix A has an inverse, then A is **invertible** (or **nonsingular**); otherwise, A is **singular.** A nonsquare matrix cannot have an inverse. Note that if A is of order $m \times n$ and B is of order $n \times m$ (where $m \neq n$), then the products AB and BA are of different orders and could therefore not be equal to each other. Not all square matrices possess inverses. If, however, a matrix does possess an inverse, then that inverse is unique.

EXAMPLE 1   The Inverse of a Matrix

Show that B is the inverse of A, where

$$A = \begin{bmatrix} -1 & 2 \\ -1 & 1 \end{bmatrix} \quad \text{and} \quad B = \begin{bmatrix} 1 & -2 \\ 1 & -1 \end{bmatrix}.$$

Solution

Using the definition of an inverse matrix, you can show that B is the inverse of A by showing that $AB = I = BA$ as follows.

$$AB = \begin{bmatrix} -1 & 2 \\ -1 & 1 \end{bmatrix}\begin{bmatrix} 1 & -2 \\ 1 & -1 \end{bmatrix} = \begin{bmatrix} -1+2 & 2-2 \\ -1+1 & 2-1 \end{bmatrix} = \begin{bmatrix} 1 & 0 \\ 0 & 1 \end{bmatrix}$$

$$BA = \begin{bmatrix} 1 & -2 \\ 1 & -1 \end{bmatrix}\begin{bmatrix} -1 & 2 \\ -1 & 1 \end{bmatrix} = \begin{bmatrix} -1+2 & 2-2 \\ -1+1 & 2-1 \end{bmatrix} = \begin{bmatrix} 1 & 0 \\ 0 & 1 \end{bmatrix}$$

REMARK   Recall that it is not always true that $AB = BA$, even if both products are defined. However, if A and B are both square matrices and $AB = I_n$, then it can be shown that $BA = I_n$. Hence, in Example 1, you needed only to check that $AB = I_2$.

The following example shows how to use a system of equations to find the inverse.

EXAMPLE 2   Finding the Inverse of a Matrix

Find the inverse of the matrix

$$A = \begin{bmatrix} 1 & 4 \\ -1 & -3 \end{bmatrix}.$$

Solution

To find the inverse of A, try to solve the matrix equation $AX = I$ for X.

$$\overset{A}{\begin{bmatrix} 1 & 4 \\ -1 & -3 \end{bmatrix}} \overset{X}{\begin{bmatrix} x_{11} & x_{12} \\ x_{21} & x_{22} \end{bmatrix}} = \overset{I}{\begin{bmatrix} 1 & 0 \\ 0 & 1 \end{bmatrix}}$$

$$\begin{bmatrix} x_{11} + 4x_{21} & x_{12} + 4x_{22} \\ -x_{11} - 3x_{21} & -x_{12} - 3x_{22} \end{bmatrix} = \begin{bmatrix} 1 & 0 \\ 0 & 1 \end{bmatrix}$$

Equating corresponding entries, you obtain the following two systems of linear equations.

$$\begin{aligned} x_{11} + 4x_{21} &= 1 & x_{12} + 4x_{22} &= 0 \\ -x_{11} - 3x_{21} &= 0 & -x_{12} - 3x_{22} &= 1 \end{aligned}$$

From the first system you find that $x_{11} = -3$ and $x_{21} = 1$, and from the second system you find that $x_{12} = -4$ and $x_{22} = 1$. Therefore, the inverse of A is

$$X = A^{-1} = \begin{bmatrix} -3 & -4 \\ 1 & 1 \end{bmatrix}.$$

Try using matrix multiplication to check this result.   ◢

Inverse Matrices and Gauss-Jordan Elimination

In Example 2, note that the two systems of linear equations have the *same coefficient matrix A*. Rather than solve the two systems represented by

$$\begin{bmatrix} 1 & 4 & \vdots & 1 \\ -1 & -3 & \vdots & 0 \end{bmatrix}$$

and

$$\begin{bmatrix} 1 & 4 & \vdots & 0 \\ -1 & -3 & \vdots & 1 \end{bmatrix}$$

separately, you can solve them *simultaneously* by **adjoining** the identity matrix to the coefficient matrix to obtain

$$\begin{bmatrix} \overset{A}{1} & 4 & \vdots & \overset{I}{1} & 0 \\ -1 & -3 & \vdots & 0 & 1 \end{bmatrix}.$$

Then, applying Gauss-Jordan elimination to this matrix, you can solve *both* systems with a single elimination process as follows.

$$\begin{bmatrix} 1 & 4 & \vdots & 1 & 0 \\ -1 & -3 & \vdots & 0 & 1 \end{bmatrix} \quad \underset{R_1 + R_2 \rightarrow}{} \begin{bmatrix} 1 & 4 & \vdots & 1 & 0 \\ 0 & 1 & \vdots & 1 & 1 \end{bmatrix}$$

$$\underset{-4R_2 + R_1 \rightarrow}{} \begin{bmatrix} 1 & 0 & \vdots & -3 & -4 \\ 0 & 1 & \vdots & 1 & 1 \end{bmatrix}$$

Thus, from the "doubly augmented" matrix $[A \vdots I]$ you obtained the matrix $[I \vdots A^{-1}]$.

$$\overset{A}{\underset{}{}}\quad\overset{I}{\underset{}{}}$$

$$\begin{bmatrix} 1 & 4 & \vdots & 1 & 0 \\ -1 & -3 & \vdots & 0 & 1 \end{bmatrix} \rightarrow \overset{I \qquad A^{-1}}{\begin{bmatrix} 1 & 0 & \vdots & -3 & -4 \\ 0 & 1 & \vdots & 1 & 1 \end{bmatrix}}$$

This procedure (or algorithm) works for an arbitrary square matrix that has an inverse.

Finding the Inverse of a Matrix by Gauss-Jordan Elimination

Let A be a square matrix of order n.

1. Write the $n \times 2n$ matrix that consists of the given matrix A on the left and the $n \times n$ identity matrix I on the right to obtain $[A \quad \vdots \quad I]$. Note that the matrices A and I are separated by a dotted line. We call this process **adjoining** the matrices A and I.
2. If possible, row reduce A to I using elementary row operations on the *entire* matrix $[A \quad \vdots \quad I]$. The result will be the matrix $[I \quad \vdots \quad A^{-1}]$. If this is not possible, then A is not invertible.
3. Check your work by multiplying to see that $AA^{-1} = I = A^{-1}A$.

EXAMPLE 3 · Finding the Inverse of a Matrix

Find the inverse of the following matrix.

$$A = \begin{bmatrix} 1 & -1 & 0 \\ 1 & 0 & -1 \\ 6 & -2 & -3 \end{bmatrix}$$

Solution

Begin by adjoining the identity matrix to A to form the matrix

$$[A \quad \vdots \quad I] = \begin{bmatrix} 1 & -1 & 0 & \vdots & 1 & 0 & 0 \\ 1 & 0 & -1 & \vdots & 0 & 1 & 0 \\ 6 & -2 & -3 & \vdots & 0 & 0 & 1 \end{bmatrix}.$$

Now, use the elementary row operations to rewrite this matrix in the form $[I \quad \vdots \quad A^{-1}]$ as follows.

$$\begin{bmatrix} 1 & -1 & 0 & \vdots & 1 & 0 & 0 \\ 1 & 0 & -1 & \vdots & 0 & 1 & 0 \\ 6 & -2 & -3 & \vdots & 0 & 0 & 1 \end{bmatrix} \begin{array}{c} \\ -R_1 + R_2 \rightarrow \\ -6R_1 + R_3 \rightarrow \end{array} \begin{bmatrix} 1 & -1 & 0 & \vdots & 1 & 0 & 0 \\ 0 & 1 & -1 & \vdots & -1 & 1 & 0 \\ 0 & 4 & -3 & \vdots & -6 & 0 & 1 \end{bmatrix}$$

$$\begin{array}{c} R_2 + R_1 \rightarrow \\ \\ -4R_2 + R_3 \rightarrow \end{array} \begin{bmatrix} 1 & 0 & -1 & \vdots & 0 & 1 & 0 \\ 0 & 1 & -1 & \vdots & -1 & 1 & 0 \\ 0 & 0 & 1 & \vdots & -2 & -4 & 1 \end{bmatrix}$$

$$\begin{array}{c} R_3 + R_1 \rightarrow \\ R_3 + R_2 \rightarrow \\ \\ \end{array} \begin{bmatrix} 1 & 0 & 0 & \vdots & -2 & -3 & 1 \\ 0 & 1 & 0 & \vdots & -3 & -3 & 1 \\ 0 & 0 & 1 & \vdots & -2 & -4 & 1 \end{bmatrix}$$

Therefore, the matrix A is invertible and its inverse is

$$A^{-1} = \begin{bmatrix} -2 & -3 & \vdots & 1 \\ -3 & -3 & \vdots & 1 \\ -2 & -4 & \vdots & 1 \end{bmatrix}.$$

Try confirming this by multiplying A and A^{-1} to obtain I.

The process shown in Example 3 applies to any $n \times n$ matrix A. If A has an inverse, this process will find it. On the other hand, if A does not have an inverse, the process will not lead to an identity matrix to the left of the dotted line.

Systems of Linear Equations

A system of linear equations can have exactly one solution, an infinite number of solutions, or no solution. If the coefficient matrix A of a *square* system (a system that has the same number of equations as variables) is invertible, the system has a unique solution, which is given as follows.

$$AX = B \qquad \text{\textit{Given equation}}$$
$$A^{-1}AX = A^{-1}B \qquad \text{\textit{Multiply both sides by A^{-1} (on the left)}}$$
$$X = A^{-1}B \qquad \text{\textit{Solution}}$$

A SYSTEM OF EQUATIONS WITH A UNIQUE SOLUTION

If A is an invertible matrix, then the system of linear equations represented by $AX = B$ has a unique solution given by

$$X = A^{-1}B.$$

Solving a system of linear equations by finding the inverse of the coefficient matrix is not very efficient. That is, it is usually more work to find A^{-1} and then multiply by B than simply to solve the system using Gaussian elimination with back-substitution. One case in which you might consider using an inverse matrix as a computational technique would be with *several* systems of linear equations, all of which have the same coefficient matrix A. In such a case, you could find the inverse matrix once and then solve each system by computing the product $A^{-1}B$. This is demonstrated in Example 4.

EXAMPLE 4 Solving a System of Equations Using an Inverse

Use an inverse matrix to solve the following systems.

a. $\begin{aligned} 2x + 3y + z &= -1 \\ 3x + 3y + z &= 1 \\ 2x + 4y + z &= -2 \end{aligned}$

b. $\begin{aligned} 2x + 3y + z &= 4 \\ 3x + 3y + z &= 8 \\ 2x + 4y + z &= 5 \end{aligned}$

Solution

First note that the coefficient matrix for each system is

$$A = \begin{bmatrix} 2 & 3 & 1 \\ 3 & 3 & 1 \\ 2 & 4 & 1 \end{bmatrix}.$$

Using Gauss-Jordan elimination, you find A^{-1} to be

$$A^{-1} = \begin{bmatrix} -1 & 1 & 0 \\ -1 & 0 & 1 \\ 6 & -2 & -3 \end{bmatrix}.$$

To solve each system, use matrix multiplication as follows.

a. $X = A^{-1}B = \begin{bmatrix} -1 & 1 & 0 \\ -1 & 0 & 1 \\ 6 & -2 & -3 \end{bmatrix}\begin{bmatrix} -1 \\ 1 \\ -2 \end{bmatrix} = \begin{bmatrix} 2 \\ -1 \\ -2 \end{bmatrix}$

The solution is $x = 2$, $y = -1$, and $z = -2$.

b. $X = A^{-1}B = \begin{bmatrix} -1 & 1 & 0 \\ -1 & 0 & 1 \\ 6 & -2 & -3 \end{bmatrix}\begin{bmatrix} 4 \\ 8 \\ 5 \end{bmatrix} = \begin{bmatrix} 4 \\ 1 \\ -7 \end{bmatrix}$

The solution is $x = 4$, $y = 1$, and $z = -7$.

DISCUSSION PROBLEM

The Factorization Principle

For *real numbers* the Factorization Principle states that if $ab = 0$, then either $a = 0$ or $b = 0$. It is possible, however, to find nonzero matrices whose product is zero. For instance,

$$\begin{bmatrix} 2 & -1 \\ -4 & 2 \end{bmatrix}\begin{bmatrix} 3 & -3 \\ 6 & -6 \end{bmatrix} = \begin{bmatrix} 0 & 0 \\ 0 & 0 \end{bmatrix}.$$

In order to obtain a factorization principle for matrices, we must restrict ourselves to a certain type of matrix. Complete the following factorization principle for matrices by determining which type of matrix will make the statement true. Then write a short paragraph that justifies your conclusion.

"Let A and B be square matrices, each of order $n \times n$. If A and B are ______ and $AB = O$ where O is the $n \times n$ zero matrix, then $A = O$ or $B = O$."

WARM UP

The following warm-up exercises involve skills that were covered in earlier sections. You will use these skills in the exercise set for this section.

In Exercises 1–8, perform the indicated matrix operations.

1. $4\begin{bmatrix} 1 & 6 \\ 0 & -4 \\ 12 & 2 \end{bmatrix}$

2. $\dfrac{1}{2}\begin{bmatrix} 11 & 10 & 48 \\ 1 & 0 & 16 \\ 0 & 2 & 8 \end{bmatrix}$

3. $\begin{bmatrix} 1 & -10 & 3 \\ 4 & 1 & 0 \end{bmatrix} - 2\begin{bmatrix} 3 & -4 & 8 \\ 0 & 7 & 1 \end{bmatrix}$

4. $\begin{bmatrix} 5 & 20 \\ -7 & 15 \end{bmatrix} - 3\begin{bmatrix} 6 & 3 \\ 4 & -2 \end{bmatrix}$

5. $\begin{bmatrix} 1 & -2 \\ -1 & 3 \end{bmatrix}\begin{bmatrix} 3 & 2 \\ 1 & 1 \end{bmatrix}$

6. $\begin{bmatrix} 1 & 0 \\ 0 & 1 \end{bmatrix}\begin{bmatrix} 6 & 5 \\ 3 & -2 \end{bmatrix}$

7. $\begin{bmatrix} 2 & 0 & 0 \\ 0 & -1 & 0 \\ 0 & 0 & 3 \end{bmatrix}\begin{bmatrix} \frac{1}{2} & 0 & 0 \\ 0 & -1 & 0 \\ 0 & 0 & \frac{1}{3} \end{bmatrix}$

8. $\begin{bmatrix} 1 & -1 & 0 \\ 1 & 0 & -1 \\ 6 & -2 & -3 \end{bmatrix}\begin{bmatrix} -2 & -3 & 1 \\ -3 & -3 & 1 \\ -2 & -4 & 1 \end{bmatrix}$

In Exercises 9 and 10, rewrite the matrices in reduced row-echelon form.

9. $\begin{bmatrix} 3 & -2 & 1 & 0 \\ 4 & -3 & 0 & 1 \end{bmatrix}$

10. $\begin{bmatrix} 1 & 1 & 2 & 1 & 0 & 0 \\ -1 & 0 & 3 & 0 & 1 & 0 \\ 1 & 2 & 8 & 0 & 0 & 1 \end{bmatrix}$

EXERCISES for Section 8.3

In Exercises 1–8, show that B is the inverse of A.

1. $A = \begin{bmatrix} 2 & 1 \\ 5 & 3 \end{bmatrix}$, $\qquad B = \begin{bmatrix} 3 & -1 \\ -5 & 2 \end{bmatrix}$

2. $A = \begin{bmatrix} 1 & -1 \\ -1 & 2 \end{bmatrix}$, $\qquad B = \begin{bmatrix} 2 & 1 \\ 1 & 1 \end{bmatrix}$

3. $A = \begin{bmatrix} 1 & 2 \\ 3 & 4 \end{bmatrix}$, $\qquad B = \begin{bmatrix} -2 & 1 \\ \frac{3}{2} & -\frac{1}{2} \end{bmatrix}$

4. $A = \begin{bmatrix} 1 & -1 \\ 2 & 3 \end{bmatrix}$, $\qquad B = \begin{bmatrix} \frac{3}{5} & \frac{1}{5} \\ -\frac{2}{5} & \frac{1}{5} \end{bmatrix}$

5. $A = \begin{bmatrix} -2 & 2 & 3 \\ 1 & -1 & 0 \\ 0 & 1 & 4 \end{bmatrix}$, $\quad B = \dfrac{1}{3}\begin{bmatrix} -4 & -5 & 3 \\ -4 & -8 & 3 \\ 1 & 2 & 0 \end{bmatrix}$

6. $A = \begin{bmatrix} 2 & -17 & 11 \\ -1 & 11 & -7 \\ 0 & 3 & -2 \end{bmatrix}$, $\quad B = \begin{bmatrix} 1 & 1 & 2 \\ 2 & 4 & -3 \\ 3 & 6 & -5 \end{bmatrix}$

7. $A = \begin{bmatrix} 2 & 0 & 1 & 1 \\ 3 & 0 & 0 & 1 \\ -1 & 1 & -2 & 1 \\ 4 & -1 & 1 & 0 \end{bmatrix}$, $B = \begin{bmatrix} -1 & 2 & -1 & -1 \\ -4 & 9 & -5 & -6 \\ 0 & 1 & -1 & -1 \\ 3 & -5 & 3 & 3 \end{bmatrix}$

8. $A = \begin{bmatrix} -1 & 1 & 0 & -1 \\ 1 & -1 & 1 & 0 \\ -1 & 1 & 2 & 0 \\ 0 & -1 & 1 & 1 \end{bmatrix}$, $B = \dfrac{1}{3}\begin{bmatrix} -3 & 1 & 1 & -3 \\ -3 & -1 & 2 & -3 \\ 0 & 1 & 1 & 0 \\ -3 & -2 & 1 & 0 \end{bmatrix}$

In Exercises 9–30, find the inverse of the matrix (if it exists).

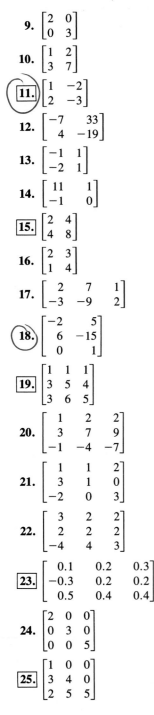

9. $\begin{bmatrix} 2 & 0 \\ 0 & 3 \end{bmatrix}$

10. $\begin{bmatrix} 1 & 2 \\ 3 & 7 \end{bmatrix}$

11. $\begin{bmatrix} 1 & -2 \\ 2 & -3 \end{bmatrix}$

12. $\begin{bmatrix} -7 & 33 \\ 4 & -19 \end{bmatrix}$

13. $\begin{bmatrix} -1 & 1 \\ -2 & 1 \end{bmatrix}$

14. $\begin{bmatrix} 11 & 1 \\ -1 & 0 \end{bmatrix}$

15. $\begin{bmatrix} 2 & 4 \\ 4 & 8 \end{bmatrix}$

16. $\begin{bmatrix} 2 & 3 \\ 1 & 4 \end{bmatrix}$

17. $\begin{bmatrix} 2 & 7 & 1 \\ -3 & -9 & 2 \end{bmatrix}$

18. $\begin{bmatrix} -2 & 5 \\ 6 & -15 \\ 0 & 1 \end{bmatrix}$

19. $\begin{bmatrix} 1 & 1 & 1 \\ 3 & 5 & 4 \\ 3 & 6 & 5 \end{bmatrix}$

20. $\begin{bmatrix} 1 & 2 & 2 \\ 3 & 7 & 9 \\ -1 & -4 & -7 \end{bmatrix}$

21. $\begin{bmatrix} 1 & 1 & 2 \\ 3 & 1 & 0 \\ -2 & 0 & 3 \end{bmatrix}$

22. $\begin{bmatrix} 3 & 2 & 2 \\ 2 & 2 & 2 \\ -4 & 4 & 3 \end{bmatrix}$

23. $\begin{bmatrix} 0.1 & 0.2 & 0.3 \\ -0.3 & 0.2 & 0.2 \\ 0.5 & 0.4 & 0.4 \end{bmatrix}$

24. $\begin{bmatrix} 2 & 0 & 0 \\ 0 & 3 & 0 \\ 0 & 0 & 5 \end{bmatrix}$

25. $\begin{bmatrix} 1 & 0 & 0 \\ 3 & 4 & 0 \\ 2 & 5 & 5 \end{bmatrix}$

26. $\begin{bmatrix} 1 & 0 & 0 \\ 3 & 0 & 0 \\ 2 & 5 & 5 \end{bmatrix}$

27. $\begin{bmatrix} 1 & 0 & 3 & 0 \\ 0 & 2 & 0 & 4 \\ 1 & 0 & 3 & 0 \\ 0 & 2 & 0 & 4 \end{bmatrix}$

28. $\begin{bmatrix} 1 & 3 & -2 & 0 \\ 0 & 2 & 4 & 6 \\ 0 & 0 & -2 & 1 \\ 0 & 0 & 0 & 5 \end{bmatrix}$

29. $\begin{bmatrix} 1 & -2 & -1 & -2 \\ 3 & -5 & -2 & -3 \\ 2 & -5 & -2 & -5 \\ -1 & 4 & 4 & 11 \end{bmatrix}$

30. $\begin{bmatrix} 4 & 8 & -7 & 14 \\ 2 & 5 & -4 & 6 \\ 0 & 2 & 1 & -7 \\ 3 & 6 & -5 & 10 \end{bmatrix}$

In Exercises 31–34, use an inverse matrix to solve the given system of linear equations. (Use the inverse matrix found in Exercise 11.)

31. $x - 2y = 5$
    $2x - 3y = 10$

32. $x - 2y = 0$
    $2x - 3y = 3$

33. $x - 2y = 4$
    $2x - 3y = 2$

34. $x - 2y = 1$
    $2x - 3y = -2$

In Exercises 35–38, use an inverse matrix to solve the given system of linear equations. (Use the inverse matrix found in Exercise 13.)

35. $-x + y = 4$
    $-2x + y = 0$

36. $-x + y = -3$
    $-2x + y = 5$

37. $-x + y = 20$
    $-2x + y = 10$

38. $-x + y = 0$
    $-2x + y = 7$

In Exercises 39 and 40, use an inverse matrix to solve the given system of linear equations. (Use the inverse matrix found in Exercise 19.)

39.
$$\begin{aligned} x + y + z &= 0 \\ 3x + 5y + 4z &= 5 \\ 3x + 6y + 5z &= 2 \end{aligned}$$

40.
$$\begin{aligned} x + y + z &= -1 \\ 3x + 5y + 4z &= 2 \\ 3x + 6y + 5z &= 0 \end{aligned}$$

In Exercises 41 and 42, use an inverse matrix to solve the given system of linear equations. (Use the inverse matrix found in Exercise 29.)

41.
$$\begin{aligned} x_1 - 2x_2 - x_3 - 2x_4 &= 0 \\ 3x_1 - 5x_2 - 2x_3 - 3x_4 &= 1 \\ 2x_1 - 5x_2 - 2x_3 - 5x_4 &= -1 \\ -x_1 + 4x_2 + 4x_3 + 11x_4 &= 2 \end{aligned}$$

42.
$$\begin{aligned} x_1 - 2x_2 - x_3 - 2x_4 &= 1 \\ 3x_1 - 5x_2 - 2x_3 - 3x_4 &= -2 \\ 2x_1 - 5x_2 - 2x_3 - 5x_4 &= 0 \\ -x_1 + 4x_2 + 4x_3 + 11x_4 &= -3 \end{aligned}$$

Bond Investments In Exercises 43–46, consider a person who invests in AAA-rated bonds, A-rated bonds, and B-rated bonds. The average yield is 6.5% on AAA-bonds, 7% on A-bonds, and 9% on B-bonds. Suppose the person invests twice as much in B-bonds as in A-bonds. A system of linear equations (where x, y, and z represent the amounts invested in AAA-, A-, and B-bonds, respectively) is as follows.

$$\begin{aligned} x + y + z &= \text{(total investment)} \\ 0.065x + 0.07y + 0.09z &= \text{(annual return)} \\ 2y - z &= 0 \end{aligned}$$

Use the inverse of the coefficient matrix of this system to find the amount invested in each type of bond, with the given total investment and annual return.

43. Total investment = \$25,000
Annual return = \$1,900

44. Total investment = \$45,000
Annual return = \$3,750

45. Total investment = \$12,000
Annual return = \$835

46. Total investment = \$500,000
Annual return = \$38,000

Circuit Analysis In Exercises 47 and 48, consider the circuit in the figure. The currents I_1, I_2, and I_3 in amperes are given by the solution to the system of linear equations

$$\begin{aligned} 2I_1 + 4I_3 &= E_1 \\ I_2 + 4I_3 &= E_2 \\ I_1 + I_2 - I_3 &= 0 \end{aligned}$$

where E_1 and E_2 are voltages. Use the inverse of the coefficient matrix of this system to find the unknown currents for the given voltages.

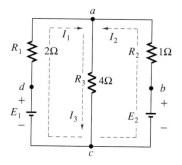

Figure for 47 and 48

47. $E_1 = 14V$, $E_2 = 28V$

48. $E_1 = 10V$, $E_2 = 10V$

SOLVING

Using Technology
to Solve Systems
of Equations

Many matrices used to solve problems in engineering, business, and science have several rows and columns. Because it is so easy to make a mistake, people who work with such matrices almost always use technology to perform the actual matrix calculations. Matrices have always been a powerful type of mathematical model—especially for dealing with problems that have a lot of data. With technology available to perform the calculations, matrices have also become a practical type of mathematical model.

EXAMPLE 1 Solving a Metallurgy Problem

Three iron alloys contain the following percents of carbon, chromium, and iron. Alloy X is a type of wrought iron, Alloy Y is a type of stainless steel, and Alloy Z is a type of cast iron. How much of the three types of alloy can you make with 15 tons of carbon, 39 tons of chromium, and 546 tons of iron?

	Alloy X	Alloy Y	Alloy Z
Carbon	1%	1%	4%
Chromium	0%	15%	3%
Iron	99%	84%	93%

Solution

Let x, y, and z represent the amount of each iron alloy. You can model the situation with the following linear system.

$$\begin{cases} 0.01x + 0.01y + 0.04z = 15 & \textit{Carbon} \\ \phantom{0.01x + {}} 0.15y + 0.03z = 39 & \textit{Chromium} \\ 0.99x + 0.84y + 0.93z = 546 & \textit{Iron} \end{cases}$$

The matrix equation that represents this system is as follows.

$$AX = B$$

$$\begin{bmatrix} 0.01 & 0.01 & 0.04 \\ 0 & 0.15 & 0.03 \\ 0.99 & 0.84 & 0.93 \end{bmatrix} \begin{bmatrix} x \\ y \\ z \end{bmatrix} = \begin{bmatrix} 15 \\ 39 \\ 546 \end{bmatrix}$$

With a graphing calculator or some other matrix operation utility, you can solve the equation as follows.

$$X = A^{-1}B = \begin{bmatrix} -25.4 & -5.4 & 1.267 \\ -6.6 & 6.733 & 0.067 \\ 33 & -0.333 & -0.333 \end{bmatrix} \begin{bmatrix} 15 \\ 39 \\ 546 \end{bmatrix} = \begin{bmatrix} 100 \\ 200 \\ 300 \end{bmatrix}$$

Thus, you can make 100 tons of Alloy X, 200 tons of Alloy Y, and 300 tons of Alloy Z. (In the solution above, we displayed the approximate values of the entries of A^{-1}. In practice, however, that is not necessary—simply instruct your calculator or computer to display $A^{-1}B$.)

EXERCISES

(*See also: Section 8.3, Exercises 39–48*)

In Exercises 1–3, use a calculator or computer (and an inverse matrix) to solve the linear system.

1. $\begin{aligned} x - y + 10z &= 2 \\ 3x \qquad + z &= 4 \\ 7x + 2y + z &= 0 \end{aligned}$

2. $\begin{aligned} 4x \qquad + 2z &= 3 \\ -x + 2y + 5z &= -1 \\ 3x + y - 7z &= 10 \end{aligned}$

3. $\begin{aligned} 2x + 6y - z &= 4 \\ 3x + y + 2z &= -4 \\ 6x - y + 3z &= 1 \end{aligned}$

4. *World Series and Super Bowl Watchers* The combined viewing audience of the 1990 World Series and Super Bowl XXV was 109.75 million. The combined female viewing audience was approximately 47.38 million. Use the bar graph shown at the right to determine the number of viewers for each sporting event. *Hint:* Let x represent the number of people who watched the 1990 World Series, and let y represent the number of people who watched Super Bowl XXV. Using the bar graph, you can write the following linear system. (*Source:* Nielsen Media Research)

$0.41x + 0.44y = 47.38$
$0.59x + 0.56y = 62.37$

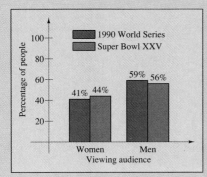

5. *Shoe Purchases* The bar graph at the right shows the percents (by age group) of the total amounts spent on three types of shoes in 1988. For instance, 18% of the total amount spent on gym shoes was spent by people in the 14–17 age group, 10% was spent by people in the 18–24 age group, and 16% was spent by people in the 25–34 age group. In 1988 the amount (in millions of dollars) spent for *all three* types of shoes is shown in the matrix. How many dollars worth of gym shoes, jogging shoes, and walking shoes were sold in 1988?

Amount (millions)	Spent on Shoes
14–17	$200.4
18–24	$154.5
25–34	$361.8

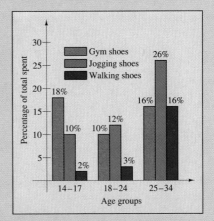

6. *Gold Alloys* "Gold" jewelry is seldom made of pure gold, because it is soft *and* it is expensive. Instead, gold is mixed with other metals to produce a harder, less expensive gold alloy. The amount of gold (by weight) in the alloy is measured in karats. A 24-karat gold mixture is 100% gold. An 18-karat gold mixture is 75% gold, and so on. Three different gold alloys contain the percents of gold, copper, and silver shown in the matrix. You have 20,144 grams of gold, 766 grams of copper, and 1,990 grams of silver. How much of the three types of alloy can you make?

Percent by Weight	Alloy X	Alloy Y	Alloy Z
Gold	94%	92%	80%
Copper	4%	2%	4%
Silver	2%	6%	16%

511

8.4 The Determinant of a Square Matrix

Introduction / Minors and Cofactors of a Square Matrix /
The Determinant of a Square Matrix / Triangular Matrices

Introduction

Every *square* matrix can be associated with a real number called its **determinant.** Determinants have many uses, and several will be discussed in this chapter.

Historically, the use of determinants arose from special number patterns that occur when solving a system of linear equations. For instance, the system

$$a_1 x + b_1 y = c_1$$
$$a_2 x + b_2 y = c_2$$

has a solution given by

$$x = \frac{c_1 b_2 - c_2 b_1}{a_1 b_2 - a_2 b_1} \quad \text{and} \quad y = \frac{a_1 c_2 - a_2 c_1}{a_1 b_2 - a_2 b_1},$$

provided $a_1 b_2 - a_2 b_1 \neq 0$. Note that the denominator of each fraction is the same, and is called the **determinant** of the coefficient matrix of the system.

Coefficient Matrix   *Determinant*

$$A = \begin{bmatrix} a_1 & b_1 \\ a_2 & b_2 \end{bmatrix} \qquad \det(A) = a_1 b_2 - a_2 b_1$$

The determinant of the matrix A can also be denoted by vertical bars on both sides of the matrix, as indicated in the following definition.

DEFINITION OF THE DETERMINANT OF A 2 × 2 MATRIX

The **determinant** of the matrix

$$A = \begin{bmatrix} a_1 & b_1 \\ a_2 & b_2 \end{bmatrix}$$

is given by

$$\det(A) = \begin{vmatrix} a_1 & b_1 \\ a_2 & b_2 \end{vmatrix} = a_1 b_2 - a_2 b_1.$$

REMARK In this text, $\det(A)$ and $|A|$ are used interchangeably to represent the determinant of A. Although vertical bars are also used to denote the absolute value of a real number, the context will show which use is intended.

A convenient method for remembering the formula for the determinant of a 2 × 2 matrix is shown in the following diagram.

$$\det(A) = \begin{vmatrix} a_1 & b_1 \\ a_2 & b_2 \end{vmatrix} = a_1 b_2 - a_2 b_1$$

Note that the determinant is given by the difference of the products of the two diagonals of the matrix.

EXAMPLE 1   The Determinant of a 2 × 2 Matrix

Find the determinant of each of the following matrices.

a. $A = \begin{bmatrix} 2 & -3 \\ 1 & 2 \end{bmatrix}$ b. $B = \begin{bmatrix} 2 & 1 \\ 4 & 2 \end{bmatrix}$ c. $C = \begin{bmatrix} 0 & 3 \\ 2 & 4 \end{bmatrix}$

Solution

a. $\det(A) = \begin{vmatrix} 2 & -3 \\ 1 & 2 \end{vmatrix} = 2(2) - 1(-3) = 4 + 3 = 7$

b. $\det(B) = \begin{vmatrix} 2 & 1 \\ 4 & 2 \end{vmatrix} = 2(2) - 4(1) = 4 - 4 = 0$

REMARK   Notice in Example 1 that the determinant of a matrix can be positive, zero, or negative.

c. $\det(C) = \begin{vmatrix} 0 & 3 \\ 2 & 4 \end{vmatrix} = 0(4) - 2(3) = 0 - 6 = -6$

The determinant of a matrix of order 1×1 is defined simply as the entry of the matrix. For instance, if $A = [-2]$, then $\det(A) = -2$.

Minors and Cofactors of a Square Matrix

To define the determinant of a square matrix of order 3×3 and higher, it is convenient to introduce the concepts of **minors** and **cofactors.**

MINORS AND COFACTORS OF A SQUARE MATRIX

If A is a square matrix, then the **minor** M_{ij} of the entry a_{ij} is the determinant of the matrix obtained by deleting the ith row and jth column of A. The **cofactor** C_{ij} of the entry a_{ij} is given by

$$C_{ij} = (-1)^{i+j} M_{ij}.$$

For example, if A is a 3×3 matrix, then the minors of a_{21} and a_{22} are shown in the following diagram.

Minor of a_{21}

$\begin{bmatrix} a_{11} & a_{12} & a_{13} \\ a_{21} & a_{22} & a_{23} \\ a_{31} & a_{32} & a_{33} \end{bmatrix}$, $M_{21} = \begin{vmatrix} a_{12} & a_{13} \\ a_{32} & a_{33} \end{vmatrix}$

Delete Row 2 and Column 1

Minor of a_{22}

$\begin{bmatrix} a_{11} & a_{12} & a_{13} \\ a_{21} & a_{22} & a_{23} \\ a_{31} & a_{32} & a_{33} \end{bmatrix}$, $M_{22} = \begin{vmatrix} a_{11} & a_{13} \\ a_{31} & a_{33} \end{vmatrix}$

Delete Row 2 and Column 2

The minors and cofactors of a matrix differ at most in sign. To obtain the cofactor C_{ij}, first find the minor M_{ij}, and then multiply by $(-1)^{i+j}$. The value of $(-1)^{i+j}$ is given by the following checkerboard pattern of $+$'s and $-$'s.

Sign Pattern for Cofactors

$$
\begin{bmatrix}
+ & - & + \\
- & + & - \\
+ & - & +
\end{bmatrix}
$$

3×3 matrix

$$
\begin{bmatrix}
+ & - & + & - \\
- & + & - & + \\
+ & - & + & - \\
- & + & - & +
\end{bmatrix}
$$

4×4 matrix

$$
\begin{bmatrix}
+ & - & + & - & + & \cdots \\
- & + & - & + & - & \cdots \\
+ & - & + & - & + & \cdots \\
- & + & - & + & - & \cdots \\
+ & - & + & - & + & \cdots \\
\vdots & \vdots & \vdots & \vdots & \vdots &
\end{bmatrix}
$$

$n \times n$ matrix

Note that *odd* positions (where $i + j$ is odd) have negative signs, and *even* positions (where $i + j$ is even) have positive signs.

EXAMPLE 2   Finding the Minors and Cofactors of a Matrix

Find all the minors and cofactors of

$$
A = \begin{bmatrix}
0 & 2 & 1 \\
3 & -1 & 2 \\
4 & 0 & 1
\end{bmatrix}.
$$

Solution

To find the minor M_{11}, delete the first row and first column of A and evaluate the determinant of the resulting matrix.

$$
\begin{bmatrix}
0 & 2 & 1 \\
3 & -1 & 2 \\
4 & 0 & 1
\end{bmatrix}, \qquad M_{11} = \begin{vmatrix} -1 & 2 \\ 0 & 1 \end{vmatrix} = -1(1) - 0(2) = -1
$$

Similarly, to find M_{21}, delete the first row and second column.

$$
\begin{bmatrix}
0 & 2 & 1 \\
3 & -1 & 2 \\
4 & 0 & 1
\end{bmatrix}, \qquad M_{12} = \begin{vmatrix} 3 & 2 \\ 4 & 1 \end{vmatrix} = 3(1) - 4(2) = -5
$$

Continuing this pattern, you obtain the following minors.

$$
\begin{array}{lll}
M_{11} = -1 & M_{12} = -5 & M_{13} = 4 \\
M_{21} = 2 & M_{22} = -4 & M_{23} = -8 \\
M_{31} = 5 & M_{32} = -3 & M_{33} = -6
\end{array}
$$

Now, to find the cofactors, combine the checkerboard pattern of signs with these minors to obtain the following.

$$
\begin{array}{lll}
C_{11} = -1 & C_{12} = 5 & C_{13} = 4 \\
C_{21} = -2 & C_{22} = -4 & C_{23} = 8 \\
C_{31} = 5 & C_{32} = 3 & C_{33} = -6
\end{array}
$$

The Determinant of a Square Matrix

Having defined the minors and cofactors of a square matrix, you are ready for a general definition of the determinant of a matrix. The definition given below is **inductive** because it uses determinants of matrices of order $n - 1$ to define the determinant of a matrix of order n.

DETERMINANT OF A SQUARE MATRIX

If A is a square matrix (of order 2×2 or greater), then the determinant of A is the sum of the entries in any row (or column) of A multiplied by their respective cofactors. For instance, expanding along the first row yields

$$|A| = a_{11}C_{11} + a_{12}C_{12} + \cdots + a_{1n}C_{1n}.$$

REMARK   Try checking that for a 2×2 matrix this definition yields $|A| = a_{11}a_{22} - a_{12}a_{21}$ as previously defined.

When you use this definition to evaluate a determinant, you are **expanding by cofactors,** as demonstrated for a 3×3 matrix in Example 3.

EXAMPLE 3   The Determinant of a Matrix of Order 3

Find the determinant of

$$A = \begin{bmatrix} 0 & 2 & 1 \\ 3 & -1 & 2 \\ 4 & 0 & 1 \end{bmatrix}.$$

Solution

Note that this matrix was given in Example 2. There you found the cofactors of the entries in the first row to be

$$C_{11} = -1, \qquad C_{12} = 5, \qquad \text{and} \qquad C_{13} = 4.$$

Therefore, by the definition of a determinant, you have the following.

$$
\begin{aligned}
|A| &= a_{11}C_{11} + a_{12}C_{12} + a_{13}C_{13} & \textit{First row expansion} \\
&= 0(-1) + 2(5) + 1(4) \\
&= 14
\end{aligned}
$$

In Example 3 you found the determinant by expanding the cofactors in the first row. You could have used any row or column. For instance, you could have expanded along the second row to obtain

$$|A| = a_{21}C_{21} + a_{22}C_{22} + a_{23}C_{23} \qquad \textit{Second row expansion}$$
$$= 3(-2) + (-1)(-4) + 2(8)$$
$$= 14$$

or along the first column to obtain

$$|A| = a_{11}C_{11} + a_{21}C_{21} + a_{31}C_{31} \qquad \textit{First column expansion}$$
$$= 0(-1) + 3(-2) + 4(5)$$
$$= 14.$$

Try some other possibilities to see that the determinant of A can be evaluated by expanding by cofactors along *any* row or column.

When expanding by cofactors you do not need to find cofactors of zero entries because zero times its cofactor is zero.

$$a_{ij}C_{ij} = (0)C_{ij} = 0$$

Thus, the row (or column) containing the most zeros is usually the best choice for expansion by cofactors. This is demonstrated in the next example.

EXAMPLE 4   The Determinant of a Matrix of Order 4

Find the determinant of

$$A = \begin{bmatrix} 1 & -2 & 3 & 0 \\ -1 & 1 & 0 & 2 \\ 0 & 2 & 0 & 3 \\ 3 & 4 & 0 & 2 \end{bmatrix}.$$

Solution

Inspecting this matrix, you see that three of the entries in the third column are zeros. Thus, you can eliminate some of the work in the expansion by using the third column.

$$|A| = 3(C_{13}) + 0(C_{23}) + 0(C_{33}) + 0(C_{43})$$

Because C_{23}, C_{33}, and C_{43} have zero coefficients, you need only find the cofactor C_{13}. To do this, delete the first row and third column of A and evaluate the determinant of the resulting matrix.

$$C_{13} = (-1)^{1+3} \begin{vmatrix} -1 & 1 & 2 \\ 0 & 2 & 3 \\ 3 & 4 & 2 \end{vmatrix}$$
$$= \begin{vmatrix} -1 & 1 & 2 \\ 0 & 2 & 3 \\ 3 & 4 & 2 \end{vmatrix}$$

Expanding by cofactors in the second row yields the following.

$$C_{13} = (0)(-1)^3 \begin{vmatrix} 1 & 2 \\ 4 & 2 \end{vmatrix} + (2)(-1)^4 \begin{vmatrix} -1 & 2 \\ 3 & 2 \end{vmatrix} + (3)(-1)^5 \begin{vmatrix} -1 & 1 \\ 3 & 4 \end{vmatrix}$$

$$= 0 + 2(1)(-8) + 3(-1)(-7)$$

$$= 5$$

Thus, you obtain

$$|A| = 3C_{13} = 3(5) = 15.$$

There is an alternative method that is commonly used for evaluating the determinant of a 3×3 matrix A. (This method works *only* for 3×3 matrices.) To apply this method, copy the first and second columns of A to form fourth and fifth columns. The determinant of A is then obtained by adding the products of the three "downward diagonals" and subtracting the products of the three "upward diagonals" as shown in the following diagram.

Subtract these three products.

Add these three products.

Thus, the determinant of the 3×3 matrix A is given by the following.

$$|A| = a_{11}a_{22}a_{33} + a_{12}a_{23}a_{31} + a_{13}a_{21}a_{32}$$
$$- a_{21}a_{22}a_{13} - a_{32}a_{23}a_{11} - a_{33}a_{21}a_{12}$$

EXAMPLE 5   The Determinant of a 3×3 Matrix

Find the determinant of

$$A = \begin{bmatrix} 0 & 2 & 1 \\ 3 & -1 & 2 \\ 4 & -4 & 1 \end{bmatrix}.$$

Solution

Since A is a 3×3 matrix, you can use the alternative procedure for finding $|A|$. Begin by recopying the first two columns and then computing the six diagonal products as follows.

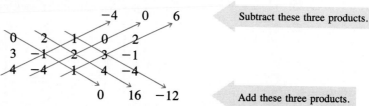

Subtract these three products.

Add these three products.

REMARK   Be sure you understand that the diagonal process illustrated in Example 5 is valid *only* for matrices of order 3×3. For matrices of higher orders, another method must be used.

Now, by adding the lower three products and subtracting the upper three products, you find the determinant of A to be

$$|A| = 0 + 16 - 12 - (-4) - 0 - 6 = 2.$$ ◢

Triangular Matrices

Evaluating determinants of matrices of order 4 or higher can be tedious. There is, however, an important exception: the determinant of a **triangular** matrix. A square matrix is **upper triangular** if it has all zero entries below its main diagonal, and **lower triangular** if it has all zero entries above its main diagonal. A matrix that is both upper and lower triangular is **diagonal.** That is, a diagonal matrix is one in which all entries above and below the main diagonal are zero.

Upper Triangular Matrix

$$\begin{bmatrix} a_{11} & a_{12} & a_{13} & \cdots & a_{1n} \\ 0 & a_{22} & a_{23} & \cdots & a_{2n} \\ 0 & 0 & a_{33} & \cdots & a_{3n} \\ \vdots & \vdots & \vdots & & \vdots \\ 0 & 0 & 0 & \cdots & a_{nn} \end{bmatrix}$$

Lower Triangular Matrix

$$\begin{bmatrix} a_{11} & 0 & 0 & \cdots & 0 \\ a_{21} & a_{22} & 0 & \cdots & 0 \\ a_{31} & a_{32} & a_{33} & \cdots & 0 \\ \vdots & \vdots & \vdots & & \vdots \\ a_{n1} & a_{n2} & a_{n3} & \cdots & a_{nn} \end{bmatrix}$$

To find the determinant of any order triangular matrix, simply form the product of the entries on the main diagonal. It is easy to see that this procedure is valid for triangular matrices of order 2×2 or 3×3. For instance, the determinant of

$$A = \begin{bmatrix} 2 & 3 & -1 \\ 0 & -1 & 2 \\ 0 & 0 & 3 \end{bmatrix}$$

can be found by expanding by the third row to obtain

$$|A| = 0 \begin{vmatrix} 3 & -1 \\ -1 & 2 \end{vmatrix} - 0 \begin{vmatrix} 2 & -1 \\ 0 & 2 \end{vmatrix} + 3 \begin{vmatrix} 2 & 3 \\ 0 & -1 \end{vmatrix} = 3(2)(-1) = -6$$

which is the product of the entries on the main diagonal.

EXAMPLE 6 The Determinant of a Triangular Matrix

Find the determinant of each of the following matrices.

a. $A = \begin{bmatrix} 2 & 0 & 0 & 0 \\ 4 & -2 & 0 & 0 \\ -5 & 6 & 1 & 0 \\ 1 & 5 & 3 & 3 \end{bmatrix}$

b. $B = \begin{bmatrix} -1 & 0 & 0 & 0 & 0 \\ 0 & 3 & 0 & 0 & 0 \\ 0 & 0 & 2 & 0 & 0 \\ 0 & 0 & 0 & 4 & 0 \\ 0 & 0 & 0 & 0 & -2 \end{bmatrix}$

Solution

a. The determinant of this triangular matrix is given by

$$|A| = (2)(-2)(1)(3) = -12.$$

b. This *diagonal* matrix is both upper and lower triangular. Therefore, its determinant is given by

$$|B| = (-1)(3)(2)(4)(-2) = 48.$$

DISCUSSION
PROBLEM

A Matrix with a Determinant of Zero

Let a, b, and c be *any* real numbers. Explain why the determinant of the following matrix is zero (regardless of the values of a, b, and c).

$$A = \begin{bmatrix} a & b & c \\ 1 & 2 & 3 \\ 2 & 4 & 6 \end{bmatrix}$$

WARM UP

The following warm-up exercises involve skills that were covered in earlier sections. You will use these skills in the exercise set for this section.

In Exercises 1–4, perform the indicated matrix operations.

1. $\begin{bmatrix} 1 & -2 \\ 0 & 3 \end{bmatrix} + \begin{bmatrix} 2 & 7 \\ 4 & -3 \end{bmatrix}$

2. $\begin{bmatrix} -2 & 5 \\ 3 & -2 \end{bmatrix} - \begin{bmatrix} 0 & -3 \\ 1 & 2 \end{bmatrix}$

3. $3\begin{bmatrix} 3 & -4 & 2 \\ 1 & 0 & -1 \\ 0 & 1 & -2 \end{bmatrix}$

4. $4\begin{bmatrix} 0 & 2 & 3 \\ -1 & 2 & 3 \\ -2 & 1 & -2 \end{bmatrix}$

In Exercises 5–10, perform the indicated arithmetic operations.

5. $[(1)(3) + (-3)(2)] - [(1)(4) + (3)(5)]$

6. $[(4)(4) + (-1)(-3)] - [(-1)(2) + (-2)(7)]$

7. $\dfrac{4(7) - 1(-2)}{(-5)(-2) - 3(4)}$

8. $\dfrac{3(6) - 2(7)}{6(-5) - 2(1)}$

9. $-5(-1)^2[6(-2) - 7(-3)]$

10. $4(-1)^3[3(6) - 2(7)]$

EXERCISES for Section 8.4

In Exercises 1–24, find the determinant of the matrix.

1. $[5]$

2. $[-8]$

3. $\begin{bmatrix} 2 & 1 \\ 3 & 4 \end{bmatrix}$

4. $\begin{bmatrix} -3 & 1 \\ 5 & 2 \end{bmatrix}$

5. $\begin{bmatrix} 5 & 2 \\ -6 & 3 \end{bmatrix}$

6. $\begin{bmatrix} 2 & -2 \\ 4 & 3 \end{bmatrix}$

7. $\begin{bmatrix} -7 & 6 \\ \frac{1}{2} & 3 \end{bmatrix}$

8. $\begin{bmatrix} 4 & -3 \\ 0 & 0 \end{bmatrix}$

9. $\begin{bmatrix} 2 & 6 \\ 0 & 3 \end{bmatrix}$

10. $\begin{bmatrix} 2 & -3 \\ -6 & 9 \end{bmatrix}$

11. $\begin{bmatrix} 2 & -1 & 0 \\ 4 & 2 & 1 \\ 4 & 2 & 1 \end{bmatrix}$

12. $\begin{bmatrix} -2 & 2 & 3 \\ 1 & -1 & 0 \\ 0 & 1 & 4 \end{bmatrix}$

13. $\begin{bmatrix} 0.3 & 0.2 & 0.2 \\ 0.2 & 0.2 & 0.2 \\ -0.4 & 0.4 & 0.3 \end{bmatrix}$

14. $\begin{bmatrix} 0.1 & 0.2 & 0.3 \\ -0.3 & 0.2 & 0.2 \\ 0.5 & 0.4 & 0.4 \end{bmatrix}$

15. $\begin{bmatrix} 1 & 4 & -2 \\ 3 & 6 & -6 \\ -2 & 1 & 4 \end{bmatrix}$

16. $\begin{bmatrix} 2 & 3 & 1 \\ 0 & 5 & -2 \\ 0 & 0 & -2 \end{bmatrix}$

17. $\begin{bmatrix} 6 & 3 & -7 \\ 0 & 0 & 0 \\ 4 & -6 & 3 \end{bmatrix}$

18. $\begin{bmatrix} 1 & 1 & 2 \\ 3 & 1 & 0 \\ -2 & 0 & 3 \end{bmatrix}$

19. $\begin{bmatrix} -1 & 2 & -5 \\ 0 & 3 & 4 \\ 0 & 0 & 3 \end{bmatrix}$

20. $\begin{bmatrix} 1 & 0 & 0 \\ -4 & -1 & 0 \\ 5 & 1 & 5 \end{bmatrix}$

21. $\begin{bmatrix} -1 & 0 & 0 & 0 \\ 2 & 3 & 0 & 0 \\ -4 & 5 & 3 & 0 \\ 1 & 0 & 2 & 2 \end{bmatrix}$

22. $\begin{bmatrix} -2 & 0 & 0 & 0 & 0 \\ 0 & 3 & 0 & 0 & 0 \\ 0 & 0 & -1 & 0 & 0 \\ 0 & 0 & 0 & 2 & 0 \\ 0 & 0 & 0 & 0 & -4 \end{bmatrix}$

23. $\begin{bmatrix} x & y & 1 \\ -2 & -2 & 1 \\ 1 & 5 & 1 \end{bmatrix}$

24. $\begin{bmatrix} 3-\lambda & 2 \\ 4 & 1-\lambda \end{bmatrix}$

In Exercises 25–28, find (a) all minors and (b) cofactors for the given matrix.

25. $\begin{bmatrix} 3 & 4 \\ 2 & -5 \end{bmatrix}$

26. $\begin{bmatrix} 11 & 0 \\ -3 & 2 \end{bmatrix}$

27. $\begin{bmatrix} 3 & -2 & 8 \\ 3 & 2 & -6 \\ -1 & 3 & 6 \end{bmatrix}$

28. $\begin{bmatrix} -2 & 9 & 4 \\ 7 & -6 & 0 \\ 6 & 7 & -6 \end{bmatrix}$

In Exercises 29–34, find the determinant of the matrix by the method of expansion by cofactors. Expand using the indicated row or column.

29. $\begin{bmatrix} -3 & 2 & 1 \\ 4 & 5 & 6 \\ 2 & -3 & 1 \end{bmatrix}$
(a) Row 1
(b) Column 2

30. $\begin{bmatrix} -3 & 4 & 2 \\ 6 & 3 & 1 \\ 4 & -7 & -8 \end{bmatrix}$
(a) Row 2
(b) Column 3

31. $\begin{bmatrix} 5 & 0 & -3 \\ 0 & 12 & 4 \\ 1 & 6 & 3 \end{bmatrix}$
(a) Row 2
(b) Column 2

32. $\begin{bmatrix} 10 & -5 & 5 \\ 30 & 0 & 10 \\ 0 & 10 & 1 \end{bmatrix}$
(a) Row 3
(b) Column 1

33. $\begin{bmatrix} 6 & 0 & -3 & 5 \\ 4 & 13 & 6 & -8 \\ -1 & 0 & 7 & 4 \\ 8 & 6 & 0 & 2 \end{bmatrix}$
(a) Row 2
(b) Column 2

34. $\begin{bmatrix} 10 & 8 & 3 & -7 \\ 4 & 0 & 5 & -6 \\ 0 & 3 & 2 & 7 \\ 1 & 0 & -3 & 2 \end{bmatrix}$
(a) Row 3
(b) Column 1

In Exercises 35–44, find the determinant of the matrix.

35. $\begin{bmatrix} 1 & 4 & -2 \\ 3 & 2 & 0 \\ -1 & 4 & 3 \end{bmatrix}$

36. $\begin{bmatrix} 2 & -1 & 3 \\ 1 & 4 & 4 \\ 1 & 0 & 2 \end{bmatrix}$

37. $\begin{bmatrix} 2 & 4 & 6 \\ 0 & 3 & 1 \\ 0 & 0 & -5 \end{bmatrix}$

38. $\begin{bmatrix} -3 & 0 & 0 \\ 7 & 11 & 0 \\ 1 & 2 & 2 \end{bmatrix}$

39. $\begin{bmatrix} 3 & 6 & -5 & 4 \\ -2 & 0 & 6 & 0 \\ 1 & 1 & 2 & 2 \\ 0 & 3 & -1 & -1 \end{bmatrix}$

40. $\begin{bmatrix} 2 & 6 & 6 & 2 \\ 2 & 7 & 3 & 6 \\ 1 & 5 & 0 & 1 \\ 3 & 7 & 0 & 7 \end{bmatrix}$

41. $\begin{bmatrix} 5 & 3 & 0 & 6 \\ 4 & 6 & 4 & 12 \\ 0 & 2 & -3 & 4 \\ 0 & 1 & -2 & 2 \end{bmatrix}$

42. $\begin{bmatrix} 1 & 4 & 3 & 2 \\ -5 & 6 & 2 & 1 \\ 0 & 0 & 0 & 0 \\ 3 & -2 & 1 & 5 \end{bmatrix}$

43.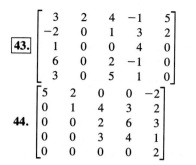
$$\begin{bmatrix} 3 & 2 & 4 & -1 & 5 \\ -2 & 0 & 1 & 3 & 2 \\ 1 & 0 & 0 & 4 & 0 \\ 6 & 0 & 2 & -1 & 0 \\ 3 & 0 & 5 & 1 & 0 \end{bmatrix}$$

44. $$\begin{bmatrix} 5 & 2 & 0 & 0 & -2 \\ 0 & 1 & 4 & 3 & 2 \\ 0 & 0 & 2 & 6 & 3 \\ 0 & 0 & 3 & 4 & 1 \\ 0 & 0 & 0 & 0 & 2 \end{bmatrix}$$

In Exercises 45 and 46, solve for x.

45. $\begin{vmatrix} x - 1 & 2 \\ 3 & x - 2 \end{vmatrix} = 0$ **46.** $\begin{vmatrix} x - 2 & -1 \\ -3 & x \end{vmatrix} = 0$

In Exercises 47–52, evaluate the determinant of the matrix where the entries are functions. Determinants of this type occur in calculus.

47. $\begin{bmatrix} 4u & -1 \\ -1 & 2v \end{bmatrix}$ **48.** $\begin{bmatrix} 3x^2 & -3y^2 \\ 1 & 1 \end{bmatrix}$

49. $\begin{bmatrix} e^{2x} & e^{3x} \\ 2e^{2x} & 3e^{3x} \end{bmatrix}$ **50.** $\begin{bmatrix} e^{-x} & xe^{-x} \\ -e^{-x} & (1 - x)e^{-x} \end{bmatrix}$

51. $\begin{bmatrix} x & \ln x \\ 1 & 1/x \end{bmatrix}$ **52.** $\begin{bmatrix} x & x \ln x \\ 1 & 1 + \ln x \end{bmatrix}$

8.5 Properties of Determinants

Elementary Row Operations / Determinants and the Inverse of a Matrix

Which of the following two determinants is easier to evaluate?

$$|A| = \begin{vmatrix} 1 & -2 & 3 & 1 \\ 4 & -6 & 3 & 2 \\ -2 & 4 & -9 & -3 \\ 3 & -6 & 9 & 2 \end{vmatrix} \quad \text{or} \quad |B| = \begin{vmatrix} 1 & -2 & 3 & 1 \\ 0 & 2 & -9 & -2 \\ 0 & 0 & -3 & -1 \\ 0 & 0 & 0 & -1 \end{vmatrix}$$

From what you now know about the determinant of a triangular matrix, it is clear that the second determinant is *much* easier to evaluate. Its determinant is simply the product of the entries on the main diagonal. That is,

$$|B| = (1)(2)(-3)(-1) = 6.$$

On the other hand, using expansion by cofactors to evaluate the first determinant is messy. For instance, if you expand by cofactors across the first row, you have the following.

$$|A| = 1 \begin{vmatrix} -6 & 3 & 2 \\ 4 & -9 & -3 \\ -6 & 9 & 2 \end{vmatrix} + 2 \begin{vmatrix} 4 & 3 & 2 \\ -2 & -9 & -3 \\ 3 & 9 & 2 \end{vmatrix} + 3 \begin{vmatrix} 4 & -6 & 2 \\ -2 & 4 & -3 \\ 3 & -6 & 2 \end{vmatrix} - 1 \begin{vmatrix} 4 & -6 & 3 \\ -2 & 4 & -9 \\ 3 & -6 & 9 \end{vmatrix}$$

Evaluating the determinants of these four 3 × 3 matrices produces

$$|A| = (1)(-60) + (2)(39) + (3)(-10) - (1)(-18) = 6.$$

Elementary Row Operations

It is not coincidental that the determinants of A and B are the same value. In fact, we obtained matrix B by performing elementary row operations on matrix A. (Try verifying this.) In this section, we consider the effect of elementary row (and column) operations on the value of a determinant. Here is an example.

> **EXAMPLE 1** **The Effect of Elementary Row Operations on a Determinant**

a. The matrix B was obtained from A by interchanging the rows of A.

$$|A| = \begin{vmatrix} 2 & -3 \\ 1 & 4 \end{vmatrix} = 11 \quad \text{and} \quad |B| = \begin{vmatrix} 1 & 4 \\ 2 & -3 \end{vmatrix} = -11$$

b. The matrix B was obtained from A by adding -2 times the first row of A to the second row of A.

$$|A| = \begin{vmatrix} 1 & -3 \\ 2 & 4 \end{vmatrix} = 10 \quad \text{and} \quad |B| = \begin{vmatrix} 1 & -3 \\ 0 & 10 \end{vmatrix} = 10$$

c. The matrix B was obtained from A by multiplying the first row of A by $\frac{1}{2}$.

$$|A| = \begin{vmatrix} 2 & -8 \\ -2 & 9 \end{vmatrix} = 2 \quad \text{and} \quad |B| = \begin{vmatrix} 1 & -4 \\ -2 & 9 \end{vmatrix} = 1$$

In Example 1, you saw that interchanging two rows of the matrix changed the sign of its determinant. Adding a multiple of one row to another did not change the determinant. Finally, multiplying a row by a nonzero constant multiplied the determinant by that same constant. The following theorem generalizes these observations.

ELEMENTARY ROW OPERATIONS AND DETERMINANTS

Let A and B be square matrices.

1. If B is obtained from A by interchanging two rows of A, then
$$|B| = -|A|.$$

2. If B is obtained from A by adding a multiple of a row of A to another row of A, then
$$|B| = |A|.$$

3. If B is obtained from A by multiplying a row of A to a nonzero constant C, then
$$|B| = c|A|.$$

REMARK Note that the third property allows us to take a common factor out of a row. For instance,

$$\begin{vmatrix} 2 & 4 \\ 1 & 3 \end{vmatrix} = 2 \begin{vmatrix} 1 & 2 \\ 1 & 3 \end{vmatrix}.$$

Factor 2 out of the first row

Moreover, the theorem is true if the word *row* is replaced by the word *column*. ◢

This theorem provides a practical way to evaluate determinants (especially determinants of large matrices). To find the determinant of a matrix A, use elementary row operations to obtain a triangular matrix B that is row equivalent to A. At each step in the elimination process, incorporate the effect of the elementary row operation on the determinant. Finally, find the determinant of B by forming the product of the entries on its main diagonal. This process is demonstrated in the next example.

EXAMPLE 2   Evaluating a Determinant

Find the determinant of

$$A = \begin{bmatrix} 2 & -3 & 10 \\ 1 & 2 & -2 \\ 0 & 1 & -3 \end{bmatrix}.$$

Solution

Using elementary row operations, rewrite A in triangular form as follows.

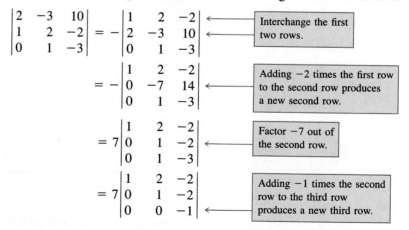

$$\begin{vmatrix} 2 & -3 & 10 \\ 1 & 2 & -2 \\ 0 & 1 & -3 \end{vmatrix} = - \begin{vmatrix} 1 & 2 & -2 \\ 2 & -3 & 10 \\ 0 & 1 & -3 \end{vmatrix} \quad \begin{array}{|l|} \hline \text{Interchange the first} \\ \text{two rows.} \\ \hline \end{array}$$

$$= - \begin{vmatrix} 1 & 2 & -2 \\ 0 & -7 & 14 \\ 0 & 1 & -3 \end{vmatrix} \quad \begin{array}{|l|} \hline \text{Adding } -2 \text{ times the first row} \\ \text{to the second row produces} \\ \text{a new second row.} \\ \hline \end{array}$$

$$= 7 \begin{vmatrix} 1 & 2 & -2 \\ 0 & 1 & -2 \\ 0 & 1 & -3 \end{vmatrix} \quad \begin{array}{|l|} \hline \text{Factor } -7 \text{ out of} \\ \text{the second row.} \\ \hline \end{array}$$

$$= 7 \begin{vmatrix} 1 & 2 & -2 \\ 0 & 1 & -2 \\ 0 & 0 & -1 \end{vmatrix} \quad \begin{array}{|l|} \hline \text{Adding } -1 \text{ times the second} \\ \text{row to the third row} \\ \text{produces a new third row.} \\ \hline \end{array}$$

Now, because the final matrix is triangular, you may conclude that the determinant is

$$|A| = 7(1)(1)(-1) = -7.$$ ◢

Conditions that Yield a Zero Determinant

If A is a square matrix and any one of the following conditions is true, then $|A| = 0$.

1. An entire row (or an entire column) is zero.
2. Two rows (or two columns) are equal.
3. One row (or column) is a multiple of another row (or column).

Recognizing the conditions listed in this theorem can make evaluating a determinant much easier. For instance, consider the following three evaluations.

$$\begin{vmatrix} 0 & 0 & 0 \\ 2 & 4 & -5 \\ 3 & -5 & 2 \end{vmatrix} = 0 \qquad \begin{vmatrix} 1 & -2 & 4 \\ 0 & 1 & 2 \\ 1 & -2 & 4 \end{vmatrix} = 0 \qquad \begin{vmatrix} 1 & 2 & -3 \\ 2 & -1 & -6 \\ -2 & 0 & 6 \end{vmatrix} = 0$$

First row has all zeros.

First and third rows are the same.

Third column is a multiple of first column.

Do not conclude that the three conditions listed in this theorem are the *only* conditions that produce a determinant of zero. The theorem is often used indirectly. That is, you can begin with a matrix that does not satisfy any of the three conditions listed in the theorem, and through elementary row or column operations obtain a matrix that does satisfy one of the conditions. Then you may conclude that the original matrix has a determinant of zero.

EXAMPLE 3   A Matrix with a Zero Determinant

Find the determinant of

$$A = \begin{bmatrix} 1 & 4 & 1 \\ 2 & -1 & 0 \\ 0 & 18 & 4 \end{bmatrix}.$$

Solution

Adding -2 times the first row to the second row produces

$$|A| = \begin{vmatrix} 1 & 4 & 1 \\ 2 & -1 & 0 \\ 0 & 18 & 4 \end{vmatrix} = \begin{vmatrix} 1 & 4 & 1 \\ 0 & -9 & -2 \\ 0 & 18 & 4 \end{vmatrix}.$$

Because the second and third rows are multiples of each other, you conclude that the determinant is zero.

Determinants and the Inverse of a Matrix

We saw in Section 8.3 that some square matrices are not invertible. However, it can be difficult to tell simply by inspection whether a matrix possesses an inverse. For instance, can you tell which of the following two matrices is invertible?

$$A = \begin{bmatrix} 0 & 2 & -1 \\ 3 & -2 & 1 \\ 3 & 2 & -1 \end{bmatrix} \quad \text{or} \quad B = \begin{bmatrix} 0 & 2 & -1 \\ 3 & 2 & 1 \\ 3 & 2 & 1 \end{bmatrix}$$

The following theorem shows how determinants can be used to classify square matrices as invertible or noninvertible.

DETERMINANT OF AN INVERTIBLE MATRIX

A square matrix A is invertible (nonsingular) if and only if

$$|A| \neq 0.$$

EXAMPLE 4   Classifying Square Matrices as Singular or Nonsingular

Which of the following matrices possesses an inverse?

a. $\begin{bmatrix} 0 & 2 & -1 \\ 3 & -2 & 1 \\ 3 & 2 & -1 \end{bmatrix}$ b. $\begin{bmatrix} 0 & 2 & -1 \\ 3 & -2 & 1 \\ 3 & 2 & 1 \end{bmatrix}$

If its equal to zero, it doesn't have an inverse

Solution

a. Because

$$\begin{vmatrix} 0 & 2 & -1 \\ 3 & -2 & 1 \\ 3 & 2 & -1 \end{vmatrix} = 0$$

this matrix has no inverse (it is singular).

b. Because

$$\begin{vmatrix} 0 & 2 & -1 \\ 3 & -2 & 1 \\ 3 & 2 & 1 \end{vmatrix} = -12 \neq 0$$

this matrix has an inverse (it is nonsingular).

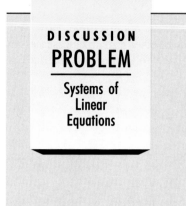

DISCUSSION
PROBLEM
Systems of Linear Equations

In Example 4, you looked at two 3×3 matrices, one that had an inverse and one that didn't. Two systems of linear equations that have these matrices as coefficient matrices are as follows.

1. $\begin{aligned} 2y - z &= -5 \\ 3x - 2y + z &= 8 \\ 3x + 2y - z &= -2 \end{aligned}$ 2. $\begin{aligned} 2y - z &= -5 \\ 3x - 2y + z &= 8 \\ 3x + 2y + z &= 0 \end{aligned}$

One of these two systems has a single solution and one has either infinitely many solutions or no solution. How can you use the result of Example 4 to determine which is which? Discuss, or write a short paragraph describing, the relationship between determinants and the number of solutions of a system of linear equations.

WARM UP

The following warm-up exercises involve skills that were covered in earlier sections. You will use these skills in the exercise set for this section.

In Exercises 1–4, write the matrices in row-echelon form.

1. $\begin{bmatrix} 2 & -6 \\ 5 & 20 \end{bmatrix}$ 2. $\begin{bmatrix} -7 & 21 \\ 3 & 5 \end{bmatrix}$

3. $\begin{bmatrix} 1 & 3 & 4 \\ 0 & 1 & 1 \\ 2 & 4 & 6 \end{bmatrix}$ 4. $\begin{bmatrix} 4 & 8 & 16 \\ 3 & -1 & 2 \\ -2 & 10 & 12 \end{bmatrix}$

In Exercises 5–10, evaluate the determinant.

5. $\begin{vmatrix} 4 & -3 \\ -2 & 1 \end{vmatrix}$ 6. $\begin{vmatrix} 10 & -20 \\ -1 & 2 \end{vmatrix}$

7. $\begin{vmatrix} 4 & 0 \\ -3 & -2 \end{vmatrix}$ 8. $\begin{vmatrix} x & x^2 \\ 1 & 2x \end{vmatrix}$

9. $\begin{vmatrix} 4 & 0 & -2 \\ 3 & 1 & 2 \\ -8 & 0 & 6 \end{vmatrix}$ 10. $\begin{vmatrix} 3 & 2 & 5 \\ 0 & 0 & -4 \\ -6 & 1 & 1 \end{vmatrix}$

EXERCISES for Section 8.5

In Exercises 1–14, state the property of determinants that verifies the equation.

1. $\begin{vmatrix} 2 & -6 \\ 1 & -3 \end{vmatrix} = 0$

2. $\begin{vmatrix} -4 & 5 \\ 12 & -15 \end{vmatrix} = 0$

3. $\begin{vmatrix} 1 & 4 & 2 \\ 0 & 0 & 0 \\ 5 & 6 & -7 \end{vmatrix} = 0$

4. $\begin{vmatrix} -4 & 3 & 2 \\ 8 & 0 & 0 \\ -4 & 3 & 2 \end{vmatrix} = 0$

5. $\begin{vmatrix} 1 & 3 & 4 \\ -7 & 2 & -5 \\ 6 & 1 & 2 \end{vmatrix} = -\begin{vmatrix} 1 & 4 & 3 \\ -7 & -5 & 2 \\ 6 & 2 & 1 \end{vmatrix}$

6. $\begin{vmatrix} 1 & 3 & 4 \\ -2 & 2 & 0 \\ 1 & 6 & 2 \end{vmatrix} = -\begin{vmatrix} 1 & 6 & 2 \\ -2 & 2 & 0 \\ 1 & 3 & 4 \end{vmatrix}$

7. $\begin{vmatrix} 5 & 10 & 15 \\ 2 & -3 & 4 \\ 2 & -7 & 1 \end{vmatrix} = 5\begin{vmatrix} 1 & 2 & 3 \\ 2 & -3 & 4 \\ 2 & -7 & 1 \end{vmatrix}$

8. $\begin{vmatrix} 1 & 8 & -3 \\ 3 & -12 & 6 \\ 7 & 4 & 9 \end{vmatrix} = 12\begin{vmatrix} 1 & 2 & -1 \\ 3 & -3 & 2 \\ 7 & 1 & 3 \end{vmatrix}$

9. $\begin{vmatrix} 5 & 0 & 10 \\ 25 & -30 & 40 \\ -15 & 5 & 20 \end{vmatrix} = 5^3\begin{vmatrix} 1 & 0 & 2 \\ 5 & -6 & 8 \\ -3 & 1 & 4 \end{vmatrix}$

10. $\begin{vmatrix} 6 & 0 & 0 \\ 0 & 6 & 0 \\ 0 & 0 & 6 \end{vmatrix} = 6^3\begin{vmatrix} 1 & 0 & 0 \\ 0 & 1 & 0 \\ 0 & 0 & 1 \end{vmatrix}$

11. $\begin{vmatrix} 2 & -3 \\ 8 & 7 \end{vmatrix} = \begin{vmatrix} 2 & -3 \\ 0 & 19 \end{vmatrix}$

12. $\begin{vmatrix} 1 & -3 \\ 5 & 2 \end{vmatrix} = \begin{vmatrix} 1 & -3 \\ 0 & 17 \end{vmatrix}$

13. $\begin{vmatrix} 3 & 2 & 4 \\ -2 & 1 & 5 \\ 5 & -7 & -20 \end{vmatrix} = \begin{vmatrix} 7 & 2 & -6 \\ 0 & 1 & 0 \\ -9 & -7 & 15 \end{vmatrix}$

14. $\begin{vmatrix} 5 & 4 & 2 \\ 2 & -3 & 4 \\ 7 & 6 & 3 \end{vmatrix} = \begin{vmatrix} 1 & 10 & -6 \\ 2 & -3 & 4 \\ 7 & 6 & 3 \end{vmatrix}$

In Exercises 15–30, use elementary row (or column) operations as aids for evaluating the determinant.

15. $\begin{vmatrix} 1 & 2 & 5 \\ 1 & 4 & 2 \\ 0 & 3 & -4 \end{vmatrix}$

16. $\begin{vmatrix} 1 & 7 & -3 \\ 1 & 3 & 1 \\ 4 & 8 & 1 \end{vmatrix}$

17. $\begin{vmatrix} 3 & -1 & -3 \\ -1 & -4 & -2 \\ 3 & -1 & -1 \end{vmatrix}$

18. $\begin{vmatrix} 4 & 3 & -2 \\ 5 & 4 & 1 \\ -2 & 3 & 4 \end{vmatrix}$

19. $\begin{vmatrix} 3 & 8 & -7 \\ 0 & -5 & 4 \\ 8 & 1 & 6 \end{vmatrix}$

20. $\begin{vmatrix} 5 & -8 & 0 \\ 9 & 7 & 4 \\ -8 & 7 & 1 \end{vmatrix}$

21. $\begin{vmatrix} 2 & -1 & 3 \\ 1 & 2 & -1 \\ 3 & -4 & 7 \end{vmatrix}$

22. $\begin{vmatrix} 2 & 0 & 1 \\ 4 & -4 & 0 \\ -1 & 5 & 2 \end{vmatrix}$

23. $\begin{vmatrix} 7 & 0 & -14 \\ -2 & 5 & 4 \\ -6 & 2 & 12 \end{vmatrix}$

24. $\begin{vmatrix} 3 & 0 & 0 \\ -2 & 5 & 0 \\ 12 & 5 & 7 \end{vmatrix}$

25. $\begin{vmatrix} 4 & -8 & 5 & 0 \\ 8 & -5 & 3 & 0 \\ 8 & 5 & 2 & 0 \\ 1 & 7 & -5 & 1 \end{vmatrix}$

26. $\begin{vmatrix} 4 & -7 & 9 & 1 \\ 6 & 2 & 7 & 0 \\ 3 & 6 & -3 & 3 \\ 0 & 7 & 4 & -1 \end{vmatrix}$

27. $\begin{vmatrix} 0 & -3 & 8 & 2 \\ 8 & 1 & -1 & 6 \\ -4 & 6 & 0 & 9 \\ -7 & 0 & 0 & 14 \end{vmatrix}$

28. $\begin{vmatrix} 1 & -1 & 8 & 4 \\ 2 & 6 & 0 & -4 \\ 2 & 0 & 2 & 6 \\ 0 & 2 & 8 & 0 \end{vmatrix}$

29. $\begin{vmatrix} 3 & -2 & 4 & 3 & 1 \\ -1 & 0 & 2 & 1 & 0 \\ 5 & -1 & 0 & 3 & 2 \\ 4 & 7 & -8 & 0 & 0 \\ 1 & 2 & 3 & 0 & 2 \end{vmatrix}$

30. $\begin{vmatrix} 4 & 2 & -1 & 0 & 3 \\ 0 & 1 & 1 & 2 & -3 \\ 0 & 0 & -2 & 8 & 12 \\ 0 & 0 & 0 & 5 & 13 \\ 0 & 0 & 0 & 0 & 3 \end{vmatrix}$

In Exercises 31–36, use a determinant to ascertain whether the matrix is invertible.

31. $\begin{bmatrix} 5 & 4 \\ 10 & 8 \end{bmatrix}$

32. $\begin{bmatrix} 3 & -6 \\ 4 & 2 \end{bmatrix}$

33. $\begin{bmatrix} 14 & 7 & 0 \\ 2 & 3 & 0 \\ 1 & -5 & 2 \end{bmatrix}$

34. $\begin{bmatrix} 1 & 0 & 4 \\ 0 & 6 & 3 \\ 2 & -1 & 4 \end{bmatrix}$

35. $\begin{bmatrix} \frac{1}{2} & \frac{3}{2} & 2 \\ \frac{2}{3} & -\frac{1}{3} & 0 \\ 1 & 1 & 1 \end{bmatrix}$

36. $\begin{bmatrix} 2 & -1 & 6 \\ 1 & -3 & 4 \\ 4 & -2 & 12 \end{bmatrix}$

In Exercises 37 and 38, find the value(s) of k such that A is singular.

37. $\begin{bmatrix} k-1 & 3 \\ 2 & k-2 \end{bmatrix}$

38. $\begin{bmatrix} 1 & 0 & 3 \\ 2 & -1 & 0 \\ 4 & 2 & k \end{bmatrix}$

In Exercises 39–42, verify the equation.

39. $\begin{vmatrix} w & x \\ y & z \end{vmatrix} = -\begin{vmatrix} y & z \\ w & x \end{vmatrix}$

40. $\begin{vmatrix} w & cx \\ y & cz \end{vmatrix} = c\begin{vmatrix} w & x \\ y & z \end{vmatrix}$

41. $\begin{vmatrix} w & x \\ y & z \end{vmatrix} = \begin{vmatrix} w & x + cw \\ y & z + cy \end{vmatrix}$

42. $\begin{vmatrix} w & x \\ cw & cx \end{vmatrix} = 0$

In Exercises 43 and 44, evaluate the determinant to verify the equation.

43. $\begin{vmatrix} 1 & x & x^2 \\ 1 & y & y^2 \\ 1 & z & z^2 \end{vmatrix} = (y - x)(z - x)(z - y)$

44. $\begin{vmatrix} a + b & a & a \\ a & a + b & a \\ a & a & a + b \end{vmatrix} = b^2(3a + b)$

In Exercises 45–48, find (a) $|A|$, (b) $|B|$, (c) AB, and (d) $|AB|$.

45. $A = \begin{bmatrix} -1 & 0 \\ 0 & 3 \end{bmatrix}$, $B = \begin{bmatrix} 2 & 0 \\ 0 & -1 \end{bmatrix}$

46. $A = \begin{bmatrix} -2 & 1 \\ 4 & -2 \end{bmatrix}$, $B = \begin{bmatrix} 1 & 2 \\ 0 & -1 \end{bmatrix}$

47. $A = \begin{bmatrix} -1 & 2 & 1 \\ 1 & 0 & 1 \\ 0 & 1 & 0 \end{bmatrix}$, $B = \begin{bmatrix} -1 & 0 & 0 \\ 0 & 2 & 0 \\ 0 & 0 & 3 \end{bmatrix}$

48. $A = \begin{bmatrix} 2 & 0 & 1 \\ 1 & -1 & 2 \\ 3 & 1 & 0 \end{bmatrix}$, $B = \begin{bmatrix} 2 & -1 & 4 \\ 0 & 1 & 3 \\ 3 & -2 & 1 \end{bmatrix}$

49. Find the square matrices A and B to demonstrate that
$$|A + B| \neq |A| + |B|.$$

8.6 Applications of Determinants and Matrices

Cramer's Rule / Area of a Triangle / Lines in the Plane / Cryptography

Cramer's Rule

So far, we have discussed three methods for solving a system of linear equations: substitution, elimination (with equations), and elimination (with matrices). We now look at one more method, **Cramer's Rule,** named after Gabriel Cramer (1704–1752). This rule uses determinants to write the solution of a system of linear equations. To see how Cramer's Rule works, take another look at the solution described at the beginning of Section 8.4. There, we pointed out that the system

$$a_1x + b_1y = c_1$$
$$a_2x + b_2y = c_2$$

has a solution given by

$$x = \frac{c_1b_2 - c_2b_1}{a_1b_2 - a_2b_1} \quad \text{and} \quad y = \frac{a_1c_2 - a_2c_1}{a_1b_2 - a_2b_1}$$

provided $a_1b_2 - a_2b_1 \neq 0$. Each numerator and denominator in this solution can be expressed as a determinant, as follows.

$$x = \frac{c_1b_2 - c_2b_1}{a_1b_2 - a_2b_1} = \frac{\begin{vmatrix} c_1 & b_1 \\ c_2 & b_2 \end{vmatrix}}{\begin{vmatrix} a_1 & b_1 \\ a_2 & b_2 \end{vmatrix}}, \quad y = \frac{a_1c_2 - a_2c_1}{a_1b_2 - a_2b_1} = \frac{\begin{vmatrix} a_1 & c_1 \\ a_2 & c_2 \end{vmatrix}}{\begin{vmatrix} a_1 & b_1 \\ a_2 & b_2 \end{vmatrix}}$$

Relative to the original system, the denominator for x and y is simply the determinant of the *coefficient* matrix of the system. We denote this determinant by D. The numerators for x and y are denoted by D_x and D_y, respectively. They are formed by using the column of constants as replacements for the coefficients of x and y, as follows.

$$
\begin{array}{cccc}
\text{Coefficient} & & & \\
\underline{\text{Matrix}} & \underline{D} & \underline{D_x} & \underline{D_y} \\
\begin{bmatrix} a_1 & b_1 \\ a_2 & b_2 \end{bmatrix} & \begin{vmatrix} a_1 & b_1 \\ a_2 & b_2 \end{vmatrix} & \begin{vmatrix} c_1 & b_1 \\ c_2 & b_2 \end{vmatrix} & \begin{vmatrix} a_1 & c_1 \\ a_2 & c_2 \end{vmatrix}
\end{array}
$$

EXAMPLE 1   Using Cramer's Rule for a 2 × 2 System

Use Cramer's Rule to solve the following system of linear equations.

$$4x - 2y = 10$$
$$3x - 5y = 11$$

Solution

To begin, find the determinant of the coefficient matrix.

$$D = \begin{vmatrix} 4 & -2 \\ 3 & -5 \end{vmatrix} = -20 - (-6) = -14$$

Because this determinant is not zero, you can apply Cramer's Rule to find the solution, as follows.

$$x = \frac{D_x}{D} = \frac{\begin{vmatrix} 10 & -2 \\ 11 & -5 \end{vmatrix}}{-14} = \frac{(-50) - (-22)}{-14} = \frac{-28}{-14} = 2$$

$$y = \frac{D_y}{D} = \frac{\begin{vmatrix} 4 & 10 \\ 3 & 11 \end{vmatrix}}{-14} = \frac{44 - 30}{-14} = \frac{14}{-14} = -1$$

Therefore, the solution is $x = 2$ and $y = -1$. Check this solution in the original system of equations.

Cramer's Rule generalizes easily to systems of n equations in n variables. The value of each variable is given as the quotient of two determinants. The denominator is the determinant of the coefficient matrix, and the numerator is the determinant of the matrix formed by replacing the column corresponding

to the variable (being solved for) with the column representing the constants. For example, the solution for x_3 in the system

$$a_{11}x_1 + a_{12}x_2 + a_{13}x_3 = b_1$$
$$a_{21}x_1 + a_{22}x_2 + a_{23}x_3 = b_2$$
$$a_{31}x_1 + a_{32}x_2 + a_{33}x_3 = b_3$$

is given by

$$x_3 = \frac{|A_3|}{|A|} = \frac{\begin{vmatrix} a_{11} & a_{12} & b_1 \\ a_{21} & a_{22} & b_2 \\ a_{31} & a_{32} & b_3 \end{vmatrix}}{\begin{vmatrix} a_{11} & a_{12} & a_{13} \\ a_{21} & a_{22} & a_{23} \\ a_{31} & a_{32} & a_{33} \end{vmatrix}}.$$

CRAMER'S RULE

If a system of n linear equations in n variables has a coefficient matrix A with a *nonzero* determinant $|A|$, then the solution of the system is given by

$$x_1 = \frac{|A_1|}{|A|}, \quad x_2 = \frac{|A_2|}{|A|}, \quad \ldots, \quad x_n = \frac{|A_n|}{|A|}$$

where the ith column of A_i is the column of constants in the system of equations. If the coefficient matrix is zero, then the system has either no solution *or* infinitely many solutions.

EXAMPLE 2 Using Cramer's Rule for a 3 × 3 System

Use Cramer's Rule to solve the following system of linear equations.

$$-x + 2y - 3z = 1$$
$$2x \qquad + z = 0$$
$$3x - 4y + 4z = 2$$

Solution

To begin, find the determinant of the coefficient matrix.

$$D = \begin{vmatrix} -1 & 2 & -3 \\ 2 & 0 & 1 \\ 3 & -4 & 4 \end{vmatrix} = -(2)\begin{vmatrix} 2 & -3 \\ -4 & 4 \end{vmatrix} + (0)\begin{vmatrix} -1 & -3 \\ 3 & 4 \end{vmatrix} - (1)\begin{vmatrix} -1 & 2 \\ 3 & -4 \end{vmatrix}$$

$$= -2(-4) + 0 - (-2)$$

$$= 10$$

Because this determinant is not zero, you can apply Cramer's Rule to find the solution, as follows.

$$x = \frac{D_x}{D} = \frac{\begin{vmatrix} 1 & 2 & -3 \\ 0 & 0 & 1 \\ 2 & -4 & 4 \end{vmatrix}}{10} = \frac{8}{10} = \frac{4}{5}$$

$$y = \frac{D_y}{D} = \frac{\begin{vmatrix} -1 & 1 & -3 \\ 2 & 0 & 1 \\ 3 & 2 & 4 \end{vmatrix}}{10} = \frac{-15}{10} = -\frac{3}{2}$$

$$z = \frac{D_z}{D} = \frac{\begin{vmatrix} -1 & 2 & 1 \\ 2 & 0 & 0 \\ 3 & -4 & 2 \end{vmatrix}}{10} = \frac{-16}{10} = -\frac{8}{5}$$

Therefore, the solution is $\left(\frac{4}{5}, -\frac{3}{2}, -\frac{8}{5}\right)$. Check this solution in the original system of equations. ◀

When using Cramer's Rule, remember that the method *does not* apply if the determinant of the coefficient matrix is zero.

Area of a Triangle

Throughout this chapter we have discussed several applications of matrices and determinants that involve a system of linear equations. In this section you will look at some *other* types of applications involving determinants and matrices.

The first gives a formula for finding the area of a triangle whose vertices are given by three points in a rectangular coordinate system.

AREA OF A TRIANGLE

The area of a triangle with vertices (x_1, y_1), (x_2, y_2), and (x_3, y_3) is given by

$$\text{Area} = \pm\frac{1}{2}\begin{vmatrix} x_1 & y_1 & 1 \\ x_2 & y_2 & 1 \\ x_3 & y_3 & 1 \end{vmatrix}$$

where the symbol $(\pm)$ indicates that the appropriate sign should be chosen to yield a positive area.

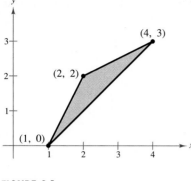

FIGURE 8.1

REMARK To see the benefit of the "determinant formula for area," you should try finding the area of the triangle in Example 3 using the standard formula: area = $\frac{1}{2}$(base)(height).

EXAMPLE 3   Finding the Area of a Triangle

Find the area of the triangle whose vertices are (1, 0), (2, 2), and (4, 3), as shown in Figure 8.1.

Solution

Choose $(x_1, y_1) = (1, 0)$, $(x_2, y_2) = (2, 2)$, and $(x_3, y_3) = (4, 3)$. Then, to find the area of the triangle, evaluate the determinant

$$
\begin{vmatrix} x_1 & y_1 & 1 \\ x_2 & y_2 & 1 \\ x_3 & y_3 & 1 \end{vmatrix} = \begin{vmatrix} 1 & 0 & 1 \\ 2 & 2 & 1 \\ 4 & 3 & 1 \end{vmatrix}
$$

$$
= 1\begin{vmatrix} 2 & 1 \\ 3 & 1 \end{vmatrix} - 0\begin{vmatrix} 2 & 1 \\ 4 & 1 \end{vmatrix} + 1\begin{vmatrix} 2 & 2 \\ 4 & 3 \end{vmatrix}
$$

$$
= 1(-1) - 0(-2) + 1(-2)
$$

$$
= -3.
$$

Using this value, you conclude that the area of the triangle is

$$
\text{Area} = -\frac{1}{2}\begin{vmatrix} 1 & 0 & 1 \\ 2 & 2 & 1 \\ 4 & 3 & 1 \end{vmatrix} = -\frac{1}{2}(-3) = \frac{3}{2}.
$$

Lines in the Plane

Suppose the three points in Example 3 had been on the same line. What would have happened had you applied the area formula to three such points? The answer is that the determinant would have been zero. Consider, for instance, the three collinear points (0, 1), (2, 2), and (4, 3), as shown in Figure 8.2. The area of the "triangle" that has these three points as vertices is

$$
\frac{1}{2}\begin{vmatrix} 0 & 1 & 1 \\ 2 & 2 & 1 \\ 4 & 3 & 1 \end{vmatrix} = \frac{1}{2}\left(0\begin{vmatrix} 2 & 1 \\ 3 & 1 \end{vmatrix} - 1\begin{vmatrix} 2 & 1 \\ 4 & 1 \end{vmatrix} + 1\begin{vmatrix} 2 & 2 \\ 4 & 3 \end{vmatrix}\right)
$$

$$
= \frac{1}{2}[0(-1) - 1(-2) + 1(-2)] = 0.
$$

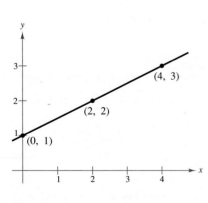

FIGURE 8.2

We generalize this result as follows.

TEST FOR COLLINEAR POINTS

Three points (x_1, y_1), (x_2, y_2), and (x_3, y_3) are collinear (lie on the same line), if and only if

$$\begin{vmatrix} x_1 & y_1 & 1 \\ x_2 & y_2 & 1 \\ x_3 & y_3 & 1 \end{vmatrix} = 0.$$

EXAMPLE 4   Testing for Collinear Points

Determine whether the points $(-2, -2)$, $(1, 1)$, and $(7, 5)$ lie on the same line. (See Figure 8.3.)

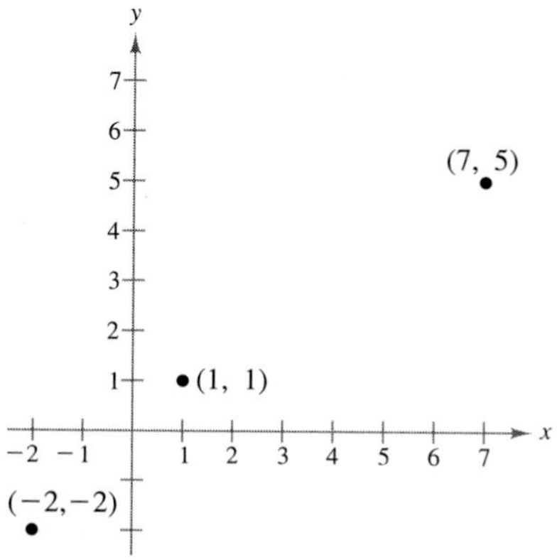

Solution

Letting $(x_1, y_1) = (-2, -2)$, $(x_2, y_2) = (1, 1)$, and $(x_3, y_3) = (7, 5)$, you have

$$\begin{vmatrix} x_1 & y_1 & 1 \\ x_2 & y_2 & 1 \\ x_3 & y_3 & 1 \end{vmatrix} = \begin{vmatrix} -2 & -2 & 1 \\ 1 & 1 & 1 \\ 7 & 5 & 1 \end{vmatrix}$$

$$= -2\begin{vmatrix} 1 & 1 \\ 5 & 1 \end{vmatrix} - (-2)\begin{vmatrix} 1 & 1 \\ 7 & 1 \end{vmatrix} + 1\begin{vmatrix} 1 & 1 \\ 7 & 5 \end{vmatrix}$$

$$= -2(-4) - (-2)(-6) + 1(-2)$$

$$= -6.$$

Because the value of this determinant is *not* zero, you conclude that the three points do not lie on the same line.

FIGURE 8.3

The test for collinear points can be adapted to another use. That is, if you are given two points in a rectangular coordinate system, then you can find the equation of the line passing through the two points as follows.

TWO-POINT FORM OF THE EQUATION OF A LINE

An equation of the line passing through the distinct points (x_1, y_1) and (x_2, y_2) is given by

$$\begin{vmatrix} x & y & 1 \\ x_1 & y_1 & 1 \\ x_2 & y_2 & 1 \end{vmatrix} = 0.$$

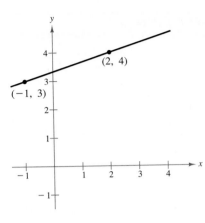

FIGURE 8.4

EXAMPLE 5 Finding an Equation of the Line Passing Through Two Points

Find an equation of the line passing through the two points (2, 4) and (−1, 3), as shown in Figure 8.4.

Solution

Applying the determinant formula for the equation of the line passing through these two points produces

$$\begin{vmatrix} x & y & 1 \\ 2 & 4 & 1 \\ -1 & 3 & 1 \end{vmatrix} = 0.$$

To evaluate this determinant, expand by cofactors along the first row to obtain the following.

$$x \begin{vmatrix} 4 & 1 \\ 3 & 1 \end{vmatrix} - y \begin{vmatrix} 2 & 1 \\ -1 & 1 \end{vmatrix} + 1 \begin{vmatrix} 2 & 4 \\ -1 & 3 \end{vmatrix} = x - 3y + 10 = 0$$

Therefore, an equation of the line is

$$x - 3y + 10 = 0.$$

Cryptography

A **cryptogram** is a message written according to a secret code. (The Greek word "kryptos" means "hidden.") Next, we describe a method for using matrix multiplication to **encode** and **decode** messages.

We begin by assigning a number to each letter in the alphabet (with 0 assigned to a blank space) as follows.

0 =	9 = I	18 = R
1 = A	10 = J	19 = S
2 = B	11 = K	20 = T
3 = C	12 = L	21 = U
4 = D	13 = M	22 = V
5 = E	14 = N	23 = W
6 = F	15 = O	24 = X
7 = G	16 = P	25 = Y
8 = H	17 = Q	26 = Z

Then the message is converted to numbers and partitioned into **uncoded row matrices,** each having n entries, as demonstrated in Example 6.

EXAMPLE 6 Forming Uncoded Row Matrices

Write the uncoded row matrices of order 1×3 for the message MEET ME MONDAY.

Solution

Partitioning the message (including blank spaces, but ignoring other punctuation) into groups of three produces the following uncoded row matrices.

[13 5 5]   [20 0 13]   [5 0 13]   [15 14 4]   [1 25 0]
 M   E E   T   M   E   M   O   N   D   A   Y

Note that a blank space is used to fill out the last uncoded row matrix.

To **encode** a message you choose an $n \times n$ invertible matrix A and multiply the uncoded row matrices (on the right) by A to obtain **coded row matrices.** This process is demonstrated in Example 7.

EXAMPLE 7   Encoding a Message

Use the following matrix to encode the message MEET ME MONDAY.

$$A = \begin{bmatrix} 1 & -2 & 2 \\ -1 & 1 & 3 \\ 1 & -1 & -4 \end{bmatrix}$$

Solution

The coded row matrices are obtained by multiplying each of the uncoded row matrices found in Example 6 by the matrix A as follows.

Uncoded Row Matrix	Encoding Matrix A	Coded Row Matrix
$[13 \quad 5 \quad 5]$	$\begin{bmatrix} 1 & -2 & 2 \\ -1 & 1 & 3 \\ 1 & -1 & -4 \end{bmatrix}$	$= [13 \quad -26 \quad 21]$
$[20 \quad 0 \quad 13]$	$\begin{bmatrix} 1 & -2 & 2 \\ -1 & 1 & 3 \\ 1 & -1 & -4 \end{bmatrix}$	$= [33 \quad -53 \quad -12]$
$[5 \quad 0 \quad 13]$	$\begin{bmatrix} 1 & -2 & 2 \\ -1 & 1 & 3 \\ 1 & -1 & -4 \end{bmatrix}$	$= [18 \quad -23 \quad -42]$
$[15 \quad 14 \quad 4]$	$\begin{bmatrix} 1 & -2 & 2 \\ -1 & 1 & 3 \\ 1 & -1 & -4 \end{bmatrix}$	$= [5 \quad -20 \quad 56]$
$[1 \quad 25 \quad 0]$	$\begin{bmatrix} 1 & -2 & 2 \\ -1 & 1 & 3 \\ 1 & -1 & -4 \end{bmatrix}$	$= [-24 \quad 23 \quad 77]$

Thus, the sequence of coded row matrices is

[13 −26 21][33 −53 −12][18 −23 −42][5 −20 56][−24 23 77].

Finally, removing the matrix notation produces the following cryptogram.

13 −26 21 33 −53 −12 18 −23 −42 5 −20 56 −24 23 77

For those who do not know the matrix A, decoding the cryptogram found in Example 7 is difficult. But for an authorized receiver who knows the matrix A, decoding is simple. The receiver need only multiply the coded row matrices by A^{-1} to retrieve the uncoded row matrices. In other words, if

$$X = [x_1 \quad x_2 \quad \cdots \quad x_n]$$

is an uncoded $1 \times n$ matrix, then $Y = XA$ is the corresponding encoded matrix. The receiver of the encoded matrix can decode Y by multiplying on the right by A^{-1} to obtain

$$YA^{-1} = (XA)A^{-1} = X.$$

This procedure is demonstrated in Example 8.

EXAMPLE 8 Decoding a Message

Use the inverse of the matrix

$$A = \begin{bmatrix} 1 & -2 & 2 \\ -1 & 1 & 3 \\ 1 & -1 & -4 \end{bmatrix}$$

to decode the cryptogram

13 −26 21 33 −53 −12 18 −23 −42 5 −20 56 −24 23 77.

Solution

Begin by using Gauss-Jordan elimination to find A^{-1}.

$$[A \; \vdots \; I]$$
$$\begin{bmatrix} 1 & -2 & 2 & \vdots & 1 & 0 & 0 \\ -1 & 1 & 3 & \vdots & 0 & 1 & 0 \\ 1 & -1 & -4 & \vdots & 0 & 0 & 1 \end{bmatrix} \rightarrow$$

$$[I \; \vdots \; A^{-1}]$$
$$\begin{bmatrix} 1 & 0 & 0 & \vdots & -1 & -10 & -8 \\ 0 & 1 & 0 & \vdots & -1 & -6 & -5 \\ 0 & 0 & 1 & \vdots & 0 & -1 & -1 \end{bmatrix}$$

Now, to decode the message, partition the message into groups of three to form the coded row matrices

[13 −26 21][33 −53 −12][18 −23 −42][5 −20 56][−24 23 77].

Then multiply each coded row matrix by A^{-1} (on the right) to obtain the decoded row matrices.

Coded Row Matrix	Decoding Matrix A^{-1}	Decoded Row Matrix

$$[13 \quad -26 \quad 21] \begin{bmatrix} -1 & -10 & -8 \\ -1 & -6 & -5 \\ 0 & -1 & -1 \end{bmatrix} = [13 \quad 5 \quad 5]$$

$$[33 \quad -53 \quad -12] \begin{bmatrix} -1 & -10 & -8 \\ -1 & -6 & -5 \\ 0 & -1 & -1 \end{bmatrix} = [20 \quad 0 \quad 13]$$

$$[18 \quad -23 \quad -42] \begin{bmatrix} -1 & -10 & -8 \\ -1 & -6 & -5 \\ 0 & -1 & -1 \end{bmatrix} = [5 \quad 0 \quad 13]$$

$$[5 \quad -20 \quad 56] \begin{bmatrix} -1 & -10 & -8 \\ -1 & -6 & -5 \\ 0 & -1 & -1 \end{bmatrix} = [15 \quad 14 \quad 4]$$

$$[-24 \quad 23 \quad 77] \begin{bmatrix} -1 & -10 & -8 \\ -1 & -6 & -5 \\ 0 & -1 & -1 \end{bmatrix} = [1 \quad 25 \quad 0]$$

Thus, the sequence of decoded row matrices is

$$[13 \quad 5 \quad 5][20 \quad 0 \quad 13][5 \quad 0 \quad 13][15 \quad 14 \quad 4][1 \quad 25 \quad 0]$$

and the message is as follows.

$$[13 \quad 5 \quad 5] \quad [20 \quad 0 \quad 13] \quad [5 \quad 0 \quad 13] \quad [15 \quad 14 \quad 4] \quad [1 \quad 25 \quad 0]$$

M E E T ▓ M E ▓ M O N D A Y ▓

DISCUSSION PROBLEM

Comparing Techniques

In Chapters 7 and 8, we have looked at several techniques for solving a system of linear equations.

1. Elimination method using equations (Section 7.3)
2. Gaussian elimination with back-substitution (Section 8.1)
3. Gauss-Jordan elimination (Section 8.1)
4. Inverse matrix method (Section 8.3)
5. Cramer's Rule (Section 8.6)

Write a short paper describing the advantages and disadvantages of each method. Include one example problem using all methods.

WARM UP

The following warm-up exercises involve skills that were covered in earlier sections. You will use these skills in the exercise set for this section.

In Exercises 1–4, solve the system of equations using Gaussian elimination with back-substitution or Gauss-Jordan elimination.

1. $x - 3y = -2$
$x + y = 2$

2. $-x + 3y = 5$
$4x - y = 2$

3. $x + 2y - z = 7$
$-y - z = 4$
$4x - z = 16$

4. $3x + 6z = 0$
$-2x + y = 5$
$y + 2z = 3$

In Exercises 5–10, evaluate the determinant.

5. $\begin{vmatrix} 10 & 8 \\ -6 & -4 \end{vmatrix}$

6. $\begin{vmatrix} -7 & 14 \\ 2 & 3 \end{vmatrix}$

7. $\begin{vmatrix} 1 & 0 & -2 \\ 0 & 1 & 0 \\ -2 & 0 & 1 \end{vmatrix}$

8. $\begin{vmatrix} 0 & 3 & 1 \\ 5 & -2 & 1 \\ 1 & 6 & 1 \end{vmatrix}$

9. $\begin{vmatrix} 0 & -2 & 1 & 0 \\ -2 & 0 & 5 & -1 \\ 1 & 5 & 0 & 2 \\ 0 & -1 & 2 & 0 \end{vmatrix}$

10. $\begin{vmatrix} 1 & 0 & -2 & 1 \\ 0 & 2 & 5 & 1 \\ 3 & -3 & 2 & 1 \\ 0 & 0 & 4 & 1 \end{vmatrix}$

EXERCISES for Section 8.6

In Exercises 1–10, use Cramer's Rule to solve (if possible) the system of equations.

1. $x + 2y = 5$
$-x + y = 1$

2. $2x - y = -10$
$3x + 2y = -1$

3. $3x + 4y = -2$
$5x + 3y = 4$

4. $18x + 12y = 13$
$30x + 24y = 23$

5. $20x + 8y = 11$
$12x - 24y = 21$

6. $13x - 6y = 17$
$26x - 12y = 8$

7. $-0.4x + 0.8y = 1.6$
$2x - 4y = 5$

8. $-0.4x + 0.8y = 1.6$
$0.2x + 0.3y = 2.2$

9. $3x + 6y = 5$
$6x + 14y = 11$

10. $3x + 2y = 1$
$2x + 10y = 6$

In Exercises 11–20, use Cramer's Rule to solve (if possible) the given system for x. You do *not* have to solve the system for the other variables.

11. $4x - y + z = -5$
$2x + 2y + 3z = 10$
$5x - 2y + 6z = 1$

12. $4x - 2y + 3z = -2$
$2x + 2y + 5z = 16$
$8x - 5y - 2z = 4$

13. $3x + 4y + 4z = 11$
$4x - 4y + 6z = 11$
$6x - 6y = 3$

14. $14x - 21y - 7z = 10$
$-4x + 2y - 2z = 4$
$56x - 21y + 7z = 5$

15. $3x + 3y + 5z = 1$
$3x + 5y + 9z = 2$
$5x + 9y + 17z = 4$

16. $2x + 3y + 5z = 4$
$3x + 5y + 9z = 7$
$5x + 9y + 17z = 13$

17. $5x - 3y + 2z = 2$
$2x + 2y - 3z = 3$
$x - 7y + 8z = -4$

18. $3x + 2y + 5z = 4$
$4x - 3y - 4z = 1$
$-8x + 2y + 3z = 0$

19. $7x - 3y \qquad + 2w = 41$
$-2x + y \qquad - w = -13$
$4x \qquad + z - 2w = 12$
$-x + y \qquad - w = -8$

20. $2x + 5y \qquad + w = 11$
$x + 4y + 2z - 2w = -7$
$2x - 2y + 5z + w = 3$
$x \qquad - 3w = 1$

21. *Circuit Analysis*   Consider the circuit in the figure. The currents I_1, I_2, and I_3 in amperes are given by the solution to the system of linear equations

$-4I_1 \qquad + 10I_3 = 5$
$\qquad 5I_2 + 10I_3 = 70$
$I_1 - I_2 + I_3 = 0.$

Use Cramer's Rule to find the three currents.

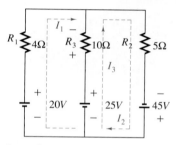

Figure for 21

22. *Circuit Analysis*   Consider the circuit in the figure. The currents I_1, I_2, and I_3 in amperes are given by the solution to the system of linear equations

$4I_1 \qquad + 8I_3 = 2$
$\qquad 2I_2 + 8I_3 = 6$
$I_1 + I_2 - I_3 = 0.$

Use Cramer's Rule to find the three currents.

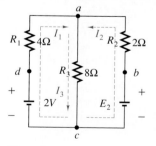

Figure for 22

23. *Maximum Social Security Contribution*   The maximum Social Security contributions for an employee between 1981 and 1989 are shown in the figure. (The figure shows the amount contributed by the *employee*. This amount is matched by the employer.) The least squares regression line $y = a + bt$ for this data is found by solving the system

$9a + 45b = 24.983$
$45a + 285b = 137.012$

where y is the contribution in 1000s of dollars and t is the calendar year with $t = 1$ corresponding to 1981. Use Cramer's Rule to solve this system, and use the result to approximate the maximum Social Security contribution in 1992. (*Source:* U.S. Social Security Administration)

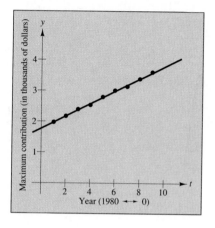

Figure for 23

24. *Pulley System*   A system of pulleys that is assumed frictionless and without mass is loaded with 192-pound and 64-pound weights (see figure). The tensions t_1 and t_2 in the ropes and the acceleration a of the 64-pound weight are found by solving the system

$t_1 - 2t_2 \qquad = 0$
$t_1 \qquad - 3a = 192$
$\qquad t_2 + 2a = 64$

where t_1 and t_2 are measured in pounds and a is in feet per second squared. Use Cramer's Rule to find the acceleration a of the system.

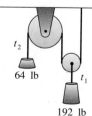

Figure for 24

In Exercises 25–34, use a determinant to find the area of the triangle with given vertices.

25. (0, 0), (3, 1), (1, 5)

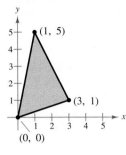

26. (0, 0), (5, −2), (4, 5)

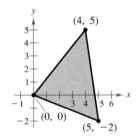

27. (−2, −3), (2, −3), (0, 4)

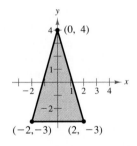

28. (−2, 1), (3, −1), (1, 6)

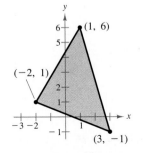

29. $\left(0, \frac{1}{2}\right)$, $\left(\frac{5}{2}, 0\right)$ (4, 3)

30. (−4, −5), (6, −1), (6, 10)

31. (−2, 4), (2, 3), (−1, 5)

32. (0, −2), (−1, 4), (3, 5)

33. (−3, 5), (2, 6), (3, −5)

34. (−2, 4), (1, 5), (3, −2)

35. *Area of a Region* A large region of forest has been infested with gypsy moths. The region is roughly triangular, as shown in the figure. From the northernmost vertex *A* of the region, the distance to vertex *B* is 25 miles south and 10 miles east, and the distance to vertex *C* is 20 miles south and 28 miles east. Approximate the number of square miles in this region.

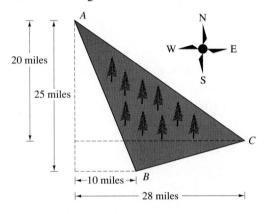

Figure for 35

36. *Area of a Region* Suppose you purchased a triangular tract of land, as shown in the figure. To estimate the number of square feet in the tract, you start at one vertex and walk 65 feet east and 50 feet north to the second vertex. Then, from the second vertex you walk 85 feet west and 30 feet north to the third vertex. How many square feet are there in the tract of land?

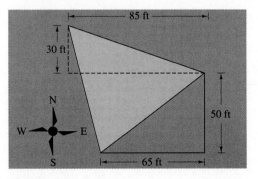

Figure for 36

In Exercises 37–42, use a determinant to ascertain if the points are collinear.

37. $(3, -1)$, $(0, -3)$, $(12, 5)$

38. $(-3, -5)$, $(6, 1)$, $(10, 2)$

39. $\left(2, -\frac{1}{2}\right)$, $(-4, 4)$, $(6, -3)$

40. $(0, 1)$, $(4, -2)$, $(-8, 7)$

41. $(0, 2)$, $(1, 2.4)$, $(-1, 1.6)$

42. $(2, 3)$, $(3, 3.5)$, $(-1, 2)$

In Exercises 43–48, use a determinant to find an equation of the line through the given points (x_1, y_1) and (x_2, y_2).

43. $(0, 0)$, $(5, 3)$

44. $(0, 0)$, $(-2, 2)$

45. $(-4, 3)$, $(2, 1)$

46. $(10, 7)$, $(-2, -7)$

47. $\left(-\frac{1}{2}, 3\right)$, $\left(\frac{5}{2}, 1\right)$

48. $\left(\frac{2}{3}, 4\right)$ $(6, 12)$

In Exercises 49–52, write a cryptogram for each message using the matrix

$$A = \begin{bmatrix} 1 & 2 & 2 \\ 3 & 7 & 9 \\ -1 & -4 & -7 \end{bmatrix}.$$

49. LANDING SUCCESSFUL

50. BEAM ME UP SCOTTY

51. HAPPY BIRTHDAY

52. OPERATION OVERLOAD

In Exercises 53 and 54, decode the cryptogram by using the inverse of matrix A from Exercises 49–52.

53. 20   17 -15 -12 -56 -104 1 -25 -65
62   143   181

54. 13 -9 -59 61   112   106 -17 -73 -131
11   24   29   65   144   172

REVIEW EXERCISES for Chapter 8

In Exercises 1–12, use matrices and elementary row operations to solve the system of equations.

1. $5x + 4y = 2$
   $-x + y = -22$

2. $2x - 5y = 2$
   $3x - 7y = 1$

3. $0.2x - 0.1y = 0.07$
   $0.4x - 0.5y = -0.01$

4. $2x + y = 0.3$
   $3x - y = -1.3$

5. $-x + y + 2z = 1$
   $2x + 3y + z = -2$
   $5x + 4y + 2z = 4$

6. $2x + 3y + z = 10$
   $2x - 3y - 3z = 22$
   $4x - 2y + 3z = -2$

7. $2x + 3y + 3z = 3$
   $6x + 6y + 12z = 13$
   $12x + 9y - z = 2$

8. $4x + 4y + 4z = 5$
   $4x - 2y - 8z = 1$
   $5x + 3y + 8z = 6$

9. $2x + y + 2z = 4$
   $2x + 2y = 5$
   $2x - y + 6z = 2$

10. $3x + 21y - 29z = -1$
   $2x + 15y - 21z = 0$

11. $x + 2y + 6z = 1$
   $2x + 5y + 15z = 4$
   $3x + y + 3z = -6$

12. $x + 2y + w = 3$
   $-3y + 3z = 0$
   $4x + 4y + z + 2w = 0$
   $2x + z = 3$

In Exercises 13–20, perform the indicated matrix operations (if possible).

13. $\begin{bmatrix} 2 & 1 & 0 \\ 0 & 5 & -4 \end{bmatrix} - 3\begin{bmatrix} 5 & 3 & -6 \\ 0 & -2 & 5 \end{bmatrix}$

14. $-2\begin{bmatrix} 1 & 2 \\ 5 & -4 \\ 6 & 0 \end{bmatrix} + 8\begin{bmatrix} 7 & 1 \\ 1 & 2 \\ 1 & 4 \end{bmatrix}$

15. $\begin{bmatrix} 1 & 2 \\ 5 & -4 \\ 6 & 0 \end{bmatrix}\begin{bmatrix} 6 & -2 & 8 \\ 4 & 0 & 0 \end{bmatrix}$

16. $\begin{bmatrix} 1 & 5 & 6 \\ 2 & -4 & 0 \end{bmatrix}\begin{bmatrix} 6 & -2 & 8 \\ 4 & 0 & 0 \end{bmatrix}$

17. $\begin{bmatrix} 1 & 5 & 6 \\ 2 & -4 & 0 \end{bmatrix}\begin{bmatrix} 6 & 4 \\ -2 & 0 \\ 8 & 0 \end{bmatrix}$

18. $\begin{bmatrix} 4 \\ 6 \end{bmatrix}\begin{bmatrix} 6 & -2 \end{bmatrix}$

19. $\begin{bmatrix} 1 & 3 & 2 \\ 0 & 2 & -4 \\ 0 & 0 & 3 \end{bmatrix}\begin{bmatrix} 4 & -3 & 2 \\ 0 & 3 & -1 \\ 0 & 0 & 2 \end{bmatrix}$

20. $\begin{bmatrix} 2 & 1 \\ 6 & 0 \end{bmatrix}\left(\begin{bmatrix} 4 & 2 \\ -3 & 1 \end{bmatrix} + \begin{bmatrix} -2 & 4 \\ 0 & 4 \end{bmatrix}\right)$

In Exercises 21–24, solve for X given

$$A = \begin{bmatrix} -4 & 0 \\ 1 & -5 \\ -3 & 2 \end{bmatrix} \quad \text{and} \quad B = \begin{bmatrix} 1 & 2 \\ -2 & 1 \\ 4 & 4 \end{bmatrix}.$$

21. $X = 3A - 2B$ **22.** $6X = 4A + 3B$

23. $3X + 2A = B$ **24.** $2A - 5B = 3X$

25. Write the system of linear equations represented by the matrix equation

$$\begin{bmatrix} 5 & 4 \\ -1 & 1 \end{bmatrix} \begin{bmatrix} x \\ y \end{bmatrix} = \begin{bmatrix} 2 \\ -22 \end{bmatrix}.$$

26. Write the matrix equation $AX = B$ for the following system of linear equations.

$$\begin{aligned} 2x + 3y + z &= 10 \\ 2x - 3y - 3z &= 22 \\ 4x - 2y + 3z &= -2 \end{aligned}$$

In Exercises 27–30, find the inverse of the matrix (if it exists).

27. $\begin{bmatrix} 2 & 6 \\ 3 & -6 \end{bmatrix}$ **28.** $\begin{bmatrix} 3 & -10 \\ 4 & 2 \end{bmatrix}$

29. $\begin{bmatrix} 2 & 0 & 3 \\ -1 & 1 & 1 \\ 2 & -2 & 1 \end{bmatrix}$ **30.** $\begin{bmatrix} 1 & 4 & 6 \\ 2 & -3 & 1 \\ -1 & 18 & 16 \end{bmatrix}$

In Exercises 31–34, evaluate the determinant.

31. $\begin{vmatrix} 50 & -30 \\ 10 & 5 \end{vmatrix}$ **32.** $\begin{vmatrix} 8 & 5 \\ 2 & -4 \end{vmatrix}$

33. $\begin{vmatrix} 3 & 0 & -4 & 0 \\ 0 & 8 & 1 & 2 \\ 6 & 1 & 8 & 2 \\ 0 & 3 & -4 & 1 \end{vmatrix}$ **34.** $\begin{vmatrix} -5 & 6 & 0 & 0 \\ 0 & 1 & -1 & 2 \\ -3 & 4 & -5 & 1 \\ 1 & 6 & 0 & 3 \end{vmatrix}$

In Exercises 35–42, solve (if possible) the system of linear equations using (a) the inverse of the coefficient matrix and (b) Cramer's Rule.

35. $\begin{aligned} x + 2y &= -1 \\ 3x + 4y &= -5 \end{aligned}$ **36.** $\begin{aligned} x + 3y &= 23 \\ -6x + 2y &= -18 \end{aligned}$

37. $\begin{aligned} -3x - 3y - 4z &= 2 \\ y + z &= -1 \\ 4x + 3y + 4z &= -1 \end{aligned}$ **38.** $\begin{aligned} x - 3y - 2z &= 8 \\ -2x + 7y + 3z &= -19 \\ x - y - 3z &= 3 \end{aligned}$

39. $\begin{aligned} x + 3y + 2z &= 2 \\ -2x - 5y - z &= 10 \\ 2x + 4y &= -12 \end{aligned}$ **40.** $\begin{aligned} 2x + 4y &= -12 \\ 3x + 4y - 2z &= -14 \\ -x + y + 2z &= -6 \end{aligned}$

41. $\begin{aligned} 2x + 3y - 4z &= 1 \\ x - y + 2z &= -4 \\ 3x + 7y - 10z &= 0 \end{aligned}$ **42.** $\begin{aligned} -x + y + z &= 6 \\ 4x - 3y + z &= 20 \\ 2x - y + 3z &= 8 \end{aligned}$

In Exercises 43–46, use a determinant to find the area of the triangle with the given vertices.

43. $(1, 0), (5, 0), (5, 8)$ **44.** $(-4, 0), (4, 0), (0, 6)$

45. $(1, 2), (4, -5), (3, 2)$ **46.** $\left(\frac{3}{2}, 1\right), \left(4, -\frac{1}{2}\right), (4, 2)$

In Exercises 47–50, use a determinant to find an equation of the line through the given points.

47. $(-4, 0), (4, 4)$ **48.** $(2, 5), (6, -1)$

49. $\left(-\frac{5}{2}, 3\right), \left(\frac{7}{2}, 1\right)$ **50.** $(-0.8, 0.2), (0.7, 3.2)$

51. *Mixture Problem*   A florist wants to arrange a dozen flowers consisting of two varieties—carnations and roses. Carnations cost \$0.75 each and roses cost \$1.50 each. How many of each should the florist use in order for the arrangement to cost \$12.00?

52. *Mixture Problem*   One hundred gallons of a 60% acid solution are obtained by mixing a 75% solution with a 50% solution. How many gallons of each must be used to obtain the desired mixture?

53. *Fitting a Parabola to Three Points*   Find an equation of the parabola $y = ax^2 + bx + c$ passing through the points $(-1, 2)$, $(0, 3)$, and $(1, 6)$.

54. *Break-Even Point*   A small business invests \$25,000 in equipment to produce a product. Each unit of the product costs \$3.75 to produce and is sold for \$5.25. How many items must be sold before the business breaks even?

55. If A is a 3×3 matrix and $|A| = 2$, then what is the value of $|4A|$? Give the reason for your answer.

56. Verify that

$$\begin{vmatrix} a_{11} & a_{12} & a_{13} \\ a_{21} & a_{22} & a_{23} \\ a_{31} + c_1 & a_{32} + c_2 & a_{33} + c_3 \end{vmatrix}$$
$$= \begin{vmatrix} a_{11} & a_{12} & a_{13} \\ a_{21} & a_{22} & a_{23} \\ a_{31} & a_{32} & a_{33} \end{vmatrix} + \begin{vmatrix} a_{11} & a_{12} & a_{13} \\ a_{21} & a_{22} & a_{23} \\ c_1 & c_2 & c_3 \end{vmatrix}.$$

C H A P T E R   9

OVERVIEW

This chapter discusses four topics related to number patterns: sequences, mathematical induction, the binomial theorem, and probability.

In the first three sections, you will learn how sequences can be used to model real-life situations. For instance, Exercise 55 on page 574 uses a geometric sequence to model the profit earned by H. J. Heinz Company from 1980 through 1989.

The last two sections contain a brief introduction to probability. Even with such a brief presentation, however, you can get an idea of the broad applicability of this topic. For instance, Example 11 on page 612 shows how probability can be used in quality control, and Exercise 52 on page 617 shows how probability is used in market surveys.

Sequences, Counting Principles, and Probability

9.1 Sequences and Summation Notation

Sequences / Factorial Notation / Summation Notation / The Sum of a Sequence / Applications

Sequences

In mathematics, the word *sequence* is used in much the same way as it is in ordinary English. When we say that a collection of objects is listed *in sequence*, we usually mean that the collection is ordered so that it has a first member, a second member, a third member, and so on.

$$1, 3, 5, 7, \ldots \qquad \text{and} \qquad \frac{1}{2}, \frac{1}{4}, \frac{1}{8}, \frac{1}{16}, \ldots$$

A sequence is a *function* whose domain is the set of positive integers. However, we usually represent a sequence by subscript notation, rather than by the standard function notation. For instance, we write the terms of the sequence

$$f(1), f(2), f(3), f(4), \ldots, f(n), \ldots$$

as

$$a_1, a_2, a_3, a_4, \ldots, a_n, \ldots.$$

Note that subscripts make up the domain of the sequence, and they serve to identify the location of a term within the sequence. For instance, a_4 is the 4th term of the sequence and a_n is the **nth term** of the sequence. The entire sequence is sometimes denoted by the short form $\{a_n\}$.

543

DEFINITION OF A SEQUENCE

An **infinite sequence** $\{a_n\}$ is a function whose domain is the set of positive integers. The function values

$$a_1, a_2, a_3, a_4, \ldots, a_n, \ldots$$

are the **terms** of the sequence. If the domain of the function consists of the first n positive integers only, then the sequence is a **finite sequence.**

On occasion it is convenient to begin subscripting a sequence with 0 instead of 1 so that the terms of the sequence become

$$a_0, a_1, a_2, a_3, a_4, \ldots, a_n, \ldots.$$

In such cases, we still call a_n the nth term of the sequence, even though it occupies the $(n + 1)$th position in the sequence.

EXAMPLE 1   Finding Terms in a Sequence

a. The first four terms of the sequence whose nth term is $a_n = 3n - 2$ are:

$$a_1 = 3(1) - 2 = 1$$
$$a_2 = 3(2) - 2 = 4$$
$$a_3 = 3(3) - 2 = 7$$
$$a_4 = 3(4) - 2 = 10.$$

b. The first four terms of the sequence whose nth term is $a_n = 3 + (-1)^n$ are:

$$a_1 = 3 + (-1)^1 = 3 - 1 = 2$$
$$a_2 = 3 + (-1)^2 = 3 + 1 = 4$$
$$a_3 = 3 + (-1)^3 = 3 - 1 = 2$$
$$a_4 = 3 + (-1)^4 = 3 + 1 = 4.$$

The terms of a sequence need not all be positive, as shown in part (b) of Example 2.

EXAMPLE 2 **Finding Terms in a Sequence**

a. The first four terms of the sequence whose nth term is $a_n = \dfrac{2n}{(1+n)}$ are:

$$a_1 = \frac{2(1)}{1+1} = \frac{2}{2} = 1$$

$$a_2 = \frac{2(2)}{1+2} = \frac{4}{3}$$

$$a_3 = \frac{2(3)}{1+3} = \frac{6}{4} = \frac{3}{2}$$

$$a_4 = \frac{2(4)}{1+4} = \frac{8}{5}.$$

b. The first four terms of the sequence whose nth term is $a_n = \dfrac{(-1)^n}{(2n-1)}$ are:

$$a_1 = \frac{(-1)^1}{2(1)-1} = \frac{-1}{2-1} = -1$$

$$a_2 = \frac{(-1)^2}{2(2)-1} = \frac{1}{4-1} = \frac{1}{3}$$

$$a_3 = \frac{(-1)^3}{2(3)-1} = \frac{-1}{6-1} = -\frac{1}{5}$$

$$a_4 = \frac{(-1)^4}{2(4)-1} = \frac{1}{8-1} = \frac{1}{7}.$$

REMARK Try finding the first four terms of the sequence whose nth term is

$$a_n = \frac{(-1)^{n+1}}{2n-1}.$$

How do they differ from the first four terms of the sequence in Example 2(b)?

It is important to realize that simply listing the first few terms is not sufficient to define a unique sequence. Consider the following sequences, both of which have the same first three terms.

$$\frac{1}{2}, \frac{1}{4}, \frac{1}{8}, \frac{1}{16}, \cdots, \frac{1}{2^n}, \cdots$$

$$\frac{1}{2}, \frac{1}{4}, \frac{1}{8}, \frac{1}{15}, \cdots, \frac{6}{(n+1)(n^2-n+6)}, \cdots$$

When given the first few terms of a sequence, the best you can do is write an *apparent* nth term for the sequence. There are likely other nth terms.

EXAMPLE 3 **Finding the nth Term of a Sequence**

Write an expression for the apparent nth term (a_n) of each of the following sequences.

a. 1, 3, 5, 7, . . .　　**b.** 2, 5, 10, 17, . . .　　**c.** $\dfrac{2}{1}, \dfrac{3}{2}, \dfrac{4}{3}, \dfrac{5}{4}, \cdots$

Solution

a. *n*: 1   2   3   4 . . . *n*
Terms: 1   3   5   7 . . . a_n
Apparent pattern: Each term is 1 less than twice *n*, which implies that

$$a_n = 2n - 1.$$

b. *n*: 1   2   3   4 . . . *n*
Terms: 2   5   10   17 . . . a_n
Apparent pattern: Each term is 1 more than the square of *n*, which implies that

$$a_n = n^2 + 1.$$

c. *n*: 1   2   3   4 . . . *n*
Terms: $\dfrac{2}{1}$ $\dfrac{3}{2}$ $\dfrac{4}{3}$ $\dfrac{5}{4}$ · · · a_n
Apparent pattern: Each term has a numerator that is greater than its denominator, which implies that

$$a_n = \frac{n+1}{n}.$$

Factorial Notation

Some very important sequences in mathematics involve terms that are defined with special types of products—**factorials.**

DEFINITION OF FACTORIAL

If *n* is a positive integer, then *n* **factorial** is defined by

$$n! = 1 \cdot 2 \cdot 3 \cdot 4 \cdots (n-1) \cdot n.$$

As a special case, we define zero factorial to be $0! = 1$.

Here are some values of *n*! for the first several nonnegative integers.

$0! = 1$

$1! = 1$

$2! = 1 \cdot 2 = 2$

$3! = 1 \cdot 2 \cdot 3 = 6$

$4! = 1 \cdot 2 \cdot 3 \cdot 4 = 24$

$5! = 1 \cdot 2 \cdot 3 \cdot 4 \cdot 5 = 120$

The value of n does not have to be very large before the value of $n!$ becomes huge. For instance, $10! = 3,628,800$. Many calculators have a factorial key, denoted by $\boxed{x!}$.

Factorials follow the same conventions for order of operations as do exponents. For instance,

$$2n! = 2(n!) = 2(1 \cdot 2 \cdot 3 \cdot 4 \cdots n)$$

whereas $(2n)! = 1 \cdot 2 \cdot 3 \cdot 4 \cdots 2n$.

EXAMPLE 4   Finding Terms of a Sequence Involving Factorials

List the first five terms of the sequence whose nth term is $a_n = 2^n/n!$. Begin with $n = 0$.

Solution

$$a_0 = \frac{2^0}{0!} = \frac{1}{1} = 1$$

$$a_1 = \frac{2^1}{1!} = \frac{2}{1} = 2$$

$$a_2 = \frac{2^2}{2!} = \frac{4}{2} = 2$$

$$a_3 = \frac{2^3}{3!} = \frac{8}{6} = \frac{4}{3}$$

$$a_4 = \frac{2^4}{4!} = \frac{16}{24} = \frac{2}{3}$$

When working with fractions involving factorials, you will often find that the fractions can be reduced.

$$\frac{n!}{(n-1)!} = \frac{1 \cdot 2 \cdot 3 \cdots (n-1) \cdot n}{1 \cdot 2 \cdot 3 \cdots (n-1)} = n$$

$$\frac{8!}{2! \cdot 6!} = \frac{1 \cdot 2 \cdot 3 \cdot 4 \cdot 5 \cdot 6 \cdot 7 \cdot 8}{1 \cdot 2 \cdot 1 \cdot 2 \cdot 3 \cdot 4 \cdot 5 \cdot 6} = \frac{7 \cdot 8}{2} = 28$$

Summation Notation

A convenient notation for the sum of the terms of a finite sequence is **summation notation** or **sigma notation** because it involves the use of the upper-case Greek letter sigma, written as Σ.

DEFINITION OF SUMMATION NOTATION

The sum of the first n terms of a sequence is represented by

$$\sum_{i=1}^{n} a_i = a_1 + a_2 + a_3 + a_4 + \cdots + a_n$$

where i is the **index of summation**, n is the **upper limit of summation**, and 1 is the **lower limit of summation**.

EXAMPLE 5   Summation Notation for Sums

a. $\displaystyle\sum_{i=1}^{5} 3i = 3(1) + 3(2) + 3(3) + 3(4) + 3(5)$

$$= 3(1 + 2 + 3 + 4 + 5)$$

$$= 3(15) = 45$$

b. $\displaystyle\sum_{i=1}^{4} 2 = 2 + 2 + 2 + 2 = 8$

c. $\displaystyle\sum_{k=3}^{6} (1 + k^2) = (1 + 3^2) + (1 + 4^2) + (1 + 5^2) + (1 + 6^2)$

$$= 10 + 17 + 26 + 37 = 90$$

d. $\displaystyle\sum_{i=0}^{8} \frac{1}{i!} = \frac{1}{0!} + \frac{1}{1!} + \frac{1}{2!} + \frac{1}{3!} + \frac{1}{4!} + \frac{1}{5!} + \frac{1}{6!} + \frac{1}{7!} + \frac{1}{8!}$

$$= 1 + 1 + \frac{1}{2} + \frac{1}{6} + \frac{1}{24} + \frac{1}{120} + \frac{1}{720} + \frac{1}{5,040} + \frac{1}{40,320}$$

$$\approx 2.71828$$

For this summation, note that the sum is very close to the irrational number $e \approx 2.718281828$. It can be shown that as more terms of the sequence whose nth term is $1/n!$ are added, the sum becomes closer and closer to e.

REMARK In Example 5, note that the lower index of a summation does not have to be 1. Also note that the index does not have to be the letter i. For instance, in part (c), the letter k is the index.

When working with summation notation, the following properties of sums are useful.

PROPERTIES OF SUMS

1. $\displaystyle\sum_{i=1}^{n} ca_i = c\sum_{i=1}^{n} a_i,$ c is any constant

2. $\displaystyle\sum_{i=1}^{n} (a_i + b_i) = \sum_{i=1}^{n} a_i + \sum_{i=1}^{n} b_i$

3. $\displaystyle\sum_{i=1}^{n} (a_i - b_i) = \sum_{i=1}^{n} a_i - \sum_{i=1}^{n} b_i$

Proof

Each of these properties follows directly from the associative property of addition, the commutative property of addition, and the distributive property of multiplication over addition. For example, note the use of the distributive property in the proof of Property 1.

$$\sum_{i=1}^{n} ca_i = ca_1 + ca_2 + ca_3 + \cdots + ca_n$$

$$= c(a_1 + a_2 + a_3 + \cdots + a_n) = c\sum_{i=1}^{n} a_i$$

Variations in the upper and lower limits of summation can produce quite different-looking summation notations for *the same sum*. For example, consider the following two sums.

$$\sum_{i=1}^{5} 3(2^i) = 3\sum_{i=1}^{5} 2^i = 3(2^1 + 2^2 + 2^3 + 2^4 + 2^5)$$

$$\sum_{i=0}^{4} 3(2^{i+1}) = 3\sum_{i=0}^{4} 2^{i+1} = 3(2^1 + 2^2 + 2^3 + 2^4 + 2^5)$$

The Sum of a Sequence

The summation of the terms of a sequence is a **series.** For a finite sequence $a_1, a_2, a_3, \ldots, a_n$, the associated series is the summation

$$\sum_{i=1}^{n} a_i = a_1 + a_2 + a_3 + \cdots + a_n.$$

The summation

$$\sum_{i=1}^{\infty} a_i = a_1 + a_2 + a_3 + a_4 + \cdots$$

is an **infinite series.** Infinite series have important uses in calculus.

EXAMPLE 6   Finding the Sum of a Sequence

Write the sum of the sequence $1, \frac{1}{2}, \frac{1}{4}, \frac{1}{8}, \frac{1}{16}, \frac{1}{32}$ in sigma notation and evaluate the associated series.

Solution

Since the denominators of the terms of the sequence are powers of 2, the apparent ith term is $a_i = 1/2^i$, starting with $i = 0$. The associated series is

$$\sum_{i=0}^{5} \frac{1}{2^i} = 1 + \frac{1}{2} + \frac{1}{4} + \frac{1}{8} + \frac{1}{16} + \frac{1}{32}$$

and its value is

$$\sum_{i=0}^{5} \frac{1}{2^i} = 1 + \frac{16}{32} + \frac{8}{32} + \frac{4}{32} + \frac{2}{32} + \frac{1}{36} = 1 + \frac{31}{32} = 1.96895.$$

Applications

Sequences have many applications in business and science.

EXAMPLE 7   Population of the United States

The resident population of the United States from 1950 to 1989 can be approximated by the model

$$a_n = \sqrt{22{,}926 + 902.5n + 2.01n^2}, \qquad n = 0, 1, \ldots, 39$$

where a_n is the population in millions and n represents the calendar year with $n = 0$ corresponding to 1950. (*Source*: U.S. Bureau of Census) Find the last five terms of this finite sequence.

Solution

The last five terms of this finite sequence are:

$$a_{35} = \sqrt{22{,}926 + 902.5(35) + 2.01(35^2)} \approx 238.7 \qquad \textit{1985 population}$$
$$a_{36} = \sqrt{22{,}926 + 902.5(36) + 2.01(36^2)} \approx 240.9 \qquad \textit{1986 population}$$
$$a_{37} = \sqrt{22{,}926 + 902.5(37) + 2.01(37^2)} \approx 243.0 \qquad \textit{1987 population}$$
$$a_{38} = \sqrt{22{,}926 + 902.5(38) + 2.01(38^2)} \approx 245.2 \qquad \textit{1988 population}$$
$$a_{39} = \sqrt{22{,}926 + 902.5(39) + 2.01(39^2)} \approx 247.3. \qquad \textit{1989 population}$$

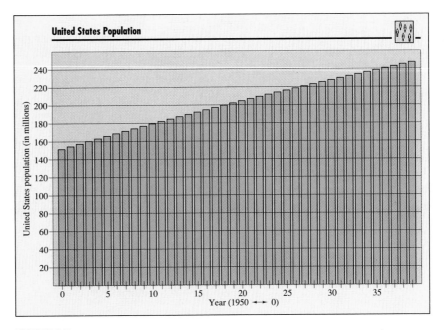

FIGURE 9.1

The bar graph in Figure 9.1 graphically represents the population given by this sequence for the entire 40-year period from 1950 to 1989.

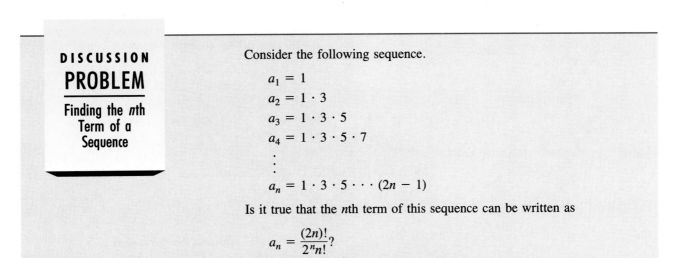

DISCUSSION
PROBLEM

Finding the *n*th Term of a Sequence

Consider the following sequence.

$$a_1 = 1$$
$$a_2 = 1 \cdot 3$$
$$a_3 = 1 \cdot 3 \cdot 5$$
$$a_4 = 1 \cdot 3 \cdot 5 \cdot 7$$
$$\vdots$$
$$a_n = 1 \cdot 3 \cdot 5 \cdots (2n - 1)$$

Is it true that the *n*th term of this sequence can be written as

$$a_n = \frac{(2n)!}{2^n n!}?$$

Write a paragraph that justifies your answer.

WARM UP

The following warm-up exercises involve skills that were covered in earlier sections. You will use these skills in the exercise set for this section.

In Exercises 1 and 2, find the required value of the function.

1. $f(n) = \dfrac{2n}{n^2 + 1}$, $f(2)$ **2.** $f(n) = \dfrac{4}{3(n + 1)}$, $f(3)$

In Exercises 3–6, factor the expression.

3. $4n^2 - 1$ **4.** $4n^2 - 8n + 3$

5. $n^2 - 3n + 2$ **6.** $n^2 + 3n + 2$

In Exercises 7–10, perform the indicated operations and/or simplify.

7. $\left(\dfrac{2}{3}\right)\left(\dfrac{3}{4}\right)\left(\dfrac{4}{5}\right)\left(\dfrac{5}{6}\right)$ **8.** $\dfrac{2 \cdot 4 \cdot 6 \cdot 8}{2^4}$

9. $\dfrac{1}{2 \cdot 2} + \dfrac{1}{2 \cdot 3} + \dfrac{1}{2 \cdot 4}$ **10.** $\dfrac{1}{1 \cdot 2} + \dfrac{1}{2 \cdot 3} + \dfrac{1}{3 \cdot 4}$

EXERCISES for Section 9.1

In Exercises 1–18, write the first five terms of the indicated sequence. (Assume n begins with 1.)

1. $a_n = 2n + 1$ **2.** $a_n = 4n - 3$

3. $a_n = 2^n$ **4.** $a_n = \left(\dfrac{1}{2}\right)^n$

5. $a_n = (-2)^n$ **6.** $a_n = \left(-\dfrac{1}{2}\right)^n$

7. $a_n = \dfrac{1 + (-1)^n}{n}$ **8.** $a_n = \dfrac{n}{n + 1}$

9. $a_n = 3 - \dfrac{1}{2^n}$ **10.** $a_n = \dfrac{3^n}{4^n}$

11. $a_n = \dfrac{1}{n^{3/2}}$ **12.** $a_n = \dfrac{3n^2 - n + 4}{2n^2 + 1}$

13. $a_n = \dfrac{3^n}{n!}$ **14.** $a_n = \dfrac{n!}{n}$

15. $a_n = \dfrac{(-1)^n}{n^2}$ **16.** $a_n = (-1)^n\left(\dfrac{n}{n + 1}\right)$

17. $a_1 = 3$ and $a_{k+1} = 2(a_k - 1)$

18. $a_1 = 4$ and $a_{k+1} = \left(\dfrac{k + 1}{2}\right)a_k$

In Exercises 19–24, simplify the ratio of factorials.

19. $\dfrac{4!}{6!}$ **20.** $\dfrac{25!}{23!}$

21. $\dfrac{(n + 1)!}{n!}$ **22.** $\dfrac{(n + 2)!}{n!}$

23. $\dfrac{(2n - 1)!}{(2n + 1)!}$ **24.** $\dfrac{(2n + 2)!}{(2n)!}$

In Exercises 25–36, write an expression for the *most apparent* nth term of the sequence. (Assume n begins with 1.)

25. $1, 4, 7, 10, 13, \ldots$ **26.** $3, 7, 11, 15, 19, \ldots$

27. $0, 3, 8, 15, 24, \ldots$ **28.** $1, \frac{1}{4}, \frac{1}{9}, \frac{1}{16}, \frac{1}{25}, \ldots$

29. $\frac{1}{2}, \frac{-1}{4}, \frac{1}{8}, \frac{-1}{16}, \ldots$ **30.** $\frac{1}{3}, \frac{2}{9}, \frac{4}{27}, \frac{8}{81}, \ldots$

31. $1 + \frac{1}{1}, 1 + \frac{1}{2}, 1 + \frac{1}{3}, 1 + \frac{1}{4}, 1 + \frac{1}{5}, \ldots$

32. $1 + \frac{1}{2}, 1 + \frac{3}{4}, 1 + \frac{7}{8}, 1 + \frac{15}{16}, 1 + \frac{31}{32}, \ldots$

33. $1, \frac{1}{2}, \frac{1}{6}, \frac{1}{24}, \frac{1}{120}, \ldots$ **34.** $2, -4, 6, -8, 10, \ldots$

35. $1, -1, 1, -1, 1, \ldots$

36. $1, 2, \dfrac{2^2}{2}, \dfrac{2^3}{6}, \dfrac{2^4}{24}, \dfrac{2^5}{120}, \cdots$

In Exercises 37–50, find the given sum.

37. $\displaystyle\sum_{i=1}^{5} (2i + 1)$ **38.** $\displaystyle\sum_{i=1}^{6} (3i - 1)$

39. $\displaystyle\sum_{k=1}^{4} 10$ **40.** $\displaystyle\sum_{k=1}^{5} 6$

41. $\displaystyle\sum_{i=0}^{4} i^2$

42. $\displaystyle\sum_{i=0}^{5} 3i^2$

43. $\displaystyle\sum_{k=0}^{3} \frac{1}{k^2 + 1}$

44. $\displaystyle\sum_{j=3}^{5} \frac{1}{j}$

45. $\displaystyle\sum_{i=1}^{4} [(i - 1)^2 + (i + 1)^3]$

46. $\displaystyle\sum_{k=2}^{5} (k + 1)(k - 3)$

47. $\displaystyle\sum_{i=1}^{4} (9 + 2i)$

48. $\displaystyle\sum_{j=0}^{4} (-2)^j$

49. $\displaystyle\sum_{k=0}^{4} \frac{(-1)^k}{k + 1}$

50. $\displaystyle\sum_{k=0}^{4} \frac{(-1)^k}{k!}$

In Exercises 51–60, use sigma notation to write the given sum.

51. $\dfrac{1}{3(1)} + \dfrac{1}{3(2)} + \dfrac{1}{3(3)} + \cdots + \dfrac{1}{3(9)}$

52. $\dfrac{5}{1 + 1} + \dfrac{5}{1 + 2} + \dfrac{5}{1 + 3} + \cdots + \dfrac{5}{1 + 15}$

53. $\left[2\left(\frac{1}{8}\right) + 3\right] + \left[2\left(\frac{2}{8}\right) + 3\right] + \cdots + \left[2\left(\frac{8}{8}\right) + 3\right]$

54. $\left[1 - \left(\frac{1}{6}\right)^2\right] + \left[1 - \left(\frac{2}{6}\right)^2\right] + \cdots + \left[1 - \left(\frac{6}{6}\right)^2\right]$

55. $3 - 9 + 27 - 81 + 243 - 729$

56. $1 - \frac{1}{2} + \frac{1}{4} - \frac{1}{8} + \cdots - \frac{1}{128}$

57. $\dfrac{1}{1^2} - \dfrac{1}{2^2} + \dfrac{1}{3^2} - \dfrac{1}{4^2} + \cdots - \dfrac{1}{20^2}$

58. $\dfrac{1}{1 \cdot 3} + \dfrac{1}{2 \cdot 4} + \dfrac{1}{3 \cdot 5} + \cdots + \dfrac{1}{10 \cdot 12}$

59. $\frac{1}{4} + \frac{3}{8} + \frac{7}{16} + \frac{15}{32} + \frac{31}{64}$

60. $\frac{1}{2} + \frac{2}{4} + \frac{6}{8} + \frac{24}{16} + \frac{120}{32} + \frac{720}{64}$

61. *Compound Interest*   A deposit of $5000 is made in an account that earns 8% interest compounded quarterly. The balance in the account after n quarters is given by

$$A_n = 5000\left(1 + \frac{0.08}{4}\right)^n, \quad n = 1, 2, 3, \ldots.$$

(a) Compute the first eight terms of this sequence.

(b) Find the balance in this account after 10 years by computing the 40th term of the sequence.

62. *Compound Interest*   A deposit of $100 is made *each* month in an account that earns 12% interest compounded monthly. The balance in the account after n months is given by

$$A_n = 100(101)[(1.01)^n - 1], \quad n = 1, 2, 3, \ldots.$$

(a) Compute the first 6 terms of this sequence.

(b) Find the balance after 5 years by computing the 60th term of the sequence.

(c) Find the balance after 20 years by computing the 240th term of the sequence.

63. *Hospital Costs*   The average cost of a day in a hospital from 1980 to 1987 is given by the model

$$a_n = 242.67 + 42.67n, \quad n = 0, 1, 2, \ldots, 7$$

where a_n is the average cost in dollars and n is the year with $n = 0$ corresponding to 1980. (*Source*: American Hospital Association) Find the terms of this finite sequence and construct a bar graph that represents the sequence.

64. *Federal Debt*   It took more than 200 years for the United States to accumulate a $1 trillion debt. Then it took just 8 years to get to $3 trillion. (*Source*: Treasury Department) The federal debt during the decade of the 1980s is approximated by the model

$$a_n = 0.1\sqrt{82 + 9n^2}, \quad n = 0, 1, 2, \ldots, 10$$

where a_n is the debt in trillions and n is the year with $n = 0$ corresponding to 1980. Find the terms of this finite sequence and construct a bar graph that represents the sequence.

65. *Corporate Dividends*   The dividends declared per share of common stock of Ameritech Corporation for the years 1985 through 1990 are shown in the figure. (*Source*: Ameritech 1990 Annual Report) These dividends can be approximated by the model

$$a_n = 0.20n + 1.17, \quad n = 5, 6, 7, 8, 9, 10$$

where a_n is the dividend in dollars and n is the year with $n = 5$ corresponding to 1985. Approximate the sum of the dividends per share of common stock for the years 1985 through 1990 by evaluating

$$\sum_{5}^{10} (0.20n + 1.17).$$

Compare this sum with the result of adding the dividends as shown in the figure.

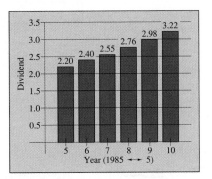

Figure for 65

66. *Total Revenue* The total annual sales for MCI Communications from 1980 through 1989 can be approximated by the model

$$a_n = 48.217n^2 + 228.1n + 311.28, \qquad n = 0, 1, 2, \ldots, 9$$

where a_n is the annual sales (in millions of dollars) and n is the year with $n = 0$ corresponding to 1980. (*Source*: MCI Communications) Find the total revenue from 1980 through 1989 by evaluating the sum

$$\sum_{0}^{9} (48.217n^2 + 228.1n + 311.28).$$

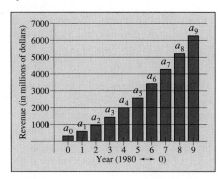

Figure for 66

In Exercises 67 and 68, use the following definition of the arithmetic mean $\bar{x}$ of a set of n measurements $x_1, x_2, x_3, \ldots, x_n$.

$$\bar{x} = \frac{1}{n} \sum_{i=1}^{n} x_i$$

67. Prove that $\sum_{i=1}^{n} (x_i - \bar{x}) = 0$.

68. Prove that $\sum_{i=1}^{n} (x_i - \bar{x})^2 = \sum_{i=1}^{n} x_i^2 - \frac{1}{n} \left(\sum_{i=1}^{n} x_i \right)^2$.

9.2 Arithmetic Sequences

Introduction / The Sum of an Arithmetic Sequence / Arithmetic Mean / Applications

Introduction

A sequence whose consecutive terms have a common difference is an **arithmetic sequence**.

DEFINITION OF AN ARITHMETIC SEQUENCE

A sequence is **arithmetic** if the differences between consecutive terms are the same. Thus, the sequence

$$a_1, a_2, a_3, a_4, \ldots, a_n, \ldots$$

is arithmetic if there is a number d such that

$$a_2 - a_1 = d, \qquad a_3 - a_2 = d, \qquad a_4 - a_3 = d,$$

and so on. The number d is the **common difference** of the arithmetic sequence.

EXAMPLE 1 **Examples of Arithmetic Sequence**

a. The sequence whose nth term is $4n + 3$ is arithmetic. For this sequence, the common difference between consecutive terms is 4.

$$7, 11, 15, 19, \ldots, 4n + 3, \ldots$$

$$11 - 7 = 4$$

b. The sequence whose nth term is $7 - 5n$ is arithmetic. For this sequence, the common difference between consecutive terms is -5.

$$2, -3, -8, -13, \ldots, 7 - 5n, \ldots$$

$$-3 - 2 = -5$$

c. The sequence whose nth term is $\frac{1}{4}(n + 3)$ is arithmetic. For this sequence, the common difference between consecutive terms is $\frac{1}{4}$.

$$1, \frac{5}{4}, \frac{3}{2}, \frac{7}{4}, \ldots, \frac{n + 3}{4}, \ldots$$

$$\frac{5}{4} - 1 = \frac{1}{4}$$

In Example 1, notice that each of the arithmetic sequences has an nth term that is of the form $dn + c$, where the common difference of the sequence is d. We summarize this result as follows.

THE nTH TERM OF AN ARITHMETIC SEQUENCE

The nth term of an arithmetic sequence has the form

$$a_n = dn + c$$

where d is the common difference between consecutive terms of the sequence and $c = a_1 - d$.

REMARK An alternative form of the nth term of an arithmetic sequence is $a_n = a_1 + (n - 1)d$.

EXAMPLE 2 **Finding the nth Term of an Arithmetic Sequence**

Find a formula for the nth term of the arithmetic sequence whose common difference is 5 and whose *second* term is 12. What is the 18th term of this sequence?

Solution

You know that the formula for the nth term is of the form $a_n = dn + c$. Moreover, because the common difference is given to be $d = 5$, the formula must have the form

$$a_n = 5n + c.$$

Using the fact that the second term is

$$a_2 = 12 = 5(2) + c$$

it follows that $c = 2$. Thus, the formula for the nth term is

$$a_n = 5n + 2.$$

The sequence, therefore, has the form

$$7, 12, 17, 22, 27, \ldots, 5n + 2, \ldots$$

and the 18th term of the sequence is

$$a_{18} = 5(18) + 2 = 92.$$

If you know the nth term of an arithmetic sequence *and* you know the common difference of the sequence, then you can find the $(n + 1)$th term by using the following **recursion formula.**

$$a_{n+1} = a_n + d$$

With such a formula, you can find any term of an arithmetic sequence, *provided* you know the previous term. For example, if you know the first term, then you can find the second term. Then, knowing the second term, you can find the third term, and so on.

EXAMPLE 3   Using a Recursion Formula

Find the ninth term of the arithmetic sequence whose first two terms are 2 and 9.

Solution

For this sequence you find that $a_1 = 2$ and $a_2 = 9$ so that the common difference is $d = 9 - 2 = 7$. There are two ways to find the ninth term. One way is to simply write out the first nine terms (by repeatedly adding 7):

$$2, 9, 16, 23, 30, 37, 44, 51, 58.$$

Another way to find the ninth term is to first find a formula for the nth term. Since the first term is 2, it follows that $c = a_1 - d = 2 - 7 = -5$. Therefore, a formula for the nth term of the sequence is $a_n = 7n - 5$, which implies that the ninth term is

$$a_9 = 7(9) - 5 = 58.$$

EXAMPLE 4   Finding the *n*th Term of an Arithmetic Sequence

The fourth term of an arithmetic sequence is 20, and the thirteenth term is 65. Write the first several terms of this sequence.

Solution

To obtain the thirteenth term from the fourth term, you would have to add the common difference d to the fourth term nine times. That is,

$$a_{13} = a_4 + 9d.$$

Since you are given $a_4 = 20$ and $a_{13} = 65$, you can solve this equation for d as follows.

$$65 = 20 + 9d$$
$$45 = 9d$$
$$5 = d$$

Now, from the formula for the *n*th term of an arithmetic sequence, you have

$$a_n = dn + c$$
$$a_4 = 5(4) + c$$
$$20 = 20 + c$$
$$0 = c.$$

Thus, $a_n = 5_n$ and the first several terms of the sequence are as follows.

$$5, 10, 15, 20, 25, 30, 35, 40, 45, 50, 55, 60, 65, \ldots$$

The Sum of an Arithmetic Sequence

Now look at a formula for finding the *sum* of a finite arithmetic sequence.

THE SUM OF A FINITE ARITHMETIC SEQUENCE

The formula for the sum of a finite arithmetic sequence with n terms is

$$S = \frac{n}{2}(a_1 + a_n).$$

Proof

Begin by generating the terms of the arithmetic sequence in two ways. In the first way, repeatedly add d to the first term to obtain

$$S = a_1 + a_2 + a_3 + \cdots + a_{n-2} + a_{n-1} + a_n$$
$$= a_1 + [a_1 + d] + [a_1 + 2d] + \cdots + [a_1 + (n-1)d].$$

In the second way, repeatedly subtract d from the nth term to obtain

$$S = a_n + a_{n-1} + a_{n-2} + \cdots + a_3 + a_2 + a_1$$
$$= a_n + [a_n - d] + [a_n - 2d] + \cdots + [a_n - (n-1)d].$$

If you add these two versions of S, the multiples of d cancel and you obtain

$$\overbrace{2S = (a_1 + a_n) + (a_1 + a_n) + (a_1 + a_n) + \cdots + (a_1 + a_n)}^{n \text{ terms}}$$
$$= n(a_1 + a_n).$$

Thus, you have

$$S = \frac{n}{2}(a_1 + a_n).$$

REMARK Be sure you see that this formula works only for *arithmetic* sequences.

EXAMPLE 5   Finding the Sum of an Arithmetic Sequence

Find the sum of the integers from 1 to 100.

Solution

The integers from 1 to 100 form an arithmetic sequence

$$1, 2, 3, 4, 5, 6, \ldots, 99, 100$$

that has 100 terms. Thus, you can use the formula for the sum of an arithmetic sequence, as follows.

$$S = 1 + 2 + 3 + 4 + 5 + 6 + \cdots + 99 + 100$$

$$= \frac{n}{2}(a_1 + a_n)$$

$$= \frac{100}{2}(1 + 100)$$

$$= 50(101)$$

$$= 5050$$

EXAMPLE 6   **Finding the Sum of an Arithmetic Sequence**

Find the sum

$$\sum_{n=1}^{150} (11n - 6).$$

Solution

From the summation

$$\sum_{n=1}^{150} (11n - 6) = 5 + 16 + 27 + 38 + \cdots$$

you can conclude that $a_1 = 5$ and $a_n = 11n - 6$. Therefore, $a_{150} = 11(150) - 6 = 1644$ and the sum of the first 150 terms is

$$S = \frac{n}{2}(a_1 + a_n) = \frac{150}{2}(5 + 1644) = 75(1649) = 123{,}675.$$

EXAMPLE 7   **Finding the Sum of an Arithmetic Sequence**

Verify the formula

$$S = 1 + 3 + 5 + \cdots + (2n - 1) = n^2.$$

Solution

Using the formula for the sum of a finite arithmetic sequence, you can write

$$S = 1 + 3 + 5 + \cdots + (2n - 1)$$

$$= \frac{n}{2}(a_1 + a_n)$$

$$= \frac{n}{2}[1 + (2n - 1)]$$

$$= \frac{n}{2}(2n)$$

$$= n^2.$$

Arithmetic Mean

Recall that $(a + b)/2$ is the midpoint between the two numbers a and b on the real number line. As a result, the terms

$$a, \frac{a + b}{2}, b$$

have a common difference. We call $(a + b)/2$ the **arithmetic mean** of the numbers a and b. We can generalize this concept by finding k numbers m_1, $m_2, m_3, \ldots, m_k$ between a and b such that the terms

$$a, m_1, m_2, m_3, \ldots, m_k, b$$

have a common difference. This process is referred to as **inserting k arithmetic means** between a and b.

EXAMPLE 8 **Inserting Arithmetic Means Between Two Numbers**

Insert three arithmetic means between 4 and 15.

Solution

You need to find three numbers m_1, m_2, and m_3 such that the terms

$$4, m_1, m_2, m_3, 15$$

have a common difference. In this case you have $a_1 = 4$, $n = 5$, and $a_5 = 15$. Therefore,

$$a_5 = 15 = a_1 + (n - 1)d = 4 + 4d.$$

Since $15 = 4 + 4d$, you find that $d = \frac{11}{4}$, and the three arithmetic means are as follows.

$$m_1 = a_1 + d = 4 + \frac{11}{4} = \frac{27}{4}$$

$$m_2 = m_1 + d = \frac{27}{4} + \frac{11}{4} = \frac{38}{4}$$

$$m_3 = m_2 + d = \frac{38}{4} + \frac{11}{4} = \frac{49}{4}$$

Applications

EXAMPLE 9 Seating Capacity

An auditorium has 20 rows of seats. There are 20 seats in the first row, 21 seats in the second row, 22 seats in the third row, and so on (see Figure 9.2). How many seats are there in all 20 rows?

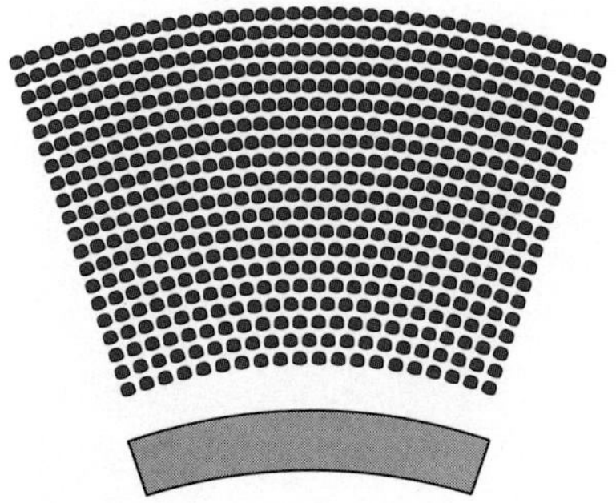

FIGURE 9.2

Solution

The number of seats in the rows forms an arithmetic sequence in which the common difference is $d = 1$. Since $c = a_1 - d = 20 - 1 = 19$, you can determine that the formula for the nth term in the sequence is $a_n = n + 19$. Therefore, the 20th term in the sequence is $a_{20} = 20 + 19 = 39$, and the total number of seats is

$$\begin{aligned} S &= 20 + 21 + 22 + \cdots + 39 \\ &= \frac{n}{2}(a_1 + a_{20}) \\ &= \frac{20}{2}(20 + 39) \\ &= 10(59) \\ &= 590. \end{aligned}$$

EXAMPLE 10 Total Sales

A small business sells \$10,000 worth of products during its first year. The owner of the business has set a goal of increasing annual sales by \$7,500 each year for 9 years. Assuming that this goal is met, find the total sales during the first 10 years this business is in operation.

Solution

The annual sales form an arithmetic sequence in which $a_1 = 10{,}000$ and $d = 7{,}500$. Thus, $c = a_1 - d = 10{,}000 - 7{,}500 = 2{,}500$ and the nth term of the sequence is

$$a_n = 7{,}500n + 2{,}500.$$

This implies that the 10th term of the sequence is $a_{10} = 77{,}500$. Therefore, the total sales for the first 10 years are as follows.

$$\begin{aligned} S &= \frac{n}{2}(a_1 + a_{10}) \\ &= \frac{10}{2}(10{,}000 + 77{,}500) \\ &= 5(87{,}500) \\ &= \$437{,}500 \end{aligned}$$

PROBLEM

Arithmetic Sequences

The first five terms of each of the indicated sequences are 1, 3, 5, 7, 9.

1. $a_n = 2n - 1$ 2. $b_n = \sqrt{4n^2 - 4n + 1}$

3. $c_n = \dfrac{2n^2 + n - 1}{n + 1}$

Does this fact *alone* mean that each sequence is arithmetic? Is each sequence arithmetic? Write a paragraph justifying your answer. Here is another sequence whose first five terms are 1, 3, 5, 7, 9. Is it arithmetic?

$$d_n = n^6 - 15n^5 + 85n^4 - 225n^3 + 274n^2 - 118n - 1$$

WARM UP

The following warm-up exercises involve skills that were covered in earlier sections. You will use these skills in the exercise set for this section.

In Exercises 1 and 2, find the sum.

1. $\displaystyle\sum_{i=1}^{6} (2i - 1)$ **2.** $\displaystyle\sum_{i=1}^{10} (4i + 2)$

In Exercises 3 and 4, find the distance between the two real numbers.

3. $\frac{5}{2}, 8$ **4.** $\frac{4}{3}, \frac{14}{3}$

In Exercises 5 and 6, find the required value of the function.

5. $f(n) = 10 + (n - 1)4$, $f(3)$ **6.** $f(n) = 1 + (n - 1)\frac{1}{3}$, $f(10)$

In Exercises 7–10, evaluate the expression.

7. $\frac{11}{2}(1 + 25)$ **8.** $\frac{16}{2}(4 + 16)$

9. $\frac{20}{2}[2(5) + (12 - 1)3]$ **10.** $\frac{8}{2}[2(-3) + (15 - 1)5]$

EXERCISES for Section 9.2

In Exercises 1–10, determine whether the sequence is arithmetic. If it is, find the common difference.

1. 4, 7, 10, 13, 16, . . .   **2.** 10, 8, 6, 4, 2, . . .

3. 1, 2, 4, 8, 16, . . .   **4.** $3, \frac{5}{2}, 2, \frac{3}{2}, 1, \ldots$

5. $\frac{9}{4}, 2, \frac{7}{4}, \frac{3}{2}, \frac{5}{4}, \ldots$ **6.** $-12, -8, -4, 0, 4, \ldots$

7. $\frac{1}{3}, \frac{2}{3}, \frac{4}{3}, \frac{8}{3}, \frac{16}{3}, \ldots$

8. ln 1, ln 2, ln 3, ln 4, ln 5, . . .

9. 5.3, 5.7, 6.1, 6.5, 6.9, . . .

10. $1^2, 2^2, 3^2, 4^2, 5^2, \ldots$

In Exercises 11–18, write the first five terms of the specified sequence. Determine whether the sequence is arithmetic, and if it is, find the common difference.

11. $a_n = 5 + 3n$ **12.** $a_n = (2^n)n$

13. $a_n = \dfrac{1}{n + 1}$ **14.** $a_n = 1 + (n - 1)4$

15. $a_n = 100 - 3n$ **16.** $a_n = 2^{n-1}$

17. $a_1 = 1, a_2 = 1, a_n = a_{n-1} + a_{n-2}, \quad n \geq 3$

18. $a_n = (-1)^n$

In Exercises 19–30, find a formula for a_n for the given arithmetic sequence.

19. $a_1 = 1, d = 3$
20. $a_1 = 15, d = 4$
21. $a_1 = 100, d = -8$
22. $a_1 = 0, d = -\frac{2}{3}$
23. $a_1 = x, d = 2x$
24. $a_1 = -y, d = 5y$
25. $4, \frac{3}{2}, -1, -\frac{7}{2}, \ldots$
26. $10, 5, 0, -5, -10, \ldots$
27. $a_1 = 5, a_4 = 15$
28. $a_1 = -4, a_5 = 16$
29. $a_3 = 94, a_6 = 85$
30. $a_5 = 190, a_{10} = 115$

In Exercises 31–40, write the first five terms of the arithmetic sequence.

31. $a_1 = 5, d = 6$
32. $a_1 = 5, d = -\frac{3}{4}$
33. $a_1 = -2.6, d = -0.4$
34. $a_1 = 16.5, d = 0.25$
35. $a_1 = \frac{3}{2}, a_{k+1} = a_k - \frac{1}{4}$
36. $a_1 = 6, a_{k+1} = a_k + 12$
37. $a_1 = 2, a_{12} = 46$
38. $a_4 = 16, a_{10} = 46$
39. $a_8 = 26, a_{12} = 42$
40. $a_3 = 19, a_{15} = -1.7$

In Exercises 41–48, find the sum of the first n terms of the arithmetic sequence.

41. $8, 20, 32, 44, \ldots, \quad n = 10$
42. $2, 8, 14, 20, \ldots, \quad n = 25$
43. $-6, -2, 2, 6, \ldots, \quad n = 50$
44. $0.5, 0.9, 1.3, 1.7, \ldots, \quad n = 10$
45. $40, 37, 34, 31, \ldots, \quad n = 10$
46. $1.50, 1.45, 1.40, 1.35, \ldots, \quad n = 20$
47. $a_1 = 100, a_{25} = 220, \quad n = 25$
48. $a_1 = 15, a_{100} = 307, \quad n = 100$

In Exercises 49–60, find the indicated sum.

49. $\displaystyle\sum_{n=1}^{50} n$
50. $\displaystyle\sum_{n=1}^{100} 2n$
51. $\displaystyle\sum_{n=1}^{100} 5n$
52. $\displaystyle\sum_{n=51}^{100} 7n$
53. $\displaystyle\sum_{n=11}^{30} n - \sum_{n=1}^{10} n$
54. $\displaystyle\sum_{n=51}^{100} n - \sum_{n=1}^{50} n$
55. $\displaystyle\sum_{n=1}^{500} (n + 3)$
56. $\displaystyle\sum_{n=1}^{250} (1000 - n)$
57. $\displaystyle\sum_{n=1}^{20} (2n + 5)$
58. $\displaystyle\sum_{n=1}^{100} \frac{n + 4}{2}$
59. $\displaystyle\sum_{n=0}^{50} (1000 - 5n)$
60. $\displaystyle\sum_{n=0}^{100} \frac{8 - 3n}{16}$

In Exercises 61–64, insert k arithmetic means between the given pair of numbers.

61. $5, 17, \quad k = 2$
62. $24, 56, \quad k = 3$
63. $3, 6, \quad k = 3$
64. $2, 5, \quad k = 4$

65. Find the sum of the first 100 odd positive integers.

66. Find the sum of the integers from -10 to 50.

67. *Job Offer* A person accepts a position with a company at a salary of \$27,500 for the first year. The person is guaranteed a raise of \$1,500 per year for the first five years.
(a) Determine the person's salary during the sixth year of employment.
(b) Determine the person's total compensation from the company through six full years of employment.

68. *Job Offer* A person accepts a position with a company at a salary of \$32,800 for the first year. The person is guaranteed a raise of \$1,750 per year for the first five years.
(a) Determine the person's salary during the sixth year of employment.
(b) Determine the person's total compensation from the company through six full years of employment.

69. *Seating Capacity* Determine the seating capacity of an auditorium with 30 rows of seats if there are 20 seats in the first row, 24 seats in the second row, 28 seats in the third row, and so on.

70. *Seating Capacity* Determine the seating capacity of an auditorium with 36 rows of seats if there are 15 seats in the first row, 18 seats in the second row, 21 seats in the third row, and so on.

71. *Brick Pattern* A brick patio is roughly the shape of a trapezoid (see figure). The patio has 20 rows of bricks. The first row has 14 bricks and the 20th row has 33 bricks. How many bricks are in the patio?

Figure for 71

72. *Falling Object* An object (with negligible air resistance) is dropped from a plane. During the first second of its fall, the object falls 4.9 meters; during the second second, it falls 14.7 meters; during the third second, it falls 24.5 meters; and during the fourth second, it falls 34.3 meters. If this arithmetic pattern continues, how many meters will the object have fallen in 10 seconds?

PROBLEM

B Y   F I N D I N G   P A T T E R N S

SOLVING

___

Finding Models
to Data by
Recognizing
Patterns

When creating a mathematical model of a real-life situation, you often do not expect the model to fit the real-life data *exactly*. For instance, in the technology feature on pages 434–435, we found linear models that approximately fit data for newspaper circulation. For that particular situation, there is no simple mathematical model that exactly fits the data.

Occasionally, however, real-life situations require mathematical models that do exactly fit the data. This often occurs in science and engineering, where you expect variables to be related by scientific or geometric principles.

EXAMPLE 1   Finding an Exact Mathematical Model

A polygon is *regular* if all its sides have the same length and all its angles have the same measure. The table gives the degree measures A_n of the interior angles of n-sided regular polygons for n equal to 3, 4, 5, 6, 7, and 8. Find a mathematical model for A_n in terms of n.

Number of Sides, n	3	4	5	6	7	8
Angle Measure, A_n	60°	90°	108°	120°	$128\frac{4}{7}°$	135°

Solution

From the table, it is not clear what pattern the angle measures are following. When hunting for a mathematical model, performing an operation of the data can help to make the pattern more evident. For instance, you might take the natural logarithm of the angle measures, or square them, or take their square roots. For this particular data collection, multiplying A_n by n produces an easily recognized pattern.

Number of Sides, n	3	4	5	6	7	8
Product, nA_n	180°	360°	540°	720°	900°	1080°

From this result, you can see that the products nA_n form an arithmetic sequence whose nth term is

$$nA_n = 180(n - 2).$$

Thus, the model for the degree measure of the interior angle of an n-sided regular polygon is

$$A_n = \frac{180(n - 2)}{n}.$$

We have verified this model for values of n from 3 through 8. To verify that the model works for larger values of n would require further justification. ◢

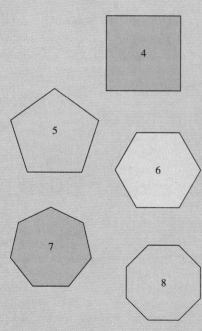

EXERCISES

(See also: Exercises 69–72, Section 9.2)

1. *Stars* One way to form an *n*-pointed star is to begin with *n* equally spaced points on a circle and connect every second point. The angle measures of the star tips for *n* = 5, 6, 7, 8, 9, 10 are shown below. Find a mathematical model for these angle measures.

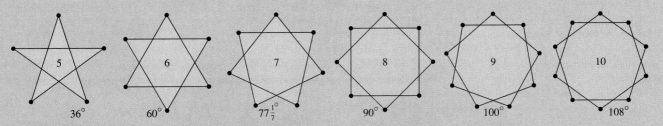

2. *More Stars* Another way to form an *n*-pointed star is to begin with *n* equally spaced points on a circle and connect every third point. The angle measures of the star tips for *n* = 7, 8, 9, 10, 11, 12 are shown below. Find a mathematical model for these angle measures.

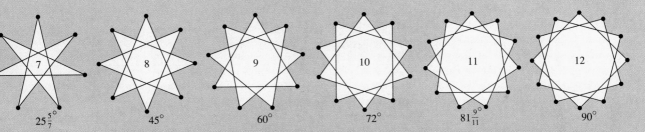

3. *Even More Stars* A regular polygon can be considered to be a "star" formed by connected adjacent points. Describe the pattern formed by the models in Example 1, Exercise 1, and Exercise 2.

> *Example 1:* Form star by connecting adjacent points.
> *Exercise 1:* Form star by connecting every second point.
> *Exercise 2:* Form star by connecting every third point.

The stars below are formed by connecting every *fourth* point on the circle. Find a model for the angle measures of these star tips. Explain your reasoning.

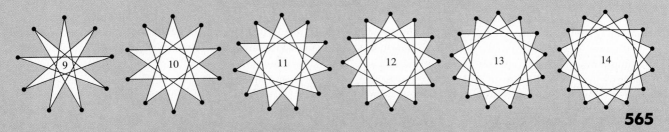

9.3 Geometric Sequences

Introduction / The Sum of a Geometric Sequence / Applications

Introduction

In Section 9.2 you saw that a sequence whose consecutive terms have a common *difference* is called an arithmetic sequence. In this section you will study another important type of sequence—a **geometric sequence.** Consecutive terms of a geometric sequence have a common *ratio*, as indicated in the following definition.

DEFINITION OF A GEOMETRIC SEQUENCE

A sequence is **geometric** if the ratios of consecutive terms are the same. Thus, the sequence

$$a_1, a_2, a_3, a_4, \ldots, a_n, \ldots$$

is geometric if there is a number r, $r \neq 0$, such that

$$\frac{a_2}{a_1} = r, \qquad \frac{a_3}{a_2} = r, \qquad \frac{a_4}{a_3} = r,$$

and so on. The number r is the **common ratio** of the geometric sequence.

EXAMPLE 1 Examples of Geometric Sequences

a. The sequence whose nth term is 2^n is geometric. For this sequence, the common ratio between consecutive terms is 2.

$$\underbrace{2, 4,}_{\frac{4}{2} = 2} 8, 16, \ldots, 2^n, \ldots$$

b. The sequence whose nth term is $4(3^n)$ is geometric. For this sequence, the common ratio between consecutive terms is 3.

$$\underbrace{12, 36,}_{\frac{36}{12} = 3} 108, 324, \ldots, 4(3^n), \ldots$$

c. The sequence whose nth term is $\left(-\frac{1}{3}\right)^n$ is geometric. For this sequence, the common ratio between consecutive terms is $-\frac{1}{3}$.

$$\underbrace{-\frac{1}{3}, \frac{1}{9},}_{\frac{1/9}{-1/3} = -\frac{1}{3}} -\frac{1}{27}, \frac{1}{81}, \ldots, \left(-\frac{1}{3}\right)^n, \ldots$$

In Example 1, notice that each of the geometric sequences has an nth term that is of the form ar^n, where the common ratio of the sequence is r. We summarize this result as follows.

THE nTH TERM OF A GEOMETRIC SEQUENCE

The nth term of a geometric sequence has the form

$$a_n = a_1 r^{n-1}$$

where r is the common ratio of consecutive terms of the sequence. Thus, every geometric sequence can be written in the following form.

$$a_1, \ a_2, \ a_3, \ a_4, \ a_5, \ \ldots, \ a_n, \ \ldots$$
$$\downarrow \ \downarrow \ \downarrow \ \downarrow \ \downarrow \ \ldots, \ \downarrow \ \ldots$$
$$a_1, \ a_1 r, \ a_1 r^2, \ a_1 r^3, \ a_1 r^4, \ \ldots, \ a_1 r^{n-1} \ \ldots$$

REMARK   If you know the nth term of a geometric sequence, then the $(n + 1)$th term can be found by multiplying by r. That is,

$$a_{n+1} = r a_n.$$

EXAMPLE 2   Finding the Terms of a Geometric Sequence

Write the first five terms of the geometric sequence whose first term is $a_1 = 3$ and whose common ratio is $r = 2$.

Solution

Starting with 3, we repeatedly multiply by 2 to obtain the following.

$$a_1 = 3$$
$$a_2 = 3(2^1) = 6$$
$$a_3 = 3(2^2) = 12$$
$$a_4 = 3(2^3) = 24$$
$$a_5 = 3(2^4) = 48$$

EXAMPLE 3   Finding a Term of a Geometric Sequence

Find the 15th term of the geometric sequence whose first term is 20 and whose common ratio is 1.05.

Solution

You can obtain the 15th term by multiplying the first term by r 14 times. Thus, since $a_1 = 20$ and $r = 1.05$, the 15th term ($n = 15$) is

$$a_{15} = a_1 r^{n-1} = 20(1.05)^{14} \approx 39.599.$$

EXAMPLE 4　Finding a Term of a Geometric Sequence

Find the 12th term of the geometric sequence

　　5, 15, 45, ...

Solution

The common ratio of this sequence is $r = \frac{15}{5} = 3$. Therefore, since the first term is $a_1 = 5$, you determine the 12th term ($n = 12$) to be

$$a_{12} = a_1 r^{n-1} = 5(3)^{11} = 5(177{,}147) = 885{,}735.$$

EXAMPLE 5　Finding a Term of a Geometric Sequence

The 4th term of a geometric sequence is 125, and the 10th term is $\frac{125}{64}$. Find the 14th term.

Solution

To obtain the 10th term from the 4th term, you can multiply the 4th term by r^6. That is, $a_{10} = a_4 r^6$. Since $a_{10} = \frac{125}{64}$ and $a_4 = 125$, you have the following.

$$\frac{125}{64} = 125r^6$$

$$\frac{1}{64} = r^6$$

$$\frac{1}{2} = r$$

Now you can obtain the 14th term by multiplying the 10th term by r^4. That is,

$$a_{14} = a_{10}r^4 = \frac{125}{64}\left(\frac{1}{2}\right)^4 = \frac{125}{1024}.$$

The Sum of a Geometric Sequence

The formula for the sum of a *finite* geometric sequence is as follows.

THE SUM OF A FINITE GEOMETRIC SEQUENCE

The sum of the finite geometric sequence

$$a_1, a_1r, a_1r^2, a_1r^3, a_1r^4, \ldots, a_1r^{n-1}$$

with common ratio $r \neq 1$ is given by

$$S = a_1\left(\frac{1 - r^n}{1 - r}\right).$$

Proof

Begin by writing out the nth partial sum.

$$S = a_1 + a_1 r + a_1 r^2 + \cdots + a_1 r^{n-2} + a_1 r^{n-1}$$

Multiplication by r yields

$$rS = a_1 r + a_1 r^2 + a_1 r^3 + \cdots + a_1 r^{n-1} + a_1 r^n.$$

Subtracting the second equation from the first yields

$$S - rS = a_1 - a_1 r^n.$$

Therefore, $S(1 - r) = a_1(1 - r^n)$, and since $r \neq 1$, you have

$$S = a_1 \left(\frac{1 - r^n}{1 - r} \right).$$

EXAMPLE 6   Finding the Sum of a Finite Geometric Sequence

Find the following sum.

$$\sum_{n=1}^{12} 4(0.3)^n$$

Solution

By writing out a few terms, you have

$$\sum_{n=1}^{12} 4(0.3)^n = 4(0.3) + 4(0.3)^2 + 4(0.3)^3 + \cdots + 4(0.3)^{12}.$$

Now, since $a_1 = 4(0.3)$, $r = 0.3$, and $n = 12$, apply the formula for the sum of a geometric sequence to obtain

$$\sum_{n=1}^{12} 4(0.3)^n = a_1 \left(\frac{1 - r^n}{1 - r} \right)$$

$$= 4(0.3) \left(\frac{1 - (0.3)^{12}}{1 - 0.3} \right)$$

$$\approx 1.714.$$

When using the formula for the sum of a finite geometric sequence, be careful to check that the index begins at $i = 1$. If the index begins at $i = 0$, you must adjust the formula for the nth partial sum, as demonstrated in the following example.

EXAMPLE 7 Finding the Sum of a Finite Geometric Sequence

Find the following sum.

$$\sum_{i=0}^{10} 10\left(-\frac{1}{2}\right)^i$$

Solution

By writing out a few terms, you have

$$\sum_{i=0}^{10} 10\left(-\frac{1}{2}\right)^i = 10 - 10\left(\frac{1}{2}\right) + 10\left(\frac{1}{2}\right)^2 - \cdots + 10\left(\frac{1}{2}\right)^{10}.$$

You can see that the *first term* is $a_1 = 10$ and $r = -\frac{1}{2}$. Moreover, by starting with $i = 0$ and ending with $i = 10$, you are adding $n = 11$ terms, which means that the partial sum is

$$\sum_{i=0}^{10} 10\left(-\frac{1}{2}\right)^i = a_1\left(\frac{1 - r^n}{1 - r}\right) = 10\left(\frac{1 - (-1/2)^{11}}{1 - (-1/2)}\right) \approx 6.670.$$

The formula for the sum of a *finite* geometric sequence can, depending on the value of r, be extended to find the sum of an *infinite* geometric sequence. Specifically, if the common ratio r has the property that $|r| < 1$, then it can be shown that r^n becomes arbitrarily close to zero as n increases without bound. Consequently,

$$S \rightarrow a_1\left(\frac{1 - 0}{1 - r}\right) = \frac{a_1}{1 - r} \qquad \text{as} \qquad n \rightarrow \infty.$$

We summarize this result as follows.

REMARK The summation $a_1 + a_1 r + a_1 r^2 + a_1 r^3 + \cdots$ is an **infinite geometric series.**

SUM OF AN INFINITE GEOMETRIC SEQUENCE

If $|r| < 1$, then the infinite geometric sequence

$$a_1, a_1 r, a_1 r^2, a_1 r^3, \ldots, a_1 r^{n-1}, \ldots$$

has the sum

$$S = \sum_{n=1}^{\infty} a_1 r^{n-1} = \frac{a_1}{1 - r}.$$

EXAMPLE 8 **Finding the Sum of an Infinite Geometric Sequence**

Find the sum of the following infinite geometric sequence.

$$4, \ 4(0.6), \ 4(0.6)^2, \ 4(0.6)^3, \ \ldots, \ 4(0.6)^{n-1}, \ \ldots$$

Solution

Since $a_1 = 4$, $r = 0.6$, and $|r| < 1$, you have

$$S = \sum_{n=1}^{\infty} 4(0.6)^{n-1}$$

$$= \frac{a_1}{1 - r}$$

$$= \frac{4}{1 - (0.6)}$$

$$= 10.$$

Applications

EXAMPLE 9 **An Application: Compound Interest**

A deposit of $50 is made the first day of each month in a savings account that pays 12% compounded monthly. What is the balance at the end of two years?

Solution

The formula for compound interest is

$$A = P\left(1 + \frac{r}{n}\right)^{tn}$$

where A is the balance of the account, P is the initial deposit, r is the annual percentage rate, n is the number of compoundings per year, and t is the time (in years). To find the balance in the account after 24 months, it is helpful to consider each of the 24 deposits separately. For example, the first deposit will gain interest for a full 24 months, and its balance will be

$$A_{24} = 50\left(1 + \frac{0.12}{12}\right)^{24} = 50(1.01)^{24}.$$

The second deposit will gain interest for 23 months, and its balance will be

$$A_{23} = 50\left(1 + \frac{0.12}{12}\right)^{23} = 50(1.01)^{23}.$$

The last (24th) deposit will gain interest for only 1 month, and its balance will be

$$A_1 = 50\left(1 + \frac{0.12}{12}\right)^1 = 50(1.01).$$

Finally, the total balance in the account will be the sum of the balances of the 24 deposits.

$$S = A_1 + A_2 + A_3 + \cdots + A_{23} + A_{24}$$

Using the formula for the sum of a geometric sequence, with $A_1 = 50(1.01)$ and $r = 1.01$, you have

$$S = 50(1.01)\left(\frac{1 - (1.01)^{24}}{1 - 1.01}\right) = \$1362.16.$$

DISCUSSION

PROBLEM

Comparing Two Sequences

The first several terms of the sequences whose nth terms are $a_n = (0.99)^{n-1}$ and $b_n = 1 - (n - 1)(0.01)$ are almost the same.

1. $a_1 = 1$
   $a_2 = 0.99$
   $a_3 = (0.99)^2 \approx 0.98$
   $a_4 = (0.99)^3 \approx 0.97$
   $a_5 = (0.99)^4 \approx 0.96$

2. $b_1 = 1$
   $b_2 = 1 - 0.01 = 0.99$
   $b_3 = 1 - 2(0.01) = 0.98$
   $b_4 = 1 - 3(0.01) = 0.97$
   $b_5 = 1 - 4(0.01) = 0.96$

Yet the two sequences have very basic differences. Discuss, or write a paragraph describing, some of the differences between the two sequences.

WARM UP

The following warm-up exercises involve skills that were covered in earlier sections. You will use these skills in the exercise set for this section.

In Exercises 1–4, evaluate the expression.

1. $\left(\frac{4}{5}\right)^3$

2. $\left(\frac{3}{4}\right)^2$

3. 2^{-4}

4. $\dfrac{5}{3^4}$

In Exercises 5–10, simplify the expression.

5. $(2n)(3n^2)$

6. $n(3n)^3$

7. $\dfrac{4n^5}{n^2}$

8. $\dfrac{(2n)^3}{8n}$

9. $[2(3)^{-4}]^n$

10. $3(4^2)^{-n}$

EXERCISES for Section 9.3

In Exercises 1–10, determine whether the sequence is geometric. If it is, find its common ratio.

1. $5, 15, 45, 135, \ldots$

2. $3, 12, 48, 192, \ldots$

3. $3, 12, 21, 30, \ldots$

4. $1, -2, 4, -8, \ldots$

5. $1, -\frac{1}{2}, \frac{1}{4}, -\frac{1}{8}, \ldots$

6. $5, 1, 0.2, 0.04, \ldots$

7. $\frac{1}{2}, \frac{2}{3}, \frac{3}{4}, \frac{4}{5}, \ldots$

8. $9, -6, 4, -\frac{8}{3}, \ldots$

9. $1, \frac{1}{2}, \frac{1}{3}, \frac{1}{4}, \ldots$

10. $\frac{1}{5}, \frac{2}{3}, \frac{3}{9}, \frac{4}{11}, \ldots$

In Exercises 11–20, write the first five terms of the geometric sequence.

11. $a_1 = 2, r = 3$

12. $a_1 = 6, r = 2$

13. $a_1 = 1, r = \frac{1}{2}$

14. $a_1 = 1, r = \frac{1}{3}$

15. $a_1 = 5, r = -\frac{1}{10}$

16. $a_1 = 6, r = -\frac{1}{4}$

17. $a_1 = 1, r = e$

18. $a_1 = 2, r = \sqrt{3}$

19. $a_1 = 3, r = \dfrac{x}{2}$

20. $a_1 = 5, r = 2x$

In Exercises 21–32, find the nth term of the geometric sequence.

21. $a_1 = 4, r = \frac{1}{2}, n = 10$

22. $a_1 = 5, r = \frac{3}{2}, n = 8$

23. $a_1 = 6, r = -\frac{1}{3}, n = 12$

24. $a_1 = 8, r = \sqrt{5}, n = 9$

25. $a_1 = 100, r = e^x, n = 9$

26. $a_1 = 1, r = -\dfrac{x}{3}, n = 7$

27. $a_1 = 500, r = 1.02, n = 40$

28. $a_1 = 1000, r = 1.005, n = 60$

29. $a_1 = 16, a_4 = \frac{27}{4}, n = 3$

30. $a_2 = 3, a_5 = \frac{3}{64}, n = 1$

31. $a_2 = -18, a_5 = \frac{2}{3}, n = 6$

32. $a_3 = \frac{16}{3}, a_5 = \frac{64}{27}, n = 7$

33. *Compound Interest* A principal of $1000 is invested at 10% interest. Find the amount after 10 years if the interest is compounded (a) annually, (b) semiannually, (c) quarterly, (d) monthly, and (e) daily.

34. *Compound Interest* A principal of $2500 is invested at 12% interest. Find the amount after 20 years if the interest is compounded (a) annually, (b) semiannually, (c) quarterly, (d) monthly, and (e) daily.

35. *Depreciation* A company buys a machine for $135,000 that depreciates at the rate of 30% per year. (In other words, at the end of each year the depreciated value is 70% of what it was at the beginning of the year.) Find the depreciated value of the machine after five full years.

36. *Population Growth* A city of 250,000 people is growing at the rate of 1.3% per year. Estimate the population of the city 30 years from now.

In Exercises 37–46, find the indicated sum.

37. $\displaystyle\sum_{n=1}^{9} 2^{n-1}$

38. $\displaystyle\sum_{n=1}^{9} (-2)^{n-1}$

39. $\displaystyle\sum_{i=1}^{7} 64\left(-\frac{1}{2}\right)^{i-1}$

40. $\displaystyle\sum_{i=1}^{6} 32\left(\frac{1}{4}\right)^{i-1}$

41. $\displaystyle\sum_{i=1}^{10} 8\left(\frac{-1}{4}\right)^{i-1}$

42. $\displaystyle\sum_{i=1}^{10} 5\left(\frac{-1}{3}\right)^{i-1}$

43. $\displaystyle\sum_{n=0}^{20} 3\left(\frac{3}{2}\right)^{n}$

44. $\displaystyle\sum_{n=0}^{15} 2\left(\frac{4}{3}\right)^{n}$

45. $\displaystyle\sum_{n=0}^{5} 300(1.06)^{n}$

46. $\displaystyle\sum_{n=0}^{6} 500(1.04)^{n}$

47. *Compound Interest* A deposit of $100 is made at the beginning of each month for five years in an account that pays 10%, compounded monthly. What is the balance A in the account at the end of five years?

$$A = 100\left(1 + \frac{0.10}{12}\right)^{1} + \cdots + 100\left(1 + \frac{0.10}{12}\right)^{60}$$

48. *Compound Interest* A deposit of $50 is made at the beginning of each month for five years in an account that pays 12%, compounded monthly. What is the balance A in the account at the end of five years?

$$A = 50\left(1 + \frac{0.12}{12}\right)^{1} + \cdots + 50\left(1 + \frac{0.12}{12}\right)^{60}$$

49. *Compound Interest* A deposit of P dollars is made at the beginning of each month in an account at an annual interest rate r, compounded monthly. The balance A after t years is

$$A = P\left(1 + \frac{r}{12}\right) + P\left(1 + \frac{r}{12}\right)^{2} + \cdots + P\left(1 + \frac{r}{12}\right)^{12t}.$$

Show that the balance is given by

$$A = P\left[\left(1 + \frac{r}{12}\right)^{12t} - 1\right]\left(1 + \frac{12}{r}\right).$$

50. *Compound Interest*   A deposit of P dollars is made each month in an account at an annual interest rate r, compounded continuously. The balance A after t years is

$$A = Pe^{r/12} + Pe^{2r/12} + \cdots + Pe^{12tr/12}.$$

Show that the balance is given by

$$A = \frac{Pe^{r/12}(e^{rt} - 1)}{e^{r/12} - 1}.$$

Compound Interest   In Exercises 51–54, consider making monthly deposits of P dollars into a savings account at an annual interest rate r. Use the results of Exercises 49 and 50 to find the balance A after t years if the interest is compounded (a) monthly and (b) continuously.

51. $P = \$50,\quad r = 7\%,\quad t = 20$ years
52. $P = \$75,\quad r = 9\%,\quad t = 25$ years
53. $P = \$100,\ r = 10\%,\ t = 40$ years
54. $P = \$20,\quad r = 6\%,\quad t = 50$ years

55. *Profit*   The annual profit for the H.J. Heinz Company from 1980 through 1989 can be approximated by the model

$$a_n = 167.5e^{0.12n},\qquad n = 0, 1, 2, \ldots, 9$$

where a_n is the annual profit in millions of dollars and n represents the year with $n = 0$ corresponding to 1980 (see figure). Use the formula for the sum of a geometric sequence to approximate the total profit earned during this 10-year period. (*Source*: H.J. Heinz Company)

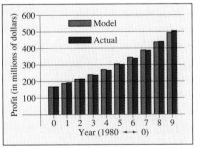

Figure for 55

56. *Would You Take This Job?*   Suppose you went to work at a company that pays $\$0.01$ for the first day, $\$0.02$ for the second day, $\$0.04$ for the third day, and so on. If the daily wage keeps doubling, what would your total income be for working (a) 29 days? (b) 30 days? (c) 31 days?

57. *Salary*   You accept a job with a salary of $\$30,000$ for the first year. Suppose that during the next 39 years you receive a 5% raise each year. What would your total compensation be over the 40-year period?

58. *Area*   The sides of a square are 16 inches long. The square is divided into nine smaller squares, and the center square is shaded dark red (see figure). Each of the eight light red squares is then divided into nine smaller squares, and the center square of each is shaded dark red. If this process is repeated four more times, determine the area of the region shaded dark red.

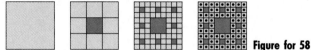

Figure for 58

In Exercises 59–68, find the sum of the infinite series.

59. $\displaystyle\sum_{n=0}^{\infty} \left(\tfrac{1}{2}\right)^n = 1 + \tfrac{1}{2} + \tfrac{1}{4} + \tfrac{1}{8} + \cdots$

60. $\displaystyle\sum_{n=0}^{\infty} 2\left(\tfrac{2}{3}\right)^n = 2 + \tfrac{4}{3} + \tfrac{8}{9} + \tfrac{16}{27} + \cdots$

61. $\displaystyle\sum_{n=0}^{\infty} \left(-\tfrac{1}{2}\right)^n = 1 - \tfrac{1}{2} + \tfrac{1}{4} - \tfrac{1}{8} + \cdots$

62. $\displaystyle\sum_{n=0}^{\infty} 2\left(-\tfrac{2}{3}\right)^n = 2 - \tfrac{4}{3} + \tfrac{8}{9} - \tfrac{16}{27} + \cdots$

63. $\displaystyle\sum_{n=0}^{\infty} 4\left(\tfrac{1}{4}\right)^n = 4 + 1 + \tfrac{1}{4} + \tfrac{1}{16} + \cdots$

64. $\displaystyle\sum_{n=0}^{\infty} \left(\tfrac{1}{10}\right)^n = 1 + 0.1 + 0.01 + 0.001 + \cdots$

65. $8 + 6 + \tfrac{9}{2} + \tfrac{27}{8} + \cdots$ **66.** $3 - 1 + \tfrac{1}{3} - \tfrac{1}{9} + \cdots$

67. $4 - 2 + 1 - \tfrac{1}{2} + \cdots$ **68.** $2 + \sqrt{2} + 1 + \dfrac{1}{\sqrt{2}} + \cdots$

69. *Distance*   A ball is dropped 16 feet. Each time it drops h feet, it rebounds $0.81h$ feet. Find the total distance it travels.

70. *Time*   The ball in Exercise 69 takes the following time for each fall.

$s_1 = -16t^2 + 16,$ $s_1 = 0$ if $t = 1$
$s_2 = -16t^2 + 16(0.81),$ $s_2 = 0$ if $t = 0.9$
$s_3 = -16t^2 + 16(0.81)^2,$ $s_3 = 0$ if $t = (0.9)^2$
$s_4 = -16t^2 + 16(0.81)^3,$ $s_4 = 0$ if $t = (0.9)^3$
$\vdots$ $\vdots$
$s_n = -16t^2 + 16(0.81)^{n-1},$ $s_n = 0$ if $t = (0.9)^{n-1}$

Beginning with s_2, the ball takes the same amount of time to bounce up as it does to fall, and thus the total time elapsed before it comes to rest is

$$t = 1 + 2\sum_{n=1}^{\infty} (0.9)^n.$$

Find this total.

9.4 Mathematical Induction

Introduction / Sums of Powers of Integers

Introduction

In this section we look at a form of mathematical proof, the principle of **mathematical induction.** It is important that you clearly see the logical need for this principle, so let's take a closer look at a problem we discussed earlier.

$$S_1 = 1 = 1^2$$
$$S_2 = 1 + 3 = 2^2$$
$$S_3 = 1 + 3 + 5 = 3^2$$
$$S_4 = 1 + 3 + 5 + 7 = 4^2$$
$$S_5 = 1 + 3 + 5 + 7 + 9 = 5^2$$

Judging from the pattern formed by these first five sums, it appears that the sum of the first n odd integers is

$$S_n = 1 + 3 + 5 + 7 + 9 + \cdots + 2n - 1 = n^2.$$

While this particular formula *is* valid, it is important for you to see that recognizing a pattern and then simply *jumping to the conclusion* that the pattern must be true for all values of n is *not* a logically valid method of proof. There are many examples in which a pattern appears to be developing for small values of n and then at some point the pattern fails. One of the most famous cases of this was the conjecture by the French mathematician Pierre de Fermat (1601–1655), who speculated that all numbers of the form

$$F_n = 2^{2^n} + 1, \quad n = 0, 1, 2, \ldots$$

are prime. For $n = 0$, 1, 2, 3, and 4, the conjecture is true.

$$F_0 = 3, \quad F_1 = 5, \quad F_2 = 17, \quad F_3 = 257, \quad F_4 = 65,537$$

The size of the next Fermat number ($F_5 = 4,294,967,297$) is so great that it was difficult for Fermat to determine whether it was prime or not. However, another well-known mathematician, Leonhard Euler (1707–1783), later found a factorization

$$F_5 = 4,294,967,297 = 641(6,700,417)$$

which proved that F_5 is not prime, and therefore Fermat's conjecture was false.

Just because a rule, pattern, or formula seems to work for several values of n, you cannot simply decide that it is valid for all values of n without going through a *legitimate proof*. Let's see how you prove such statements by the principle of **mathematical induction.**

THE PRINCIPLE OF MATHEMATICAL INDUCTION

Let P_n be a statement involving the positive integer n. If

1. P_1 is true, and
2. the truth of P_k implies the truth of P_{k+1}, for every positive integer k,

then P_n must be true for all positive integers n.

REMARK It is important to recognize that both parts of the Principle of Mathematical Induction are necessary.

To apply the Principle of Mathematical Induction you need to be able to determine the statement P_{k+1} for a given statement P_k.

EXAMPLE 1 A Preliminary Example

Find P_{k+1} for the following.

a. P_k: $S_k = \dfrac{k^2(k+1)^2}{4}$

b. P_k: $S_k = 1 + 5 + 9 + \cdots + [4(k-1) - 3] + (4k - 3)$

c. P_k: $3^k \geq 2k + 1$

Solution

a. Substituting $k + 1$ for k, you have

$$P_{k+1}: S_{k+1} = \frac{(k+1)^2(k+1+1)^2}{4} \qquad \textit{Replace k by k + 1}$$

$$= \frac{(k+1)^2(k+2)^2}{4}. \qquad \textit{Simplify}$$

b. In this case you have

$$P_{k+1}: S_{k+1} = 1 + 5 + 9 + \cdots + (4[(k+1) - 1] - 3) + [4(k+1) - 3]$$

$$= 1 + 5 + 9 + \cdots + (4k - 3) + (4k + 1).$$

c. Replacing k by $k + 1$ in the statement $3^k \geq 2k + 1$, you have

$$P_{k+1}: 3^{k+1} \geq 2(k+1) + 1$$

$$3^{k+1} \geq 2k + 3.$$

EXAMPLE 2 Using Mathematical Induction

Use mathematical induction to prove the following formula.

$$S_n = 1 + 3 + 5 + 7 + \cdots + (2n - 1) = n^2$$

Solution

Mathematical induction consists of two distinct parts. First, you must show that the formula is true when $n = 1$.

1. When $n = 1$, the formula is valid, since

$$S_1 = 1 = 1^2.$$

The second part of mathematical induction has two steps. The first step is to assume that the formula is valid for *some* integer k. The second step is to use this assumption to prove that the formula is valid for the next integer, $k + 1$.

2. Assuming that the formula

$$S_k = 1 + 3 + 5 + 7 + \cdots + (2k - 1) = k^2$$

is true, you must show that the formula $S_{k+1} = (k + 1)^2$ is true.

$$
\begin{aligned}
S_{k+1} &= 1 + 3 + 5 + 7 + \cdots + (2k - 1) + [2(k + 1) - 1] \\
&= [1 + 3 + 5 + 7 + \cdots + (2k - 1)] + (2k + 2 - 1) \\
&= S_k + (2k + 1) \\
&= k^2 + 2k + 1 \\
&= (k + 1)^2
\end{aligned}
$$

REMARK When using mathematical induction to prove a *summation* formula (like the one in Example 2), it is helpful to think of S_{k+1} as $S_{k+1} = S_k + a_{k+1}$, where a_{k+1} is the $(k + 1)$ term of the original sum.

Combining the results of parts (1) and (2), you conclude by mathematical induction that the formula is valid for *all* positive integer values of n. ◢

A well-known illustration used to explain why the principle of mathematical induction works is the unending line of dominoes shown in Figure 9.3. If the line actually contains infinitely many dominoes, then it is clear that you could not knock the entire line down by knocking down only *one domino* at a time. However, suppose it were true that each domino would knock down the next one as it fell. Then you could knock them all down simply by pushing the first one and starting a chain reaction. Mathematical induction works in the same way. If the truth of P_k implies the truth of P_{k+1} and if P_1 is true, then the chain reaction proceeds as follows:

P_1 implies P_2

P_2 implies P_3

P_3 implies P_4

and so on.

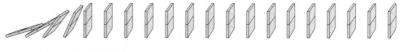

FIGURE 9.3

The first domino knocks over the second which knocks over the third which knocks over the fourth and so on.

It occasionally happens that a statement involving natural numbers is not true for the first $k - 1$ positive integers but is true for all values of $n \geq k$. In these instances, you use a slight variation of the principle of mathematical induction in which you verify P_k rather than P_1. This variation is called the **extended principle of mathematical induction.** To see the validity of this, note from Figure 9.3 that all but the first $k - 1$ dominoes can be knocked down by knocking over the kth domino. This suggests that you can prove a statement P_n to be true for $n \geq k$ by showing that P_k is true and that P_k implies P_{k+1}. In Exercises 29–32 of this section you are asked to apply this extension of mathematical induction.

EXAMPLE 3   Using Mathematical Induction

Use mathematical induction to prove the following formula.

$$S_n = 1^2 + 2^2 + 3^2 + 4^2 + \cdots + n^2 = \frac{n(n + 1)(2n + 1)}{6}$$

Solution

1. When $n = 1$, the formula is valid, because

$$S_1 = 1^2 = \frac{1(2)(3)}{6}.$$

2. Assuming that

$$S_k = 1^2 + 2^2 + 3^2 + 4^2 + \cdots + k^2 = \frac{k(k + 1)(2k + 1)}{6}$$

you must show that

$$S_{k+1} = \frac{(k + 1)(k + 2)(2k + 3)}{6}.$$

To do this, write the following.

$$
\begin{aligned}
S_{k+1} &= S_k + a_{k+1} \\
&= (1^2 + 2^2 + 3^2 + 4^2 + \cdots + k^2) + (k + 1)^2 \\
&= \frac{k(k + 1)(2k + 1)}{6} + (k + 1)^2 \\
&= \frac{k(k + 1)(2k + 1) + 6(k + 1)^2}{6} \\
&= \frac{(k + 1)[k(2k + 1) + 6(k + 1)]}{6} \\
&= \frac{(k + 1)[2k^2 + 7k + 6]}{6} \\
&= \frac{(k + 1)(k + 2)(2k + 3)}{6}
\end{aligned}
$$

Combining the results of parts (1) and (2), you conclude by mathematical induction that the formula is valid for *all n* ≥ 1.   ◢

Sums of Powers of Integers

The formula in Example 3 is one of a collection of useful summation formulas. We summarize this and other formulas dealing with the sum of various powers of the first *n* positive integers as follows.

SUMS OF POWERS OF INTEGERS

1. $1 + 2 + 3 + 4 + \cdots + n = \dfrac{n(n + 1)}{2}$

2. $1^2 + 2^2 + 3^2 + 4^2 + \cdots + n^2 = \dfrac{n(n + 1)(2n + 1)}{6}$

3. $1^3 + 2^3 + 3^3 + 4^3 + \cdots + n^3 = \dfrac{n^2(n + 1)^2}{4}$

4. $1^4 + 2^4 + 3^4 + 4^4 + \cdots + n^4$
$$= \dfrac{n(n + 1)(2n + 1)(3n^2 + 3n - 1)}{30}$$

5. $1^5 + 2^5 + 3^5 + 4^5 + \cdots + n^5 = \dfrac{n^2(n + 1)^2(2n^2 + 2n - 1)}{12}$

REMARK   Each of these formulas for sums can be proved by mathematical induction. (See Exercises 21–23, 25, 26.)

EXAMPLE 4   Finding a Sum of Powers of Integers

Find the following sum.

$$1^3 + 2^3 + 3^3 + 4^3 + 5^3 + 6^3 + 7^3$$

Solution

Using the formula for the sum of the cubes of the first *n* positive integers, you obtain the following.

$$1^3 + 2^3 + 3^3 + 4^3 + 5^3 + 6^3 + 7^3 = \frac{7^2(7 + 1)^2}{4}$$
$$= \frac{49(64)}{4}$$
$$= 784$$

Check this sum by adding the numbers 1, 8, 27, 64, 125, 216, and 343.   ◢

EXAMPLE 5   Proving an Inequality by Mathematical Induction

Prove that $n < 2^n$ for all positive integers n.

Solution

1. For $n = 1$, the formula is true, since

$$1 < 2^1.$$

2. Assuming that

$$k < 2^k$$

you need to show that $k + 1 < 2^{k+1}$. For $n = k$, you have

$$2^{k+1} = 2(2^k) > 2(k) = 2k. \qquad \textit{By assumption}$$

Since $2k = k + k > k + 1$ for all $k > 1$, it follows that

$$2^{k+1} > 2k > k + 1$$

or

$$k + 1 < 2^{k+1}.$$

Hence, $n < 2^n$ for all integers $n \geq 1$.

DISCUSSION

PROBLEM

The Sum of the Angles of a Regular Polygon

A *regular n*-sided polygon is a polygon that has n equal sides and n equal angles. For instance, an equilateral triangle is a regular three-sided polygon. Each angle of an equilateral triangle measures 60°, and the sum of all three angles is 180°. Similarly, the sum of the four angles of a regular four-sided polygon (a square) is 360°. From the following four regular polygons (see Figure 9.4), find a formula for the sum of the angles of a regular *n*-sided polygon. Describe how you could *prove* that your formula is valid. Do you think that mathematical induction would be an appropriate technique to prove the validity of your formula?

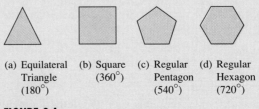

(a) Equilateral   (b) Square   (c) Regular   (d) Regular
    Triangle   (360°)   Pentagon   Hexagon
    (180°)   (540°)   (720°)

FIGURE 9.4

WARM UP

The following warm-up exercises involve skills that were covered in earlier sections. You will use these skills in the exercise set for this section.

In Exercises 1–4, find the required sum.

1. $\sum\limits_{k=3}^{6} (2k - 3)$

2. $\sum\limits_{j=1}^{5} (j^2 - j)$

3. $\sum\limits_{k=2}^{5} \dfrac{1}{k}$

4. $\sum\limits_{i=1}^{2} \left(1 + \dfrac{1}{i}\right)$

In Exercises 5–10, simplify the expression.

5. $\dfrac{2(k + 1) + 3}{5}$

6. $\dfrac{3(k + 1) - 2}{6}$

7. $2 \cdot 2^{2(k+1)}$

8. $\dfrac{3^{2k}}{3^{2(k+1)}}$

9. $\dfrac{k + 1}{k^2 + k}$

10. $\dfrac{\sqrt{32}}{\sqrt{50}}$

EXERCISES for Section 9.4

In Exercises 1–10, find the indicated sum using the formulas for the sums of powers of integers.

1. $\sum\limits_{n=1}^{20} n$

2. $\sum\limits_{n=1}^{50} n$

3. $\sum\limits_{n=1}^{6} n^2$

4. $\sum\limits_{n=1}^{10} n^2$

5. $\sum\limits_{n=1}^{5} n^3$

6. $\sum\limits_{n=1}^{8} n^3$

7. $\sum\limits_{n=1}^{6} n^4$

8. $\sum\limits_{n=1}^{4} n^5$

9. $\sum\limits_{n=1}^{6} (n^2 - n)$

10. $\sum\limits_{n=1}^{10} (n^3 - n^2)$

In Exercises 11–14, find S_{k+1} for the given S_k.

11. $S_k = \dfrac{5}{k(k + 1)}$

12. $S_k = \dfrac{1}{(k + 1)(k + 3)}$

13. $S_k = \dfrac{k^2(k + 1)^2}{4}$

14. $S_k = \dfrac{k}{2}(3k - 1)$

In Exercises 15–28, use mathematical induction to prove the given formula for every positive integer n.

15. $2 + 4 + 6 + 8 + \cdots + 2n = n(n + 1)$

16. $3 + 7 + 11 + 15 + \cdots + (4n - 1) = n(2n + 1)$

17. $2 + 7 + 12 + 17 + \cdots + (5n - 3) = \dfrac{n}{2}(5n - 1)$

18. $1 + 4 + 7 + 10 + \cdots + (3n - 2) = \dfrac{n}{2}(3n - 1)$

19. $1 + 2 + 2^2 + 2^3 + \cdots + 2^{n-1} = 2^n - 1$

20. $2(1 + 3 + 3^2 + 3^3 + \cdots + 3^{n-1}) = 3^n - 1$

21. $1 + 2 + 3 + 4 + \cdots + n = \dfrac{n(n + 1)}{2}$

22. $1^2 + 2^2 + 3^2 + 4^2 + \cdots + n^2 = \dfrac{n(n + 1)(2n + 1)}{6}$

23. $1^3 + 2^3 + 3^3 + 4^3 + \cdots + n^3 = \dfrac{n^2(n + 1)^2}{4}$

24. $\left(1 + \dfrac{1}{1}\right)\left(1 + \dfrac{1}{2}\right)\left(1 + \dfrac{1}{3}\right) \cdots \left(1 + \dfrac{1}{n}\right) = n + 1$

25. $\sum\limits_{i=1}^{n} i^5 = \dfrac{n^2(n + 1)^2(2n^2 + 2n - 1)}{12}$

26. $\displaystyle\sum_{i=1}^{n} i^4 = \frac{n(n+1)(2n+1)(3n^2+3n-1)}{30}$

27. $\displaystyle\sum_{i=1}^{n} i(i+1) = \frac{n(n+1)(n+2)}{3}$

28. $\displaystyle\sum_{i=1}^{n} \frac{1}{(2i-1)(2i+1)} = \frac{n}{2n+1}$

In Exercises 29–32, use mathematical induction to prove the given inequality for the indicated integer values of n.

29. $\left(\dfrac{4}{3}\right)^n > n, \quad n \geq 7$

30. $\dfrac{1}{\sqrt{1}} + \dfrac{1}{\sqrt{2}} + \dfrac{1}{\sqrt{3}} + \cdots + \dfrac{1}{\sqrt{n}} > \sqrt{n}, \quad n \geq 2$

31. $n! > 2^n, \quad n \geq 4$

32. $\left(\dfrac{x}{y}\right)^{n+1} < \left(\dfrac{x}{y}\right)^n, \quad$ if $n \geq 1$ and $0 < x < y$

In Exercises 33–42, use mathematical induction to prove the given property for all positive integers n.

33. $(ab)^n = a^n b^n$

34. $\left(\dfrac{a}{b}\right)^n = \dfrac{a^n}{b^n}$

35. If $x_1 \neq 0, x_2 \neq 0, \ldots, x_n \neq 0$, then

$$(x_1 x_2 x_3 \cdots x_n)^{-1} = x_1^{-1} x_2^{-1} x_3^{-1} \cdots x_n^{-1}.$$

36. If $x_1 > 0, x_2 > 0, \ldots, x_n > 0$, then

$$\ln(x_1 x_2 x_3 \cdots x_n)$$
$$= \ln x_1 + \ln x_2 + \ln x_3 + \cdots + \ln x_n.$$

37. Generalized Distributive Law:

$$x(y_1 + y_2 + \cdots + y_n) = xy_1 + xy_2 + \cdots + xy_n$$

38. $x^n - y^n = (x - y)(x^{n-1} + x^{n-2}y + \cdots + xy^{n-2} + y^{n-1})$

[*Hint*: $x^{n+1} - y^{n+1} = x^n(x - y) + y(x^n - y^n)$]

39. $(a + bi)^n$ and $(a - bi)^n$ are complex conjugates for all $n \geq 1$.

40. A factor of $(n^3 + 3n^2 + 2n)$ is 3.

41. A factor of $(2^{2n-1} + 3^{2n-1})$ is 5.

42.
$$\begin{vmatrix} a_{11} & 0 & 0 & \cdots & 0 \\ a_{21} & a_{22} & 0 & \cdots & 0 \\ \vdots & \vdots & & & \vdots \\ a_{n1} & a_{n2} & a_{n3} & \cdots & a_{nn} \end{vmatrix} = a_{11} a_{22} a_{33} \cdots a_{nn}$$

9.5 The Binomial Theorem

Binomial Coefficients / Pascal's Triangle / Binomial Expansions

Binomial Coefficients

Recall that a **binomial** is a polynomial that has two terms. In this section, you will look at a formula that gives a quick method of raising a binomial to a power. To begin, look at the expansion of $(x + y)^n$ for several values of n.

$$(x + y)^0 = 1$$
$$(x + y)^1 = x + y$$
$$(x + y)^2 = x^2 + 2xy + y^2$$
$$(x + y)^3 = x^3 + 3x^2y + 3xy^2 + y^3$$
$$(x + y)^4 = x^4 + 4x^3y + 6x^2y^2 + 4xy^3 + y^4$$
$$(x + y)^5 = x^5 + 5x^4y + 10x^3y^2 + 10x^2y^3 + 5xy^4 + y^5$$

There are several observations you can make about these expansions of $(x + y)^n$.

1. In each expansion, there are $n + 1$ terms.
2. In each expansion, x and y have symmetrical roles. The powers of x decrease by 1 in successive terms, while the powers of y increase by 1.
3. The sum of the powers of each term in a binomial expansion is n. For example, in the expansion of $(x + y)^5$, the sum of the powers of each term is 5, as follows.

$$\overset{4 + 1 = 5 \qquad 3 + 2 = 5}{(x + y)^5 = x^5 + 5\,\overbrace{x^4y^1} + \;10\,\overbrace{x^3y^2} + 10x^2y^3 + 5xy^4 + y^5}$$

4. The first term is x^n, the last term is y^n, and each of these terms has a coefficient of 1.
5. The coefficients increase and then decrease in a symmetrical pattern. For $(x + y)^5$, the pattern is

$$1 \qquad 5 \qquad 10 \qquad 10 \qquad 5 \qquad 1.$$

The most difficult part of a binomial expansion is finding the coefficients of the interior terms. To find these **binomial coefficients,** we use a well-known theorem called the **Binomial Theorem.**

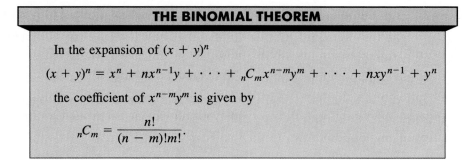

THE BINOMIAL THEOREM

In the expansion of $(x + y)^n$

$$(x + y)^n = x^n + nx^{n-1}y + \cdots + {}_nC_m x^{n-m}y^m + \cdots + nxy^{n-1} + y^n$$

the coefficient of $x^{n-m}y^m$ is given by

$$_nC_m = \frac{n!}{(n - m)!m!}.$$

Proof

The Binomial Theorem can be proved quite nicely using mathematical induction. The steps are straightforward but look a little messy, so we will present only an outline on the proof.

1. If $n = 1$, then you have

$$(x + y)^1 = x^1 + y^1 = {}_1C_0 x + {}_1C_1 y$$

and the formula is valid.

2. Assuming the formula is true for $n = k$, then the coefficient of $x^{k-m}y^m$ is given by

$$_kC_m = \frac{k!}{(k - m)!m!} = \frac{k(k - 1)(k - 2) \cdots (k - m + 1)}{m!}.$$

To show that the formula is true for $n = k + 1$, look at the coefficient of $x^{k+1-m}y^m$ in the expansion of

$$(x + y)^{k+1} = (x + y)^k(x + y).$$

From the right-hand side, you can determine that the term involving $x^{k+1-m}y^m$ is the sum of two products.

$$({}_kC_m x^{k-m}y^m)(x) + ({}_kC_{m-1} x^{k+1-m}y^{m-1})(y)$$

$$= \left[\frac{k!}{(k - m)!m!} + \frac{k!}{(k - m + 1)!(m - 1)!} \right] x^{k+1-m}y^m$$

$$= \left[\frac{(k + 1 - m)k!}{(k + 1 - m)!m!} + \frac{k!m}{(k + 1 - m)!m!} \right] x^{k+1-m}y^m$$

$$= \left[\frac{k!(k + 1 - m + m)}{(k + 1 - m)!m!} \right] x^{k+1-m}y^m$$

$$= \left[\frac{(k + 1)!}{(k + 1 - m)!m!} \right] x^{k+1-m}y^m$$

$$= {}_{k+1}C_m x^{k+1-m}y^m$$

REMARK The symbol $\binom{n}{m}$ is often used in place of ${}_nC_m$ to denote binomial coefficients.

Thus, by mathematical induction, the Binomial Theorem is valid for all positive integers n. ◢

EXAMPLE 1 Finding Binomial Coefficients

Find the following binomial coefficients.

a. $_8C_2$ **b.** $_{10}C_3$
c. $_7C_3$ **d.** $_7C_4$

Solution

Note in parts (a) and (b) how the numerator factorial can be factored in order to cancel one of the denominator factorials.

a. $_8C_2 = \dfrac{8!}{6!2!} = \dfrac{(8 \cdot 7) \cdot 6!}{6! \cdot 2!} = \dfrac{8 \cdot 7}{2 \cdot 1} = 28$

b. $_{10}C_3 = \dfrac{10!}{7!3!} = \dfrac{(10 \cdot 9 \cdot 8) \cdot 7!}{7! \cdot 3!} = \dfrac{10 \cdot 9 \cdot 8}{3 \cdot 2 \cdot 1} = 120$

c. $_7C_3 = \dfrac{7 \cdot 6 \cdot 5}{3 \cdot 2 \cdot 1} = 35$

d. $_7C_4 = \dfrac{7 \cdot 6 \cdot 5 \cdot 4}{4 \cdot 3 \cdot 2 \cdot 1} = 35$

When $m \neq 0$ or $m \neq n$, as in the above example, there is a simple pattern for evaluating binomial coefficients.

$$_8C_2 = \overbrace{\dfrac{8 \cdot 7}{\underbrace{2 \cdot 1}_{2 \text{ factorial}}}}^{2 \text{ factors}} \quad \text{and} \quad _{10}C_3 = \overbrace{\dfrac{10 \cdot 9 \cdot 8}{\underbrace{3 \cdot 2 \cdot 1}_{3 \text{ factorial}}}}^{3 \text{ factors}}$$

In general, you have the following.

$$_nC_m = \dfrac{\overbrace{n(n-1)(n-2) \cdots}^{m \text{ factors}}}{m!}, \qquad 0 < m < n$$

It is not a coincidence that the results in parts (c) and (d) of Example 1 are the same. In general, it is true that $_nC_m = {_nC_{n-m}}$. For instance,

$$_6C_0 = {_6C_6} = 1, \qquad _6C_1 = {_6C_5} = 6, \qquad \text{and} \qquad _6C_2 = {_6C_4} = 15.$$

This shows the symmetric property of binomial coefficients that was identified earlier. By calculating $_6C_1$ and $_6C_5$, you can see that the simpler of the two symmetric coefficients to calculate is the one with the smaller number on the right.

Pascal's Triangle

There is a convenient way to remember a pattern for binomial coefficients. By arranging the coefficients in a triangular pattern, you obtain the following array, which is **Pascal's Triangle.** This triangle is named after the famous French mathematician Blaise Pascal (1623–1662).

$$
\begin{array}{ccccccccccccccc}
&&&&&&&&1\\
&&&&&&&1&&1\\
&&&&&&1&&2&&1\\
&&&&&1&&3&&3&&1\\
&&&&1&&4&&6&&4&&1\\
&&&1&&5&&10&&10&&5&&1\\
&&1&&6&&15&&20&&15&&6&&1\\
&1&&7&&21&&35&&35&&21&&7&&1
\end{array}
$$

The first and last number in each row of Pascal's Triangle is 1. Every other number in each row is formed by adding the two numbers immediately above the number. For example, the two numbers above 35 are 15 and 20.

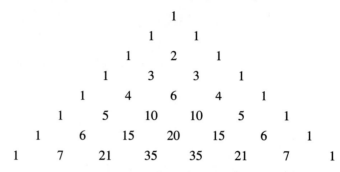

$15 + 20 = 35$

Pascal noticed that numbers in this triangle are precisely the same numbers that are the coefficients of binomial expansions, as follows.

$$(x + y)^0 = 1$$
$$(x + y)^1 = 1x + 1y$$
$$(x + y)^2 = 1x^2 + 2xy + 1y^2$$
$$(x + y)^3 = 1x^3 + 3x^2y + 3xy^2 + 1y^3$$
$$(x + y)^4 = 1x^4 + 4x^3y + 6x^2y^2 + 4xy^3 + 1y^4$$
$$(x + y)^5 = 1x^5 + 5x^4y + 10x^3y^2 + 10x^2y^3 + 5xy^4 + 1y^5$$
$$(x + y)^6 = 1x^6 + 6x^5y + 15x^4y^2 + 20x^3y^3 + 15x^2y^4 + 6xy^5 + 1y^6$$
$$(x + y)^7 = 1x^7 + 7x^6y + 21x^5y^2 + 35x^4y^3 + 35x^3y^4 + 21x^2y^5 + 7xy^6 + 1y^6$$

Because the top row in Pascal's Triangle corresponds to the binomial expansion $(x + y)^0 = 1$, it is called the **zero row.** Similarly, the next row corresponds to the binomial expansion $(x + y)^1 = 1(x) + 1(y)$, and it is called the **first row.** In general, the *n*th row in Pascal's Triangle gives the coefficients of $(x + y)^n$.

EXAMPLE 2   Using Pascal's Triangle

Use Pascal's Triangle to find the following binomial coefficients.

$$_8C_0, \quad _8C_1, \quad _8C_2, \quad _8C_3, \quad _8C_4, \quad _8C_5, \quad _8C_6, \quad _8C_7, \quad _8C_8$$

Solution

These nine binomial coefficients represent the eighth row of Pascal's Triangle. Thus, using the seventh row of the triangle, you can calculate the numbers in the eighth row, as follows.

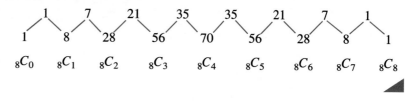

Binomial Expansions

As mentioned at the beginning of this section, when you write out the coefficients for a binomial that is raised to a power, you are **expanding a binomial.** The formulas for binomial coefficients give you an easy way to expand binomials, as demonstrated in the next three examples.

EXAMPLE 3   Expanding a Binomial

Write the expansion for the following expression.

$$(x + 1)^3$$

Solution

The binomial coefficients from the third row of Pascal's Triangle are 1, 3, 3, 1. Therefore, the expansion is as follows.

$$(x + 1)^3 = (1)x^3 + (3)x^2(1) + (3)x(1^2) + (1)(1^3)$$
$$= x^3 + 3x^2 + 3x + 1$$

To expand binomials representing *differences*, rather than sums, alternate signs.

$$(x - 1)^3 = x^3 - 3x^2 + 3x - 1$$
$$(x - 1)^4 = x^4 - 4x^3 + 6x^2 - 4x + 1$$

EXAMPLE 4   Expanding a Binomial

Write the expansion for the following expression.

$$(x + 3)^4$$

Solution

The binomial coefficients from the fourth row of Pascal's Triangle are 1, 4, 6, 4, 1. Therefore, the expansion is as follows.

$$(x + 3)^4 = (1)x^4 + (4)x^3(3) + (6)x^2(3^2) + (4)x(3^3) + (1)(3^4)$$
$$= x^4 + 12x^3 + 54x^2 + 108x + 81$$

EXAMPLE 5   Expanding a Binomial

Write the expansion for the following expression.

$$(x - 2y)^4$$

Solution

The binomial coefficients from the fourth row of Pascal's Triangle are 1, 4, 6, 4, 1. Therefore, the expansion is as follows.

$$(x - 2y)^4 = (1)x^4 - (4)x^3(2y) + (6)x^2(2y)^2 - (4)x(2y)^3 + (1)(2y)^4$$
$$= x^4 - 8x^3y + 24x^2y^2 - 32xy^3 + 16y^4$$

EXAMPLE 6   Finding a Specified Term in a Binomial Expansion

Find the sixth term in the expansion of $(3a + 2b)^{12}$.

Solution

Using the Binomial Theorem, let $x = 3a$ and $y = 2b$ and note that in the *sixth* term, the exponent of y is $m = 5$ and the exponent of x is $n - m = 12 - 5 = 7$. Consequently, the sixth term of the expansion is

$$_{12}C_5 x^7 y^5 = \frac{12 \cdot 11 \cdot 10 \cdot 9 \cdot 8}{5!}(3a)^7(2b)^5.$$

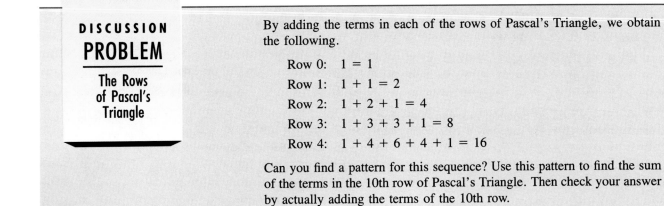

DISCUSSION PROBLEM

The Rows of Pascal's Triangle

By adding the terms in each of the rows of Pascal's Triangle, we obtain the following.

Row 0: $1 = 1$

Row 1: $1 + 1 = 2$

Row 2: $1 + 2 + 1 = 4$

Row 3: $1 + 3 + 3 + 1 = 8$

Row 4: $1 + 4 + 6 + 4 + 1 = 16$

Can you find a pattern for this sequence? Use this pattern to find the sum of the terms in the 10th row of Pascal's Triangle. Then check your answer by actually adding the terms of the 10th row.

WARM UP

The following warm-up exercises involve skills that were covered in earlier sections. You will use these skills in the exercise set for this section.

In Exercises 1–6, perform the indicated operations and/or simplify.

1. $5x^2(x^3 + 3)$ **2.** $(x + 5)(x^2 - 3)$

3. $(x + 4)^2$ **4.** $(2x - 3)^2$

5. $x^2y(3xy^{-2})$ **6.** $(-2z)^5$

In Exercises 7–10, evaluate the expression.

7. $5!$ **8.** $\dfrac{8!}{5!}$

9. $\dfrac{10!}{7!}$ **10.** $\dfrac{6!}{3!3!}$

EXERCISES for Section 9.5

In Exercises 1–10, evaluate $_nC_m$.

1. $_5C_3$ **2.** $_8C_6$

3. $_{12}C_0$ **4.** $_{20}C_{20}$

5. $_{20}C_{15}$ **6.** $_{12}C_5$

7. $_{100}C_{98}$ **8.** $_{10}C_4$

9. $_{100}C_2$ **10.** $_{10}C_6$

In Exercises 11–30, use the Binomial Theorem to expand and simplify the expression.

11. $(x + 1)^4$ **12.** $(x + 1)^6$

13. $(a + 2)^3$ **14.** $(a + 3)^4$

15. $(y - 2)^4$ **16.** $(y - 2)^5$

17. $(x + y)^5$ **18.** $(x + y)^6$

19. $(r + 3s)^6$ **20.** $(x + 2y)^4$

21. $(x - y)^5$ **22.** $(2x - y)^5$

23. $(1 - 2x)^3$

24. $(5 - 3y)^3$

25. $(x^2 + 5)^4$

26. $(x^2 + y^2)^6$

27. $\left(\dfrac{1}{x} + y\right)^5$

28. $\left(\dfrac{1}{x} + 2y\right)^6$

29. $2(x - 3)^4 + 5(x - 3)^2$

30. $3(x + 1)^5 - 4(x + 1)^3$

In Exercises 31–36, use the Binomial Theorem to expand the complex number. Simplify your answer by recalling that $i^2 = -1$.

31. $(1 + i)^4$

32. $(2 - i)^5$

33. $(2 - 3i)^6$

34. $(5 + \sqrt{-9})^3$

35. $\left(\dfrac{-1}{2} + \dfrac{\sqrt{3}}{2}i\right)^3$

36. $(5 - \sqrt{3}i)^4$

In Exercises 37–40, expand the binomial using Pascal's Triangle to determine the coefficients.

37. $(2t - s)^5$

38. $(x + 2y)^5$

39. $(3 - 2z)^4$

40. $(3y + 2)^5$

In Exercises 41–48, find the coefficient a of the given term in the expansion of the binomial.

Binomial	*Term*
41. $(x + 3)^{12}$	ax^5
42. $(x^2 + 3)^{12}$	ax^8
43. $(x - 2y)^{10}$	$ax^8 y^2$
44. $(4x - y)^{10}$	$ax^2 y^8$
45. $(3x - 2y)^9$	$ax^4 y^5$
46. $(2x - 3y)^8$	$ax^6 y^2$
47. $(x^2 + y)^{10}$	$ax^8 y^6$
48. $(z^2 - 1)^{12}$	az^6

In Exercises 49–54, use the Binomial Theorem to expand the given expression. In the study of probability, it is sometimes necessary to use the expansion of $(p + q)^n$, where $p + q = 1$.

49. $\left(\dfrac{1}{2} + \dfrac{1}{2}\right)^7$

50. $\left(\dfrac{1}{4} + \dfrac{3}{4}\right)^{10}$

51. $\left(\dfrac{1}{3} + \dfrac{2}{3}\right)^8$

52. $(0.3 + 0.7)^{12}$

53. $(0.6 + 0.4)^5$

54. $(0.35 + 0.65)^6$

In Exercises 55–58, use the Binomial Theorem to approximate the given quantity accurate to three decimal places. For example, in Exercise 55 you have

$$(1.02)^8 = (1 + 0.02)^8 = 1 + 8(0.02) + 28(0.02)^2 + \ldots.$$

55. $(1.02)^8$

56. $(2.005)^{10}$

57. $(2.99)^{12}$

58. $(1.98)^9$

In Exercises 59–62, shift the graph of f to the right or left as indicated to form the graph of g. Then write the polynomial function g in standard form.

59. $f(x) = -x^2 + 3x + 2$

60. $f(x) = 2x^2 - 4x + 1$

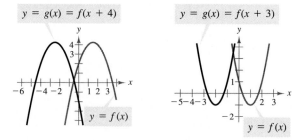

61. $f(x) = x^3 - 4x$

62. $f(x) = -x^4 + 4x^2 - 1$

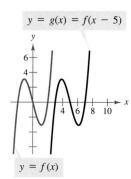

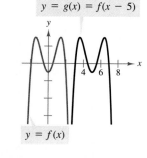

63. *Life Insurance* The average amount of life insurance per household (in households that carry life insurance) from 1970 through 1988 can be approximated by the model

$$f(t) = 0.2187t^2 + 0.6715t + 26.67, \qquad 0 \le t \le 18.$$

In this model, $f(t)$ represents the amount of life insurance (in 1000s of dollars) and t represents the calendar year with $t = 0$ corresponding to 1970 (see figure). You want to adjust this model so that $t = 0$ corresponds to 1980 rather than 1970. To do this, you shift the graph of f 10 units *to the left* and obtain

$$g(t) = f(t + 10)$$
$$= 0.2187(t + 10)^2 + 0.6715(t + 10) + 26.67.$$

Write this new polynomial function in standard form. (*Source*: American Council of Life Insurance)

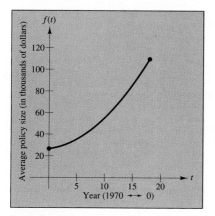

Figure for 63

64. *Health Maintenance Organizations* The number of people enrolled in health maintenance organizations (HMOs) in the United States from 1976 through 1988 can be approximated by the model

$$f(t) = 215t^2 - 470t + 6700, \qquad 0 \le t \le 12.$$

In this model, $f(t)$ represents the number of people (in 1000s) and t represents the calendar year with $t = 0$ corresponding to 1976 (see figure). You want to adjust this model so that $t = 0$ corresponds to 1980 rather than 1976.

To do this, you shift the graph of f 4 units *to the left* and obtain

$$g(t) = f(t + 4)$$
$$= 215(t + 4)^2 - 470(t + 4) + 6700.$$

Write this new polynomial function in standard form. (*Source*: Group Health Insurance Association of America)

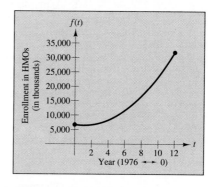

Figure for 64

In Exercises 65–68, prove the given property for all integers m and n where $0 \le m \le n$.

65. $_nC_m = {_nC_{n-m}}$

66. $_nC_0 - {_nC_1} + {_nC_2} - \cdots \pm {_nC_n} = 0$

67. $_{n+1}C_m = {_nC_m} + {_nC_{m-1}}$

68. $_{2n}C_n = ({_nC_0})^2 + ({_nC_1})^2 + ({_nC_2})^2 + \cdots + ({_nC_n})^2$

69. Prove that the sum of the numbers in the nth row of Pascal's Triangle is 2^n. (*Hint*: Consider $2^n = (1 + 1)^n$.)

9.6 Counting Principles, Permutations, Combinations

Simple Counting Problems / Counting Principles / Permutations / Combinations

Simple Counting Problems

In the last two sections of this chapter, we give a brief introduction to some of the basic counting principles and their application to probability. In the next section, you will see that much of probability has to do with counting the number of ways an event can occur. Examples 1, 2, and 3 describe some simple counting problems.

EXAMPLE 1   A Random Number Generator

A random number generator (on a computer) selects an integer from 1 to 40. Find the number of ways the following events can occur.

a. An even integer is selected.
b. A number less than 10 is selected.
c. A prime number is selected.

Solution

a. Since half of the numbers between 1 and 40 are even, this event can occur in 20 different ways.
b. The integers between 1 and 40 that are less than 10 are given in the following set.

$$\{1, 2, 3, 4, 5, 6, 7, 8, 9\}$$

Since this set has nine members, there are nine different ways this event can happen.
c. The prime numbers between 1 and 40 are given in the following set.

$$\{2, 3, 5, 7, 11, 13, 17, 19, 23, 29, 31, 37\}$$

Since this set has 12 members, there are 12 different ways this event can happen.

EXAMPLE 2   Selecting Pairs of Numbers at Random

Eight pieces of paper are numbered from 1 to 8 and placed in a box. One piece of paper is drawn from the box, its number is written down, and the piece of paper is replaced in the box. Then a second piece of paper is drawn from the box, and its number is written down. Finally, the two numbers are added together. How many different ways can a total of 12 be obtained?

Solution

To solve this problem, you count the different ways that a total of 12 can be obtained using two numbers between 1 and 8.

$$\boxed{\text{First number}} \; + \; \boxed{\text{Second number}} \; = \; \boxed{12}$$

After considering the various possibilities, you see that this equation can be solved in the following five ways.

First Number	Second Number
4	8
5	7
6	6
7	5
8	4

Thus, a total of 12 can be obtained in five different ways.

Solving counting problems can be tricky. Often, seemingly minor changes in the statement of a problem can affect the answer. For instance, compare the counting problem in the next example with that given in Example 2.

EXAMPLE 3   Selecting Pairs of Numbers at Random

Eight pieces of paper are numbered from 1 to 8 and placed in a box. Two pieces of paper are drawn from the box, the number on each piece of paper is written down, and the two are totaled. How many different ways can a total of 12 be obtained?

Solution

To solve this problem, you count the different ways that a total of 12 can be obtained *using two different numbers* between 1 and 8.

REMARK   The difference between the counting problems in Examples 2 and 3 is that the random selection in Example 2 occurs **with replacement,** whereas the random selection in Example 3 occurs **without replacement,** which eliminates the possibility of choosing two 6's.

First Number	Second Number
4	8
5	7
7	5
8	4

Thus, a total of 12 can be obtained in four different ways.

Counting Principles

In the first three examples, we looked at simple counting problems in which we can *list* each possible way that an event can occur. When it is possible, this is always the best way to solve a counting problem. However, some events can occur in so many different ways that it is not feasible to write out the entire list. In such cases, we must rely on formulas and counting principles. The most important of these is the **Fundamental Counting Principle.**

FUNDAMENTAL COUNTING PRINCIPLE

Let E_1 and E_2 be two events. The first event E_1 can occur in m_1 different ways. After E_1 has occurred, E_2 can occur in m_2 different ways. The number of ways that the two events can occur is

$$m_1 \cdot m_2.$$

REMARK The Fundamental Counting Principle can be extended to three or more events. For instance, the number of ways that three events E_1, E_2, and E_3 can occur is $m_1 \cdot m_2 \cdot m_3$.

EXAMPLE 4 Applying the Fundamental Counting Principle

How many different pairs of letters from the English alphabet are possible? (Disregard the difference between uppercase and lowercase letters.)

Solution

This experiment has two events. The first event is the choice of the first letter, and the second event is the choice of the second letter. Since the English alphabet contains 26 letters, it follows that each event can occur in 26 ways.

Two-letter "words"

26   26

Thus, using the Fundamental Counting Principle, it follows that the number of two-letter "words" is

$$26 \cdot 26 = 676.$$

EXAMPLE 5 Applying the Fundamental Counting Principle

Telephone numbers in the United States have ten digits. The first three are the *area code* and the next seven are the *local telephone number*. How many different telephone numbers are possible within each area code? (Note that a local telephone number cannot begin with 0 or 1.)

Solution

Since the first digit cannot be 0 or 1, there are only eight choices for the first digit. For each of the other six digits, there are ten choices.

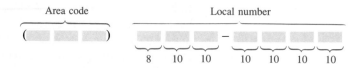

Area code   Local number

8 10 10   10 10 10 10

Thus, using the Fundamental Counting Principle, the number of local telephone numbers that are possible within each area code is

$$8 \cdot 10 \cdot 10 \cdot 10 \cdot 10 \cdot 10 \cdot 10 = 8{,}000{,}000.$$

Permutations

One important application of the Fundamental Counting Principle is in determining the number of ways that n elements can be arranged (in order). We call an ordering of n elements a **permutation** of the elements.

DEFINITION OF PERMUTATION

A **permutation** of n different elements is an ordering of the elements such that one element is first, one is second, one is third, and so on.

EXAMPLE 6   Listing Permutations

Write the different permutations of the letters A, B, and C.

Solution

These three letters can be arranged in the following six different ways.

A, B, C,   B, A, C,   C, A, B

A, C, B,   B, C, A,   C, B, A

Thus, you see that these three letters have six different permutations. ◢

In Example 6, you were able to list the different permutations of three letters. However, you could also have used the Fundamental Counting Principle. To do this, you could reason that there are three choices for the first letter, two choices for the second, and only one choice for the third, as follows.

Permutations of three letters

3   2   1

Thus, the number of permutations of three letters is

$3 \cdot 2 \cdot 1 = 3! = 6.$

EXAMPLE 7   Finding the Number of Permutations of *n* Elements

How many different permutations are possible for the letters A, B, C, D, E, and F?

Solution

There are too many different permutations to list, so you use the following reasoning.

1st position:	Any of the *six* letters.
2nd position:	Any of the remaining *five* letters.
3rd position:	Any of the remaining *four* letters.
4th position:	Any of the remaining *three* letters.
5th position:	Any of the remaining *two* letters.
6th position:	The *one* remaining letter.

Thus, the number of choices for the six positions is as follows.

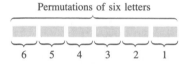

Permutations of six letters

Using the Fundamental Counting Principle, you find that the total number of permutations of the six letters is

$$6 \cdot 5 \cdot 4 \cdot 3 \cdot 2 \cdot 1 = 6! = 720.$$

The results obtained in Examples 6 and 7 can be generalized to conclude that the number of permutations of *n* different elements is *n*!.

NUMBER OF PERMUTATIONS OF *n* ELEMENTS

The number of permutations of *n* elements is given by

$$n \cdot (n - 1) \cdots 4 \cdot 3 \cdot 2 \cdot 1 = n!.$$

In other words, there are *n*! different ways that *n* elements can be ordered.

EXAMPLE 8   Finding the Number of Permutations

Suppose that you are a supervisor for 11 different employees. One of your responsibilities is to perform an annual evaluation for each employee, and then rank the 11 different performances. (In other words, one employee must be ranked as having the best performance, one must be ranked second, and so on.) How many different rankings are possible?

Solution

Since there are 11 different employees, you have 11 choices for first ranking. After choosing the first ranking, you can choose any of the remaining 10 for second ranking, and so on.

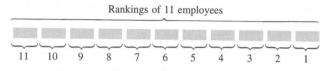

Rankings of 11 employees

11 10 9 8 7 6 5 4 3 2 1

Thus, the number of different rankings is

$$11! = 39,916,800.$$

Occasionally, you may be interested in ordering a *subset* of a collection of elements rather than the entire collection. For example, you might want to choose (and order) *m* elements out of a collection of *n* elements. Such an ordering is a **permutation of *n* elements taken *m* at a time.**

EXAMPLE 9   Permutations of *n* Elements Taken *m* at a Time

Eight horses are running in a race. In how many different ways can these horses come in first, second, and third? (Assume that there are no ties.)

Solution

You have the following possibilities.

Win   (1st position):   *Eight* choices
Place (2nd position):   *Seven* choices
Show (3rd position):   *Six* choices

Using the Fundamental Counting Principle, multiply these three numbers together to obtain the following.

Different orders of horses

8 7 6

Thus, there are $8 \cdot 7 \cdot 6 = 336$ different orders.

The result of Example 9 can be generalized as follows.

PERMUTATIONS OF *n* ELEMENTS TAKEN *m* AT A TIME

The number of permutations of *n* elements taken *m* at a time is

$$_nP_m = \frac{n!}{(n-m)!} = n(n-1)(n-2)\cdots(n-m+1).$$

Using this formula, you can rework Example 9 to find that the number of permutations of eight horses taken three at a time is

$$_8P_3 = \frac{8!}{(8-3)!} = \frac{8!}{5!} = \frac{8 \cdot 7 \cdot 6 \cdot 5!}{5!} = 8 \cdot 7 \cdot 6 = 336$$

which is the same answer you obtained in the solution of Example 9.

Remember that for permutations, order is important. Thus, if you are looking at the possible permutations of the letters A, B, C, and D taken three at a time, the permutations (A, B, D) and (B, A, D) would be different (since the *order* of the elements is different).

Suppose, however, that you are asked to find the possible permutations of the letters A, A, B, and C. The total number of permutations of the four letters would be $_4P_4 = 4!$. However, not all of these arrangements would be *distinguishable* because there are two A's in the list. To find the number of distinguishable permutations, you can use the following formula.

DISTINGUISHABLE PERMUTATIONS

Suppose a set of *n* objects has n_1 of one kind of object, n_2 of a second kind, n_3 of a third kind, and so on, with $n = n_1 + n_2 + n_3 + \cdots + n_k$. Then the number of **distinguishable permutations** of the *n* objects is

$$\frac{n!}{n_1! \cdot n_2! \cdot n_3! \cdots n_k!}.$$

EXAMPLE 10   Distinguishable Permutations

In how many distinguishable ways can the letters in BANANA be written?

Solution

This word has six letters, of which three are A's, two are N's and one is a B. Thus, the number of distinguishable ways the letters can be written is

$$\frac{6!}{3! \cdot 2! \cdot 1!} = \frac{6 \cdot 5 \cdot 4 \cdot 3!}{3! \cdot 2!} = 60.$$

The 60 different "words" are as follows.

AAABNN	AAANBN	AAANNB	AABANN	AABNAN	AABNNA
AANABN	AANANB	AANBAN	AANBNA	AANNAB	AANNBA
ABAANN	ABANAN	ABANNA	ABNAAN	ABNANA	ABNNAA
ANAABN	ANAANB	ANABAN	ANABNA	ANANAB	ANANBA
ANBAAN	ANBANA	ANBNAA	ANNAAB	ANNABA	ANNBAA
BAAANN	BAANAN	BAANNA	BANAAN	BANANA	BANNAA
BNAAAN	BNAANA	BNANAA	BNNAAA	NAAABN	NAAANB
NAABAN	NAABNA	NAANAB	NAANBA	NABAAN	NABANA
NABNAA	NANAAB	NANABA	NANBAA	NBAAAN	NBAANA
NBANAA	NBNAAA	NNAAAB	NNAABA	NNABAA	NNBAAA

Combinations

When counting the number of possible permutations of a set of elements, *order* is important. As a final topic in this section, we look at a method of selecting subsets of a larger set in which order is *not important*. Such subsets are **combinations of *n* elements taken *m* at a time.** For instance, the combinations

$$\{A, B, C\} \quad \text{and} \quad \{B, A, C\}$$

are equivalent because both sets contain the same three elements, and the order in which the elements are listed is *not important*. Hence, we would count only one of the two sets. A common example of how a combination occurs is a card game in which the player is free to reorder the cards after they have been dealt.

EXAMPLE 11   Combination of *n* Elements Taken *m* at a Time

In how many different ways can three letters be chosen from the letters A, B, C, D, and E? (The order of the three letters is not important.)

Solution

The following subsets represent the different combinations of three letters that can be chosen from five letters.

$$\{A, B, C\} \quad \{A, B, D\}$$
$$\{A, B, E\} \quad \{A, C, D\}$$
$$\{A, C, E\} \quad \{A, D, E\}$$
$$\{B, C, D\} \quad \{B, C, E\}$$
$$\{B, D, E\} \quad \{C, D, E\}$$

From this list, you conclude that there are ten different ways that three letters can be chosen from five letters.

The formula for the number of *combinations* of *n* elements taken *m* at a time is as follows.

COMBINATIONS OF *n* ELEMENTS TAKEN *m* AT A TIME

The number of combinations of n elements taken m at a time is

$$_nC_m = \frac{n!}{(n-m)!m!}.$$

Note that the formula for $_nC_m$ is the same as the one given for binomial coefficients. To see how this formula is used, solve the counting problem given in Example 11. In that problem, you must find the number of combinations of five elements taken three at a time. Thus, $n = 5$, $m = 3$, and the number of combinations is

$$_5C_3 = \frac{5!}{2!3!} = \frac{5 \cdot 4 \cdot 3}{3 \cdot 2 \cdot 1} = 10$$

which is the same answer you obtained in Example 11.

EXAMPLE 12 Combinations of *n* Elements Taken *m* at a Time

A standard poker hand consists of five cards dealt from a deck of 52. How many different poker hands are possible? (After the cards are dealt, the player may reorder them, and therefore order is not important.)

Solution

Use the formula for the number of combinations of 52 elements taken five at a time, as follows.

$$_{52}C_5 = \frac{52!}{47!5!} = \frac{52 \cdot 51 \cdot 50 \cdot 49 \cdot 48}{5 \cdot 4 \cdot 3 \cdot 2 \cdot 1} = 2{,}598{,}960 \text{ different hands}$$

EXAMPLE 13 Combinations and the Fundamental Counting Principle

The traveling squad for a college basketball team consists of two centers, five forwards, and four guards. In how many ways can the coach select a starting team of one center, two forwards, and two guards?

Solution

The number of ways to select one center is

$$_2C_1 = \frac{2!}{1!(1!)} = 2.$$

The number of ways to select two forwards from among five is

$$_5C_2 = \frac{5!}{3!(2!)} = 10.$$

The number of ways to select two guards from among four is

$$_4C_2 = \frac{4!}{2!(2!)} = 6.$$

Therefore, the total number of ways to select a starting team is

$$_2C_1 \cdot {}_5C_2 \cdot {}_4C_2 = 2 \cdot 10 \cdot 6 = 120.$$

◀

DISCUSSION

PROBLEM

You Be
the
Instructor

Suppose you are teaching an algebra class and are writing a test for this chapter. Create two word problems that you think are appropriate for the test. One of the word problems should deal with permutations and the other should deal with combinations. (Assume that your students will have only five minutes to solve each problem.)

WARM UP

The following warm-up exercises involve skills that were covered in earlier sections. You will use these skills in the exercise set for this section.

In Exercises 1–4, evaluate the expression.

1. $13 \cdot 8^2 \cdot 2^3$

2. $10^2 \cdot 9^3 \cdot 4$

3. $\dfrac{12!}{2!(7!)(3!)}$

4. $\dfrac{25!}{22!}$

In Exercises 5 and 6, find the binomial coefficient.

5. $_{12}C_7$

6. $_{25}C_{22}$

In Exercises 7–10, simplify the expression.

7. $\dfrac{n!}{(n-4)!}$

8. $\dfrac{(2n)!}{4(2n-3)!}$

9. $\dfrac{2 \cdot 4 \cdot 6 \cdot 8 \cdots (2n)}{2^n}$

10. $\dfrac{3 \cdot 6 \cdot 9 \cdot 12 \cdots (3n)}{3^n}$

EXERCISES for Section 9.6

1. A bag contains 10 marbles numbered 1 through 10. A marble is selected, its number is recorded, and the marble is *replaced* in the bag. Then a second marble is drawn and its number recorded. Finally, the recorded numbers are added. How many different ways can a sum of 8 be obtained?

2. A bag contains 10 marbles numbered 1 through 10. Two marbles are selected and their numbers are recorded. Then the recorded numbers are added. How many different ways can a sum of 8 be obtained?

3. *Job Applicants*   A small college needs two additional faculty members: a chemist and a statistician. In how many ways can these positions be filled if there are three applicants for the chemistry position and four for the position in statistics?

4. *Computer Systems*   A customer in a computer store can choose one of three monitors, one of two keyboards, and one of four computers. If all the choices are compatible, how many different systems could be chosen?

5. *Toboggan Ride*   Four people are lining up for a ride on a toboggan, but only two of the four are willing to take the first position. With that constraint, in how many ways can the four people be seated on the toboggan?

6. *Course Schedule*   A college student is preparing a course schedule for the next semester. The student may select one of two mathematics courses, one of three science courses, and one of five courses from the social sciences and humanities. How many schedules are possible?

7. *License Plate Numbers*   In a certain state the automobile license plates consist of two letters followed by a four-digit number. How many distinct license plate numbers can be formed?

8. *License Plate Numbers*   In a certain state the automobile license plates consist of two letters followed by a four-digit number. To avoid confusion between "O" and "zero" and "I" and "one," the letters "O" and "I" are not used. How many distinct license plate numbers can be formed?

9. *True-False Exam*   In how many ways can a six-question true-false exam be answered? (Assume that no questions are omitted.)

10. *Multiple Choice*   In how many ways can a 10-question multiple choice exam be answered if there are four choices of answers for each question? (Assume that no questions are omitted.)

11. *Three-Digit Numbers*   How many three-digit numbers can be formed under the following conditions?
(a) The leading digit cannot be zero.
(b) The leading digit cannot be zero and no repetition of digits is allowed.
(c) The leading digit cannot be zero and the number must be a multiple of 5.
(d) The number is at least 400.

12. *Four-Digit Numbers*   How many four-digit numbers can be formed under the following conditions?
(a) The leading digit cannot be zero.
(b) The leading digit cannot be zero and no repetition of digits is allowed.

(c) The leading digit cannot be zero and the number must be less than 5000.
(d) The leading digit cannot be zero and the number must be even.

13. *Combination Lock*   A combination lock will open when the right choice of three numbers (from 1 to 40, inclusive) is selected. How many different lock combinations are possible?

14. *Combination Lock*   A combination lock will open when the right choice of three numbers (from 1 to 50, inclusive) is selected. How many different lock combinations are possible?

15. *Concert Seats*   Three couples have reserved seats in a given row for a concert. In how many different ways can they be seated if
(a) there are no seating restrictions?
(b) the two members of each couple wish to sit together?

16. *Single File*   In how many orders can three girls and two boys walk through a doorway single-file if
(a) there are no restrictions?
(b) the boys go before the girls?
(c) the girls go before the boys?

In Exercises 17–26, evaluate $_nP_m$.

17. $_4P_4$
18. $_5P_5$
19. $_8P_3$
20. $_{20}P_2$
21. $_{20}P_5$
22. $_{100}P_1$
23. $_{100}P_2$
24. $_{10}P_2$
25. $_5P_4$
26. $_7P_4$

27. Write all the permutations of the letters A, B, C, and D.

28. Write all the permutations of the letters A, B, C, and D if the letters B and C must remain between the letters A and D.

29. *Posing for a Photograph*   In how many ways can five children line up in one row to have their picture taken?

30. *Riding in a Car*   In how many ways can six people sit in a six-passenger car?

31. *Choosing Officers*   From a pool of 12 candidates, the offices of president, vice-president, secretary, and treasurer will be filled. In how many different ways can the offices be filled, if each of the 12 candidates can hold any office?

32. *Assembly Line Production*   There are four processes involved in assembling a certain product, and these can be performed in any order. The management wants to test each order to determine which is the least time-consuming. How many different orders will have to be tested?

In Exercises 33–38, find the number of distinguishable permutations of the given group of letters.

33. A, A, G, E, E, E, M

34. B, B, B, T, T, T, T, T

35. A, A, Y, Y, Y, Y, X, X, X

36. K, K, M, M, M, L, L, N, N

37. A, L, G, E, B, R, A

38. M, I, S, S, I, S, S, I, P, P, I

39. Write all the possible selections of two letters that can be formed from the letters A, B, C, D, E, and F. (The order of the two letters is not important.)

40. Write all the possible selections of three letters that can be formed from the letters A, B, C, D, E, and F. (The order of the three letters is not important.)

41. *Forming an Experimental Group*   In order to conduct a certain experiment, four students are randomly selected from a class of 20. How many different groups of four students are possible?

42. *Test Questions*   A student may answer any 10 questions from a total of 12 questions on an exam. In how many different ways can the student select the questions?

43. *Lottery Choices*   There are 40 numbers in a particular state lottery. In how many ways can a player select six of the numbers? (The order of selection is not important.)

44. *Lottery Choices*   There are 50 numbers in a particular state lottery. In how many ways can a player select six of the numbers? (The order of selection is not important.)

45. *Number of Subsets*   How many subsets of four elements can be formed from a set of 100 elements?

46. *Number of Subsets*   How many subsets of five elements can be formed from a set of 80 elements?

47. *Forming a Committee*   A committee composed of three graduate students and two undergraduate students is to be selected from a group of eight graduates and five undergraduates. How many different committees can be formed?

48. *Defective Units*   A shipment of 12 microwave ovens contains three defective units. In how many ways can a vending company purchase four of these units and receive (a) all good units, (b) two good units, and (c) at least two good units?

49. *Job Applicants*   An employer interviews eight people for four openings in the company. Three of the eight people are women. If all eight are qualified, in how many ways could the employer fill the four positions if (a) the selection is random and (b) exactly two are women?

50. *Poker Hand*   Five cards are selected from an ordinary deck of 52 playing cards. In how many ways can you get a full house? (A full house consists of three of one kind and two of another. For example, A-A-A-5-5 and K-K-K-10-10 are full houses.)

51. *Forming a Committee*   Four people are to be selected at random from a group of four couples. In how many ways can this be done, given the following conditions?
(a) There are no restrictions.
(b) There is to be at least one couple in the group of four.
(c) The selection must include one member from each couple.

52. *Interpersonal Relationships*   The complexity of the interpersonal relationships increases dramatically as the size of a group increases. Determine the number of two-person relationships in a group of people of size (a) 3, (b) 8, (c) 12, and (d) 20.

In Exercises 53–56, find the number of diagonals of the given polygon. (A line segment connecting any two non-adjacent vertices is called a *diagonal* of the polygon.)

53. Pentagon

54. Hexagon

55. Octagon

56. Decagon (10 sides)

In Exercises 57 and 58, solve for n.

57. $14 \cdot {}_nP_3 = {}_{n+2}P_4$

58. ${}_nP_5 = 18 \, {}_{n-2}P_4$

In Exercises 59–63, prove the identity.

59. ${}_nP_{n-1} = {}_nP_n$

60. ${}_nP_1 = {}_nC_1$

61. ${}_nC_{n-1} = {}_nC_1$

62. ${}_nC_n = {}_nC_0$

63. ${}_nC_m = \dfrac{{}_nP_m}{m!}$

9.7 Probability

Sample Spaces / The Probability of an Event / Mutually Exclusive Events /
Independent Events / The Complement of an Event

Sample Spaces

As a member of a complex society, you are used to living with varying amounts
of uncertainty. For example, you may be questioning the likelihood of getting
a good job after graduation, of winning a state lottery, of having an accident
on your next trip home, or of any of several other possibilities.

In assigning measurements to uncertainties in everyday life, we often use
ambiguous terminology, such as *fairly certain*, *probable*, or *highly unlikely*.
In mathematics, we attempt to remove this ambiguity by assigning a number
to the likelihood of the occurrence of an event. We call this measurement the
probability that the event will occur. For example, if we toss a fair coin, we
say that the probability that it will land heads up is one-half, or 50%.

In the study of probability, any happening whose result is uncertain is
an **experiment.** The various possible results of the experiment are **outcomes,**
and the collection of all possible outcomes of an experiment is the **sample
space** of the experiment. Finally, any subcollection of a sample space is an
event. In this section we will deal only with sample spaces in which each
outcome is equally likely, such as flipping a fair coin or tossing a fair die.

EXAMPLE 1 Finding the Sample Space

An experiment consists of tossing a six-sided die.

a. What is the sample space?
b. Describe the event corresponding to a number greater than 2 turning up.

Solution

a. The sample space consists of six outcomes, which you represent by the
numbers 1 through 6. That is,

$$S = \{1, 2, 3, 4, 5, 6\}.$$

Note that each of the outcomes in the sample space is equally likely
(assuming the die is balanced).
b. The *event* corresponding to a number greater than 2 turning up is the
following subset of S.

$$A = \{3, 4, 5, 6\}$$

To describe sample spaces in such a way that each outcome is equally
likely, we must sometimes distinguish between various outcomes in ways that
appear artificial. The next example illustrates such a situation.

EXAMPLE 2   Finding the Sample Space

Find the sample spaces for the following.

a. One coin is tossed.
b. Two coins are tossed.
c. Three coins are tossed.

Solution

a. Since the coin will land either heads up (denoted by H) or tails up (denoted by T), the sample space is

$$S = \{H, T\}.$$

b. Since either coin can land heads up or tails up, the possible outcomes are as follows.

HH = heads up on both coins

HT = heads up on first coin and tails up on second coin

TH = tails up on first coin and heads up on second coin

TT = tails up on both coins

Thus, the sample space is

$$S = \{HH, HT, TH, TT\}.$$

Note that you must distinguish between the two cases HT and TH, even though these two outcomes appear to be similar.

c. Following the notation of part (b), the sample space is

$$S = \{HHH, HHT, HTH, HTT, THH, THT, TTH, TTT\}.$$

The Probability of an Event

To calculate the probability of an event, you count the number of outcomes in the event and in the sample space. The *number of outcomes* in event E is denoted by $n(E)$, and the number of outcomes in the sample space S is denoted by $n(S)$.

THE PROBABILITY OF AN EVENT

If an event E has $n(E)$ equally likely outcomes and its sample space S has $n(S)$ equally likely outcomes, then the **probability** of event E is

$$P(E) = \frac{n(E)}{n(S)}.$$

Because the number of outcomes in an event must be less than or equal to the number of outcomes in the sample space, you can see that the probability of an event must be a number between 0 and 1. That is, for any event E, it must be true that $0 \le P(E) \le 1$.

PROPERTIES OF THE PROBABILITY OF AN EVENT

Let E be an event that is a subset of a finite sample space S.

1. $0 \le P(E) \le 1$
2. If $P(E) = 0$, then the event E *cannot occur*, and E is an **impossible event.**
3. If $P(E) = 1$, then the event E *must occur*, and E is a **certain event.**

EXAMPLE 3   Finding the Probability of an Event

Find the probability of the following events.

a. Two coins are tossed. What is the probability that both land heads up?

b. A card is drawn from a standard deck of playing cards. What is the probability that it is an ace?

Solution

a. Following the procedure in Example 2(b), let

$$E = \{HH\} \quad \text{and} \quad S = \{HH, HT, TH, TT\}.$$

The probability of getting two heads is

$$P(E) = \frac{n(E)}{n(S)} = \frac{1}{4}.$$

b. Since there are 52 cards in a standard deck of playing cards and there are four aces (one in each suit), the probability of drawing an ace is

$$P(E) = \frac{n(E)}{n(S)} = \frac{4}{52} = \frac{1}{13}.$$

FIGURE 9.5

EXAMPLE 4   Finding the Probability of an Event

Two six-sided dice are tossed. What is the probability that the total of the two dice is 7? (See Figure 9.5.)

Solution

Since there are six possible outcomes on each die, you use the Fundamental Counting Principle to conclude that there are

$6 \cdot 6 = 36$ different outcomes

when two dice are tossed. To find the probability of rolling a total of 7, you must first count the number of ways this can occur.

	Total of 7					
First die	1	2	3	4	5	6
Second die	6	5	4	3	2	1

Thus, a total of 7 can be rolled in six ways, which means that the probability of rolling a 7 is

$$P(E) = \frac{n(E)}{n(S)} = \frac{6}{36} = \frac{1}{6}.$$

You could have written out each sample space in Examples 3 and 4 and simply counted the outcomes in the desired events. For larger sample spaces, however, you must make more use of the counting principles discussed in the previous section.

EXAMPLE 5   Finding the Probability of an Event

Twelve-sided dice can be constructed (in the shape of regular dodecahedrons) so that each of the numbers from 1 to 6 appears twice on each die, as shown in Figure 9.6. Prove that these dice can be used in any game requiring ordinary six-sided dice without changing the probability of different outcomes.

Solution

For an ordinary six-sided die, each of the numbers 1, 2, 3, 4, 5, and 6 occurs only once, so the probability of any particular number coming up is

$$P(E) = \frac{n(E)}{n(S)} = \frac{1}{6}.$$

For one of the twelve-sided dice, each number occurs twice, so the probability of any particular number coming up is

$$P(E) = \frac{n(E)}{n(S)} = \frac{2}{12} = \frac{1}{6}.$$

Thus, the twelve-sided dice can be used in place of the six-sided dice without changing the probabilities.

FIGURE 9.6

EXAMPLE 6   The Probability of Winning a Lottery

A state lottery is set up so that each player chooses six different numbers from 1 to 40. If these six numbers match the six numbers drawn by the lottery commission, the player wins (or shares) the top prize. What is the probability of winning the top prize in this game?

Solution

Since the order of the numbers is not important, use the formula for the number of combinations of 40 elements taken six at a time to determine the size of the sample space.

$$n(S) = {}_{40}C_6 = \frac{40 \cdot 39 \cdot 38 \cdot 37 \cdot 36 \cdot 35}{6 \cdot 5 \cdot 4 \cdot 3 \cdot 2 \cdot 1} = 3{,}838{,}380$$

If a person buys only one ticket, the probability of winning is

$$P(E) = \frac{n(E)}{n(S)} = \frac{1}{3{,}838{,}380}.$$

EXAMPLE 7   Random Selection

The total number of colleges and universities in the United States in 1987 is shown in Figure 9.7. (*Source*: U.S. National Center for Education Statistics) Suppose one institution is selected at random. What is the probability that the institution is in one of three southern regions?

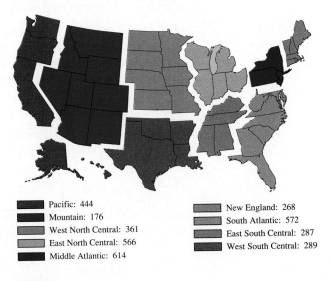

Pacific: 444
Mountain: 176
West North Central: 361
East North Central: 566
Middle Atlantic: 614

New England: 268
South Atlantic: 572
East South Central: 287
West South Central: 289

FIGURE 9.7

Solution

Begin by finding the total number of colleges and universities.

Total = 268 + 614 + 566 + 361 + 572 + 287 + 289 + 176 + 444 = 3577

Since there are 572 + 287 + 289 = 1148 colleges and universities in the three southern regions, the probability that the institution is from one of these regions is

$$P(E) = \frac{n(E)}{n(S)} = \frac{1148}{3577} \approx 0.321.$$

◢

Mutually Exclusive Events

Two events A and B (from the same sample space) are **mutually exclusive** if A and B have no outcomes in common. In the terminology of sets, we say that the **intersection of A and B** is the empty set, which implies that

$$P(A \cap B) = 0.$$

For instance, if two dice are tossed, the event A of rolling a total of six and the event B of rolling a total of nine are mutually exclusive. To find the probability that one or the other of two mutually exclusive events will occur, *add* their individual probabilities.

PROBABILITY OF THE UNION OF TWO EVENTS

If A and B are events in the same sample space, then the probability of A or B occurring is given by

$$P(A \cup B) = P(A) + P(B) - P(A \cap B).$$

If A and B are mutually exclusive, then $P(A \cap B) = 0$ and it follows that

$$P(A \cup B) = P(A) + P(B).$$

EXAMPLE 8   Finding the Probability of the Union of Two Events

One card is selected from a standard deck of 52 playing cards. What is the probability that the card is either a heart or a face card?

Solution

Since the deck has 13 hearts, the probability of selecting a heart (event A) is

$$P(A) = \frac{13}{52}.$$

Selecting a heart

Similarly, since the deck has 12 face cards, the probability of selecting a face card (event B) is

$$P(B) = \frac{12}{52}.$$

Selecting a face card

Now, because three of the cards are hearts and face cards (see Figure 9.8), it follows that

$$P(A \cap B) = \frac{3}{52}.$$

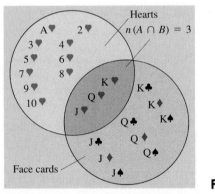

FIGURE 9.8

Finally, applying the formula for the probability of the union of two events, you conclude that the probability of selecting a heart or a face card is

$$P(A \cup B) = P(A) + P(B) - P(A \cap B)$$
$$= \frac{13}{52} + \frac{12}{52} - \frac{3}{52}$$
$$= \frac{22}{52} \approx 0.423.$$

Independent Events

Two events are **independent** if the occurrence of one has no effect on the occurrence of the other. To find the probability that two independent events will occur, *multiply* the probabilities of each. For instance, rolling a total of 12 with two-sided dice has no effect on the outcome for future rolls of the dice.

> ### PROBABILITY OF INDEPENDENT EVENTS
>
> If A and B are independent events, then the probability that both A and B will occur is
>
> $$P(A \text{ and } B) = P(A) \cdot P(B).$$

EXAMPLE 9 Probability of Independent Events

A random number generator on a computer selects three integers from 1 to 20. What is the probability that all three numbers are less than or equal to 5?

Solution

If the random number generator is truly random, then you can conclude that the selection of any given number will not affect the selection of the next number. This means that the three choices represent independent events. Furthermore, since the probability of selecting a number from 1 to 5 is

$$P(A) = \frac{5}{20} = \frac{1}{4}$$

you can conclude that the probability of selecting all three numbers less than or equal to 5 is

$$P(A) \cdot P(A) \cdot P(A) = \left(\frac{1}{4}\right)\left(\frac{1}{4}\right)\left(\frac{1}{4}\right) = \frac{1}{64}.$$

EXAMPLE 10 Probability of Independent Events

In 1988, 54.2% of the population of the United States were 30 years old or older. Suppose that in a survey, 10 people were chosen at random from the population. What is the probability that all 10 are 30 years old or older?

Solution

Let A represent choosing a person who is 30 years old or older. Since the probability of choosing a person who is 30 years old or older is 0.542, you conclude that the probability that all 10 people are 30 years old or older is

$$P(A)^{10} = (0.542)^{10} \approx 0.0022.$$

The Complement of an Event

The **complement of an event** A is the collection of all outcomes in the sample space that are not in A. We denote the complement of event A by A'. Since $P(A \text{ or } A') = 1$ and since A and A' are mutually exclusive, we have $P(A) + P(A') = 1$. Therefore, the probability of A' is given by

$$P(A') = 1 - P(A).$$

For instance, if the probability of *winning* a certain game is

$$P(A) = \frac{1}{4}$$

then the probability of *losing* the game is

$$P(A') = 1 - \frac{1}{4} = \frac{3}{4}.$$

EXAMPLE 11   **Finding the Probability of the Complement of an Event**

A manufacturer has determined that a certain machine averages one faulty unit for every 1000 it produces. What is the probability that an order of 200 units will have one or more faulty units?

Solution

To solve this problem as stated, you would need to find the probability of having exactly one faulty unit, exactly two faulty units, exactly three faulty units, and so on. However, using complements, you can simply find the probability that all units are perfect and then subtract this value from 1. Since the probability that any given unit is perfect is $999/1000$, the probability that all 200 units are perfect is

$$P(A) = \left(\frac{999}{1000}\right)^{200} \approx 0.8186.$$

Therefore, the probability that at least one unit is faulty is

$$P(A') = 1 - P(A) \approx 0.1814.$$

DISCUSSION

PROBLEM

An Experiment
in
Probability

In this section you have been finding probabilities from a *theoretical* point of view. Another way to find probabilities is from an *experimental* point of view. For instance, suppose you want to find the probability of obtaining a given total when two six-sided dice are tossed. The following BASIC program simulates the tossing of a pair of dice 5,000 times.

```
10   RANDOMIZE
20   DIM TALLY(12)
30   FOR I=1 TO 5000
40   ROLLONE=INT(6*RND)+1
50   ROLLTWO=INT(6*RND)+1
60   DICETOTAL=ROLLONE+ROLLTWO
70   TALLY(DICETOTAL)=TALLY(DICETOTAL)+1
80   NEXT
90   FOR I=2 TO 12
100  PRINT "TOTAL OF",I,"OCCURRED",TALLY(I),
     "TIMES"
110  NEXT
120  END
```

When you run this program, the printout is as follows.

```
TOTAL OF  2 OCCURRED 139 TIMES
TOTAL OF  3 OCCURRED 264 TIMES
TOTAL OF  4 OCCURRED 443 TIMES
TOTAL OF  5 OCCURRED 553 TIMES
TOTAL OF  6 OCCURRED 691 TIMES
TOTAL OF  7 OCCURRED 810 TIMES
TOTAL OF  8 OCCURRED 715 TIMES
TOTAL OF  9 OCCURRED 557 TIMES
TOTAL OF 10 OCCURRED 398 TIMES
TOTAL OF 11 OCCURRED 270 TIMES
TOTAL OF 12 OCCURRED 160 TIMES
```

In Example 4 you found that the theoretical probability of tossing a total of 7 on a pair of dice is $\frac{1}{6} \approx 0.167$. From this experiment, you find that the experimental probability of tossing a total of 7 on a pair of dice is $810/5,000 \approx 0.162$. Try this experiment on a computer that has the BASIC language. By increasing the number of trials from 5,000 to 10,000, does your experimental result get closer to the theoretical result?

WARM UP

The following warm-up exercises involve skills that were covered in earlier sections. You will use these skills in the exercise set for this section.

In Exercises 1–8, evaluate the expression.

1. $\dfrac{1}{4} + \dfrac{5}{8} - \dfrac{5}{16}$

2. $\dfrac{4}{15} + \dfrac{3}{5} - \dfrac{1}{3}$

3. $\dfrac{5 \cdot 4}{5!}$

4. $\dfrac{5!22!}{27!}$

5. $\dfrac{4!8!}{12!}$

6. $\dfrac{9 \cdot 8 \cdot 7 \cdot 6 \cdot 5}{9!}$

7. $\dfrac{{}_5C_3}{{}_{10}C_3}$

8. $\dfrac{{}_{10}C_2 \cdot {}_{10}C_2}{{}_{20}C_4}$

In Exercises 9 and 10, evaluate the expression. (Round to three decimal places.)

9. $\left(\dfrac{99}{100}\right)^{100}$

10. $1 - \left(\dfrac{89}{100}\right)^{50}$

EXERCISES for Section 9.7

In Exercises 1–6, determine the sample space for the given experiment.

1. A coin and a die are tossed.

2. A die is tossed twice and the sum of the points is recorded.

3. A taste tester has to rank three varieties of yogurt, A, B, and C, according to preference.

4. Two marbles are selected from a sack containing two red marbles, two blue marbles, and one black marble. The color of each marble is recorded.

5. Two county supervisors are selected from five supervisors, A, B, C, D, and E, to study a recycling plan.

6. A salesperson makes a presentation about a product in three homes per day. In each home there may be a sale (denote by *S*) or there may be no sale (denote by *F*).

Heads or Tails   In Exercises 7–10, find the required probability in the experiment of tossing a coin three times. Use the sample space $S = \{HHH, HHT, HTH, HTT, THH, THT, TTH, TTT\}$.

7. The probability of getting exactly one tail.

8. The probability of getting a head on the first toss.

9. The probability of getting at least one head.

10. The probability of getting at least two heads.

Drawing a Card   In Exercises 11–14, find the required probability in the experiment of selecting one card from a standard deck of 52 playing cards.

11. The probability of getting a face card.

12. The probability of not getting a face card.

13. The probability of getting a black card that is not a face card.

14. The probability that the card will be a 6 or less.

Tossing a Die   In Exercises 15–20, find the required probability in the experiment of tossing a six-sided die twice.

15. The probability that the sum is 4.

16. The probability that the sum is less than 11.

17. The probability that the sum is at least 7.

18. The probability that the total is 2, 3, or 12.

19. The probability that the sum is odd and no more than 7.

20. The probability that the sum is odd or a prime.

Drawing Marbles In Exercises 21–24, find the required probability in the experiment of drawing two marbles (the first is *not* replaced before the second is drawn) from a bag containing one green, two yellow, and three red marbles.

21. The probability of drawing two red marbles.

22. The probability of drawing two yellow marbles.

23. The probability of drawing neither yellow marble.

24. The probability of drawing marbles of different colors.

In Exercises 25 and 26, you are given the probability that an event *will* happen. Find the probability that the event *will not* happen.

25. $p = 0.7$

26. $p = 0.36$

In Exercises 27 and 28, you are given the probability that an event *will not* happen. Find the probability that the event *will* happen.

27. $p = 0.15$

28. $p = 0.84$

29. *Alumni Association* The alumni office of a college is sending a survey to selected members of the class of 1990. Of the 1254 people who graduated that year, 672 were women, 124 of whom went on to graduate school. Of the 582 male graduates, 198 went on to graduate school. If an alumni member is selected at random, what is the probability that the person is (a) female, (b) male, and (c) female and did not attend graduate school?

30. *Post High-School Education* In a high school graduating class of 72 students, 28 are on the honor roll. Of these 28, 18 are going on to college. Of the other 44 students, 12 are going on to college. If a student is selected at random from the class, what is the probability that the person chosen is (a) going to college, (b) not going to college, and (c) on the honor roll but not going to college?

31. *Winning an Election* Taylor, Moore, and Jenkins are candidates for public office. It is estimated that Moore and Jenkins have about the same probability of winning, and Taylor is believed to be twice as likely to win as either of the others. Find the probability of each candidate winning the election.

32. *Winning an Election* Three people have been nominated for president of a college class. From a small poll, it is estimated that the probability of the first candidate winning the election is 0.37, and the probability of the second candidate winning the election is 0.44. What is the probability that the third candidate will win?

33. *Preparing for a Test* An instructor gives her class a list of 20 study problems, from which she will select 10 to be answered on an exam. If a given student knows how to solve 15 of the problems, find the probability that the student will be able to answer (a) all 10 questions on the exam, (b) exactly eight questions on the exam, and (c) at least nine questions on the exam.

34. *Preparing for a Test* An instructor gives his class a list of eight study problems, from which he will select five to be answered on an exam. If a given student knows how to solve six of the problems, find the probability that the student will be able to answer (a) all five questions on the exam, (b) exactly four questions on the exam, and (c) at least four questions on the exam.

35. *Letter Mix-Up* Four letters and envelopes are addressed to four different people. If the letters are randomly inserted into the envelopes, what is the probability that (a) exactly one will be inserted in the correct envelope and (b) at least one will be inserted in the correct envelope?

36. *Payroll Mix-Up* Five paychecks and envelopes are addressed to five different people. If the paychecks are randomly inserted into the envelopes, what is the probability that (a) exactly one will be inserted in the correct envelope and (b) at least one will be inserted in the correct envelope?

37. *Game Show* On a game show you are given five digits to arrange in the proper order to give the price of a car. If you are correct, you win the car. What is the probability of winning, given the following conditions?
(a) You guess the position of each digit.
(b) You know the first digit, but must guess the remaining four.

38. *Game Show* On a game show you are given four digits to arrange in the proper order to give the price of a car. If you are correct, you win the car. What is the probability of winning, given the following conditions?
(a) You guess the position of each digit.
(b) You know the first digit, but must guess the remaining three.

39. *Drawing Cards from a Deck* Two cards are selected at random from an ordinary deck of 52 playing cards. Find the probability that two aces are selected, given the following conditions.
(a) The cards are drawn in sequence, with the first card being replaced and the deck reshuffled prior to the second drawing.
(b) The two cards are drawn consecutively, without replacement.

40. *Poker Hand* Five cards are drawn from an ordinary deck of 52 playing cards. What is the probability of getting a full house?

41. *Defective Units* A shipment of 12 microwave ovens contains three defective units. A vending company has ordered four of these 12 units, and since each is identically packaged, the selection will be random. What is the probability that (a) all four units are good, (b) exactly two units are good, and (c) at least two units are good?

42. *Defective Units* A shipment of 20 compact disc players contains four defective units. A retail outlet has ordered five of these units. What is the probability that (a) all five players are good, (b) exactly four of the players are good, and (c) at least one of the players is defective?

43. *Random Number Generator* Two integers (between 1 and 30 inclusive) are chosen by a random number generator on a computer. What is the probability that (a) the numbers are both even, (b) one number is even and one is odd, (c) both numbers are less than 10, and (d) the same number is chosen twice?

44. *Random Number Generator* Two integers (between 1 and 40 inclusive) are chosen by a random number generator on a computer. What is the probability that (a) the numbers are both even, (b) one number is even and one is odd, (c) both numbers are no more than 30, and (d) the same number is chosen twice?

45. *Backup System* A space vehicle has an independent backup system for one of its communication networks. The probability that either system will function satisfactorily for the duration of a flight is 0.985. What is the probability that during a given flight (a) both systems function satisfactorily, (b) at least one system functions satisfactorily, and (c) both systems fail?

46. *Backup Vehicle* A fire company keeps two rescue vehicles to serve the community. Because of the demand on the company's time and the chance of mechanical failure, the probability that a specific vehicle is available when needed is 90%. If the availability of one vehicle is *independent* of the other, find the probability that (a) both vehicles are available at a given time, (b) neither vehicle is available at a given time, and (c) at least one vehicle is available at a given time.

47. *Making a Sale* A sales representative makes a sale at a rate of approximately one-fourth of all calls. If, on a given day, the representative contacts five potential clients, what is the probability that a sale will be made with (a) all five contacts, (b) none of the contacts, and (c) at least one contact?

48. *Making a Sale* A sales representative makes a sale at a rate of approximately one-third of all calls. If, on a given day, the representative contacts four potential clients, what is the probability that a sale will be made with (a) all four contacts, (b) none of the contacts, and (c) at least one contact?

49. *A Boy or a Girl?* Assume that the probability of the birth of a child of a particular sex is 50%. In a family with four children, what is the probability that (a) all the children are boys, (b) all the children are the same sex, and (c) there is at least one boy?

50. *A Boy or a Girl?* Assume that the probability of the birth of a child of a particular sex is 50%. In a family with six children, what is the probability that (a) all the children are girls, (b) all the children are the same sex, and (c) there is at least one girl?

51. *Is That Cash or Charge?* According to a survey by *USA Today*, the method used by Christmas shoppers to pay for gifts is as shown in the pie chart. Suppose two Christmas shoppers are chosen at random. What is the probability that both shoppers paid for their gifts only in cash?

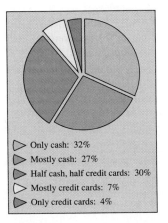

- ▷ Only cash: 32%
- ▷ Mostly cash: 27%
- ▷ Half cash, half credit cards: 30%
- ▷ Mostly credit cards: 7%
- ▷ Only credit cards: 4%

Figure for 51

52. *Flexible Work Hours*   In a survey by *Robert Hall International*, people were asked if they would prefer to work flexible hours—even if it meant slower career advancement—so they could spend more time with their family. The results of the survey are shown in the figure. Suppose three people from the survey were chosen at random. What is the probability that all three people would prefer flexible work hours?

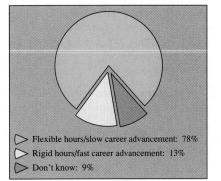

▷ Flexible hours/slow career advancement: 78%
▷ Rigid hours/fast career advancement: 13%
▷ Don't know: 9%

Figure for 52

REVIEW EXERCISES for Chapter 9

In Exercises 1–4, use sigma notation to write the given sum.

1. $\dfrac{1}{2(1)} + \dfrac{1}{2(2)} + \dfrac{1}{2(3)} + \cdots + \dfrac{1}{2(20)}$

2. $2(1^2) + 2(2^2) + 2(3^2) + \cdots + 2(9^2)$

3. $\frac{1}{2} + \frac{2}{3} + \frac{3}{4} + \cdots + \frac{9}{10}$

4. $1 - \frac{1}{3} + \frac{1}{9} - \frac{1}{27} + \cdots$

In Exercises 5–14, find the sum.

5. $\displaystyle\sum_{i=1}^{6} 5$

6. $\displaystyle\sum_{j=1}^{8} (20 - 3j)$

7. $\displaystyle\sum_{i=0}^{6} 2^i$

8. $\displaystyle\sum_{i=0}^{4} 3^i$

9. $\displaystyle\sum_{k=0}^{\infty} 4\left(\frac{2}{3}\right)^k$

10. $\displaystyle\sum_{k=0}^{\infty} 1.3\left(\frac{1}{10}\right)^k$

11. $\displaystyle\sum_{k=1}^{11} \left(\frac{2}{3}k + 4\right)$

12. $\displaystyle\sum_{k=1}^{25} \left(\frac{3k+1}{4}\right)$

13. $\displaystyle\sum_{n=0}^{10} (n^2 + 3)$

14. $\displaystyle\sum_{n=1}^{100} \left(\frac{1}{n} - \frac{1}{n+1}\right)$

In Exercises 15–18, write the first five terms of the arithmetic sequence.

15. $a_1 = 3,\ d = 4$

16. $a_1 = 8,\ d = -2$

17. $a_4 = 10,\ a_{10} = 28$

18. $a_2 = 14,\ a_6 = 22$

In Exercises 19 and 20, write an expression for the *n*th term of the specified arithmetic sequence and find the sum of the first 20 terms of the sequence.

19. $a_1 = 100,\ d = -3$

20. $a_1 = 10,\ a_3 = 28$

21. Find the sum of the first 100 positive multiples of 5.

22. Find the sum of the integers from 20 to 80 (inclusive).

In Exercises 23–26, write the first five terms of the geometric sequence.

23. $a_1 = 4,\ r = -\frac{1}{4}$

24. $a_1 = 2,\ r = 2$

25. $a_1 = 9,\ a_3 = 4$

26. $a_1 = 2,\ a_3 = 12$

In Exercises 27 and 28, write an expression for the *n*th term of the specified geometric sequence and find the sum of the first 20 terms of the sequence.

27. $a_1 = 16,\ a_2 = -8$

28. $a_1 = 100,\ r = 1.05$

29. *Depreciation*   A company buys a machine for $120,000. During the next five years it will depreciate at the rate of 30% per year. (That is, at the end of each year, the depreciated value will be 70% of what it was at the beginning of the year.)
   (a) Find the formula for the *t*th term of a geometric sequence that gives the value of the machine *t* full years after it was purchased.
   (b) Find the depreciated value of the machine at the end of five full years.

30. *Total Compensation*   Suppose you accept a job that pays a salary of $32,000 the first year, and that you will receive a 5.5% raise for each of the next 39 years. What will your total salary be over the 40-year period?

31. *Compound Interest*   A deposit of $200 is made at the beginning of each month for two years into an account that pays 6%, compounded monthly. What is the balance in the account at the end of two years?

32. *Compound Interest* A deposit of $100 is made at the beginning of each month for 10 years into an account that pays 6.5%, compounded monthly. What is the balance in the account at the end of 10 years?

In Exercises 33–36, use mathematical induction to prove the given formula for every positive integer n.

33. $1 + 4 + \cdots + (3n - 2) = \dfrac{n}{2}(3n - 1)$

34. $1 + \dfrac{3}{2} + 2 + \dfrac{5}{2} + \cdots + \dfrac{1}{2}(n + 1) = \dfrac{n}{4}(n + 3)$

35. $\displaystyle\sum_{i=0}^{n-1} ar^i = \dfrac{a(1 - r^n)}{1 - r}$

36. $\displaystyle\sum_{k=0}^{n-1} (a + kd) = \dfrac{n}{2}[2a + (n - 1)d]$

In Exercises 37–40, evaluate the given expression.

37. $_6C_4$

38. $_{10}C_7$

39. $_8P_5$

40. $_{12}P_3$

In Exercises 41–46, use the Binomial Theorem to expand the binomial. Simplify your answer. (Remember that $i = \sqrt{-1}$.)

41. $\left(\dfrac{x}{2} + y\right)^4$

42. $(a - 3b)^5$

43. $\left(\dfrac{2}{x} - 3x\right)^6$

44. $(3x + y^2)^7$

45. $(5 + 2i)^4$

46. $(4 - 5i)^3$

47. *Interpersonal Relationships* The complexity of interpersonal relationships increases dramatically as the size of a group increases. Determine the number of different two-person relationships in a family of (a) 2, (b) 4, and (c) 6.

48. *Morse Code* In Morse Code, all characters are transmitted using a sequence of dits and dahs. How many different characters can be formed by a sequence of three dits and dahs? (These can be repeated. For example, dit-dit-dit represents the letter *s*.)

49. *Amateur Radio* A Novice Amateur Radio license consists of two letters, one digit, and then three more letters. How many different licenses can be issued if no restrictions are placed on the letters or digits?

50. *Connecting Points with Lines* How many different straight-line segments are determined by (a) five noncollinear points and (b) ten noncollinear points?

51. *Matching Socks* A man has five pairs of socks (no two pairs are the same color). If he randomly selects two socks from the drawer, what is the probability that he gets a matched pair?

52. *Bookshelf Order* A child carries a five-volume set of books to a bookshelf. The child is not able to read, and hence cannot distinguish one volume from another. What is the probability that the books are shelved in the correct order?

53. *Roll of the Dice* Are the chances of rolling a 3 with one die the same as the chances of rolling a total of 6 with two dice? If not, which has the higher probability?

54. *Roll of the Dice* A die is rolled six times. What is the probability that each side will appear exactly once?

55. *Tossing a Coin* Find the probability of obtaining at least one tail when a coin is tossed five times.

56. *Parental Independence* In the *Marriott Seniors' Attitudes Survey*, senior citizens were asked if they would live with their children when they reached the point of not being able to live alone. The results are shown in the figure. Suppose three senior citizens who could not live alone are randomly selected. What is the probability that all three are not living with their children?

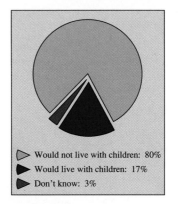

▷ Would not live with children: 80%
▶ Would live with children: 17%
▶ Don't know: 3%

Figure for 56

57. *Card Game* Five cards are drawn from an ordinary deck of 52 playing cards. Find the probability of getting two pairs. (For example, the hand could be A-A-5-5-Q or 4-4-7-7-K.)

58. *Birthday Problem*
(a) What is the probability that, in a group of 10 people, at least two have the same birthday? (Assume there are 365 different birthdays in a year.)
(b) How large must the group be before the probability that two have the same birthday is at least 50%?

CUMULATIVE TEST for Chapters 7–9

Take this test as you would take a test in class. After you are done, check your work with the answers given in the back of the book.

1. Solve the system by the method of substitution.

$$y = 3 - x^2$$
$$2(y - 2) = x - 1$$

2. Solve the linear system by the method of elimination.

$$x + 3y = -1$$
$$2x + 4y = 0$$

3. Solve the linear system by the method of elimination.

$$-2x + 4y - z = 3$$
$$x - 2y + 2z = -6$$
$$x - 3y - z = 1$$

4. Use Gauss-Jordan elimination to solve the linear system.

$$x + 3y - 2z = -7$$
$$-2x + y - z = -5$$
$$4x + y + z = 3$$

5. Find the value of a such that the system is inconsistent.

$$ax - 8y = 9$$
$$3x + 4y = 0$$

6. Sketch a graph of the solution of the system of inequalities.

$$3x + 4y \geq 16$$
$$3x - 4y \leq 8$$
$$y \leq 4$$

7. Find $2A - B$, given the matrices.

$$A = \begin{bmatrix} 6 & -1 \\ 2 & 4 \\ -3 & 5 \end{bmatrix}, \quad B = \begin{bmatrix} 1 & 4 \\ -1 & 5 \\ 1 & 10 \end{bmatrix}$$

8. Find AB, if possible.

$$A = \begin{bmatrix} 4 & -3 \\ 2 & 1 \\ 5 & 0 \end{bmatrix}, \quad B = \begin{bmatrix} 3 & -2 \\ 1 & -3 \end{bmatrix}$$

9. Find the inverse (if it exists) of the matrix.

$$\begin{bmatrix} 1 & 2 & -1 \\ 3 & 7 & -10 \\ -5 & -7 & -15 \end{bmatrix}$$

10. Evaluate the determinant.

$$\begin{vmatrix} 1 & 1 & 1 \\ 2 & -1 & -2 \\ 1 & -2 & -1 \end{vmatrix}$$

11. Simplify: $\dfrac{10!}{8!}$

12. Find the sum of the first 20 terms of the arithmetic sequence
8, 12, 16, 20, ...

13. Write the first five terms of the geometric sequence given $a_1 = 54$ and $r = -\frac{1}{3}$.

14. Find the sum of the infinite geometric series: $\sum\limits_{i=0}^{\infty} 3\left(\frac{1}{2}\right)^i$

15. Use mathematical induction to prove the formula
$$3 + 7 + 11 + 15 + \cdots + (4n - 1) = n(2n + 1).$$

16. Use the Binomial Theorem to expand and simplify $(z - 3)^4$.

17. Find the equation of the circle $x^2 + y^2 + Dx + Ey + F = 0$ that passes through the points $(0, 0)$, $(0, 4)$, and $(3, -4)$.

18. Maximize the objective function $z = 3x + 2y$ subject to the following constraints:
$$\begin{aligned} x &\geq 0 \\ y &\geq 0 \\ x + 4y &\leq 20 \\ 2x + y &\leq 12 \end{aligned}$$

19. Use determinants to find the area of the triangle with vertices $(0, 0)$, $(6, 2)$, and $(8, 10)$.

20. You accept a job with a salary of $32,000 for the first year. Suppose that during the next nine years you receive a 5% raise each year. What would your total compensation be over the ten-year period?

21. A personnel manager has ten applicants to fill three different positions in a corporation. In how many ways can this be done, assuming all the applicants are qualified for any of the three positions?

22. On a game show, the digits 3, 4, and 5 are given to arrange in the proper order to show the price of an appliance. If you are correct, you win the appliance. What is the probability of winning if
(a) you have no idea of the price of the appliance?
(b) you know the price of the appliance is at least $400?

Appendixes

Appendix A
Graphing Utilities

Introduction

In Section 3.2, you studied the point-plotting method for sketching the graph of an equation. One of the disadvantages of the point-plotting method is that in order to get a good idea about the shape of a graph, you need to plot *many* points. With only a few points, you could badly misrepresent the graph. For instance, consider the equation

$$y = \frac{1}{30}x(39 - 10x^2 + x^4).$$

Suppose you plotted only five points: $(-3, -3)$, $(-1, -1)$, $(0, 0)$, $(1, 1)$, and $(3, 3)$, as shown in Figure A.1. From these five points, you might assume that the graph of the equation is a straight line. That, however, is not correct. By plotting several more points you can see that the actual graph is not straight at all! (See Figure A.2.)

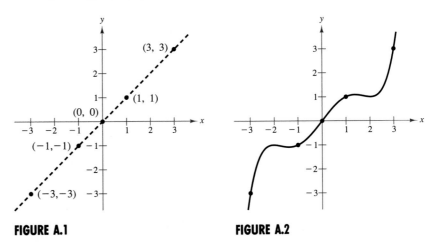

FIGURE A.1   **FIGURE A.2**

Thus, the point-plotting method leaves us with a dilemma. On the one hand, the method can be very inaccurate if only a few points are plotted. But, on the other hand, it is very time consuming to plot a dozen (or more) points. Technology can help us solve this dilemma. Plotting several (even several hundred) points in a rectangular coordinate system is something that a graphing utility can do easily.

A5

The point-plotting method is the method used by *all* graphing packages for computers and *all* graphing calculators. Each computer or calculator screen is made up of a grid of hundreds or thousands of small areas called **pixels.** Screens that have many pixels per inch are said to have a higher **resolution** than screens that don't have as many. For instance, the screen shown in Figure A.3(a) has a higher resolution than the screen shown in Figure A.3(b). Note that the "graph" of the line on the first screen looks more like a line than the "graph" on the second screen.

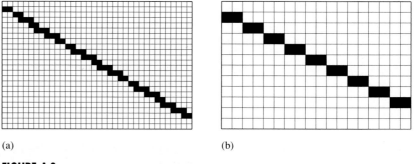

(a)   (b)

FIGURE A.3

Screens on most graphing calculators have 48 pixels per inch. Screens on computer monitors typically have between 32 and 100 pixels per inch.

EXAMPLE 1   Using Pixels to Sketch a Graph

Use the grid shown in Figure A.4 to sketch a graph of $y = \frac{1}{2}x^2$. Each pixel on the grid must be either on (shaded black) or off (unshaded).

Solution

To shade the grid, we use the following rule. If a pixel contains a plotted point of the graph, then it will be "on"; otherwise, the pixel will be "off." Using this rule, the graph of the curve looks like that shown in Figure A.5.

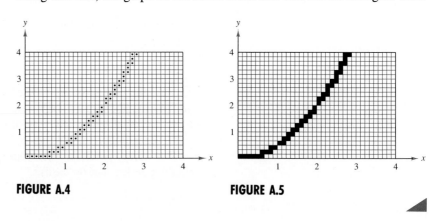

FIGURE A.4   **FIGURE A.5**

Basic Graphing

There are many different types of graphing utilities—graphing calculators and software packages for computers. The procedures used to draw a graph are similar with most of these utilities.

Basic Graphing Steps for a Graphing Utility

To draw the graph of an equation involving x and y with a graphing utility, use the following steps.

1. Rewrite the equation so that y is isolated on the left side of the equation.
2. Set the boundaries of the viewing rectangle by entering the minimum and maximum x-values and the minimum and maximum y-values.
3. Enter the equation in the form y = (expression involving x). Read the user's guide that accompanies your graphing utility to see how the equation should be entered.
4. Activate the graphing utility.

EXAMPLE 2   Sketching the Graph of an Equation

Sketch the graph of $2y + x^3 = 4x$.

Solution

To begin, solve the given equation for y in terms of x.

$$2y + x^3 = 4x \qquad \text{\textit{Given equation}}$$
$$2y = -x^3 + 4x \qquad \text{\textit{Subtract } x^3 \text{ from both sides}}$$
$$y = -\frac{1}{2}x^3 + 2x \qquad \text{\textit{Divide both sides by 2}}$$

Set the viewing rectangle so that $-10 \le x \le 10$ and $-10 \le y \le 10$. (On some graphing utilities, this is the default setting.) Next, enter the equation into the graphing utility.

$$Y = -X \wedge 3/2 + 2 * X$$

Finally, activate the graphing utility. The display screen should look like that shown in Figure A.6.

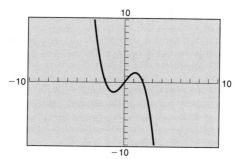

FIGURE A.6

In Figure A.6, notice that the calculator screen does not label the tick marks on the *x*-axis or the *y*-axis. To see what the tick marks represent, check the values in the utility's "range."

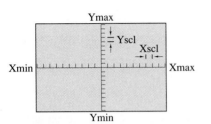

FIGURE A.7

Range

Xmin = −10	*The minimum x-value is −10.*
Xmax = 10	*The maximum x-value is 10.*
Xscl = 1	*The x-scale is 1 unit per tick mark.*
Ymin = −10	*The minimum y-value is −10.*
Ymax = 10	*The maximum y-value is 10.*
Yscl = 1	*The y-scale is 1 unit per tick mark.*
Xres = 1	*The x-resolution is 1 plotted point per 1 pixel.*

These settings are summarized visually in Figure A.7.

EXAMPLE 3   Graphing an Equation Involving Absolute Value

Sketch the graph of $y = |x - 3|$.

Solution

This equation is already written so that *y* is isolated on the left side of the equation, so you can enter the equation as follows.

$$Y = \text{abs}(X - 3)$$

After activating the graphing utility, its screen should look like the one shown in Figure A.8.

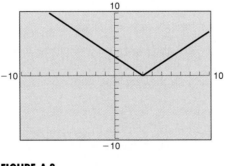

FIGURE A.8

Special Features

In order to be able to use your graphing calculator to its best advantage, you must be able to determine a proper viewing rectangle and use the zoom feature. The next two examples show how this is done.

EXAMPLE 4   Determining a Viewing Rectangle

Sketch the graph of $y = x^2 + 12$.

Solution

Begin as usual by entering the equation.

$Y = X \wedge 2 + 12$

Activate the graphing utility. If you used a viewing rectangle in which $-10 \leq x \leq 10$ and $-10 \leq y \leq 10$, then no part of the graph will appear on the screen, as shown in Figure A.9(a). The reason for this is that the lowest point on the graph of $y = x^2 + 12$ occurs at the point (0, 12). With the viewing rectangle in Figure A.9(a), the largest y-value is 10. In other words, none of the graph is visible on a screen whose y-values range between -10 and 10.

To be able to see the graph, change Ymax=10 to Ymax=30, Yscl=1 to Yscl=5. Now activate the graphing utility and you will obtain the graph shown in Figure A.9(b). On this graph, note that each tick mark on the y-axis represents 5 units because you changed the y-scale to 5. Also note that the highest point on the y-axis is now 30 because you changed the maximum value of y to 30.

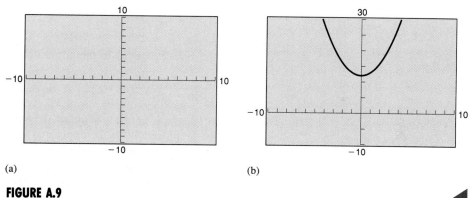

(a)   (b)

FIGURE A.9

EXAMPLE 5   Using the Zoom Feature

Sketch the graph of $y = x^3 - x^2 - x$. How many x-intercepts does this graph have?

Solution

Begin by drawing the graph on a "standard" viewing rectangle as shown in Figure A.10(a). From the display screen, it is clear that the graph has at least one intercept (just to the left of $x = 2$), but it is difficult to determine whether the graph has other intercepts. To obtain a better view of the graph near $x = -1$, you can use the zoom feature of the graphing utility. The redrawn screen is shown in Figure A.10(b). From this screen you can tell that the graph has three x-intercepts whose x-coordinates are approximately -0.6, 0, and 1.6.

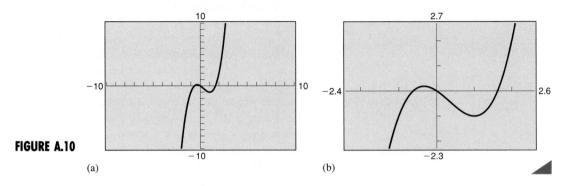

FIGURE A.10

(a)   (b)

EXAMPLE 6   Sketching More Than One Graph on the Same Screen

Sketch the graphs of $y = -\sqrt{36 - x^2}$ and $y = \sqrt{36 - x^2}$ on the same screen.

Solution

To begin, enter both equations in the graphing utility.

$$Y = \sqrt{(36 - X \wedge 2)}$$
$$Y = -\sqrt{(36 - X \wedge 2)}$$

Then, activate the graphing utility to obtain the graph shown in Figure A.11(a). Notice that the graph should be the upper and lower parts of the circle given by $x^2 + y^2 = 6^2$. The reason it doesn't look like a circle is that, with the standard settings, the tick marks on the x-axis are farther apart than the tick marks on the y-axis. To correct this, change the viewing rectangle so that $-15 \leq x \leq 15$. The redrawn screen is shown in Figure A.11(b). Notice that in this screen the graph appears to be more circular.

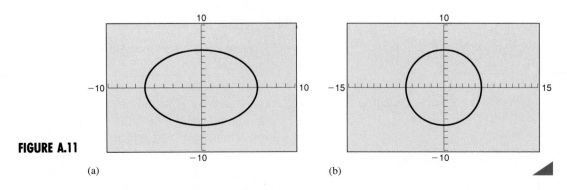

FIGURE A.11

(a)   (b)

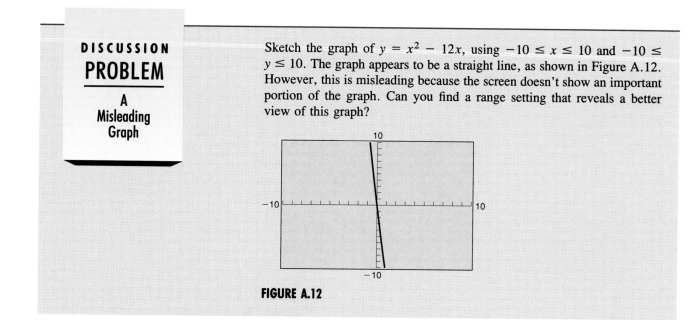

Sketch the graph of $y = x^2 - 12x$, using $-10 \le x \le 10$ and $-10 \le y \le 10$. The graph appears to be a straight line, as shown in Figure A.12. However, this is misleading because the screen doesn't show an important portion of the graph. Can you find a range setting that reveals a better view of this graph?

FIGURE A.12

WARM UP

In Exercises 1–10, solve for y in terms of x.

1. $3x + y = 4$
2. $x - y = 0$
3. $2x + 3y = 2$
4. $4x - 5y = -2$
5. $3x + 4y - 5 = 0$
6. $-2x - 3y + 6 = 0$
7. $x^2 + y - 4 = 0$
8. $-2x^2 + 3y + 2 = 0$
9. $x^2 + y^2 = 4$
10. $x^2 - y^2 = 9$

EXERCISES for Appendix A

In Exercises 1–20, use a graphing utility to sketch the graph of the equation. Use a setting on each graph of $-10 \le x \le 10$ and $-10 \le y \le 10$.

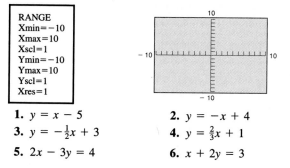

```
RANGE
Xmin=-10
Xmax=10
Xscl=1
Ymin=-10
Ymax=10
Yscl=1
Xres=1
```

1. $y = x - 5$
2. $y = -x + 4$
3. $y = -\frac{1}{2}x + 3$
4. $y = \frac{2}{3}x + 1$
5. $2x - 3y = 4$
6. $x + 2y = 3$

7. $y = \frac{1}{2}x^2 - 1$
8. $y = -x^2 + 6$
9. $y = x^2 - 4x - 5$
10. $y = x^2 - 3x + 2$
11. $y = -x^2 + 2x + 1$
12. $y = -x^2 + 4x - 1$
13. $2y = x^2 + 2x - 3$
14. $3y = -x^2 - 4x + 5$
15. $y = |x + 5|$
16. $y = \frac{1}{2}|x - 6|$
17. $y = \sqrt{x^2 + 1}$
18. $y = 2\sqrt{x^2 + 2} - 4$
19. $y = \frac{1}{5}(-x^3 + 16x)$
20. $y = \frac{1}{8}(x^3 + 8x^2)$

In Exercises 21–30, use a graphing utility to match the equation with its graph. [The graphs are labeled (a), (b), (c), (d), (e), (f), (g), (h), (i), and (j).]

21. $y = x$
22. $y = -x$
23. $y = x^2$
24. $y = -x^2$

25. $y = x^3$

26. $y = -x^3$

27. $y = |x|$

28. $y = -|x|$

29. $y = \sqrt{x}$

30. $y = -\sqrt{x}$

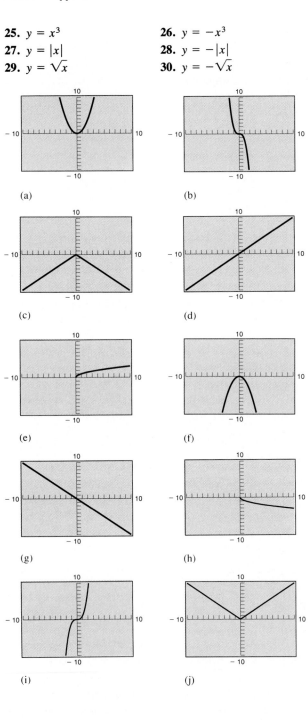

(a)

(b)

(c)

(d)

(e)

(f)

(g)

(h)

(i)

(j)

In Exercises 31–34, use a graphing utility to sketch the graph of the equation. Use the indicated setting.

31. $y = -2x^2 + 12x + 14$

32. $y = -x^2 + 5x + 6$

```
RANGE
Xmin=-5
Xmax=10
Xscl=1
Ymin=-5
Ymax=35
Yscl=5
Xres=1
```

```
RANGE
Xmin=-8
Xmax=4
Xscl=1
Ymin=-5
Ymax=15
Yscl=5
Xres=1
```

33. $y = x^3 + 6x^2$

34. $y = -x^3 + 16x$

```
RANGE
Xmin=-10
Xmax=5
Xscl=1
Ymin=-4
Ymax=36
Yscl=3
Xres=1
```

```
RANGE
Xmin=-6
Xmax=6
Xscl=1
Ymin=-25
Ymax=25
Yscl=5
Xres=1
```

In Exercises 35–38, find a setting on a graphing utility so that the graph of the equation agrees with the graph shown.

35. $y = -x^2 - 4x + 20$

36. $y = x^2 + 12x - 8$

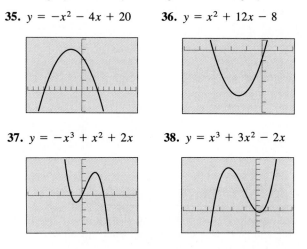

37. $y = -x^3 + x^2 + 2x$

38. $y = x^3 + 3x^2 - 2x$

In Exercises 39–42, use a graphing utility to find the number of x-intercepts of the equation.

39. $y = \frac{1}{8}(4x^2 - 32x + 65)$

40. $y = \frac{1}{4}(-4x^2 + 16x - 15)$

41. $y = 4x^3 - 20x^2 - 4x + 61$

42. $y = \frac{1}{4}(2x^3 + 6x^2 - 4x + 1)$

In Exercises 43–46, use a graphing utility to sketch the graphs of the equations on the same screen. Using a "square setting," what geometrical shape is bounded by the graphs?

43. $y = |x| - 4$
$y = -|x| + 4$

44. $y = x + |x| - 4$
$y = x - |x| + 4$

45. $y = -\sqrt{25 - x^2}$
$y = \sqrt{25 - x^2}$

46. $y = 6$
$y = -\sqrt{3}x - 4$
$y = \sqrt{3}x - 4$

Ever Been Married?   In Exercises 47–50, use the following models, which relate ages to the percentages of American males and females who have never been married.

$$y = \frac{0.36 - 0.0056x}{1 - 0.0817x + 0.00226x^2}, \quad \begin{matrix} \text{Males} \\ 20 \le x \le 50 \end{matrix}$$

$$y = \frac{100}{8.944 - 0.886x + 0.249x^2}, \quad \begin{matrix} \text{Females} \\ 20 \le x \le 50 \end{matrix}$$

In these models, y is the percent of the population (in decimal form) who have never been married and x is the age of the person. (*Source:* U.S. Bureau of the Census)

47. Use a graphing utility to sketch the graph of both equations giving the percentages of American males and females who have never been married. Use the following range settings.

```
RANGE
Xmin=20
Xmax=50
Xscl=5
Ymin=0
Ymax=1
Yscl=0.1
Xres=1
```

48. Write a short paragraph describing the relationship between the two graphs that were plotted in Exercise 47.

49. Suppose an American male is chosen at random from the population. If the person is 25 years old, what is the probability that he has never been married?

50. Suppose an American female is chosen at random from the population. If the person is 25 years old, what is the probability that she has never been married?

Earnings and Dividends   In Exercises 51–54, use the following model, which approximates the relationship between dividends per share and earnings per share for the Pall Corporation between 1982 and 1989.

$$y = -0.166 + 0.502x - 0.0953x^2, \quad 0.25 \le x \le 2$$

In this model, y is the dividends per share (in dollars) and x is the earnings per share (in dollars). (*Source:* Standard ASE Stock Reports)

51. Use a graphing utility to sketch the graph of the model that gives the dividend per share in terms of the earnings per share. Use the following range settings.

```
RANGE
Xmin=0
Xmax=2
Xscl=0.25
Ymin=0
Ymax=0.5
Yscl=0.1
Xres=1
```

52. According to the given model, what size dividend would the Pall Corporation pay if the earnings per share were $1.30?

53. Use a trace feature on your graphing utility to estimate the earnings per share that would produce a dividend per share of $0.25. The choices are labeled (a), (b), (c), and (d). (Find the y-value that is as close to 0.25 as possible. The x-value that is displayed will then be the approximate earnings per share that would produce a dividend per share of $0.25.)

(a) $1.00   (b) $1.03   (c) $1.06   (d) $1.09

54. The *payout ratio* for a stock is the ratio of the dividend per share to earnings per share. Use the model to find the payout ratio for an earnings per share of (a) $0.75, (b) $1.00, and (c) $1.25.

Appendix B

Using Logarithmic Tables

Although it is more efficient to use calculators than tables in computations with logarithmic functions, it is instructive to see how to work with tables. Using base 10, note first that every positive real number can be written as a product $c \times 10^k$, where $1 \leq c < 10$ and k is an integer. For example, $1989 = 1.989 \times 10^3$, $5.37 = 5.37 \times 10^0$, and $0.0439 = 4.39 \times 10^{-2}$. Suppose you apply the properties of logarithms to the number 1989.

$$1989 = 1.989 \times 10^3$$
$$\log_{10} 1989 = \log_{10}(1.989 \times 10^3)$$
$$= \log_{10}(1.989) + \log_{10}(10^3)$$
$$= \log_{10}(1.989) + 3 \log_{10}(10)$$
$$= \log_{10}(1.989) + 3$$

In general, for any positive real number x (expressible as $x = c \times 10^k$), its common logarithm has the **standard form**

$$\log_{10} x = \log_{10} c + \log_{10}(10^k) = \log_{10} c + k$$

where $1 \leq c < 10$. We call $\log_{10} c$ the **mantissa** and k the **characteristic** of $\log_{10} x$. Since the function $f(x) = \log_{10} x$ increases as x increases, and since $1 \leq c < 10$, it follows that

$$\log_{10} 1 \leq \log_{10} c < \log_{10} 10$$
$$0 \leq \log_{10} c < 1$$

which means that the *mantissa* of $\log_{10} x$ lies between 0 and 1.

The common logarithm table in Appendix E gives four-decimal-place approximations of the *mantissa* for the logarithm of every three-digit number between 1.00 and 9.99. The next example shows how to use the table in Appendix E to approximate common logarithms.

EXAMPLE 1   Approximating Common Logarithms with Tables

Use the tables in Appendix E to approximate the following.

a. $\log_{10} 85.6$ **b.** $\log_{10} 0.000329$

A15

Solution

a. Since $85.6 = 8.56 \times 10^1$, the characteristic is 1. Using the common logarithm table, you can see that the mantissa is $\log_{10} 8.56 \approx 0.9325$. Therefore,

$$
\begin{aligned}
\log_{10} 85.6 &= \text{(mantissa)} + \text{(characteristic)} \\
&= \log_{10} 8.56 + 1 \\
&\approx 0.9325 + 1 \\
&= 1.9325.
\end{aligned}
$$

b. Since $0.000329 = 3.29 \times 10^{-4}$, the characteristic is -4. From the common logarithm table for the mantissa 3.29, you obtain

$$
\begin{aligned}
\log_{10} 0.000329 &= \log_{10} 3.29 + (-4) \\
&\approx 0.3598 - 4 \\
&= -3.6402.
\end{aligned}
$$

The next example shows how to combine the use of properties of logarithms with tables to evaluate logarithms.

EXAMPLE 2   Combining Properties of Logarithms with Tables

Use the tables in Appendix E to approximate $\log_{10} \sqrt[3]{38.6}$.

Solution

$$
\begin{aligned}
\log_{10} \sqrt[3]{38.6} &= \frac{1}{3} \log_{10} 38.6 \\
&= \frac{1}{3}(\log_{10} 3.86 + 1) \\
&\approx \frac{1}{3}(0.5866 + 1) \\
&\approx 0.5289
\end{aligned}
$$

The table for common logarithms can be used in the *reverse* manner to find the number (called an **antilogarithm**) that has a given logarithm. We demonstrate this procedure in Example 3.

EXAMPLE 3   Finding the Antilogarithm of a Number

Use the tables in Appendix E to approximate the value of x in each of the following.

a. $\log_{10} x = 2.6571$ **b.** $x = 10^{-3.6364}$

Solution

a. You know that $\log_{10} x = 2.6571 = 0.6571 + 2$. Thus, the mantissa is 0.6571 and the characteristic is 2. From the table, you find that the mantissa 0.6571 corresponds approximately to $\log_{10} 4.54$. Since the characteristic is 2, it follows that x is given by

$$x \approx 4.54 \times 10^2 = 454.$$

b. In logarithmic form, this exponential equation can be written as $\log_{10} x = -3.6364$. To obtain the standard form, add and subtract 4 to obtain

$$\log_{10} x = (4 - 3.6364) - 4 = 0.3636 - 4.$$

Thus, the mantissa is 0.3636 and the characteristic is -4. From the table, you find that

$$x \approx 2.31 \times 10^{-4} = 0.000231.$$

For numbers with more than three nonzero digits, you can still use the common logarithm tables by applying a procedure called **linear interpolation.** This procedure is based on the fact that changes in $\log_{10} x$ are approximately proportional to the corresponding changes in x. We demonstrate the procedure in Example 4.

EXAMPLE 4   Linear Interpolation

Use linear interpolation and the tables in Appendix E to approximate the value of $\log_{10} 5.382$.

Solution

The three-digit x-values in the table that are closest to 5.382 are $x = 5.38$ and $x = 5.39$. Use the logarithms of these two values in the following arrangement.

$$0.01 \left\{ 0.002 \left\{ \begin{matrix} \log_{10} 5.38 \approx 0.7308 \\ \log_{10} 5.382 \approx ? \end{matrix} \right\} d \atop \log_{10} 5.39 \approx 0.7316 \right\} 0.0008$$

Note that the differences between the x-values are beside the left braces and the differences between the corresponding logarithms are beside the right braces. From this arrangement, you can write the following proportion.

$$\frac{d}{0.0008} = \frac{0.002}{0.01}$$

$$d = (0.0008)\left(\frac{0.002}{0.01}\right) = 0.00016 \approx 0.0002$$

Therefore,

$$\log_{10} 5.382 \approx \log_{10} 5.38 + d \approx 0.7308 + 0.002 = 0.7310.$$

The next example shows how to perform numerical computations with logarithms.

EXAMPLE 5   Using Logarithms to Perform Numerical Computations

Use the tables in Appendix E to approximate the value of

$$x = \frac{(1.9)^3}{\sqrt{82.7}}.$$

Solution

Using the properties of logarithms, you can write

$$\log_{10} x = 3 \log_{10} 1.9 - \frac{1}{2} \log_{10} 82.7$$

$$\approx 3(0.2788) - \frac{1}{2}(1.9175) \approx -0.12235.$$

By adding and subtracting 1, you obtain the standard form

$$\log_{10} x \approx (1 - 0.12235) - 1 = 0.87765 - 1.$$

Finally, since the antilogarithm of 0.87765 is approximately 7.54, we find that

$$x \approx 7.54 \times 10^{-1} = 0.754.$$

EXERCISES for Appendix B

In Exercises 1–4, approximate the common logarithm of the given number by using the table in Appendix E.

1. (a) 417   (b) 0.0417
2. (a) 985   (b) 9.85
3. (a) 6300   (b) 1000
4. (a) 0.0001   (b) 41.3

In Exercises 5–8, approximate the common logarithm of the given quantity by using the table in Appendix E.

5. (a) $\dfrac{5.30}{21.5}$ (b) $(30500)(0.258)$

6. (a) $\sqrt[3]{5.33}$ (b) $(1.02)^{36}$

7. (a) $\sqrt[5]{7200}$ (b) $(3.4)^8$

8. (a) $\dfrac{(3.6)^6}{500}$ (b) $(0.245)^4(8.7)^3$

In Exercises 9–12, find N (antilogarithm) by using the table in Appendix E.

9. (a) $N = 10^{4.3979}$ (b) $N = 10^{-1.6021}$
10. (a) $\log_{10} N = 3.6702$ (b) $\log_{10} N = -2.3298$
11. (a) $\log_{10} N = 6.1335$ (b) $\log_{10} N = 8.1335 - 10$
12. (a) $\log_{10} N = 4.8420$ (b) $\log_{10} N = 7.8420 - 10$

In Exercises 13 and 14, use linear interpolation to approximate the common logarithm of the given number by using the table in Appendix E.

13. (a) 4385   (b) 0.6058
14. (a) 125.2   (b) 0.08675

In Exercises 15 and 16, use linear interpolation to approximate the antilogarithm N by using the table in Appendix E.

15. (a) $\log_{10} N = 5.6175$ (b) $\log_{10} N = -2.1503$

16. (a) $N = 10^{0.5743}$ (b) $N = 10^{9.9317}$

In Exercises 17–22, approximate the given quantity by using common logarithms.

17. $\dfrac{(86.4)(8.09)}{38.6}$ **18.** $\dfrac{1243}{(42.8)(67.9)}$

19. $\sqrt[3]{86.5}$ **20.** $\sqrt[4]{(4.705)(18.86)}$

21. $500(1.03)^{20}$ **22.** $(0.2313)^6$

In Exercises 23 and 24, approximate the natural logarithm of the given number by using the table in Appendix D.

23. (a) 6.24   (b) 9.55

24. (a) 2.605   (b) 3.005

In Exercises 25 and 26, approximate the antilogarithm N by using the table in Appendix D.

25. (a) $\ln N = 2.0096$ (b) $\ln N = 1.4422$

26. (a) $\ln N = 1.1233$ (b) $\ln N = 0.2271$

In Exercises 27 and 28, approximate the exponential by using the table in Appendix C.

27. (a) $e^{3.5}$ (b) $e^{-3.5}$

28. (a) $e^{6.2}$ (b) $e^{-6.2}$

Appendix C

Exponential Tables

x	e^x	e^{-x}
0.0	1.0000	1.0000
0.1	1.1052	0.9048
0.2	1.2214	0.8187
0.3	1.3499	0.7408
0.4	1.4918	0.6703
0.5	1.6487	0.6065
0.6	1.8221	0.5488
0.7	2.0138	0.4966
0.8	2.2255	0.4493
0.9	2.4596	0.4066
1.0	2.7183	0.3679
1.1	3.0042	0.3329
1.2	3.3201	0.3012
1.3	3.6693	0.2725
1.4	4.0552	0.2466
1.5	4.4817	0.2231
1.6	4.9530	0.2019
1.7	5.4739	0.1827
1.8	6.0496	0.1653
1.9	6.6859	0.1496
2.0	7.3891	0.1353
2.1	8.1662	0.1225
2.2	9.0250	0.1108
2.3	9.9742	0.1003
2.4	11.023	0.0907
2.5	12.182	0.0821
2.6	13.464	0.0743
2.7	14.880	0.0672
2.8	16.445	0.0608
2.9	18.174	0.0550
3.0	20.086	0.0498
3.1	22.198	0.0450
3.2	24.533	0.0408
3.3	27.113	0.0369
3.4	29.964	0.0334

x	e^x	e^{-x}
3.5	33.115	0.0302
3.6	36.598	0.0273
3.7	40.447	0.0247
3.8	44.701	0.0224
3.9	49.402	0.0202
4.0	54.598	0.0183
4.1	60.340	0.0166
4.2	66.686	0.0150
4.3	73.700	0.0136
4.4	81.451	0.0123
4.5	90.017	0.0111
4.6	99.484	0.0101
4.7	109.95	0.0091
4.8	121.51	0.0082
4.9	134.29	0.0074
5.0	148.41	0.0067
5.1	164.02	0.0061
5.2	181.27	0.0055
5.3	200.34	0.0050
5.4	221.41	0.0045
5.5	244.69	0.0041
5.6	270.43	0.0037
5.7	298.87	0.0033
5.8	330.30	0.0030
5.9	365.04	0.0027
6.0	403.43	0.0025
6.1	445.86	0.0022
6.2	492.75	0.0020
6.3	544.57	0.0018
6.4	601.85	0.0017
6.5	665.14	0.0015
6.6	735.10	0.0014
6.7	812.41	0.0012
6.8	897.85	0.0011
6.9	992.27	0.0010

x	e^x	e^{-x}
7.0	1096.63	0.0009
7.1	1211.97	0.0008
7.2	1339.43	0.0007
7.3	1480.30	0.0007
7.4	1635.98	0.0006
7.5	1808.04	0.0006
7.6	1998.20	0.0005
7.7	2208.35	0.0005
7.8	2440.60	0.0004
7.9	2697.28	0.0004
8.0	2980.96	0.0003
8.1	3294.47	0.0003
8.2	3640.95	0.0003
8.3	4023.87	0.0002
8.4	4447.07	0.0002
8.5	4914.77	0.0002
8.6	5431.66	0.0002
8.7	6002.91	0.0002
8.8	6634.24	0.0002
8.9	7331.97	0.0001
9.0	8103.08	0.0001
9.1	8955.29	0.0001
9.2	9897.13	0.0001
9.3	10938.02	0.0001
9.4	12088.38	0.0001
9.5	13359.73	0.0001
9.6	14764.78	0.0001
9.7	16317.61	0.0001
9.8	18033.74	0.0001
9.9	19930.37	0.0001
10.0	22026.47	0.0000

Appendix D

Natural Logarithmic Tables

	0.00	0.01	0.02	0.03	0.04	0.05	0.06	0.07	0.08	0.09
1.0	0.0000	0.0100	0.0198	0.0296	0.0392	0.0488	0.0583	0.0677	0.0770	0.0862
1.1	0.0953	0.1044	0.1133	0.1222	0.1310	0.1398	0.1484	0.1570	0.1655	0.1740
1.2	0.1823	0.1906	0.1989	0.2070	0.2151	0.2231	0.2311	0.2390	0.2469	0.2546
1.3	0.2624	0.2700	0.2776	0.2852	0.2927	0.3001	0.3075	0.3148	0.3221	0.3293
1.4	0.3365	0.3436	0.3507	0.3577	0.3646	0.3716	0.3784	0.3853	0.3920	0.3988
1.5	0.4055	0.4121	0.4187	0.4253	0.4318	0.4383	0.4447	0.4511	0.4574	0.4637
1.6	0.4700	0.4762	0.4824	0.4886	0.4947	0.5008	0.5068	0.5128	0.5188	0.5247
1.7	0.5306	0.5365	0.5423	0.5481	0.5539	0.5596	0.5653	0.5710	0.5766	0.5822
1.8	0.5878	0.5933	0.5988	0.6043	0.6098	0.6152	0.6206	0.6259	0.6313	0.6366
1.9	0.6419	0.6471	0.6523	0.6575	0.6627	0.6678	0.6729	0.6780	0.6831	0.6881
2.0	0.6931	0.6981	0.7031	0.7080	0.7129	0.7178	0.7227	0.7275	0.7324	0.7372
2.1	0.7419	0.7467	0.7514	0.7561	0.7608	0.7655	0.7701	0.7747	0.7793	0.7839
2.2	0.7885	0.7930	0.7975	0.8020	0.8065	0.8109	0.8154	0.8198	0.8242	0.8286
2.3	0.8329	0.8372	0.8416	0.8459	0.8502	0.8544	0.8587	0.8629	0.8671	0.8713
2.4	0.8755	0.8796	0.8838	0.8879	0.8920	0.8961	0.9002	0.9042	0.9083	0.9123
2.5	0.9163	0.9203	0.9243	0.9282	0.9322	0.9361	0.9400	0.9439	0.9478	0.9517
2.6	0.9555	0.9594	0.9632	0.9670	0.9708	0.9746	0.9783	0.9821	0.9858	0.9895
2.7	0.9933	0.9969	1.0006	1.0043	1.0080	1.0116	1.0152	1.0188	1.0225	1.0260
2.8	1.0296	1.0332	1.0367	1.0403	1.0438	1.0473	1.0508	1.0543	1.0578	1.0613
2.9	1.0647	1.0682	1.0716	1.0750	1.0784	1.0818	1.0852	1.0886	1.0919	1.0953
3.0	1.0986	1.1019	1.1053	1.1086	1.1119	1.1151	1.1184	1.1217	1.1249	1.1282
3.1	1.1314	1.1346	1.1378	1.1410	1.1442	1.1474	1.1506	1.1537	1.1569	1.1600
3.2	1.1632	1.1663	1.1694	1.1725	1.1756	1.1787	1.1817	1.1848	1.1878	1.1909
3.3	1.1939	1.1969	1.2000	1.2030	1.2060	1.2090	1.2119	1.2149	1.2179	1.2208
3.4	1.2238	1.2267	1.2296	1.2326	1.2355	1.2384	1.2413	1.2442	1.2470	1.2499
3.5	1.2528	1.2556	1.2585	1.2613	1.2641	1.2669	1.2698	1.2726	1.2754	1.2782
3.6	1.2809	1.2837	1.2865	1.2892	1.2920	1.2947	1.2975	1.3002	1.3029	1.3056
3.7	1.3083	1.3110	1.3137	1.3164	1.3191	1.3218	1.3244	1.3271	1.3297	1.3324
3.8	1.3350	1.3376	1.3403	1.3429	1.3455	1.3481	1.3507	1.3533	1.3558	1.3584
3.9	1.3610	1.3635	1.3661	1.3686	1.3712	1.3737	1.3762	1.3788	1.3813	1.3838
4.0	1.3863	1.3888	1.3913	1.3938	1.3962	1.3987	1.4012	1.4036	1.4061	1.4085
4.1	1.4110	1.4134	1.4159	1.4183	1.4207	1.4231	1.4255	1.4279	1.4303	1.4327
4.2	1.4351	1.4375	1.4398	1.4422	1.4446	1.4469	1.4493	1.4516	1.4540	1.4563
4.3	1.4586	1.4609	1.4633	1.4656	1.4679	1.4702	1.4725	1.4748	1.4770	1.4793
4.4	1.4816	1.4839	1.4861	1.4884	1.4907	1.4929	1.4951	1.4974	1.4996	1.5019
4.5	1.5041	1.5063	1.5085	1.5107	1.5129	1.5151	1.5173	1.5195	1.5217	1.5239
4.6	1.5261	1.5282	1.5304	1.5326	1.5347	1.5369	1.5390	1.5412	1.5433	1.5454
4.7	1.5476	1.5497	1.5518	1.5539	1.5560	1.5581	1.5602	1.5623	1.5644	1.5665
4.8	1.5686	1.5707	1.5728	1.5748	1.5769	1.5790	1.5810	1.5831	1.5851	1.5872
4.9	1.5892	1.5913	1.5933	1.5953	1.5974	1.5994	1.6014	1.6034	1.6054	1.6074
5.0	1.6094	1.6114	1.6134	1.6154	1.6174	1.6194	1.6214	1.6233	1.6253	1.6273
5.1	1.6292	1.6312	1.6332	1.6351	1.6371	1.6390	1.6409	1.6429	1.6448	1.6467
5.2	1.6487	1.6506	1.6525	1.6544	1.6563	1.6582	1.6601	1.6620	1.6639	1.6658
5.3	1.6677	1.6696	1.6715	1.6734	1.6752	1.6771	1.6790	1.6808	1.6827	1.6845
5.4	1.6864	1.6882	1.6901	1.6919	1.6938	1.6956	1.6974	1.6993	1.7011	1.7029

Natural Logarithmic Tables (Continued)

	0.00	0.01	0.02	0.03	0.04	0.05	0.06	0.07	0.08	0.09
5.5	1.7047	1.7066	1.7084	1.7102	1.7120	1.7138	1.7156	1.7174	1.7192	1.7210
5.6	1.7228	1.7246	1.7263	1.7281	1.7299	1.7317	1.7334	1.7352	1.7370	1.7387
5.7	1.7405	1.7422	1.7440	1.7457	1.7475	1.7492	1.7509	1.7527	1.7544	1.7561
5.8	1.7579	1.7596	1.7613	1.7630	1.7647	1.7664	1.7681	1.7699	1.7716	1.7733
5.9	1.7750	1.7766	1.7783	1.7800	1.7817	1.7834	1.7851	1.7867	1.7884	1.7901
6.0	1.7918	1.7934	1.7951	1.7967	1.7984	1.8001	1.8017	1.8034	1.8050	1.8066
6.1	1.8083	1.8099	1.8116	1.8132	1.8148	1.8165	1.8181	1.8197	1.8213	1.8229
6.2	1.8245	1.8262	1.8278	1.8294	1.8310	1.8326	1.8342	1.8358	1.8374	1.8390
6.3	1.8405	1.8421	1.8437	1.8453	1.8469	1.8485	1.8500	1.8516	1.8532	1.8547
6.4	1.8563	1.8579	1.8594	1.8610	1.8625	1.8641	1.8656	1.8672	1.8687	1.8703
6.5	1.8718	1.8733	1.8749	1.8764	1.8779	1.8795	1.8810	1.8825	1.8840	1.8856
6.6	1.8871	1.8886	1.8901	1.8916	1.8931	1.8946	1.8961	1.8976	1.8991	1.9006
6.7	1.9021	1.9036	1.9051	1.9066	1.9081	1.9095	1.9110	1.9125	1.9140	1.9155
6.8	1.9169	1.9184	1.9199	1.9213	1.9228	1.9242	1.9257	1.9272	1.9286	1.9301
6.9	1.9315	1.9330	1.9344	1.9359	1.9373	1.9387	1.9402	1.9416	1.9430	1.9445
7.0	1.9459	1.9473	1.9488	1.9502	1.9516	1.9530	1.9544	1.9559	1.9573	1.9587
7.1	1.9601	1.9615	1.9629	1.9643	1.9657	1.9671	1.9685	1.9699	1.9713	1.9727
7.2	1.9741	1.9755	1.9769	1.9782	1.9796	1.9810	1.9824	1.9838	1.9851	1.9865
7.3	1.9879	1.9892	1.9906	1.9920	1.9933	1.9947	1.9961	1.9974	1.9988	2.0001
7.4	2.0015	2.0028	2.0042	2.0055	2.0069	2.0082	2.0096	2.0109	2.0122	2.0136
7.5	2.0149	2.0162	2.0176	2.0189	2.0202	2.0215	2.0229	2.0242	2.0255	2.0268
7.6	2.0281	2.0295	2.0308	2.0321	2.0334	2.0347	2.0360	2.0373	2.0386	2.0399
7.7	2.0412	2.0425	2.0438	2.0451	2.0464	2.0477	2.0490	2.0503	2.0516	2.0528
7.8	2.0541	2.0554	2.0567	2.0580	2.0592	2.0605	2.0618	2.0631	2.0643	2.0656
7.9	2.0669	2.0681	2.0694	2.0707	2.0719	2.0732	2.0744	2.0757	2.0769	2.0782
8.0	2.0794	2.0807	2.0819	2.0832	2.0844	2.0857	2.0869	2.0882	2.0894	2.0906
8.1	2.0919	2.0931	2.0943	2.0956	2.0968	2.0980	2.0992	2.1005	2.1017	2.1029
8.2	2.1041	2.1054	2.1066	2.1078	2.1090	2.1102	2.1114	2.1126	2.1138	2.1150
8.3	2.1163	2.1175	2.1187	2.1199	2.1211	2.1223	2.1235	2.1247	2.1258	2.1270
8.4	2.1282	2.1294	2.1306	2.1318	2.1330	2.1342	2.1353	2.1365	2.1377	2.1389
8.5	2.1401	2.1412	2.1424	2.1436	2.1448	2.1459	2.1471	2.1483	2.1494	2.1506
8.6	2.1518	2.1529	2.1541	2.1552	2.1564	2.1576	2.1587	2.1599	2.1610	2.1622
8.7	2.1633	2.1645	2.1656	2.1668	2.1679	2.1691	2.1702	2.1713	2.1725	2.1736
8.8	2.1748	2.1759	2.1770	2.1782	2.1793	2.1804	2.1815	2.1827	2.1838	2.1849
8.9	2.1861	2.1872	2.1883	2.1894	2.1905	2.1917	2.1928	2.1939	2.1950	2.1961
9.0	2.1972	2.1983	2.1994	2.2006	2.2017	2.2028	2.2039	2.2050	2.2061	2.2072
9.1	2.2083	2.2094	2.2105	2.2116	2.2127	2.2138	2.2148	2.2159	2.2170	2.2181
9.2	2.2192	2.2203	2.2214	2.2225	2.2235	2.2246	2.2257	2.2268	2.2279	2.2289
9.3	2.2300	2.2311	2.2322	2.2332	2.2343	2.2354	2.2364	2.2375	2.2386	2.2396
9.4	2.2407	2.2418	2.2428	2.2439	2.2450	2.2460	2.2471	2.2481	2.2492	2.2502
9.5	2.2513	2.2523	2.2534	2.2544	2.2555	2.2565	2.2576	2.2586	2.2597	2.2607
9.6	2.2618	2.2628	2.2638	2.2649	2.2659	2.2670	2.2680	2.2690	2.2701	2.2711
9.7	2.2721	2.2732	2.2742	2.2752	2.2762	2.2773	2.2783	2.2793	2.2803	2.2814
9.8	2.2824	2.2834	2.2844	2.2854	2.2865	2.2875	2.2885	2.2895	2.2905	2.2915
9.9	2.2925	2.2935	2.2946	2.2956	2.2966	2.2976	2.2986	2.2996	2.3006	2.3016

Appendix E

Common Logarithmic Tables

	0.00	0.01	0.02	0.03	0.04	0.05	0.06	0.07	0.08	0.09
1.0	0.0000	0.0043	0.0086	0.0128	0.0170	0.0212	0.0253	0.0294	0.0334	0.0374
1.1	0.0414	0.0453	0.0492	0.0531	0.0569	0.0607	0.0645	0.0682	0.0719	0.0755
1.2	0.0792	0.0828	0.0864	0.0899	0.0934	0.0969	0.1004	0.1038	0.1072	0.1106
1.3	0.1139	0.1173	0.1206	0.1239	0.1271	0.1303	0.1335	0.1367	0.1399	0.1430
1.4	0.1461	0.1492	0.1523	0.1553	0.1584	0.1614	0.1644	0.1673	0.1703	0.1732
1.5	0.1761	0.1790	0.1818	0.1847	0.1875	0.1903	0.1931	0.1959	0.1987	0.2014
1.6	0.2041	0.2068	0.2095	0.2122	0.2148	0.2175	0.2201	0.2227	0.2253	0.2279
1.7	0.2304	0.2330	0.2355	0.2380	0.2405	0.2430	0.2455	0.2480	0.2504	0.2529
1.8	0.2553	0.2577	0.2601	0.2625	0.2648	0.2672	0.2695	0.2718	0.2742	0.2765
1.9	0.2788	0.2810	0.2833	0.2856	0.2878	0.2900	0.2923	0.2945	0.2967	0.2989
2.0	0.3010	0.3032	0.3054	0.3075	0.3096	0.3118	0.3139	0.3160	0.3181	0.3201
2.1	0.3222	0.3243	0.3263	0.3284	0.3304	0.3324	0.3345	0.3365	0.3385	0.3404
2.2	0.3424	0.3444	0.3464	0.3483	0.3502	0.3522	0.3541	0.3560	0.3579	0.3598
2.3	0.3617	0.3636	0.3655	0.3674	0.3692	0.3711	0.3729	0.3747	0.3766	0.3784
2.4	0.3802	0.3820	0.3838	0.3856	0.3874	0.3892	0.3909	0.3927	0.3945	0.3962
2.5	0.3979	0.3997	0.4014	0.4031	0.4048	0.4065	0.4082	0.4099	0.4116	0.4133
2.6	0.4150	0.4166	0.4183	0.4200	0.4216	0.4232	0.4249	0.4265	0.4281	0.4298
2.7	0.4314	0.4330	0.4346	0.4362	0.4378	0.4393	0.4409	0.4425	0.4440	0.4456
2.8	0.4472	0.4487	0.4502	0.4518	0.4533	0.4548	0.4564	0.4579	0.4594	0.4609
2.9	0.4624	0.4639	0.4654	0.4669	0.4683	0.4698	0.4713	0.4728	0.4742	0.4757
3.0	0.4771	0.4786	0.4800	0.4814	0.4829	0.4843	0.4857	0.4871	0.4886	0.4900
3.1	0.4914	0.4928	0.4942	0.4955	0.4969	0.4983	0.4997	0.5011	0.5024	0.5038
3.2	0.5052	0.5065	0.5079	0.5092	0.5105	0.5119	0.5132	0.5145	0.5159	0.5172
3.3	0.5185	0.5198	0.5211	0.5224	0.5237	0.5250	0.5263	0.5276	0.5289	0.5302
3.4	0.5315	0.5328	0.5340	0.5353	0.5366	0.5378	0.5391	0.5403	0.5416	0.5428
3.5	0.5441	0.5453	0.5465	0.5478	0.5490	0.5502	0.5514	0.5527	0.5539	0.5551
3.6	0.5563	0.5575	0.5587	0.5599	0.5611	0.5623	0.5635	0.5647	0.5658	0.5670
3.7	0.5682	0.5694	0.5705	0.5717	0.5729	0.5740	0.5752	0.5763	0.5775	0.5786
3.8	0.5798	0.5809	0.5821	0.5832	0.5843	0.5855	0.5866	0.5877	0.5888	0.5899
3.9	0.5911	0.5922	0.5933	0.5944	0.5955	0.5966	0.5977	0.5988	0.5999	0.6010
4.0	0.6021	0.6031	0.6042	0.6053	0.6064	0.6075	0.6085	0.6096	0.6107	0.6117
4.1	0.6128	0.6138	0.6149	0.6160	0.6170	0.6180	0.6191	0.6201	0.6212	0.6222
4.2	0.6232	0.6243	0.6253	0.6263	0.6274	0.6284	0.6294	0.6304	0.6314	0.6325
4.3	0.6335	0.6345	0.6355	0.6365	0.6375	0.6385	0.6395	0.6405	0.6415	0.6425
4.4	0.6435	0.6444	0.6454	0.6464	0.6474	0.6484	0.6493	0.6503	0.6513	0.6522
4.5	0.6532	0.6542	0.6551	0.6561	0.6571	0.6580	0.6590	0.6599	0.6609	0.6618
4.6	0.6628	0.6637	0.6646	0.6656	0.6665	0.6675	0.6684	0.6693	0.6702	0.6712
4.7	0.6721	0.6730	0.6739	0.6749	0.6758	0.6767	0.6776	0.6785	0.6794	0.6803
4.8	0.6812	0.6821	0.6830	0.6839	0.6848	0.6857	0.6866	0.6875	0.6884	0.6893
4.9	0.6902	0.6911	0.6920	0.6928	0.6937	0.6946	0.6955	0.6964	0.6972	0.6981
5.0	0.6990	0.6998	0.7007	0.7016	0.7024	0.7033	0.7042	0.7050	0.7059	0.7067
5.1	0.7076	0.7084	0.7093	0.7101	0.7110	0.7118	0.7126	0.7135	0.7143	0.7152
5.2	0.7160	0.7168	0.7177	0.7185	0.7193	0.7202	0.7210	0.7218	0.7226	0.7235
5.3	0.7243	0.7251	0.7259	0.7267	0.7275	0.7284	0.7292	0.7300	0.7308	0.7316
5.4	0.7324	0.7332	0.7340	0.7348	0.7356	0.7364	0.7372	0.7380	0.7388	0.7396

Common Logarithmic Tables (Continued)

	0.00	0.01	0.02	0.03	0.04	0.05	0.06	0.07	0.08	0.09
5.5	0.7404	0.7412	0.7419	0.7427	0.7435	0.7443	0.7451	0.7459	0.7466	0.7474
5.6	0.7482	0.7490	0.7497	0.7505	0.7513	0.7520	0.7528	0.7536	0.7543	0.7551
5.7	0.7559	0.7566	0.7574	0.7582	0.7589	0.7597	0.7604	0.7612	0.7619	0.7627
5.8	0.7634	0.7642	0.7649	0.7657	0.7664	0.7672	0.7679	0.7686	0.7694	0.7701
5.9	0.7709	0.7716	0.7723	0.7731	0.7738	0.7745	0.7752	0.7760	0.7767	0.7774
6.0	0.7782	0.7789	0.7796	0.7803	0.7810	0.7818	0.7825	0.7832	0.7839	0.7846
6.1	0.7853	0.7860	0.7868	0.7875	0.7882	0.7889	0.7896	0.7903	0.7910	0.7917
6.2	0.7924	0.7931	0.7938	0.7945	0.7952	0.7959	0.7966	0.7973	0.7980	0.7987
6.3	0.7993	0.8000	0.8007	0.8014	0.8021	0.8028	0.8035	0.8041	0.8048	0.8055
6.4	0.8062	0.8069	0.8075	0.8082	0.8089	0.8096	0.8102	0.8109	0.8116	0.8122
6.5	0.8129	0.8136	0.8142	0.8149	0.8156	0.8162	0.8169	0.8176	0.8182	0.8189
6.6	0.8195	0.8202	0.8209	0.8215	0.8222	0.8228	0.8235	0.8241	0.8248	0.8254
6.7	0.8261	0.8267	0.8274	0.8280	0.8287	0.8293	0.8299	0.8306	0.8312	0.8319
6.8	0.8325	0.8331	0.8338	0.8344	0.8351	0.8357	0.8363	0.8370	0.8376	0.8382
6.9	0.8388	0.8395	0.8401	0.8407	0.8414	0.8420	0.8426	0.8432	0.8439	0.8445
7.0	0.8451	0.8457	0.8463	0.8470	0.8476	0.8482	0.8488	0.8494	0.8500	0.8506
7.1	0.8513	0.8519	0.8525	0.8531	0.8537	0.8543	0.8549	0.8555	0.8561	0.8567
7.2	0.8573	0.8579	0.8585	0.8591	0.8597	0.8603	0.8609	0.8615	0.8621	0.8627
7.3	0.8633	0.8639	0.8645	0.8651	0.8657	0.8663	0.8669	0.8675	0.8681	0.8686
7.4	0.8692	0.8698	0.8704	0.8710	0.8716	0.8722	0.8727	0.8733	0.8739	0.8745
7.5	0.8751	0.8756	0.8762	0.8768	0.8774	0.8779	0.8785	0.8791	0.8797	0.8802
7.6	0.8808	0.8814	0.8820	0.8825	0.8831	0.8837	0.8842	0.8848	0.8854	0.8859
7.7	0.8865	0.8871	0.8876	0.8882	0.8887	0.8893	0.8899	0.8904	0.8910	0.8915
7.8	0.8921	0.8927	0.8932	0.8938	0.8943	0.8949	0.8954	0.8960	0.8965	0.8971
7.9	0.8976	0.8982	0.8987	0.8993	0.8998	0.9004	0.9009	0.9015	0.9020	0.9025
8.0	0.9031	0.9036	0.9042	0.9047	0.9053	0.9058	0.9063	0.9069	0.9074	0.9079
8.1	0.9085	0.9090	0.9096	0.9101	0.9106	0.9112	0.9117	0.9122	0.9128	0.9133
8.2	0.9138	0.9143	0.9149	0.9154	0.9159	0.9165	0.9170	0.9175	0.9180	0.9186
8.3	0.9191	0.9196	0.9201	0.9206	0.9212	0.9217	0.9222	0.9227	0.9232	0.9238
8.4	0.9243	0.9248	0.9253	0.9258	0.9263	0.9269	0.9274	0.9279	0.9284	0.9289
8.5	0.9294	0.9299	0.9304	0.9309	0.9315	0.9320	0.9325	0.9330	0.9335	0.9340
8.6	0.9345	0.9350	0.9355	0.9360	0.9365	0.9370	0.9375	0.9380	0.9385	0.9390
8.7	0.9395	0.9400	0.9405	0.9410	0.9415	0.9420	0.9425	0.9430	0.9435	0.9440
8.8	0.9445	0.9450	0.9455	0.9460	0.9465	0.9469	0.9474	0.9479	0.9484	0.9489
8.9	0.9494	0.9499	0.9504	0.9509	0.9513	0.9518	0.9523	0.9528	0.9533	0.9538
9.0	0.9542	0.9547	0.9552	0.9557	0.9562	0.9566	0.9571	0.9576	0.9581	0.9586
9.1	0.9590	0.9595	0.9600	0.9605	0.9609	0.9614	0.9619	0.9624	0.9628	0.9633
9.2	0.9638	0.9643	0.9647	0.9652	0.9657	0.9661	0.9666	0.9671	0.9675	0.9680
9.3	0.9685	0.9689	0.9694	0.9699	0.9703	0.9708	0.9713	0.9717	0.9722	0.9727
9.4	0.9731	0.9736	0.9741	0.9745	0.9750	0.9754	0.9759	0.9764	0.9768	0.9773
9.5	0.9777	0.9782	0.9786	0.9791	0.9795	0.9800	0.9805	0.9809	0.9814	0.9818
9.6	0.9823	0.9827	0.9832	0.9836	0.9841	0.9845	0.9850	0.9854	0.9859	0.9863
9.7	0.9868	0.9872	0.9877	0.9881	0.9886	0.9890	0.9894	0.9899	0.9903	0.9908
9.8	0.9912	0.9917	0.9921	0.9926	0.9930	0.9934	0.9939	0.9943	0.9948	0.9952
9.9	0.9956	0.9961	0.9965	0.9969	0.9974	0.9978	0.9983	0.9987	0.9991	0.9996

Appendix F
Programs

To illustrate the power and versatility of programmable calculators, this appendix offers a variety of types of programs, including a simulation program (Inverse Functions), a tutorial program (Shifting, Stretching, and Reflecting Graphs), and programs for solving an equation or systems of equations. Programs are given for the following graphics calculators from Texas Instruments, Casio, and Sharp.

- TI-80
- TI-81
- TI-82
- TI-85
- Sharp EL 9200
- Sharp EL 9300
- Casio 6300
- Casio 7700
- Casio fx-9700GE
- Casio cfx-9800G

Similar programs can be written for other brands and models of graphics calculators.

Enter a program in your calculator, then refer to the text discussion and apply the program as appropriate. Section references are provided to help you locate the text discussion of the mathematical concepts demonstrated.

Functions   (Section 3.3)

Most graphing utilities can be used to evaluate a function at several values of x.

TI-80

```
PROGRAM:EVALUATE
:Lbl A
:Input "ENTER X ",X
:Disp Y₁
:Goto A
```

To use this program, enter a function in Y_1. Functions may also be evaluated directly on the TI-80's home screen.

TI-81

```
Prgm1: EVALUATE
:Lbl 1
:Disp "ENTER X"
:Input X
:Disp Y₁
:Goto 1
```

To use this program, enter a function in Y_1. Functions may also be evaluated on the TI-81's home screen.

TI-82

```
PROGRAM:EVALUATE
:Lbl A
:Input "ENTER X ",X
:Disp Y₁
:Goto A
```

To use this program, enter a function in Y_1. Functions may also be evaluated directly on the TI-82's home screen.

TI-85

```
 PROGRAM:
:Lbl A
:Input "Enter x ",x
:Disp y1
:Goto A
```

To use this program, enter a function in $y1$.

Sharp EL 9200
Sharp EL 9300

```
Evaluate
-----------REAL
Goto top
Label eqtn
Y=f(X)
Return
Label top
Input X
Gosub equation
Print Y
Goto top
End
```

To use this program, replace $f(X)$ with your expression in X.

Casio 6300

EVALUATE:
Lbl 1:
"X="?→X:
"F(X)=" ◢ Prog 0 ◢
Goto 1

To use this program, write the function as Prog 0.

Casio 7700

EVALUATE
Lbl 1
"X="?→X
"F(X)=":f₁ ◢
Goto 1

To use this program, enter a function in f_1.

Casio fx-9700GE
Casio cfx-9800G

EVALUATE↵
Lbl 1↵
"X="?→X↵
"F(X)=":Y1 ◢
Goto 1

To use this program, enter a function in Y1. Functions may also be evaluated in a table from the Casio calculator's TABLE&GRAPH MENU.

Shifting, Reflecting, and Stretching Graphs   (Section 3.5)

These programs will sketch a graph of the function $y = R(x + H)^2 + V$ where $R = \pm 1$, H is an integer between -6 and 6, and V is an integer between -3 and 3. These programs give you practice working with reflections, horizontal shifts, and vertical shifts.

TI-80

```
PROGRAM:PARABOLA
:-6 + int (12rand)→H
:-3 + int (6rand)→V
:rand→R
:If R < .5
:Then
:-1→R
:Else
:1→R
:End
:"R(X + H)² + V"→Y₁
:-9→Xmin
:9→Xmax
:1→Xscl
:-6→Ymin
:6→Ymax
:1→Yscl
:DispGraph
:Pause
:Disp "R = ",R
:Disp "H = ",H
:Disp "V = ",V
```

Press **ENTER** after the graph to display the coordinates.

TI-81

```
Prgm2: PARABOLA
:Rand→H
:-6+Int (12H) →H
:Rand→V
:-3+Int (6V) →V
:Rand→R
:If R<.5
:-1→R
:If R>.49
:1→R
:"R(X+H)²+V" →Y₁
:-9 →Xmin
:9 →Xmax
:1 →Xscl
:-6 →Ymin
:6 →Ymax
:1 →Yscl
:DispGraph
:Pause
:Disp "Y=R(X+H)²+V"
:Disp "R="
:Disp R
:Disp "H="
:Disp H
:Disp "V="
:Disp V
:End
```

Press ENTER after the graph to display the coordinates.

TI-82

```
PROGRAM:PARABOLA
:-6 + int (12rand)→H
:-3 + int (6rand)→V
:rand→R
:If R < .5
:Then
:-1→R
:Else
:1→R
:End
:"R(X + H)² + V"→Y₁
:-9→Xmin
:9→Xmax
:1→Xscl
:-6→Ymin
:6→Ymax
:1→Yscl
:DispGraph
:Pause
:Disp "R = ",R
:Disp "H = ",H
:Disp "V = ",V
```

Press **ENTER** after the graph to display the coordinates.

TI-85

```
 PROGRAM:
: rand→H
: -6+int(12H)→H
: rand→V
: -3+int(6V)→V
: rand→R
: If R<.5
: -1→R
: If R>.49
: 1→R
: y1=R(x+H)²+V
: -9→xMin
: 9→xMax
: 1→xScl
: -6→yMin
: 6→yMax
: 1→yScl
: DispG
: Pause
: Disp "Y=R(X+H)²+V"
: Disp "R=",R
: Disp "H=",H
: Disp "V=",V
```

Sharp EL 9200
Sharp EL 9300

```
Parabola
------------REAL
H=int (random*12)-6
V=int (random*6)-3
S=(random*2)-1
R=S/abs S
Range -9,9,1,-6,6,1
Graph R(X+H)²+V
Wait
Print "Y=R(X+H)²+V
Print R
Print H
Print V
End
```

Pressing Enter after the graph will display the coefficients.

Casio 6300

```
R(X+H)²+V:
-6+Int (12Ran#)→H:
-3+Int (6Ran#)→V:
Ran#<.5⇒-1→R:1→R:
Range -9,9,1,-6,6,1:
Graph Y=R(X+H)²+V◢
"Y=R(X+H)²+V"◢
"R="◢R◢
"H="◢H◢
"V="◢V
```

Casio 7700

```
R(X+H)²+V
-6+Int (12Ran#)→H
-3+Int (6Ran#)→V
Ran#<.5⇒-1→R:1→R
Range -9,9,1,-6,6,1
Graph Y=R(X+H)²+V◢
"Y=R(X+H)²+V"
"R=":R◢
"H=":H◢
"V=":V
```

Casio fx-9700GE
Casio cfx-9800G

```
R(X+H)²+V↵
-6+Int (12Ran#)→H↵
-3+Int (6Ran#)→V↵
Ran#→R↵
R<.5⇒-1→R↵
R≥.5⇒1→R↵
Range -9,9,1,-6,6,1↵
Graph Y=R(X+H)²+V◢
"Y=R(X+H)²+V"↵
"R=":R◢
"H=":H◢
"V=":V
```

Press **ENTER** after the graph to display the coefficients.

Inverse Functions   (Section 3.7)

These programs graph a function *f and* its reflection in the line $y = x$.

TI-80

```
PROGRAM:REFLECT
:47Xmin/63→Ymin
:47Xmax/63→Ymax
:Xscl→Yscl
:"X"→Y₂
:DispGraph
:(Xmax−Xmin)/62→I
:Xmin→X
:Lbl A
:Pt-On(Y₁,X)
:X+I→X
:If X>Xmax
:Stop
:Goto A
```

To use this program, enter the function in Y_1 and set a viewing rectangle.

TI-81

```
Prgm3: REFLECT
:2Xmin/3 → Ymin
:2Xmax/3 → Ymax
:Xscl → Yscl
:"X" → Y₂
:DispGraph
:(Xmax-Xmin)/95 → I
:Xmin → X
:Lbl 1
:PT-On(Y₁,X)
:X+I → X
:If X>Xmax
:End
:Goto 1
```

To use this program, enter the function in Y_1 and set a viewing rectangle.

TI-82

```
PROGRAM:REFLECT
:63Xmin/95→Ymin
:63Xmax/95→Ymax
:Xscl→Yscl
:"X"→Y₂
:DispGraph
:(Xmax−Xmin)/94→I
:Xmin→X
:While X≤Xmax
:Pt-On(Y₁,X)
:X+I→X
:End
```

To use this program, enter the function in Y_1 and set a viewing rectangle.

TI-85

```
 PROGRAM:
:63×Min/127→yMin
:63×Max/127→yMax
:xScl→yScl
:y2=x
:DispG
:(xMax-xMin)/126→I
:xMin→x
:Lbl A
:PtOn(y1,x)
:x+I→x
:If x>xMax
:Stop
:Goto A
```

Sharp EL 9200
Sharp EL 9300

```
Reflection
------------REAL
Goto top
Label eqtn
Y=X^3+X+1
Return
Label rng
xmin=-10
xmax=10
xstp=(xmax-xmin)/10
ymin=2xmin/3
ymax=2xmax/3
ystp=xstp
Range xmin,xmax,xstp,ymin,ymax,ystp
Return
Label top
Gosub rng
Graph X
step=(xmax-xmin)/(94*2)
X=xmin
Label 1
Gosub eqtn
Plot X,Y
Plot Y,X
X=X+step
If X<=xmax Goto 1
End
```

To use this program, enter a function in X in the third line.

Casio 7700

```
REFLECTION
"GRAPH -A TO A"
"A="?→A
Range -A,A,1,-2A+3,2A+3,1
Graph Y=f₁
-A→B
Lbl 1
B→X
Plot f₁,B
B+A÷32→B
B≤A⇒Goto 1:Graph Y=X
```

To use this program, enter the function in f_1 and set a viewing rectangle.

Casio 6300

```
REFLECTION:
"-A TO A"◢
"A="?→A:
Range -A,A,1,-2A+3,2A+3,1:
-A→B:
Lbl 1:
B→X:
Prog 0:
Ans→Y:
Plot B,Y:
B+A÷24→B:
B≤A⇒Goto 1:
-A→B:
Lbl 2:
B→X:
Prog 0:
Ans→Y:
Plot Y,B:
B+A÷24→B:
B≤A⇒Goto 2:Graph Y=X
```

To use this program, write the function as Prog 0 and set a viewing rectangle.

Casio cfx-9800G

```
REFLECTION↵
63Xmin÷95→A↵
63Xmax÷95→B↵
Xscl→C↵
Range ,,,A,B,C↵
(Xmax−Xmin)÷94→I↵
Xmax→M↵
Xmin→D↵
Graph Y=f₁↵
Lbl 1↵
D→X↵
Plot f₁,D↵
D+I→D↵
D≤M⇒Goto 1:Graph Y=X
```

Casio fx-9700GE

```
REFLECTION↵
63Xmin÷127→A↵
63Xmax÷127→B↵
Xscl→C↵
Range ,,,A,B,C↵
(Xmax−Xmin)÷126→I↵
Xmax→M↵
Xmin→D↵
Graph Y=f₁↵
Lbl 1↵
D→X↵
Plot f₁,D↵
D+I→D↵
D≤M⇒Goto 1:Graph Y=X
```

To use either program, enter a function in f_1 and set a viewing rectangle.

The Quadratic Formula   (Section 2.4)

These programs will display solutions to quadratic equations or the words "No Real Solution." To use the program, write the equation in standard form and then enter the values of a, b, and c.

TI-80

```
PROGRAM:QUADRAT
:Disp "AX² +BX +C = 0"
:Input "ENTER A ", A
:Input "ENTER B ", B
:Input "ENTER C ", C
:B²-4AC→D
:If D≥0
:Then
:(-B+√D)/(2A)→M
:Disp M
:(-B-√D)/(2A)→N
:Disp N
:Else
:Disp "NO REAL SOLUTION"
:End
```

TI-81

```
Prgm4: QUADRAT
:Disp "ENTER A"
:Input A
:Disp "ENTER B"
:Input B
:Disp "ENTER C"
:Input C
:B²-4AC →D
:If D<0
:Goto 1
:((-B+√D)/(2A)) →S
:Disp S
:((-B-√D)/(2A)) →S
:Disp S
:End
:Lbl 1
:Disp "NO REAL"
:Disp "SOLUTION"
:End
```

TI-82

```
PROGRAM:QUADRAT
:Disp "AX² +BX +C = 0"
:Input "ENTER A ", A
:Input "ENTER B ", B
:Input "ENTER C ", C
:B²-4AC→D
:If D≥0
:Then
:(-B+√D)/(2A)→M
:Disp M
:(-B-√D)/(2A)→N
:Disp N
:Else
:Disp "NO REAL SOLUTION"
:End
```

TI-85

```
PROGRAM:
:Input "ENTER A ",A
:Input "ENTER B ",B
:Input "ENTER C ",C
:B²-4*A*C→D
:Disp (-B+√D)/(2A)
:Disp (-B-√D)/(2A)
```

Solutions to quadratic equations are also available directly by using the TI-85 POLY function.

Sharp EL 9200
Sharp EL 9300

```
Quadratic
----------COMPLEX
Input A
Input B
Input C
D=B²-4AC
x1=(-B+√D)/(2A)
x2=(-B-√D)/(2A)
Print x1
Print x2
X=x1
Y=x2
End
```

This program is written in the program's complex mode, so both real and complex answers are given. The answers are also stored under variables X and Y so they can be used in the calculator mode.

Casio 6300

```
QUADRATICS:
"AX²+BX+C=0"◢
"A="?→A:
"B="?→B:
"C="?→C:
B²-4AC→D:
D<0⟹Goto 1:
"X="◢(-B+√D)÷2A◢
"OR X="◢(-B-√D)÷2A:
Goto 2:
Lbl 1:
"NO REAL SOLUTION"
Lbl 2
```

Casio 7700

```
QUADRATICS
"AX²+BX+C=0"
"A="?→A
"B="?→B
"C="?→C
B²-4AC→D
D<0⟹Goto 1
"X=":(-B+√D)÷2A◢
"OR X=":(-B-√D)÷2A
Goto 2
Lbl 1
"NO REAL SOLUTION"
Lbl 2
```

Casio fx-9700GE
Casio cfx-9800G

```
QUADRATIC↵
"AX²+BX+C=0"↵
"A="?→A↵
"B="?→B↵
"C="?→C↵
B²-4AC→D↵
(-B+√D)÷2A◢
(-B-√D)÷2A
```

Both real and complex answers are given. Solutions to quadratic equations are also available directly from the Casio calculator's EQUATION MENU.

Systems of Linear Equations in Two Variables   (Section 7.2)

The general solution of the linear system

$$ax + by = c$$
$$dx + ey = f$$

is $x = (ce - bf)/(ae - db)$ and $y = (af - cd)/(ae - db)$. If $ae - db = 0$, then the system does not have a unique solution. Graphing utility programs for solving such a system are given below.

TI-80

```
PROGRAM:SOLVE
:Disp "AX+BY=C"
:Input "ENTER A ", A
:Input "ENTER B ", B
:Input "ENTER C ", C
:Disp "DX+EY=F"
:Input "ENTER D ", D
:Input "ENTER E ", E
:Input "ENTER F ", F
:If AE-DB=0
:Then
:Disp "NO SOLUTION"
:Else
:(CE-BF)/(AE-DB)→X
:(AF-CD)/(AE-DB)→Y
:Disp X
:Disp Y
:End
```

TI-81

```
Prgm5: SOLVE
:Disp "ENTER A,B,C,D,E,F"
:Input A
:Input B
:Input C
:Input D
:Input E
:Input F
:If AE-DB=0
:Goto 1
:(CE-BF)/(AE-DB)→X
:(AF-CD)/(AE-DB)→Y
:Disp X
:Disp Y
:Lbl 1
:End
```

TI-82

```
PROGRAM:SOLVE
:Disp "AX+BY=C"
:Prompt A
:Prompt B
:Prompt C
:Disp "DX+EY=F"
:Prompt D
:Prompt E
:Prompt F
:If AE-DB=0
:Then
:Disp "NO SOLUTION"
:Else
:(CE-BF)/(AE-DB)→X
:(AF-CD)/(AE-DB)→Y
:Disp X
:Disp Y
:End
```

TI-85

```
PROGRAM:
:Disp "ax+by=c"
:Input "Enter a ",A
:Input "Enter b ",B
:Input "Enter c ",C
:Disp "dx+ey=f"
:Input "Enter d ",D
:Input "Enter e ",E
:Input "Enter f ",F
:If A*E-D*B=0
:Goto A
:(C*E-B*F)/(A*E-D*B)→X
:(A*F-C*D)/(A*E-D*B)→Y
:Disp X
:Disp Y
:Lbl A
```

Sharp EL 9200
Sharp EL 9300

```
Solve
----------REAL
Input A
Input B
Input C
Input D
Input E
Input F
If A*E-D*B=0 Goto 1
X=(C*E-B*F)/(A*E-D*B)
Y=(A*F-C*D)/(A*E-D*B)
Print X
Print Y
End
Label 1
Print "No solution
End
```

Equations must be entered in the form: $Ax + By = C$; $Dx + Ey = F$. Uppercase letters are used so that the values can be accessed in the calculation mode of the calculator.

Casio 6300

```
SOLVE:
"A="?→A:
"B="?→B:
"C="?→C:
"D="?→D:
"E="?→E:
"F="?→F:
AE-DB=0⇒Goto 1:
"X=" ◢(CE-BF)÷(AE-DB)◢
"Y=" ◢(AF-CD)÷(AE-DB):
Goto 2:
Lbl 1:
"NO UNIQUE SOLUTION":
Lbl 2
```

Casio 7700

```
SOLVE
"ENTER A,B,C,D,E,F"
"A="?→A
"B="?→B
"C="?→C
"D="?→D
"E="?→E
"F="?→F
AE-DB=0⇒Goto 1
"X=":(CE-BF)÷(AE-DB)◢
"Y=":(AF-CD)÷(AE-DB)
Goto 2
Lbl 1
"NO UNIQUE SOLUTION"
Lbl 2
```

Casio fx-9700GE
Casio cfx-9800G

```
SOLVE↵
"ENTER A,B,C,D,E,F"↵
"A":?→A↵
"B":?→B↵
"C":?→C↵
"D":?→D↵
"E":?→E↵
"F":?→F↵
AE-DB=0⇨Goto 1↵
"X=":(CE-BF)÷(AE-DB)◢
"Y=":(AF-CD)÷(AE-DB)↵
Goto 2↵
Lbl 1↵
"NO UNIQUE SOLUTION"↵
Lbl 2
```

Solutions to systems of linear equations are also available directly from the Casio calculator's EQUATION MENU.

Sequences and Summation Notation   (Section 9.1)

These programs find the sum of a finite sequence.

TI-80

PROGRAM:SUM
:Input "LOWER LIMIT ", M
:Input "UPPER LIMIT ", N
:sum seq(Y_1,X,M,N,1)→S
:Disp "SUM = ",S

To use this program, store the formula for the nth term as Y_1. You may also find the sum of a sequence directly on the TI-80's home screen.

TI-81

Prgm6: SUM
:Disp "ENTER M"
:Input M
:Disp "ENTER N"
:Input N
:0 → S
:M → X
:Lbl 1
:S+Y_1 → S
:If X=N
:Goto 2
:X+1 → X
:Goto 1
:Lbl 2
:Disp S

To use this program, store the formula for the nth term as Y_1.

TI-82

PROGRAM:SUM
:Input "LOWER LIMIT ", M
:Input "UPPER LIMIT ", N
:sum seq(Y_1,X,M,N,1)→S
:Disp "SUM = ",S

To use this program, store the formula for the nth term as Y_1. You may also find the sum of a sequence directly on the TI-82's home screen.

TI-85

PROGRAM:
:Input "ENTER M ",M
:Input "ENTER N ",N
:0→S
:M→x
:Lbl A
:S+y1→S
:If x=N
:Goto B
:x+1→x
:Goto A
:Lbl B
:Disp S

To use this program, store the formula for the nth term as $y1$.

Sharp EL 9260
Sharp EL 9300

```
Sum
----------REAL
Goto 1
Label eqtn
y=√ (22926+902.5x+2.01x²)
Return
Label 1
Print "Enter start
Input m
Print "Enter end
Input n
If n<m Goto 1
sum=0
x=m
Label 2
Gosub eqtn
sum=sum+y
x=x+1
If x<=n Goto 2
Print sum
End
```

To use this program, enter the formula in x for the nth term into the y-line at the beginning of the program.

Casio 6300

```
SUM:
"LOWER LIMIT"?→M:
"UPPER LIMIT"?→N:
0→S:
M→X:
Lbl 1:
Prog 0:
Ans+S→S:
X+1→X:
X≤N⇒Goto 1:
"SUM="◢S
```

To use this program, store the formula for the nth term as Prog 0.

Casio 7700

```
SUM
"LOWER LIMIT"?→M
"UPPER LIMIT"?→N
0→S
M→X
Lbl 1
S+f₁→S
X+1→X
X≤N⇒Goto 1
"SUM=":S
```

To use this program, store the formula for the nth term in f_1.

Casio fx-9700GE
Casio cfx-9800G

```
SUM↵
"LOWER LIMIT":?→M↵
"UPPER LIMIT":?→N↵
0→S↵
M→X↵
Lbl 1↵
S+f₁→S↵
X+1→X↵
X≤N⇒Goto 1↵
"SUM=":S
```

To use this program, store the formula for the nth term in f_1. The sum of a series is also available directly from the Casio calculator's TABLE&GRAPH MENU.

Appendix G
Additional Problem Solving with Technology

This appendix offers 300 additional exercises that take advantage of the capabilities and versatility of graphing calculators as a tool for problem solving. Keyed to the text by section number and title, each exercise set contains problems for which graphing technology is required. All of the exercises were written to accommodate any graphing calculator or computer graphing utility.

These exercises encourage the use of graphing technology as an efficient and effective problem–solving tool in a variety of ways, including visualization, exploration and discovery, and verification. A wide range of problem types are represented, offering opportunities for solving multi-part problems, writing exercises, real-life applications, and exercises that require graphical interpretation.

EXERCISES for Appendix G

SECTION 3.2   Graphs of Equations

In Exercises 1–6, use a graphing utility to graph the equation. (Use the standard viewing rectangle.) Determine the number of times (if any) the graph intersects each coordinate axis.

1. $y = 3x^4 - 6x^2$

2. $y = \frac{1}{27}\left(x^4 + 4x^3\right)$

3. $y = x\sqrt{4 - x}$

4. $y = x\sqrt{4 - x^2}$

5. $y = \dfrac{10x}{x^2 + 1}$

6. $y = \dfrac{10}{x^2 + 1}$

In Exercises 7–12, use a graphing utility to graph the equation. Use the specified viewing rectangle. Determine the number of times the graph intersects each coordinate axis.

7. $y = x^4 - 4x^3 + 16x$

Xmin = −5
Xmax = 5
Xscl = 1
Ymin = −15
Ymax = 30
Yscl = 5

8. $y = 4x^3 - x^4$

Xmin = −2
Xmax = 6
Xscl = 1
Ymin = −2
Ymax = 30
Yscl = 2

9. $y = 100x\sqrt{25 - x}$

Xmin = −30
Xmax = 30
Xscl = 5
Ymin = −5000
Ymax = 5000
Yscl = 1000

10. $y = 100x\sqrt{25 - x^2}$

Xmin = −8
Xmax = 8
Xscl = 1
Ymin = −2000
Ymax = 2000
Yscl = 500

11. $x^2 - 100y - 1000 = 0$

Xmin = −100
Xmax = 100
Xscl = 10
Ymin = −10
Ymax = 10
Yscl = 1

12. $2x^3 - 100x - 15{,}625 + 250y = 0$

Xmin = −20
Xmax = 25
Xscl = 2
Ymin = −2
Ymax = 100
Yscl = 5

In Exercises 13–16, solve for y and use a graphing utility to graph each of the resulting equations on the same viewing rectangle. Adjust the viewing rectangle so a circle really does appear circular. (Your graphing utility may have a *square* setting that does this automatically.)

13. Circle: $x^2 + y^2 = 64$

14. Circle: $x^2 + y^2 = 49$

15. Ellipse: $6x^2 + y^2 = 72$

16. Ellipse: $x^2 + 9y^2 = 81$

In Exercises 17 and 18, describe the given viewing rectangle.

17. $y = -(x - 5)^2(x - 15)$

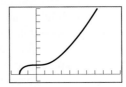

18. $y = \sqrt{x^3 + 8}$

In Exercises 19 and 20, use a graphing utility to graph the equation using each of the suggested viewing rectangles. Assume that the equation gives the profit y when x units of a product are sold. Note that a graph can distort the information presented simply by changing the viewing rectangle. Which viewing rectangle would be selected by a person who wishes to argue that profits will increase dramatically with increased sales?

19. $y = 0.25x - 50$

Xmin = −3		Xmin = −3	
Xmax = 800		Xmax = 1000	
Xscl = 50		Xscl = 100	
Ymin = −20		Ymin = −100	
Ymax = 100		Ymax = 500	
Yscl = 10		Yscl = 40	

20. $y = 2.44x - \dfrac{x^2}{20{,}000} - 5000$

Xmin = −5000		Xmin = −5000	
Xmax = 22000		Xmax = 22000	
Xscl = 5000		Xscl = 5000	
Ymin = −20000		Ymin = −5000	
Ymax = 60000		Ymax = 24000	
Yscl = 10000		Yscl = 5000	

SECTION 3.3   Lines in the Plane

1. *Exploration*   Use a graphing utility to compare the slopes of the lines given by $y = ax$ with $a = 0.5, 1, 2,$ and 4. What do you observe about the slopes of the lines? Repeat the experiment with $a = -0.5, -1, -2,$ and -4. What do you observe about the slopes of these lines? (*Hint:* Use a square setting to guarantee a true geometric perspective.)

In Exercises 2 and 3, use a graphing utility to graph the equation using each of the suggested viewing rectangles. Note that the viewing rectangle selected will alter the appearance of the slope.

2. $y = 0.5x - 3$

Xmin = -5
Xmax = 10
Xscl = 1
Ymin = -10
Ymax = 5
Yscl = 1

Xmin = -2
Xmax = 10
Xscl = 1
Ymin = -4
Ymax = 1
Yscl = 1

3. $y = -8x + 5$

Xmin = -5
Xmax = 5
Xscl = 1
Ymin = -10
Ymax = 10
Yscl = 1

Xmin = -5
Xmax = 10
Xscl = 1
Ymin = -80
Ymax = 80
Yscl = 20

In Exercises 4–7, use a graphing utility to graph the three equations on the same viewing rectangle. Adjust the viewing rectangle so the slope appears visually correct. (Your calculator may have a *square* setting that does this automatically.)

4. $y = 2x$ $y = -2x$ $y = \frac{1}{2}x$

5. $y = \frac{2}{3}x$ $y = -\frac{3}{2}x$ $y = \frac{2}{3}x + 2$

6. $y = -\frac{1}{2}x$ $y = -\frac{1}{2}x + 3$ $y = 2x - 4$

7. $y = x - 8$ $y = x + 1$ $y = -x + 3$

SECTION 3.4   Functions

1. *Exploration*   Use a graphing utility to graph $x^2 + y = 1$. Then use the graph to write a convincing argument that each x-value has at most one y-value.
    Use a graphing utility to graph $-x + y^2 = 1$. (*Hint:* You will need to use two equations.) Then use the graph to find an x-value that corresponds to two y-values. Why does the graph not represent y as a function of x?

In Exercises 2–7, use a graphing utility to graph the function. Use the graph to approximate the domain of the function.

2. $f(x) = \sqrt{5 - x}$ 3. $g(x) = \sqrt{25 - x^2}$

4. $g(x) = \sqrt{x^2 - 4}$ 5. $f(x) = \frac{2}{3}\sqrt{x^2 - 9}$

6. $h(x) = \dfrac{2}{x - 3}$ 7. $y = \dfrac{4x}{x^2 - 9}$

In Exercises 8–11, use a graphing utility to graph the functions. Use the graph to approximate any points of intersection.

8. $f(x) = x^2 - 1,$ $g(x) = 3$

9. $f(x) = x^2 + 1,$ $g(x) = 2x$

10. $f(x) = x^3 - 3x^2 + 3,$ $g(x) = 3 - 2x$

11. $f(x) = \frac{1}{3}(x + 4),$ $g(x) = x^{2/3}$

12. *Athletics*   A baseball is hit 3 feet above the ground. The path of the baseball is given by the function

$$h(x) = -\tfrac{1}{200}x^2 + x + 3$$

where the height h and the horizontal distance x are measured in feet.

(a) Use a graphing utility to graph the path of the baseball. Select a viewing rectangle which shows the entire path.

(b) Suppose the ball was caught by an outfielder when the ball was at the height of 7 feet. Use the graph to estimate the distance between home-plate and the outfielder.

SECTION 3.5 Graphs of Functions

In Exercises 1–4, select the viewing rectangle on a graphing utility that shows the most complete graph of the function.

1. $f(x) = -0.2x^2 + 3x + 32$

(a)

Xmin = -2
Xmax = 20
Xscl = 1
Ymin = -10
Ymax = 30
Yscl = 4

(b)

Xmin = -10
Xmax = 30
Xscl = 5
Ymin = -5
Ymax = 50
Yscl = 5

(c)

Xmin = 0
Xmax = 10
Xscl = 0.5
Ymin = 0
Ymax = 200
Yscl = 25

2. $f(x) = 6[x - (0.1x)^5]$

(a)

Xmin = -500
Xmax = 500
Xscl = 50
Ymin = -500
Ymax = 500
Yscl = 50

(b)

Xmin = -25
Xmax = 25
Xscl = 5
Ymin = -25
Ymax = 25
Yscl = 5

(c)

Xmin = -20
Xmax = 20
Xscl = 5
Ymin = -100
Ymax = 100
Yscl = 20

3. $f(x) = 4x^3 - x^4$

(a)

Xmin = -2
Xmax = 6
Xscl = 1
Ymin = -10
Ymax = 30
Yscl = 4

(b)

Xmin = -50
Xmax = 50
Xscl = 5
Ymin = -50
Ymax = 50
Yscl = 5

(c)

Xmin = 0
Xmax = 2
Xscl = 0.2
Ymin = -2
Ymax = 2
Yscl = 0.5

4. $f(x) = 10x\sqrt{400 - x^2}$

(a)

Xmin = -5
Xmax = 50
Xscl = 5
Ymin = -5000
Ymax = 5000
Yscl = 500

(b)

Xmin = -20
Xmax = 20
Xscl = 2
Ymin = -500
Ymax = 500
Yscl = 50

(c)

Xmin = -25
Xmax = 25
Xscl = 5
Ymin = -2000
Ymax = 2000
Yscl = 200

In Exercises 5–12, use a graphing utility to graph the function. Approximate the intervals over which the function is increasing, decreasing, or constant.

5. $f(x) = 2x$

6. $f(x) = x^2 - 2x$

7. $f(x) = x^3 - 3x^2$

8. $f(x) = \sqrt{x^2 - 4}$

9. $f(x) = 3x^4 - 6x^2$

10. $f(x) = x^{2/3} = (x^2)^{1/3}$

11. $f(x) = x\sqrt{x + 3}$

12. $f(x) = |x + 1| + |x - 1|$

In Exercises 13–18, use a graphing utility to approximate (to two decimal place accuracy) any relative minimum or maximum values of the function.

13. $f(x) = x^2 - 6x$

14. $f(x) = (x - 1)^2(x + 2)$

15. $g(x) = 2x^3 + 3x^2 - 12x$

16. $g(x) = x^3 - 6x^2 + 15$

17. $h(x) = (x - 1)\sqrt{x}$

18. $h(x) = x\sqrt{4 - x}$

In Exercises 19–22, use a graphing utility to graph the three functions on the same set of coordinate axes. Describe the graphs of g and h relative to the graph of f.

19. $f(x) = x^3 - 3x^2$

$g(x) = f(x + 2)$

$h(x) = f\left(\frac{1}{2}x\right)$

20. $f(x) = x^3 - 3x^2 + 2$

$g(x) = f(x - 1)$

$h(x) = f(2x)$

21. $f(x) = x^3 - 3x^2$

$g(x) = -\frac{1}{3}f(x)$

$h(x) = f(-x)$

22. $f(x) = x^3 - 3x^2 + 2$

$g(x) = -f(x)$

$h(x) = f(-x)$

23. *Minimum Cost* A power station is on one side of a river that is $\frac{1}{2}$ mile wide, and a factory is 6 miles downstream on the other side (see figure). It costs $6 per foot to run power lines overland and $8 per foot to run them underwater.

(a) Show that the total cost for the power line is given by

$$T = 8(5280)\sqrt{x^2 + \tfrac{1}{4}} + 6(5280)(6 - x).$$

(b) Use a graphing utility to graph the cost function.

(c) To determine the most economical path for the power line, approximate (to two decimal place accuracy) the value of x that minimizes the cost function.

Factory

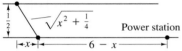 Power station

Figure for 23

In Exercises 6–9, use the results of Exercises 2–5 and your imagination to guess the shape of the graph of the function. Then use a graphing utility to graph the function, and compare the result with the graph drawn by hand.

6. $f(x) = x^2(x - 6)^2$

7. $f(x) = x^3(x - 6)^2$

8. $f(x) = x^2(x - 6)^3$

9. $f(x) = x^3(x - 6)^3$

10. Use a graphing utility to graph the functions $f(x) = x$ and $g(x) = x^3/3$. Use the graphs to identify the following graphs for $f + g$ and $f - g$. Give reasons for your answers.

(a)   (b)

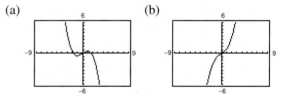

SECTION 3.6 Combinations of Functions

1. *Exploration* Display the graphs of $y = x^2 + c$ where $c = -2, 0, 2,$ and 4. Use the result to describe the effect that c has on the graph.

Display the graphs of $y = (x + c)^2$ where $c = -2, 0, 2$ and 4. Use the result to describe the effect that c has on the graph.

2. Use a graphing utility to graph $f(x) = x^2$, $g(x) = x^4$, and $h(x) = x^6$ in the same viewing rectangle. Describe any similarities and differences you observe among the graphs.

3. Use a graphing utility to graph $f(x) = x$, $g(x) = x^3$, and $h(x) = x^5$ in the same viewing rectangle. Describe any similarities and differences you observe among the graphs.

4. Use the results of Exercises 2 and 3 to sketch the graphs of $f(x) = x^{10}$ and $g(x) = x^{11}$, without the aid of a graphing utility.

5. Use the results of Exercises 2 and 3 to sketch the graphs of $f(x) = (x - 3)^3$, $g(x) = (x + 1)^2$, and $h(x) = (x - 4)^5$, without the aid of a graphing utility.

SECTION 3.7 Inverse Functions

1. *Exploration* Sketch the graphs of the functions given in Example 2, along with the graph of $y = x$.

$$f(x) = x^3 - 1$$
$$g(x) = \sqrt[3]{x + 1}$$

(Use the viewing rectangle $-6 \le x \le 6$ and $-4 \le y \le 4$.) Describe how the two graphs are related.

In Exercises 2–7, match the function with the graph of its inverse. Verify your answers by finding the inverse of the function and using a graphing utility to graph the function and its inverse. [The graphs are labeled (a), (b), (c), (d), (e), and (f).]

2. $f(x) = \frac{1}{4}x$

3. $f(x) = 2(x - 1)$

4. $f(x) = \sqrt{x - 3}$

5. $f(x) = x^2 - 1, \ x \ge 0$

6. $f(x) = \dfrac{x^3}{3}$

7. $f(x) = \dfrac{x}{x + 1}$

(a)

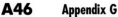

(b)

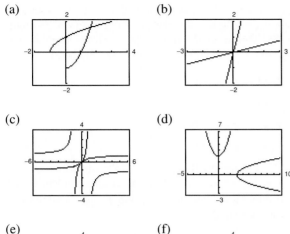

(c)

(d)

(e)

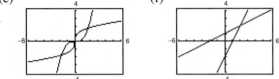

(f)

SECTION 4.1   Quadratic Functions

1. *Exploration*   Use a graphing utility to graph $y = ax^2$ with $a = -2, -1, 1,$ and 2. How does the value of a affect the graph?

    Use a graphing utility to graph $y = (x - h)^2$ with $h = -4, -2, 2,$ and 4. How does the value of h affect the graph?

    Use a graphing utility to graph $y = x^2 + k$ with $k = -4, -2, 2,$ and 4. How does the value of k affect the graph?

In Exercises 2–5, use a graphing utility to compare the graph of each function with the graph of $y = x^2$.

2. (a) $f(x) = \frac{1}{6}x^2$

    (b) $g(x) = -3x^2$

3. (a) $f(x) = -\frac{1}{2}x^2 + 4$

    (b) $g(x) = 2x^2 - 5$

4. (a) $f(x) = -(x + 3)^2 + 3$

    (b) $g(x) = \frac{4}{3}(x - 4)^2 - 3$

5. (a) $f(x) = 0.75(x - 4.5)^2 + 5.2$

    (b) $g(x) = -0.2(x + 1.4)^2 + 3.8$

In Exercises 6–9, use a graphing utility to approximate (accurate to three decimal places) the solutions of the equation.

6. $5.1x^2 - 1.7x - 3.2 = 0$

7. $-0.005x^2 + 0.101x - 0.193 = 0$

8. $422x^2 - 506x - 347 = 0$

9. $2x^2 - 2.50x - 0.42 = 0$

10. *Recreation Spending*   The total number of dollars spent on recreation in the United States from 1980 to 1988 can be approximated by the model

$$S = 116,289 + 9506.7t + 841.45t^2,$$

where the spending is measured in millions of dollars and the time t represents the year, with $t = 0$ corresponding to 1980.

    (a) Use a graphing utility to graph the recreational spending function.

    (b) Move the cursor along the graph to predict the year when total recreational spending will reach $300,000,000,000. (*Source:* U. S. Bureau of Economic Analysis)

SECTION 4.2   Polynomial Functions of Higher Degree

1. *Exploration*   Use a graphing utility to investigate the behavior of the graph of

$$y = x^3 - 105x^2 + 21.$$

First use a viewing rectangle in which $-2 \le x \le 2$ and $-10 \le y \le 30$. How complete a view of the graph does this viewing rectangle show? Does the graph move down as x increases indefinitely? Find a viewing rectangle that gives a good view of the basic characteristics of the graph.

In Exercises 2–5, use a graphing utility to graph the functions f and g in the same viewing rectangle. Zoom out sufficiently far to show that the right-hand and left-hand behavior of f and g are identical.

2. $f(x) = x^3 - 9x + 1$ $g(x) = x^3$

3. $f(x) = -\frac{1}{3}(x^3 - 3x + 2)$ $g(x) = -\frac{1}{3}x^3$

4. $f(x) = -(x^4 - 4x^3 + 16x)$ $g(x) = -x^4$

5. $f(x) = 3x^4 - 6x^2$ $g(x) = 3x^4$

In Exercises 6–9, use the Intermediate Value Theorem and a graphing utility to find the intervals of length 1 in which the polynomial function is guaranteed to have a zero. (See Example 8.)

6. $f(x) = x^3 - 3x^2 + 3$

7. $f(x) = 0.11x^3 - 2.07x^2 + 9.81x - 6.88$

8. $g(x) = 3x^4 + 4x^3 - 3$

9. $h(x) = x^4 - 10x^2 + 2$

10. *Tree Growth* The growth of a red oak tree is approximated by the function

$$G = -0.003t^3 + 0.137t^2 + 0.458t - 0.839,$$

$$2 \le t \le 34,$$

where G is the height of the tree in feet and t is its age in years. Use a graphing utility to obtain a graph of this function. Estimate the age of the tree when it is growing most rapidly. This point is called the **point of diminishing returns** because the increase in yield will be less with each additional year. (*Hint:* Use a viewing rectangle in which $-10 \le x \le 45$ and $-5 \le y \le 60$.)

SECTION 4.4   Real Zeros of Polynomial Functions

1. *Exploration* Use a graphing utility to sketch the graph of $y = 24x^3 - 36x + 17$. Describe a viewing rectangle that allows you to determine the number of real solutions of the equation $24x^3 - 36x + 17 = 0$.

Use the same technique to determine the number of real solutions of $97x^3 - 102x^2 - 200x - 63 = 0$.

In Exercises 2–9, use a graphing utility to approximate any solutions (accurate to three decimal places) of the equation. [Remember to write the equation in the form $f(x) = 0$.]

2. $\frac{1}{4}(x^2 - 10x + 17) = 0$

3. $-2(x^2 - 6x + 6) = 0$

4. $x^3 + x + 4 = 0$

5. $\frac{1}{9}x^3 + x + 4 = 0$

6. $2x^3 - x^2 - 18x + 9 = 0$

7. $4x^3 + 12x^2 - 26x - 24 = 0$

8. $x^4 = 2x^3 + 1$

9. $x^5 = 3 + 2x^3$

In Exercises 10–19, use a graphing utility to approximate any points of intersection (accurate to three decimal places) of the graphs of the equations.

10. $y = 2 - x$
$y = 2x - 1$

11. $y = 7 - x$
$y = \frac{3}{2} - \frac{11}{2}x$

12. $y = 4 - x^2$
$y = 2x - 1$

13. $y = x^3 - 3$
$y = 5 - 2x$

14. $y = 8$
$y = 3x^2 + 2x$

15. $y = 32$
$y = x^5 - x^2$

16. $y = 3(x + 1)$
$y = x^2 + 2x + 1$

17. $y = x + 2$
$y = -x^2 + 4x + 2$

18. $y = 2x^2$
$y = x^4 - 2x^2$

19. $y = -x$
$y = 2x - x^2$

20. *Mixture Problem* A 55-gallon barrel contains a mixture with a concentration of 33%. You are instructed to remove x gallons of this mixture and replace it with 100% concentrate.

(a) Write the amount of concentrate in the final mixture as a function of x.

(b) Use a graphing utility to graph the concentration function. What is the domain of the function?

(c) Approximate (accurate to one decimal place) the value of x if the final mixture is 60% concentrate.

SECTION 4.5   The Fundamental Theorem of Algebra

In Exercises 1 and 2, use the graph of the function to determine the number of real solutions of the equation $f(x) = 0$.

1. $f(x) = x^4 - 2x^3 + 4x - 4$

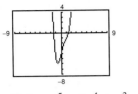

2. $f(x) = x^5 - 3x^4 + x^3 - x^2 + 2$

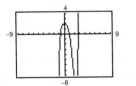

In Exercises 3–10, (a) use a graphing utility to verify that there are real zeros of f, (b) move the cursor along the graph to approximate the real zero(s) of the function accurate to two decimal places, and (c) find the complex zeros of f.

3. $f(x) = x^3 - 6x^2 + 13x - 10$

4. $f(x) = 3x^3 - 4x^2 + 8x + 8$

5. $f(x) = x^4 - 2x^3 - 18x^2 + 32x + 32$

6. $f(x) = 4x^4 + 4x^3 - 23x^2 - 20x + 15$

7. $f(x) = x^4 - 4x^3 + 6x^2 - 4x + 5$

8. $f(x) = 2x^4 - 2x^3 + 9x^2 - 8x + 4$

9. $f(x) = 9x^5 - 54x^4 + 82x^3 - 6x^2 + 9x$

10. $f(x) = 2x^5 - 2x^4 - 3x^3 + 3x^2 - 5x + 5$

11. *Dimensions of a Region*   A rectangular region with perimeter 230 feet has length x.

(a) Express the area of the rectangle as a function of x.

(b) Use a graphing utility to graph the area function. Because area is nonnegative, what is the domain of the function?

(c) Approximate (accurate to one decimal place) the dimensions of the region if its area is 2000 square feet. (Note that there are two values of x such that the area of the region is 2000 square feet.) Which x-value is appropriate in this problem? Explain.

SECTION 5.1   Rational Functions

In Exercises 1–4, (a) determine the domain of f and g, (b) find any vertical asymptotes of f, (c) use a graphing utility to obtain the graphs of f and g in the same viewing rectangle, and (d) explain why the graphing utility does or does not show the difference in the domains of f and g.

1. $f(x) = \dfrac{x^2 - 4}{x + 2}$

$g(x) = x - 2$

2. $f(x) = \dfrac{x^2(x - 3)}{x^2 - 3x}$

$g(x) = x$

3. $f(x) = \dfrac{x - 3}{x^2 - 3x}$

$g(x) = \dfrac{1}{x}$

4. $f(x) = \dfrac{2x - 8}{x^2 - 9x + 20}$

$g(x) = \dfrac{2}{x - 5}$

In Exercises 5–12, use a graphing utility to graph the function, and give its domain and range.

5. $f(x) = \dfrac{2 + x}{1 - x}$

6. $f(x) = \dfrac{3 - x}{2 - x}$

7. $f(t) = \dfrac{3t + 1}{t}$

8. $h(x) = \dfrac{1}{x - 3} + 1$

9. $h(t) = \dfrac{4}{t^2 + 1}$

10. $g(x) = -\dfrac{x}{(x - 2)^2}$

11. $f(x) = \dfrac{20x}{x^2 + 1} - \dfrac{1}{x}$

12. $f(x) = 5\left(\dfrac{1}{x - 4} - \dfrac{1}{x + 2} \right)$

In Exercises 13 and 14, use a graphing utility to graph the function and note that a graph may have two horizontal asymptotes.

13. $h(x) = \dfrac{6x}{\sqrt{x^2 + 1}}$

14. $g(x) = \dfrac{4|x - 2|}{x + 1}$

In Exercises 15 and 16, use a graphing utility to graph the function and note that a graph may cross its horizontal asymptote.

15. $f(x) = \dfrac{4(x-1)^2}{x^2 - 4x + 5}$ **16.** $g(x) = \dfrac{3x^4 - 5x + 3}{x^4 + 1}$

In Exercises 17 and 18, use a graphing utility to graph the function. Explain why there is no vertical asymptote when a superficial examination of the function may indicate that there should be one.

17. $h(x) = \dfrac{6 - 2x}{3 - x}$ **18.** $g(x) = \dfrac{x^2 + x - 2}{x - 1}$

In Exercises 19 and 20, use a graphing utility to graph the function and locate any relative maximum or minimum points on the graph.

19. $f(x) = \dfrac{3(x+1)}{x^2 + x + 1}$ **20.** $C(x) = x + \dfrac{16}{x}$

21. *Page Design*   A page that is x inches wide and y inches high contains 30 square inches of print. The margins at the top and bottom are 2 inches, and the margins on each side are 1 inch (see figure).

(a) Show here that the total area A of the page is given by
$$A = \frac{2x(2x + 11)}{x - 2}.$$

(b) Determine the domain of the function based on the physical constraints of the problem.

(c) Use a graphing utility to graph the area function and approximate the page size so the least amount of paper will be used.

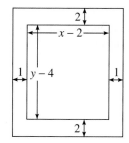

Figure for 21

22. *Exploration*   Use a graphing utility to graph $y = (x^2 - x)/(x + 1)$ and $y = x - 2$ on the same viewing rectangle. What do you observe as x increases or decreases without bound? Write a convincing argument that the line $y = x - 2$ is an asymptote of the graph of the rational function. (*Hint:* Use long division as part of your argument.)

SECTION 5.3   Conic Sections

In Exercises 1–6, match the equation with its graph. Use a graphing utility to confirm your selections. [The graphs are labeled (a), (b), (c), (d), (e), and (f).]

1. $x^2 - 2y^2 = 4$ **2.** $x^2 - y^2 = 0$

3. $x^2 + 2y^2 = 4$ **4.** $x^2 + y^2 = 4$

5. $x^2 + 2y = 4$ **6.** $2x^2 + y^2 = 4$

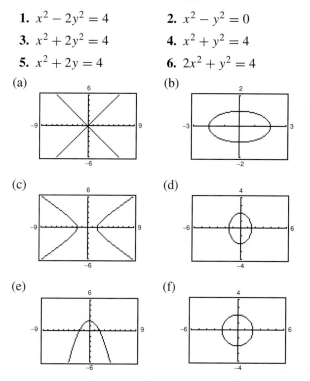

In Exercises 7–10, use a graphing utility to graph the conic. (Recall that it is first necessary to solve the equation for y and obtain two functions.)

7. $6x^2 - 3y^2 = 4$ **8.** $6x^2 + 3y^2 = 10$

9. $6x + 3y^2 = 10$ **10.** $2x^2 + \sqrt{15}y^2 = 25$

11. Use a graphing utility to sketch the graph of the equation

$$\frac{x^2}{9} + \frac{y^2}{9(1 - e^2)} = 1$$

for $e = 0.25$, $e = 0.50$, $e = 0.75$, 0.9, and $e = 0.95$. Discuss the change in the shape of the graph as e approaches 1.

SECTION 5.4   Conic Sections and Translations

In Exercises 1–6, match the equation with its graph. Use a graphing utility to confirm your selections. [The graphs are labeled (a), (b), (c), (d), (e), and (f).]

1. $\dfrac{(x - 2)^2}{4} + \dfrac{(y + 1)^2}{4} = 1$

2. $\dfrac{(x - 2)^2}{4} - \dfrac{(y + 1)^2}{4} = 1$

3. $\dfrac{(x - 2)^2}{4} + (y + 1)^2 = 1$

4. $\dfrac{(x - 2)^2}{4} + (y + 1) = 0$

5. $\dfrac{(x - 2)^2}{4} + \dfrac{(y + 1)^2}{9} = 1$

6. $\dfrac{(x - 2)^2}{4} - (y + 1)^2 = 0$

(a)   (b)

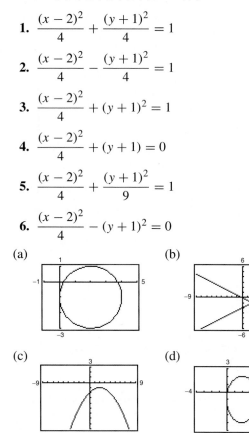

(c)   (d)

(e)   (f)

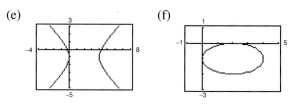

SECTION 6.1   Exponential Functions

1. *Exploration*   Use a graphing utility to graph $y = a^x$ with $a = 2, 3$, and 5 on the same viewing rectangle. (Use a viewing rectangle in which $-2 \le x \le 1$ and $0 \le y \le 2$.) How do the graphs compare with each other? Which graph is on the top in the interval $(-\infty, 0)$? Which is on the bottom? Which graph is on the top in the interval $(0, \infty)$? Which is on the bottom?

Repeat this experiment with the graphs of $y = a^{-x}$ with $a = 2, 3$, and 5.

2. Use a graphing utility to graph $y = 3^x$ and $y = 4^x$ in the same viewing rectangle. Use the graphs to solve the inequality $4^x < 3^x$.

3. Use a graphing utility to graph $y = \left(\frac{1}{2}\right)^x$ and $y = \left(\frac{1}{4}\right)^x$ in the same viewing rectangle. Use the graphs to solve the inequality $\left(\frac{1}{4}\right)^x < \left(\frac{1}{2}\right)^x$.

In Exercises 4–9, use a graphing utility to graph the exponential function.

4. $y = 1.08^{-5x}$ **5.** $y = 1.08^{5x}$

6. $s(t) = 2e^{0.12t}$ **7.** $s(t) = 3e^{-0.2t}$

8. $g(x) = 1 + e^{-x}$ **9.** $h(x) = e^{x-2}$

10. Graph the following functions in the same viewing rectangle of a graphing utility.

(a) $f(x) = 3^x$

(b) $g(x) = f(x - 2) = 3^{x-2}$

(c) $h(x) = -\frac{1}{2}f(x) = -\frac{1}{2}3^x$

(d) $q(x) = f(-x) + 3 = 3^{-x} + 3$

11. Use a graphing utility to graph each of the following functions. Use the graphs to determine any asymptotes of the functions.

(a) $f(x) = \dfrac{8}{1 + e^{-0.5x}}$ (b) $g(x) = \dfrac{8}{1 + e^{-0.5/x}}$

12. Use a graphing utility to graph each of the following functions. Use the graphs to determine where each function is increasing and decreasing, and approximate any relative maximum or minimum values of each function.

(a) $f(x) = x^2 e^{-x}$ (b) $g(x) = x2^{3-x}$

13. Use a graphing utility to demonstrate that

$$\left(1 + \frac{0.5}{x}\right)^x \rightarrow e^{0.5}$$

as x increases without bound.

SECTION 6.2   Logarithmic Functions

1. *Exploration*   Use a graphing utility to graph $y = \log_{10} x$ and $y = 8$ on the same viewing rectangle. Do the graphs intersect? If so, find a viewing rectangle that shows the point of intersection. What is the point of intersection?

In Exercises 2 and 3, use a graphing utility to graph the function. Use the graph to determine the intervals in which the function is increasing and decreasing and approximate any relative maximum or minimum values of the function.

2. $f(x) = \dfrac{x}{2} - \ln\dfrac{x}{4}$ **3.** $g(x) = \dfrac{12 \ln x}{x}$

4. Use a graphing utility to graph f and g on the same screen. Then determine which is increasing at the greater rate for "large" values of x. What can you conclude about the rate of growth of the natural logarithmic function?

(a) $f(x) = \ln x$, $g(x) = \sqrt{x}$
(b) $f(x) = \ln x$, $g(x) = \sqrt[4]{x}$

SECTION 6.3   Properties of Logarithms

1. *Exploration*   Use a graphing utility to graph $y = \log_a x$ and $y = (\ln x)/(\ln a)$ with $a = 2, 3$, and 5 on the same viewing rectangle. (Use a viewing rectangle in which $0 \le x \le 10$ and $-4 \le y \le 4$.) On the interval $(0, 1)$, which graph is on top? Which is on the bottom? On the interval $(1, \infty)$, which graph is on top? Which is on the bottom?

2. Use a graphing utility to graph

$$f(x) = \ln\frac{x}{2}, \quad g(x) = \frac{\ln x}{\ln 2}, \quad h(x) = \ln x - \ln 2$$

in the same viewing rectangle. Which two functions have identical graphs?

SECTION 6.4   Solving Exponential and Logarithmic Equations

1. *Exploration*   Use a graphing utility to graph $y = \ln x$ and $y = x^2 - 2$ on the same viewing rectangle. (Use a viewing rectangle in which $0 \le x \le 3$ and $-5 \le y \le 5$.) Then zoom in to approximate the two points of intersection.

In Exercises 2–7, use a graphing utility to solve the equation. (Round your result to three decimal places.)

2. $250e^{-0.01x} = 100$ **3.** $12e^{x/2} = 225$

4. $50(200 - e^{0.12t}) = 6000$

5. $\ln x + 2\ln(x + 1) = 4$

6. $\dfrac{25 \log_{10}(x + 1)}{x} = 2$

7. $\dfrac{1000}{1 + 2e^{-0.04t}} = 750$

In Exercises 8 and 9, use a graphing utility to graph the functions and approximate any points of intersection.

8. $f(x) = \sqrt{x}$, $g(x) = 5e^{-x/2}$

9. $f(x) = \dfrac{4}{x}$, $g(x) = \ln x$

10. *Population Growth* The growth of a population is modeled by the function

$$P(t) = \frac{500}{1 + 4e^{-0.09t}}$$

where P is the population in hundreds and t is time in years.

(a) Use a graphing utility to graph the model. Use the model to estimate the upper asymptote and interpret its meaning in the context of the model.

(b) Approximate (accurate to one decimal place) the times when $P = 200$, $P = 300$, and $P = 400$.

SECTION 6.5 Exponential and Logarithmic Applications

1. *Exploration* A person deposits $1000 in an account that pays 9.5% per year, compounded quarterly. The balance after t years is

$$A = 1000 \left(1 + \frac{0.095}{4} \right)^{4t}.$$

Another person deposits $1000 in an account that pays 9.5% per year, compounded continuously. The balance after t years is $A = 1000e^{0.095t}$. Use a graphing utility to graph both equations on the same viewing rectangle. (Use a viewing rectangle in which $0 \le x \le 8$ and $1000 \le y \le 2200$.) What do you conclude? Which of the two accounts will reach a balance of $2000 first? Justify your answer by zooming in near $x = 7$.

Profit In Exercises 2 and 3, match the profit function with its graph where P is profit is thousands of dollars after spending x hundred dollars in advertising. Use a graphing utility to verify your selection. [The graphs are labeled (a) and (b).]

2. $P(x) = 50 \left(1 - e^{-0.04x} \right)$

3. $P(x) = \dfrac{50}{1 + 4e^{-0.04x}}$

(a)   (b)

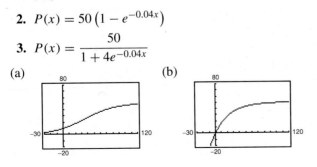

4. *Investing for Retirement* Many people put off saving for retirement until a higher interest rate is available or until they feel financially able to put aside a larger sum of money. Consider the following three options assuming the interest is compounded continuously. Use a graphing utility to graph the exponential functions showing the growth of each investment. Which option is best and why? Make a generalization on retirement savings strategy based on your findings.

(a) Putting aside $2000 (lump sum) at 6.6% 30 years before retirement.

(b) Putting aside $3000 (lump sum) at 7.8% 20 years before retirement.

(c) Putting aside $5000 (lump sum) at 10.2% 10 years before retirement.

SECTION 7.1 Systems of Equations

In Exercises 1–12, use a graphing utility to graph the equations and approximate any solutions of the system of equations.

1. $-2x + y = 1$
$\quad\quad x - 3y = 2$

2. $5x - 6y = -30$
$\quad\quad 5x + 4y = 20$

3. $2x - 5y = 20$
$\quad\quad 4x - 5y = 40$

4. $5x + 3y = 24$
$\quad\quad x - 2y = 10$

5. $y = x^2$
$\quad\quad y = 4x - x^2$

6. $y = 8 - x^2$
$\quad\quad y = 6 - x$

7. $x^2 - y^2 = 12$
$\quad\quad x - 2y = 0$

8. $x^2 + y^2 = 20$
$\quad\quad x + 3y = 10$

9. $\sqrt{x} - y = 0$
$\quad\quad x - 5y = -6$

10. $x^2 - y^2 = 1$
$\quad\quad \dfrac{x^2}{2} + y^2 = 1$

11. $y = x^3$
$\quad\quad y = x^3 - 3x^2 + 3x$

12. $y = \frac{1}{5}(24 - x)$
$\quad\quad y = \sqrt{64 - x^2}$

In Exercises 13 and 14, use a calculator to graph each equation in the system. The graphs appear parallel. Yet, from the slope-intercept form of the line, you find that the slopes are not equal and thus, the graphs intersect. Find the point of intersection of the two lines.

13. $x - 100y = -200$

$3x - 275y = 198$

14. $35x - 33y = 0$

$12x - 11y = 92$

Break-Even Analysis In Exercises 15 and 16, use a graphing utility to graph the equations for cost C and revenue R. Use the graphs to approximate the sales, x, necessary to break even. (Round your answer up to the nearest whole unit.)

15. $C = 0.85x + 35,000, \ R = 1.55x$

16. $C = 6x + 500,000, \ R = 35x$

6. *Break-Even Analysis* A certain truck model costs $24,500 with a gasoline engine and $25,750 with a diesel engine. The number of miles per gallon of fuel for trucks with these two engines are 17 and 25, respectively. Assume that the price of each type of fuel is $1.269 per gallon.

(a) Show that the cost C_g of driving the gasoline-powered truck x miles is

$$C_g = 24,500 + \frac{1.269x}{17}$$

and the cost C_d of driving the diesel model x miles is

$$C_d = 25,750 + \frac{1.269x}{25}.$$

(b) Use a graphing utility to graph the cost functions and use the graphs to approximate the mileage at which the diesel-powered truck becomes more economical than the gasoline-powered truck.

SECTION 7.2 Systems of Linear Equations in Two Variables

1. *Exploration* Rewrite the system of equations in slope-intercept form and graph using a graphing utility. What is the relationship between the slopes of the two lines and the number of points of intersection?

(a) $2x + 4y = 8$ (b) $-x + 5y = 15$

$\quad 4x - 3y = -6$ $2x - 10y = 7$

(c) $x - y = 9$

$\quad 2x - 2y = 18$

In Exercises 2–5, use a graphing utility to graph the equations and approximate any solutions of the system of equations.

2. $5x + 4y = 35$ **3.** $5x - 4y = 0$

$\quad -x + 3y = 12$ $-3x + 8y = 14$

4. $4x - y = 3$ **5.** $x - 6y = 2$

$\quad 6x + 2y = 1$ $2x + 3y = 9$

SECTION 7.4 Systems of Inequalities

In Exercises 1–4, match the system of inequalities with the graph of its solution. Use a graphing utility to verify your answer. [The graphs are labeled (a), (b), (c), and (d).]

1. $2x + 5y \le 10$ **2.** $2x + 5y \le 20$

$\quad\quad\quad x \ge 0$ $2x - 5y \le 0$

$\quad\quad\quad y \ge 0$ $x \ge 0$

3. $y < 5\ln x$ **4.** $x^2 + y^2 \le 16$

$\quad y \ge 0$ $y \ge 3$

$\quad x \ge 1$

$\quad x \le e$

(a)

(b)

(c)

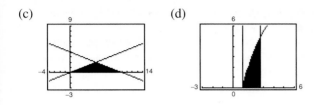

(d)

In Exercises 5–10, use a graphing utility to graph the solution of the system of inequalities. (*Note:* If your graphing utility has *shading* capabilities, shade the region representing the solution.)

5. $x - 2y \le 6$
$x + 2y \ge 6$
$x \quad\quad \le 6$

6. $-x + 3y \le 9$
$x + 5y \le 9$
$y \ge 1$

7. $y < 4x - x^2$
$y > x$

8. $x^2 + y^2 \le 36$
$x^2 + y^2 > 16$

9. $y < 8e^{-x/2}$
$y \ge 0$
$0 \le x \le 4$

10. $xy \le 8$
$y \ge 1$
$x \ge 2$

SECTION 8.2   Operations with Matrices

In Exercises 1–4, use a graphing utility to find (a) $A + 3B$, and (b) $2A - B$. If it is not possible, state the reason.

1. $A = \begin{bmatrix} 4 & 6 & 1 \\ -2 & 2 & 5 \end{bmatrix}$ $B = \begin{bmatrix} 3 & 2 & 6 \\ 2 & 3 & -3 \end{bmatrix}$

2. $A = \begin{bmatrix} 1 & 1 & 0 & 5 \\ -2 & -1 & 2 & -10 \\ 3 & 6 & 7 & 14 \end{bmatrix}$

$B = \begin{bmatrix} 1 & 2 & -1 & 3 \\ 3 & 7 & -5 & 14 \\ -2 & -1 & -3 & 8 \end{bmatrix}$

3. $A = \begin{bmatrix} 1 & -1 & -1 & 1 \\ 4 & -4 & 1 & 8 \\ -6 & 8 & 18 & 0 \end{bmatrix}$

$B = \begin{bmatrix} 1 & -3 & 0 & -7 \\ -3 & 10 & 1 & 23 \\ 4 & -10 & 2 & -24 \end{bmatrix}$

4. $A = \begin{bmatrix} 3 & -2 & 0 & 5 \\ -4 & 7 & 11 & 1 \end{bmatrix}$

$B = \begin{bmatrix} -2 & 2 & 7 \\ 9 & 8 & 3 \\ -12 & 5 & 20 \end{bmatrix}$

In Exercises 5–10, use a graphing utility to find (a) AB and (b) A^2. If not possible, state the reason.

5. $A = \begin{bmatrix} 3 & -1 \\ -2 & 0 \end{bmatrix}$ $B = \begin{bmatrix} -1 & 4 \\ 1 & 4 \end{bmatrix}$

6. $A = \begin{bmatrix} 5 & 4 & 2 \\ 1 & 1 & 3 \end{bmatrix}$ $B = \begin{bmatrix} -5 & 8 \\ 0 & 2 \\ 7 & -3 \end{bmatrix}$

7. $A = \begin{bmatrix} 3 & -4 \\ -5 & 10 \end{bmatrix}$ $B = \begin{bmatrix} 12 & 8 & 0 \\ -6 & 4 & 10 \end{bmatrix}$

8. $A = \begin{bmatrix} 6 & 13 & 8 \\ \frac{1}{2} & \frac{3}{2} & 1 \\ 4 & 6 & 8 \end{bmatrix}$ $B = \begin{bmatrix} 3 & 4 & 2 \\ -2 & 5 & 3 \end{bmatrix}$

9. $A = \begin{bmatrix} 3 & 0 & 2 & 8 \end{bmatrix}$ $B = \begin{bmatrix} -5 \\ 10 \\ -12 \\ 25 \end{bmatrix}$

10. $A = \begin{bmatrix} 3 & 1 & 2 \\ 0 & -5 & 6 \\ 0 & 0 & 1 \end{bmatrix}$ $B = \begin{bmatrix} 12 & 0 & 0 \\ 10 & 8 & 0 \\ -5 & 4 & 2 \end{bmatrix}$

SECTION 8.3   The Inverse of a Square Matrix

In Exercises 1–2, use a graphing utility to show that B is the inverse of A.

1. $A = \begin{bmatrix} 3 & -2 \\ 1 & 1 \end{bmatrix}$ $B = \begin{bmatrix} \frac{1}{5} & \frac{2}{5} \\ -\frac{1}{5} & \frac{3}{5} \end{bmatrix}$

2. $A = \begin{bmatrix} 1 & -2 & -1 \\ 2 & 0 & 1 \\ 3 & -2 & 1 \end{bmatrix}$ $B = \begin{bmatrix} \frac{1}{2} & 1 & -\frac{1}{2} \\ \frac{1}{4} & 1 & -\frac{3}{4} \\ -1 & -1 & 1 \end{bmatrix}$

In Exercises 3–8, use a graphing utility to find the inverse of the matrix (if it exists).

3. $\begin{bmatrix} 2 & 3 \\ 1 & 2 \end{bmatrix}$

4. $\begin{bmatrix} 5 & 7 \\ 7 & 10 \end{bmatrix}$

5. $\begin{bmatrix} 5 & -4 \\ 2 & 1 \end{bmatrix}$

6. $\begin{bmatrix} -4 & 10 \\ 6 & 15 \end{bmatrix}$

7. $\begin{bmatrix} 10 & 12 & 6 \\ 5 & 6 & 1 \\ 1 & 1 & 2 \end{bmatrix}$

8. $\begin{bmatrix} 2 & 1 & 1 & 0 \\ 3 & 6 & 2 & 1 \\ -2 & -2 & 0 & 1 \\ 1 & 0 & 4 & 4 \end{bmatrix}$

In Exercises 9–14, use a graphing utility and the inverse matrix to solve (if possible) the system of equations.

9. $5x + 2y = -8$
$2x - 3y = -7$

10. $1.5x + 3.2y = 22$
$-x + y = 1$

11. $7x + 2y + 3z = 15$
$5x - 3y \quad\quad = 13$
$4x + y + z = 8$

12. $3x - 7y + z = 5$
$x + y - 3z = 1$
$5x - 5y - 5z = 3$

13. $-x + 4y + 2z = 3$
$11x + 3y + z = 40$
$5x - y + 3z = -4$

14. $2x - 8y + 3z = 52$
$5x \quad\quad + 4z = 23$
$x - 3y - 6z = 6$

SECTION 8.6   Applications of Determinants and Matrices

In Exercises 1–4, use a graphing utility to evaluate the determinant of the matrix.

1. $\begin{bmatrix} 35 & 15 & 70 \\ -8 & 20 & 3 \\ -5 & 6 & 20 \end{bmatrix}$

2. $\begin{bmatrix} 3 & -1 & 2 \\ 1 & -1 & 2 \\ -2 & 3 & 10 \end{bmatrix}$

3. $\begin{bmatrix} 0.3 & -0.2 & 0.5 \\ 0.6 & 0.4 & -0.3 \\ 1.2 & 0 & 0.7 \end{bmatrix}$

4. $\begin{bmatrix} \frac{3}{2} & -\frac{3}{4} & 1 \\ 10 & 8 & 7 \\ 12 & -4 & 12 \end{bmatrix}$

In Exercises 5–8, use a graphing utility and Cramer's Rule to solve the system of equations.

5. $4x - y = -2$
$-2x + y = 3$

6. $4x + 8y = 6$
$8x + 26y = 19$

7. $3x + y + z = 6$
$x - 4y + 2z = -1$
$x - 3y + z = 0$

8. $3x - 2y + 3z = 8$
$x + 3y + 6z = -3$
$x + 2y + 9z = -5$

Geometry In Exercises 9 and 10, use a graphing utility and a determinant to find the area of the triangle with the given vertices.

9. $(-2, 1)$, $(3, -1)$, $(1, 6)$

10. $(0, \frac{1}{2})$, $(\frac{5}{2}, 0)$, $(4, 3)$

Collinear Points In Exercises 11 and 12, use a graphing utility and a determinant to determine whether the given points are collinear.

11. $(-1, 11)$, $(0, 8)$, $(2, 2)$
12. $(-1, -1)$, $(1, 9)$, $(2, 13)$

Curve Fitting In Exercises 13 and 14, use a graphing utility and Cramer's Rule to find the equation of the parabola $y = ax^2 + bx + c$ that passes through the given points.

13. $(0, 1)$, $(1, -3)$, $(-2, 21)$

14. $(2, 3)$, $(-1, \frac{9}{2})$, $(-2, 9)$

SECTION 9.1   Sequences and Summation Notation

In Exercises 1–4, match the sequence with the graph of its first ten terms.

1. $a_n = \dfrac{8}{n}$

2. $a_n = \dfrac{5n}{n+1}$

3. $a_n = 4 - \dfrac{1}{2^n}$

4. $a_n = \left(-\dfrac{3}{2}\right)^n$

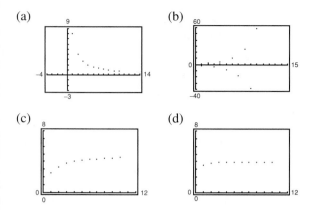

(a)

(b)

(c)

(d)

In Exercises 5–8, use a graphing utility to graph the first ten terms of the sequence. (Assume n begins with 1.)

5. $a_n = 8\left(-\dfrac{3}{4}\right)^n$

6. $a_n = \dfrac{2^n}{n!}$

7. $a_n = \dfrac{4}{n^{3/2}}$

8. $a_n = \dfrac{n^2}{n+1}$

In Exercises 9–12, use a graphing utility to find the sum.

9. $\displaystyle\sum_{i=0}^{8} (2i+3)$

10. $\displaystyle\sum_{i=1}^{20} \left(\tfrac{1}{4}i - 1\right)$

11. $\displaystyle\sum_{j=1}^{12} \dfrac{(-)^{j+1}6}{j}$

12. $\displaystyle\sum_{n=1}^{8} \dfrac{2}{n^2}$

13. *Compound Interest* A deposit of $500 is made in an account that earns 6% interest compounded yearly. The balance in the account after n years is given by

$$A_n = 500(1 + 0.06)^n, \quad n = 1, 2, 3, \ldots$$

(a) Compute the first eight terms of this sequence.

(b) Use a graphing utility to graph the first 40 terms of the sequence. Explain why the points are increasing at a more rapid rate as time passes.

SECTION 9.3 Geometric Sequences

In Exercises 1–4, match the geometric sequence with the graph of its first ten terms.

1. $a_n = 60\left(-\dfrac{1}{3}\right)n$

2. $a_n = 2\left(\dfrac{4}{3}\right)^n$

3. $a_n = 100(1.05)^n$

4. $a_n = 500(0.80)^n$

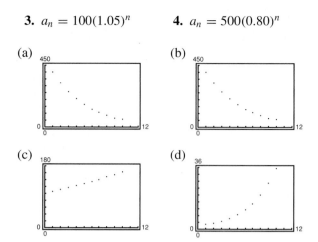

(a)

(b)

(c)

(d)

In Exercises 5–8, use a graphing utility to graph the first ten terms of the sequence. (Assume n begins with 1.)

5. $a_n = 12\left(\dfrac{2}{3}\right)^{n-1}$

6. $a_n = 12\left(-\dfrac{2}{3}\right)^{n-1}$

7. $a_n = 4\left(\dfrac{3}{2}\right)^{n-1}$

8. $a_n = 4\left(-\dfrac{3}{2}\right)^{n-1}$

In Exercises 9–14, use a graphing utility to find the sum.

9. $\displaystyle\sum_{i=1}^{10} 2^{i-1}$

10. $\displaystyle\sum_{i=1}^{20} 8\left(-\dfrac{1}{4}\right)^{i-1}$

11. $\displaystyle\sum_{i=1}^{20} 100(1.1)^{i-1}$

12. $\displaystyle\sum_{i=1}^{40} 50(1.07)^{i-1}$

13. $\displaystyle\sum_{i=1}^{8} (-3)^{i-1}$

14. $\displaystyle\sum_{i=1}^{30} 18\left(\dfrac{2}{3}\right)^{i-1}$

15. *Cooling* The temperature of an item is 70° Fahrenheit when it is placed in a freezer. Its temperature n hours after being placed in the freezer is 20% less than 1 hour earlier.

(a) Find a formula for the nth term of the geometric sequence that gives the temperature of the item n hours after being placed in the freezer.

(b) Use a graphing utility to estimate the time when the item freezes.

Answers to Warm Ups, Odd-Numbered Exercises, and Cumulative Tests

CHAPTER 1

Section 1.1 *(page 6)*

1. **(a)** 5, 1 **(b)** $-9, 5, 0, 1$ **(c)** $-9, -\frac{7}{2}, 5, \frac{2}{3}, 0, 1$
   (d) $\sqrt{2}$
3. **(a)** None **(b)** -13 **(c)** $2.01, 0.666\ldots, -13$
   (d) $0.010110111\ldots$
5. **(a)** $\frac{6}{3}$ **(b)** $\frac{6}{3}$ **(c)** $-\frac{1}{3}, \frac{6}{3}, -7.5$ **(d)** $-\pi, \frac{1}{2}\sqrt{2}$
7. $\frac{3}{2} < 7$ 9. $-4 > -8$

11. $\frac{5}{6} > \frac{2}{3}$

13. $x \le 5$ is the set of all real numbers less than or equal to 5.

15. $x < 0$ is the set of all negative real numbers.

17. $x \ge 4$ is the set of all real numbers greater than or equal to 4.

19. $-2 < x < 2$ is the set of all real numbers greater than -2 and less than 2.

21. $-1 \le x < 0$ is the set of all negative real numbers greater than or equal to -1.

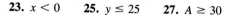

23. $x < 0$ 25. $y \le 25$ 27. $A \ge 30$

29. $3.5\% \le r \le 6\%$ 31. 10   33. $\pi - 3 \approx 0.1416$
35. -1 37. -9 39. 3.75   41. $|-3| > -|-3|$
43. $-5 = -|5|$ 45. $-|-2| = -|2|$ 47. 4
49. $\frac{5}{2}$ 51. 51   53. 14.99   55. $|x - 5| \le 3$
57. $|7 - 18| = 11$ mi   59. $|y| \ge 6$
61. $|\$113,356.52 - \$112,700.00| = \$656.52 > \$500.00$
   $0.05(\$112,700.00) = \$5,635.00$
   Since the actual expenses differ from the budget by more than $500.00, there is failure to meet the budget variance test.
63. $|\$37,335.80 - \$37,640.00| = \$304.20 < \$500.00$
   $0.05(\$37,640.00) = \$1,882.00$
   Since the difference between actual expenses and the budget is less than $500.00 and less than 5% of the budgeted amount, there is compliance with the budget variance test.
65. $|77.8 - 92.2| = 14.4$
   There was a deficit of $14.4 billion.
67. $|520.0 - 590.2| = 70.2$
   There was a deficit of $70.2 billion.
69. $\frac{127}{90}, \frac{584}{413}, \frac{7071}{5000}, \sqrt{2}, \frac{47}{33}$
71. 0.625   73. $0.123123\ldots$

75.

n	10	1	0.5	0.01	0.0001	0.000001
$5/n$	0.5	5	10	500	50,000	5,000,000

77. True   79. True

Section 1.2 *(page 16)*

WARM UP   1. $-4 < -2$ 2. $0 > -3$
3. $\sqrt{3} > 1.73$ 4. $-\pi < -3$ 5. 2   6. 4
7. 5   8. 4   9. 14   10. -2

1. $7x, 4$ 3. $x^2, -4x, 8$ 5. $4x^3, x, -5$
7. **(a)** -10 **(b)** -6 9. **(a)** 14   **(b)** 2
11. **(a)** Division by 0 is undefined.   **(b)** 0
13. Commutative (addition)   15. Inverse (multiplication)

17. Distributive property   **19.** Identity (multiplication)
21. Associative and Commutative (multiplication)
23. 0   **25.** Division by 0 is undefined.   **27.** 6
29. $\frac{1}{2}$ **31.** $\frac{3}{8}$ **33.** $\frac{3}{10}$ **35.** 48   **37.** -2.57
39. 1.56   **41.** 125   **43.** 729   **45.** 5184   **47.** $\frac{16}{3}$
49. -24 **51.** 5   **53.** $-125z^3$ **55.** $5x^6$

57. $24y^{10}$ **59.** $3x^2$ **61.** $\frac{7}{x}$ **63.** $\frac{4}{3}(x+y)^2$

65. 1   **67.** $\frac{1}{4x^4}$ **69.** $-2x^3$ **71.** $\frac{10}{x}$ **73.** $\frac{a^6}{64b^9}$

75. $\frac{1}{625x^8y^8}$ **77.** 3^{3n} **79.** $\frac{b^5}{a^5}$ **81.** 5.75×10^7

83. 8.99×10^{-5} **85.** 524,000,000
87. 0.00000000048
89. (a) 954.448   (b) 3.077×10^{10} **91.** $8\frac{1}{3}$ min
93. State highway   20,728
       Municipal street   9,885
       Local street   8,473
       Interstate highway   4,965
       Other/unknown   1,503
95. $5(2.7 - 9.4)$

Section 1.3   (page 27)

WARM UP   **1.** $\frac{4}{27}$ **2.** 48   **3.** $-8x^3$ **4.** $6x^7$

5. $28x^6$ **6.** $\frac{1}{5}x^2$ **7.** $3z^4$ **8.** $\frac{25}{4x^2}$ **9.** 1

10. $(x+2)^{10}$

1. $9^{1/2} = 3$ **3.** $\sqrt[5]{32} = 2$ **5.** $\sqrt{196} = 14$
7. $(-216)^{1/3} = -6$ **9.** $\sqrt[3]{27^2} = 9$ **11.** $81^{3/4} = 27$
13. 3   **15.** 2   **17.** 3   **19.** $\frac{1}{2}$ **21.** -125
23. 216   **25.** $\frac{1}{8}$ **27.** $\frac{27}{8}$ **29.** -4 **31.** $2\sqrt{2}$

33. 3×10^{-2} **35.** $6x\sqrt{2x}$ **37.** $\frac{3|x|}{z}\sqrt{\frac{2}{z}}$

39. $2x\sqrt[3]{2x^2}$ **41.** $\frac{5|x|\sqrt{3}}{y^2}$ **43.** $\frac{\sqrt{3}}{3}$ **45.** $4\sqrt[3]{4}$

47. $\frac{x(5+\sqrt{3})}{11}$ **49.** $3(\sqrt{6}-\sqrt{5})$ **51.** $\frac{2}{\sqrt{2}}$

53. $\frac{2}{3(\sqrt{5}-\sqrt{3})}$ **55.** $-\frac{1}{2(\sqrt{7}+3)}$ **57.** $3^{1/2} = \sqrt{3}$
59. $(x+1)^{2/3} = \sqrt[3]{(x+1)^2}$ **61.** $2\sqrt[4]{2}$ **63.** $\sqrt[8]{2x}$
65. $2\sqrt{x}$ **67.** $34\sqrt{2}$ **69.** $4\sqrt[4]{y}$ **71.** 625

73. $\frac{2}{x}$ **75.** $\frac{1}{x^3}$ **77.** 7.550   **79.** 2.236

81. 9.137   **83.** $\sqrt{5} + \sqrt{3} > \sqrt{5+3}$
85. $5 > \sqrt{3^2 + 2^2}$ **87.** $\sqrt{3} \cdot \sqrt[4]{3} > \sqrt[3]{3}$
89. $0.215 = 21.5\%$ **91.** 24 in. $\times$ 24 in. $\times$ 24 in.

93. $\frac{\pi}{2} \approx 1.57$ sec   **95.** 1

97. When any positive integer is squared, the unit digit is 0, 1, 4, 5, 6, or 9. Therefore, $\sqrt{5233}$ is not an integer.

Section 1.4   (page 35)

WARM UP   **1.** $42x^3$ **2.** $-20z^2$ **3.** $-27x^6$

4. $-3x^6$ **5.** $\frac{9}{4}z^3$ **6.** $4\sqrt{3}$ **7.** $\frac{9}{4x^2}$ **8.** 8

9. $\sqrt{2}$ **10.** $-3x$

1. Degree: 2, Leading coefficient: 2
3. Degree: 5, Leading coefficient: 1
5. Degree: 5, Leading coefficient: 4
7. Polynomial: $-3x^3 + 2x + 8$
9. Not a polynomial because of the operation of division
11. Polynomial: $-y^4 + y^3 + y^2$
13. $-2x - 10$ **15.** $3x^3 - 2x + 2$
17. $8x^3 + 29x^2 + 11$ **19.** $12z + 8$
21. $3x^3 - 6x^2 + 3x$ **23.** $-15z^2 + 5z$
25. $30x^3 + 12x^2$ **27.** $x^2 + 7x + 12$
29. $6x^2 - 7x - 5$ **31.** $4x^2 + 12x + 9$
33. $4x^2 - 20xy + 25y^2$
35. $x^2 + 2xy + y^2 - 6x - 6y + 9$ **37.** $x^2 - 100$
39. $x^2 - 4y^2$ **41.** $m^2 - n^2 - 6m + 9$ **43.** $4r^4 - 25$
45. $x^3 + 3x^2 + 3x + 1$ **47.** $8x^3 - 12x^2y + 6xy^2 - y^3$
49. $x - y$ **51.** $16x^6 - 24x^3 + 9$
53. $x^4 - x^3 + 5x^2 - 9x - 36$ **55.** $x^4 + x^2 + 1$
57. $2x^2 + 2x$ **59.** $x^3 + 4x^2 - 5x - 20$
61. $500r^2 + 1000r + 500$
63. $V = x(26 - 2x)(18 - 2x)$
       $= 4x(x - 13)(x - 9)$

x (in.)	1	2	3
V (cu in.)	384	616	720

65. $3x^2 + 7x$
67. Total $= 0.14x^2 - 3.33x + 58.40$

x (mph)	Total
30	84.5 ft
40	149.2 ft
55	298.75 ft

69. $m + n$
71. Instead of squaring the binomial, the student squared each term. The correct method is as follows: $x^2 + 2(x)(-3) + (-3)^2 = x^2 - 6x + 9$.

Section 1.5   *(page 45)*

WARM UP   **1.** $15x^2 - 6x$ **2.** $-2y^2 - 2y$
3. $4x^2 + 12x + 9$ **4.** $9x^2 - 48x + 64$
5. $2x^2 + 13x - 24$ **6.** $-5z^2 - z + 4$ **7.** $4y^2 - 1$
8. $x^2 - a^2$ **9.** $x^3 + 12x^2 + 48x + 64$
10. $8x^3 - 36x^2 + 54x - 27$

1. $3(x + 2)$ **3.** $2x(x^2 - 3)$ **5.** $(x - 1)(x + 5)$
7. $(x + 6)(x - 6)$ **9.** $(4y + 3)(4y - 3)$
11. $(x + 1)(x - 3)$ **13.** $(x - 2)^2$ **15.** $(2t + 1)^2$
17. $(5y - 1)^2$ **19.** $(x + 2)(x - 1)$
21. $(s - 3)(s - 2)$ **23.** $(y + 5)(y - 4)$
25. $(x - 20)(x - 10)$ **27.** $(3x - 2)(x - 1)$
29. $(3z + 1)(3z - 2)$ **31.** $(5x + 1)(x + 5)$
33. $(x - 2)(x^2 + 2x + 4)$ **35.** $(y + 4)(y^2 - 4y + 16)$
37. $(2t - 1)(4t^2 + 2t + 1)$ **39.** $(x - 1)(x^2 + 2)$
41. $(2x - 1)(x^2 - 3)$ **43.** $(3 + x)(2 - x^3)$
45. $(x + 2)(3x + 4)$ **47.** $(2x - 1)(3x + 2)$
49. $(3x - 1)(5x - 2)$ **51.** $x^2(x - 4)$ **53.** $(1 - 2x)^2$
55. $2x(x + 1)(x - 2)$ **57.** $(9x + 1)(x + 1)$
59. $(3x + 1)(x^2 + 5)$ **61.** $x(x - 4)(x^2 + 1)$
63. $-(x + 2)(x + 6)$ **65.** $(x + 1)^2(x - 1)^2$
67. $2(t - 2)(t^2 + 2t + 4)$ **69.** $(2x - 1)(6x - 1)$
71. $-(x + 1)(x - 3)(x + 9)$ **73.** $7(x^2 + 1)(3x^2 - 1)$

75. $-2x(x - 5)^3(x + 5)$ **77.** $-(x^2 + 1)^4\left(\dfrac{x^2}{2} + 1\right)$

79. $\dfrac{-1}{(x^2 + 1)^5}$

81. Two possible values of c are 2 and -12.
83. b
85. a
87.

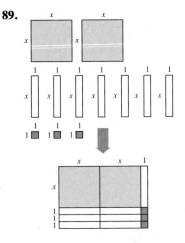

89.

91. (a) $V = \pi h(R + r)(R - r)$ **(b)** $V = 2\pi\left(\dfrac{R + r}{2}\right)(R - r)h$

Section 1.6   *(page 54)*

WARM UP   **1.** $5x^2(1 - 3x)$ **2.** $(4x + 3)(4x - 3)$
3. $(3x - 1)^2$ **4.** $(2y + 3)^2$ **5.** $(z + 3)(z + 1)$
6. $(x - 5)(x - 10)$ **7.** $(3 - x)(1 + 3x)$
8. $(3x - 1)(x - 15)$ **9.** $(s + 1)(s + 2)(s - 2)$
10. $(y + 4)(y^2 - 4y + 16)$

1. All real numbers   **3.** All nonnegative real numbers
5. All real numbers x such that $x \neq 2$
7. All real numbers x such that $x \neq 0$ and $x \neq 4$
9. All real numbers x such that $x \geq -1$
11. $3x, x \neq 0$ **13.** $x - 2, x \neq 2$

15. $x + 2, x \neq -2$ **17.** $\dfrac{3x}{2}, x \neq 0$

19. $\dfrac{3y}{y + 1}, x \neq 0$ **21.** $-\frac{1}{2}, x \neq 5$

23. $\dfrac{x(x + 3)}{x - 2}, x \neq -2$ **25.** $\dfrac{y - 4}{y + 6}, y \neq 3$

27. $-(x^2 + 1), x \neq 2$ **29.** $z - 2$

31. $\dfrac{1}{5(x - 2)}, x \neq 1$ **33.** $-\dfrac{x(x + 7)}{x + 1}, x \neq 9$

35. $\dfrac{r + 1}{r}, r \neq 1$ **37.** $\dfrac{t - 3}{(t + 3)(t - 2)}, t \neq -2$

39. $\dfrac{x - y}{x(x + y)^2}, x \neq -2y$ **41.** $\frac{3}{2}, x \neq -y$

43. $x(x + 1), x \neq 0, -1$ **45.** $\dfrac{x + 5}{x - 1}$ **47.** $\dfrac{6x + 13}{x + 3}$

49. $-\dfrac{2}{x - 2}$ **51.** $\dfrac{x - 4}{(x + 2)(x - 2)(x - 1)}$

53. $-\dfrac{x^2 + 3}{(x + 1)(x - 2)(x - 3)}$ **55.** $\dfrac{2 - x}{x^2 + 1}, x \neq 0$

57. $\frac{1}{2}, x \neq 2$ **59.** $\dfrac{1}{x}, x \neq -1$ **61.** $\dfrac{(x + 3)^3}{2x(x - 3)}$

63. $-\dfrac{2x + h}{x^2(x + h)^2}, h \neq 0$ **65.** $\dfrac{2x - 1}{2x}, x > 0$

67. $-\dfrac{1}{t^2\sqrt{t^2 + 1}}$ **69.** $-\dfrac{1}{x^2(x + 1)^{3/4}}$

71. $\dfrac{1}{\sqrt{x + 2} + \sqrt{x}}$ **73. (a)** $\frac{1}{16}$ min   **(b)** $\dfrac{x}{16}$ min

(c) $\frac{60}{16} = \frac{15}{4}$ min   **75.** $\dfrac{11x}{30}$ **77. (a)** 12.65%

(b) $\dfrac{288(MN - P)}{N(MN + 12P)}$ **79.** $\dfrac{R_1R_2}{R_1 + R_2}$

Section 1.7   (page 62)

1. $2x - (3y + 4) = 2x - 3y - 4$
3. $5z + 3(x - 2) = 5z + 3x - 6$
5. $-\dfrac{x - 3}{x - 1} = \dfrac{3 - x}{x - 1}$ **7.** $a\left(\dfrac{x}{y}\right) = \dfrac{ax}{y}$
9. $(4x)^2 = 16x^2$ **11.** $\sqrt{x + 9}$ cannot be simplified.
13. $\dfrac{6x + y}{6x - y}$ cannot be simplified.   **15.** $\dfrac{1}{x + y^{-1}} = \dfrac{y}{xy + 1}$
17. $x(2x - 1)^2 = x(4x^2 - 4x + 1)$
19. $\sqrt[3]{x^3 + 7x^2}$ cannot be simplified.
21. $\dfrac{3}{x} + \dfrac{4}{y} = \dfrac{3y + 4x}{xy}$
23. $\dfrac{1}{2y} = \dfrac{1}{2} \cdot \dfrac{1}{y}$ **25.** $3x + 2$ **27.** $\frac{1}{3}$ **29.** $-\frac{1}{4}$
31. 2   **33.** $\frac{1}{2}$ **35.** $\dfrac{1}{2x^2}$ **37.** 1, 2   **39.** $1 + x$
41. -1 **43.** $3x - 1$ **45.** $16x^{-1} - 5 - x$
47. $4x^{8/3} - 7x^{5/3} + x^{-1/3}$ **49.** $3x^{-1/2} - 5x^{3/2} - x^{7/2}$
51. $\dfrac{-2(21x^2 + 2x - 27)}{(x^2 - 3)^3(6x + 1)^4}$ **53.** $\dfrac{27x^2 - 24x + 2}{(6x + 1)^4}$
55. $\dfrac{-5}{(3x + 2)^{7/4}(2x + 3)^{2/3}}$ **57.** $\dfrac{4x - 3}{(3x - 1)^{4/3}}$

Chapter 1   Review Exercises   (page 63)

1. (a) 11   **(b)** $11, -14$ **(c)** $11, -14, -\frac{8}{9}, \frac{5}{2}, 0.4$
     (d) $\sqrt{6}$
3. $-4 < -3$

5. The set consists of all real numbers less than or equal to 7.

7. -14 **9.** $|x - 7| \geq 4$ **11.** $|y + 30| < 5$
13. -11 **15.** 25   **17.** -144 **19.** $\frac{15,625}{729}$

21. 18   **23.** 9×10^8 **25.** 2.74×10^6
27. 483,300,000   **29. (a)** 11,414.125
(b) 18,380.160   **31.** 280   **33.** $\frac{12}{7}$ **35.** -1
37. $16x^4$ **39.** 3^{16} **41.** 5
43. $\sqrt{10x}$ cannot be simplified.   **45.** $16^{1/2} = 4$
47. $2x^2$ **49.** $2\sqrt{2}$ **51.** $2 + \sqrt{3}$
53. $4x^2 - 12x + 9$ **55.** $x^5 - 2x^4 + x^3 - x^2 + 2x - 1$
57. $y^6 + y^4 - y^3 - y$ **59.** $x(x + 1)(x - 1)$
61. $(x + 10)(2x + 1)$ **63.** $(x - 1)(x^2 + 2)$
65. $9x^2 - 10x + 48$ **67.** -1 **69.** $\dfrac{1}{x^2}$
71. $\dfrac{x(5x - 6)}{5}, x \neq -\frac{3}{2}, 0$ **73.** $\dfrac{x^3 - x + 3}{(x + 2)(x - 1)}$
75. $\dfrac{x + 1}{x(x^2 + 1)}$ **77.** $\dfrac{2x^2 - 3x + 2}{(x - 2)^2(x + 2)}$
79. $-\dfrac{1}{xy(x + y)}, x \neq y$ **81.** $\dfrac{3ax^2}{(a^2 - x)(a - x)}, x \neq 0, a, a^2$

83.

n	1	10	10^2	10^4	10^6	10^{10}
$\dfrac{5}{\sqrt{n}}$	5	1.5811	0.5	0.05	0.005	0.00005

$5/\sqrt{n}$ approaches 0 as n increases without bound.
85. Proof
87. $\pi R^2 - \pi r^2 = \pi(R + r)(R - r)$

CHAPTER 2

Section 2.1   (page 74)

WARM UP   **1.** $-3x - 10$ **2.** $5x - 12$ **3.** x

4. $x + 26$ **5.** $\dfrac{8x}{15}$ **6.** $\dfrac{3x}{4}$ **7.** $-\dfrac{1}{x(x + 1)}$ **8.** $\dfrac{5}{x}$

9. $\dfrac{7x - 8}{x(x - 2)}$ **10.** $-\dfrac{2}{x^2 - 1}$

1. Identity   **3.** Conditional   **5.** Identity
7. Identity   **9.** Conditional
11. (a) No   **(b)** No   **(c)** Yes   **(d)** No
13. (a) Yes   **(b)** Yes   **(c)** No   **(d)** No
15. (a) Yes   **(b)** No   **(c)** No   **(d)** No
17. (a) No   **(b)** No   **(c)** Yes   **(d)** Yes
19. 5   **21.** -4 **23.** 3   **25.** 9   **27.** -26
29. -4 **31.** $-\frac{6}{5}$ **33.** 9   **35.** No solution
37. 10   **39.** 4   **41.** 3   **43.** 5   **45.** No solution
47. $\frac{11}{6}$ **49.** $\frac{5}{3}$ **51.** No solution   **53.** 0
55. All real numbers   **57.** $\dfrac{1}{3 - a}, a \neq 3$
59. $\dfrac{1 + 4b}{2 + a}, a \neq -2$ **61.** $x \approx 138.889$
63. $x \approx 62.372$

65. One equation of this form is $x + 4 = 3x$.

67. (a) 6.46   **(b)** $\frac{1.73}{0.27} \approx 6.41$ **69. (a)** 1.00

(b) $\frac{6.01}{5.98} \approx 1.01$ **71.** 61.2 in.

73. $S = \begin{cases} \frac{1}{2}x + 8000, & 0 \le x \le 16{,}000 \\ x & x > 16{,}000 \end{cases}$ **75.** \$7600

77. 10 in.

Section 2.2   (page 85)

WARM UP   **1.** 14   **2.** 4   **3.** -3 **5.** -2 **7.** $\frac{2}{5}$
9. 6   **10.** $-\frac{11}{5}$

1. $S = n + (n + 1) = 2n + 1$ **3.** $d = 50t$
5. $A = 0.2x$ **7.** $P = 2x + 2(2x) = 6x$
9. $C = 1200 + 25x$ **11.** 262, 263   **13.** 37, 185
15. $-5, -4$ **17.** 13.5   **19.** 1192.5   **21.** 135%
23. 175
25. Income Taxes: \$446 billion; Corporation Taxes: \$103 billion; Social Security Taxes: \$360 billion; Other: \$82 billion
27. \$22,316.98
29. January: \$71,590, February: \$85,908   **31.** 1700%
33. 860%   **35.** $15' \times 22\frac{1}{2}'$ **37.** 97   **39.** 3 hr
41. $\frac{1}{3}$ hr   **43. (a)** ≈ 3.8 hr, 3.2 hr   **(b)** ≈ 1.1 hr
(c) 25.6 mi   **45.** $66\frac{2}{3}$ mph   **47.** 1.29 sec
49. 57.1 ft   **51.** 4.36 ft
53. \$10,000 at 11%; \$15,000 at $12\frac{1}{2}$%
55. First three quarters: 11.5%, Last quarter: 10%
57. ≈ 32.1 gal   **59.** 50 lb of each kind

61. 8064 units   **63.** 6 ft   **65.** $\frac{2A}{b}$ **67.** $\frac{S}{1 + R}$

69. $\frac{A - P}{Pt}$ **71.** $\frac{2A - ah}{h}$ **73.** $\frac{3V + \pi h^3}{3\pi h^2}$

75. $\frac{L - L_0}{L_0 \Delta t}$ **77.** $\frac{Fr^2}{\alpha m_1}$ **79.** $\frac{(n - 1)fR_2}{R_2 + (n - 1)f}$

81. $\frac{L - a + d}{d}$ **83.** $\frac{S - a}{S - L}$

Section 2.3   (page 98)

WARM UP   **1.** $\dfrac{\sqrt{14}}{10}$ **2.** $4\sqrt{2}$ **3.** 14   **4.** $\dfrac{\sqrt{10}}{4}$

5. $x(3x + 7)$ **6.** $(2x + 5)(2x - 5)$
7. $-(x - 7)(x - 15)$ **8.** $(x - 2)(x + 9)$
9. $(5x - 1)(2x + 3)$ **10.** $(6x - 1)(x - 12)$

1. $2x^2 + 5x - 3 = 0$ **3.** $x^2 - 25x = 0$
5. $x^2 - 6x + 7 = 0$ **7.** $2x^2 - 2x + 1 = 0$
9. $3x^2 - 60x - 10 = 0$ **11.** $0, -\frac{1}{2}$ **13.** $4, -2$
15. -5 **17.** $3, -\frac{1}{2}$ **19.** $2, -6$ **21.** $-a$
23. $\pm 4; \pm 4.00$ **25.** $\pm\sqrt{7}; \pm 2.65$

27. $\pm 2\sqrt{3}; \pm 3.46$ **29.** $12 \pm 3\sqrt{2}; 16.24, 7.76$
31. $-2 \pm 2\sqrt{3}; 1.46, -5.46$ **33.** $2; 2.00$ **35.** $0, 2$

37. $4, -8$ **39.** $-3 \pm \sqrt{7}$ **41.** $1 \pm \dfrac{\sqrt{6}}{3}$

43. $2 \pm 2\sqrt{3}$ **45.** ± 8 **47.** $1 \pm \sqrt{2}$ **49.** $\pm\frac{3}{4}$
51. $\frac{3}{2}$ **53.** $6, -12$ **55.** $\frac{3}{2}, -\frac{1}{2}$ **57.** $5, -\frac{10}{3}$

59. $9, 3$ **61.** $\frac{1}{2} \pm \sqrt{3}$ **63.** $\dfrac{3}{5} \pm \dfrac{\sqrt{2}}{2}$ **65.** $-1, -5$

67. $-\frac{1}{2}$ **69.** $\dfrac{1}{(x - 2)^2 - 16}$ **71.** $\dfrac{1}{\sqrt{9 - (x - 3)^2}}$

73. $34' \times 48'$ **75.** $\dfrac{\sqrt{1821}}{4} \approx 10.67$ sec

77. $\dfrac{5\sqrt{2}}{2} \approx 3.54$ cm   **79.** $1500\sqrt{2} \approx 2121.32$ ft

81. 50,000 units   **83.** 1990; yes; yes   **85.** 0, 1

Section 2.4   (page 108)

WARM UP   **1.** $3\sqrt{17}$ **2.** $2\sqrt{3}$ **3.** $4\sqrt{6}$
4. $3\sqrt{73}$ **5.** $2, -1$ **6.** $\frac{3}{2}, -3$ **7.** $5, -1$
8. $\frac{1}{2}, -7$ **9.** $3, 2$ **10.** $4, -1$

1. One real solution   **3.** No real solutions
5. Two real solutions   **7.** $\frac{1}{2}, -1$ **9.** $\frac{1}{4}, -\frac{3}{4}$
11. $1 \pm \sqrt{3}$ **13.** $-7 \pm \sqrt{5}$ **15.** $-4 \pm 2\sqrt{5}$
17. $\dfrac{2}{3} \pm \dfrac{\sqrt{7}}{3}$ **19.** $-\dfrac{1}{3} \pm \dfrac{\sqrt{11}}{6}$ **21.** $-\frac{1}{2} \pm \sqrt{2}$

23. $\frac{2}{7}$ **25.** $-\dfrac{8}{5} \pm \dfrac{\sqrt{3}}{5}$ **27.** $6 \pm \sqrt{11}$

29. $x \approx 0.976, -0.643$ **31.** $x \approx 1.687, -0.488$
33. -11 **35.** $\pm\sqrt{10}$ **37.** $\dfrac{-3 \pm \sqrt{5}}{2}$ **39.** $-2, 4$

41. ± 2 **43.** 50, 50   **45.** 7, 8   **47.** 258 units
49. 653 units
51. $x = 35$ ft, $y = 20$ ft or $x = 15$ ft, $y = \frac{140}{3}$ ft
53. 19.098 ft, 9.5 times   **55.** 6 in. $\times$ 6 in.   **57.** 1990
59. 550 mph, 600 mph   **61.** $30\sqrt{6}$ ft ≈ 73.5 ft

Section 2.5   (page 117)

WARM UP   **1.** $2\sqrt{3}$ **2.** $10\sqrt{5}$ **3.** $\sqrt{5}$
4. $-6\sqrt{3}$ **5.** 12   **6.** 48   **7.** $\dfrac{\sqrt{3}}{3}$ **8.** $\sqrt{2}$

9. $\dfrac{1}{2} \pm \dfrac{\sqrt{5}}{2}$ **10.** $-1 \pm \sqrt{2}$

1. $a = -10, b = 6$ **3.** $a = 6, b = 5$ **5.** $4 + 3i$
7. $2 - 3\sqrt{3}i$ **9.** $5\sqrt{3}i$ **11.** $-1 - 6i$ **13.** 8
15. $0.3i$ **17.** $11 - i$ **19.** 4   **21.** $3 - 3\sqrt{2}i$

23. $-14 + 20i$ **25.** $\frac{1}{6} + \frac{7}{6}i$ **27.** $5 - 3i, 34$
29. $-2 + \sqrt{5}i, 9$ **31.** $-20i, 400$ **33.** $\sqrt{8}, 8$
35. $-2\sqrt{3}$ **37.** -10 **39.** $5 + i$ **41.** $12 + 30i$
43. 24 **45.** $-9 + 40i$ **47.** -10 **49.** $\frac{16}{41} + \frac{20}{41}i$
51. $\frac{3}{5} + \frac{4}{5}i$ **53.** $-7 - 6i$ **55.** $-\frac{9}{1681} + \frac{40}{1681}i$
57. $1 \pm i$ **59.** $-2 \pm \frac{1}{2}i$ **61.** $-\frac{3}{2}, -\frac{5}{2}$

63. $\frac{1}{8} \pm \frac{\sqrt{11}}{8}i$ **65.** $i, -1, -i, 1, i, -1, -i, 1, i,$

$-1, -i, 1, i, -1, -i, 1$ **67.** $-1 + 6i$ **69.** $-5i$
71. $-375\sqrt{3}i$ **73.** i **75.** $8, 8, 8$

Section 2.6 (page 127)

WARM UP **1.** 11 **2.** 20, -3 **3.** 5, -45
4. 0, $-\frac{1}{5}$ **5.** $\frac{2}{3}, -2$ **6.** $\frac{11}{6}, -\frac{5}{2}$ **7.** 1, -5

8. $\frac{3}{2}, -\frac{5}{2}$ **9.** $\frac{3 \pm \sqrt{5}}{2}$ **10.** $2 \pm \sqrt{2}$

1. $0, \pm \frac{3\sqrt{2}}{2}$ **3.** 3, -1, 0 **5.** $\pm 3, \pm 3i$

7. $-3, 0$ **9.** 3, 1, -1 **11.** $\pm 1, \frac{1}{2} \pm \frac{\sqrt{3}}{2}i$

13. $\pm 3, \pm 1$ **15.** $\pm 2, \pm 3i$ **17.** $\pm \frac{1}{2}, \pm 4$

19. $1, -2, 1 \pm \sqrt{3}i, -\frac{1}{2} \pm \frac{\sqrt{3}i}{2}$ **21.** $-\frac{1}{5}, -\frac{1}{3}$

23. $\frac{1}{4}$ **25.** $1, -\frac{125}{8}$ **27.** 50 **29.** 26 **31.** -16
33. 6, 5 **35.** 2, -5 **37.** 0 **39.** 36 **41.** $\frac{101}{4}$
43. 0, 4 **45.** -59, 69 **47.** 78
49. $\pm \sqrt{69}, \pm \sqrt{59}i$ **51.** $1, \frac{2}{5}$ **53.** 4, -5

55. $\frac{-3 \pm \sqrt{21}}{6}$ **57.** $2, -\frac{3}{2}$ **59.** -1 **61.** 1, -3

63. 1, -3 **65.** 3, -2 **67.** $\sqrt{3}, -3$ **69.** 10, -1
71. $x \approx \pm 1.038$ **73.** $x \approx 16.756$
75. The original group consisted of 34 students.
77. 400 mph **79.** 7% **81.** (a) $211.6°$
(b) 24.725 psi **83.** 500 units **85.** 0.26 mi or 1 mi

87. $h = \frac{1}{\pi r}\sqrt{S^2 - \pi^2 r^4}$

89. One correct answer out of many is $a = 4$ and $b = 24$.

Section 2.7 (page 137)

WARM UP **1.** $-\frac{1}{2}$ **2.** $-\frac{1}{6}$ **3.** -3 **4.** $\frac{13}{2}$
5. $x \geq 0$ **6.** $-3 < z < 10$ **7.** $P \leq 2$
8. $W \geq 200$ **9.** 2, 7 **10.** 0, 1

1. $-1 \leq x \leq 3$; Bounded **3.** $10 < x$; Unbounded
5. (a) Yes **(b)** No **(c)** Yes **(d)** No
7. (a) Yes **(b)** No **(c)** No **(d)** Yes
9. c **11.** f **13.** g **15.** b

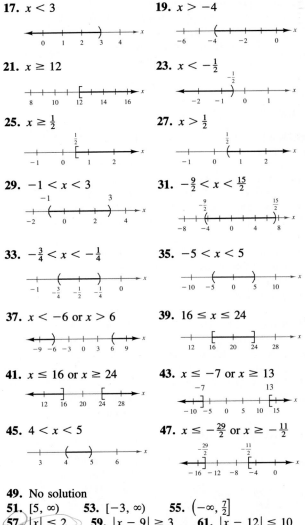

17. $x < 3$ **19.** $x > -4$

21. $x \geq 12$ **23.** $x < -\frac{1}{2}$

25. $x \geq \frac{1}{2}$ **27.** $x > \frac{1}{2}$

29. $-1 < x < 3$ **31.** $-\frac{9}{2} < x < \frac{15}{2}$

33. $-\frac{3}{4} < x < -\frac{1}{4}$ **35.** $-5 < x < 5$

37. $x < -6$ or $x > 6$ **39.** $16 \leq x \leq 24$

41. $x \leq 16$ or $x \geq 24$ **43.** $x \leq -7$ or $x \geq 13$

45. $4 < x < 5$ **47.** $x \leq -\frac{29}{2}$ or $x \geq -\frac{11}{2}$

49. No solution
51. $[5, \infty)$ **53.** $[-3, \infty)$ **55.** $\left(-\infty, \frac{7}{2}\right]$
57. $|x| \leq 2$ **59.** $|x - 9| \geq 3$ **61.** $|x - 12| \leq 10$
63. $|x + 3| > 5$ **65.** More than 400 miles
67. $r > 12.5\%$ **69.** $x \geq 36$ units
71. $133 \approx \frac{400}{3} \leq x \leq \frac{700}{3} \approx 233$
73. $106.864 \leq \text{area} \leq 109.464$ **75.** $65.8 \leq h \leq 71.2$

Section 2.8 (page 148)

WARM UP **1.** $y < -6$ **2.** $z > -\frac{9}{2}$
3. $-3 \leq x < 1$ **4.** $x \leq -5$ **5.** $-3 < x$
6. $5 < x < 7$ **7.** $-\frac{7}{2} \leq x \leq \frac{7}{2}$ **8.** $x < 2, x > 4$
9. $x < -6, x > -2$ **10.** $-2 \leq x \leq 6$

1. $[-3, 3]$ **3.** $(-\infty, -2) \cup (2, \infty)$ **5.** $(-7, 3)$
7. $(-\infty, -5] \cup [1, \infty)$ **9.** $(-3, 2)$
11. $(-\infty, -1) \cup (1, \infty)$ **13.** $(-3, 1)$

15. $(-\infty, 0) \cup \left(0, \frac{3}{2}\right)$ **17.** $[-2, 0] \cup [2, \infty)$
19. $[-2, \infty)$ **21.** $(-\infty, -1) \cup (0, 1)$
23. $(-\infty, -1) \cup (4, \infty)$ **25.** $(5, 15)$
27. $\left(-5, -\frac{3}{2}\right) \cup (-1, \infty)$ **29.** $\left(-\frac{3}{4}, 3\right) \cup [6, \infty)$
31. $(-3, -2) \cup [0, 3)$
33. $(-\infty, -2) \cup (-1, 1) \cup (3, \infty)$ **35.** $[-2, 2]$
37. $(-\infty, 3] \cup [4, \infty)$ **39.** $[-4, 3]$
41. $(-0.13, 25.13)$ **43. (a)** 10 sec
(b) 4 sec $< t <$ 6 sec
45. Between 13.8 meters and 36.2 meters   **47.** $R_1 \geq 2$
49. 1993

Chapter 2   Review Exercises   *(page 150)*

1. Identity   **3. (a)** No   **(b)** Yes   **(c)** Yes
(d) No   **5.** 20   **7.** $-\frac{1}{2}$ **9.** $\frac{1}{5}$ **11.** 0, 2
13. $\frac{4}{3}, -\frac{1}{2}$ **15.** $-4 \pm 3\sqrt{2}$ **17.** $6 \pm \sqrt{6}$

19. $0, \frac{12}{5}$ **21.** 0, 1, 2   **23.** $\pm\dfrac{\sqrt{2}}{2}$ **25.** 2, 6

27. $\pm\dfrac{\sqrt{2}}{2}$ **29.** 5   **31.** $\frac{25}{4}$ **33.** No solution

35. $-124, 126$ **37.** $-4, -2 \pm \dfrac{\sqrt{95}}{5}$ **39.** $-5, 15$

41. 1, 3   **43.** $r = \sqrt{\dfrac{3V}{\pi h}}$ **45.** $p = \dfrac{k}{3\pi r^2 L}$

47. $\left(-\frac{5}{3}, \infty\right)$ **49.** $[-2, 2]$ **51.** $(-\infty, 3), (5, \infty)$
53. $(1, 3)$ **55.** $(-\infty, 0], [3, \infty)$ **57.** $[5, \infty)$
59. $3 + 7i$ **61.** $-\sqrt{2}i$ **63.** $40 + 65i$
65. $-4 - 46i$ **67.** $1 - 6i$ **69.** $\frac{4}{3}i$
71. September: \$325,000; October: \$364,000
73. 2.857 qt   **75.** $1000\sqrt{2} \approx 1{,}414.21$ ft
77. 4 farmers   **79.** 143,203 units   **81.** $x \geq 36$

CHAPTER 3

Section 3.1   *(page 162)*

WARM UP   **1.** 5   **2.** $3\sqrt{2}$ **3.** 1   **4.** -2
5. $3(\sqrt{2} + \sqrt{5})$ **6.** $2(\sqrt{3} + \sqrt{11})$ **7.** $-3, 11$
8. 9, 1   **9.** 11   **10.** 4

1.
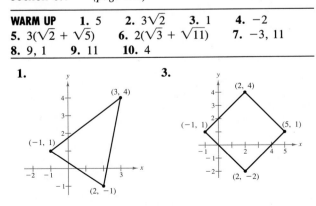

3.

5. $(-1, 1), (3, 2), (0, 4)$ **7.** 8   **9.** 5
11. (a) $a = 4, b = 3, c = 5$ **(b)** 5
13. (a) $a = 10, b = 3, c = \sqrt{109}$ **(b)** $\sqrt{109}$
15. (a)   **17. (a)**

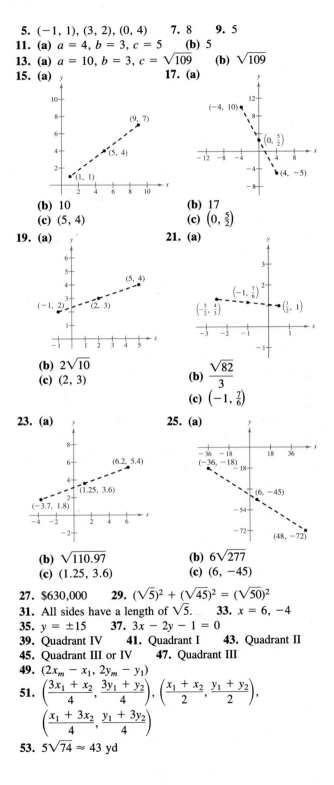

(b) 10   **(b)** 17
(c) $(5, 4)$ **(c)** $\left(0, \frac{5}{2}\right)$

19. (a)   **21. (a)**

(b) $2\sqrt{10}$ **(b)** $\dfrac{\sqrt{82}}{3}$
(c) $(2, 3)$
    (c) $\left(-1, \frac{7}{6}\right)$

23. (a)   **25. (a)**

(b) $\sqrt{110.97}$ **(b)** $6\sqrt{277}$
(c) $(1.25, 3.6)$ **(c)** $(6, -45)$

27. \$630,000   **29.** $(\sqrt{5})^2 + (\sqrt{45})^2 = (\sqrt{50})^2$
31. All sides have a length of $\sqrt{5}$.   **33.** $x = 6, -4$
35. $y = \pm 15$ **37.** $3x - 2y - 1 = 0$
39. Quadrant IV   **41.** Quadrant I   **43.** Quadrant II
45. Quadrant III or IV   **47.** Quadrant III
49. $(2x_m - x_1, 2y_m - y_1)$
51. $\left(\dfrac{3x_1 + x_2}{4}, \dfrac{3y_1 + y_2}{4}\right), \left(\dfrac{x_1 + x_2}{2}, \dfrac{y_1 + y_2}{2}\right),$
$\left(\dfrac{x_1 + 3x_2}{4}, \dfrac{y_1 + 3y_2}{4}\right)$
53. $5\sqrt{74} \approx 43$ yd

55.

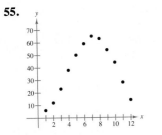

57. $14 per 100 lb of milk in 1989   **59.** $\approx 1782\%$

61.

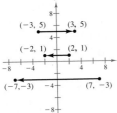

The points are reflected through the y-axis.

Section 3.2   *(page 173)*

WARM UP   **1.** $y = \dfrac{3x - 2}{5}$ **2.** $y = -\dfrac{(x - 5)(x + 1)}{2}$

3. 2   **5.** 0, ± 3 **4.** 1, -5 **5.** 0, ± 3 **6.** ± 2
7. $y = x^3 + 4x$ **8.** $x^2 + y^2 = 4$ **9.** $y = 4x^2 + 8$
10. $y^2 = -3x + 4$

1. (a) Yes   **(b)** Yes   **3. (a)** No   **(b)** Yes
5. (a) Yes   **(b)** Yes   **7.** 2   **9.** 4

11.

x	-4	-2	0	2	4
y	11	7	3	-1	-5
(x, y)	$(-4, 11)$	$(-2, 7)$	$(0, 3)$	$(2, -1)$	$(4, -5)$

13. $(5, 0), (0, -5)$ **15.** $(-2, 0), (1, 0), (0, -2)$
17. $(0, 0), (-2, 0)$ **19.** $(1, 0), \left(0, \frac{1}{2}\right)$
21. y-axis symmetry   **23.** x-axis symmetry
25. Origin symmetry   **27.** Origin symmetry

29.

31.

33. c   **35.** d   **37.** e
39. Intercepts: $\left(\frac{2}{3}, 0\right), (0, 2)$
No symmetry

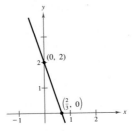

41. Intercepts: $(-1, 0), (1, 0), (0, 1)$
Symmetry: y-axis

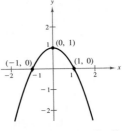

43. Intercepts: $(3, 0), (1, 0), (0, 3)$
No symmetry

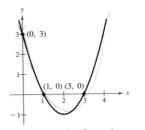

45. Intercepts: $\left(-\sqrt[3]{2}, 0\right), (0, 2)$
No symmetry

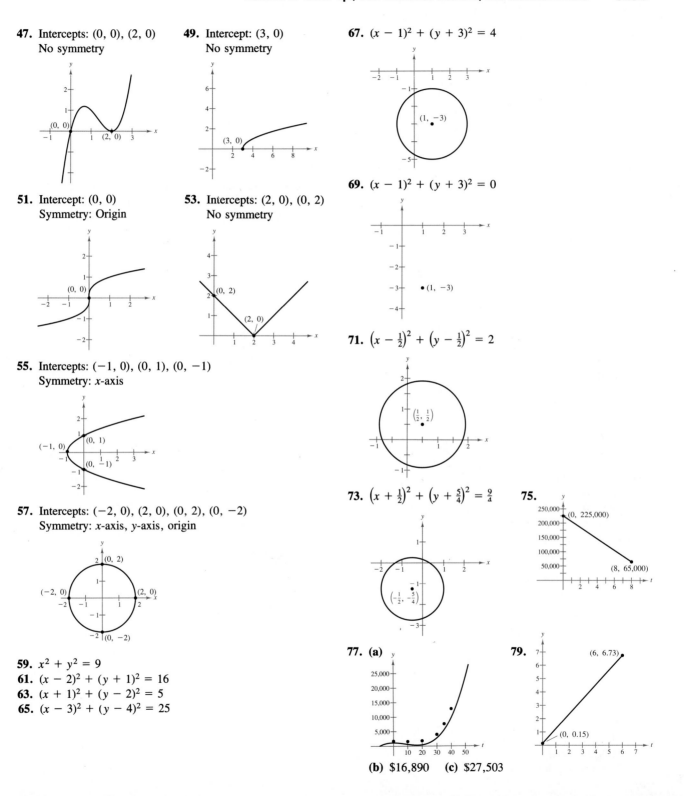

47. Intercepts: (0, 0), (2, 0)
No symmetry

49. Intercept: (3, 0)
No symmetry

67. $(x - 1)^2 + (y + 3)^2 = 4$

51. Intercept: (0, 0)
Symmetry: Origin

53. Intercepts: (2, 0), (0, 2)
No symmetry

69. $(x - 1)^2 + (y + 3)^2 = 0$

55. Intercepts: (−1, 0), (0, 1), (0, −1)
Symmetry: x-axis

71. $\left(x - \frac{1}{2}\right)^2 + \left(y - \frac{1}{2}\right)^2 = 2$

57. Intercepts: (−2, 0), (2, 0), (0, 2), (0, −2)
Symmetry: x-axis, y-axis, origin

73. $\left(x + \frac{1}{2}\right)^2 + \left(y + \frac{5}{4}\right)^2 = \frac{9}{4}$

75.

59. $x^2 + y^2 = 9$
61. $(x - 2)^2 + (y + 1)^2 = 16$
63. $(x + 1)^2 + (y - 2)^2 = 5$
65. $(x - 3)^2 + (y - 4)^2 = 25$

77. (a)

79.

(b) $16,890   **(c)** $27,503

81. There is an unlimited number of correct answers, one of which is $a = 1$ and $b = -5$.

Section 3.3 *(page 185)*

WARM UP **1.** $-\frac{9}{2}$ **2.** $-\frac{13}{3}$ **3.** $-\frac{5}{4}$ **4.** $\frac{1}{2}$
5. $y = \frac{2}{3}x - \frac{5}{3}$ **6.** $y = -2x$ **7.** $y = 3x - 1$
8. $y = \frac{2}{3}x + 5$ **9.** $y = -2x + 7$ **10.** $y = x + 3$

1. $\frac{6}{5}$ **3.** 0 **5.** -3

7.

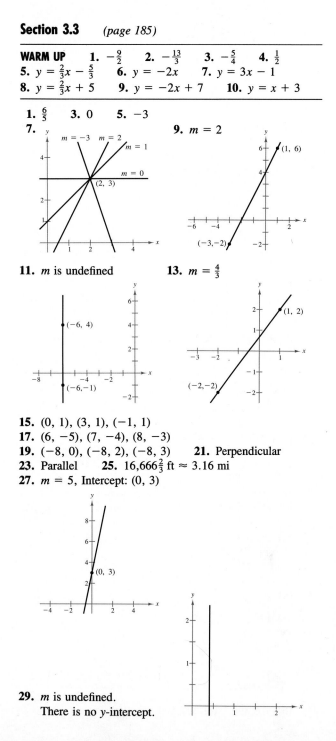

9. $m = 2$

11. m is undefined

13. $m = \frac{4}{3}$

15. $(0, 1)$, $(3, 1)$, $(-1, 1)$
17. $(6, -5)$, $(7, -4)$, $(8, -3)$
19. $(-8, 0)$, $(-8, 2)$, $(-8, 3)$ **21.** Perpendicular
23. Parallel **25.** $16{,}666\frac{2}{3}$ ft ≈ 3.16 mi
27. $m = 5$, Intercept: $(0, 3)$

29. m is undefined.
There is no y-intercept.

31. $m = -\frac{7}{6}$, Intercept: $(0, 5)$

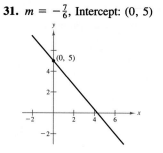

33. $3x + 5y - 10 = 0$ **35.** $x + 2y - 3 = 0$
37. $x + 8 = 0$ **39.** $2x - 5y + 1 = 0$
41. $3x - y - 2 = 0$

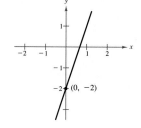

43. $2x + y = 0$

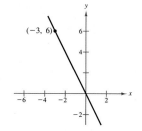

45. $x + 3y - 4 = 0$

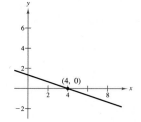

47. $x - 6 = 0$

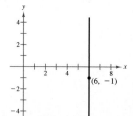

49. $2y - 5 = 0$

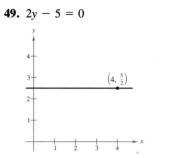

51. $3x + 2y - 6 = 0$ **53.** $12x + 3y + 2 = 0$
55. $x + y - 3 = 0$ **57.** (a) $2x - y - 3 = 0$
(b) $x + 2y - 4 = 0$ **59.** (a) $3x + 4y + 2 = 0$
(b) $4x - 3y + 36 = 0$ **61.** (a) $y = 0$
(b) $x + 1 = 0$ **63.** $V = 125t + 2,540$
65. $V = 2,000t + 20,400$ **67.** b   **69.** a
71. $F = \frac{9}{5}C + 32$ **73.** $39,500
75. $V = -175t + 875$ **77.** $S = 0.85L$
79. $W = 0.07S + 2500$ **81.** (a) $C = 16.75t + 36,500$
(b) $R = 27t$ (c) $P = 10.25t - 36,500$
(d) $t \approx 3,561$ hr   **83.** $y = 100 + 30t$

Section 3.4   *(page 197)*

WARM UP   **1.** -73 **2.** 13   **3.** $2(x + 2)$
4. $-8(x - 2)$ **5.** $y = \frac{7}{5} - \frac{2}{5}x$ **6.** $y = \pm x$
7. $x \le -2, x \ge 2$ **8.** $-3 \le x \le 3$
9. All real numbers   **10.** $x \le 1, x \ge 2$

1. (a) Function
(b) Not a function, since the element 1 in A is matched
with two elements, -2 and 1, in B
(c) Function
(d) Not a function, since not all elements of A are
matched with an element in B
3. Not a function   **5.** Function   **7.** Function
9. Not a function
11. (a) $\dfrac{1}{4 + 1}$ (b) $\dfrac{1}{0 + 1}$ (c) $\dfrac{1}{4x + 1}$
(d) $\dfrac{1}{(x + h) + 1}$
13. (a) -1 (b) -9 (c) $2x - 5$
15. (a) 0   (b) -0.75 (c) $x^2 + 2x$
17. (a) 1   (b) 2.5   (c) $3 - 2|x|$
19. (a) $-\frac{1}{9}$ (b) Undefined   (c) $\dfrac{1}{y^2 + 6y}$
21. (a) 1   (b) -1 (c) $\dfrac{|x - 1|}{x - 1}$
23. (a) -1 (b) 2   (c) 6   **25.** 5   **27.** ± 3
29. All real numbers x

31. All real numbers except $t = 0$
33. $y \ge 10$ **35.** $-1 \le x \le 1$
37. All real numbers except $x = 0, -2$
39. $(-2, 4), (-1, 1), (0, 0), (1, 1), (2, 4)$
41. $(-2, 0), (-1, 1), (0, \sqrt{2}), (1, \sqrt{3}), (2, 2)$
43. $2, -1$ **45.** $3, 0$ **47.** $3 + h$
49. $3x\Delta x + 3x^2 + (\Delta x)^2$ **51.** 3   **53.** $A = \dfrac{C^2}{4\pi}$
55. $A = \dfrac{x^2}{x - 1}, x > 1$
57. $V = 4x(6 - x)^2, 0 < x < 6$
59. $h = \sqrt{d^2 - 2000^2}, d \ge 2000$
61. (a) $C = 12.30x + 98,000$ (b) $R = 17.98x$
(c) $P = 5.68x - 98,000$
63. (a) $R = \dfrac{240n - n^2}{20}$

(b)

n	90	100	110	120	130	140	150
$R(n)$	$675	$700	$715	$720	$715	$700	$675

Section 3.5   *(page 213)*

WARM UP   **1.** 2   **2.** 0   **3.** $-\dfrac{3}{x}$ **4.** $x^2 + 3$
5. $0, \pm 4$ **6.** $\frac{1}{2}, 1$ **7.** All real numbers except $x = 4$
8. All real numbers except $x = 4, 5$ **9.** $t \le \frac{5}{3}$
10. All real numbers

1. Domain: $[1, \infty)$, Range: $[0, \infty)$
3. Domain: $(-\infty, -2], [2, \infty)$, Range: $[0, \infty)$
5. Domain: $[-5, 5]$, Range: $[0, 5]$
7. Function   **9.** Not a function   **11.** Function
13. (a) Increasing on $(-\infty, \infty)$ (b) Odd function
15. (a) Increasing on $(-\infty, 0), (2, \infty)$, Decreasing on $(0, 2)$
(b) Neither even nor odd
17. (a) Increasing on $(-1, 0), (1, \infty)$,
Decreasing on $(-\infty, -1), (0, 1)$
(b) Even function
19. (a) Increasing on $(-2, \infty)$, Decreasing on $(-3, -2)$
(b) Neither even nor odd
21. Even   **23.** Odd   **25.** Neither even nor odd
27. Even   **29.** Neither even nor odd

31. Odd

33. Neither even nor odd

35. Neither even nor odd

37. Neither even nor odd

39. Neither even nor odd

41. $(-\infty, 4]$

43. $(-\infty, -3], [3, \infty)$

45. $[-1, 1]$

47. $(-\infty, \infty)$

49. $f(x) < 0$ for all x

51. (a) (b) (c)

53. (a) (b) (c) (d) (e) (f)

55. (a) $g(x) = (x - 1)^2 + 1$

(b) $g(x) = -(x + 1)^2$

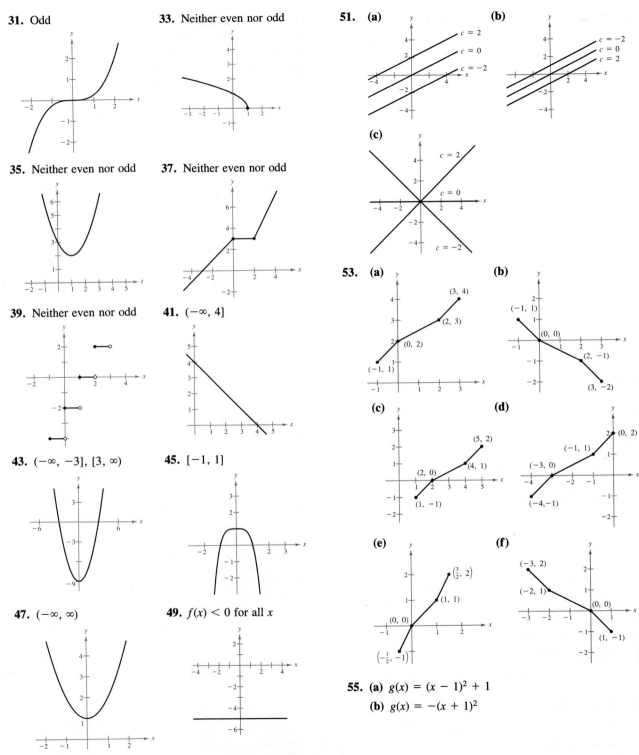

57. $C = 0.65 + 0.42[\![t]\!]$ **59.** 350,000 units

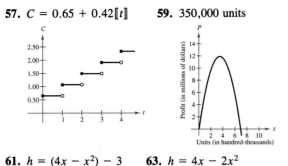

61. $h = (4x - x^2) - 3$ **63.** $h = 4x - 2x^2$
65. $L = (4 - y^2) - (y + 2)$
67. $f(-x) = a_{2n+1}(-x)^{2n+1} + a_{2n-1}(-x)^{2n-1} + \cdots +$
$$a_3(-x)^3 + a_1(-x)$$
$$= -(a_{2n+1}x^{2n+1} + a_{2n-1}x^{2n-1} + \cdots + a_3x^3 +$$
$$a_1x)$$
$$= -f(x)$$

69.

Interval	Intake Pipe	Drain Pipe 1	Drain Pipe 2
[0, 5]	Open	Closed	Closed
[5, 10]	Open	Open	Closed
[10, 20]	Closed	Closed	Closed
[20, 30]	Closed	Closed	Open
[30, 40]	Open	Open	Open
[40, 45]	Open	Closed	Open
[45, 50]	Open	Open	Open
[50, 60]	Open	Open	Closed

Section 3.6   (page 222)

WARM UP   **1.** $\dfrac{1}{x(1 - x)}$ **2.** $-\dfrac{12}{(x + 3)(x - 3)}$

3. $\dfrac{3x - 2}{x(x - 2)}$ **4.** $\dfrac{4x - 5}{3(x - 5)}$ **5.** $\sqrt{\dfrac{x - 1}{x + 1}}$

6. $\dfrac{x + 1}{x(x + 2)}$ **7.** $5(x - 2)$ **8.** $\dfrac{x + 1}{(x - 2)(x + 3)}$

9. $\dfrac{1 + 5x}{3x - 1}$ **10.** $\dfrac{x + 4}{4x}$

1. (a) $2x$ **(b)** 2 **(c)** $x^2 - 1$ **(d)** $\dfrac{x + 1}{x - 1}, x \neq 1$

3. (a) $x^2 - x + 1$ **(b)** $x^2 + x - 1$ **(c)** $x^2 - x^3$
(d) $\dfrac{x^2}{1 - x}, x \neq 1$

5. (a) $x^2 + 5 + \sqrt{1 - x}$ **(b)** $x^2 + 5 - \sqrt{1 - x}$
(c) $(x^2 + 5)\sqrt{1 - x}$ **(d)** $\dfrac{x^2 + 5}{\sqrt{1 - x}}, x < 1$

7. (a) $\dfrac{x + 1}{x^2}$ **(b)** $\dfrac{x - 1}{x^2}$ **(c)** $\dfrac{1}{x^3}$ **(d)** $x, x \neq 0$

9. 9   **11.** 5   **13.** $4t^2 - 2t + 5$ **15.** 0   **17.** 26
19. $\frac{3}{5}$ **21. (a)** $(x - 1)^2$ **(b)** $x^2 - 1$ **(c)** x^4
23. (a) $20 - 3x$ **(b)** $-3x$ **(c)** $9x + 20$
25. (a) $\sqrt{x^2 + 4}$ **(b)** $x + 4$
27. (a) $x - \frac{8}{3}$ **(b)** $x - 8$ **29. (a)** $\sqrt[4]{x}$ **(b)** $\sqrt[4]{x}$
31. (a) $|x + 6|$ **(b)** $|x| + 6$
33. (a) 3   **(b)** 0   **35. (a)** 0   **(b)** 4
37. $f(x) = x^2, g(x) = 2x + 1$
39. $f(x) = \sqrt[3]{x}, g(x) = x^2 - 4$
41. $f(x) = \dfrac{1}{x}, g(x) = x + 2$
43. $f(x) = x^2 + 2x, g(x) = x + 4$
45. (a) $x \geq 0$ **(b)** All real numbers
(c) All real numbers
47. (a) All real numbers except $x = \pm 1$
(b) All real numbers
(c) All real numbers except $x = -2, 0$
49. $T = \frac{3}{4}x + \frac{1}{15}x^2$

51. $(A \circ r)(t) = 0.36\pi t^2$
$A \circ r$ represents the area of the circle at time t.
53. $(C \circ x)(t) = 3000t + 750$
$C \circ x$ represents the cost after t production hours.

Section 3.7   (page 232)

WARM UP   **1.** All real numbers   **2.** $[-1, \infty)$
3. All real numbers except $x = 0, 2$
4. All real numbers except $x = -\frac{5}{3}$ **5.** x **6.** x

7. x **8.** x **9.** $x = \frac{3}{2}y + 3$ **10.** $x = \dfrac{y^3}{2} + 2$

1. $f^{-1}(x) = \frac{1}{8}x$ **3.** $f^{-1}(x) = x - 10$
5. $f^{-1}(x) = x^3$

7. (a) $f(g(x)) = f\left(\dfrac{x}{2}\right) = 2\left(\dfrac{x}{2}\right) = x$ **(b)**
$$g(f(x)) = g(2x) = \dfrac{(2x)}{2} = x$$

9. (a) $f(g(x)) = f\left(\dfrac{x-1}{5}\right) = 5\left(\dfrac{x-1}{5}\right) + 1 = x$

$g(f(x)) = g(5x+1) = \dfrac{(5x+1) - 1}{5} = x$

(b)

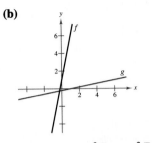

11. (a) $f(g(x)) = f(\sqrt[3]{x}) = (\sqrt[3]{x})^3 = x$
$g(f(x)) = g(x^3) = \sqrt[3]{x^3} = x$

(b)

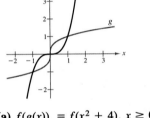

13. (a) $f(g(x)) = f(x^2 + 4), \ x \geq 0$
$\qquad = \sqrt{(x^2 + 4) - 4} = x$
$g(f(x)) = g(\sqrt{x-4})$
$\qquad = (\sqrt{x-4})^2 + 4 = x$

(b)

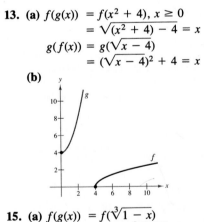

15. (a) $f(g(x)) = f(\sqrt[3]{1-x})$
$\qquad = 1 - (\sqrt[3]{1-x})^3 = x$
$g(f(x)) = g(1 - x^3)$
$\qquad = \sqrt[3]{1 - (1 - x^3)} = x$

(b)

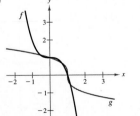

17. One-to-one   **19.** Not one-to-one   **21.** One-to-one
23. Not one-to-one   **25.** Not one-to-one

27. $f^{-1}(x) = \dfrac{x+3}{2}$ **29.** $f^{-1}(x) = \sqrt[5]{x}$

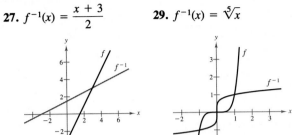

31. $f^{-1}(x) = x^2, \ x \geq 0$

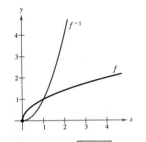

33. $f^{-1}(x) = \sqrt{4 - x^2}, \ 0 \leq x \leq 2$

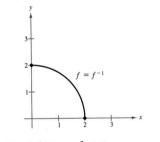

35. $f^{-1}(x) = x^3 + 1$

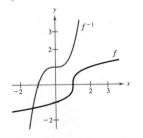

37. Not one-to-one   **39.** $g^{-1}(x) = 8x$
41. Not one-to-one   **43.** $f^{-1}(x) = \sqrt{x} - 3, \ x \geq 0$
45. $h^{-1}(x) = \dfrac{1}{x}$ **47.** $f^{-1}(x) = \dfrac{x^2 - 3}{2}, \ x \geq 0$
49. Not one-to-one   **51.** $f^{-1}(x) = -\sqrt{25 - x}, \ x \leq 25$

53. $f^{-1}(x) = \sqrt{x} + 3, x \geq 0$ **55.** $f^{-1}(x) = x - 3, x \geq 0$

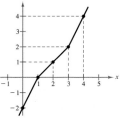

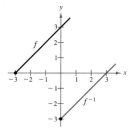

57.

x	0	1	2	3	4
$f^{-1}(x)$	-2	0	1	2	4

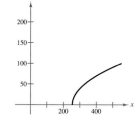

59. 32   **61.** 600   **63.** $2\sqrt[3]{x} + 3$

65. (a) $y = \dfrac{5\sqrt{6}}{3}\sqrt{2x - 509}$

y: Percentage load; x: Exhaust temperature

(b)

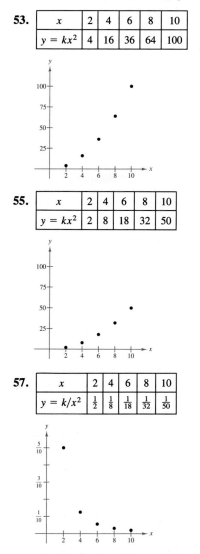

67. False   **69.** True

Section 3.8   (page 240)

WARM UP   **1.** $\frac{1}{3}$ **2.** $\frac{9}{16}$ **3.** $\frac{128}{3}$ **4.** 75   **5.** $\frac{275}{27}$
6. $\frac{105}{128}$ **7.** 27   **8.** $\frac{2}{9}$ **9.** $\frac{28}{13}$ **10.** 157.5

1. $A = kr^2$ **3.** $y = \dfrac{k}{x^2}$ **5.** $z = k\sqrt[3]{u}$ **7.** $z = kuv$

9. $F = \dfrac{kg}{r^2}$ **11.** $P = \dfrac{k}{V}$ **13.** $F = \dfrac{km_1m_2}{r^2}$

15. The area of a triangle is jointly proportional to the magnitude of the base and the height.

17. The volume of a sphere varies directly as the cube of its radius.

19. Average speed is directly proportional to the distance and inversely proportional to the time.

21. $y = \frac{5}{2}x$ **23.** $A = \pi r^2$ **25.** $y = \dfrac{75}{x}$

27. $h = \dfrac{12}{t^3}$ **29.** $z = 2xy$ **31.** $F = 14rs^3$

33. $z = \dfrac{2x^2}{3y}$ **35.** $S = \dfrac{4L}{3(L - S)}$ **37. (a)** 2 in.

(b) 15 lb   **39.** 39.47 lb   **41.** 0.61 mph   **43.** 506 ft
45. 400 ft   **47.** No. The largest size is the best buy.
49. The illumination is one-fourth the original.
51. The velocity is increased by $\frac{1}{3}$.

53.

x	2	4	6	8	10
$y = kx^2$	4	16	36	64	100

55.

x	2	4	6	8	10
$y = kx^2$	2	8	18	32	50

57.

x	2	4	6	8	10
$y = k/x^2$	$\frac{1}{2}$	$\frac{1}{8}$	$\frac{1}{18}$	$\frac{1}{32}$	$\frac{1}{50}$

59.

x	2	4	6	8	10
$y = k/x^2$	$\frac{5}{2}$	$\frac{5}{8}$	$\frac{5}{18}$	$\frac{5}{32}$	$\frac{1}{10}$

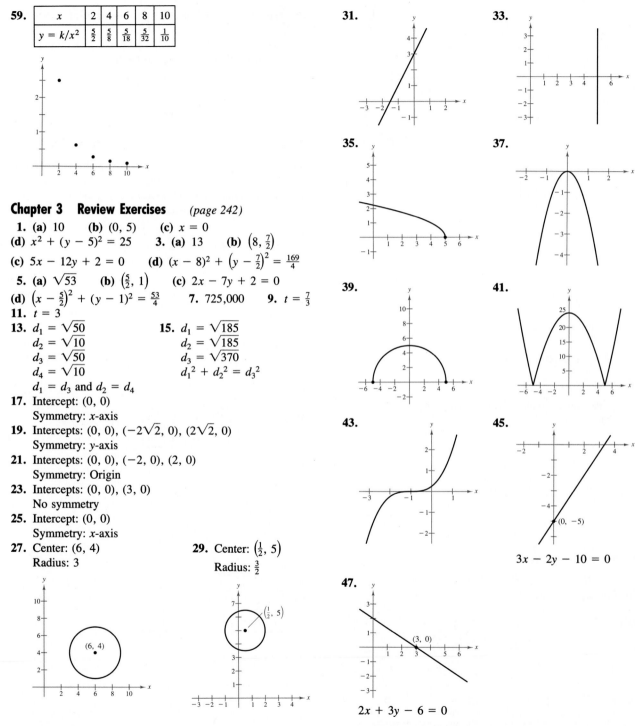

Chapter 3 Review Exercises *(page 242)*

1. (a) 10 **(b)** $(0, 5)$ **(c)** $x = 0$
(d) $x^2 + (y - 5)^2 = 25$ **3. (a)** 13 **(b)** $\left(8, \frac{7}{2}\right)$
(c) $5x - 12y + 2 = 0$ **(d)** $(x - 8)^2 + \left(y - \frac{7}{2}\right)^2 = \frac{169}{4}$
5. (a) $\sqrt{53}$ **(b)** $\left(\frac{5}{2}, 1\right)$ **(c)** $2x - 7y + 2 = 0$
(d) $\left(x - \frac{5}{2}\right)^2 + (y - 1)^2 = \frac{53}{4}$ **7.** 725,000 **9.** $t = \frac{7}{3}$
11. $t = 3$
13. $d_1 = \sqrt{50}$ **15.** $d_1 = \sqrt{185}$
    $d_2 = \sqrt{10}$ $d_2 = \sqrt{185}$
    $d_3 = \sqrt{50}$ $d_3 = \sqrt{370}$
    $d_4 = \sqrt{10}$ $d_1^2 + d_2^2 = d_3^2$
    $d_1 = d_3$ and $d_2 = d_4$
17. Intercept: $(0, 0)$
  Symmetry: x-axis
19. Intercepts: $(0, 0)$, $(-2\sqrt{2}, 0)$, $(2\sqrt{2}, 0)$
  Symmetry: y-axis
21. Intercepts: $(0, 0)$, $(-2, 0)$, $(2, 0)$
  Symmetry: Origin
23. Intercepts: $(0, 0)$, $(3, 0)$
  No symmetry
25. Intercept: $(0, 0)$
  Symmetry: x-axis
27. Center: $(6, 4)$ **29.** Center: $\left(\frac{1}{2}, 5\right)$
  Radius: 3   Radius: $\frac{3}{2}$

31.

33.

35.

37.

39.

41.

43.

45.

$3x - 2y - 10 = 0$

47.

$2x + 3y - 6 = 0$

49. (a) $5x - 4y - 23 = 0$ **(b)** $4x + 5y - 2 = 0$

51. $210,000 **53. (a)** 5 **(b)** 17 **(c)** $t^4 + 1$
(d) $-x^2 - 1$ **55. (a)** -14 **(b)** $-5x^2 - 30x - 39$
(c) -30 **(d)** $-10x - 5\Delta x$ **57.** $[-5, 5]$
59. All real numbers except $s = 3$
61. All real numbers except $x = -2, 3$
63. (a) $f^{-1}(x) = 2x + 6$

(b)
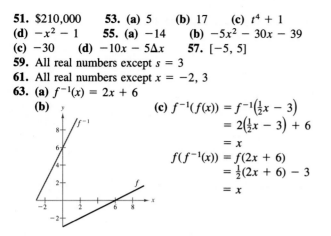

(c) $f^{-1}(f(x)) = f^{-1}\left(\frac{1}{2}x - 3\right)$
$= 2\left(\frac{1}{2}x - 3\right) + 6$
$= x$
$f(f^{-1}(x)) = f(2x + 6)$
$= \frac{1}{2}(2x + 6) - 3$
$= x$

65. (a) $f^{-1}(x) = x^2 - 1, x \geq 0$

(b)
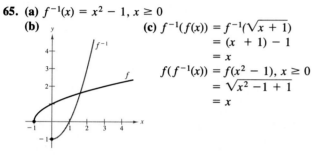

(c) $f^{-1}(f(x)) = f^{-1}(\sqrt{x + 1})$
$= (x + 1) - 1$
$= x$
$f(f^{-1}(x)) = f(x^2 - 1), x \geq 0$
$= \sqrt{x^2 - 1 + 1}$
$= x$

67. (a) $f^{-1}(x) = \sqrt{x + 5}, x \geq -5$

(b)

(c) $f^{-1}(f(x)) = f^{-1}(x^2 - 5), x \geq 0$
$= \sqrt{x^2 - 5 + 5}$
$= x$
$f(f^{-1}(x)) = f(\sqrt{x + 5})$
$= (x + 5) - 5$
$= x$

69. $x \geq 4, f^{-1}(x) = \sqrt{\frac{x}{2}} + 4, x \geq 0$

71. $x \geq 2, f^{-1}(x) = \sqrt{x^2 + 4}, x \geq 0$ **73.** -7
75. 5 **77.** 23 **79.** 9 **81. (a)** 16 ft/sec
(b) 1.5 sec **(c)** -16 ft/sec
83. $A = x(12 - x), (0, 6]$

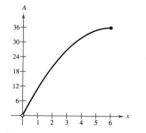

85. $F = \frac{1}{3}x\sqrt{y}$ **87.** $z = \dfrac{32x^2}{25y}$ **89.** 2,438.7 kilowatts

CUMULATIVE TEST FOR CHAPTERS 1–3 *(page 246)*

1. $\frac{2}{45}$ **3.** $\dfrac{4x^3}{15y^5}$ **5.** $x^3 - x^2 - 5x + 6$

7. $\dfrac{s - 1}{(s + 3)(s + 1)}$ **9.** $\frac{2}{3}$ **11.** $-2 \pm i$

13. $-1 < x < 5$ **15.**

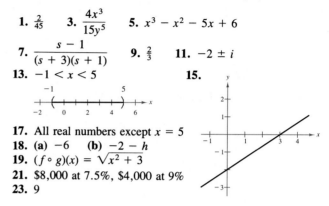

17. All real numbers except $x = 5$
18. (a) -6 **(b)** $-2 - h$
19. $(f \circ g)(x) = \sqrt{x^2 + 3}$
21. $8,000 at 7.5%, $4,000 at 9%
23. 9

CHAPTER 4

Section 4.1 *(page 254)*

WARM UP **1.** $\frac{1}{2}, -6$ **2.** $-\frac{3}{5}, 3$ **3.** $\frac{3}{2}, -1$
4. -10 **5.** $3 \pm \sqrt{5}$ **6.** $-2 \pm \sqrt{3}$ **7.** $4 \pm \dfrac{\sqrt{14}}{2}$
8. $-5 \pm \dfrac{\sqrt{3}}{3}$ **9.** $-\dfrac{3}{2} \pm \dfrac{\sqrt{3}}{2}i$ **10.** $-\dfrac{3}{2} \pm \dfrac{\sqrt{21}}{2}$

1. f **3.** c **5.** b **7.** $f(x) = (x - 2)^2$
9. $f(x) = -(x + 2)^2 + 4$ **11.** $f(x) = -2(x + 3)^2 + 3$
13. Vertex: $(0, -5)$, Intercepts: $(\pm\sqrt{5}, 0), (0, -5)$

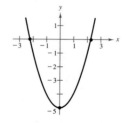

15. Vertex: $(0, 16)$, Intercepts: $(\pm4, 0), (0, 16)$

17. Vertex: $(-5, -6)$, Intercepts: $(-5 \pm \sqrt{6}, 0)$, $(0, 19)$

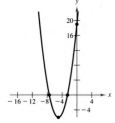

19. Vertex: $(4, 0)$, Intercepts: $(4, 0)$, $(0, 16)$

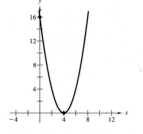

21. Vertex: $(-1, 4)$, Intercepts: $(1, 0)$, $(-3, 0)$, $(0, 3)$

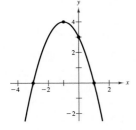

23. Vertex: $\left(\frac{1}{2}, 1\right)$, Intercept: $\left(0, \frac{5}{4}\right)$

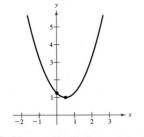

25. Vertex: $(1, 6)$, Intercepts: $(1 \pm \sqrt{6}, 0)$, $(0, 5)$

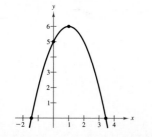

27. Vertex: $\left(\frac{1}{2}, 20\right)$, Intercept: $(0, 21)$

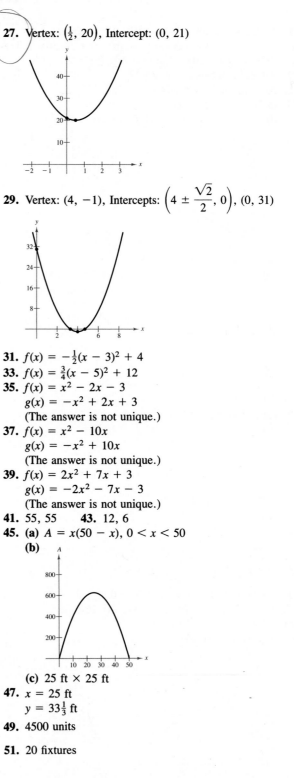

29. Vertex: $(4, -1)$, Intercepts: $\left(4 \pm \dfrac{\sqrt{2}}{2}, 0\right)$, $(0, 31)$

31. $f(x) = -\frac{1}{2}(x - 3)^2 + 4$
33. $f(x) = \frac{3}{4}(x - 5)^2 + 12$
35. $f(x) = x^2 - 2x - 3$
  $g(x) = -x^2 + 2x + 3$
  (The answer is not unique.)
37. $f(x) = x^2 - 10x$
  $g(x) = -x^2 + 10x$
  (The answer is not unique.)
39. $f(x) = 2x^2 + 7x + 3$
  $g(x) = -2x^2 - 7x - 3$
  (The answer is not unique.)
41. 55, 55   **43.** 12, 6
45. **(a)** $A = x(50 - x)$, $0 < x < 50$
  (b)

(c) 25 ft $\times$ 25 ft
47. $x = 25$ ft
  $y = 33\frac{1}{3}$ ft
49. 4500 units

51. 20 fixtures

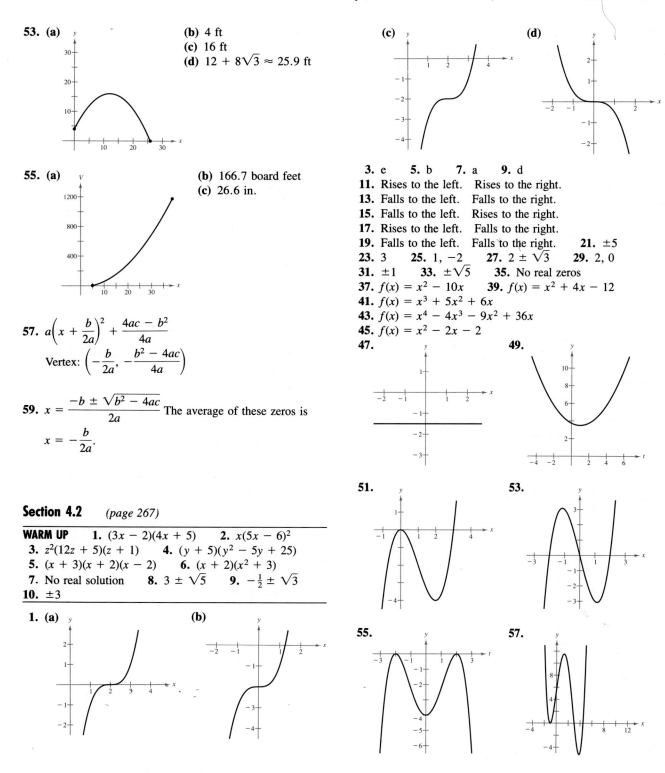

53. (a)

(b) 4 ft
(c) 16 ft
(d) $12 + 8\sqrt{3} \approx 25.9$ ft

(c)

(d)

55. (a)

(b) 166.7 board feet
(c) 26.6 in.

3. e   **5.** b   **7.** a   **9.** d
11. Rises to the left.   Rises to the right.
13. Falls to the left.   Falls to the right.
15. Falls to the left.   Rises to the right.
17. Rises to the left.   Falls to the right.
19. Falls to the left.   Falls to the right.   **21.** ± 5
23. 3   **25.** 1, -2 **27.** $2 \pm \sqrt{3}$ **29.** 2, 0
31. ± 1 **33.** $\pm\sqrt{5}$ **35.** No real zeros
37. $f(x) = x^2 - 10x$ **39.** $f(x) = x^2 + 4x - 12$
41. $f(x) = x^3 + 5x^2 + 6x$
43. $f(x) = x^4 - 4x^3 - 9x^2 + 36x$
45. $f(x) = x^2 - 2x - 2$
47.   **49.**

57. $a\left(x + \dfrac{b}{2a}\right)^2 + \dfrac{4ac - b^2}{4a}$

Vertex: $\left(-\dfrac{b}{2a}, -\dfrac{b^2 - 4ac}{4a}\right)$

59. $x = \dfrac{-b \pm \sqrt{b^2 - 4ac}}{2a}$ The average of these zeros is

$x = -\dfrac{b}{2a}$.

51.   **53.**

Section 4.2   (page 267)

WARM UP   **1.** $(3x - 2)(4x + 5)$ **2.** $x(5x - 6)^2$
3. $z^2(12z + 5)(z + 1)$ **4.** $(y + 5)(y^2 - 5y + 25)$
5. $(x + 3)(x + 2)(x - 2)$ **6.** $(x + 2)(x^2 + 3)$
7. No real solution   **8.** $3 \pm \sqrt{5}$ **9.** $-\frac{1}{2} \pm \sqrt{3}$
10. ± 3

55.   **57.**

1. (a)   **(b)**

59.

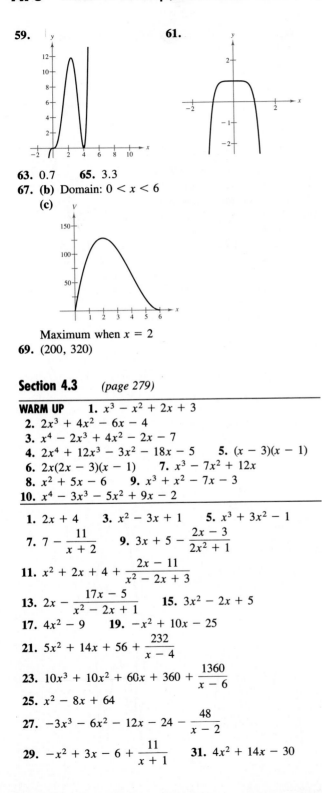

61.

63. 0.7   **65.** 3.3

67. (b) Domain: $0 < x < 6$

(c)

Maximum when $x = 2$

69. (200, 320)

Section 4.3   *(page 279)*

WARM UP   **1.** $x^3 - x^2 + 2x + 3$

2. $2x^3 + 4x^2 - 6x - 4$

3. $x^4 - 2x^3 + 4x^2 - 2x - 7$

4. $2x^4 + 12x^3 - 3x^2 - 18x - 5$ **5.** $(x - 3)(x - 1)$

6. $2x(2x - 3)(x - 1)$ **7.** $x^3 - 7x^2 + 12x$

8. $x^2 + 5x - 6$ **9.** $x^3 + x^2 - 7x - 3$

10. $x^4 - 3x^3 - 5x^2 + 9x - 2$

1. $2x + 4$ **3.** $x^2 - 3x + 1$ **5.** $x^3 + 3x^2 - 1$

7. $7 - \dfrac{11}{x + 2}$ **9.** $3x + 5 - \dfrac{2x - 3}{2x^2 + 1}$

11. $x^2 + 2x + 4 + \dfrac{2x - 11}{x^2 - 2x + 3}$

13. $2x - \dfrac{17x - 5}{x^2 - 2x + 1}$ **15.** $3x^2 - 2x + 5$

17. $4x^2 - 9$ **19.** $-x^2 + 10x - 25$

21. $5x^2 + 14x + 56 + \dfrac{232}{x - 4}$

23. $10x^3 + 10x^2 + 60x + 360 + \dfrac{1360}{x - 6}$

25. $x^2 - 8x + 64$

27. $-3x^3 - 6x^2 - 12x - 24 - \dfrac{48}{x - 2}$

29. $-x^2 + 3x - 6 + \dfrac{11}{x + 1}$ **31.** $4x^2 + 14x - 30$

33. $(x - 2)(x + 3)(x - 1)$ **35.** $(2x - 1)(x - 5)(x - 2)$

37. $(x + \sqrt{3})(x - \sqrt{3})(x + 2)$

39. $(x - 1)(x - 1 - \sqrt{3})(x - 1 + \sqrt{3})$

41. $f(x) = (x - 4)(x^2 + 3x - 2) + 3, f(4) = 3$

43. $f(x) = (x - \sqrt{2})[x^2 + (3 + \sqrt{2})x + 3\sqrt{2}] - 8$,
  $f(\sqrt{2}) = -8$

45. (a) 1   **(b)** 4   **(c)** 4   **(d)** 1954   **47. (a)** 97

(b) $-\frac{5}{3}$ **(c)** 17   **(d)** -199 **49. (a)** 72   **(b)** 0

(c) 37.648   **(d)** 30   **51.** $2x^2 - x - 1$

53. $x^2 + 2x - 3$ **55.** $x^2 + 3x$ **57.** 3300 rpm

59. (Answers are not unique.)

(a) $f(x) = (x - 2)x^2 + 5 = x^3 - 2x^2 + 5$

(b) $f(x) = -(x + 3)x^2 + 1 = -x^3 - 3x^2 + 1$

Section 4.4   *(page 289)*

WARM UP   **1.** $f(x) = 3x^3 - 8x^2 - 5x + 6$

2. $f(x) = 4x^4 - 3x^3 - 16x^2 + 12x$

3. $x^4 - 3x^3 + 5 + \dfrac{3}{x + 3}$ **4.** $3x^3 + 15x^2 - 9 - \dfrac{2}{x + \frac{2}{3}}$

5. $\frac{1}{2}, -3 \pm \sqrt{5}$ **6.** $10, -\frac{2}{3}, -\frac{3}{2}$ **7.** $-\frac{3}{4}, 2 \pm \sqrt{2}$

8. $\frac{2}{5}, -\frac{7}{2}, -2$ **9.** $\pm\sqrt{2}, \pm 1$ **10.** $\pm 2, \pm\sqrt{3}$

1. One negative zero   **3.** No real zeros

5. One positive zero   **7.** One or three positive zeros

9. Two or no positive zeros   **11.** $\pm 1, \pm 2, \pm 4$

13. $\pm 1, \pm 3, \pm\frac{1}{2}, \pm\frac{3}{2}, \pm\frac{1}{4}, \pm\frac{3}{4}$

15. $\pm 1, \pm 2, \pm 4, \pm 8, \pm\frac{1}{2}$ **17. (a)** Upper bound

(b) Lower bound   **(c)** Neither   **19. (a)** Neither

(b) Lower bound   **(c)** Upper bound   **21.** 1, 2, 3

23. $1, -1, 4$ **25.** $-1, -10$ **27.** 1, 2   **29.** $\frac{1}{2}, -1$

31. $1, -\frac{1}{2}$ **33.** $-\frac{3}{4}$ **35.** $\pm 1, \pm\sqrt{2}$ **37.** $-1, 2$

39. $0, -1, -3, 4$ **41.** $-2, 4, -\frac{1}{2}$

43. $0, 3, 4, \pm\sqrt{2}$

45. (a) $\pm 1, \pm 3, \pm\frac{1}{2}, \pm\frac{3}{2}, \pm\frac{1}{4}, \pm\frac{3}{4}, \pm\frac{1}{8}, \pm\frac{3}{8}, \pm\frac{1}{16}, \pm\frac{3}{16},$
  $\pm\frac{1}{32}, \pm\frac{3}{32}$

(b)

(c) $1, \frac{3}{4}, -\frac{1}{8}$

47. (a) $\pm1, \pm2, \pm3, \pm6, \pm9, \pm\frac{1}{2}, \pm\frac{3}{2}, \pm\frac{9}{2}, \pm\frac{1}{4}, \pm\frac{3}{4},$
$\pm\frac{9}{4}, \pm18$

(b)

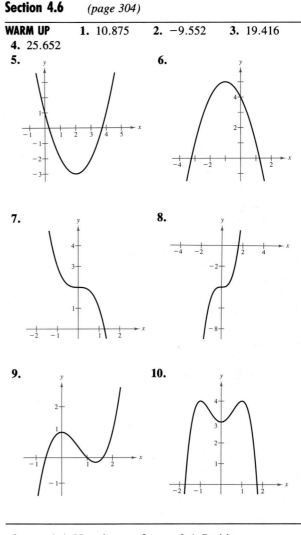

(c) $-2, \frac{1}{8} \pm \frac{\sqrt{145}}{8}$

49. $\pm2, \pm\frac{3}{2}$ **51.** $\pm1, \frac{1}{4}$ **53.** d   **55.** b
57. 3.77 in. $\times$ 7.77 in. $\times$ 0.614 in.
59. 18 in. $\times$ 18 in. $\times$ 36 in.

Section 4.5   *(page 299)*

WARM UP   **1.** $4 - \sqrt{29}i, 4 + \sqrt{29}i$
2. $-5 - 12i, -5 + 12i$ **3.** $-1 + 4\sqrt{2}i, -1 - 4\sqrt{2}i$
4. $6 + \frac{1}{2}i, 6 - \frac{1}{2}i$ **5.** $-13 + 9i$ **6.** $12 + 16i$
7. $26 + 22i$ **8.** 29 **9.** i **10.** $-9 + 46i$

1. $\pm5i, (x + 5i)(x - 5i)$
3. $2 \pm \sqrt{3}, (x - 2 - \sqrt{3})(x - 2 + \sqrt{3})$
5. $\pm3, \pm3i, (x + 3)(x - 3)(x + 3i)(x - 3i)$
7. $1 \pm i, (z - 1 + i)(z - 1 - i)$
9. $2, 2 \pm i, (x - 2)(x - 2 + i)(x - 2 - i)$
11. $-5, 4 \pm 3i, (t + 5)(t - 4 + 3i)(t - 4 - 3i)$
13. $-10, -7 \pm 5i, (x + 10)(x + 7 - 5i)(x + 7 + 5i)$
15. $-\frac{3}{4}, 1 \pm \frac{1}{2}i, (4x + 3)(2x - 2 + i)(2x - 2 - i)$
17. $-2, 1 \pm \sqrt{2}i, (x + 2)(x - 1 + \sqrt{2}i)(x - 1 - \sqrt{2}i)$
19. $-\frac{1}{5}, 1 \pm \sqrt{5}i, (5x + 1)(x - 1 + \sqrt{5}i)(x - 1 - \sqrt{5}i)$
21. $2, \pm2i, (x - 2)^2(x + 2i)(x - 2i)$
23. $\pm i, \pm3i, (x + i)(x - i)(x + 3i)(x - 3i)$
25. $-2, -\frac{1}{2}, \pm i, (x + 2)(2x + 1)(x + i)(x - i)$
27. $x^3 - x^2 + 25x - 25$ **29.** $x^3 - 10x^2 + 33x - 34$
31. $x^4 + 37x^2 + 36$ **33.** $x^4 + 8x^3 + 9x^2 - 10x + 100$
35. $16x^4 + 36x^3 + 16x^2 + x - 30$
37. (a) $(x^2 + 9)(x^2 - 3)$
(b) $(x^2 + 9)(x + \sqrt{3})(x - \sqrt{3})$
(c) $(x + 3i)(x - 3i)(x + \sqrt{3})(x - \sqrt{3})$
39. (a) $(x^2 - 2x - 2)(x^2 - 2x + 3)$
(b) $(x - 1 + \sqrt{3})(x - 1 - \sqrt{3})(x^2 - 2x + 3)$
(c) $(x - 1 + \sqrt{3})(x - 1 - \sqrt{3})(x - 1 + \sqrt{2}i) \cdot$
$(x - 1 - \sqrt{2}i)$ **41.** $-\frac{3}{2}, \pm5i$ **43.** $\pm2i, 1, -\frac{1}{2}$
45. $-3 \pm i, \frac{1}{4}$ **47.** $2, -3 \pm \sqrt{2}i$ **49.** $\frac{3}{4}, \frac{1}{2} \pm \frac{\sqrt{5}}{2}i$

51. Setting $h = 64$ and solving the resulting equation yields imaginary roots.
53. $x^2 + b$

Section 4.6   *(page 304)*

WARM UP   **1.** 10.875   **2.** -9.552 **3.** 19.416
4. 25.652
5.   **6.**

7.   **8.**

9.   **10.**

1. $x = 1.4$: Negative   **3.** $x = 2.4$: Positive
$x = 1.5$: Positive $x = 2.5$: Negative
5. 0.68   **7.** 0.21   **9.** 2.77   **11.** $-1.16, 1.45$
13. $-4.60, -1.04, 5.64$ **15.** 0.53   **17.** 4.97
19. 1.56   **21.** \$384,000   **23.** 4.5 hr
25. $f(x) = x^4 - 16x^3 + 83x^2 - 156x + 66$

Chapter 4 Review Exercises *(page 307)*

1.

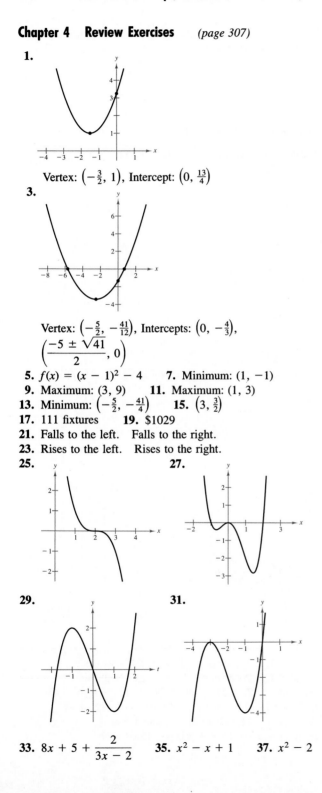

Vertex: $\left(-\frac{3}{2}, 1\right)$, Intercept: $\left(0, \frac{13}{4}\right)$

3.

Vertex: $\left(-\frac{5}{2}, -\frac{41}{12}\right)$, Intercepts: $\left(0, -\frac{4}{3}\right)$, $\left(\dfrac{-5 \pm \sqrt{41}}{2}, 0\right)$

5. $f(x) = (x - 1)^2 - 4$ **7.** Minimum: $(1, -1)$

9. Maximum: $(3, 9)$ **11.** Maximum: $(1, 3)$

13. Minimum: $\left(-\frac{5}{2}, -\frac{41}{4}\right)$ **15.** $\left(3, \frac{3}{2}\right)$

17. 111 fixtures **19.** \$1029

21. Falls to the left. Falls to the right.

23. Rises to the left. Rises to the right.

25.

27.

29.

31.

33. $8x + 5 + \dfrac{2}{3x - 2}$ **35.** $x^2 - x + 1$ **37.** $x^2 - 2$

39. $x^2 - 3x + 2 - \dfrac{1}{x^2 + 2}$

41. $0.25x^3 - 3.5x^2 - 7x - 14 - \dfrac{28}{x - 2}$

43. $6x^3 - 27x$ **45.** $2x^2 - (3 - 4i)x + (1 - 2i)$

47. (a) No **(b)** Yes **(c)** Yes **(d)** No

49. (a) No **(b)** Yes **(c)** Yes **(d)** No

51. (a) 580 **(b)** 0 **53. (a)** -421 **(b)** 96

55. $f(x) = 6x^4 + 13x^3 + 7x^2 - x - 1$

57. $f(x) = 3x^4 - 14x^3 + 17x^2 - 42x + 24$

59. Two or no positive zeros and one negative zero

61. ± 1, ± 3, ± 5, ± 15, $\pm \frac{1}{2}$, $\pm \frac{3}{2}$, $\pm \frac{5}{2}$, $\pm \frac{15}{2}$, $\pm \frac{1}{4}$, $\pm \frac{3}{4}$, $\pm \frac{5}{4}$, $\pm \frac{15}{4}$ **63.** $1, \frac{3}{4}$ **65.** $\frac{5}{6}, \pm 2i$ **67.** $-1, \frac{3}{2}, 3, \frac{2}{3}$

69. 0.47 **71.** 3.26 **73.** 11.30 ft

CHAPTER 5

Section 5.1 *(page 321)*

WARM UP **1.** $(x - 5)(x + 2)$ **2.** $(x - 5)(x - 2)$
3. $x(x + 1)(x + 3)$ **4.** $(x^2 - 2)(x - 4)$
5. **6.**

7. **8.**

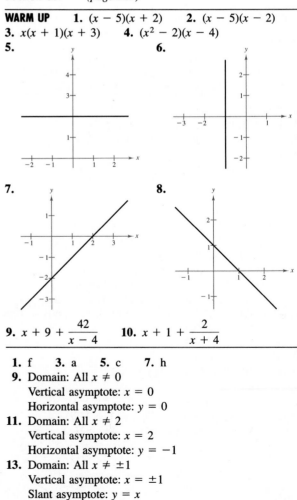

9. $x + 9 + \dfrac{42}{x - 4}$ **10.** $x + 1 + \dfrac{2}{x + 4}$

1. f **3.** a **5.** c **7.** h

9. Domain: All $x \neq 0$
Vertical asymptote: $x = 0$
Horizontal asymptote: $y = 0$

11. Domain: All $x \neq 2$
Vertical asymptote: $x = 2$
Horizontal asymptote: $y = -1$

13. Domain: All $x \neq \pm 1$
Vertical asymptote: $x = \pm 1$
Slant asymptote: $y = x$

15. Domain: All reals
Horizontal asymptote: $y = 3$
17. Domain: All reals
19.   **21.**

23.   **25.**

27.   **29.**

31.   **33.**

35.   **37.**

39.   **41.**

43.   **45.**

47.   **49.**

51.   **53.**

55.   **57.**

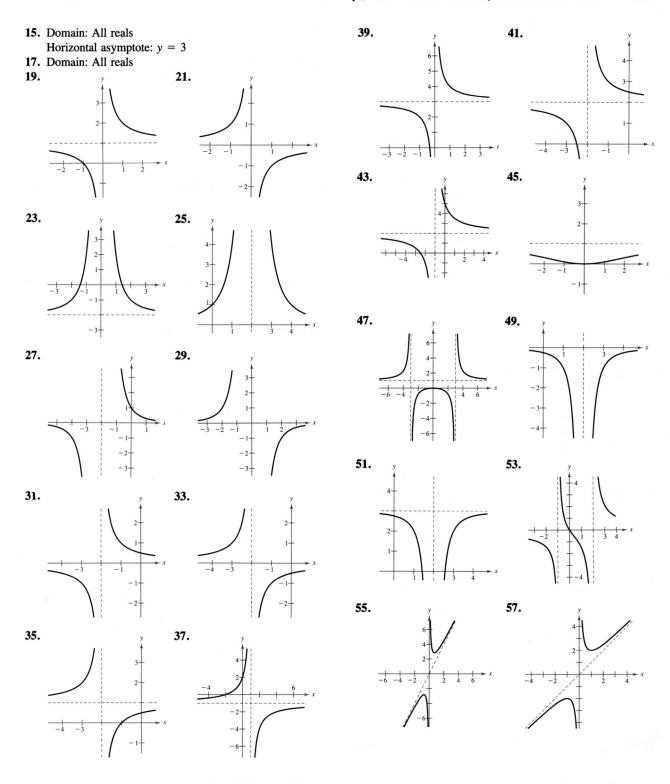

59.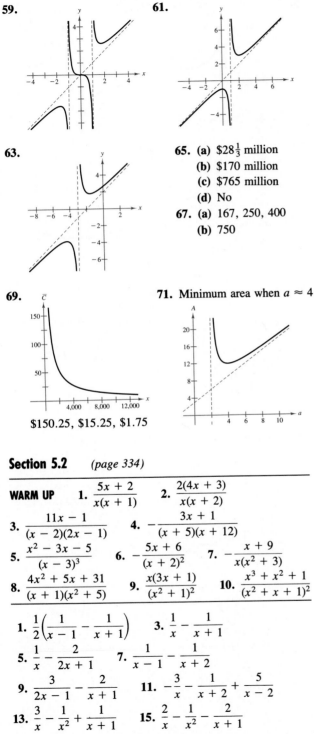

61.

63.

65. (a) $\$28\frac{1}{3}$ million
 (b) $\$170$ million
 (c) $\$765$ million
 (d) No

67. (a) 167, 250, 400
 (b) 750

69.

71. Minimum area when $a \approx 4$

$\$150.25, \$15.25, \$1.75$

17. $\dfrac{3}{x-3} + \dfrac{9}{(x-3)^2}$ **19.** $-\dfrac{1}{x} + \dfrac{2x}{x^2+1}$

21. $\dfrac{1}{3(x^2+2)} - \dfrac{1}{6(x+2)} + \dfrac{1}{6(x-2)}$

23. $\dfrac{1}{8(2x+1)} + \dfrac{1}{8(2x-1)} - \dfrac{x}{2(4x^2+1)}$

25. $\dfrac{1}{x^2+2} + \dfrac{x}{(x^2+2)^2}$

27. $\dfrac{1}{x+1} + \dfrac{2}{x^2-2x+3}$

29. $2x + \dfrac{1}{2}\left(\dfrac{3}{x-4} - \dfrac{1}{x+2}\right)$

31. $x + 3 + \dfrac{6}{x-1} + \dfrac{4}{(x-1)^2} + \dfrac{1}{(x-1)^3}$

33. $\dfrac{1}{2a}\left(\dfrac{1}{a+x} + \dfrac{1}{a-x}\right)$ **35.** $\dfrac{1}{L}\left(\dfrac{1}{y} + \dfrac{1}{L-y}\right)$

Section 5.3 *(page 344)*

WARM UP **1.** $9x^2 + 16y^2 = 144$ **2.** $x^2 + 4y^2 = 32$
3. $16x^2 - y^2 = 4$ **4.** $243x^2 + 4y^2 = 9$
5. $c = 2\sqrt{2}$ **6.** $c = \sqrt{13}$ **7.** $c = 2\sqrt{3}$
8. $c = \sqrt{5}$ **9.** $d = 4$ **10.** $d = 2$

1. a **3.** d **5.** g **7.** e
9. Vertex: $(0, 0)$, Focus: $\left(0, \frac{1}{16}\right)$

11. Vertex: $(0, 0)$, Focus: $\left(-\frac{3}{2}, 0\right)$

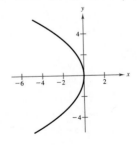

Section 5.2 *(page 334)*

WARM UP **1.** $\dfrac{5x+2}{x(x+1)}$ **2.** $\dfrac{2(4x+3)}{x(x+2)}$

3. $\dfrac{11x-1}{(x-2)(2x-1)}$ **4.** $-\dfrac{3x+1}{(x+5)(x+12)}$

5. $\dfrac{x^2-3x-5}{(x-3)^3}$ **6.** $-\dfrac{5x+6}{(x+2)^2}$ **7.** $-\dfrac{x+9}{x(x^2+3)}$

8. $\dfrac{4x^2+5x+31}{(x+1)(x^2+5)}$ **9.** $\dfrac{x(3x+1)}{(x^2+1)^2}$ **10.** $\dfrac{x^3+x^2+1}{(x^2+x+1)^2}$

1. $\dfrac{1}{2}\left(\dfrac{1}{x-1} - \dfrac{1}{x+1}\right)$ **3.** $\dfrac{1}{x} - \dfrac{1}{x+1}$

5. $\dfrac{1}{x} - \dfrac{2}{2x+1}$ **7.** $\dfrac{1}{x-1} - \dfrac{1}{x+2}$

9. $\dfrac{3}{2x-1} - \dfrac{2}{x+1}$ **11.** $-\dfrac{3}{x} - \dfrac{1}{x+2} + \dfrac{5}{x-2}$

13. $\dfrac{3}{x} - \dfrac{1}{x^2} + \dfrac{1}{x+1}$ **15.** $\dfrac{2}{x} - \dfrac{1}{x^2} - \dfrac{2}{x+1}$

13. Vertex: $(0, 0)$, Focus: $(0, -2)$

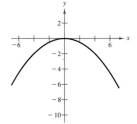

15. Vertex: $(0, 0)$, Focus: $(2, 0)$

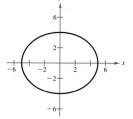

17. $x^2 = -6y$

19. $y^2 = -8x$

21. $x^2 = 4y$

23. $x^2 = -8y$

25. $y^2 = 9x$

27. Center: $(0, 0)$, Vertices: $(\pm 5, 0)$

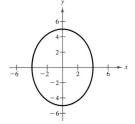

29. Center: $(0, 0)$, Vertices: $(0, \pm 5)$

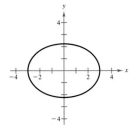

31. Center: $(0, 0)$, Vertices: $(\pm 3, 0)$

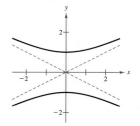

33. Center: $(0, 0)$, Vertices: $(0, \pm\sqrt{5})$

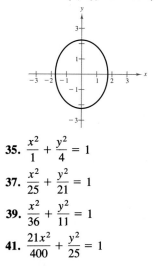

35. $\dfrac{x^2}{1} + \dfrac{y^2}{4} = 1$

37. $\dfrac{x^2}{25} + \dfrac{y^2}{21} = 1$

39. $\dfrac{x^2}{36} + \dfrac{y^2}{11} = 1$

41. $\dfrac{21x^2}{400} + \dfrac{y^2}{25} = 1$

43. Center: $(0, 0)$, Vertices: $(\pm 1, 0)$

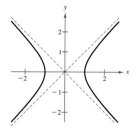

45. Center: $(0, 0)$, Vertices: $(0, \pm 1)$

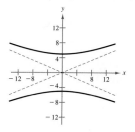

47. Center: $(0, 0)$, Vertices: $(0, \pm 5)$

49. Center: $(0, 0)$, Vertices: $(\pm\sqrt{3}, 0)$

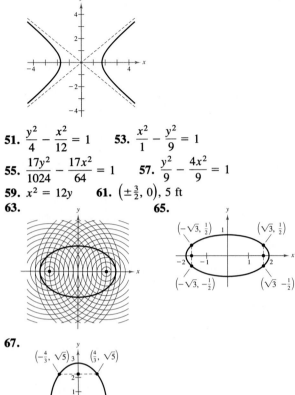

51. $\dfrac{y^2}{4} - \dfrac{x^2}{12} = 1$ **53.** $\dfrac{x^2}{1} - \dfrac{y^2}{9} = 1$

55. $\dfrac{17y^2}{1024} - \dfrac{17x^2}{64} = 1$ **57.** $\dfrac{y^2}{9} - \dfrac{4x^2}{9} = 1$

59. $x^2 = 12y$ **61.** $\left(\pm\frac{3}{2}, 0\right)$, 5 ft

63.

65.

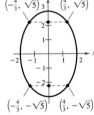

67.

69. $x \approx 110.3$ mi

73. The shape continuously changes from an ellipse with a vertical major axis of length 8 and minor axis length 2 to a circle with a diameter of 8 and then to an ellipse with a horizontal major axis of length 16 and minor axis length 8.

Section 5.4   *(page 353)*

WARM UP	**1.** Hyperbola	**2.** Ellipse	**3.** Parabola
4. Hyperbola	**5.** Ellipse	**6.** Circle	**7.** Hyperbola
8. Parabola	**9.** Parabola	**10.** Ellipse	

1. Vertex: $(1, -2)$, Focus: $(1, -4)$, Directrix: $y = 0$

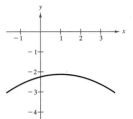

3. Vertex: $\left(5, -\frac{1}{2}\right)$, Focus: $\left(\frac{11}{2}, -\frac{1}{2}\right)$, Directrix: $x = \frac{9}{2}$

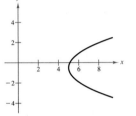

5. Vertex: $(1, 1)$, Focus: $(1, 2)$, Directrix: $y = 0$

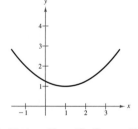

7. Vertex: $(8, -1)$, Focus: $(9, -1)$, Directrix: $x = 7$

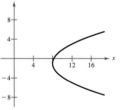

9. Vertex: $(-2, -3)$, Focus: $(-4, -3)$, Directrix: $x = 0$

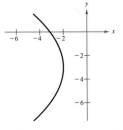

11. Vertex: $(-1, 2)$, Focus: $(0, 2)$, Directrix: $x = -2$

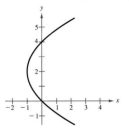

13. $(y - 2)^2 = -8(x - 3)$ **15.** $x^2 = 8(y - 4)$
17. $(y - 2)^2 = 8x$ **19.** $x^2 = -(y - 4)$
21. Center: $(1, 5)$, Vertices: $(1, 10)$, $(1, 0)$, Foci: $(1, 9)$, $(1, 1)$

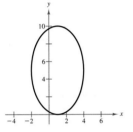

23. Center: $(-2, 3)$, Vertices: $(-2, 6)$, $(-2, 0)$, Foci: $(-2, 3 \pm \sqrt{5})$

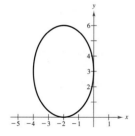

25. Center: $(1, -1)$, Vertices: $\left(\frac{9}{4}, -1\right)$, $\left(-\frac{1}{4}, -1\right)$, Foci: $\left(\frac{7}{4}, -1\right)$, $\left(\frac{1}{4}, -1\right)$

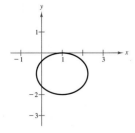

27. Center: $\left(\frac{1}{2}, -1\right)$, Vertices: $\left(\frac{1}{2} \pm \sqrt{5}, -1\right)$, Foci: $\left(\frac{1}{2} \pm \sqrt{2}, -1\right)$

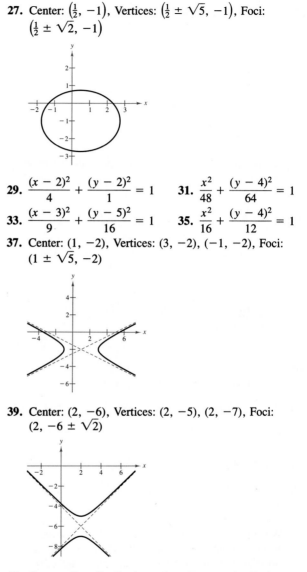

29. $\dfrac{(x - 2)^2}{4} + \dfrac{(y - 2)^2}{1} = 1$ **31.** $\dfrac{x^2}{48} + \dfrac{(y - 4)^2}{64} = 1$

33. $\dfrac{(x - 3)^2}{9} + \dfrac{(y - 5)^2}{16} = 1$ **35.** $\dfrac{x^2}{16} + \dfrac{(y - 4)^2}{12} = 1$

37. Center: $(1, -2)$, Vertices: $(3, -2)$, $(-1, -2)$, Foci: $(1 \pm \sqrt{5}, -2)$

39. Center: $(2, -6)$, Vertices: $(2, -5)$, $(2, -7)$, Foci: $(2, -6 \pm \sqrt{2})$

41. Center: $(2, -3)$, Vertices: $(3, -3)$, $(1, -3)$, Foci: $(2 \pm \sqrt{10}, -3)$

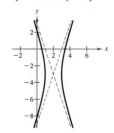

43. Center: $(1, -3)$, Vertices: $(1, -3 \pm \sqrt{2})$, Foci: $(1, -3 \pm 2\sqrt{5})$

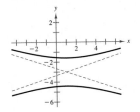

9.

11.

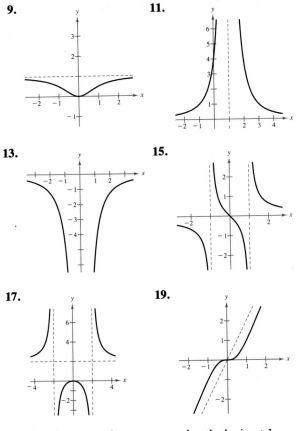

45. The graph of this equation is two lines intersecting at $(-1, -3)$.

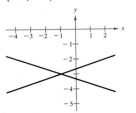

13.

15.

47. $\dfrac{(x-4)^2}{4} - \dfrac{y^2}{12} = 1$ **49.** $\dfrac{(y-5)^2}{16} - \dfrac{(x-4)^2}{9} = 1$

51. $\dfrac{y^2}{9} - \dfrac{4(x-2)^2}{9} = 1$ **53.** $\dfrac{(x-3)^2}{9} - \dfrac{(y-2)^2}{4} = 1$

55. Circle   **57.** Hyperbola   **59.** Ellipse

61. Parabola   **63. (a)** $17,500\sqrt{2}$ mph

(b) $x^2 = -16,400(y - 4,100)$ **65.** $\dfrac{x^2}{25} + \dfrac{y^2}{16} = 1$

67. 91,376,731 mi;   94,537,269 mi   **69.** 0.0543

17.

19.

Chapter 5   Review Exercises   *(page 356)*

1. Domain: All reals such that $x \neq -3$
Vertical asymptote: $x = -3$
Horizontal asymptote: $y = 0$

3. Domain: All reals such that $x \neq \pm 2$
Vertical asymptote: $x = 2$, $x = -2$
Horizontal asymptote: $y = 1$

5.

7.

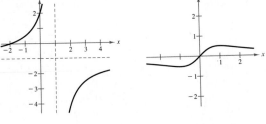

21. As x increases, the cost approaches the horizontal asymptote $\overline{C} = 0.5$.

23. (a) \$176 million   **(b)** \$528 million
(c) \$1584 million   **(d)** No

25.

80.3 mg/dc²/hr

27. $\dfrac{3}{x+2} - \dfrac{4}{x+4}$

29. $1 - \dfrac{25}{8(x+5)} + \dfrac{9}{8(x-3)}$

31. $\dfrac{1}{2}\left(\dfrac{3}{x-1} - \dfrac{x-3}{x^2+1}\right)$

33. $\dfrac{3x}{x^2+1} + \dfrac{x}{(x^2+1)^2}$

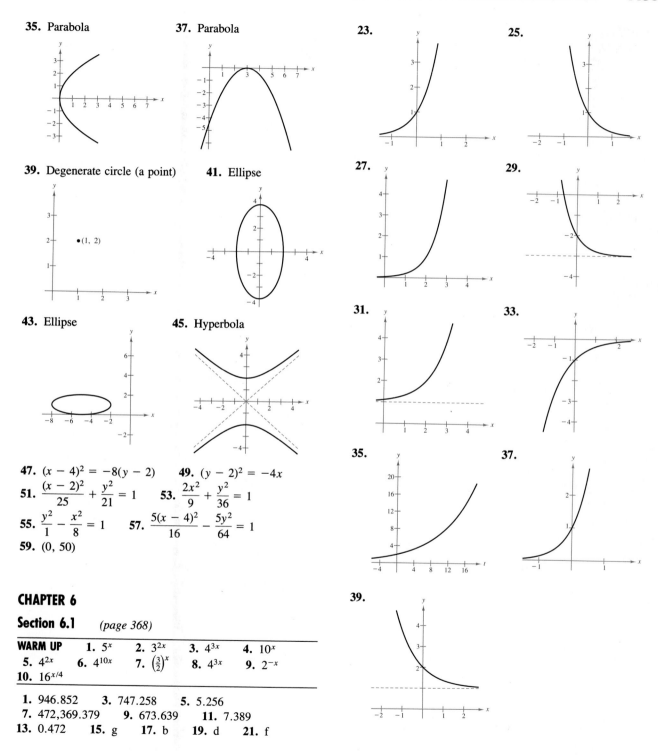

35. Parabola

37. Parabola

23.

25.

39. Degenerate circle (a point)

41. Ellipse

27.

29.

43. Ellipse

45. Hyperbola

31.

33.

47. $(x - 4)^2 = -8(y - 2)$ **49.** $(y - 2)^2 = -4x$

51. $\dfrac{(x - 2)^2}{25} + \dfrac{y^2}{21} = 1$ **53.** $\dfrac{2x^2}{9} + \dfrac{y^2}{36} = 1$

55. $\dfrac{y^2}{1} - \dfrac{x^2}{8} = 1$ **57.** $\dfrac{5(x - 4)^2}{16} - \dfrac{5y^2}{64} = 1$

59. $(0, 50)$

35.

37.

39.

CHAPTER 6

Section 6.1   *(page 368)*

WARM UP   **1.** 5^x **2.** 3^{2x} **3.** 4^{3x} **4.** 10^x
5. 4^{2x} **6.** 4^{10x} **7.** $\left(\frac{3}{2}\right)^x$ **8.** 4^{3x} **9.** 2^{-x}
10. $16^{x/4}$

1. 946.852   **3.** 747.258   **5.** 5.256
7. 472,369.379   **9.** 673.639   **11.** 7.389
13. 0.472   **15.** g   **17.** b   **19.** d   **21.** f

41.

n	1	2	4
A	$7,764.62	$8,017.84	$8,155.09

n	12	365	Continuous compounding
A	$8,250.97	$8,298.66	$8,300.29

43.

n	1	2	4
A	$24,115.73	$25,714.29	$26,602.23

n	12	365	Continuous compounding
A	$27,231.38	$27,547.07	$27,557.94

45.

t	1	10	20
P	$91,393.12	$40,656.97	$16,529.89

t	30	40	50
P	$6,720.55	$2,732.37	$1,110.90

47.

t	1	10	20
P	$90,521.24	$36,940.70	$13,646.15

t	30	40	50
P	$5,040.98	$1,862.17	$687.90

49. $222,822.57 **51. (a)** $472.70 **(b)** $298.29
53. (a) 100 **(b)** 300 **(c)** 900
55. (a) 25 units **(b)** 16.297 units
(c)

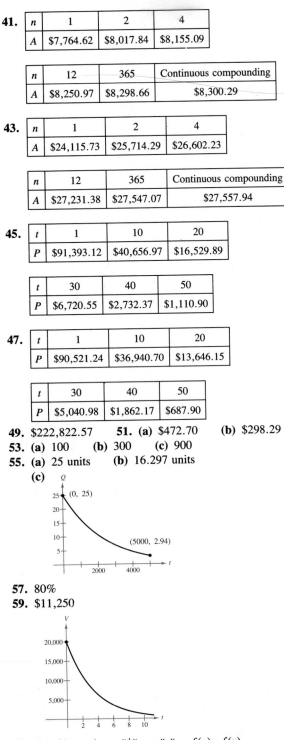

57. 80%
59. $11,250

61. (a) $f(u + v) = a^{u+v} = a^u a^v = f(u) \cdot f(v)$
(b) $f(2x) = a^{2x} = (a^x)^2 = [f(x)]^2$

Section 6.2 *(page 379)*

WARM UP **1.** 3 **2.** 0 **3.** −1 **4.** 1
5. 7.389 **6.** 0.368
7. Graph is shifted 2 units to the left.
8. Graph is reflected about the x-axis.
9. Graph is shifted down 1 unit.
10. Graph is reflected about the y-axis.

1. 4 **3.** −2 **5.** $\frac{1}{2}$ **7.** 0 **9.** −2 **11.** 3
13. −2 **15.** 2 **17.** $\log_5 125 = 3$ **19.** $\log_{81} 3 = \frac{1}{4}$
21. $\log_6 \frac{1}{36} = -2$ **23.** $\ln 20.0855 \ldots = 3$ **25.** $\ln 4 = x$
27. 2.538 **29.** −0.319 **31.** 2.913 **33.** 1.005
35. **37.**

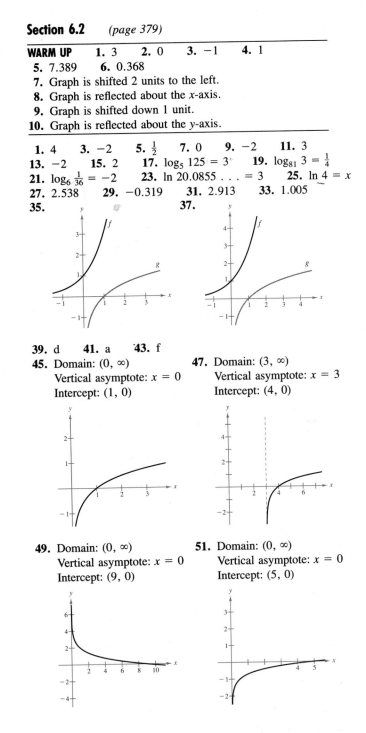

39. d **41.** a **43.** f
45. Domain: $(0, \infty)$ **47.** Domain: $(3, \infty)$
Vertical asymptote: $x = 0$ Vertical asymptote: $x = 3$
Intercept: $(1, 0)$ Intercept: $(4, 0)$

49. Domain: $(0, \infty)$ **51.** Domain: $(0, \infty)$
Vertical asymptote: $x = 0$ Vertical asymptote: $x = 0$
Intercept: $(9, 0)$ Intercept: $(5, 0)$

53. Domain: $(2, \infty)$
Vertical asymptote: $x = 2$
Intercept: $(3, 0)$

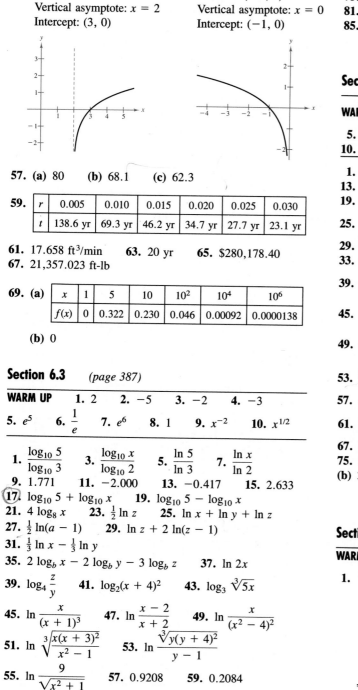

55. Domain: $(-\infty, 0)$
Vertical asymptote: $x = 0$
Intercept: $(-1, 0)$

75. 4.5   **77.** $\frac{3}{2}$ **79.** $\frac{1}{2} + \frac{1}{2} \log_7 10$
81. $-3 - \log_5 2$ **83.** $6 + \ln 5$
85. $\beta = 10(\log_{10} I + 16)$,   60 db

57. (a) 80   (b) 68.1   (c) 62.3

59.

r	0.005	0.010	0.015	0.020	0.025	0.030
t	138.6 yr	69.3 yr	46.2 yr	34.7 yr	27.7 yr	23.1 yr

61. 17.658 ft^3/min   **63.** 20 yr   **65.** \$280,178.40
67. 21,357.023 ft-lb

69. (a)

x	1	5	10	10^2	10^4	10^6
$f(x)$	0	0.322	0.230	0.046	0.00092	0.0000138

(b) 0

Section 6.3   (page 387)

WARM UP   **1.** 2   **2.** -5 **3.** -2 **4.** -3
5. e^5 **6.** $\dfrac{1}{e}$ **7.** e^6 **8.** 1   **9.** x^{-2} **10.** $x^{1/2}$

1. $\dfrac{\log_{10} 5}{\log_{10} 3}$ **3.** $\dfrac{\log_{10} x}{\log_{10} 2}$ **5.** $\dfrac{\ln 5}{\ln 3}$ **7.** $\dfrac{\ln x}{\ln 2}$
9. 1.771   **11.** -2.000 **13.** -0.417 **15.** 2.633
17. $\log_{10} 5 + \log_{10} x$ **19.** $\log_{10} 5 - \log_{10} x$
21. $4 \log_8 x$ **23.** $\frac{1}{2} \ln z$ **25.** $\ln x + \ln y + \ln z$
27. $\frac{1}{2} \ln(a - 1)$ **29.** $\ln z + 2 \ln(z - 1)$
31. $\frac{1}{3} \ln x - \frac{1}{3} \ln y$
35. $2 \log_b x - 2 \log_b y - 3 \log_b z$ **37.** $\ln 2x$
39. $\log_4 \dfrac{z}{y}$ **41.** $\log_2(x + 4)^2$ **43.** $\log_3 \sqrt[3]{5x}$
45. $\ln \dfrac{x}{(x + 1)^3}$ **47.** $\ln \dfrac{x - 2}{x + 2}$ **49.** $\ln \dfrac{x}{(x^2 - 4)^2}$
51. $\ln \sqrt[3]{\dfrac{x(x + 3)^2}{x^2 - 1}}$ **53.** $\ln \dfrac{\sqrt[3]{y(y + 4)^2}}{y - 1}$
55. $\ln \dfrac{9}{\sqrt{x^2 + 1}}$ **57.** 0.9208   **59.** 0.2084
61. 1.6542   **63.** 0.1781   **65.** -0.7124
67. 0.91355   **69.** 2.0367   **71.** 2   **73.** 2.4

Section 6.4   (page 395)

WARM UP   **1.** $\dfrac{\ln 3}{\ln 2}$ **2.** $1 + \dfrac{2}{\ln 4}$ **3.** $\dfrac{e}{2}$ **4.** $2e$
5. $2 \pm i$ **6.** $\frac{1}{2}, 1$ **7.** $2x$ **8.** $3x$ **9.** $2x$
10. $-x^2$

1. 2   **3.** -2 **5.** 3   **7.** 64   **9.** $\frac{1}{10}$ **11.** x^2
13. $5x + 2$ **15.** x^2 **17.** $\ln 10 \approx 2.303$
19. $\ln \frac{39}{2} \approx 2.970$ **21.** $\ln 15 \approx 2.708$ **23.** 0
25. $\dfrac{\ln 12}{3} \approx 0.828$ **27.** $\ln \frac{5}{3} \approx 0.511$
29. $\frac{2}{3} \ln \frac{962}{3} \approx 3.847$ **31.** $\ln 5 \approx 1.609$
33. $\frac{1}{2} \ln \frac{1}{3} \approx -0.549$ **35.** 0   **37.** $\log_{10} 42 \approx 1.623$
39. $\dfrac{\ln 80}{2 \ln 3} \approx 1.994$ **41.** 2   **43.** $\frac{1}{2} \log_{10} 36 \approx 0.778$
45. $1 + \dfrac{\ln 7}{\ln 5} \approx 2.209$ **47.** $\dfrac{\ln 25 + \ln 3}{\ln 9 - \ln 5} \approx 7.345$
49. $\dfrac{\ln 2}{12 \ln\left(1 + \frac{0.10}{12}\right)} \approx 6.960$ **51.** $e^5 \approx 148.413$
53. $\dfrac{e^{2.4}}{2} \approx 5.512$ **55.** $\frac{1}{4} = 0.250$
57. $e^2 - 2 \approx 5.389$ **59.** $1 + \sqrt{1 + e} \approx 2.928$
61. 103   **63.** $\dfrac{-1 + \sqrt{17}}{2} \approx 1.562$ **65.** 4
67. No solution   **69.** 3   **71.** 8.2 yr   **73.** 12.9 yr
75. (a) 1426 units   (b) 1498 units   **77.** (a) 29.3 yr
(b) 39.8 yr   **79.** Males: 69.71 in.;   Females: 64.51 in.

Section 6.5   (page 404)

WARM UP

1.
2.

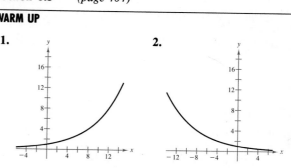

3.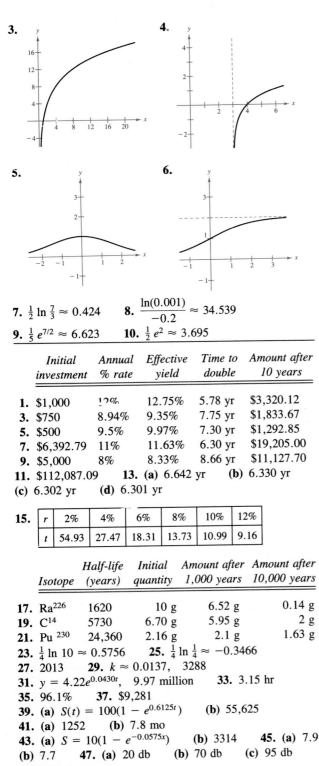

4.

5.

6.

7. $\frac{1}{2} \ln \frac{7}{3} \approx 0.424$ **8.** $\dfrac{\ln(0.001)}{-0.2} \approx 34.539$

9. $\frac{1}{5} e^{7/2} \approx 6.623$ **10.** $\frac{1}{2} e^2 \approx 3.695$

	Initial investment	Annual % rate	Effective yield	Time to double	Amount after 10 years
1.	$1,000	12%	12.75%	5.78 yr	$3,320.12
3.	$750	8.94%	9.35%	7.75 yr	$1,833.67
5.	$500	9.5%	9.97%	7.30 yr	$1,292.85
7.	$6,392.79	11%	11.63%	6.30 yr	$19,205.00
9.	$5,000	8%	8.33%	8.66 yr	$11,127.70

11. $112,087.09   **13. (a)** 6.642 yr   **(b)** 6.330 yr
(c) 6.302 yr   **(d)** 6.301 yr

15.

r	2%	4%	6%	8%	10%	12%
t	54.93	27.47	18.31	13.73	10.99	9.16

Isotope	Half-life (years)	Initial quantity	Amount after 1,000 years	Amount after 10,000 years
17. Ra^{226}	1620	10 g	6.52 g	0.14 g
19. C^{14}	5730	6.70 g	5.95 g	2 g
21. Pu^{230}	24,360	2.16 g	2.1 g	1.63 g

23. $\frac{1}{4} \ln 10 \approx 0.5756$ **25.** $\frac{1}{4} \ln \frac{1}{4} \approx -0.3466$
27. 2013   **29.** $k \approx 0.0137$, 3288
31. $y = 4.22e^{0.0430t}$, 9.97 million   **33.** 3.15 hr
35. 96.1%   **37.** $9,281
39. (a) $S(t) = 100(1 - e^{0.6125t})$ **(b)** 55,625
41. (a) 1252   **(b)** 7.8 mo
43. (a) $S = 10(1 - e^{-0.0575x})$ **(b)** 3314   **45. (a)** 7.9
(b) 7.7   **47. (a)** 20 db   **(b)** 70 db   **(c)** 95 db

(d) 120 db   **49.** 95%   **51.** 4.64
53. 1.6×10^{-6} moles per liter   **55.** 10^7 **57.** 7:30 A.M.

Chapter 6   Review Exercises   *(page 408)*

1. d   **3.** a   **5.** c
7.

9.

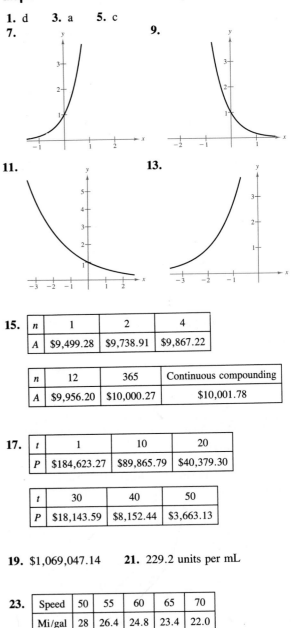

11.

13.

15.

n	1	2	4
A	$9,499.28	$9,738.91	$9,867.22

n	12	365	Continuous compounding
A	$9,956.20	$10,000.27	$10,001.78

17.

t	1	10	20
P	$184,623.27	$89,865.79	$40,379.30

t	30	40	50
P	$18,143.59	$8,152.44	$3,663.13

19. $1,069,047.14   **21.** 229.2 units per mL

23.

Speed	50	55	60	65	70
Mi/gal	28	26.4	24.8	23.4	22.0

25.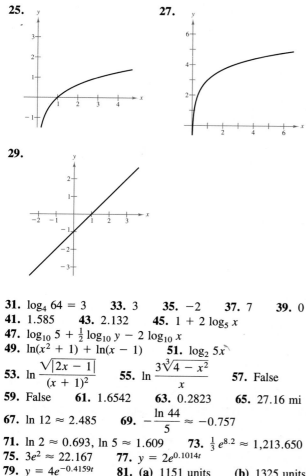

27.

29.

9.

11. $(x - 3)^2 = \frac{3}{2}(y + 2)$

13.

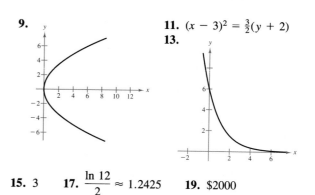

15. 3   **17.** $\dfrac{\ln 12}{2} \approx 1.2425$ **19.** $2000

CHAPTER 7

Section 7.1   (page 420)

WARM UP

1.

2.

3.

4.

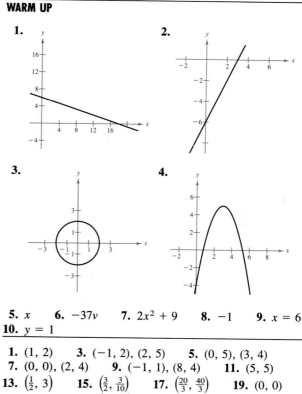

31. $\log_4 64 = 3$ **33.** 3   **35.** -2 **37.** 7   **39.** 0
41. 1.585   **43.** 2.132   **45.** $1 + 2 \log_5 x$
47. $\log_{10} 5 + \frac{1}{2} \log_{10} y - 2 \log_{10} x$
49. $\ln(x^2 + 1) + \ln(x - 1)$ **51.** $\log_2 5x$
53. $\ln \dfrac{\sqrt{|2x - 1|}}{(x + 1)^2}$ **55.** $\ln \dfrac{3\sqrt[3]{4 - x^2}}{x}$ **57.** False
59. False   **61.** 1.6542   **63.** 0.2823   **65.** 27.16 mi
67. $\ln 12 \approx 2.485$ **69.** $-\dfrac{\ln 44}{5} \approx -0.757$
71. $\ln 2 \approx 0.693$, $\ln 5 \approx 1.609$ **73.** $\frac{1}{3} e^{8.2} \approx 1{,}213.650$
75. $3e^2 \approx 22.167$ **77.** $y = 2e^{0.1014t}$
79. $y = 4e^{-0.4159t}$ **81.** (a) 1151 units   (b) 1325 units
83. (a) 8.94%   (b) $1833.67   (c) 9.35%
85. $10^{-3.5}$

CUMULATIVE TEST FOR CHAPTERS 4–6   (page 411)

1.

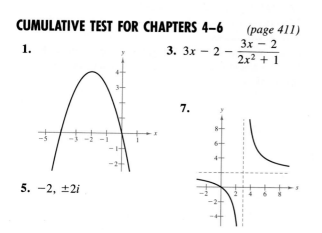

3. $3x - 2 - \dfrac{3x - 2}{2x^2 + 1}$

7.

5. $-2, \pm 2i$

5. x **6.** $-37v$ **7.** $2x^2 + 9$ **8.** -1 **9.** $x = 6$
10. $y = 1$

1. $(1, 2)$ **3.** $(-1, 2), (2, 5)$ **5.** $(0, 5), (3, 4)$
7. $(0, 0), (2, 4)$ **9.** $(-1, 1), (8, 4)$ **11.** $(5, 5)$
13. $\left(\frac{1}{2}, 3\right)$ **15.** $\left(\frac{3}{2}, \frac{3}{10}\right)$ **17.** $\left(\frac{20}{3}, \frac{40}{3}\right)$ **19.** $(0, 0)$
21. $(1, 2)$ **23.** $\left(\frac{29}{10}, \frac{21}{10}\right), (-2, 0)$
25. $(-1, -2), (2, 1)$ **27.** $(-1, 0), (0, 1), (1, 0)$
29. $\left(\frac{1}{2}, 2\right), \left(-4, -\frac{1}{4}\right)$ **31.** $(2, 2), (4, 0)$
33. No points of intersection   **35.** $(3, \pm 4)$ **37.** $(0, 1)$
39. $(0, 0), (1, 1)$ **41.** $(0, -13), (\pm 12, 5)$
43. 193 units   **45.** 233,334 units   **47.** 6,400 units

49. $13,000 at 8%, $12,000 at 8.25%
51. More than $8,333.33   **53. (a)** 24.7 in
(b) Doyle Log Rule   **55.** 8 mi × 12 mi
57. (a) $y = 2x$ **(b)** $y = 0$ **(c)** $y = x - 2$

Section 7.2   *(page 431)*

WARM UP

1.
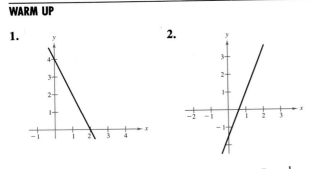
2.

3. $x - y + 4 = 0$ **4.** $5x + 3y - 28 = 0$ **5.** $-\frac{1}{2}$
6. $\frac{7}{4}$ **7.** Perpendicular   **8.** Parallel
9. Neither parallel nor perpendicular   **10.** Perpendicular

1. (2, 0)   **3.** (−1, −1)

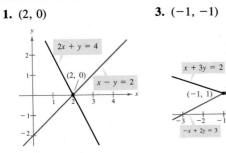

5. Inconsistent   **7.** All points (x, y) lying on
                                          the line $3x - 2y = 6$

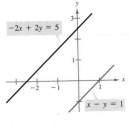
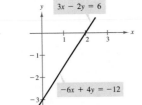

9. $\left(-\frac{1}{3}, -\frac{2}{3}\right)$

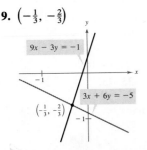

11. $\left(\frac{5}{2}, \frac{3}{4}\right)$ **13.** (3, 4)   **15.** (4, −1)   **17.** (40, 40)
19. Inconsistent   **21.** $\left(\frac{18}{5}, \frac{3}{5}\right)$ **23.** (5, −2)
25. All points (x, y) lying on the line $x - 2y = 5$.
27. $\left(\frac{90}{31}, -\frac{67}{31}\right)$ **29.** $\left(-\frac{6}{35}, \frac{43}{35}\right)$
31. (79,400, 398)   It is necessary to change the scale on the axes to see the point of intersection of the lines.
33. 550 mph, 50 mph
35. $\frac{20}{3}$ gal of 20% solution, $\frac{10}{3}$ gal of 50% solution
37. $4,000 at 10.5%, $8,000 at 12%
39. 375 adults, 125 children   **41.** (80, 10)
43. (100, 200)   **45.** (2,000,000, 100)
47. 75 mi, 225 mi   **49.** $y = 0.97x + 2.10$
51. $y = 0.318\,x + 4.061$ **53.** $y = \frac{3}{4}x + \frac{4}{3}$
55. $y = -2x + 4$ **57.** $y = -240x + 685$, 349 units
59. $x + 2y = 8$
      $x + 4y = 13$

Section 7.3   *(page 447)*

WARM UP   **1.** (15, 10)   **2.** $\left(-2, -\frac{8}{3}\right)$ **3.** (28, 4)
4. (4, 3)   **5.** Not a solution   **6.** Not a solution
7. Solution   **8.** Solution   **9.** $5a + 2$ **10.** $a + 13$

1. (1, 2, 3)   **3.** (2, −3, −2)   **5.** (5, −2, 0)
7. Inconsistent   **9.** $\left(1, -\frac{3}{2}, \frac{1}{2}\right)$
11. $(-3a + 10, 5a - 7, a)$ **13.** $\left(13 - 4a, \frac{45}{2} - \frac{15}{2}a, a\right)$
15. $(-a, 2a - 1, a)$ **17.** $\left(\frac{1}{2} - \frac{3}{2}a, 1 - \frac{2}{3}a, a\right)$
19. (1, 1, 1, 1)   **21.** Inconsistent   **23.** (0, 0, 0)
25. $\left(-\frac{3}{2}a, \frac{4}{3}a, a\right)$ **27.** $y = 2x^2 + 3x - 4$
29. $y = x^2 - 4x + 3$ **31.** $x^2 + y^2 - 4x = 0$
33. $x^2 + y^2 - 6x - 8y = 0$ **35.** $s = -16t^2 + 144$
37. $s = -16t^2 - 32t + 500$
39. $4,000 at 5%, $5,000 at 6%, $7,000 at 7%
41. $300,000 at 8%, $400,000 at 9%, $75,000 at 10%
43. $250,000 - \frac{1}{2}s$ in certificates of deposit
      $125,000 + \frac{1}{2}s$ in municipal bonds
      $125,000 - s$ in blue-chip stocks
      s in growth stocks
45. 20 gal of spray X, 18 gal of spray Y, 16 gal of spray Z
47. Use 4 medium trucks or use 2 large, 1 medium, and 2 small trucks
49. $t_1 = 96$ lb
      $t_2 = 48$ lb
      $a = -16$ ft/sec^2

51. $\frac{1}{2}\left(-\frac{2}{x} + \frac{1}{x - 1} + \frac{1}{x + 1}\right)$
53. $\frac{1}{2}\left(\frac{1}{x} - \frac{1}{x - 2} + \frac{2}{x + 3}\right)$
55. $y = -\frac{5}{24}x^2 - \frac{3}{10}x + \frac{41}{6}$ **57.** $y = x^2 - x$

59. $y = 0.141x^2 - 4.427x + 58.400$

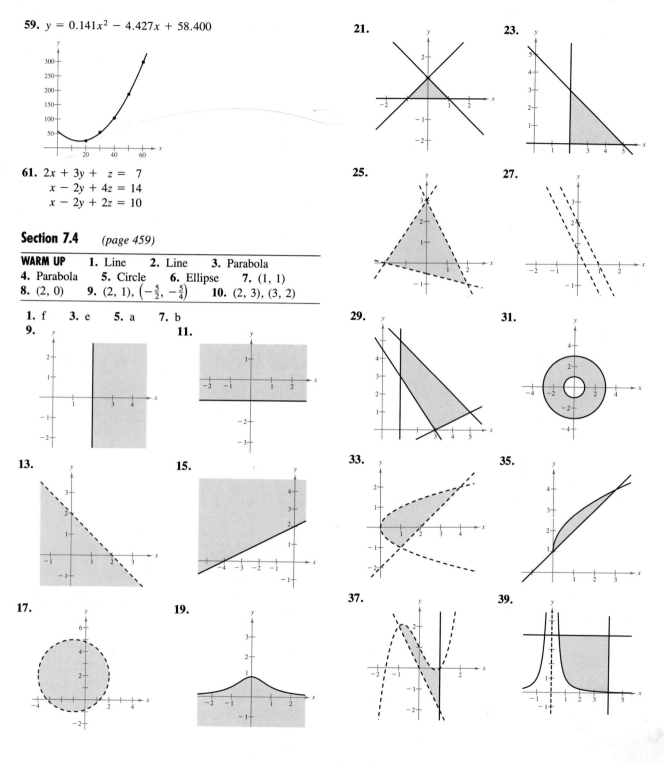

61. $2x + 3y + z = 7$
$x - 2y + 4z = 14$
$x - 2y + 2z = 10$

Section 7.4 *(page 459)*

WARM UP   **1.** Line   **2.** Line   **3.** Parabola
4. Parabola   **5.** Circle   **6.** Ellipse   **7.** (1, 1)
8. (2, 0)   **9.** (2, 1), $\left(-\frac{5}{2}, -\frac{5}{4}\right)$ **10.** (2, 3), (3, 2)

1. f   **3.** e   **5.** a   **7.** b
9.
11.
13.
15.
17.
19.
21.
23.
25.
27.
29.
31.
33.
35.
37.
39.

41. $2 \leq x \leq 5$
$1 \leq y \leq 7$

43. $y \leq \frac{3}{2}x$
$y \leq -x + 5$
$y \geq 0$

45. $x^2 + y^2 \leq 16$
$x \geq 0$
$y \geq 0$

47. $x + \frac{3}{2}y \leq 12$
$\frac{4}{3}x + \frac{3}{2}y \leq 15$
$x \qquad \geq 0$
$y \geq 0$

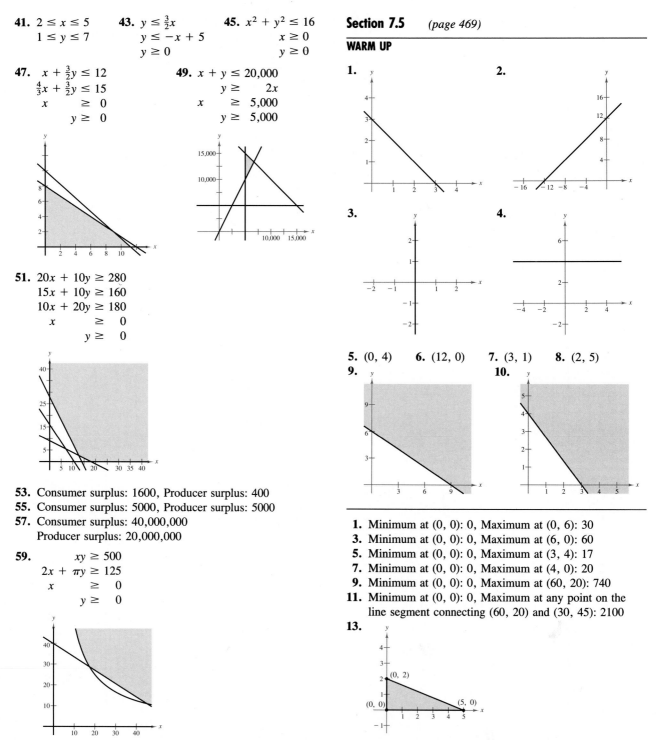

49. $x + y \leq 20{,}000$
$y \geq 2x$
$x \geq 5{,}000$
$y \geq 5{,}000$

51. $20x + 10y \geq 280$
$15x + 10y \geq 160$
$10x + 20y \geq 180$
$x \qquad \geq 0$
$y \geq 0$

53. Consumer surplus: 1600, Producer surplus: 400
55. Consumer surplus: 5000, Producer surplus: 5000
57. Consumer surplus: 40,000,000
Producer surplus: 20,000,000

59.
$xy \geq 500$
$2x + \pi y \geq 125$
$x \qquad \geq 0$
$y \geq 0$

Section 7.5 *(page 469)*

WARM UP

1.

2.

3.

4.

5. $(0, 4)$ **6.** $(12, 0)$ **7.** $(3, 1)$ **8.** $(2, 5)$
9. **10.**

1. Minimum at $(0, 0)$: 0, Maximum at $(0, 6)$: 30
3. Minimum at $(0, 0)$: 0, Maximum at $(6, 0)$: 60
5. Minimum at $(0, 0)$: 0, Maximum at $(3, 4)$: 17
7. Minimum at $(0, 0)$: 0, Maximum at $(4, 0)$: 20
9. Minimum at $(0, 0)$: 0, Maximum at $(60, 20)$: 740
11. Minimum at $(0, 0)$: 0, Maximum at any point on the line segment connecting $(60, 20)$ and $(30, 45)$: 2100
13.

$(0, 2)$
$(0, 0)$ $(5, 0)$

Minimum at $(0, 0)$: 0, Maximum at $(5, 0)$: 30

15.

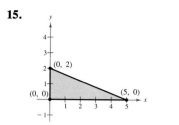

Minimum at (0, 0): 0, Maximum at (0, 2): 48

17.

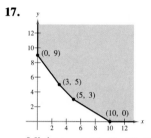

Minimum at (5, 3): 35, No maximum

19.

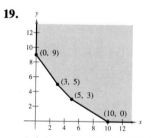

Minimum at (10, 0): 20, No maximum

21.

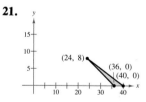

Minimum at (24, 8): 104, Maximum at (40, 0): 160

23.

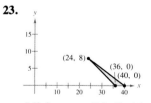

Minimum at (36, 0): 36, Maximum at (24, 8): 56

25. Maximum at (3, 6): 12   **27.** Maximum at (0, 10): 10

29. Maximum at (0, 5): 25   **31.** Maximum at (4, 4): 36

33. 200 units of the $250 model, 50 units of the $400 model

35. 3 bags of Brand X, 6 bags of Brand Y

37. 750 units of Model A, 1,000 units of Model B, Maximum profit: $83,750

39. 8 audits, 8 tax returns, Maximum revenue: $18,400

41.

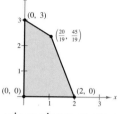

z is maximum at any point on the line segment between (2, 0) and $\left(\frac{20}{19}, \frac{45}{19}\right)$.

43.

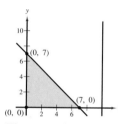

The constraint $x \le 10$ is extraneous.
Maximum 14 at (0, 7)

45.

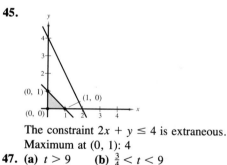

The constraint $2x + y \le 4$ is extraneous.
Maximum at (0, 1): 4

47. (a) $t > 9$ (b) $\frac{3}{4} < t < 9$

Chapter 7   Review Exercises   *(page 472)*

1. (1, 1)   **3.** (5, 4)   **5.** (0, 0), (2, 8), (−2, 8)

7. (0, 0), (−3, 3)   **9.** 4762 units   **11.** $\left(\frac{5}{2}, 3\right)$

13. (−0.5, 0.8)   **15.** (0, 0)

17. All points (x, y) lying on the line $5x − 8y = 14$.

19. 40 gal of the 75% solution, 60 gal of the 50% solution

21. 218.75 mi/hr, 193.75 mi/hr   **23.** $\left(\frac{500,000}{7}, \frac{159}{7}\right)$

25. $\left(\frac{24}{5}, \frac{22}{5}, -\frac{8}{5}\right)$ **27.** (3a + 4, 2a + 5, a)

29. (−3a + 2, 5a + 6, a)   **31.** $y = 2x^2 + x − 6$

33. $x^2 + y^2 − 4x + 2y − 4 = 0$

35. 10 gal of spray X, 5 gal of spray Y, 12 gal of spray Z

37. $y = 1.01x + 1.54$

39.

41.

43.

45.

47. $-x + y \leq 4$
$2x + y \leq 22$
$-x + y \geq -2$
$2x + y \geq 7$

49. $x + y \leq 1500$
$x \geq 400$
$y \geq 600$

51. Consumer surplus: 4,500,000
Producer surplus: 9,000,000

53. Maximum at (5, 8): 47

55. Minimum at (15, 0): 26.25

57. 5 units of A, 2 units of B, Maximum profit: $138

59. 3 bags of Brand X, 2 bags of Brand Y

CHAPTER 8

Section 8.1   (page 485)

WARM UP   **1.** -3 **2.** 30   **3.** 6   **4.** $-\frac{1}{9}$
5. Solution   **6.** Not a solution   **7.** (5, 2)
8. $\left(\frac{12}{5}, -3\right)$ **9.** (40, 14, 2)   **10.** $\left(\frac{15}{2}, 4, 1\right)$

1. 3×2 **3.** 5×1 **5.** 2×2
7. Reduced row-echelon form

9. Not in row-echelon form

11. $\begin{bmatrix} 1 & 4 & 3 \\ 0 & 2 & -1 \end{bmatrix}$

13. $\begin{bmatrix} 1 & 1 & 4 & -1 \\ 0 & 5 & -2 & 6 \\ 0 & 3 & 20 & 4 \end{bmatrix}$ $\begin{bmatrix} 1 & 1 & 4 & -1 \\ 0 & 1 & -\frac{2}{5} & \frac{6}{5} \\ 0 & 3 & 20 & 4 \end{bmatrix}$

15. (a) $\begin{bmatrix} 1 & 2 & 3 \\ 0 & -5 & -10 \\ 3 & 1 & -1 \end{bmatrix}$ **(b)** $\begin{bmatrix} 1 & 2 & 3 \\ 0 & -5 & -10 \\ 0 & -5 & -10 \end{bmatrix}$

(c) $\begin{bmatrix} 1 & 2 & 3 \\ 0 & -5 & -10 \\ 0 & 0 & 0 \end{bmatrix}$ **(d)** $\begin{bmatrix} 1 & 2 & 3 \\ 0 & 1 & 2 \\ 0 & 0 & 0 \end{bmatrix}$

(e) $\begin{bmatrix} 1 & 0 & -1 \\ 0 & 1 & 2 \\ 0 & 0 & 0 \end{bmatrix}$

17. $\begin{bmatrix} 1 & 1 & 0 & 5 \\ 0 & 1 & 2 & 0 \\ 0 & 0 & 1 & -1 \end{bmatrix}$ **19.** $\begin{bmatrix} 1 & -1 & -1 & 1 \\ 0 & 1 & 6 & 3 \\ 0 & 0 & 0 & 0 \end{bmatrix}$

21. $\begin{bmatrix} 1 & 0 & 0 \\ 0 & 1 & 0 \\ 0 & 0 & 1 \end{bmatrix}$ **23.** $\begin{bmatrix} 1 & 2 & 0 & 0 \\ 0 & 0 & 1 & 0 \\ 0 & 0 & 0 & 1 \\ 0 & 0 & 0 & 0 \end{bmatrix}$

25. $4x + 3y = 8$
$x - 2y = 3$

27. $x \qquad + 2z = -10$
$3y - z = 5$
$4x + 2y \qquad = 3$

29. $x - 2y = 4$
$y = -3$
$(-2, -3)$

31. $x - y + 2z = 4$
$y - z = 2$
$z = -2$
$(8, 0, -2)$

33. $(7, -5)$ **35.** $(-4, -8, 2)$ **37.** $(3, 2)$
39. $(4, -2)$ **41.** $\left(\frac{1}{2}, -\frac{3}{4}\right)$ **43.** Inconsistent
45. $(4, -3, 2)$ **47.** $(2a + 1, 3a + 2, a)$
49. $(5a + 4, -3a + 2, a)$ **51.** $(0, 2 - 4a, a)$
53. $(1, 0, 4, -2)$ **55.** $(0, 0)$ **57.** $(-2a, a, a)$
59. \$800,000 at 8%, \$500,000 at 9%, \$200,000 at 12%
61. $y = x^2 + 2x + 4$ **63.** $x^2 + y^2 - 5x - 3y + 6 = 0$
65. $x + y + 7z = -1$
$2x + y + 10z = -3$
$3x + 2y + 17z = -4$

Section 8.2   (page 498)

WARM UP   **1.** -5 **2.** -7
3. Not in reduced row-echelon form
4. Not in reduced row-echelon form
5. $\begin{bmatrix} -5 & 10 & : & 12 \\ 7 & -3 & : & 0 \\ -1 & 7 & : & 25 \end{bmatrix}$ **6.** $\begin{bmatrix} 10 & 15 & -9 & : & 42 \\ 6 & -5 & 0 & : & 0 \end{bmatrix}$

7. $(0, 2)$ **8.** $(2 + a, 3 - a, a)$ **9.** $(1 - 2a, a, -1)$
10. $(2, -1, -1)$

---

1. $x = -4,\ y = 22$ **3.** $x = 2,\ y = 3$

5. (a) $\begin{bmatrix} 3 & -2 \\ 1 & 7 \end{bmatrix}$ **(b)** $\begin{bmatrix} -1 & 0 \\ 3 & -9 \end{bmatrix}$ **(c)** $\begin{bmatrix} 3 & -3 \\ 6 & -3 \end{bmatrix}$

(d) $\begin{bmatrix} -1 & -1 \\ 8 & -19 \end{bmatrix}$ **7. (a)** $\begin{bmatrix} 7 & 3 \\ 1 & 9 \\ -2 & 15 \end{bmatrix}$ **(b)** $\begin{bmatrix} 5 & -5 \\ 3 & -1 \\ -4 & -5 \end{bmatrix}$

(c) $\begin{bmatrix} 18 & -3 \\ 6 & 12 \\ -9 & 15 \end{bmatrix}$ **(d)** $\begin{bmatrix} 16 & -11 \\ 8 & 2 \\ -11 & -5 \end{bmatrix}$

9. (a) $\begin{bmatrix} 3 & 3 & -2 & 1 & 1 \\ -2 & 5 & 7 & -6 & -8 \end{bmatrix}$

(b) $\begin{bmatrix} 1 & 1 & 0 & -1 & 1 \\ 4 & -3 & -11 & 6 & 6 \end{bmatrix}$

(c) $\begin{bmatrix} 6 & 6 & -3 & 0 & 3 \\ 3 & 3 & -6 & 0 & -3 \end{bmatrix}$

(d) $\begin{bmatrix} 4 & 4 & -1 & -2 & 3 \\ 9 & -5 & -24 & 12 & 11 \end{bmatrix}$ **11. (a)** $\begin{bmatrix} 0 & 15 \\ 6 & 12 \end{bmatrix}$

(b) $\begin{bmatrix} -2 & 2 \\ 31 & 14 \end{bmatrix}$ **(c)** $\begin{bmatrix} 9 & 6 \\ 12 & 12 \end{bmatrix}$ **13. (a)** $\begin{bmatrix} 0 & -10 \\ 10 & 0 \end{bmatrix}$

(b) $\begin{bmatrix} 0 & -10 \\ 10 & 0 \end{bmatrix}$ **(c)** $\begin{bmatrix} 8 & -6 \\ 6 & 8 \end{bmatrix}$

15. (a) $\begin{bmatrix} 6 & -21 & 15 \\ 8 & -23 & 19 \\ 4 & 7 & 5 \end{bmatrix}$ **(b)** $\begin{bmatrix} 9 & 0 & 13 \\ 7 & -2 & 21 \\ 1 & 4 & -19 \end{bmatrix}$

(c) $\begin{bmatrix} 20 & 7 & -8 \\ 24 & 7 & -2 \\ 2 & -5 & 30 \end{bmatrix}$ **17.** Not possible

19. $\begin{bmatrix} -1 & 19 \\ 4 & -27 \\ 0 & 14 \end{bmatrix}$ **21.** $\begin{bmatrix} 1 & 0 & 0 \\ 0 & 1 & 0 \\ 0 & 0 & \frac{7}{2} \end{bmatrix}$ **23.** $\begin{bmatrix} 60 & 72 \\ -20 & -24 \\ 10 & 12 \\ 60 & 72 \end{bmatrix}$

25. $\begin{bmatrix} -6 & -9 \\ -1 & 0 \\ 17 & -10 \end{bmatrix}$ **27.** $\begin{bmatrix} 3 & 3 \\ -\frac{1}{2} & 0 \\ -\frac{13}{2} & \frac{11}{2} \end{bmatrix}$

29. $A = \begin{bmatrix} -1 & 1 \\ -2 & 1 \end{bmatrix}$ **31.** $A = \begin{bmatrix} 1 & -2 & 3 \\ -1 & 3 & -1 \\ 2 & -5 & 5 \end{bmatrix}$

$X = \begin{bmatrix} x \\ y \end{bmatrix}$

$B = \begin{bmatrix} 4 \\ 0 \end{bmatrix}$ $X = \begin{bmatrix} x \\ y \\ z \end{bmatrix}$

$x = 4,\ y = 8$ $B = \begin{bmatrix} 9 \\ -6 \\ 17 \end{bmatrix}$

$x = 1,\ y = -1,\ z = 2$

33. $\begin{bmatrix} -4 & 0 \\ 8 & 2 \end{bmatrix}$ **35.** $\begin{bmatrix} 0 & 0 & 0 \\ 0 & 0 & 0 \\ 0 & 0 & 0 \end{bmatrix}$

37. $AC = BC = \begin{bmatrix} 12 & -6 & 9 \\ 16 & -8 & 12 \\ 4 & -2 & 3 \end{bmatrix}$ **39.** $\begin{bmatrix} 72 & 48 & 24 \\ 36 & 108 & 72 \end{bmatrix}$

41. $AB = [\$1250 \quad \$1331.25 \quad \$981.25]$
The entries represent the profit from the two products at each of the three outlets.

43. (a) $\$18,300$ **(b)** $\$21,260$

(c) $\begin{bmatrix} \$15,770 & \$18,300 \\ \$26,500 & \$29,250 \\ \$21,260 & \$24,150 \end{bmatrix}$

The entries are the wholesale and retail price of the inventory at each outlet.

45. $\begin{bmatrix} 0.40 & 0.15 & 0.15 \\ 0.28 & 0.53 & 0.17 \\ 0.32 & 0.32 & 0.68 \end{bmatrix}$

Section 8.3   *(page 507)*

WARM UP   **1.** $\begin{bmatrix} 4 & 24 \\ 0 & -16 \\ 48 & 8 \end{bmatrix}$ **2.** $\begin{bmatrix} \frac{11}{2} & 5 & 24 \\ \frac{1}{2} & 0 & 8 \\ 0 & 1 & 4 \end{bmatrix}$

3. $\begin{bmatrix} -5 & -2 & -13 \\ 4 & -13 & -2 \end{bmatrix}$ **4.** $\begin{bmatrix} -13 & 11 \\ -19 & 21 \end{bmatrix}$ **5.** $\begin{bmatrix} 1 & 0 \\ 0 & 1 \end{bmatrix}$

6. $\begin{bmatrix} 6 & 5 \\ 3 & -2 \end{bmatrix}$ **7.** $\begin{bmatrix} 1 & 0 & 0 \\ 0 & 1 & 0 \\ 0 & 0 & 1 \end{bmatrix}$ **8.** $\begin{bmatrix} 1 & 0 & 0 \\ 0 & 1 & 0 \\ 0 & 0 & 1 \end{bmatrix}$

9. $\begin{bmatrix} 1 & 0 & 3 & -2 \\ 0 & 1 & 4 & -3 \end{bmatrix}$

10. $\begin{bmatrix} 1 & 0 & 0 & -6 & -4 & 3 \\ 0 & 1 & 0 & 11 & 6 & -5 \\ 0 & 0 & 1 & -2 & -1 & 1 \end{bmatrix}$

1–8. Proofs

9. $\begin{bmatrix} \frac{1}{2} & 0 \\ 0 & \frac{1}{3} \end{bmatrix}$ **11.** $\begin{bmatrix} -3 & 2 \\ -2 & 1 \end{bmatrix}$ **13.** $\begin{bmatrix} 1 & -1 \\ 2 & -1 \end{bmatrix}$

15. Does not exist   **17.** Does not exist

19. $\begin{bmatrix} 1 & 1 & -1 \\ -3 & 2 & -1 \\ 3 & -3 & 2 \end{bmatrix}$ **21.** $\frac{1}{2}\begin{bmatrix} -3 & 3 & 2 \\ 9 & -7 & -6 \\ -2 & 2 & 2 \end{bmatrix}$

23. $\frac{5}{11}\begin{bmatrix} 0 & -4 & 2 \\ -22 & 11 & 11 \\ 22 & -6 & -8 \end{bmatrix}$ **25.** $\begin{bmatrix} 1 & 0 & 0 \\ -0.75 & 0.25 & 0 \\ 0.35 & -0.25 & 0.2 \end{bmatrix}$

27. Does not exist   **29.** $\begin{bmatrix} -24 & 7 & 1 & -2 \\ -10 & 3 & 0 & -1 \\ -29 & 7 & 3 & -2 \\ 12 & -3 & -1 & 1 \end{bmatrix}$

31. $(5, 0)$ **33.** $(-8, -6)$ **35.** $(4, 8)$ **37.** $(10, 30)$
39. $(3, 8, -11)$ **41.** $(2, 1, 0, 0)$
43. $\$10,000$ in AAA rated bonds, $\$5,000$ in A rated bonds, $\$10,000$ in B rated bonds

45. $9,000 in AAA rated bonds, $1,000 in A rated bonds, $2,000 in B rated bonds

47. $I_1 = -3$ amps
$I_2 = 8$ amps
$I_3 = 5$ amps

Section 8.4 *(page 520)*

WARM UP **1.** $\begin{bmatrix} 3 & 5 \\ 4 & 0 \end{bmatrix}$ **2.** $\begin{bmatrix} -2 & 8 \\ 2 & -4 \end{bmatrix}$

3. $\begin{bmatrix} 9 & -12 & 6 \\ 3 & 0 & -3 \\ 0 & 3 & -6 \end{bmatrix}$ **4.** $\begin{bmatrix} 0 & 8 & 12 \\ -4 & 8 & 12 \\ -8 & 4 & -8 \end{bmatrix}$ **5.** -22

6. 35 **7.** -15 **8.** $-\frac{1}{8}$ **9.** -45 **10.** -16

1. 5 **3.** 5 **5.** 27 **7.** -24 **9.** 6 **11.** 0
13. -0.002 **15.** 0 **17.** 0 **19.** -9 **21.** -18
23. $-7x + 3y - 8$
25. (a) $M_{11} = -5$, $M_{12} = 2$, $M_{21} = 4$, $M_{22} = 3$
  (b) $C_{11} = -5$, $C_{12} = -2$, $C_{21} = -4$, $C_{22} = 3$
27. (a) $M_{11} = 30$, $M_{12} = 12$, $M_{13} = 11$, $M_{21} = -36$,
  $M_{22} = 26$, $M_{23} = 7$, $M_{31} = -4$, $M_{32} = -42$,
  $M_{33} = 12$
  (b) $C_{11} = 30$, $C_{12} = -12$, $C_{13} = 11$, $C_{21} = 36$,
  $C_{22} = 26$, $C_{23} = -7$, $C_{31} = -4$, $C_{32} = 42$, $C_{33} = 12$
29. -75 **31.** 96 **33.** 170 **35.** -58 **37.** -30
39. -108 **41.** 0 **43.** 412 **45.** $x = -1$, $x = 4$
47. $8uv - 1$ **49.** e^{5x} **51.** $1 - \ln x$

Section 8.5 *(page 527)*

WARM UP **1.** $\begin{bmatrix} 1 & -3 \\ 0 & 1 \end{bmatrix}$ **2.** $\begin{bmatrix} 1 & -3 \\ 0 & 1 \end{bmatrix}$

3. $\begin{bmatrix} 1 & 3 & 4 \\ 0 & 1 & 1 \\ 0 & 0 & 0 \end{bmatrix}$ **4.** $\begin{bmatrix} 1 & 2 & 4 \\ 0 & 1 & \frac{10}{7} \\ 0 & 0 & 0 \end{bmatrix}$ **5.** -2 **6.** 0

7. -8 **8.** x^2 **9.** 8 **10.** 60

1. Column 2 is a multiple of Column 1.
3. Row 2 has only zero entries.
5. The interchange of Columns 2 and 3 results in a change of sign of the determinant.
7. Multiplying any row by a constant multiplies the value of the determinant by that constant.
9. Multiplying the entries of all three rows by 5 multiplies the value of the determinant by 5^3.
11. Adding -4 times the entries of Row 1 to the elements of Row 2 leaves the determinant unchanged.
13. Adding multiples of Column 2 to Columns 1 and 3 leaves the determinant unchanged.

15. 1 **17.** -26 **19.** -126 **21.** 0 **23.** 0
25. 236 **27.** 7441 **29.** 410 **31.** Not invertible
33. Invertible **35.** Invertible **37.** $k = -1$, $k = 4$
39. Proof **41.** Proof **43.** Proof
45. (a) -3 **(b)** -2 **(c)** $\begin{bmatrix} -2 & 0 \\ 0 & -3 \end{bmatrix}$ **(d)** 6

47. (a) 2 **(b)** -6 **(c)** $\begin{bmatrix} 1 & 4 & 3 \\ -1 & 0 & 3 \\ 0 & 2 & 0 \end{bmatrix}$ **(d)** -12

Section 8.6 *(page 538)*

WARM UP **1.** $(1, 1)$ **2.** $(1, 2)$ **3.** $(3, 0, -4)$
4. $(-2, 1, 1)$ **5.** 8 **6.** -49 **7.** -3 **8.** 20
9. 9 **10.** 35

1. $(1, 2)$ **3.** $(2, -2)$ **5.** $\left(\frac{3}{4}, -\frac{1}{2}\right)$
7. Cramer's Rule does not apply. **9.** $\left(\frac{2}{3}, \frac{1}{2}\right)$ **11.** -1
13. 1 **15.** 0 **17.** Cramer's Rule does not apply.
19. 5
21. $I_1 = \frac{125}{22}$ amps
  $I_2 = \frac{93}{11}$ amps
  $I_3 = \frac{61}{22}$ amps
23. $y = 1.768 + 0.202t$; Maximum contribution is about $4200. **25.** 7 **27.** 14 **29.** $\frac{33}{8}$ **31.** $\frac{5}{2}$
33. 28 **35.** 250 sq mi **37.** Collinear
39. Not collinear **41.** Collinear **43.** $3x - 5y = 0$
45. $x + 3y - 5 = 0$ **47.** $2x + 3y - 8 = 0$
49. 1 -25 -65 17 15 -9 -12 -62 -119 27 51 48 43 67 48 57 111 117
51. -5 -41 -87 91 207 257 11 -5 -41 40 80 84 76 177 227
53. SEND PLANES

Chapter 8 Review Exercises *(page 541)*

1. $(10, -12)$ **3.** $(0.6, 0.5)$ **5.** $(2, -3, 3)$
7. $\left(\frac{1}{2}, -\frac{1}{3}, 1\right)$ **9.** $\left(-2a + \frac{3}{2}, 2a + 1, a\right)$
11. Inconsistent **13.** $\begin{bmatrix} -13 & -8 & 18 \\ 0 & 11 & -19 \end{bmatrix}$

15. $\begin{bmatrix} 14 & -2 & 8 \\ 14 & -10 & 40 \\ 36 & -12 & 48 \end{bmatrix}$ **17.** $\begin{bmatrix} 44 & 4 \\ 20 & 8 \end{bmatrix}$

19. $\begin{bmatrix} 4 & 6 & 3 \\ 0 & 6 & -10 \\ 0 & 0 & 6 \end{bmatrix}$ **21.** $\begin{bmatrix} -14 & -4 \\ 7 & -17 \\ -17 & -2 \end{bmatrix}$

23. $\frac{1}{3}\begin{bmatrix} 9 & 2 \\ -4 & 11 \\ 10 & 0 \end{bmatrix}$

25. $5x + 4y = 2$
  $-x + y = -22$

27. $\begin{bmatrix} \frac{1}{5} & \frac{1}{5} \\ \frac{1}{10} & -\frac{1}{15} \end{bmatrix}$ **29.** $\begin{bmatrix} \frac{1}{2} & -1 & -\frac{1}{2} \\ \frac{1}{2} & -\frac{2}{3} & -\frac{5}{6} \\ 0 & \frac{2}{3} & \frac{1}{3} \end{bmatrix}$

31. 550 **33.** 279 **35.** $(-3, 1)$ **37.** $(1, 1, -2)$
39. $(2, -4, 6)$ **41.** Inconsistent **43.** 16 **45.** 7
47. $x - 2y + 4 = 0$ **49.** $2x + 6y - 13 = 0$
51. 8 carnations, 4 roses **53.** $y = x^2 + 2x + 3$
55. 128; Each of the three rows is multiplied by 4.

CHAPTER 9

Section 9.1 *(page 552)*

WARM UP **1.** $\frac{4}{5}$ **2.** $\frac{1}{3}$ **3.** $(2n + 1)(2n - 1)$
4. $(2n - 1)(2n - 3)$ **5.** $(n - 1)(n - 2)$
6. $(n + 1)(n + 2)$ **7.** $\frac{1}{3}$ **8.** 24 **9.** $\frac{13}{24}$ **10.** $\frac{3}{4}$

1. 3, 5, 7, 9, 11 **3.** 2, 4, 8, 16, 32
5. $-2, 4, -8, 16, -32$ **7.** $0, 1, 0, \frac{1}{2}, 0$
9. $\frac{5}{2}, \frac{11}{4}, \frac{23}{8}, \frac{47}{16}, \frac{95}{32}$ **11.** $1, \frac{1}{2^{3/2}}, \frac{1}{3^{3/2}}, \frac{1}{4^{3/2}}, \frac{1}{5^{3/2}}$
13. $3, \frac{9}{2}, \frac{9}{2}, \frac{27}{8}, \frac{81}{40}$ **15.** $-1, \frac{1}{4}, -\frac{1}{9}, \frac{1}{16}, -\frac{1}{25}$
17. 3, 4, 6, 10, 18 **19.** $\frac{1}{30}$ **21.** $n + 1$
23. $\dfrac{1}{2n(2n + 1)}$ **25.** $a_n = 3n - 2$ **27.** $a_n = n^2 - 1$
29. $a_n = \dfrac{(-1)^{n+1}}{2^n}$ **31.** $a_n = 1 + \dfrac{1}{n}$ **33.** $a_n = \dfrac{1}{n!}$
35. $a_n = (-1)^{n+1}$ **37.** 35 **39.** 40 **41.** 30
43. $\frac{9}{5}$ **45.** 238 **47.** 56 **49.** $\frac{47}{60}$ **51.** $\displaystyle\sum_{i=1}^{9} \frac{1}{3i}$
53. $\displaystyle\sum_{i=1}^{8} \left[2\left(\frac{i}{8}\right) + 3\right]$ **55.** $\displaystyle\sum_{i=1}^{6}(-1)^{i+1}3^i$
57. $\displaystyle\sum_{i=1}^{20} \frac{(-1)^{i+1}}{i^2}$ **59.** $\displaystyle\sum_{i=1}^{5} \frac{2^i - 1}{2^{i+1}}$
61. (a) $A_1 = \$5,100.00$, $A_2 = \$5,202.00$, $A_3 = \$5,306.04$,
$A_4 = \$5,412.16$, $A_5 = \$5,520.40$, $A_6 = \$5,630.81$,
$A_7 = \$5,743.43$, $A_8 = \$5,858.30$
(b) $\$11,040.20$
63. $a_0 = 242.67$, $a_1 = 285.34$, $a_2 = 328.01$,
$a_3 = 370.68$, $a_4 = 413.35$, $a_5 = 456.02$,
$a_6 = 498.69$, $a_7 = 541.36$

65. 16.02; Result of adding dividends in the figure is 16.11.

Section 9.2 *(page 562)*

WARM UP **1.** 36 **2.** 240 **3.** $\frac{11}{2}$ **4.** $\frac{10}{3}$
5. 18 **6.** 4 **7.** 143 **8.** 160 **9.** 430
10. 256

1. Arithmetic sequence, $d = 3$
3. Not an arithmetic sequence
5. Arithmetic sequence, $d = -\frac{1}{4}$
7. Not an arithmetic sequence
9. Arithmetic sequence, $d = 0.4$
11. 8, 11, 14, 17, 20; Arithmetic sequence, $d = 3$
13. $\frac{1}{2}, \frac{1}{3}, \frac{1}{4}, \frac{1}{5}, \frac{1}{6}$; Not an arithmetic sequence
15. 97, 94, 91, 88, 85; Arithmetic sequence, $d = -3$
17. 1, 1, 2, 3, 5; Not an arithmetic sequence
19. $a_n = 3n - 2$ **21.** $a_n = -8n + 108$
23. $a_n = 2xn - x$ **25.** $a_n = -\frac{5}{2}n + \frac{13}{2}$
27. $a_n = \frac{10}{3}n + \frac{5}{3}$ **29.** $a_n = -3n + 103$
31. 5, 11, 17, 23, 29
33. $-2.6, -3.0, -3.4, -3.8, -4.2$ **35.** $\frac{3}{2}, \frac{5}{4}, 1, \frac{3}{4}, \frac{1}{2}$
37. 2, 6, 10, 14, 18 **39.** $-2, 2, 6, 10, 14$ **41.** 620
43. 4600 **45.** 265 **47.** 4000 **49.** 1275
51. 25,250 **53.** 355 **55.** 126,750 **57.** 520
59. 44,625 **61.** 9, 13 **63.** $\frac{15}{4}, \frac{9}{2}, \frac{21}{4}$ **65.** 10,000
67. (a) $\$35,000$ **(b)** $\$187,500$ **69.** 2340
71. 470 bricks

Section 9.3 *(page 573)*

WARM UP **1.** $\frac{64}{125}$ **2.** $\frac{9}{16}$ **3.** $\frac{1}{16}$ **4.** $\frac{5}{81}$ **5.** $6n^3$
6. $27n^4$ **7.** $4n^3$ **8.** n^2 **9.** $\dfrac{2^n}{81^n}$ **10.** $\dfrac{3}{16^n}$

1. Geometric sequence, $r = 3$
3. Not a geometric sequence
5. Geometric sequence, $r = -\frac{1}{2}$
7. Not a geometric sequence
9. Not a geometric sequence **11.** 2, 6, 18, 54, 162
13. $1, \frac{1}{2}, \frac{1}{4}, \frac{1}{8}, \frac{1}{16}$ **15.** $5, -\frac{1}{2}, \frac{1}{20}, -\frac{1}{200}, \frac{1}{2000}$
17. $1, e, e^2, e^3, e^4$ **19.** $3, \dfrac{3x}{2}, \dfrac{3x^2}{4}, \dfrac{3x^3}{8}, \dfrac{3x^4}{16}$
21. $\left(\frac{1}{2}\right)^7$ **23.** $-\dfrac{2}{3^{10}}$ **25.** $100e^{8x}$ **27.** $500(1.02)^{39}$
29. 9 **31.** $-\frac{2}{9}$ **33. (a)** $\$2,593.74$ **(b)** $\$2,653.30$
(c) $\$2,685.06$ **(d)** $\$2,707.04$ **(e)** $\$2,717.91$
35. $\$22,689.45$ **37.** 511 **39.** 43 **41.** ≈ 6.4
43. $\approx 29,921.31$ **45.** ≈ 2092.60 **47.** $\$7,808.24$
49. Proof **51. (a)** $\$26,198.27$ **(b)** $\$26,263.88$

53. (a) $637,678.02 **(b)** $645,861.43
55. $3,048.1 million **57.** $3,623,993.23 **59.** 2
61. $\frac{2}{3}$ **63.** $\frac{16}{3}$ **65.** 32 **67.** $\frac{8}{3}$ **69.** 152.42 ft

Section 9.4 (page 581)

WARM UP **1.** 24 **2.** 40 **3.** $\frac{77}{60}$ **4.** $\frac{7}{2}$
5. $\frac{2k+5}{5}$ **6.** $\frac{3k+1}{6}$ **7.** $8 \cdot 2^{2k} = 2^{2k+3}$ **8.** $\frac{1}{9}$
9. $\frac{1}{k}$ **10.** $\frac{4}{5}$

1. 210 **3.** 91 **5.** 225 **7.** 2275 **9.** 70
11. $\frac{5}{(k+1)(k+2)}$ **13.** $\frac{(k+1)^2(k+2)^2}{4}$

Section 9.5 (page 589)

WARM UP **1.** $5x^5 + 15x^2$ **2.** $x^3 + 5x^2 - 3x - 15$
3. $x^2 + 8x + 16$ **4.** $4x^2 - 12x + 9$ **5.** $\frac{3x^3}{y}$
6. $-32z^5$ **7.** 120 **8.** 336 **9.** 720 **10.** 20

1. 10 **3.** 1 **5.** 15,504 **7.** 4950 **9.** 4950
11. $x^4 + 4x^3 + 6x^2 + 4x + 1$
13. $a^3 + 6a^2 + 12a + 8$
15. $y^4 - 8y^3 + 24y^2 - 32y + 16$
17. $x^5 + 5x^4y + 10x^3y^2 + 10x^2y^3 + 5xy^4 + y^5$
19. $r^6 + 18r^5s + 135r^4s^2 + 540r^3s^3 + 1215r^2s^4 + 1458rs^5 + 729s^6$
21. $x^5 - 5x^4y + 10x^3y^2 - 10x^2y^3 + 5xy^4 - y^5$
23. $1 - 6x + 12x^2 - 8x^3$
25. $x^8 + 20x^6 + 150x^4 + 500x^2 + 625$
27. $\frac{1}{x^5} + \frac{5y}{x^4} + \frac{10y^2}{x^3} + \frac{10y^3}{x^2} + \frac{5y^4}{x} + y^5$
29. $2x^4 - 24x^3 + 113x^2 - 246x + 207$ **31.** -4
33. $2035 + 828i$ **35.** 1
37. $32t^5 - 80t^4s + 80t^3s^2 - 40t^2s^3 + 10ts^4 - s^5$
39. $81 - 216z + 216z^2 - 96z^3 + 16z^4$ **41.** 1,732,104
43. 180 **45.** $-326,592$ **47.** 210
49. $\frac{21}{128} + \frac{7}{128} + \frac{21}{128} + \frac{35}{128} + \frac{35}{128} + \frac{21}{128} + \frac{7}{128} + \frac{1}{128}$
51. $\frac{1}{6561} + \frac{16}{6561} + \frac{112}{6561} + \frac{448}{6561} + \frac{1120}{6561} + \frac{1792}{6561} + \frac{1792}{6561} + \frac{1024}{6561} + \frac{256}{6561}$
53. $0.07776 + 0.25920 + 0.34560 + 0.23040 + 0.07680 + 0.01024$
55. 1.172 **57.** 510,568.785
59. $g(x) = -x^2 - 5x - 2$
61. $g(x) = x^3 - 15x^2 + 71x - 105$
63. $g(t) = 0.2187t^2 + 5.0455t + 55.255$

Section 9.6 (page 601)

WARM UP **1.** 6656 **2.** 291,600 **3.** 7920
4. 13,800 **5.** 792 **6.** 2300
7. $n(n-1)(n-2)(n-3)$ **8.** $n(n-1)(2n-1)$
9. $n!$ **10.** $n!$

1. 7 **3.** 12 **5.** 12 **7.** 6,760,000 **9.** 64
11. (a) 900 **(b)** 648 **(c)** 180 **(d)** 600
13. 64,000 **15. (a)** 720 **(b)** 48 **17.** 24
19. 336 **21.** 1,860,480 **23.** 9900 **25.** 120
27. ABCD, ABDC, ACBD, ACDB, ADBC, ADCB, BACD, BADC, CABD, CADB, DABC, DACB, BCAD, BDAC, CBAD, CDAB, DBAC, DCAB, BCDA, BDCA, CBDA, CDBA, DBCA, DCBA
29. 120 **31.** 11,880 **33.** 420 **35.** 1260
37. 2520
39. AB, AC, AD, AE, AF, BC, BD, BE, BF, CD, CE, CF, DE, DF, EF
41. 4845 **43.** 3,838,380 **45.** 3,921,225 **47.** 560
49. (a) 70 **(b)** 30 **51. (a)** 70 **(b)** 54 **(c)** 16
53. 5 **55.** 20 **57.** $n = 5$ or $n = 6$

Section 9.7 (page 614)

WARM UP **1.** $\frac{9}{16}$ **2.** $\frac{8}{15}$ **3.** $\frac{1}{6}$ **4.** $\frac{1}{80,730}$
5. $\frac{1}{495}$ **6.** $\frac{1}{24}$ **7.** $\frac{1}{12}$ **8.** $\frac{135}{323}$ **9.** 0.366
10. 0.997

1. $\{(h, 1), (h, 2), (h, 3), (h, 4), (h, 5), (h, 6), (t, 1), (t, 2), (t, 3), (t, 4), (t, 5), (t, 6)\}$
3. $\{ABC, ACB, BAC, BCA, CAB, CBA\}$
5. $\{AB, AC, AD, AE, BC, BD, BE, CD, CE, DE\}$
7. $\frac{3}{8}$ **9.** $\frac{7}{8}$ **11.** $\frac{3}{13}$ **13.** $\frac{5}{13}$ **15.** $\frac{1}{12}$ **17.** $\frac{7}{12}$
19. $\frac{1}{3}$ **21.** $\frac{1}{5}$ **23.** $\frac{2}{5}$ **25.** 0.3 **27.** 0.85
29. (a) $\frac{112}{209}$ **(b)** $\frac{97}{209}$ **(c)** $\frac{274}{627}$
31. $P(\{\text{Taylor wins}\}) = 0.50$, $P(\{\text{Moore wins}\}) = P(\{\text{Jenkins wins}\}) = 0.25$
33. (a) $\frac{21}{1292} = 0.016$ **(b)** $\frac{225}{646} \approx 0.348$
(c) $\frac{49}{323} \approx 0.152$ **35. (a)** $\frac{1}{3}$ **(b)** $\frac{5}{8}$ **37. (a)** $\frac{1}{120}$
(b) $\frac{1}{24}$ **39. (a)** $\frac{1}{169}$ **(b)** $\frac{1}{221}$ **41. (a)** $\frac{14}{55}$ **(b)** $\frac{12}{55}$
(c) $\frac{54}{55}$ **43. (a)** $\frac{1}{4}$ **(b)** $\frac{1}{2}$ **(c)** $\frac{9}{100}$ **(d)** $\frac{1}{30}$
45. (a) ≈ 0.9702 **(b)** ≈ 0.9998 **(c)** ≈ 0.0002
47. (a) $\frac{1}{1024}$ **(b)** $\frac{243}{1024}$ **(c)** $\frac{781}{1024}$ **49. (a)** $\frac{1}{16}$
(b) $\frac{1}{8}$ **(c)** $\frac{15}{16}$ **51.** 0.1024

Chapter 9 Review Exercises *(page 617)*

1. $\sum_{k=1}^{20} \frac{1}{2k}$ **3.** $\sum_{k=1}^{9} \frac{k}{k+1}$ **5.** 30 **7.** 127

9. 12 **11.** 88 **13.** 418 **15.** 3, 7, 11, 15, 19

17. 1, 4, 7, 10, 13 **19.** $a_n = -3n + 103$; 1430

21. 25,250 **23.** 4, -1, $\frac{1}{4}$, $-\frac{1}{16}$, $\frac{1}{64}$

25. 9, 6, 4, $\frac{8}{3}$, $\frac{16}{9}$ or 9, -6, 4, $-\frac{8}{3}$, $\frac{16}{9}$

27. $a_n = 16\left(-\frac{1}{2}\right)^{n-1}$, ≈ 10.667 **29. (a)** $a_t = 120,000(0.7)^t$

(b) \$20,168.40 **31.** \$5,111.82 **33.** Proof

35. Proof **37.** 15 **39.** 6720

41. $\frac{x^4}{16} + \frac{x^3 y}{2} + \frac{3x^2 y^2}{2} + 2xy^3 + y^4$

43. $\frac{64}{x^6} - \frac{576}{x^4} + \frac{2160}{x^2} - 4320 + 4860x^2 - 2916x^4 + 729x^6$

45. $41 + 840i$ **47. (a)** 1 **(b)** 6 **(c)** 15

49. 118,813,760 **51.** $\frac{1}{9}$

53. $P(\{3\}) = \frac{1}{6}$, $P(\{(1, 5), (5, 1), (2, 4), (4, 2), (3, 3)\}) = \frac{5}{36}$

55. $\frac{31}{32}$ **57.** 0.0475

CUMULATIVE TEST FOR CHAPTERS 7-9 *(page 619)*

1. $(1, 2)$, $\left(-\frac{3}{2}, \frac{3}{4}\right)$ **3.** $(4, 2, -3)$ **5.** -6

7. $\begin{bmatrix} 11 & -6 \\ 5 & 3 \\ -7 & 0 \end{bmatrix}$ **9.** $\begin{bmatrix} -175 & 37 & -13 \\ 95 & -20 & 7 \\ 14 & -3 & 1 \end{bmatrix}$ **11.** 90

13. 54, -18, 6, -2, $\frac{2}{3}$ **15.** Proof

17. $x^2 + y^2 - \frac{41}{3}x - 4y = 0$ **19.** 22 **21.** 120

APPENDIX A *(page A11)*

WARM UP **1.** $y = 4 - 3x$ **2.** $y = x$

3. $y = \frac{2}{3}(1 - x)$ **4.** $y = \frac{2}{5}(2x + 1)$

5. $y = \frac{1}{4}(5 - 3x)$ **6.** $y = \frac{2}{3}(-x + 3)$

7. $y = 4 - x^2$ **8.** $y = \frac{2}{3}(x^2 - 1)$

9. $y = \pm\sqrt{4 - x^2}$ **10.** $y = \pm\sqrt{x^2 - 9}$

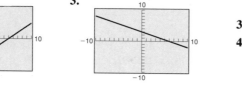

1. **3.**

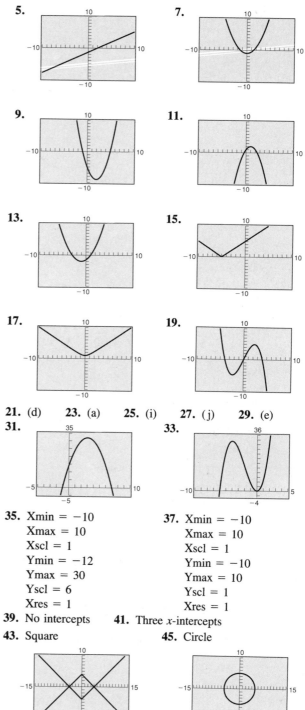

5. **7.**

9. **11.**

13. **15.**

17. **19.**

21. (d) **23. (a)** **25. (i)** **27. (j)** **29. (e)**

31. **33.**

35. Xmin $= -10$ **37.** Xmin $= -10$
Xmax $= 10$ Xmax $= 10$
Xscl $= 1$ Xscl $= 1$
Ymin $= -12$ Ymin $= -10$
Ymax $= 30$ Ymax $= 10$
Yscl $= 6$ Yscl $= 1$
Xres $= 1$ Xres $= 1$

39. No intercepts **41.** Three x-intercepts

43. Square **45.** Circle

47.
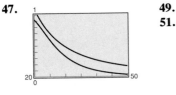

49. 0.59
51. 0.5
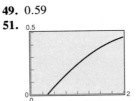

53. (b)

Appendix B   *(page A18)*

1. (a) 2.6201   **(b)** −1.3799   **3. (a)** 3.7993
(b) 3   **5. (a)** −0.6081   **(b)** 3.8959
7. (a) 0.7715   **(b)** 4.2520   **9. (a)** 25,000
(b) 0.025   **11. (a)** 1,360,000   **(b)** 0.0136
13. (a) 3.6420   **(b)** −0.2176   **15. (a)** 414,500
(b) 0.007075   **17.** 18.10   **19.** 4.42   **21.** 901.5
23. (a) 1.8310   **(b)** 2.2565   **25. (a)** 7.46
(b) 4.23   **27. (a)** 33.115   **(b)** 0.0302

Appendix G   *(page A41)*

Section 3.2

1.
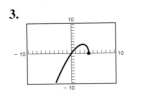

The graph intersects the *x*-axis three times.
The graph intersects the *y*-axis once.

3.
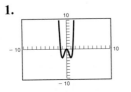

The graph intersects the *x*-axis twice.
The graph intersects the *y*-axis once.

5.
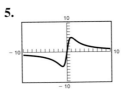

The graph intersects the *x*-axis once.
The graph intersects the *y*-axis once.

7.
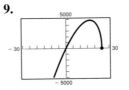

The graph intersects the *x*-axis twice.
The graph intersects the *y*-axis once.

9.
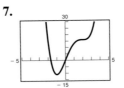

The graph intersects the *x*-axis twice.
The graph intersects the *y*-axis once.

11.
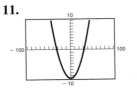

The graph intersects the *x*-axis twice.
The graph intersects the *y*-axis once.

13. $y = \sqrt{64 - x^2}$, $y = -\sqrt{64 - x^2}$

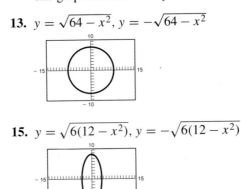

15. $y = \sqrt{6(12 - x^2)}$, $y = -\sqrt{6(12 - x^2)}$

17.

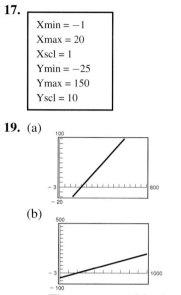

```
Xmin = -1
Xmax = 20
Xscl = 1
Ymin = -25
Ymax = 150
Yscl = 10
```

19. (a)

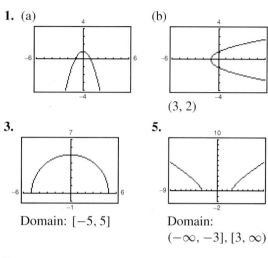

(b)

The person would select the first viewing rectangle.

Section 3.3

1. (a) $y = 2x$ $y = x$ **(b)** $y = -x$ $y = -2x$
$y = 4x$ $y = -4x$
$y = 0.5x$ $y = -0.5x$

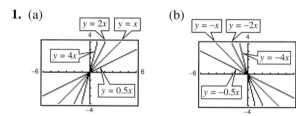

The lines become steeper as the magnitude of the slope increases.

3. (a)   **(b)**

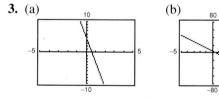

5. $y = \frac{2}{3}x + 2$ **7.** $y = x + 1$
$y = -\frac{3}{2}x$ $y = x - 8$
$y = \frac{2}{3}x$ $y = -x + 3$

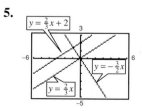
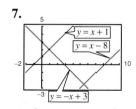

Section 3.4

1. (a)   **(b)**

(3, 2)

3.   **5.**

Domain: $[-5, 5]$ Domain:
$(-\infty, -3], [3, \infty)$

7.

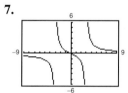

Domain: $(-\infty, -3), (-3, 3), (3, \infty)$

9.   **11.**

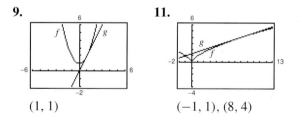

(1, 1) $(-1, 1), (8, 4)$

Section 3.5

1. b   **3.** a

5.

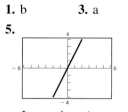

Increasing: $(-\infty, \infty)$

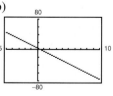

7.

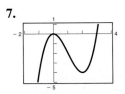

Increasing: $(-\infty, 0), (2, \infty)$
Decreasing: $(0, 2)$

9.

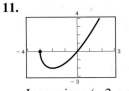

Increasing: $(-1, 0), (1, \infty)$
Decreasing: $(-\infty, -1), (0, 1)$

11.

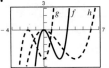

Increasing: $(-2, \infty)$
Decreasing: $(-3, -2)$

13. Relative minimum: $(3, -9)$

15. Relative minimum: $(1, -7)$
Relative maximum: $(-2, 20)$

17. Relative minimum: $(0.33, -0.38)$

19.

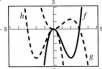

$g(x)$ is shifted two units to the left of $f(x)$.
$h(x)$ is a vertical shrink of $f(x)$.

21.

$g(x)$ is a vertical shrink of $f(x)$ and a reflection across the x-axis.

$h(x)$ is a reflection across the y-axis.

23. (b)

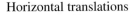

(c) $x \approx 0.57$ mi

Section 3.6

1. (a)   (b)

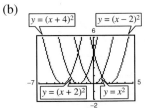

Vertical translations   Horizontal translations

3.

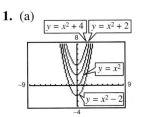

The three functions are odd. As the exponents increase, the graphs become flatter in the interval $(-1, 1)$.

5.

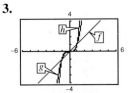

7.

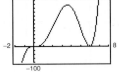

9.

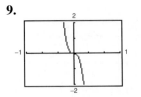

Section 3.7

1.

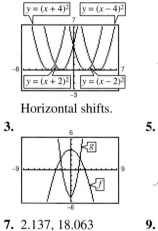

The graph of g is a reflection of f across the line $y = x$.

3. f   **5.** a   **7.** c

Section 4.1

1.

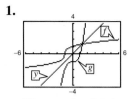

Vertical stretch and reflections across the x-axis.

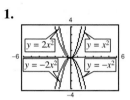

Horizontal shifts.   Vertical shifts

3.   **5.**

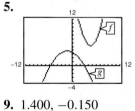

7. 2.137, 18.063   **9.** 1.400, −0.150

Section 4.2

1.

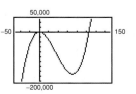

3.

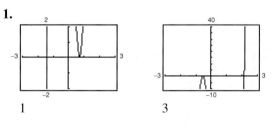

5.

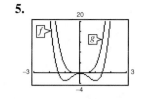

7. [0, 1], [6, 7], [11, 12]

9. [−4, −3], [−1, 0], [0, 1], [3, 4]

Section 4.4

1.

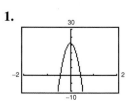

1   3

3. 1.268, 4.732   **5.** −2.422

7. −4.206, −0.735, 1.941

9. 1.638   **11.** (−1.222, 8.222)

13. (1.670, 1.660)   **15.** (2.050, 32)

17. (0, 2), (3, 5)   **19.** (0, 0), (3, −3)

Section 4.5

1. 2

3. (b) 2   **5.** (b) ±4, −0.73, 2.73

   (c) $2 \pm i$ (c) No complex zeros

7. (b) No real zeros   **9.** (b) 0, 3

   (c) $\pm i, 2 \pm i$ (c) $\pm \frac{1}{3}i$

11. (a) $A = x(115 - x)$

   (b)   (c) 93.6 ft

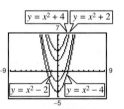

$0 < x < 115$

Section 5.1

1. (a) Domain of f: $(-\infty, -2), (-2, \infty)$
   Domain of g: $(-\infty, \infty)$
   (b) No vertical asymptote
   (c)

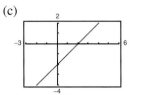

   (d) The graph differs at only one point and this cannot be seen on the display of the graphing utility.

3. (a) Domain of f: $(-\infty, 0), (0, 3), (3, \infty)$
   Domain of g: $(-\infty, 0), (0, \infty)$
   (b) $x = 0$
   (c)

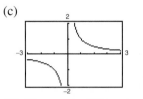

   (d) The graph differs at only one point and this cannot be seen on the display of the graphing utility.

5.

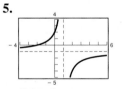

Domain:
all real numbers $x \neq 1$

Range:
all real numbers $y \neq -1$

7.

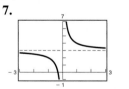

Domain:
all real numbers $t \neq 0$

Range:
all real numbers $y \neq 3$

9.

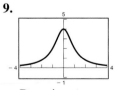

Domain:
all real numbers

Range: $0 < y \leq 4$

11.

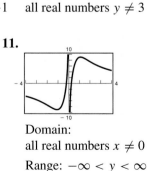

Domain:
all real numbers $x \neq 0$

Range: $-\infty < y < \infty$

13.

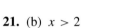

15.

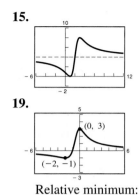

17.

The fraction is not re-
duced to lowest terms.

19.

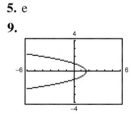

Relative minimum:
$(-2, -1)$
Relative maximum:
$(0, 3)$

21. (b) $x > 2$
    (c)

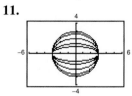

5.87 in. × 11.75 in.

Section 5.3

1. c   3. b   5. e
7.   9.

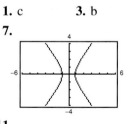
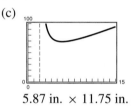

11.

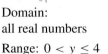

The graph becomes flatter as e approaches 1.

Section 5.4

1. a   3. f   5. d

Section 6.1

1.

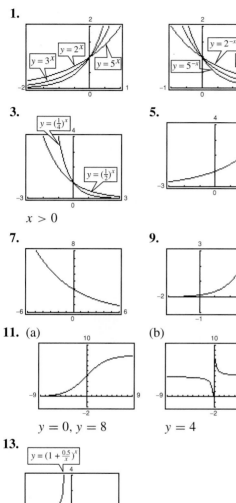

3.

$x > 0$

5.

7.

9.

11. (a)

(b)

$y = 0, y = 8$ $y = 4$

13.

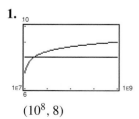

Section 6.2

1.

$(10^8, 8)$

3.

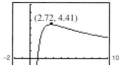

Increasing: $(0, e)$
Decreasing: (e, ∞)
Relative maximum: $(e, 4.415)$

Section 6.3

1.

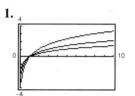

For a given a the graphs are identical.

Section 6.4

1. $(0.138, -1.981)$ **3.** 5.862
    $(1.564, 0.448)$

5. 3.158   **7.** 27.465

9. $(3.327, 1.202)$

Section 6.5

1.

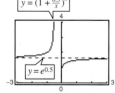

3. a

Balance increases more
rapidly when the in-
terest is compounded
continuously.

Section 7.1

1.

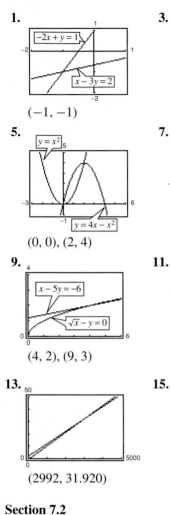

$(-1, -1)$

3.

$(10, 0)$

5.

$(0, 0), (2, 4)$

7.

$(4, 2), (-4, -2)$

9.

$(4, 2), (9, 3)$

11.

$(0, 0), (1, 1)$

13.

$(2992, 31.920)$

15.

$50{,}000$ units

Section 7.2

1. (a)

Slopes unequal;
One solution

(b)

Slopes equal;
No solution

(c)

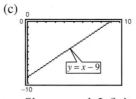

Slopes equal; Infinite number of solutions

3.

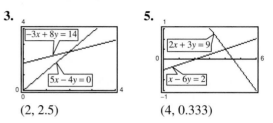

$(2, 2.5)$

5.

$(4, 0.333)$

Section 7.4

1. b   **3.** d

5.

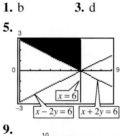

7.

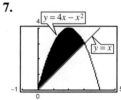

9.

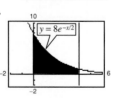

Section 8.2

1. (a) $\begin{bmatrix} 13 & 12 & 19 \\ 4 & 11 & -4 \end{bmatrix}$ (b) $\begin{bmatrix} 5 & 10 & -4 \\ -6 & 1 & 13 \end{bmatrix}$

3. (a) $\begin{bmatrix} 4 & -10 & -1 & -20 \\ -5 & 26 & 4 & 77 \\ 6 & -22 & 24 & -72 \end{bmatrix}$

(b) $\begin{bmatrix} 1 & 1 & -2 & 9 \\ 11 & -18 & 1 & -7 \\ -17 & 26 & 34 & 24 \end{bmatrix}$

5. (a) $\begin{bmatrix} -4 & 8 \\ 2 & -8 \end{bmatrix}$ (b) $\begin{bmatrix} 11 & -3 \\ -6 & 2 \end{bmatrix}$

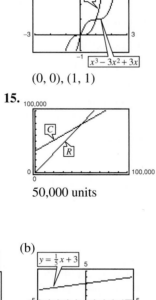

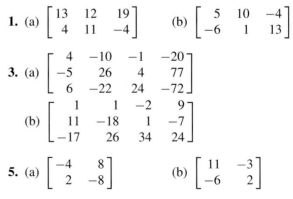

7. (a) $\begin{bmatrix} 60 & 8 & -40 \\ -120 & 0 & 100 \end{bmatrix}$ (b) $\begin{bmatrix} 29 & 52 \\ -65 & 120 \end{bmatrix}$

9. (a) [161] (b) Not possible; incorrect order

Section 8.3

1. Exercise 1 is a proof.

3. $\begin{bmatrix} 2 & -3 \\ -1 & 2 \end{bmatrix}$

5. $\begin{bmatrix} \frac{1}{13} & \frac{4}{13} \\ -\frac{2}{13} & \frac{5}{13} \end{bmatrix}$

7. $\begin{bmatrix} -\frac{11}{4} & \frac{9}{2} & 6 \\ \frac{9}{4} & -\frac{7}{2} & -5 \\ \frac{1}{4} & -\frac{1}{2} & 0 \end{bmatrix}$

9. $(-2, 1)$

11. $(2, -1, 1)$

13. $(3, 4, -5)$

Section 8.6

1. 19,185

3. 0

5. $\left(\frac{1}{2}, 4\right)$

7. $\left(2, \frac{1}{2}, -\frac{1}{2}\right)$

9. $\frac{31}{2}$

11. Collinear

13. $y = 2x^2 - 6x + 1$

Section 9.1

1. a

3. d

5.

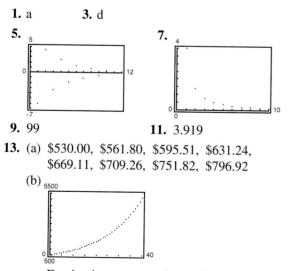

9. 99

11. 3.919

13. (a) $530.00, $561.80, $595.51, $631.24, $669.11, $709.26, $751.82, $796.92

(b)

Earning interest on an increasing principal.

Section 9.3

1. b

3. c

5.

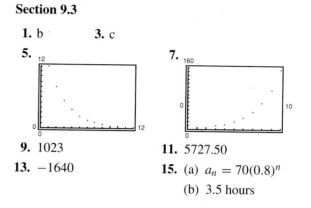

9. 1023

11. 5727.50

13. -1640

15. (a) $a_n = 70(0.8)^n$

(b) 3.5 hours

Index of Applications

Biology and Life Science Applications

Business Applications

Index

FORMULAS FROM GEOMETRY

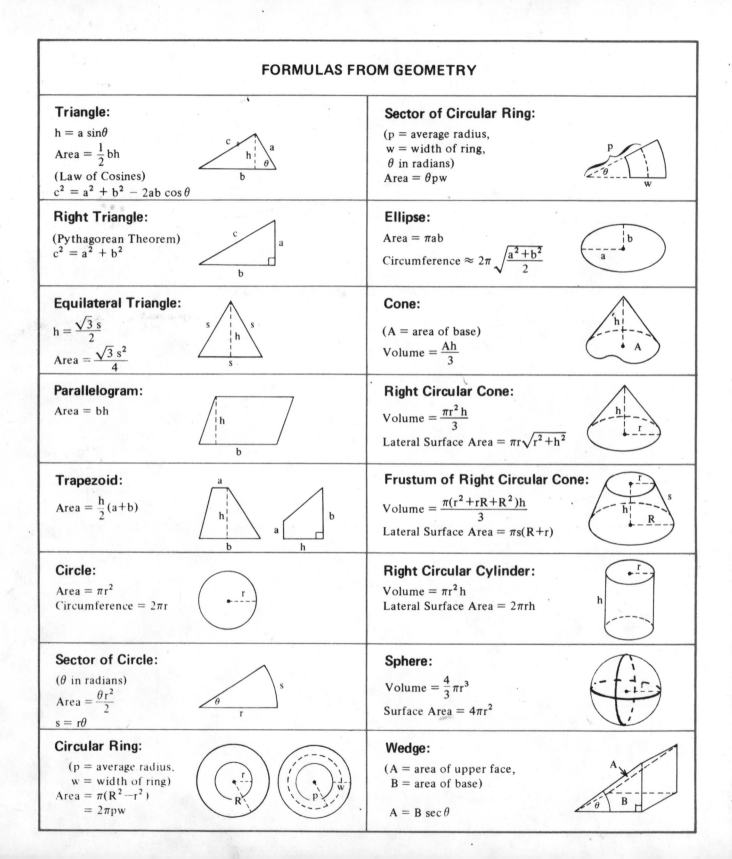

Triangle:

$h = a \sin\theta$

$\text{Area} = \frac{1}{2}bh$

(Law of Cosines)

$c^2 = a^2 + b^2 - 2ab \cos\theta$

Right Triangle:

(Pythagorean Theorem)

$c^2 = a^2 + b^2$

Equilateral Triangle:

$h = \frac{\sqrt{3}\,s}{2}$

$\text{Area} = \frac{\sqrt{3}\,s^2}{4}$

Parallelogram:

$\text{Area} = bh$

Trapezoid:

$\text{Area} = \frac{h}{2}(a+b)$

Circle:

$\text{Area} = \pi r^2$

$\text{Circumference} = 2\pi r$

Sector of Circle:

(θ in radians)

$\text{Area} = \frac{\theta r^2}{2}$

$s = r\theta$

Circular Ring:

(p = average radius,
w = width of ring)

$\text{Area} = \pi(R^2 - r^2)$

$\quad\quad = 2\pi pw$

Sector of Circular Ring:

(p = average radius,
w = width of ring,
θ in radians)

$\text{Area} = \theta pw$

Ellipse:

$\text{Area} = \pi ab$

$\text{Circumference} \approx 2\pi \sqrt{\dfrac{a^2 + b^2}{2}}$

Cone:

(A = area of base)

$\text{Volume} = \frac{Ah}{3}$

Right Circular Cone:

$\text{Volume} = \frac{\pi r^2 h}{3}$

$\text{Lateral Surface Area} = \pi r\sqrt{r^2 + h^2}$

Frustum of Right Circular Cone:

$\text{Volume} = \frac{\pi(r^2 + rR + R^2)h}{3}$

$\text{Lateral Surface Area} = \pi s(R + r)$

Right Circular Cylinder:

$\text{Volume} = \pi r^2 h$

$\text{Lateral Surface Area} = 2\pi rh$

Sphere:

$\text{Volume} = \frac{4}{3}\pi r^3$

$\text{Surface Area} = 4\pi r^2$

Wedge:

(A = area of upper face,
B = area of base)

$A = B \sec\theta$